Mechanical Engineering Series

Frederick F. Ling
Editor-in-Chief

Mechanical Engineering Series

J. Angeles, **Fundamentals of Robotic Mechanical Systems: Theory, Methods, and Algorithms, 2nd ed.**

P. Basu, C. Kefa, and L. Jestin, **Boilers and Burners: Design and Theory**

J.M. Berthelot, **Composite Materials: Mechanical Behavior and Structural Analysis**

I.J. Busch-Vishniac, **Electromechanical Sensors and Actuators**

J. Chakrabarty, **Applied Plasticity**

K.K. Choi and N.H. Kim, **Structural Sensitivity Analysis and Optimization 1: Linear Systems**

K.K. Choi and N.H. Kim, **Structural Sensitivity Analysis and Optimization 2: Nonlinear Systems and Applications**

G. Chryssolouris, **Laser Machining: Theory and Practice**

V.N. Constantinescu, **Laminar Viscous Flow**

G.A. Costello, **Theory of Wire Rope, 2nd Ed.**

K. Czolczynski, **Rotordynamics of Gas-Lubricated Journal Bearing Systems**

M.S. Darlow, **Balancing of High-Speed Machinery**

J.F. Doyle, **Nonlinear Analysis of Thin-Walled Structures: Statics, Dynamics, and Stability**

J.F. Doyle, **Wave Propagation in Structures: Spectral Analysis Using Fast Discrete Fourier Transforms, 2nd ed.**

P.A. Engel, **Structural Analysis of Printed Circuit Board Systems**

A.C. Fischer-Cripps, **Introduction to Contact Mechanics**

A.C. Fischer-Cripps, **Nanoindentations, 2nd ed.**

J. García de Jalón and E. Bayo, **Kinematic and Dynamic Simulation of Multibody Systems: The Real-Time Challenge**

W.K. Gawronski, **Advanced Structural Dynamics and Active Control of Structures**

W.K. Gawronski, **Dynamics and Control of Structures: A Modal Approach**

G. Genta, **Dynamics of Rotating Systems**

(continued after index)

R. A. Howland

Intermediate Dynamics: A Linear Algebraic Approach

Springer

R. A. Howland
University of Notre Dame

Editor-in-Chief
Frederick F. Ling
Earnest F. Gloyna Regents Chair Emeritus in Engineering
Department of Mechanical Engineering
The University of Texas at Austin
Austin, TX 78712-1063, USA
and
Distinguished William Howard Hart
Professor Emeritus
Department of Mechanical Engineering,
Aeronautical Engineering and Mechanics
Rensselaer Polytechnic Institute
Troy, NY 12180-3590, USA

Intermediate Dynamics: A Linear Algebraic Approach

e-ISBN 0-387-28316-1
ISBN: 978-1-4419-3920-3 e-ISBN: 978-0-387-28316-6

Printed on acid-free paper.

springeronline.com

Dedicated to My Folks

Mechanical Engineering Series

Frederick F. Ling
Editor-in-Chief

The Mechanical Engineering Series features graduate texts and research monographs to address the need for information in contemporary mechanical engineering, including areas of concentration of applied mechanics, biomechanics, computational mechanics, dynamical systems and control, energetics, mechanics of materials, processing, production systems, thermal science, and tribology.

Series Preface

Mechanical engineering, and engineering discipline born of the needs of the industrial revolution, is once again asked to do its substantial share in the call for industrial renewal. The general call is urgent as we face profound issues of productivity and competitiveness that require engineering solutions, among others. The Mechanical Engineering Series is a series featuring graduate texts and research monographs intended to address the need for information in contemporary areas of mechanical engineering.

The series is conceived as a comprehensive one that covers a broad range of concentrations important to mechanical engineering graduate education and research. We are fortunate to have a distinguished roster of consulting editors, each an expert in one of the areas of concentration. The names of the consulting editors are listed on page vi of this volume. The areas of concentration are applied mechanics, biomechanics, computational mechanics, dynamic systems and control, energetics, mechanics of materials, processing, thermal science, and tribology.

Preface

A number of colleges and universities offer an upper-level undergraduate course usually going under the rubric of "Intermediate" or "Advanced" Dynamics—a successor to the first dynamics offering generally required of all students. Typically common to such courses is coverage of 3-D rigid body dynamics and Lagrangian mechanics, with other topics locally discretionary. While there are a small number of texts available for such offerings, there is a notable paucity aimed at "mainstream" undergraduates, and instructors often resort to utilizing sections in the first Mechanics text not covered in the introductory course, at least for the 3-D rigid body dynamics. Though closely allied to its planar counterpart, this topic is far more complex than its predecessor: in kinematics, one must account for possible change in direction of the angular velocity; and the kinetic "moment of inertia," a simple scalar in the planar formulation, must be replaced by a tensor quantity. If elementary texts' presentation of planar dynamics is adequate, their treatment of three-dimensional dynamics is rather less satisfactory: It is common to expand vector equations of motion in *components*—in a *particular* choice of axes—and consider only a few special instances of their application (*e.g.* fixed-axis rotation in the Euler equations of motion). The presentation of principal coordinates is typically somewhat *ad hoc*, either merely stating the procedure to find such axes in general, or even more commonly invoking the "It can be shown that..." mantra. Machines seem not to *exist* in 3-D! And equations of motion for the gyroscope are derived independently of the more general ones—a practice lending a certain air of mystery to this important topic.

Such an approach can be frustrating to the student with any degree of curiosity and is counterproductive pedagogically: the component-wise expression of vector quantities has long since disappeared from even Sophomore-level courses in Mechanics, in good part because the complexity of notation obscures the relative simplicity of the concepts involved. But the Euler equations can be expressed both succinctly and generally through the introduction of matrices. The typical exposition of principal axes overlooks the fact that this is precisely the *same* device used to find the *same* "principal axes" in solid mechanics (explicitly through a rotation); few students recognize this fact, and, unfortunately, few instructors take the opportunity to point this out and unify the concepts. And principal axes themselves are, in fact, merely an application of an even more general technique utilized in linear algebra leading to the diagonalization of matrices (at least the real, symmetric ones encountered in both solid mechanics and dynamics). These facts alone suggest a linear algebraic approach to the subject.

A knowledge of linear algebra is, however, more beneficial to the scientist and engineer than merely to be able to diagonalize matrices: Eigenvectors and eigenvalues pervade both fields; yet, while students can typically *find* these quantities and use them to whatever end they have been instructed in, few can answer the simple question "What is an eigenvector?" As the field of robotics becomes ever more mainstream, a facility with [3-D] rotation matrices becomes increasingly important. Even the mundane issue of solving linear equations is often incomplete or, worse still, inaccurate: "All you need is as many equations as unknowns. If you have fewer than that, there is no solution." (The first of these statements is incomplete, the second downright wrong!) Such fallacies are likely not altogether the students' fault: few curricula allow the time to devote a full, formal course to the field, and knowledge of the material is typically gleaned piecemeal on an "as-need" basis. The result is a fractionated view with the intellectual gaps alluded to.

Yet a full course may not be necessary: For the past several years, the Intermediate Dynamics course at Notre Dame has started with an only 2–3 week presentation of linear algebra, both as a prelude to the three-dimensional dynamics to follow, and for its intrinsic pedagogical merit—to organize the bits and pieces of concepts into some organic whole. However successful the latter goal has been, the former has proven beneficial.

With regard to the other topic of Lagrangian mechanics, the situation is perhaps even more critical. At a time when the analysis of large-scale systems has become increasingly important, the presentation of energy-based dynamical techniques has been surprisingly absent from most undergraduate texts altogether. These approaches are founded on virtual work (not typically the undergraduate's favorite topic!) and not only eliminate the need to consider the forces at interconnecting pins [assumed frictionless], but also free the designer from the relatively small number of vector coordinate systems available to describe a problem: he can select a set of coordinate ideally suited to the one at hand.

With all this in mind, the following text commits to paper a course which has gradually developed at Notre Dame as *its* "Intermediate Dynamics" offering. It starts with a relatively short, but rigorous, exposition of linear systems, culminating in the diagonalization (where possible) of matrices—the foundation of principal coordinates. There is even an [optional] section dealing with Jordan normal form, rarely presented to students at this level. In order to understand this process fully, it is necessary that the student be familiar with how the [matrix] *representation* of a linear operator (or of a vector itself) changes with a transformation of basis, as well as how the eigenvectors—in fact the new axes themselves—affect this particular choice of basis. That, at least in the case of real, symmetric, square inertia matrices, this corresponds to a *rotation* of axes requires knowledge of axis rotation and the matrices which generate such rotations. This, in turn, demands an appreciation of bases themselves and, particularly, the idea of linear independence (which many students feel deals exclusively with the Wronskian) and partitioned matrix multiplication. By the time this is done, little more effort is required to deal with vector spaces in general.

This text in fact grew out of the need to dispatch a [perceived] responsibility to rigor (*i.e.proofs* of theorems) without bogging down class presentation with such details. Yet the overall approach to even the mathematical material of linear algebra is a "minimalist" one: rather than a large number of arcane theorems and ideas, the theoretical underpinning of the subject is provided by, and unified through, the basic theme of *linear independence*—the *echelon form* for vectors and [subsequently] matrices, and the *rank* of the latter. It can be argued that these are the concepts the engineer and scientist can—*should*—appreciate anyhow. Partitioning establishes the connection between vectors and [the rows/columns of] matrices, and rank provides the criterion for the solution of linear systems (which, in turn, fold back onto eigenvectors). In order to avoid the student's becoming fixated too early on square matrices, this fundamental theory is developed in the context of linear transformations between spaces of *arbitrary* dimension. It is only after this has been done that we specialize to *square* matrices, where the inverse, eigenvectors, and even properties of determinants follow naturally. Throughout, the distinction between vectors and tensors, and their *representations*—one which is generally blurred in the student's mind because it is so rarely stressed in presentation—is heavily emphasized.

Theory, such as the conditions under which systems of linear equations have a solution, is actually important in application. But this linear algebra Part is more than mere theory: Linear independence, for example, leads to the concept of matrix rank, which then becomes a criterion for predicting the number of solutions to a set of linear equations; when the cross product is shown to be equivalent to a *matrix* product of rank 2, indeterminacy of angular velocity and acceleration from the rotating coordinate system equations in the next Part becomes a natural consequence. Similarly, rotation matrices first appear as an example of orthogonal matrices, which then are used in the diagonalization of real symmetric matrices culminating the entire first Part; though this returns in the next Part in the guise of principal axes, its inverse—the rotation *from* principal axes to arbitrary ones—becomes a fundamental technique for the determination of the inertia tensor.

Given the audience for which this course is intended, the approach has been surprisingly successful: one still recalls the delight of one student who said that she had decided to attack a particular problem with rotation matrices, and "It worked!" Admittedly, such appreciation is often delayed until the part on rigid body dynamics has been covered—yet another reason for trying to have some of the more technical detail in the text rather than being presented in class.

This next Part on 3-D dynamics starts with a relatively routine exposition of kinematics, though rather more detail than usual is given to constraints on the motion resulting from interconnections, and there is a perhaps unique demonstration that the fundamental relation $d\boldsymbol{r} = d\boldsymbol{\theta} \times \boldsymbol{r}$ results from nothing more than the fixed distance between points in a rigid body. Theory from the first

Part becomes integrated into the presentation in a discussion of the indeterminacy of angular velocity and acceleration *without* such constraints. Kinetics is preceded by a review of particle and system-of-particles kinetics; this is done to stress the *particle* foundation on which even rigid body kinetics is based as much as to make the text self-contained. The derivation of the Euler equations is also relatively routine, but here the similarity to most existing texts ends: these equations are presented in *matrix* form, and principal coordinates dispatched with reference to diagonalization covered in the previous Part. The flexibility afforded by this allows an *arbitrary* choice of coordinates in terms of which to represent the relevant equations and, again, admits of a more transparent comprehension than the norm. 3-D rigid body *machine kinetics*, almost universally ignored in elementary presentations, is also covered: the emphasis is on integrating the kinematics and kinetics into a *system of linear equations*, making the difference between single rigid bodies and "machines" quantitative rather than qualitative. There is also a rather careful discussion of the *forces* at [ball-and-socket] connections in machines; this is a topic often misunderstood in *Statics*, let alone Dynamics! While the gyroscope equations of motion are developed *de nuvo* as in most texts, they are also obtained by direct application of the Euler equations; this is to overcome the stigma possibly resulting from independent derivation—the misconception that gyroscopes are somehow "special," not covered by the allegedly more general theory. There is a brief section detailing the use of the general equations of motion to obtain the variational equations necessary to analyze stability. Finally, a more or less routine treatment of energy and momentum methods is presented, though the implementation of kinetic constants and the conditions under which vanishing force or moment lead to such constants are presented and utilized; this is to set the scene for the next Part, where Analytical Dynamics employs special techniques to uncover such "integrals of the motion."

That next Part treats Analytical Mechanics. Lagrangian dynamics is developed first, based on Newton's equations of motion rather than functional minimization. Though the latter is mentioned as an alternative approach, it seems a major investment of effort—and time—to develop one relatively abstruse concept (however important intrinsically), just to derive another almost as bad by appeal to a "principle" (Least Action) whose origin is more teleological than physical! Rather more care than usual, however, is taken to relate the concepts of kinematic "constraints" and the associated kinematical equations among the coordinates; this is done better to explain Lagrange multipliers. Also included is a section on the use of non-holonomic constraints, in good part to introduce Lagrange multipliers as a means of dealing with them; while the latter is typically the *only* methodology presented in connection with this topic, here it is actually preceded by a discussion of purely algebraic elimination of redundant constraints, in the hopes of demonstrating the fundamental issue itself. Withal, the "spin" put on this section emphasizes the freedom to select *arbitrary* coordinates instead of being "shackled" by the standard vector coordinate systems developed to deal with Newton's laws. The discussion of the kinetic constants of energy and momentum in the previous Part is complemented by a discussion of "integrals of the motion" in this one; this is the springboard to introduce Jacobi's integral, and ignorable coordinates become a means of uncovering other such constants of the motion in Lagrangian systems—at least, if one can find the Right Coordinates! The Part concludes with a chapter on Hamiltonian dynamics. Though this topic is almost universally ignored by engineers, it is as universal in application as Conservation of Energy and is the *lingua franca* of Dynamical Systems, with which every modern-day practitioner must have some familiarity. Unlike most introductions to the field, which stop after having obtained the equations of motion, this presentation includes a discussion of canonical transformations, separability, and the Hamilton-Jacobi Equation: the fact that there is a *systematic* means of obtaining those variables—"ignorable" coordinates *and* momenta—in which the Hamiltonian system is completely soluble is, after all, the *raison d'etre* for invoking a system in this form, as opposed to the previous Lagrangian formulation, in the first place. Somewhat more attention to time-dependent Hamiltonian systems is given than usual. Additionally, like the Lagrangian presentation, there is a discussion of casting time as an ordinary variable; though this is often touched upon by more advanced texts on analytical mechanics, the matter seems to be dropped almost immediately, without benefit of examples to demonstrate exactly what this perspective entails; one, demonstrating whatever benefits it might enjoy, is included in this chapter.

A Note on Notation. We shall occasionally have the need to distinguish between "unordered" and "ordered" sets. *Un*ordered sets will be denoted by braces: $\{a, b\} = \{b, a\}$, while *ordered* sets will be denoted by parentheses: $(a, b) \neq (b, a)$. This convention is totally consistent with the shorthand notation for matrices: $\mathsf{A} = (a_{ij})$.

Throughout, vectors are distinguished by boldface: $\boldsymbol{V}$; unit vectors are distinguished with a "hat" and are generally lower-case letters: $\hat{\boldsymbol{\imath}}$. Tensor/matrix quantities are written in a boldface sans-serif font, typically in upper-case: I (though the matrix *representation* of a vector otherwise denoted with a lower-case boldface will retain the case: $\mathsf{v} \sim \boldsymbol{v}$).

Material not considered essential to the later material is presented in a smaller type face; this is not meant to denigrate the material so much as to provide a visual map to the overall picture. Students interested more in applications are typically impatient with a mathematical Theorem/Proof format, yet it surely is necessary to feel that a field has been developed logically. For this reason, each section concludes with a brief guide of what results are used merely to demonstrate later, more important results, and what are important intrinsically; it is hoped this will provide some "topography" of the material presented.

A Note on Style. There have been universally two comments by students regarding this text: "It's all there," and "We can hear you talking." In retrospect, this is less a "text" than *lectures* on the various topics. Examples are unusually—perhaps overwhelmingly!—complete, with all steps motivated and virtually all intermediate calculations presented; though this breaks the "one-page format" currently in favor, it counters student frustration with "terseness." And the style is more narrative than expository. It is hoped that lecturers do not find this jarring.

R.A. Howland
South Bend, Indiana

Contents

Part I

Linear Algebra

Prologue

The primary motivation for this part is to lay the foundation for the next one, dealing with 3-D rigid body dynamics. It will be seen there that the "inertia," I, a quantity which is a simple *scalar* in planar problems, blossoms into a "tensor" in three-dimensional ones. But in the same way that vectors can be *represented* in terms of "basis" vectors $\hat{\imath}$, $\hat{\jmath}$, and $\hat{k}$, tensors can be *represented* as 3×3 matrices, and formulating the various kinetic quantities in terms of these matrices makes the fundamental equations of 3-D dynamics far more transparent and comprehensible than, for example, simply writing out the components of the equations of motion. (Anyone who has seen older publications in which the separate components of moments and/or angular momentum written out doggedly, again and again, will appreciate the visual economy and conceptual clarity of simply writing out the cross product!) More importantly, however, the more general notation allows a freedom of choice of coordinate system.

While it is obvious that the representation of vectors will change when a different basis is used, it is not clear that the same holds for matrix representations. But it turns out that there is always a choice of basis in which the inertia tensor can be represented by a particularly simple matrix—one which is *diagonal* in form. Such a choice of basis—tantamount to a choice of coordinates axes—is referred to in that context as "principle axes." These happen to be precisely the same "principle axes" the student my have encountered in a course in Solid Mechanics; both, in turn, rely on techniques in the mathematical field of "linear algebra" aimed at generating a standard, "canonical" form for a given matrix.

Thus the following three chapters comprising this Part culminate in a discussion of such canonical forms. But to understand this it is also necessary to see how a change of basis affects the representation of a matrix (or even a vector, for that matter). And, since it turns out that the new principle axes are nothing more than the eigenvectors of the inertia matrix—and the inertia matrix in these axes nothing more than that with the eigenvalues arrayed along the diagonal—it is necessary to understand what these are. (If one were to walk up to you on the street as ask the question "Yes, but just what *is* an eigenvector?", could you answer [at least after recovering from the initial shock]?) To find these, in turn, it is necessary to solve a "homogeneous" system of linear equations; under what conditions can a solution to such a system—or any other linear system, for that matter—be found? The answer lies in the number of "linearly independent" equations available. And if this diagonalization depends on the "basis" chosen to represent it, just what is all this basis business about anyhow?

In fact, at the end of the day, just what is a *vector*?

The ideas of basis and linear independence are fundamental and will pervade all of the following Part. But they are concepts from the field of *vector spaces*, so these will be considered first. That "vectors" are more general than merely the directed line segments—"arrows"—that most scientists and engineers are familiar with will be pointed out; indeed, a broad class of objects satisfy exactly the same properties that position, velocity, forces, *et al* (and the way we add and multiply them) do. While starting from this point is probably only reasonable for mathematicians, students at this level from other disciplines have developed the mathematical sophistication to appreciate this fact, if only in retrospect. Thus the initial chapter adopts the perspective viewing vectors as those *objects* (and the operations on them) *satisfying certain properties*. It is hoped that this approach will allow engineers and scientists to be conversant with their more "mathematical" colleagues who, in the end, do themselves deal with the same problems they do!

The second chapter then discusses these objects called tensors—"*linear transformations*" between vector spaces—and how at least the "second-order" tensors we are primarily interested in can be represented as matrices of arbitrary (non-square) dimension. The latter occupy the bulk of this chapter, with a discussion of how they depend on the very same basis the vectors do, and how a change of this basis will change the representations of both the vectors and the tensors operating on them. If such transformations "map" one vector *to* another, the inverse operation finds where a given vector has been transformed *from*; this is precisely the problem of determining the *solution* of a [linear] equation.

Though the above has been done in the context of transformations between *arbitrary* vector spaces, we then specialize to mappings between spaces of the *same* dimension—represented as *square* matrices. Eigenvectors and eigenvalues, inverse, and determinants are treated—more for sake of completeness than anything else, since most students will likely already have encountered them at this stage. The criterion to predict, for a given matrix, how many linearly independent eigenvectors there are is developed. Finally all this will be brought together in a presentation of normal forms—diagonalized matrices when we have available a full set of linearly independent eigenvectors, and a block-diagonal form when we don't. These are simply the [matrix] representations of linear transformations in a distinguishing basis, and actually what the entire Part has been aiming at: the "principle axes" used so routinely in the following Part on 3-D rigid-body dynamics are nothing more than those which make the "inertial tensor" assume a diagonalized form.

Chapter 1

Vector Spaces

Introduction

When "vectors" are first introduced to the student, they are almost universally motivated by physical quantities such as force and position. It is observed that these have a magnitude and a direction, suggesting that we might define objects having these characteristics as "vectors." Thus, for most scientists and engineers, vectors are *directed line segments*—"arrows," as it were. The same motivation then makes more palatable a curious "arithmetic of arrows": we can find the "sum" of two arrows by placing them together head-to-tail, and we can "multiply them by numbers" through a scaling, and possibly reversal of direction, of the arrow. It is possible to show from nothing more than this that these two operations have certain properties, enumerated in Table 1.1 (see page 7).

At this stage one typically sees the introduction of a set of mutually perpendicular "unit vectors" (vectors having a magnitude of 1), $\hat{\boldsymbol{\imath}}$, $\hat{\boldsymbol{\jmath}}$, and $\hat{\boldsymbol{k}}$—*themselves directed line segments.* One then *represents* each vector $\boldsymbol{A}$ in terms of its [scalar] "components" along each of these unit vectors: $\boldsymbol{A} = A_x\hat{\boldsymbol{\imath}} + A_y\hat{\boldsymbol{\jmath}} + A_z\hat{\boldsymbol{k}}$. (One might choose to "order" these vectors $\hat{\boldsymbol{\imath}}$, $\hat{\boldsymbol{\jmath}}$, and $\hat{\boldsymbol{k}}$; then the same vector can be written in the form of an "ordered triple" $\boldsymbol{A} = (A_x, A_y, A_z)$.) Then the very same operations of "vector addition" and "scalar multiplication" defined for directed line segments can be reformulated and expressed in the Cartesian (or ordered triple) form. Presently other coordinate systems may be introduced—in polar coordinates or path-dependent tangential-normal coordinates for example, one might use unit vectors which move with the point of interest, but all quantities are still referred to a set of fundamental vectors.

[The observant reader will note that nothing has been said above regarding the "dot product" or "cross product" typically also defined for vectors. There is a reason for this omission; suffice it to point out that these operations are defined *in terms of* the magnitudes and (relative) directions of the two vectors, so hold *only* for directed line segments.]

The unit vectors introduced are an example of a "*basis*"; the expression of $\boldsymbol{A}$ in terms of its components is referred to as its *representation* in terms of that basis. Although the facts are obvious, it is rarely noted that a) the *representation* of $\boldsymbol{A}$ will change for a different choice of $\hat{\imath}$, $\hat{\jmath}$, and $\hat{k}$, but *the vector* $\boldsymbol{A}$ *remains the same*—*it* has the same magnitude and direction regardless of the representation; and b) that, in effect, we are representing the directed line segment $\boldsymbol{A}$ in terms of *real numbers* A_x, A_y, and A_z. The latter is particularly important because it allows us to deal with even three-dimensional vectors in terms of real numbers instead of having to utilize trigonometry (*spherical* trigonometry, at that!) to find, say, the sum of two vectors.

In the third chapter, we shall see how to predict the representation of a vector given in one basis when we change to another one utilizing matrix products. While talking about matrices, however, we shall also be interested in other applications, such as how they can be used to solve a system of linear equations. A fundamental idea which will run through these chapters—which pervades *all* of the field dealing with such operations, "linear algebra"—is that of "linear independence." But this concept, like that of basis, is one characterizing *vectors.* Thus the goal if this chapter is to introduce these ideas for later application. Along the way, we shall see precisely what characterizes a "basis"—what is required of a set of vectors to qualify as a basis. It will be seen to be far more general than merely a set of mutually perpendicular unit vectors; indeed, the entire concept of "vector" will be seen to be far broader than simply the directed line segments formulation the reader is likely familiar with.

To do this we are going to turn the first presentation of vectors on its ear: rather than defining the vectors and operations and obtaining the properties in Table 1.1 these operations satisfy, we shall instead say that *any* set of objects, on which there are two operations satisfying the properties in Table 1.1, are vectors. Objects never thought of as being "vectors" will suddenly emerge as, in fact, being vectors, and many of the properties of such objects, rather than being properties only of these objects, will become the common to all objects (and operations) we define as a "*vector space.*" It is a new perspective, but one which, with past experience, it is possible to appreciate in retrospect. Along the way we shall allude to some of the concepts of "algebra" from the field of mathematics. Though not the sort of thing which necessarily leads to the solution of problems, it does provide the non-mathematician with some background and jargon used by "applied mathematicians" in their description of problems engineers and scientists actually *do* deal with, particularly in the field of non-linear dynamics—an area at which the interests of both intersect.

1.1 Vectors

Consider an arbitrary set of elements $\mathcal{V}$. Given a pair of operations [corresponding to] "vector addition," $\oplus$, and "scalar multiplication," $\odot$, $\mathcal{V}$ and its two operations are said to form a *vector space* if, for $\boldsymbol{v}$, $\boldsymbol{v_1}$, $\boldsymbol{v_2}$, and $\boldsymbol{v_3}$ in $\mathcal{V}$, and real numbers a and b, these operations satisfy the properties in Table 1.1.

vector addition:

(G1) $v_1 \oplus v_2 \in \mathcal{V}$ (closure)
(G2) $(v_1 \oplus v_2) \oplus v_3 = v_1 \oplus (v_2 \oplus v_3)$ (associativity)
(G3) $\exists \mathbf{0} \in \mathcal{V} : v \oplus \mathbf{0} = v$ for all $v \in \mathcal{V}$ (additive identity element)
(G4) for each $v \in \mathcal{V}, \exists \overline{v} \in \mathcal{V} : v \oplus \overline{v} = \mathbf{0}$ (additive inverse)
(G5) $v_1 \oplus v_2 = v_2 \oplus v_1$ (commutativity)

scalar multiplication:

(V1) $a \odot v \in \mathcal{V}$ (closure)
(V2) $a \odot (v_1 \oplus v_2) = (a \odot v_1) \oplus (a \odot v_2)$ (distribution)
(V3) $(a + b) \odot v = (a \odot v) \oplus (b \odot v)$ (distribution)
(V4) $(a \cdot b) \odot v = a \odot (b \odot v)$ (associativity)
(V5) $1 \odot v = v$

Table 1.1: Properties of Vector Space Operations

1.1.1 The "Algebra" of Vector Spaces[1]

There are several things to be noted about this table. The first is that the five "G" properties deal only with the "$\oplus$" operation. The latter "V" properties relate that operation to the "$\odot$" one.

"Closure" is generally regarded by non-mathematicians as so "obvious" that it hardly seems worth mentioning. Yet, depending on how the set $\mathcal{V}$ is defined, reasonably non-pathological cases in which this is violated can be found. But in properties (V2) for example, $(v_1 \oplus v_2)$ is an element of $\mathcal{V}$ by (G1), while $(a \odot v)$ and $(b \odot v)$ are by (V1); thus we know that the "$\odot$" operation on the left-hand side of this equation is defined (and the result is known to be contained in $\mathcal{V}$), while the two terms on the right-hand side not only can be added with $\oplus$, but their "vector sum" is also in $\mathcal{V}$. It seems a little fussy, but all legal (and perhaps somewhat legalistic).

In the same way, to make a big thing about "associative" laws always seems a little strange. But remember that the operations we are dealing with here are "binary": they operate on only *two* elements at a time. The associative laws group elements in pairs (so the operations are defined in the first place), and then show that it makes no difference how the grouping occurs.

It is important to note that there are actually *four* types of operations used in this table: in addition to the $\odot$ operation between numbers and vectors, and the $\oplus$ between vectors themselves, there appear "+" and "·" between the real numbers; again, we are just being careful about which what operations apply to what elements. Actually, what all of the equalities in the Table imply is not just that the operations on each side are defined, but that they are, in fact, *equal*!

[1] Note (see page xiii in the Foreword) that "optional" material is set off by using a smaller typeface.

Property (V5) appears a little strange at first. What is significant is that the number "1" is that real number which, when it [real-]multiplies any *other* real number, gives the second number back—the "multiplicative identity element" for real numbers. That property states that "1" will also give $\boldsymbol{v}$ back when it *scalar multiplies* $\boldsymbol{v}$.

It all seems a little fussy to the applied scientist and engineer, but what is really being done is to "abstract" the *structure* of a given set of numbers, to see the essence of what makes these numbers combine the way they do. This is really what the term "algebra" (and thus "linear algebra") means to the mathematician. While it doesn't actually help to solve any problems, it does free one's perspective from the prejudice of experience with one or another type of arithmetic. This, in turn, allows one to analyze the essence of a given system, and perhaps discover unexpected results unencumbered by that prejudice.

In fact, the "G" in the identifying letters actually refers to the word "group"—itself a one-operation algebraic structure *defined* by these properties. The simplest example is the integers: they form a group under addition (though not multiplication: there is no multiplicative inverse; fractions—*rational numbers*—are required to allow this). In this regard, the last property, (G5), defined the group as a "commutative" or "Abelian" group. We will shortly be dealing with the classic instance of an operation which forms a group with appropriate elements but which is *not* Abelian: the set of $n \times n$ matrices under multiplication. The [matrix] product of two matrices **A** and **B** is another square matrix [of the same dimension, so *closed*], but generally **AB** $\neq$ **BA**.

One last note is in order: All of the above has been couched in terms of "real numbers" because this is the application in which we are primarily interested. But, in keeping with the perspective of this section, it should be noted that the actual definition of a vector space is not limited to this set of elements to define scalar multiplication. Rather, the elements multiplying vectors are supposed to come from a "field"—yet another algebraic structure. Unfortunately, a field is defined in terms of still *another* algebraic structure, the "*ring*":

Recall that a vector space as defined above really dealt with *two* sets of elements: elements of the set $\mathcal{V}$ itself, and the [set of] real numbers; in addition, there were two operations on *each* of these sets: "addition," "$\oplus$" and "+" respectively, and "multiplication," "$\cdot$" on the reals and "$\odot$" *between* the reals and elements of $\mathcal{V}$. In view of the fact that we are examining the elements which "scalar multiply" the vectors, it makes sense that we focus on the two operations "+" and "$\cdot$" on those elements:

Definition 1.1.1. A *ring* is a set of elements $\mathcal{R}$ and two operations "+" and "$\cdot$" such that elements of $\mathcal{R}$ form an Abelian group under "+," and, for arbitrary $\boldsymbol{r_1}$, $\boldsymbol{r_2}$, and $\boldsymbol{r_3}$ in $\mathcal{R}$, "$\cdot$" satisfies the three properties in Table 1.2:

(RM1)	$r_1 \cdot r_2 \in \mathcal{R}$ (closure)
(RM2)	$(r_1 \cdot r_2) \cdot r_3 = r_1 \cdot (r_2 \cdot r_3)$ (associativity)
(RM3)	$(r_1 + r_2) \cdot r_3 = (r_1 \cdot r_3) + (r_2 \cdot r_3)$ (distribution)
(RM4)	$r_1 \cdot (r_2 + r_3) = (r_1 \cdot r_2) + (r_1 \cdot r_3)$ (distribution)

Table 1.2: Properties of Multiplication in Rings

Note that the "$\cdot$" operation is *not commutative* (though the "+" operation is, by the ring definition that the elements and "+" form an Abelian group); that's why there are two "distributive" laws. Again we return to the example of square matrices: we can define the "+" operation to be an element-wise sum, and the "·" operation to be ordinary matrix multiplication; then matrices, under these two operations, do in fact form a ring (in which, as noted above, the multiplication operation is non-commutative).

Now that we have dispatched the ring, we can finally define the *field*:

Definition 1.1.2. A *field* is a ring whose non-0 elements form an Abelian group under multiplication ("·").

Recall that "0" is the *additive* identity element—that which, when "added" to any element returns that same element; thus, in effect, this definition is meant to ensure that the "reciprocal" of an element exists while precluding division by 0. Clearly, real numbers (under the usual definitions of addition and multiplication) form a field; so do the rational numbers (fractions). On the other hand, the set of integers (positive *and* negative ones—otherwise there is no "additive inverse" for each) form a ring under the usual operations, but *not* a field: there is no multiplicative inverse without fractions.

Having finally sorted out what a "field" is, we are ready to view vector spaces from the most general perspective: The elements used to scalar multiply element of the vector space must arise from a *field*, $\mathcal{F}$. In this context, we refer to the vector space as consisting of elements from one set, $\mathcal{V}$, on which vector addition is defined, in combination with another set, $\mathcal{F}$, from which the "scalars" come in order to define scalar multiplication; the resulting vector space is referred to as a *vector space over the field* $\mathcal{F}$ when we wish to emphasize the identity of those elements which scalar multiply the "vectors." In all the below examples, and almost exclusively in this book, we shall be talking about vector spaces over the *reals*. Again, while a vector space consists technically of the set $\mathcal{V}$ of "vectors," the field $\mathcal{F}$ of "scalars" [and the operations of "addition" and "multiplication" defined on *it*!] *and* the two operations, "$\oplus$" and "$\odot$" of "vector addition" on $\mathcal{V}$ and "scalar multiplication" between $\mathcal{F}$ and $\mathcal{V}$—which really comprises another set of two spaces and two operations $(\mathcal{V}, \mathcal{F}, \oplus, \odot)$—we shall typically employ the more economical [and perhaps abusive] description of simply "the vector space $\mathcal{V}$" where the field and operations are all implicitly understood. This is a reasonably common practice and exactly what was done in the examples below. One must keep in mind, however, that there are really *four* types of elements required to define a vector space completely, and that the various operations on the sets $\mathcal{V}$ and $\mathcal{F}$ are distinct.

Now having this general definition of a vector space, we are armed to see how many familiar mathematical quantities actually qualify as such a space. Recall that in order to define the space in question, we are obliged to define 1) the elements and 2) the [two] operations; if they satisfy the properties in Table 1.1, they are, in fact, a vector space. Though these examples will be identified by the *elements* comprising the space, the actual vector space involves both the elements *and* the operations on it.

Example 1.1.1. *Directed Line Segments:* under the operations of head-to-tail "addition" and "scalar multiplication" by the reals. This is the classic case, characterized by position, velocity, force, *etc.* |[2]

[2]To set off the Examples, they are terminated by the symbol "|."

Example 1.1.2. *Expressions of the form* $a_1\boldsymbol{u}_1 + a_2\boldsymbol{u}_2 + \ldots + a_n\boldsymbol{u}_n$*:* [Note that we are not calling the $\boldsymbol{u}_i$ *vectors*!] If we define "addition" of such forms

$$(a_1\boldsymbol{u}_1 + a_2\boldsymbol{u}_2 + \ldots + a_n\boldsymbol{u}_n) \oplus (b_1\boldsymbol{u}_1 + b_2\boldsymbol{u}_2 + \ldots + b_n\boldsymbol{u}_n) \equiv (a_1 + b_1)\boldsymbol{u}_1 + (a_2 + b_2)\boldsymbol{u}_2 + \cdots + (a_n + b_n)\boldsymbol{u}_n,$$

and "scalar multiplication"

$$c \odot (a_1\boldsymbol{u}_1 + a_2\boldsymbol{u}_2 + \ldots + a_n\boldsymbol{u}_n) \equiv (c \cdot a_1)\boldsymbol{u}_1 + (c \cdot a_2)\boldsymbol{u}_2 + \cdots + (c \cdot a_n)\boldsymbol{u}_n,$$

the properties defining a vector space are preserved. (Note that closure results from the definitions sending one *form* into the *same form*.) |

The $\boldsymbol{u}_1$ here act merely as "place keepers." This is effectively the same role played by the ordered nature of the next example:

Example 1.1.3. *[Ordered] Row/Column n-tuples:* Our "vectors" consist of arrays (here columns) of real numbers. We define (almost obviously) addition

$$\begin{pmatrix} x_1 \\ x_2 \\ \vdots \\ x_n \end{pmatrix} \oplus \begin{pmatrix} y_1 \\ y_2 \\ \vdots \\ y_n \end{pmatrix} \equiv \begin{pmatrix} x_1+y_1 \\ x_2+y_2 \\ \vdots \\ x_n+y_n \end{pmatrix}$$

(where the simple "+" is that for real numbers) and scalar multiplication

$$c \odot \begin{pmatrix} x_1 \\ x_2 \\ \vdots \\ x_n \end{pmatrix} \equiv \begin{pmatrix} c \cdot x_1 \\ c \cdot x_2 \\ \vdots \\ c \cdot x_n \end{pmatrix}.$$

By the corresponding properties of real numbers, all the vector properties are satisfied. |

Example 1.1.4. $m \times n$ *Matrices:* Simply generalizing the above definitions for addition—note that these arrays are "ordered" just like their row or column counterparts—

$$\begin{pmatrix} a_{11} & a_{12} & \ldots & a_{1n} \\ a_{21} & a_{22} & \ldots & a_{2n} \\ \vdots & \vdots & & \vdots \\ a_{m1} & a_{m2} & \ldots & a_{mn} \end{pmatrix} \oplus \begin{pmatrix} b_{11} & b_{12} & \ldots & b_{1n} \\ b_{21} & b_{22} & \ldots & b_{2n} \\ \vdots & \vdots & & \vdots \\ b_{m1} & b_{m2} & \ldots & b_{mn} \end{pmatrix} \equiv \begin{pmatrix} a_{11}+b_{11} & a_{12}+b_{12} & \ldots & a_{1n}+b_{1n} \\ a_{21}+b_{21} & a_{22}+b_{22} & \ldots & a_{2n}+b_{2n} \\ \vdots & \vdots & & \vdots \\ a_{m1}+b_{m1} & a_{m2}+b_{m2} & \ldots & a_{mn}+b_{mn} \end{pmatrix}$$

and scalar multiplication

$$c \odot \begin{pmatrix} a_{11} & a_{12} & \dots & a_{1n} \\ a_{21} & a_{22} & \dots & a_{2n} \\ \vdots & \vdots & & \vdots \\ a_{m1} & a_{m2} & \dots & a_{mn} \end{pmatrix} \equiv \begin{pmatrix} c \cdot a_{11} & c \cdot a_{12} & \dots & c \cdot a_{1n} \\ c \cdot a_{21} & c \cdot a_{22} & \dots & c \cdot a_{2n} \\ \vdots & \vdots & & \vdots \\ c \cdot a_{m1} & c \cdot a_{m2} & \dots & c \cdot a_{mn} \end{pmatrix}$$

—we see that such objects, too, form a vector space. |

The above examples are almost mundane. But now we shall see several more examples of "function spaces" showing that even certain *functions* can be viewed as being vector spaces.

Example 1.1.5. *n^{th}-degree Polynomials:* We regard polynomials of [at most] n^{th}-degree to be the elements:

$$f(x) = a_0 + a_1 x + a_2 x^2 + \ldots + a_n x^n$$

and define, again obviously, our addition:

$$f_1(x) \oplus f_2(x) \equiv (a_0 + b_0) + (a_1 + b_1)x + (a_2 + b_2)x^2 + \ldots + (a_n + b_n)x^n$$

and scalar multiplication

$$c \odot f(x) \equiv c \cdot a_0 + c \cdot a_1 x + c \cdot a_2 x^2 + \ldots + c \cdot a_n x^n.$$

Again, we have a vector space. (Note that both operations give n^{th}-order polynomials—*i.e.* they are *closed.*)

Once again, the x^n effectively play the role of mere place keepers. |

Example 1.1.6. *Solutions to Homogeneous n^{th}-order Linear Differential Equations:* Given the nth-order linear ordinary differential equation in independent variable t

$$x^{[n]} + a_{n-1}(t)x^{[n-1]} + \cdots + a_o(t)x = 0,$$

because of the linearity, any two solutions $x_1(t)$ and $x_2(t)$, under the above definitions of [functional] "vector addition" and "scalar multiplication" will satisfy this equation. Thus these operations satisfy closure and we can regard the set of all such solutions [to this system] as comprising a vector space. This important topic will return in Section 2.4. |

Example 1.1.7. *Periodic Functions with Period T:* Without belaboring the point any further, such functions are closed under the type of functional addition and multiplication by scalars defined above—they are still periodic with period T—and thus form a vector space. |

The last example will comprise the entire next chapter:

Example 1.1.8. *Linear Transformations from One Vector Space to Another:* In order to discuss this, we must first define what we mean by a "linear transformation":

Definition 1.1.3. A *linear transformation* from a vector space $\mathcal{V}_1$ to another vector space $\mathcal{V}_2$ is a function, L, on $\mathcal{V}_1$ taking $\boldsymbol{v}_1$ and $\boldsymbol{v}_2 \in \mathcal{V}_1$:

$$L(a\boldsymbol{v}_1 + b\boldsymbol{v}_2) = aL(\boldsymbol{v}_1) + bL(\boldsymbol{v}_2) \tag{1.1-1}$$

(Note that the the $\boldsymbol{v}_i$ and "+" on the left-hand side of this equation refer to $\mathcal{V}_1$; in particular, the "+" is "vector addition" on $\mathcal{V}_1$. On the right-hand side, on the other hand, $L(\boldsymbol{v}_i)$ are in $\mathcal{V}_2$ and "+" is "vector addition" defined in *that* space. Similarly, the "scalar products" one each side are those appropriate to the two spaces. [To be technical in the spirit of the previous discussion of "fields," we must have both vector spaces defined over the *same* field!])

For any such linear transformations L, L_1 and L_2, we can define vector addition of two such transformations

$$(L_1 \oplus L_2)(\boldsymbol{v}) \equiv L_1(\boldsymbol{v}) + L_2(\boldsymbol{v})$$

and scalar multiplication

$$(c \odot L) \equiv cL(\boldsymbol{v})$$

(Again, the "addition" and "scalar multiplication" on the right-hand side of the above two defining equations are those in $\mathcal{V}_2$.) By the definition of linear transformations (1.1-1), $(L_1 \oplus L_2)$ and $(c \odot L)$ will also satisfy this definition and be linear transformations; thus the above definitions of vector addition and scalar multiplication are closed, and satisfy the defining properties of a vector space. |

Homework:

1. Show the set of linear transformations comprises a vector space.

Many of the above examples might appear at first to be somewhat artificial. The point is, however, that they all, at least formally, satisfy the criteria of being a vector space, and we shall return to them later to exemplify the concept of basis.

1.1.2 Subspaces of a Vector Space

We have seen that a vector space consists of several components: a set of elements comprising the space itself, a "field" of elements (along with its own operations) with which we can "scalar multiply" elements of the vector space, and a pair of operations defining "vector addition" and "scalar multiplication" which satisfy certain criteria. Typically, we concentrate on the first, assuming the others implicitly. We shall briefly focus on the identity of that set of elements here, too.

Given the field and operations, all elements of a vector space satisfy the properties of Table 1.1; in particular, application of the operations to elements of the vector space result in other elements *in* the vector space—the *closure* properties (G1) and (V1) in the Table. But it might happen that a certain *subset* of elements in the space, identified by some criterion to be a "proper subset" of that space (so forming a "smaller" set not equal to the original space) will, when operated upon by the same operations, give back elements *in the subset.* A brief example might be in order to illustrate this idea:

Example 1.1.9. Consider the space of row 4-tuples—elements of the form (v_1, v_2, v_3, v_4)—under the operations defined for such elements in Example 1.1.3. Elements of the form $(v_1, v_2, 0, v_4)$ are clearly elements of this space; but the results of vector addition and scalar multiplication on such elements return an element of *exactly the same form.* |

Clearly such elements satisfy properties (G2)-(G5) and (V2)-(V5) in the Table: they *inherit* these properties by virtue of being elements of the vector space in the large. But they also enjoy a unique *closure* property in that they *remain* in the subset under the general operations. Such elements then, through this closure property, *form a vector space in their own right*: they satisfy *all* the properties of the Table *including closure.* Elements which retain their identity under the general operations seem special enough to warrant their own category:

Definition 1.1.4. A [proper] subset of elements of a vector space which remains closed under the operations of that vector space are said to form a [vector] *subspace* of that space.

Though at this stage such elements and the "subspace" they comprise might seem merely a curiosity, they are relatively important, and it is useful to have a criterion to identify them when they occur. Since it is the closure property which characterizes them, it is no surprise that the test depends only on closure:

Theorem 1.1.1. A subset of a vector space forms a subspace if and only if $((a \otimes \boldsymbol{v}_1) \oplus (b \otimes \boldsymbol{v}_2))$, for all "scalars" a and b and all $\boldsymbol{v}_1$ and $\boldsymbol{v}_2$ in the subspace, is itself an element of the subspace.

Homework:

1. Prove this.

1.2 The Basis of a Vector Space

This chapter has been motivated by appeal to our experience with "Cartesian" vectors, in which we introduce the fundamental set of vectors $\hat{\boldsymbol{\imath}}$, $\hat{\boldsymbol{\jmath}}$, and $\hat{\boldsymbol{k}}$ in terms of which to represent all *other* vectors in the form $\boldsymbol{A} = A_x\hat{\boldsymbol{\imath}} + A_y\hat{\boldsymbol{\jmath}} + A_z\hat{\boldsymbol{k}}$. [Note that after all the care exercised in the previous section to differentiate between the *scalar* "+" and "·" and *vector* "$\oplus$" and "$\odot$," we are reverting to a more common, albeit sloppier, notation. The distinction between these should be clear from context.] In terms of the "algebra" discussed in the previous section, we see that each part of this sum is the scalar products of a real with a vector (so a vector by Property (V1) in Table 1.1), and the resulting vectors are summed using *vector* addition, so yielding a vector (by Property (G1) of the same Table); this is the reason the two defining operations on vectors are so important. Such a vector sum of scalar products is referred to as a *linear combination* of the vectors.

While these fundamental vectors $\hat{\imath}$, $\hat{\jmath}$, and $\hat{k}$ form an "orthonormal set"—mutually perpendicular *unit* vectors—we shall examine in this section the question of how much we might be able to relax this convention: Do the vectors *have* to be unit vectors? Or perpendicular? And how many such vectors are required in the first place? All of these questions revolve around the issue of a *basis* for a vector space.

In essence, a basis is a *minimal* set of vectors, *from the original vector space*, which can be used to represent *any* other vector from that space as a linear combination. Rather like Goldilocks, we must have just the right number of vectors, neither too few (in order to express *any* vector), nor too many (so that it is a *minimal* set). The first of these criteria demands that the basis vectors (in reasonably descriptive mathematical jargon) *span* the set; the second that they be *linearly independent*. The latter concept is going to return continually in the next chapter and is likely one of the most fundamental in all of linear algebra.

1.2.1 Spanning Sets

Though we have already introduced the term, just to keep things neat and tidy, let us formally define the concept of linear combination:

Definition 1.2.1. A *linear combination* of a set of vectors $\{\boldsymbol{v}_1, \boldsymbol{v}_2, \ldots, \boldsymbol{v}_n\}$ is a vector sum of scalar products of the vectors: $a_1\boldsymbol{v}_1 + a_2\boldsymbol{v}_2 + \ldots + a_n\boldsymbol{v}_n$.

We note three special cases of this definition given the set $\{\boldsymbol{v}_1, \boldsymbol{v}_2, \ldots, \boldsymbol{v}_n\}$:

1. *Multiplication of a [single] vector by a scalar:* $\boldsymbol{v}_i$ can be multiplied by a scalar c by simply forming the linear combination with $a_i \equiv c$ and $a_j \equiv 0$ for $j \neq i$.

2. *Addition of two vectors:* $\boldsymbol{v}_i + \boldsymbol{v}_j$ is just the linear combination in which $a_i = a_j = 1$ and $a_k \equiv 0$ for $k \neq i, j$.

   Thus far, there's nothing new here: the above two are just the fundamental operations on vector spaces, cast in the language of linear combinations. But:

3. *Interchange of two vectors in an ordered set:* If the set is an *ordered* set of vectors $(\boldsymbol{v}_1, \ldots, \boldsymbol{v}_i, \ldots, \boldsymbol{v}_j, \ldots)$. The claim is that we can put this into the

form $(\boldsymbol{v}_1, \ldots, \boldsymbol{v}_j, \ldots, \boldsymbol{v}_i, \ldots)$ through a sequence of linear combinations:

$$\begin{aligned}(\boldsymbol{v}_1, \ldots, \boldsymbol{v}_i, \ldots, \boldsymbol{v}_j, \ldots) &\rightarrow (\boldsymbol{v}_1, \ldots, \boldsymbol{v}_i, \ldots, \boldsymbol{v}_j + \boldsymbol{v}_i, \ldots)\\ &\rightarrow (\boldsymbol{v}_1, \ldots, -\boldsymbol{v}_j, \ldots, \boldsymbol{v}_j + \boldsymbol{v}_i, \ldots)\\ &\rightarrow (\boldsymbol{v}_1, \ldots, -\boldsymbol{v}_j, \ldots, \boldsymbol{v}_i, \ldots)\\ &\rightarrow (\boldsymbol{v}_1, \ldots, \boldsymbol{v}_j, \ldots, \boldsymbol{v}_i, \ldots).\end{aligned}$$

Thus the concept of "linear combination" is more general that the definition alone might suggest.

With this, it is easy to formalize the concept of spanning:

Definition 1.2.2. A set of vectors $\{\boldsymbol{u}_1, \boldsymbol{u}_2, \ldots, \boldsymbol{u}_n\}$ in a vector space $\mathcal{V}$ is said to *span* the space iff every $\boldsymbol{v} \in \mathcal{V}$ can be written as a linear combination of the $\boldsymbol{u}$'s: $\boldsymbol{v} = v_1\boldsymbol{u}_1 + v_2\boldsymbol{u}_2 + \ldots + v_n\boldsymbol{u}_n$.

We give two examples of this. The first is drawn from the space of directed line segments. We purposely consider non-orthonormal vectors:

Example 1.2.1 *Directed Line Segments:* Consider, in the plane containing them, the directed line segment vectors $\boldsymbol{u}_1$, $\boldsymbol{u}_2$ with lengths 1 and 2 respectively and making an angle of 60° with each other, and a vector $\boldsymbol{A}$ of length 2 making an angle of 135° with $\boldsymbol{u}_1$. We are asked to express $\boldsymbol{A}$ as a linear combination of $\boldsymbol{u}_1$ and $\boldsymbol{u}_2$.

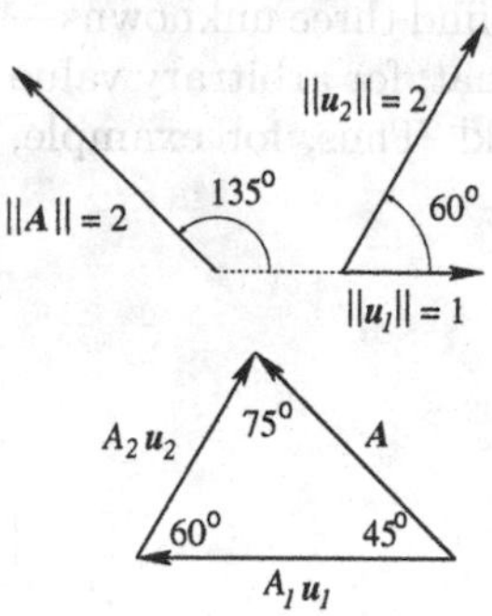

We desire A_1 and A_2 such that $\boldsymbol{A} = A_1\boldsymbol{u}_1 + A_2\boldsymbol{u}_2$. Since we are dealing with directed line segments, it is necessary to use the "head-to-tail" definition of vector addition, shown at right in the diagram above. Since we know all the angles in the resulting triangle, trigonometry tells us that

$$\frac{\|\boldsymbol{A}\|}{\sin 60°} = \frac{\|A_1\boldsymbol{u}_1\|}{\sin 75°} = \frac{\|A_2\boldsymbol{u}_2\|}{\sin 135°}$$

or, since $\|A_i\boldsymbol{u}_i\| = |A_i|\,\|\boldsymbol{u}_i\|$,[3] [by the definition of scalar multiplication for this vector space],

$$\frac{2}{\sin 60°} = \frac{|A_1|(1)}{\sin 75°} = \frac{|A_2|(2)}{\sin 45°}.$$

Solving, we get that $|A_1| = 2.231$ and $|A_2| = 0.816$. Taking into account that $A_1\boldsymbol{u}_1$ reverses the direction of $\boldsymbol{u}_1$ so that A_1 must be negative, we obtain finally that $\boldsymbol{A} = -2.231\boldsymbol{u}_1 + 0.816\,\boldsymbol{u}_2$. |

Note in the above example that the **0**-vector—just a point—can only be expressed as the linear combination $0\ \boldsymbol{u}_1 + 0\ \boldsymbol{u}_2$: there is no way that the two $\boldsymbol{u}$'s can be added together with non-0 coefficients to close the vector triangle.

The next example considers column 2-tuples to be the vector space:

[3] Note the distinction between the *norm*—the magnitude—of a *vector*, $\|.\|$, and the *absolute value*, $|.|$, of a *scalar*.

Example 1.2.2. *n-tuples (Over-spanned Space):* Consider the vectors

$$u_1 \equiv \begin{pmatrix}1\\0\end{pmatrix} \quad u_2 \equiv \begin{pmatrix}0\\1\end{pmatrix} \quad u_3 \equiv \begin{pmatrix}1\\1\end{pmatrix}$$

and the vector in this space

$$v \equiv \begin{pmatrix}3\\2\end{pmatrix}.$$

Express v as a linear combination of the u's.

Solution. Again, we seek the scalars v_1, v_2, and v_3 such that

$$v \equiv \begin{pmatrix}3\\2\end{pmatrix} = v_1\begin{pmatrix}1\\0\end{pmatrix} + v_2\begin{pmatrix}0\\1\end{pmatrix} + v_3\begin{pmatrix}1\\1\end{pmatrix}$$

or, by the definitions of both scalar multiplication and vector addition for this space

$$\begin{pmatrix}v_1+v_3\\v_2+v_3\end{pmatrix} = \begin{pmatrix}3\\2\end{pmatrix}.$$

In this case—we have only two equation through which to find three unknowns—there is no unique solution[4]; indeed, *any* v_1 and v_2 such that, for arbitrary value of v_3, satisfy $v_1 = 3 - v_3$ and $v_2 = 2 - v_3$ will work! Thus, for example, $v = 2u_1 + u_2 + u_3 = 4u_1 + 3u_2 - u_3 = \ldots etc.$

In this case, the $\mathbf{0}$-vector, simply the 2-tuple

$$\begin{pmatrix}0\\0\end{pmatrix}$$

can be expressed using any $v_1 = -v_3$ and $v_2 = -v_3$. |

In both the examples above, the given vectors spanned the respective spaces; in one case the trigonometry yielded a unique linear combination, while in the other solution of the linear system did not. The problem with the second case is obvious: there were "too many" vectors; indeed, for this example u_3 was superfluous, in the sense that $u_3 = u_1 + u_2$. But this is not an issue of simply having "the same number of spanning vectors as the 'dimension' of the space" (whatever the latter means) since, for example, had one of the two vectors in the first example been the $\mathbf{0}$, they wouldn't have spanned the space at all! Both of these ideas are unified in the next section.

1.2.2 Linear Independence

In the second example above, we observed that u_3 could be expressed in terms of u_1 and u_2; *i.e.*, there were c_1 and c_2 such that we could write

$$u_3 = c_1 u_1 + c_2 u_2.$$

[4]See Section 2.3

But why select $\mathbf{u}_3$, when we could just as easily have selected one of the others to be solved for in terms of the remaining two? In order to avoid distinguishing one vector from the rest, note that we could write the above equation in a slightly more symmetric form:

$$c_1\mathbf{u}_1 + c_2\mathbf{u}_2 - \mathbf{u}_3 = \mathbf{0}$$

—one which more equitably treats the three vectors, yet is true if and only if one can be solved for in terms of the others. This leads to the formal definition of *linear independence*:

Definition 1.2.3. A set of vectors $\{\mathbf{v}_1, \mathbf{v}_2, \ldots, \mathbf{v}_n\}$ is *linearly dependent* iff there are $\{c_1, c_2, \ldots, c_n\}$ *not all zero* such that $c_1\mathbf{v}_1 + c_2\mathbf{v}_2 + \ldots + c_n\mathbf{v}_n = \mathbf{0}$.

Clearly, if all the c_i in the above linear combination *are* 0, each term in the vector sum will vanish and so will the sum itself; this is the reason for the stricture that they not all vanish. Again, we see that this is equivalent to being able to solve for one vector in terms of the others: if, in particular, $c_{i_o} \neq 0$, the above allows us to solve for $\mathbf{v}_{i_o}$ in terms of the other $\mathbf{v}_i$:

$$\mathbf{v}_{i_o} = -\frac{1}{c_{i_o}} \sum_{i \neq i_0} c_i \mathbf{v}_i.$$

A *linearly independent* set is, quite simply (and logically), one that is not linearly *de*pendent! But, in application, it is useful to state the contrapositive of the above definition explicitly:

Definition 1.2.4. A set of vectors $\{\mathbf{v}_1, \mathbf{v}_2, \ldots, \mathbf{v}_n\}$ is *linearly independent* iff $c_1\mathbf{v}_1 + c_2\mathbf{v}_2 + \ldots + c_n\mathbf{v}_n = \mathbf{0}$ implies that $c_i = 0$ *for all i*.

Now we see the importance of the observation in the Examples of the previous section that $\mathbf{0}$ could not, or could, respectively, be written as a linear combination of the constituent vectors: this is precisely the test for linear independence! Using the above definition, it is easy to show the following useful results:

Theorem 1.2.1. Any set of vectors $\{\mathbf{v}_1, \mathbf{v}_2, \ldots, \mathbf{v}_n\}$ containing the zero vector $\mathbf{0}$ is linearly dependent.

Theorem 1.2.2. Any *subset* of a set $\{\mathbf{v}_1, \mathbf{v}_2, \ldots, \mathbf{v}_n\}$ of linearly *in*dependent vectors is also linearly independent.

Homework:

1. Prove these two Theorems.

Warning: Many students may first have seen "linear dependence" in a course on differential equations, in connection with the "Wronskian" in that field. Reliance on that criterion alone as a "quick-and-dirty" criterion for linear independence is shortsighted however: In the first place, it evaluates the determinant of *numbers*, useful only in the case of row or column n-tuples, Example 1.1.3 above; how, for example, could we have used this to show the linear independence of directed line segments, Example 1.2.1? Secondly, even if our vector space *does* consist of such n-tuples with numerical components, the determinant is only defined for a *square* matrix; what if one is examining the linear independence of *fewer* (or *more*) than n vectors? The above definitions, however, work in both cases.

We shall now consider another criterion for linear independence. Unlike the above definition, however, it will be applicable only to the n-tuple row or column vectors, and thus be apparently limited in scope. Actually, as we shall see, it can be applied to *any* vectors (Theorem 1.3.2); and in the discussion of matrices [of real numbers], it is of fundamental utility.

A Test for Linear Independence of n-tuples: Reduction to Echelon Form

The rationale for this test is to consider a set of n-tuple vectors and determine the c_i in the definition, 1.2.4, through a systematic process rather than by solving a set of linear equations as done in Example 1.2.2. This process, in turn, reduces the set of vectors to a standard form, *echelon form*, whose individual vectors reflect precisely the linear independence of the original set. The technique rests on the following:

Theorem 1.2.3. The set of m n-tuples

$$\boldsymbol{v}_1 = \begin{pmatrix} 1 \\ 0 \\ \vdots \\ 0 \\ v_{1,m+1} \\ \vdots \\ v_{1n} \end{pmatrix}, \ \boldsymbol{v}_2 = \begin{pmatrix} 0 \\ 1 \\ \vdots \\ 0 \\ v_{2,m+1} \\ \vdots \\ v_{2n} \end{pmatrix}, \ \ldots, \ \boldsymbol{v}_m = \begin{pmatrix} 0 \\ 0 \\ \vdots \\ 1 \\ v_{m,m+1} \\ \vdots \\ v_{mn} \end{pmatrix} \tag{1.2-1}$$

is linearly independent.

Homework:

1. Prove this.

The idea is to examine the linear independence of the set of vectors by trying to reduce it, through a sequence of linear combinations of the vectors among themselves, to the above "*echelon*"[5] form. This can best be seen by an example:

Example 1.2.3. *Reduction to Echelon Form:* Investigate the linear independence of the vectors

$$\mathbf{v}_1 = \begin{pmatrix} 3 \\ 2 \\ -5 \\ 1 \end{pmatrix},\ \mathbf{v}_2 = \begin{pmatrix} 9 \\ -4 \\ 15 \\ 4 \end{pmatrix},\ \mathbf{v}_3 = \begin{pmatrix} 3 \\ -8 \\ 25 \\ 2 \end{pmatrix},\ \mathbf{v}_4 = \begin{pmatrix} -6 \\ 8 \\ -10 \\ 7 \end{pmatrix}.$$

Solution. Noting the "standard form" of the above vectors (1.2-1), we get a 1 in the first entry in $\mathbf{v}_1$ and 0 in that entry in the rest by performing the following operations—linear combinations of the vectors with $\mathbf{v}_1$:

$$\mathbf{v}_1' \equiv \mathbf{v}_1/3: \quad \mathbf{v}_2' \equiv \mathbf{v}_2 - 3\mathbf{v}_1: \quad \mathbf{v}_3' \equiv \mathbf{v}_3 - \mathbf{v}_1: \quad \mathbf{v}_4' \equiv \mathbf{v}_4 + 2\mathbf{v}_1:$$

$$\begin{pmatrix} 1 \\ 2/3 \\ -5/3 \\ 1/3 \end{pmatrix} \quad \begin{pmatrix} 0 \\ -10 \\ 30 \\ 1 \end{pmatrix} \quad \begin{pmatrix} 0 \\ -10 \\ 30 \\ 1 \end{pmatrix} \quad \begin{pmatrix} 0 \\ 12 \\ -20 \\ 9 \end{pmatrix}$$

—then get 1 in the second entry of the second vector and corresponding 0's in the rest with $\mathbf{v}_2'$:

$$\mathbf{v}_1'' \equiv \mathbf{v}_1' + 2\mathbf{v}_2'/30: \quad \mathbf{v}_2'' \equiv \mathbf{v}_2'/(-10): \quad \mathbf{v}_3'' \equiv \mathbf{v}_3' - \mathbf{v}_2': \quad \mathbf{v}_4'' \equiv \mathbf{v}_4' + 6\mathbf{v}_2'/5:$$

$$\begin{pmatrix} 1 \\ 0 \\ 1/3 \\ 2/5 \end{pmatrix} \quad \begin{pmatrix} 0 \\ 1 \\ -3 \\ -1/10 \end{pmatrix} \quad \begin{pmatrix} 0 \\ 0 \\ 0 \\ 0 \end{pmatrix} \quad \begin{pmatrix} 0 \\ 0 \\ 16 \\ 51/6 \end{pmatrix}$$

—and finally, since $\mathbf{v}_3'' \equiv \mathbf{0}$, eliminate the third entry in the others with $\mathbf{v}_4''$:

$$\mathbf{v}_1''' \equiv \mathbf{v}_1'' - \mathbf{v}_4''/48: \quad \mathbf{v}_2''' \equiv \mathbf{v}_2'' + 3\mathbf{v}_4''/16: \quad \mathbf{v}_3''' \equiv \mathbf{v}_3'': \quad \mathbf{v}_4''' \equiv \mathbf{v}_4''/16:$$

$$\begin{pmatrix} 1 \\ 0 \\ 0 \\ 107/480 \end{pmatrix} \quad \begin{pmatrix} 0 \\ 1 \\ 0 \\ 239/160 \end{pmatrix} \quad \begin{pmatrix} 0 \\ 0 \\ 0 \\ 0 \end{pmatrix} \quad \begin{pmatrix} 0 \\ 0 \\ 1 \\ 51/96 \end{pmatrix}.$$

This demonstrates that only three of the original four vectors were linearly independent, but it actually generates a linear combination proving the linear dependence by Definition 1.2.3:

$$\mathbf{0} = \mathbf{v}_3'' \equiv \mathbf{v}_3' - \mathbf{v}_2' \equiv (\mathbf{v}_3 - \mathbf{v}_1) - (\mathbf{v}_2 - 3\mathbf{v}_1) = \mathbf{v}_3 - \mathbf{v}_2 + 2\mathbf{v}_1. \tag{1.2-2}$$

[5]from the French *échelon*, literally "rung of a ladder." Note that many use the term "*reduced* echelon" form to refer to that in which the leading term in normalized. We invariably normalize so do not make such a distinction.

Note that even if the order of the $\boldsymbol{v}_i$ had been different—say that $\boldsymbol{v}_3$ and $\boldsymbol{v}_4$ had been switched—the same [now *fourth*] vector would have vanished, giving the correspondingly same linear combination. Furthermore, the "echelon form" at the end would be exactly the same except for the ordering. |

Now having exemplified the process, we must justify it. The technique is somewhat indirect, forming linear combinations of the vectors in the set to yield either [linearly independent] vectors in the form of Theorem 1.2.3 or $\mathbf{0}$'s. Further, we are concluding that the vectors in the *original* set *corresponding* to the new, linearly combined, vectors, are precisely those which are linearly independent. Clearly what is necessary first is a theorem like the following:

Theorem 1.2.4. A set of vectors $\boldsymbol{v}_1, \boldsymbol{v}_2, \ldots, \boldsymbol{v}_n$ are linearly independent iff there exist n linearly independent linear combinations $\boldsymbol{V}_i = \sum_j V_{ij}\boldsymbol{v}_j$ of the vectors.

In fact, this is exactly what we shall prove. But before doing so, we need another fundamental result:

Theorem 1.2.5. In a space spanned by the n linearly independent vectors $\boldsymbol{v}_i$, $i = 1, \ldots, n$, there exist at most n linearly independent linear combinations $\boldsymbol{V}_i = \sum_{j=1}^n V_{ij}\boldsymbol{v}_j$ of them. (*Note:* this does *not* say the $\boldsymbol{V}_i$ are unique—there can be any number of different $\{\boldsymbol{V}_i\}$; only that each such set contains at most n linearly independent vectors!)

Proof. We know at least one set of n linear combinations—the $\{\boldsymbol{v}_i,\ i = 1, \ldots, n\}$ themselves—are linearly independent; assume [they or any other linear combinations] $\{\boldsymbol{V}_i = \sum_{j=1}^n V_{ij}\boldsymbol{v}_j, i = 1 \ldots n\}$ are linearly independent and consider the set comprising another $\boldsymbol{V}_{n+1} = \sum_{j=1}^n V_{(n+1)j}\boldsymbol{v}_j$ amended to them. Is this new set linearly dependent or independent; *i.e.*, under what circumstances does $\sum_{i=1}^{n+1} c_i\boldsymbol{V}_i$ vanish? Note that we can assume $\boldsymbol{V}_{n+1} \neq \mathbf{0}$; otherwise the amended set would be linearly dependent already, by Theorem 1.2.1. In the same spirit, we can also assume $c_{n+1} \neq 0$ in this sum: if it did, the linear independence of the original $\boldsymbol{V}_i$ would force their c_i to vanish. Thus we can "normalize" the linear independence criterion, $\sum_{i=1}^{n+1} c_i\boldsymbol{V}_i = \mathbf{0}$, to $\sum_{i=1}^n c_i\boldsymbol{V}_i + \boldsymbol{V}_{n+1} = \mathbf{0}$ by dividing the first by $c_{n+1} \neq 0$.

Now if this last form vanishes,

$$\mathbf{0} = \sum_{i=1}^n c_i\boldsymbol{V}_i + \boldsymbol{V}_{n+1} = \sum_{i=1}^n c_i \sum_{j=1}^n V_{ij}\boldsymbol{v}_j + \sum_{j=1}^n V_{(n+1)j}\boldsymbol{v}_j$$
$$= \sum_{j=1}^n \left(\sum_{i=1}^n c_i V_{ij} + V_{(n+1)j} \right) \boldsymbol{v}_j.$$

We see immediately that the c_i cannot vanish for all i; if they did, the sum in parentheses would vanish, collapsing the entire last sum to $\boldsymbol{V}_{n+1}$ and so forcing $\boldsymbol{V}_{n+1} \equiv \mathbf{0}$ and violating our assumption that it isn't. Thus there are $c_i \neq 0$ such that the first sum in the series vanishes, making $\{\boldsymbol{V}_i,\ i = 1, \ldots, n+1\}$

linearly dependent, and proving that the original n are the most which can be linearly independent. □

Note: This is a fundamental result—one which will return in our discussion of "dimension" of a vector space in Section 1.2.3.

We are now ready to prove the above theorem:

Proof of Theorem 1.2.4. The "implication" is trivial: Assume the $\{\mathbf{v}_i\}$ are linearly independent; then the linear combinations

$$\begin{aligned} \mathbf{V}_1 &\equiv (1)\mathbf{v}_1 + (0)\mathbf{v}_2 + \cdots + (0)\mathbf{v}_n \\ \mathbf{V}_2 &\equiv (0)\mathbf{v}_1 + (1)\mathbf{v}_2 + \cdots + (0)\mathbf{v}_n \\ &\vdots \\ \mathbf{V}_n &\equiv (0)\mathbf{v}_1 + (0)\mathbf{v}_2 + \cdots + (1)\mathbf{v}_n \end{aligned}$$

are linearly independent.

Conversely, assume the set $\{\mathbf{V}_i \equiv \sum_j V_{ij}\mathbf{v}_j, i = 1 \ldots n\}$ are linearly independent, and consider the two spaces

$$\begin{aligned} \mathcal{v} &\equiv \{\text{all vectors spanned by the } \mathbf{v}_i,\ i = 1, \ldots, n\} \\ \mathcal{V} &\equiv \{\text{all vectors spanned by the } \mathbf{V}_i,\ i = 1, \ldots, n\}. \end{aligned}$$

While all n of the $\mathbf{V}_i$ are linearly independent by hypothesis, *at most* n of the $\mathbf{v}_i$ are; whatever number are is also the maximum number of linearly independent linear combinations $\mathcal{v}$ can support, by the above Theorem. But the vectors $\{\mathbf{V}_i,\ i = 1, \ldots, n\} \subseteq \mathcal{V}$; and linear combinations *of* linear combinations of the $\mathbf{v}_i$ are themselves linear combinations of the $\mathbf{v}_i$, so $\mathcal{V} \subseteq \mathcal{v}$. Thus the n $\mathbf{V}_i \in \mathcal{v}$, so *at least* n of the spanning $\mathbf{v}_i$ must be linearly independent, or Theorem 1.2.5 is violated. Thus *all* n of the $\mathbf{v}_i$ must be linearly independent. □

This shows that, in the Example 1.2.3 above, having obtained three linearly independent linear combinations (the $\mathbf{v}_i'''$) of the original vectors $\mathbf{v}_i$, three of the latter were, in fact, linearly independent. But it *doesn't* show that the first, second, and fourth were the linearly independent ones. This is a result of the systematic way the echelon form was obtained:

Recall that we seek to have the ith vector have a 1 in its ith row, with 0's above and below. But the 0's *above* the ith row result from linear combinations of vectors *preceding* the ith vector, while those below come from linear combinations with those *following* it. If a vector is "cleared" in the process of trying to zero out entries above the ith, such as the the third vector was in the second iteration in Example 1.2.3 was, it is the result of linear combinations with vectors *only preceding it*; we have yet to operate on entries *below* the ith using the following ones. This is shown explicitly in (1.2-2) in the example. Thus $\mathbf{v}_3$ is linearly dependent on $\mathbf{v}_1$ and $\mathbf{v}_2$ but *not* on $\mathbf{v}_4$ [yet].

[5] Following common practice, proofs are set off by terminating them with a symbol, here "□."

But if we are not interested in *which* vectors are linearly independent (and typically, in application, we aren't), all that is necessary is to obtain the *number* of linearly independent linear combinations, regardless of the order in which the original set was presented. This means that we can "switch" the order of vectors at will in order to standardize that order. (Recall the remark on page 14: such an interchange is *itself* a linear combination!) In particular, this "standardized order" in the present context is one in which the ith vector is reduced to a "1" in its ith row, with 0's below, precisely that in Theorem 1.2.3. This is the approach that will be followed in the sequel.

Having exemplified the procedure and stated the theory on which it rests, we can now organize it into a systematic technique for examining the linear independence of a set of m vectors. This will be done inductively, assuming that we have already reduced i vectors in the set to the form (1.2-1) through the ith row:

$$\underbrace{\underbrace{\begin{pmatrix}1\\0\\\vdots\\0\\v_{1,i+1}\\v_{1,i+2}\\\vdots\\v_{1n}\end{pmatrix},\begin{pmatrix}0\\1\\\vdots\\0\\v_{2,i+1}\\v_{2,i+2}\\\vdots\\v_{2n}\end{pmatrix},\ldots,\begin{pmatrix}0\\0\\\vdots\\1\\v_{i,i+1}\\v_{i,i+2}\\\vdots\\v_{in}\end{pmatrix}}_{i \text{ in standard form}},\begin{pmatrix}0\\0\\\vdots\\0\\v_{i+1,i+1}\\v_{i+1,i+2}\\\vdots\\v_{i+1,n}\end{pmatrix}}_{i+1 \text{ linearly independent}},\ldots,\begin{pmatrix}0\\0\\\vdots\\0\\v_{m,i+1}\\v_{m,i+2}\\\vdots\\v_{mn}\end{pmatrix}$$

(in which the jth component of $\mathbf{v}_i$ is v_{ij}), and are dealing with putting the $(i+1)^{st}$ vector in standard form. (Initially, $i \equiv 0$.) If $v_{i+1,i+1} \neq 0$, we proceed as usual. By performing linear combinations of the preceding i vectors, this set gets reduced to an echelon form

$$\begin{pmatrix}1\\0\\\vdots\\0\\0\\v'_{1,i+2}\\\vdots\\v'_{1n}\end{pmatrix},\begin{pmatrix}0\\1\\\vdots\\0\\0\\v'_{2,i+2}\\\vdots\\v'_{2n}\end{pmatrix},\ldots,\begin{pmatrix}0\\0\\\vdots\\1\\0\\v'_{i,i+2}\\\vdots\\v'_{in}\end{pmatrix},\begin{pmatrix}0\\0\\\vdots\\0\\1\\v'_{i+1,i+2}\\\vdots\\v'_{i+1,n}\end{pmatrix},\ldots,\begin{pmatrix}0\\0\\\vdots\\0\\0\\v'_{m,i+2}\\\vdots\\v'_{mn}\end{pmatrix}$$

It might be the case that the $(i+1)^{st}$ component of $\mathbf{v}_{i+1}$, $v_{i+1,i+1}$, is, in fact, 0; in fact, it might be the case that $v_{i+1,j} = 0$ for *all* $j > i$—*i.e.*, that $\mathbf{v}_{i+1} = \mathbf{0}$—as happened with $\mathbf{v}''_3$ in Example 1.2.3 above, in which case $\mathbf{v}_{i+1}$ is a linear combination of the $\mathbf{v}_j, j \leq i$ to the left of $\mathbf{v}_{i+1}$, and we have shown that only i

of the $i+1$ vectors thus far put in standard form are linearly independent. In any event, if we are only interested in *how many* linearly independent vectors there were originally, we are free to reorder the set by switching $\boldsymbol{v}_{i+1}$ with one of the $\boldsymbol{v}_j$ in which $v_{j,i+1} \neq 0$. (Note that the two vectors being transposed have, to this point, been operated on *only by those to the left of* $\boldsymbol{v}_{i+1}$.) We then normalize that entry and use it to eliminate the $(i+1)^{st}$ entries in the other vectors.

But it is conceivable that, in fact, there *is* no other such vector with a non-0 $(i+1)^{st}$ entry; that, at this stage in the procedure, the vectors have assumed the form

$$\underbrace{\underbrace{\begin{pmatrix}1\\0\\\vdots\\0\\v_{1,i+1}\\v_{1,i+2}\\\vdots\\v_{1n}\end{pmatrix}, \begin{pmatrix}0\\1\\\vdots\\0\\v_{2,i+1}\\v_{2,i+2}\\\vdots\\v_{2n}\end{pmatrix}, \ldots, \begin{pmatrix}0\\0\\\vdots\\1\\v_{i,i+1}\\v_{i,i+2}\\\vdots\\v_{in}\end{pmatrix}}_{\textbf{\textit{i} in standard form}}, \begin{pmatrix}0\\0\\\vdots\\0\\0\\v_{i+1,i+2}\\\vdots\\v_{i+1,n}\end{pmatrix}}_{\textbf{\textit{i}+1 linearly independent \textit{using row} \textit{i}+2!}}, \ldots, \begin{pmatrix}0\\0\\\vdots\\0\\0\\v_{m,i+2}\\\vdots\\v_{mn}\end{pmatrix},$$

with a *0* in the $(i+1)^{st}$ row of not only the $(i+1)^{st}$ vector, but also in all others to its right. We briefly sketch an example of such a case:

Example 1.2.4. *"Skip" case:* Reduce the vectors

$$\boldsymbol{v}_1 = \begin{pmatrix}1\\1\\-1\\0\\11\\5\end{pmatrix}, \boldsymbol{v}_2 = \begin{pmatrix}1\\4\\-7\\-3\\14\\8\end{pmatrix}, \boldsymbol{v}_3 = \begin{pmatrix}1\\0\\1\\4\\34\\7\end{pmatrix}, \boldsymbol{v}_4 = \begin{pmatrix}1\\3\\-5\\-4\\-3\\5\end{pmatrix}, \boldsymbol{v}_5 = \begin{pmatrix}1\\3\\-5\\0\\29\\9\end{pmatrix};$$

to [reduced] echelon form.

Results of the obvious operations using, respectively,

$\underline{\boldsymbol{v}_1}$:

$$\boldsymbol{v}_1' = \begin{pmatrix}1\\1\\-1\\0\\11\\5\end{pmatrix}, \boldsymbol{v}_2' = \begin{pmatrix}0\\3\\-6\\-3\\3\\3\end{pmatrix}, \boldsymbol{v}_3' = \begin{pmatrix}0\\-1\\2\\4\\23\\2\end{pmatrix}, \boldsymbol{v}_4' = \begin{pmatrix}0\\2\\-4\\-4\\-14\\0\end{pmatrix} \boldsymbol{v}_5' = \begin{pmatrix}0\\2\\-4\\0\\18\\4\end{pmatrix}$$

then $\underline{\boldsymbol{v}_2'}$:

$$\boldsymbol{v}_1''=\begin{pmatrix}1\\0\\1\\1\\10\\4\end{pmatrix},\boldsymbol{v}_2''=\begin{pmatrix}0\\1\\-2\\-1\\1\\1\end{pmatrix},\boldsymbol{v}_3''=\begin{pmatrix}0\\0\\0\\3\\24\\3\end{pmatrix},\boldsymbol{v}_4''=\begin{pmatrix}0\\0\\0\\-2\\-16\\-2\end{pmatrix},\boldsymbol{v}_5''=\begin{pmatrix}0\\0\\0\\2\\16\\2\end{pmatrix}$$

yield a "pathological" case: though $\boldsymbol{v}_3''$ is not voided out altogether, neither it *nor* $\boldsymbol{v}_4''$ *nor* $\boldsymbol{v}_5''$ have any entry in their third rows; thus we cannot eliminate terms of this row in the vectors to the left. To continue at all, we must go to the fourth: now operating on $\underline{\boldsymbol{v}_3''}$, it turns out that the last two vectors both void out altogether:

$$\boldsymbol{v}_1'''=\begin{pmatrix}1\\0\\1\\0\\2\\3\end{pmatrix},\boldsymbol{v}_2'''=\begin{pmatrix}0\\1\\-2\\0\\9\\2\end{pmatrix},\boldsymbol{v}_3'''=\begin{pmatrix}0\\0\\0\\1\\8\\1\end{pmatrix},\boldsymbol{v}_4'''=\begin{pmatrix}0\\0\\0\\0\\0\\0\end{pmatrix},\boldsymbol{v}_5'''=\begin{pmatrix}0\\0\\0\\0\\0\\0\end{pmatrix}.$$

We thus have that only three of these column 6-vectors are linearly independent; more importantly (for purposes of this example, in any event), we note the signature "skip form" for these vectors. ∎

In this case we cannot eliminate those components in the $(i+1)^{st}$ entry to the left, because we don't have anything to form linear combinations with! Nevertheless, we can look to the $(i+2)^{nd}$ "row" of entries for a non-0 value and, assuming we find one, continue the process from there. This introduces a "skip" in the final form at the end of this step, yielding a set

$$\begin{pmatrix}1\\0\\\vdots\\0\\v_{1,i+1}\\0\\\vdots\\v_{1n}'\end{pmatrix},\begin{pmatrix}0\\1\\\vdots\\0\\v_{2,i+1}\\0\\\vdots\\v_{2n}'\end{pmatrix},\ldots,\begin{pmatrix}0\\0\\\vdots\\1\\v_{i,i+1}\\0\\\vdots\\v_{in}'\end{pmatrix},\begin{pmatrix}0\\0\\\vdots\\0\\0\\1\\\vdots\\v_{i+1,n}'\end{pmatrix},\ldots,\begin{pmatrix}0\\0\\\vdots\\0\\0\\0\\\vdots\\v_{mn}'\end{pmatrix},$$

but the resulting $i+1$ vectors are still linearly independent. Note that such a skip manifests itself in these *column* n-tuples by having a 0 instead of a 1 in the ith entry of the ith vector along with generally *non*-0 i^{th} entries to the *left* (though still 0's to the *right*).

The process continues until we either a) run out of vectors, b) run out of non-0 "rows," or c) run out of rows altogether. The final set assumes a form in which $\boldsymbol{v}'_i$ has either a 1 or a 0 in its ith entry. In the former case, all *other* vectors have a 0 in this entry; in the latter, either $\boldsymbol{v}'_i \equiv 0$ (showing that the original $\boldsymbol{v}_i$ is a linear combination of those to its left), or $v_{j,i} = 0$ for $j \geq i$ (though with possibly *non*-0 entries for $j < i$)—the "skip case" above. In either event, *the number of linearly independent vectors is the number of "diagonal 1's"*.

Thus, given a set of m n-tuple vectors, there are three possible relations between the two numbers:

1. $m < n$: If there are fewer vectors than the number of entries in each, the "best" we can hope for—that we avoid the "0 row syndrome" mentioned above—is that the final product of this process will be of the form (1.2-1):

$$\begin{pmatrix} 1 \\ 0 \\ \vdots \\ 0 \\ v_{1,m+1} \\ \vdots \\ v_{1n} \end{pmatrix}, \begin{pmatrix} 0 \\ 1 \\ \vdots \\ 0 \\ v_{2,m+1} \\ \vdots \\ v_{2n} \end{pmatrix}, \ldots, \begin{pmatrix} 0 \\ 0 \\ \vdots \\ 1 \\ v_{m,m+1} \\ \vdots \\ v_{in} \end{pmatrix};$$

   i.e., that all m are linearly independent. This is case "a)" above. Even if there are the "skips" mentioned above, there is still the possibility that the original set is linearly independent; but if there are too many skips, we might run out of "rows" first, giving even fewer than m linearly independent combinations of the original ones. In the pathological limit of such a situation, there is *no* non-0 element in *any* of the lower rows from $\boldsymbol{v}_{i+1}$ on; then *all* the remaining vectors are linearly dependent on the preceding ones (case "b)" above).

2. $m = n$: The "best case scenario" here is that we end up with a set of vectors

$$\begin{pmatrix} 1 \\ 0 \\ \vdots \\ 0 \end{pmatrix}, \begin{pmatrix} 0 \\ 1 \\ \vdots \\ 0 \end{pmatrix}, \ldots, \begin{pmatrix} 0 \\ 0 \\ \vdots \\ 1 \end{pmatrix};$$

   i.e., that all m vectors are linearly independent. But "skips" would reduce this number to less than m.

3. $m > n$: Now the best we can hope for is a final set of the form

$$\begin{pmatrix} 1 \\ 0 \\ \vdots \\ 0 \end{pmatrix}, \begin{pmatrix} 0 \\ 1 \\ \vdots \\ 0 \end{pmatrix}, \ldots, \begin{pmatrix} 0 \\ 0 \\ \vdots \\ 1 \end{pmatrix}, \begin{pmatrix} 0 \\ 0 \\ \vdots \\ 0 \end{pmatrix}, \ldots, \begin{pmatrix} 0 \\ 0 \\ \vdots \\ 0 \end{pmatrix};$$

the maximum number of linearly independent columns is n (case "c)"). "Skips" would reduce this number even further, since we would run out of rows earlier.

While all of the above has been couched in terms of column vectors, analogous results are obtained for row vectors: in particular, the above three "best" cases of final form become:

1. $m < n$:
$$\begin{gathered}(1,0,\ldots,0,v_{1,m+1},\ldots,v_{1n})\ ,\\(0,1,\ldots,0,v_{2,m+1},\ldots,v_{2n})\ ,\\\vdots\\(0,0,\ldots,1,v_{m,m+1},\ldots,v_{in})\ ,\end{gathered}$$

2. $m = n$:
$$\begin{gathered}(1,0,\ldots,0)\ ,\\(0,1,\ldots,0)\ ,\\\vdots\\(0,0,\ldots,1,)\ ,\end{gathered}$$

and

3. $m > n$:
$$\begin{gathered}(1,0,\ldots,0)\ ,\\(0,1,\ldots,0)\ ,\\\vdots\\(0,0,\ldots,1)\ ,\\(0,0,\ldots,0)\ ,\\\vdots\\(0,0,\ldots,0)\ ,\end{gathered}$$

respectively. We observe that, in the case of *row* n-tuples, "skips" manifest themselves again by having a 0 in the ith entry in the ith row and generally non-0 entries *above* this; those *below* still vanish, however.

There is one last comment which must be made regarding this procedure for determining the linear dependence/independence of a set of vectors: In forming the intermediate linear combinations $c_i\boldsymbol{v}_i + c_j\boldsymbol{v}_j$ in the ith vectors, *we clearly cannot allow* $c_i = 0$; to do so would eliminate $\boldsymbol{v}_i$ from the set altogether, replacing it by $\mathbf{0}$ which, by Theorem 1.2.1, would make the resulting set linearly *de*pendent, even if the original set wasn't. However trivial this observation might seem, it is important to make it (with a view of later application of the criterion in what follows).

This section appears long, primarily because of the detailed discussion regarding the "echelon test." In point of fact, however, there have been only two fundamental ideas: that of linear dependence and independence, and the test itself for linear independence among a set of row/column vectors; the major theorem in this section merely supports the latter. The key points are what linear independence *is* (and what it *isn't*—the "Wronskian"!), and the fact that linear independence of a set of row/column vectors is equivalent to reduction of that set of vectors to "echelon form." Despite its apparently rather narrow application, this criterion will become fundamental when applied to matrix algebra in the next chapter. But the importance of linear independence cannot be overstated: the concept will pervade all of the sequel. It should be firmly in mind.

Homework:

1. Theorem 1.2.5 emphasizes the fact that there are generally *many* sets of linearly independent linear combinations of the original linearly independent $\mathbf{v}_i$. Show, for example, that the set $\{\mathbf{V}_i \equiv \sum_{j=1}^{i} \mathbf{v}_i\}$ (*e.g.*, $\mathbf{V}_1 \equiv \mathbf{v}_1, \mathbf{V}_2 \equiv \mathbf{v}_1 + \mathbf{v}_2$, *etc.*) form a linearly independent set.
2. Show that if a vector $\mathbf{v} \notin \{$vectors spanned by linearly *in*dependent $\mathbf{v}_i$, $i = 1, \ldots, n\}$, then the set $\{\mathbf{v}_1, \mathbf{v}_2, \ldots, \mathbf{v}_n, \mathbf{v}\}$ are linearly independent.
3. Show that
$$\begin{pmatrix} 3 \\ 2 \end{pmatrix} \text{ and } \begin{pmatrix} -1 \\ 4 \end{pmatrix}$$
are linearly independent.
4. Show that
$$\begin{pmatrix} 3 \\ 2 \end{pmatrix} \text{ and } \begin{pmatrix} -6 \\ -4 \end{pmatrix}$$
are linearly dependent.
5. Show that
$$\begin{pmatrix} 1 \\ -3 \\ 2 \end{pmatrix}, \begin{pmatrix} -1 \\ -11 \\ 4 \end{pmatrix}, \begin{pmatrix} 2 \\ 1 \\ 1 \end{pmatrix}$$
are linearly dependent.
6. Investigate the linear independence of the vectors
$$\begin{pmatrix} 0 \\ 1 \\ -1 \\ 3 \end{pmatrix}, \begin{pmatrix} 3 \\ 5 \\ 2 \\ 2 \end{pmatrix}, \text{ and } \begin{pmatrix} 6 \\ 7 \\ 7 \\ -5 \end{pmatrix}$$
using
   (a) the *definition* of linear independence, and
   (b) reduction to "echelon form."
   (c) If it turns out the above vectors are linearly *de*pendent, find a linear combination of them which vanishes.

7. Show that the *matrices* (viewed as a vector space)

$$\begin{pmatrix} 1 & -3 \\ 4 & 2 \end{pmatrix}, \begin{pmatrix} 2 & -3 \\ 6 & -2 \end{pmatrix}$$

are linearly independent.

1.2.3 Bases and the Dimension of a Vector Space

Knowing now the criteria that a given set of vectors is large enough (that it is a spanning set) and that it is not too large (that vectors in the set are linearly independent) we are prepared to define a basis:

Definition 1.2.5. A *basis* for a vector space $\mathcal{V}$ is a linearly independent set of vectors *from* $\mathcal{V}$, $\{\boldsymbol{u}_1, \boldsymbol{u}_2, \ldots, \boldsymbol{u}_n\}$ which span $\mathcal{V}$; *i.e.*, for any $\boldsymbol{v} \in \mathcal{V}$, $\boldsymbol{v}$ can be written as a linear combination of the $\boldsymbol{u}_i$: $\boldsymbol{v} = v_1\boldsymbol{u}_1 + v_2\boldsymbol{u}_2 + \ldots + v_n\boldsymbol{u}_n$.

The vectors need be neither "normalized" to unit magnitude nor "orthogonal" like most systems with which the reader is familiar (though this choice does have certain conveniences); in fact, the concept of orthogonality only has meaning for vectors of the class of directed line segments (though a related criterion, the "inner product," can be used more generally). In this regard, the set of *non*-orthonormal vectors in Example 1.2.1, since they both spanned the planar space considered and were linearly independent, constitute a basis for that space. A point not often emphasized, however, is that the basis vectors must be *from the space itself.*

With the concept of basis in hand, we are ready to define the concept of *dimension* of a vector space:

Definition 1.2.6. The *dimension* of a vector space is the number of vectors in a basis for that space.

The concept of dimension is important, in that it characterizes the vector space in a sense we shall discuss below. For the time being, however, we shall obtain some important results regarding dimension; in particular, we must explain why the above definition states merely "*a* basis." Though we mention one example of a vector space of infinite dimension, the following will characterize finite-dimensional vector spaces:

Theorems on Dimension

The first of these is really just a restatement of Theorem 1.2.5 in the present context. In that theorem we considered "a space spanned by the n linearly independent vectors"—precisely what we now define to be a *basis.* Though we could leave it at that, we will give an alternate proof:

Theorem 1.2.6. The dimension of a vector space is the maximum number of linearly independent vectors in that space.

Proof. Assume the space has dimension n—*i.e.*, it has a basis $\{\boldsymbol{u}_1, \boldsymbol{u}_2, \ldots, \boldsymbol{u}_n\}$ of linearly independent vectors which spans it. We claim that there is no *other* vector $\boldsymbol{v}$ independent of the $\{\boldsymbol{u}_i\}$.

For assume there were a $\boldsymbol{v}$ such that $\{\boldsymbol{u}_1, \boldsymbol{u}_2, \ldots, \boldsymbol{u}_n, \boldsymbol{v}\}$ were linearly independent. We must take $\boldsymbol{v} \neq \boldsymbol{0}$ (for otherwise, by Theorem 1.2.1, the set would be linearly dependent from the start). Since the $\{\boldsymbol{u}_i\}$ span the space already, we can write $\boldsymbol{v} = \sum v_i \boldsymbol{u}_i$—where the v_i are not all zero, since otherwise $\boldsymbol{v}$ is! But this means that the linear combination $(1)\boldsymbol{v} - \sum v_i \boldsymbol{u}_i = \boldsymbol{0}$, violating the linear independence assumed. □

Theorem 1.2.7. The dimension of a vector space is unique; *i.e.*, given two bases, $\{\boldsymbol{u}_1, \boldsymbol{u}_2, \ldots, \boldsymbol{u}_m\}$ and $\{\boldsymbol{v}_1, \boldsymbol{v}_2, \ldots, \boldsymbol{v}_n\}$, then $m = n$.

Proof. If $m \neq n$, then either $m < n$ or $m > n$, both violating the above maximality of the number of linearly independent vectors! □

Theorem 1.2.8. *Any* set of n linearly independent vectors forms a basis for a vector space of that dimension.

Proof. Assume $\{\boldsymbol{u}_1, \boldsymbol{u}_2, \ldots, \boldsymbol{u}_n\}$ are linearly independent (so none is $\boldsymbol{0}$!) Then, for any $\boldsymbol{v}$ in the space, the set of $n+1$ vectors $\{\boldsymbol{u}_1, \boldsymbol{u}_2, \ldots, \boldsymbol{u}_n, \boldsymbol{v}\}$ must be linearly *de*pendent, since there are only n linearly *in*dependent ones. Thus $\sum c_i \boldsymbol{u}_i + c_{n+1}\boldsymbol{v} = \boldsymbol{0}$, where not all the $c_i = 0$. In particular, $c_{n+1} \neq 0$: if it were, the $\sum c_i \boldsymbol{u}_i = \boldsymbol{0}$, meaning (by the linear independence of the $\{\boldsymbol{u}_i\}$) that all the $c_i = 0$ for $i \leq n$; but then *all* the c_i vanish. Thus we can solve for $\boldsymbol{v}$:

$$\boldsymbol{v} = -\frac{1}{c_{n+1}} \sum c_i \boldsymbol{u}_i,$$

showing that the $\{\boldsymbol{u}_i\}$ span the space. Since they are already linearly independent, they form a basis. □

This last theorem explains why, in the definition of dimension above, we can use any basis whatsoever.

Now let us return to the examples of vector spaces given in Section 1.1:

Example 1.2.5. *Directed Line Segments:* Any pair of non-parallel directed line segments in 2-D, or any three non-coplanar, non-parallel directed line segments in 3-D form a basis. The dimensions of these two spaces are (not surprisingly) 2 and 3 respectively! |

Example 1.2.6. *Expressions of the form $a_1\boldsymbol{u}_1 + a_2\boldsymbol{u}_2 + \ldots + a_n\boldsymbol{u}_n$:* The vectors $\{\boldsymbol{u}_1, \boldsymbol{u}_2, \ldots, \boldsymbol{u}_n\}$ form a basis. These are elements of the space: $\boldsymbol{u}_i \equiv 1 \cdot \boldsymbol{u}_i$. The space is of dimension n. |

Example 1.2.7. *[Ordered] Row/Column n-tuples:* In keeping with this example earlier, we will consider columns. One convenient basis for these is the so-called "*natural basis*" :

$$\mathbf{e}_1 \equiv \begin{pmatrix} 1 \\ 0 \\ \vdots \\ 0 \end{pmatrix}, \quad \mathbf{e}_2 \equiv \begin{pmatrix} 0 \\ 1 \\ \vdots \\ 0 \end{pmatrix}, \ldots, \mathbf{e}_n \equiv \begin{pmatrix} 0 \\ 0 \\ \vdots \\ 1 \end{pmatrix}; \tag{1.2-3}$$

any $\mathbf{v} = (c_1, c_2, \ldots, c_n)^{\mathsf{T}} = c_1\mathbf{e}_1 + c_2\mathbf{e}_2 + \ldots + c_n\mathbf{e}_n$.[6] Once again, to specify a basis for a vector space, that basis must come *from* the space; here our space is of columns, so the basis must be, too. And, once again, this space is of dimension n. |

Example 1.2.8. *$m \times n$ Matrices:* We can form a "natural basis," analogous to the above, for these, too—[matrix] elements of the original space

$$\mathsf{E}_{11} \equiv \begin{pmatrix} 1 & 0 & \ldots & 0 \\ 0 & 0 & \ldots & 0 \\ \vdots & \vdots & \ldots & \vdots \\ 0 & 0 & \ldots & 0 \end{pmatrix}, \quad \mathsf{E}_{12} \equiv \begin{pmatrix} 0 & 1 & \ldots & 0 \\ 0 & 0 & \ldots & 0 \\ \vdots & \vdots & \ldots & \vdots \\ 0 & 0 & \ldots & 0 \end{pmatrix}, \ldots, \quad \mathsf{E}_{1n} \equiv \begin{pmatrix} 0 & 0 & \ldots & 1 \\ 0 & 0 & \ldots & 0 \\ \vdots & \vdots & \ldots & \vdots \\ 0 & 0 & \ldots & 0 \end{pmatrix}$$

$$\vdots \qquad\qquad\qquad \vdots \qquad\qquad\qquad \vdots \tag{1.2-4}$$

$$\mathsf{E}_{m1} \equiv \begin{pmatrix} 0 & 0 & \ldots & 0 \\ 0 & 0 & \ldots & 0 \\ \vdots & \vdots & \ldots & \vdots \\ 1 & 0 & \ldots & 0 \end{pmatrix}, \quad \mathsf{E}_{m2} \equiv \begin{pmatrix} 0 & 1 & \ldots & 0 \\ 0 & 0 & \ldots & 0 \\ \vdots & \vdots & \ldots & \vdots \\ 0 & 1 & \ldots & 0 \end{pmatrix}, \ldots, \quad \mathsf{E}_{mn} \equiv \begin{pmatrix} 0 & 0 & \ldots & 1 \\ 0 & 0 & \ldots & 0 \\ \vdots & \vdots & \ldots & \vdots \\ 0 & 0 & \ldots & 1 \end{pmatrix}.$$

Here the space is of dimension mn. |

Example 1.2.9. *n^{th}-degree Polynomials:* Here the basis is $\{x^0, x^1, \ldots, x^n\}$, of dimension $n + 1$. |

Example 1.2.10. *Solutions to Homogeneous n^{th}-order Linear Differential Equations:* We have already seen (1.1.6) that the solutions of a given n^{th}-order linear differential equation

$$x^{(n)} + a_{n-1}(t)x^{(n-1)} + \cdots + a_o(t)x = 0$$

constitute a vector space; it is rather less obvious what the dimension of that space will be. Observe, however, that we can write the above equation as one involving a column n-tuple $\mathbf{x} = (x^{(n-1)} \; x^{(n-2)} \; \ldots \; x)^{\mathsf{T}}$:

$$\frac{d\mathbf{x}}{dt} \equiv \frac{d}{dt}\begin{pmatrix} x^{(n-1)} \\ x^{(n-2)} \\ \vdots \\ x \end{pmatrix} = \begin{pmatrix} -a_{n-1}(t)x^{(n-1)} - \cdots - a_o(t)x \\ x^{(n-2)} \\ \vdots \\ x^{(1)} \end{pmatrix} \equiv \mathbf{f}(\mathbf{x}; t);$$

[6] The superscript "T" on a row/column vector $\mathbf{v}$, $\mathbf{v}^{\mathsf{T}}$, denotes the *transpose* of $\mathbf{v}$: if $\mathbf{v}$ is a row vector, $\mathbf{v}^{\mathsf{T}}$ is that *column* vector with the same components and conversely. See Definition 2.1.1 below. It is introduced here just for compactness.

thus, in fact, the solution to this equation is itself an n-tuple—*i.e.*, it lies in a space of dimension n. The *basis* for this space would then consist of *any* set of n linearly independent solutions of the associated equation.

This construction might, at first, appear somewhat contrived and artificial; such reservation likely results from viewing the original differential equation as one in the *scalar* "unknown" x. But recall that, in addition to the differential equation that x satisfies, the solution also invokes the *initial conditions* on not only x but *all its derivatives* (up to the nth, which, once the values of the lower derivatives are specified, is itself known). In effect, we can conceive of a point $\boldsymbol{x}$ in the n-dimensional space of x and its derivatives, evolving from the point $\boldsymbol{x}(0)$ (given by its initial conditions) along the solution $\boldsymbol{x}(t)$. In this context, the n-dimensional space of $\boldsymbol{x}$ is referred to as the *phase space.*

This perspective is actually quite fruitful: the theory of ordinary differential equations—even *non*-linear ones—rests on the foundation that, as long as the function $\boldsymbol{f}$ is *continuous*[7], the *first*-order equation $\dot{\boldsymbol{x}} = \boldsymbol{f}(\boldsymbol{x})$ not only *has* a solution satisfying its initial conditions ("existence"), but that solution is *unique.* These results are immediately applicable to the case in which the right-hand side in the above is independent of t—*i.e.*, that the system is "*autonomous.*" But we can treat the more general *non*-autonomous case easily by simply regarding t itself as one component of the unknown vector $\boldsymbol{x}' \equiv (\boldsymbol{x}\ t)^{\mathsf{T}}$: It is trivially the case that

$$\frac{d\boldsymbol{x}'}{dt} \equiv \frac{d}{dt}\begin{pmatrix} x^{(n-1)} \\ x^{(n-2)} \\ \vdots \\ x \\ t \end{pmatrix} = \begin{pmatrix} -a_{n-1}(t)x^{(n-1)} - \cdots - a_o(t)x \\ x^{(n-2)} \\ \vdots \\ x^{(1)} \\ 1 \end{pmatrix} \equiv \boldsymbol{f}'(\boldsymbol{x}')$$

satisfies both the original differential equation and the hypotheses of the fundamental theorem: if $\boldsymbol{f}(\boldsymbol{x};t)$ is continuous in $\boldsymbol{x}$ and $f(t)$ in t, then the above $\boldsymbol{f}'(\boldsymbol{x}')$ is clearly continuous in $\boldsymbol{x}'$. The $(n+1)$-dimensional space of [the original] $\boldsymbol{x}$ and t is referred to as the *extended* phase space. |

Example 1.2.11. *Periodic Functions with Period T:* This example is a rather less prosaic than the preceding ones. By Fourier's theorem, we know that any periodic function can be expanded in a generally infinite series of sines and cosines in the argument $2\pi x/T$; thus the set of all such functions—themselves periodic of period T—form an *infinite*-dimensional space! |

The observant reader will note that the last example in Section 1.1, Example 1.1.8, has been omitted. This is not a oversight. While the first two "function" vector spaces above are somewhat "concrete" with regard to establishing a basis, the last is more general and difficult to pin down. And since

[7]Actually, the result is shown for a somewhat less restrictive condition that $\boldsymbol{f}$ satisfy a "*Lipschitz condition,*" but continuous or *piecewise* continuous functions are the norm in most physical applications.

such transformations are dignified by their own chapter, it seems appropriate to forestall discussion of their basis until that point.

We note in passing that all the above, which has been applied to "vector spaces" will clearly also hold for *sub*spaces of a vector space: as vector spaces in their own right, subspaces will have their own bases and dimensions, with the latter generally less than the space in which they are contained. Note, however, that the basis of the full space will have, among its elements, basis elements of its subspaces; conversely, knowing the existence of a subspace and its basis, it is possible to build up a basis for the *full* space by augmenting the subspace basis with other, linearly independent, vectors.

Summary. This section has been, admittedly, a little technical. The key points, however, are terribly important: A *basis* is *any* set of *linearly independent* vectors *from the space* which *span* it; *i.e.*, any vector can be expressed as a *linear combination* of the basis vectors. The *dimension* of the space is the number of vectors in *a* basis, or, equivalently, the *maximum* number of linearly independent vectors in that space. While the basis may not be unique, the dimension is.

As mentioned in the Prolog to this Part, the concept of linear independence is important and has already assumed a central position in the development. This section also presents a procedure, the echelon test, to ascertain whether a given set of vectors are linearly independent or not. This, too, will return in the sequel.

1.3 The Representation of Vectors

We have earlier observed that writing vectors in the form

$$\boldsymbol{v} = v_1\boldsymbol{u}_1 + v_2\boldsymbol{u}_2 + \ldots + v_n\boldsymbol{u}_n \tag{1.3-1}$$

is more than a mere convenience: it allows us to operate on the $\boldsymbol{v}$'s *through manipulation of the* v_i rather than the original vectors or even the $\boldsymbol{u}_i$. If the v_i are real (or even complex) scalars, such manipulation becomes effectively trivial compared with, say, the original operations on the original vectors. Thus, as a matter of almost necessary convenience, this is the methodology employed. In order to discuss this approach, it is necessary to introduce a little jargon:

Definition 1.3.1. We call the right-hand side of (1.3-1) the *representation of* $\boldsymbol{v}$ *relative to the* $\{\boldsymbol{u}_i\}$; the $\{v_i\}$ are called the *components of* $\boldsymbol{v}$ *relative to the* $\{\boldsymbol{u}_i\}$.

Note that in order to be able to write *any* vector in the form (1.3-1), it is necessary that the $\boldsymbol{u}_i$ *span* the space—see Section 1.2.1. But in Example 1.2.2 of that section, we saw that such a representation might not be unique; this fact would certainly make it more difficult to identify vectors in a given representation, greatly complicating the very process we are trying to simplify!

But we also showed in the latter example that the $\boldsymbol{u}_i$ were *linearly dependent*, rather than the "minimal" set required for a *basis*. We might then suspect that this problem of non-uniqueness would not arise for a proper basis, and we would, in fact, be right:

Theorem 1.3.1. The representation of a vector relative to a basis is unique.

Proof. Assume there *were* two representations for a vector $\boldsymbol{v} \in \mathcal{V}$ with basis $\{\boldsymbol{u}_i\}$:

$$\boldsymbol{v} = v_1\boldsymbol{u}_1 + v_2\boldsymbol{u}_2 + \ldots + v_n\boldsymbol{u}_n = v'_1\boldsymbol{u}_1 + v'_2\boldsymbol{u}_2 + \ldots + v'_n\boldsymbol{u}_n.$$

But then

$$(v_1 - v'_1)\boldsymbol{u}_1 + (v_2 - v'_2)\boldsymbol{u}_2 + \cdots + (v_n - v'_n)\boldsymbol{u}_n = \boldsymbol{0}$$

—which means, by the linear independence of the basis, that $(v_i - v'_i) = 0$, or $v_i = v'_i$, for all i. □

This is precisely the reason that, if a vector $\boldsymbol{v} = v_1\boldsymbol{u}_1 + v_2\boldsymbol{u}_2 + \ldots + v_n\boldsymbol{u}_n = v'_1\boldsymbol{u}_1 + v'_2\boldsymbol{u}_2 + \ldots + v'_n\boldsymbol{u}_n$ *relative to the same basis*, we can conclude that $v'_i = v_i$: the representation is *unique*. Thus henceforth we shall assume that all representations are being made relative to a *basis*, precisely so that there is a *single* representation for each vector in the space under consideration.

But if the representation relative to a *particular* basis is unique, recall that the choice of basis itself is *not*: *any* set of linearly independent vectors spanning the space will qualify. Thus it becomes important to know not only the representation, but also *which basis* is referenced in that representation:

Example 1.3.1 *Non-unique basis:* We again return to the space of directed line segments, and consider the vector $\boldsymbol{A}$ of length 2 in Example 1.2.1, as well as three possible choices of basis: a) those given in the original example; and two pairs of *orthonormal* vectors, b) with $\hat{\boldsymbol{u}}_1$ parallel to the original $\boldsymbol{u}_1$ and the c) rotated by 45° from that:

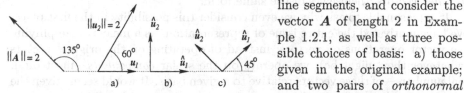

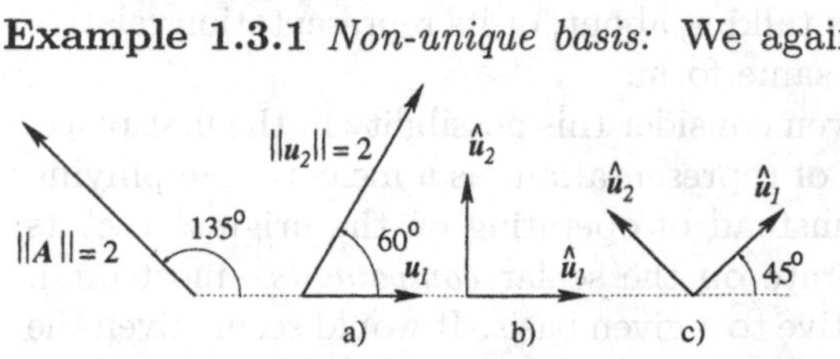

Relative to the first basis, $\boldsymbol{A} = -2.231\boldsymbol{u}_1 + 0.816\boldsymbol{u}_2$ (see Example 1.2.1), relative to the second $\boldsymbol{A} = \sqrt{2}(-\boldsymbol{u}_1 + \boldsymbol{u}_2)$, while relative to the third, $\boldsymbol{A} = 2\boldsymbol{u}_2$. The *representations* are different, but $\boldsymbol{A}$ is the same in all three cases. |

Throughout, however, it must be kept in mind that the *original* vector, regardless of what vectors it might be expressed in terms of, remains the *same*: Though the *representation* of the vector might change dramatically depending on which basis vectors it is referred to, the original vector does not. And, since representations relative to different bases represent the same vector, different representations can, in fact, be equated to one another (though we cannot equate *components* of that vector in the different bases).

1.3.1 n-tuple Representations of Vectors

Heretofore we have considered the basis to be an arbitrary collection of vectors $\{\boldsymbol{u}_i\}$. But if we choose to *order* the set and form $(\boldsymbol{u}_1, \boldsymbol{u}_2, \ldots, \boldsymbol{u}_n)$ so that $\boldsymbol{u}_1$ corresponds to the first element of the set, $\boldsymbol{u}_2$ corresponds to the second, *etc.*, it is possible to express the representation much more economically by simply expressing the *components* in a similarly-ordered set—*i.e.*, an *n-tuple.* Thus, relative to the *ordered* set $(\boldsymbol{u}_1, \boldsymbol{u}_2, \ldots, \boldsymbol{u}_n)$, the vector $\boldsymbol{v} = v_1\boldsymbol{u}_1 + v_2\boldsymbol{u}_2 + \ldots + v_n\boldsymbol{u}_n$ can be expressed simply as

$$\begin{pmatrix} v_1 \\ v_2 \\ \vdots \\ v_n \end{pmatrix}.$$

(For reasons which will become clearer in the next chapter, we generally express vectors as *column* n-tuples rather than as rows.) Thus, for example, in the above Example 1.3.1, the three representations of $\boldsymbol{A}$ could *themselves* be represented as

$$\begin{pmatrix} -2.231 \\ 0.816 \end{pmatrix}, \begin{pmatrix} -\sqrt{2} \\ \sqrt{2} \end{pmatrix}, \text{ and} \begin{pmatrix} 0 \\ 2 \end{pmatrix},$$

respectively. Again, keep in mind that all three *represent* the *same* vector, even though they "look" different, and could be equated (though not component-wise).

This is all well and good for most vectors, but if the original vector space *itself* consisted of n-tuples, some confusion can arise: one n-tuple becomes represented as yet *another* n-tuple! Thus it becomes critical to know whether, in such a space, it is the *original* vector one is talking about, or its representation relative to *other* vectors, themselves of the same form.

It is reasonable to ask why we even consider this possibility in the first place: We have motivated the entire issue of representations as a means of simplifying the vector operations on vectors; instead of operating on the original vectors (whatever they might be), we operate on the scalar *components*—most often real numbers—of those vectors relative to a given basis. It would seem, given the inherent simplicity of operation on n-tuples, that it would hardly be necessary to introduce any basis different from the "natural basis" (see (1.2-3)) in terms of which such n-tuples are all implicitly defined in the first place. But there are problems in the next chapter in which this is done, just to engrain the issue of which basis a given vector is referred to. We illustrate this idea with an example:

Example 1.3.2. Consider, in the space of 2-tuples, the vector

$$\boldsymbol{v} = \begin{pmatrix} 2 \\ 0 \end{pmatrix}$$

Note that this is *already* represented in terms of the "natural basis" [of 2-tuples]

$$\boldsymbol{e}_1 \equiv \begin{pmatrix} 1 \\ 0 \end{pmatrix}, \; \boldsymbol{e}_2 \equiv \begin{pmatrix} 0 \\ 1 \end{pmatrix};$$

indeed,

$$\boldsymbol{v} = 2\boldsymbol{e}_1 + 0\boldsymbol{e}_2 \text{ or, as a 2-tuple: } \begin{pmatrix} 2 \\ 0 \end{pmatrix}.$$

Now consider this same vector in, as it were, an *un*natural basis

$$\boldsymbol{u}_1 \equiv \begin{pmatrix} 1 \\ 1 \end{pmatrix}, \; \boldsymbol{u}_2 \equiv \begin{pmatrix} 1 \\ 0 \end{pmatrix}.$$

[These are, indeed, linearly independent. Using the definition 1.2.4:

$$c_1\boldsymbol{u}_1 + c_2\boldsymbol{u}_2 \equiv \begin{pmatrix} c_1 + c_2 \\ c_1 \end{pmatrix} = \mathbf{0}$$

iff c_1, and therefore c_2, both vanish.]

To represent $\boldsymbol{v}$ in terms of $\boldsymbol{u}_1$ and $\boldsymbol{u}_2$, we must have

$$\boldsymbol{v} = \begin{pmatrix} 2 \\ 0 \end{pmatrix} = c_1\boldsymbol{u}_1 + c_2\boldsymbol{u}_2 \equiv c_1 \begin{pmatrix} 1 \\ 1 \end{pmatrix} + c_2 \begin{pmatrix} 1 \\ 0 \end{pmatrix} = \begin{pmatrix} c_1 + c_2 \\ c_1 \end{pmatrix}$$

Solving, again $c_1 = 0$ but now $c_2 = 2$; thus, in terms of $\boldsymbol{u}_1$ and $\boldsymbol{u}_2$,

$$\boldsymbol{v} = 0\boldsymbol{u}_1 + 2\boldsymbol{u}_2 \text{ or, as a 2-tuple: } \begin{pmatrix} 0 \\ 2 \end{pmatrix}$$

Note the dramatically different representations of the vector relative to the two bases! |

A rather superficial reading of this example might suggest that

$$\begin{pmatrix} 2 \\ 0 \end{pmatrix} = \begin{pmatrix} 0 \\ 2 \end{pmatrix}, \tag{1.3-2}$$

but this would be wrong: the first is the *original* vector $\boldsymbol{v}$ [in the *natural* basis], while the second is its *representation* in terms of $\boldsymbol{u}_1$ and $\boldsymbol{u}_2$. The way out of this confusion could be notation to suggest what signifies what. To the extent it is possible, the distinction between a *vector* (of whatever ilk) and its [n-tuple] *representation* will be to denote the first with Roman boldface and the latter with sans-serif bold; thus, in the above example,

$$\boldsymbol{v} = \begin{pmatrix} 2 \\ 0 \end{pmatrix}; \mathsf{v} = \begin{pmatrix} 0 \\ 2 \end{pmatrix}. \tag{1.3-3}$$

This doesn't completely obviate the confusion: strictly speaking, the representation would also have to specify the particular *basis* (itself specified in terms of the *natural* one!) in which that representation was valid, but doing so would soon become cumbersome. Thus we will *finesse* this fussy practice, being satisfied with merely indicating when a given n-tuple *is* a representation.

A simple but important instance of these ideas—one unfortunately liable to misinterpretation—is the representation of the *basis* vectors themselves. Since each basis vector in the [*ordered*] set $(\boldsymbol{u}_1, \boldsymbol{u}_2, \ldots, \boldsymbol{u}_n)$ can be represented $\boldsymbol{u}_i = (0)\boldsymbol{u}_1 + (0)\boldsymbol{u}_2 + \cdots + (1)\boldsymbol{u}_i + \cdots + (0)\boldsymbol{u}_n$, that vector's n-tuple representation $\mathbf{u}_i$ [note notation] in terms of *that* basis will *always* be of the form

$$i \rightarrow \begin{pmatrix} 0 \\ 0 \\ \vdots \\ 0 \\ 1 \\ 0 \\ \vdots \\ 0 \end{pmatrix} = \mathbf{u}_i, \tag{1.3-4}$$

regardless of the nature of the $\boldsymbol{u}_i$. The confusion can arise because of the identity of form with the *natural basis* of n-tuples, equation (1.2-3): that equation is a *basis* for *any* n-tuple, the above is the *representation* of *any* basis vector in *terms* of such n-tuples. In any event, *neither* is [necessarily] the *Cartesian* basis $(\hat{\boldsymbol{\imath}}, \hat{\boldsymbol{\jmath}}, \hat{\boldsymbol{k}})$ (though (1.3-4) is assuredly the n-tuple *representation* of these vectors). As if all that weren't bad enough, one, can certainly write, in comparison with equations (1.3-2) and (1.3-3), that $\mathbf{e}_i = \boldsymbol{e}_i$!

The student should be certain the above distinctions are understood.

Representations allow us to deal with n-tuples rather than the original vectors. They also allows application of the "echelon test" in Section 1.2.2 to *any* class of vectors—to the extent that vectors whose [n-tuple] *representations* are linearly independent are *themselves* linearly independent. We now show that:

Theorem 1.3.2. A set of vectors $\{\boldsymbol{v}_1, \boldsymbol{v}_2, \ldots, \boldsymbol{v}_m\}$ is linearly independent iff their n-tuple representations relative to an arbitrary basis are.

Proof. Assume the $\{\boldsymbol{v}_1, \boldsymbol{v}_2, \ldots, \boldsymbol{v}_m\}$ are in a space with basis $(\boldsymbol{u}_1, \boldsymbol{u}_2, \ldots, \boldsymbol{u}_n)$; *i.e.*, we can represent each

$$\boldsymbol{v}_i = \sum_j v_{ji} \boldsymbol{u}_j$$

and consider the column representation of each:

$$\mathbf{v}_i \equiv \begin{pmatrix} v_{1i} \\ v_{2i} \\ \vdots \\ v_{ni} \end{pmatrix}.$$

We show the "implication" by assuming the $\{\boldsymbol{v}_i\}$ are linearly independent. Now if a linear combination of their representations were to vanish,

$$\sum_i c_i \mathbf{v}_i = \begin{pmatrix} \sum c_i v_{1i} \\ \sum c_i v_{2i} \\ \vdots \\ \sum c_i v_{ni} \end{pmatrix} \equiv \begin{pmatrix} C_1 \\ C_2 \\ \vdots \\ C_n \end{pmatrix} = \mathbf{0}, \tag{1.3-5}$$

then the *same* linear combination of the $\{\boldsymbol{v}_i\}$ would:

$$\mathbf{0} = \sum_j C_j \boldsymbol{u}_j \equiv \sum_j \left(\sum_i c_i v_{ji} \right) \boldsymbol{u}_j = \sum_i c_i \left(\sum_j v_{ji} \boldsymbol{u}_j \right) = \sum_i c_i \boldsymbol{v}_i.$$

But, by the assumed linear independence of the $\{\boldsymbol{v}_i\}$, this means that each $c_i = 0$ here and in (1.3-5), showing that the $\{\mathsf{v}_i\}$ in the latter are linearly independent.

Conversely, assume the $\{\mathsf{v}_i\}$ are linearly independent. Were a linear combination of the $\{\boldsymbol{v}_i\}$ to vanish,

$$\mathbf{0} = \sum_i c_i \boldsymbol{v}_i = \sum_i \left(\sum_j v_{ji} \boldsymbol{u}_j \right) = \sum_j \left(\sum_i c_i v_{ji} \right) \boldsymbol{u}_j \equiv \sum_j C_j \boldsymbol{u}_j, \qquad (1.3\text{-}6)$$

then, by the linear independence of the basis vectors $(\boldsymbol{u}_j)$, each C_j would have to vanish. But then

$$\mathbf{0} \equiv \begin{pmatrix} C_1 \\ C_2 \\ \vdots \\ C_n \end{pmatrix} \equiv \begin{pmatrix} \sum c_i v_{1i} \\ \sum c_i v_{2i} \\ \vdots \\ \sum c_i v_{ni} \end{pmatrix} = \sum_i c_i \begin{pmatrix} v_{1i} \\ v_{2i} \\ \vdots \\ v_{ni} \end{pmatrix} \equiv \sum_i c_i \mathsf{v}_i,$$

which would mean the $c_i = 0$ here and in (1.3-6), by the assumed linear independence of the $\{\mathsf{v}_i\}$. Thus the $\{\boldsymbol{v}_i\}$ are linearly independent. □

1.3.2 Representations and Units

Most physical quantities are not purely numerical; they require one or another set of *units* in terms of which to measure them. Thus units must somehow enter the representation of vector quantities demanding units. There is a somewhat subtle point, rarely mentioned, regarding just how this is to be done.

Consider and arbitrary $\boldsymbol{v} = v_1\boldsymbol{u}_1 + v_2\boldsymbol{u}_2 + \ldots + v_n\boldsymbol{u}_n$. We have two options regarding where the units for this vector are to be introduced: we can either

- attach units to the *components*, v_i, leaving the basis vectors *unitless*, OR
- attach units to the *basis* vectors, leaving the components unitless.

The former is almost invariably the tack taken: it allows vectors of various physical type (position, velocity, *etc.*) to be expressed in a single vector space, laying the burden of unit identification on the components. Thus, for example, the familiar Cartesian 3-D basis ($\hat{\boldsymbol{\imath}}$, $\hat{\boldsymbol{\jmath}}$, and $\hat{\boldsymbol{k}}$) consists of *unitless* unit vectors.

The alternative, attaching units to the basis vectors, would identify the vector space with a certain physical quantity; thus we could speak of the "position space," the "angular velocity space," *etc.*, in which the respective quantities would have *unitless* components. But implementing this would soon lead to chaos: were one to attach, say, length units to a basis, the components of position vectors would be unitless, while those of velocity would have units $[T^{-1}]$, while moments' components would have units of *force* and forces units of $[M/T^2]$—a situation lacking any aesthetic appeal whatsoever. And that doesn't even begin to address the issue of what happens if we choose to *change* our length units!

The point is that basis vectors will here universally be chosen to be unitless. And any change of basis will thus require no reference to whatever units are chosen to measure any physical quantity expressed in terms of that basis.

1.3.3 Isomorphisms among Vector Spaces of the Same Dimension

We have now seen how any vector in an n-dimensional vector space (one with n basis vectors, recall) can be represented as an n-tuple if we order the basis by simply arraying the components of a given vector relative to that basis in the same order. In effect, then, to each vector in the original space there is a representation—a *unique* representation by Theorem 1.3.1—in terms of the basis, and thus in its n-tuple representation. But this means that there is a one-to-one relation between vectors in an n-dimensional space and the n-tuples. Further, relating the vector operations in the [original] vector space to those of the n-tuple space using Example 1.1.2 (where now we *do* mean the $\boldsymbol{u}_i$ to be vectors!) and Example 1.1.3, we see that the the result of operations in the original space correspond to those in the n-tuple space. Such a relationship between the elements and operations on two [different] vectors spaces is called an *isomorphism* between them. What we have observed, then, is that any n-dimensional vector space is isomorphic to that of n-tuples. But that means that any *two* different n-dimensional vector spaces are isomorphic to [the same] space of n-tuples, and thus to each other. Thus we have all but proved the following:

Theorem 1.3.3. All vector spaces of the same dimension are isomorphic to each other.

In a highfalutin mathematical fashion, this simply says in regard to vector spaces of a given dimension, that "once you've see one vector space you've seen them all"—that you can operate in *any* vector space of the same dimension and get the same results (at least with regard to the fundamental operations on such spaces). This is the reason one often sees vector properties "proven" in *Cartesian* form, with results extended to vectors in, say, *polar* form. And it is the sense in which the dimension of a vector space characterizes it, as mentioned on page 28.

The isomorphisms among all the finite-dimensional vector spaces exemplified in Section 1.1 can be exhibited explicitly, again with reference to the space of n-tuples: The directed line segments in terms of their basis $(\hat{\boldsymbol{\imath}},\hat{\boldsymbol{\jmath}},\hat{\boldsymbol{k}})$ as $\boldsymbol{v}=v_x\hat{\boldsymbol{\imath}} + v_y\hat{\boldsymbol{\jmath}} + v_z\hat{\boldsymbol{k}}$, or as a row 3-tuple as (v_x, v_y, v_z), the "forms" in Example 1.1.2 as $(a_1, a_2, \ldots, a_n)$, the $m \times n$ matrices as a [*long*!] mn-tuple $(a_{11}, a_{12}, \ldots, a_{1m}, a_{21}, a_{22}, \ldots, a_{mn})$, and n^{th}-degree polynomials as $(a_0, a_1, \ldots, a_n)$—all n-tuples of various lengths. In particular, if the various lengths happen to be the same, they are all isomorphic to each other!

It is important to recognize that it is really just this isomorphism which allows us to operate on the n-tuple representation in the first place: we can identify each vector with its n-tuple representation, and then the *result* of *n-tuple* vector addition and scalar multiplication with the corresponding result in the *original* space. This is really the thrust of Theorem 1.3.2.

Summary. The scalars v_i in writing $\boldsymbol{v} = v_1\boldsymbol{u}_1 + v_2\boldsymbol{u}_2 + \ldots + v_n\boldsymbol{u}_n$ relative to a basis $\{\boldsymbol{u}_i\}$ are called the *components* of $\boldsymbol{v}$ relative to that basis, and are *unique* for a given $\boldsymbol{v}$. If we *order* the basis, $(\boldsymbol{u}_i)$, those scalars can themselves be arrayed in a [ordered] n-tuple v, the *representation* of $\boldsymbol{v}$; this representation is then also unique *relative to that basis.* In particular, the basis vectors themselves can be represented by elements of the *natural basis.* But if we represent $\boldsymbol{v}$ relative to a *different* basis, its *representation* will change, even though the original vector remains the same. This distinction must be kept in mind, particularly when the original vectors are *themselves* n-tuples!

The fact that we can represent vectors of any ilk as unique n-tuples becomes significant because the linear independence of the *original* vectors is reflected precisely in that of it *n-tuples.* Thus the echelon test, originally developed only for n-tuples, now can be applied to the n-tuple representations of *arbitrary* vectors.

The units of physical quantities can be carried by either the *basis* vectors or the *components* of vectors relative to that basis. The latter is far less confusing, and it is the practice followed explicitly in this book (and implicitly in most others).

Since the n-tuple representation of any vector in an n-dimensional space is unique relative to its basis, there is a one-to-one relationship between arbitrary n-vectors and n-tuple vectors. In particular, the n-tuples provide an identification—an *isomorphism*—among the vectors of *arbitrary* n-dimensional spaces.

Summary

In view of the fact that the student allegedly "knows vectors," there has been quite a bit of new information—at least a new perspective—contained in this chapter. The key points are that:

- *vectors* are *any* elements which, along with the *defined* operations of "vector addition" and "scalar multiplication," satisfy the properties in Table 1.1
- for any vector space there is a *basis* of vectors in that space which
  - *span* the space
  - are *linearly independent*
  - can *uniquely represent* any vector as a linear combination of the basis vectors, whose scalar factors are *components* of the vector relative to the given basis
  - will generally have a *different* representation relative to a different basis, even though the original vector remains the *same*
- the *dimension* of a vector space is the number of vectors in *any* basis
- vector spaces of the same dimension are *equivalent* ("isomorphic")

- we can represent vectors as *n-tuples* of the components, so that
- linear independence of a set of vectors can be determined by putting their representations [in the *same* basis!] in "*echelon form*"

Chapter 2

Linear Transformations on Vector Spaces

Introduction

Despite the somewhat daunting title of this chapter, we are going to be primarily concerned with *matrices*. But in the same way that we saw in the last chapter that we could *represent* vectors as n-tuples, it will turn out that we can also *represent* linear transformations as matrices. And in the same way that we can easily manipulate a vector of any type by operating on its n-tuple representation of real numbers, we can similarly deal most easily with linear transformations on an arbitrary set of vectors by using their matrix representations. In fact, at the end of the day, even an n-tuple is really a matrix—a *column* matrix (or *row* matrix, if we choose).

We have already seen that the representation of a vector changes with the choice of basis; in some way, this is almost obvious. What is less obvious is that, when we change the basis in terms of which we express *vectors*, we also change the representation of linear *transformations* operating on those vectors. Here is the central issue in this chapter: how do the *representations* of both vectors *and* linear transformations change with a change of basis?

This becomes important in the next Part, where we deal with 3-D rigid body motion. It turns out there that the angular momentum of a rigid body is, effectively, a linear transformation on the angular velocity, one which [linearly] *transforms* the angular velocity to an angular momentum; thus, from what we said above, it can be represented as a matrix—the *inertia* matrix. But it also happens that, because of the special form of this matrix, there is always a particular set of axes—in fact, *orthogonal* axes—in terms of which this matrix representation is particularly simple: a *diagonal* matrix. This choice of axes is called the "principle axes" and is, in the language employed heretofore, simply a change of basis resulting in this simple representation. Thus the next chapter culminates in the process by which we can diagonalize matrices, the conditions under which it can be done in the first place, and what can be done if these conditions are not met.

That process is bound up with the eigenvectors and eigenvalues of the matrix, and whether those eigenvectors are linearly independent; this is why vector independence was emphasized so much in the last chapter. But the determination of eigenvectors hinges on the solution to a "homogeneous" vector equation; under what conditions can such a solution be found? Indeed, under what conditions can the solution to a "*non*-homogeneous" equation (whatever either of these terms means) be found? This question itself is of intrinsic interest even outside the present context and, once again, rests on the idea of linear independence.

So how is the material organized?

In the last chapter, we presumed sufficient experience with "vectors" that we could immediately consider them from a general perspective; expecting that the student is likely to have had rather less exposure to matrices, however, we shall start this chapter with a brief review of the operations on matrices: Though we have seen that matrices themselves form a vector space under obvious definitions of vector addition and scalar multiplication, it is a new operation, peculiar to matrices, which really gives them their utility in application: matrix multiplication. We discuss operations employing it, and introduce a perspective which will return throughout the chapter, "partitioning": how matrices, originally simply an array of numbers, can be viewed as an array of *vectors*, and how we quantify the linear independence of such vectors through the *rank* of the matrix they comprise. Determination of the rank will return, under this interpretation, to the echelon test for linear independence of vectors presented in the last chapter, where now the "vectors" are simply the rows or columns of the matrix. Using the idea of partitioning, we also demonstrate the *block operations* on matrices, in which entire sub-blocks of a matrix are treated as if they were scalars.

We then move onto the contact point with "linear transformations" around which this chapter is organized—what they are and how they can be represented in terms of matrices. We have seen in the previous chapter [Example 1.1.8] that such transformations—*functions*, if you will—themselves form a vector space and thus must posses a basis; in this chapter we will explicitly exhibit that basis [of *arbitrary* linear transformations]. We shall see there the reason that such transformations can be represented as matrices, and the first inklings of how the basis of the space on which the transformations act affects the basis of the transformations themselves. Just as any other functions, linear transformations have a domain and a range; we will see that these form vector spaces in their own right, examine a particular *sub*space of the domain, the *null space*, and obtain an important result regarding the dimensions of all these [vector] spaces.

With an eye toward the determination of eigenvectors and eigenvalues, we then discuss the essentially inverse operation to "transformation," solution of linear systems of equations, both homogeneous and non-homogeneous. Once again the ideas of linear dependence and independence become central, as do the echelon test for linear independence and the null space we have regarded earlier.

Finally, all this material is brought together in a somewhat surprising guise: linear differential equations, used to exemplify a topic inexorably intertwined with linear transformations, *linear operators*. Here the "vector space" is that of functions and their derivatives, the "operator" can be *represented* as a matrix, the "null space" becomes the solution to the *homogeneous* differential equation, and solution becomes tantamount to that of a "system of linear equations," with corresponding results. This topic concludes the chapter.

2.1 Matrices

A *matrix* is, at its core, merely a rectangular array of numbers:

$$\mathsf{A} = \begin{pmatrix} a_{11} & a_{12} & \dots & a_{1n} \\ a_{21} & a_{22} & \dots & a_{2n} \\ \vdots & \vdots & & \vdots \\ a_{m1} & a_{m2} & \dots & a_{mn} \end{pmatrix};$$

such a matrix with m rows and n columns is, when one wishes to emphasize these dimensions, referred to as an $m \times n$ *matrix* [note the order—*row* dimension first], occasionally written $\mathsf{A}_{m\times n}$. Rather than writing out the entire matrix as done above, a more economical notation is commonly employed, simply writing out a "typical element" of the matrix:

$$\mathsf{A}_{m\times n} = (a_{ij})_{m\times n} \text{ or simply } \mathsf{A} = (a_{ij})$$

This briefer notation will be employed extensively in the sequel, as, for example, in the following definitions.

Frequently it becomes useful to interchange the rows and columns of a matrix or, equivalently, to "reflect" the matrix across its "main diagonal"—the one running from the upper left to the lower right: $(\cdot_{\cdot_{\cdot}})$. The matrix resulting from this operation on A is called the *transpose* of A, A^T:

Definition 2.1.1. The *transpose* of a matrix $\mathsf{A} = (a_{ij})$

$$\mathsf{A}^\mathsf{T} \equiv (a_{ji}).$$

Though the "rectangular" nature of matrices has been stated above, "matrices" are actually more ubiquitous than than might be inferred from this definition. Indeed, we note in passing that the n-tuple representation of *vectors*

$$\mathsf{v} = \begin{pmatrix} a_1 \\ a_2 \\ \vdots \\ a_n \end{pmatrix},$$

dealt with extensively in the first chapter, can itself be regarded as a special form of matrix, an $n \times 1$ *column matrix*. Notwithstanding the fact that we have continually pointed out that matrices themselves comprise a vector space (see Example 1.1.4), we shall generally reserve the term "vector" [representation] for just such a column matrix throughout this text.

We now introduce a perspective which will be utilized extensively in the sequel.

2.1.1 The "Partitioning" and Rank of Matrices

We have frequently employed the convention that vectors can be represented as column n-tuples. With this in mind, it is not too much of a stretch to view a matrix as a catenation of such n-tuples:

$$\mathbf{A} = \begin{pmatrix} a_{11} & a_{12} & \dots & a_{1n} \\ a_{21} & a_{22} & \dots & a_{2n} \\ \vdots & \vdots & & \vdots \\ a_{m1} & a_{m2} & \dots & a_{mn} \end{pmatrix} \equiv (\mathbf{a}_1, \mathbf{a}_2, \dots \mathbf{a}_n) \text{ for } \mathbf{a}_j \equiv \begin{pmatrix} a_{1j} \\ a_{2j} \\ \vdots \\ a_{mj} \end{pmatrix}, \qquad (2.1\text{-}1)$$

or

$$\mathbf{A} = \begin{pmatrix} a_{11} & a_{12} & \dots & a_{1n} \\ a_{21} & a_{22} & \dots & a_{2n} \\ \vdots & \vdots & & \vdots \\ a_{m1} & a_{m2} & \dots & a_{mn} \end{pmatrix} \equiv \begin{pmatrix} \mathbf{a}_1^\mathsf{T} \\ \mathbf{a}_2^\mathsf{T} \\ \vdots \\ \mathbf{a}_m^\mathsf{T} \end{pmatrix} \text{ for } \mathbf{a}_i \equiv \begin{pmatrix} a_{i1} \\ a_{i2} \\ \vdots \\ a_{in} \end{pmatrix}. \qquad (2.1\text{-}2)$$

Note that the identity of the "$\mathbf{a}_i$" is different in the above two equations.

Thus, rather than considering the matrix as a whole, we can view it as being made up of a set of row or column n-tuples—"vectors" in their own right. But this means that we can consider *linear combinations* of these "vectors." In the context of matrices, we refer to such linear combinations as *elementary row* (or *column*) *operations.*

From this perspective, it makes sense to consider the linear independence of the constituent "vectors" in a matrix. We characterize this particular aspect of a matrix through a measure called the *rank*:

The Rank of a Matrix

Definition 2.1.2. The *column rank* of a matrix is the number of linearly independent columns [or, more exactly, n-tuple column *vectors*] in the matrix.

Definition 2.1.3. The *row rank* of a matrix is the number of linearly independent rows [or, more exactly, n-tuple row vectors] in the matrix.

These definitions reference, respectively, the columns and rows of a matrix; it is perfectly reasonable to expect that, in a rectangular matrix whose row and column dimensions are different, the number of linearly independent *rows* [n-tuple vectors] might be different from the number of linearly independent columns. In fact, however, this is *not* the case: the number of linearly independent rows and columns in a matrix are exactly the *same*! This rather remarkable result is contained in the following theorem:

Theorem 2.1.1. The *row* rank and *column* rank of a matrix are the same.

Proof. Assume the *row* rank of an $m \times n$ $\mathbf{A} = (a_{ij})$ is r; thus there are r linearly independent rows of $\mathbf{A}$,

$$\begin{gathered}(a^o_{11}, a^o_{12}, \ldots a^o_{1n})\\ \vdots \\ (a^o_{r1}, a^o_{r2}, \ldots a^o_{rn})\end{gathered}$$

—where each of these equals one of the original $m \geq r$ rows of $\mathbf{A}$. Thus we can express *all* m rows of $\mathbf{A}$ in terms of—as *linear combinations of*—these r rows:

$$\begin{aligned}(a_{11}, a_{12}, \ldots a_{1n}) &= c_{11}(a^o_{11}, a^o_{12}, \ldots a^o_{1n}) + \quad \ldots \quad + c_{1r}(a^o_{r1}, a^o_{r2}, \ldots a^o_{rn})\\ &\vdots\\ (a_{m1}, a_{m2}, \ldots a_{mn}) &= c_{m1}(a^o_{11}, a^o_{12}, \ldots a^o_{1n}) + \quad \ldots \quad + c_{mr}(a^o_{r1}, a^o_{r2}, \ldots a^o_{rn})\end{aligned}$$

Now consider [any] k^{th} *column* of these equations:

$$\begin{pmatrix} a_{1k}\\ a_{2k}\\ \vdots\\ a_{mk}\end{pmatrix} = a^o_{1k}\begin{pmatrix} c_{11}\\ c_{21}\\ \vdots\\ c_{m1}\end{pmatrix} + a^o_{2k}\begin{pmatrix} c_{12}\\ c_{22}\\ \vdots\\ c_{m2}\end{pmatrix} + \cdots + a^o_{rk}\begin{pmatrix} c_{1r}\\ c_{2r}\\ \vdots\\ c_{mr}\end{pmatrix}.$$

This says the n columns of $\mathbf{A}$ are *spanned* by the r columns

$$\begin{pmatrix} c_{1j}\\ c_{2j}\\ \vdots\\ c_{mj}\end{pmatrix}, \quad j = 1, \ldots, r.$$

Thus at most r of these columns are linearly independent; *i.e.*, the number of linearly independent *columns*, the *column rank* of $\mathbf{A}$, $\leq r$, the *row* rank; *i.e.*, column rank $\leq$ row rank.

Now do the *same* thing to the rows of $\mathbf{A}^\mathsf{T}$, whose rows are the *columns* of $\mathbf{A}$; we will get the result that row rank $\leq$ column rank. Putting these final results together, we conclude that column rank $=$ row rank. □

Thus it is no longer necessary to make any distinction between the "row" and "column" ranks; we merely refer to *the rank* of the matrix $\mathbf{A}$. That is all well and good, but, assuming that rank is important (it is), how are we to determine either one?

Recall that in Section 1.2.2 we presented a procedure (page 18) which could be used to test whether a given set of n-tuples were linearly independent: if the n-tuples could be reduced, through a sequence of linear combinations among themselves, to a set of the form (1.2-1), they were linearly independent. The results of that procedure carry over directly to matrices: if the columns of an $m \times n$ matrix—"vectors" according the perspective of (2.1-1)—can be put in the form

$$\begin{pmatrix} 1 & 0 & \cdots & 0 & 0 & \cdots & 0 \\ 0 & 1 & \cdots & 0 & 0 & \cdots & 0 \\ \vdots & \vdots & \cdots & \vdots & \vdots & \cdots & \vdots \\ 0 & 0 & \cdots & 1 & 0 & \cdots & 0 \\ v_{r+1,1} & v_{r+1,2} & \cdots & v_{r+1,r} & 0 & \cdots & 0 \\ \vdots & \vdots & \cdots & \vdots & \vdots & \cdots & \vdots \\ v_{m,1} & v_{m,2} & \cdots & v_{m,r} & 0 & \cdots & 0 \end{pmatrix} \qquad (2.1\text{-}3)$$

through linear combinations—*i.e.* "elementary column operations"—giving r columns in this form, then the rank of the matrix is r. Similarly, if we can, through linear combinations of the *rows* of the matrix—"elementary row operations" —put this in the form

$$\begin{pmatrix} 1 & 0 & \cdots & 0 & 0 & v_{1,r+1} & \cdots & v_{1,n} \\ 0 & 1 & \cdots & 0 & 0 & v_{2,r+1} & \cdots & v_{2,n} \\ \vdots & \vdots & \cdots & \vdots & \vdots & \vdots & \cdots & \vdots \\ 0 & 0 & \cdots & 0 & 1 & v_{r,r+1} & \cdots & v_{r,n} \\ 0 & 0 & \cdots & 0 & 0 & 0 & \cdots & 0 \\ \vdots & \vdots & \cdots & \vdots & \vdots & \vdots & \cdots & \vdots \\ 0 & 0 & \cdots & 0 & 0 & 0 & \cdots & 0 \end{pmatrix}, \qquad (2.1\text{-}4)$$

the matrix, once again, has rank r. Though it is unlikely that one would choose to do so in determination of the rank, recall (page 26) that we must explicitly *avoid* multiplying a row or column by 0 when adding another to it; doing so would effectively *force* that row (column) to vanish, lowering the rank of the matrix from what it originally was.

In this context, the reduced form of the matrix is referred to as its *echelon form.* This criterion will become important in application, with the rank itself bearing directly on the solution of linear equations, in particular, in cases in which one desires to determine eigenvalues and eigenvectors. Rank will even become important when we consider, in the context of general linear systems, the solution to [linear] ordinary differential equations. However curious and arcane it might seem at present, it will return!

In systematizing the echelon procedure, we also mentioned (page 24) the somewhat pathological possibility that there might be "skips" introduced into the final echelon form of the vectors—cases in which there is a 0 in the ith entry of the ith vector with possibly non-0 values in those entries for the $v'_j, j < i$. In terms of matrices, this would evidence itself with an echelon form resulting from *column* operations

$$
\text{“skip row”} \longrightarrow \quad
\begin{pmatrix}
1 & 0 & \cdots & 0 & 0 & 0 & \cdots & 0\\
0 & 1 & \cdots & 0 & 0 & 0 & \cdots & 0\\
\vdots & \vdots & \ddots & \vdots & \vdots & \vdots & \ddots & \vdots\\
0 & 0 & \cdots & 1 & 0 & 0 & \cdots & 0\\
v_{k+1,1} & v_{k+1,2} & \cdots & v_{k+1,k} & 0 & 0 & \cdots & 0\\
0 & 0 & \cdots & 0 & 1 & 0 & \cdots & 0\\
v_{k+3,1} & v_{k+3,2} & \cdots & & v_{k+3,k+1} & 0 & \cdots & 0\\
\vdots & \vdots & & & \vdots & \vdots & & \vdots\\
v_{m,1} & v_{m,2} & & & v_{m,k+1} & 0 & \cdots & 0
\end{pmatrix}
$$

As was noted then, the $k+1$ vectors—the number of *columns* with a 1 in them—are linearly independent. In the present context regarding the rank of the matrix, this makes sense: though the "1" appears in the $(k+2)^{nd}$ *row* of the matrix, only $k+1$ of these rows are linearly independent, since the $(k+1)^{st}$ row is clearly a linear combination of those above it. Equivalently, considering only the $k \times (k+1)$ *sub*matrix in the upper left, the *maximum* rank this can have is k—the [maximum] number of rows. However this result is arrived at, though, we can view the rows of the matrix as "vectors," and follow the discussion of Chapter 1 [assuming no reordering], to conclude that the $(k+1)^{st}$ row in the *original* (pre-echelon) matrix must also have been a linear combination of the rows above *it*. Similarly, had *row* operations be used to determine the rank, the "skip signature" would have been a matrix of the form

$$
\begin{array}{c}\text{“skip column”}\\ \downarrow\end{array}
$$

$$
\begin{pmatrix}
1 & 0 & \cdots & 0 & v_{1,k+1} & 0 & v_{1,k+3} & \cdots & v_{1,n}\\
0 & 1 & \cdots & 0 & v_{2,k+1} & 0 & v_{2,k+3} & \cdots & v_{2,n}\\
\vdots & \vdots & \cdots & \vdots & \vdots & \vdots & \vdots & \cdots & \vdots\\
0 & 0 & \cdots & 1 & v_{k,k+1} & 0 & \vdots & \cdots & \vdots\\
0 & 0 & \cdots & 0 & 0 & 1 & v_{k+1,k+3} & \cdots & v_{k+1,n}\\
0 & 0 & \cdots & 0 & 0 & 0 & 0 & \cdots & 0\\
0 & 0 & \cdots & 0 & 0 & 0 & 0 & \cdots & 0\\
\vdots & \vdots & \cdots & \vdots & \vdots & \vdots & 0 & \cdots & \vdots\\
0 & 0 & \cdots & 0 & 0 & 0 & 0 & \cdots & 0
\end{pmatrix}
$$

—in which now the $(k+1)^{st}$ *column* would be a linear combination of those to its left.

Thus we have "skips" resulting from *column* operations manifest themselves with "skip *rows*," and those due to *row* operations evidenced through "skip *columns*"; in each case, be they rows or columns, they are linear combinations of those preceding them and thus *skip rows/columns do not increase the rank of the echelon form.* (Though the result has been demonstrated for only a *single* such skip, in fact it clearly would hold for *any* number.) This observation will prove useful when we discuss the solution of systems of linear equations in Section 2.3.

We note in passing that, in order to determine the rank of a matrix, such skips can be eliminated by using *both* row *and* column operations simultaneously; in this regard, recall Theorem 2.1.1. We adopt the convention, however, of using only *one* type of operation in a given case.

2.1.2 Operations on Matrices

Though we have already discussed a means of defining "vector addition" and "scalar multiplication" on matrices (Example 1.1.4), we restate them here, both to get all the relevant material in this chapter, and to illustrate again the above shorthand notation:

Definition 2.1.4. Given two arbitrary matrices $\mathbf{A} = (a_{ij})$ and $\mathbf{B} = (b_{ij})$ *of the same dimension*, we define the *[vector] sum* of $\mathbf{A}$ and $\mathbf{B}$

$$\mathbf{A} \oplus \mathbf{B} \equiv (a_{ij} + b_{ij})$$

and the the *scalar product* of a scalar c with $\mathbf{A}$

$$c \odot \mathbf{A} \equiv (c \cdot a_{ij}).$$

We have already pointed out that such definitions allow the set of $m \times n$ matrices to be regarded as a vector space; we have even displayed a *basis* for that space, (1.2-4) in Example 1.2.8, of dimension mn. If this were all there is to matrices, little would be made of them. But the importance of matrices in application rests on a third operation on matrices, *matrix multiplication*:

Definition 2.1.5. Consider an $m \times k$ matrix $\mathbf{A} = (a_{i\ell})$ and a $k \times n$ matrix $\mathbf{B} = (b_{\ell j})$. We define the *matrix product* $\mathbf{C} \equiv \mathbf{AB}$ to be the $m \times n$ matrix

$$(c_{ij}) \equiv \left(\sum_{\ell=1}^{k} a_{i\ell} b_{\ell j} \right). \tag{2.1-5}$$

Note that each term in the sum for c_{ij} is a product of terms whose "outer indices" are i and j; summation is on the "inner" ones. This observation is of more than incidental importance, as we shall see presently when we consider the transpose of a matrix product.

In computation, few people use (2.1-5) directly; rather they consider the rows of the first matrix (holding i fixed) and columns of the second (holding j fixed) and think "$(\rightarrow)(\downarrow)$." However one thinks about it, however, it is important to note that matrix multiplication is defined *only for matrices whose number of columns of the first equals the number of rows of the second.* Thus, although we have $\mathbf{AB}$ above, $\mathbf{BA}$ is generally not even defined! But, even in the case of *square* matrices (which we will consider in some detail in the next chapter), in which multiplication on the reverse order *is* defined, in general $\mathbf{BA} \neq \mathbf{AB}$.

Homework:

1. For two square matrices show by example that $\mathsf{BA} \neq \mathsf{AB}$.
2. Assume two compatible matrices A and B are functions of a *scalar* variable s. Show that $\frac{d\mathsf{AB}}{ds} = \frac{d\mathsf{A}}{ds}\mathsf{B} + \mathsf{A}\frac{d\mathsf{B}}{ds}$, where we define, for $\mathsf{A} \equiv (a_{ij})$, $\frac{d\mathsf{A}}{ds} \equiv (\frac{da_{ij}(s)}{ds})$.
3. (*Eigenvectors and Eigenvalues*) Consider the matrix

$$\mathsf{A} = \begin{pmatrix} 7 & -2 & -2 \\ -2 & 1 & 4 \\ -2 & 4 & 1 \end{pmatrix}$$

   (a) Show that the product $\mathsf{A}\boldsymbol{v}$, for $\boldsymbol{v} = (\mathbf{0}, \mathbf{1}, \mathbf{-1})^{\mathsf{T}}$, yields a vector along $\boldsymbol{v}$.

   (b) Now [magic trick!] pick [*almost!*] any vector $\boldsymbol{v}'$ *other* than $\boldsymbol{v}$ and show that $\mathsf{A}\boldsymbol{v}'$ *isn't* along that vector.

   Clearly, vectors which, when multiplied by a matrix, give a vector along (parallel or antiparallel to) themselves are distinguished and called *eigenvectors* of the corresponding matrix. Since the resulting vector *is* parallel or antiparallel to the original, say $\boldsymbol{\xi}$, it can be represented in the form $\lambda\boldsymbol{\xi}$, where λ is a [positive or negative] *scalar*; such λ is called the *eigenvalue* associated with $\boldsymbol{\xi}$. These topics will form the subject of the whole of Section 3.5; their applications in science and engineering are numerous!

Now having defined, in particular, matrix multiplication, we shall discuss some applications of the concept:

Inner Product

The first application of matrix multiplication gives us an operation *reminiscent* of the "dot product" familiar from vectors:

Definition 2.1.6. Given two $n \times 1$ column matrices v_1 and v_2 consisting of *real*[1] numbers, we define the *inner product*

$$\langle \mathsf{v}_1, \mathsf{v}_2 \rangle \equiv \mathsf{v}_1{}^{\mathsf{T}}\mathsf{v}_2 \tag{2.1-6}$$

Several remarks are in order about this definition: In the first place, it is obvious that the inner product results in a scalar:

$$\mathsf{v}_1{}^{\mathsf{T}}\mathsf{v}_2 = (v_{11}, v_{12}, \ldots v_{1n}) \begin{pmatrix} v_{21} \\ v_{22} \\ \vdots \\ v_{2n} \end{pmatrix} = \sum_{i=1}^{n} v_{1i}v_{2i}. \tag{2.1-7}$$

[1] The definition changes for complex numbers; see equation (3.5-3).

But this means that the inner product *commutes*:

$$\mathbf{v_2}^\mathsf{T}\mathbf{v_1} = (v_{21}, v_{22}, \ldots v_{2n})\begin{pmatrix} v_{11} \\ v_{12} \\ \vdots \\ v_{1n} \end{pmatrix} = \sum_{i=1}^{n} v_{2i}v_{1i} = \sum_{i=1}^{n} v_{1i}v_{2i} \equiv \mathbf{v_1}^\mathsf{T}\mathbf{v_2}. \tag{2.1-8}$$

It also means that the the inner product is *linear* in its two entries; *e.g.*,

$$\begin{aligned}(\mathbf{v_{11}} + \mathbf{v_{12}})^\mathsf{T}\mathbf{v_2} &= (v_{11_1} + v_{12_1}, v_{11_2} + v_{12_2}, \ldots, v_{11_n} + v_{12_n})\begin{pmatrix} v_{21} \\ v_{22} \\ \vdots \\ v_{2n} \end{pmatrix} \\ &= \sum_{i=1}^{n}(v_{11_i} + v_{12_i})v_{2i} = \sum_{i=1}^{n} v_{11_i}v_{2i} + \sum_{i=1}^{n} v_{12_i}v_{2i} \\ &\equiv \mathbf{v_{11}}^\mathsf{T}\mathbf{v_2} + \mathbf{v_{12}}^\mathsf{T}\mathbf{v_2},\end{aligned} \tag{2.1-9}$$

and similarly for $\mathbf{v_1}^\mathsf{T}(\mathbf{v_{21}} + \mathbf{v_{22}})$. (This is sometimes called a "*bilinear mapping*," since it is linear in *both* its arguments).

The second point is that the product of two *matrices* can be cast in terms of inner products: referring to equations (2.1-1) and (2.1-2), we see that we can write

$$\mathbf{AB} = \begin{pmatrix} \mathbf{a}_1^\mathsf{T} \\ \mathbf{a}_2^\mathsf{T} \\ \vdots \\ \mathbf{a}_n^\mathsf{T} \end{pmatrix}(\mathbf{b}_1, \mathbf{b}_2, \ldots \mathbf{b}_n) = (\langle \mathbf{a}_i, \mathbf{b}_j \rangle) \tag{2.1-10}$$

—the *matrix* of inner products (recall the shorthand notation indicated on page 43). Thus, in effect, the matrix product $\mathbf{AB}$ is nothing more than the aggregation of inner products of the column vectors of $\mathbf{A}^\mathsf{T}$ and $\mathbf{B}$. [Note that the curious product above is *defined*: the first "matrix" has the same number of "columns," 1, as the second has "rows"! This anticipates the next subsection.]

Finally, inner product was described above as being *reminiscent* of the dot product; it is important to realize that the two operations are *different*: In the first place, the dot product $\boldsymbol{A}\cdot\boldsymbol{B} \equiv \|\boldsymbol{A}\|\,\|\boldsymbol{B}\|\cos\theta$, as mentioned in Section 1.1, is defined only for directed line segments, since it references the *magnitudes* and *directions* [through "the angle between them"] of the two vectors; the inner product, on the other hand, is defined only for [column] n-tuples. And of the former, it is always true that $\|\boldsymbol{A}\|^2 = \boldsymbol{A}\cdot\boldsymbol{A}$; but, in general, it is *not* true, even for the *representations* $\mathbf{A}$ of directed line segment vectors $\boldsymbol{A}$, that $\|\boldsymbol{A}\|^2 = \langle \mathbf{A}, \mathbf{A} \rangle$.

Notwithstanding this *caveat*, we can at least borrow a term, "orthogonal," which is normally associated with [the perpendicularity of] directed line segments, and apply it to matrices (and thus, by extension to the *representations* of vectors, to *any* vector space):

Definition 2.1.7. Two column matrices $\mathbf{v}_1$ and $\mathbf{v}_2$, of the same dimension, are said to be *orthogonal* if their inner product vanishes: $\langle \mathbf{v}_1, \mathbf{v}_2 \rangle = 0$.

In the same fashion, it is no greater abuse to use the inner product to *define* a length:

Definition 2.1.8. The *length* $\|\mathbf{v}\|$ of a column matrix $\mathbf{v}$ is the square root of its inner product with itself: $\|\mathbf{v}\|^2 \equiv \langle \mathbf{v}, \mathbf{v} \rangle$.

Homework:

1. Consider two [directed line segment] basis vectors, $\boldsymbol{u}_1$ and $\boldsymbol{u}_2$, *not necessarily perpendicular or even unit vectors!* For any vector $\boldsymbol{v}$ lying in their plane, we can find the components, v_i, of $\boldsymbol{v}$ relative to that basis:

$$\boldsymbol{v} = v_1\boldsymbol{u}_1 + v_2\boldsymbol{u}_2$$

by forming a vector triangle. (It is helpful to draw the triangle, noting that the lengths of its sides parallel to the $\boldsymbol{u}_i$ are $\|c_i\boldsymbol{u}_i\| = |c_i|\,\|\boldsymbol{u}_i\|$.) Show that

   (a) $\|\boldsymbol{v}\|^2$ is, indeed, equal to the *dot product* $\boldsymbol{v} \cdot \boldsymbol{v}$,

   (b) the result of (a) gives just the Law of Cosines (careful of the definition of the angle here!), and that

   (c) $\|\boldsymbol{v}\|^2$ is *not* the *inner product* of the component column vector $(c_1 c_2)^{\mathsf{T}}$ *unless*

      i. the $\boldsymbol{u}_i$ are perpendicular *and*

      ii. the $\|\boldsymbol{u}_i\|$ are both 1.

2. Show, by example, that the dot and inner product don't generally give the same result.

Transpose of a Matrix Product

This entire topic consists of relating the transpose of a matrix product to the product of its transposed constituents:

Theorem 2.1.2. Given two multiplication compatible matrices $\mathbf{A}$ and $\mathbf{B}$,

$$(\mathbf{AB})^{\mathsf{T}} = \mathbf{B}^{\mathsf{T}}\mathbf{A}^{\mathsf{T}} \tag{2.1-11}$$

The reversal of order seems curious initially, but makes sense when we consider the dimensions of the two matrices. If $\mathbf{A}$ is $m \times k$ and $\mathbf{B}$ is $k \times n$, their transposes will be $k \times m$ and $n \times k$ respectively; thus the reversal of order not only makes the product on the right-hand side above defined, it is *necessary* to make the second product compatible!

Proof. Initially keeping track of both the dimensions and the ranges of the indices, to show how these carry through the transpose.

$$\begin{aligned}
(\mathbf{AB})^{\mathsf{T}} &\equiv \left(\left(\sum_{\ell=1}^{k} a_{i\ell}b_{\ell j}\right)_{m\times n}\right)^{\mathsf{T}} && i=1,\ldots,m;\ j=1,\ldots,n\\
&\equiv \left(\sum_{\ell=1}^{k} a_{j\ell}b_{\ell i}\right)_{n\times m} && i=1,\ldots,n;\ j=1,\ldots,m\\
&= \left(\sum_{\ell=1}^{k} a^{\mathsf{t}}_{\ell j}b^{\mathsf{t}}_{i\ell}\right) && (a^t_{ij})\equiv\mathbf{A}^{\mathsf{T}},\ (b^t_{ij})\equiv\mathbf{B}^{\mathsf{T}}\\
&= \left(\sum_{\ell=1}^{k} b^{\mathsf{t}}_{i\ell}a^{\mathsf{t}}_{\ell j}\right)\\
&\equiv \mathbf{B}^{\mathsf{T}}\mathbf{A}^{\mathsf{T}}
\end{aligned}$$

The fourth equation appears arbitrary until one recalls the remark following the equation (2.1-5) defining matrix multiplication: summation is on the *inner* indices; thus we must write the previous equation in this form to conclude the last. □

Block Multiplication of Partitioned Matrices

In Section 2.1.1, we pointed out that we could "partition" a matrix into row or column "vectors." We now extend that device to a larger granularity: Rather than dividing the matrix into individual rows or columns, we can divide it into rectangular *blocks*:

$$\mathbf{A} = \begin{pmatrix} \mathbf{A}_{11} & \mathbf{A}_{12} \\ \mathbf{A}_{21} & \mathbf{A}_{22} \end{pmatrix}$$

—where $\mathbf{A}_{11}$ is $m_1 \times k_1$, $\mathbf{A}_{12}$ is $m_1 \times k_2$, $\mathbf{A}_{21}$ is $m_2 \times k_1$ and $\mathbf{A}_{22}$ is $m_2 \times k_2$. We note parenthetically that the *transpose* of such a block-partitioned matrix is the transpose of the transposed blocks; *i.e.*,

$$\mathbf{A}^{\mathsf{T}} = \begin{pmatrix} \mathbf{A}_{11}^{\mathsf{T}} & \mathbf{A}_{21}^{\mathsf{T}} \\ \mathbf{A}_{12}^{\mathsf{T}} & \mathbf{A}_{22}^{\mathsf{T}} \end{pmatrix} \tag{2.1-12}$$

If, now, we do the same to another matrix $\mathbf{B}$:

$$\mathbf{B} = \begin{pmatrix} \mathbf{B}_{11} & \mathbf{B}_{12} \\ \mathbf{B}_{21} & \mathbf{B}_{22} \end{pmatrix}$$

—where $\mathbf{B}_{11}$ is $k_1 \times n_1$, $\mathbf{B}_{12}$ is $k_1 \times n_2$, $\mathbf{B}_{21}$ is $k_2 \times n_1$ and $\mathbf{B}_{22}$ is $k_2 \times n_2$ (so that the number of columns in each submatrix of $\mathbf{A}$ is "compatible" with the corresponding number of rows in each of $\mathbf{B}$), it is easy (if somewhat inelegant) to show, by keeping track of each of the indices in the respective products in (2.1-5), that,

$$\mathbf{AB} = \begin{pmatrix} \mathbf{A}_{11}\mathbf{B}_{11} + \mathbf{A}_{12}\mathbf{B}_{21} & \mathbf{A}_{11}\mathbf{B}_{12} + \mathbf{A}_{12}\mathbf{B}_{22} \\ \mathbf{A}_{21}\mathbf{B}_{11} + \mathbf{A}_{22}\mathbf{B}_{21} & \mathbf{A}_{21}\mathbf{B}_{12} + \mathbf{A}_{22}\mathbf{B}_{22} \end{pmatrix} \tag{2.1-13}$$

—exactly as if the matrix blocks were *scalars*! Note that, just as for scalars, the "outside indices" of each product correspond to its entry in the "product." And the dimensions of the block products are of dimension

$$\begin{pmatrix} m_1 \times n_1 & m_1 \times n_2 \\ m_2 \times n_1 & m_2 \times n_2 \end{pmatrix},$$

since each of the separate block products in the sums are of corresponding dimension.

Once established, this approach becomes a powerful tool for the analysis of operations involving matrix multiplication. Thus, for example, we can prove the *associativity of matrix multiplication* (matrix multiplication being *defined* initially as a *binary* operation between only *two* matrices): assuming that $\mathbf{A}$, $\mathbf{B}$, and $\mathbf{C}$ are "multiplication compatible"—*i.e.*, $\mathbf{B}$ has the same number of rows as $\mathbf{A}$ has columns, with a similar relation between $\mathbf{C}$ and $\mathbf{B}$—we obtain, with remarkable efficiency, that matrix products are *associative*—

$$(\mathbf{AB})\mathbf{C} = \mathbf{A}(\mathbf{BC}) \tag{2.1-14}$$

—by simply treating the matrices as "scalars." Similarly, we obtain the *distributive laws of matrix multiplication over matrix addition*: if $\mathbf{A}$ and $\mathbf{B}$ are of the same dimension, and both are multiplication compatible with $\mathbf{C}$,

$$(\mathbf{A} + \mathbf{B})\mathbf{C} = \mathbf{AC} + \mathbf{BC}, \tag{2.1-15}$$

and, again assuming appropriate dimensioning,

$$\mathbf{A}(\mathbf{B} + \mathbf{C}) = \mathbf{AB} + \mathbf{AC}. \tag{2.1-16}$$

There are limits to this device, however; in particular, one must retain the *order* of multiplication and not try to infer that matrix multiplication is commutative just because the multiplication of real numbers is!

Homework:

1. Prove equation (2.1-13).
2. Consider the 3×2 and 2×4 matrices $\mathbf{A}$ and $\mathbf{B}$:

$$\mathbf{A} = \begin{pmatrix} 3 & \vdots & 4 \\ \cdots & & \cdots \\ 1 & \vdots & -1 \\ -2 & \vdots & 3 \end{pmatrix} \text{ and } \mathbf{B} = \begin{pmatrix} 1 & \vdots & 0 & 0 & 5 \\ \cdots & & \cdots & & \\ 2 & \vdots & 2 & 0 & 1 \end{pmatrix}.$$

(a) Ignoring, for the time being, the partitioning indicated, simply find the product matrix **AB**.

(b) Assume the partitioning indicated defines four submatrices for each **A** and **B**. Multiplying the two matrices using *block partitioning*, show the result is the same as for the above.

(c) Now consider the entire matrix **A** but consider **B** to be partitioned into four column matrices: $\mathbf{B} = (\mathbf{b}_1\mathbf{b}_2\mathbf{b}_3\mathbf{b}_4)$. Then show that that $\mathbf{AB} = (\mathbf{Ab}_1\mathbf{Ab}_2\mathbf{Ab}_3\mathbf{Ab}_4)$.

Elementary Operations through Matrix Products

Though marginally interesting in application, the results of this section will provide the theoretical underpinning for a great deal of this chapter and the next. Recall that, in Section 2.1.1, we pointed out that the "vectors" formed by "partitioning" a matrix could be combined in linear combinations, effecting "elementary row/column operations" on the matrix. At the end of that section we translated the vector echelon test (page 18) to a similar test on matrices to examine the linear independence of these rows or columns. It was pointed out that this test corresponds to applying such elementary operations to the matrix. We shall now show how such elementary operations can, in fact, be represented as a *product* of that matrix with another, mundanely rudimentary one. Before considering matrices, however, we shall first examine what happens when we pre- or post-multiply a matrix by certain elementary *vectors*.

Back in Section 1.2.3, Example 1.2.7 exhibited a basis for n-tuple vectors, the *natural basis*, with 0's everywhere except for a "1" in a single row/column:

$$i \rightarrow \begin{pmatrix} 0 \\ 0 \\ \vdots \\ 0 \\ 1 \\ 0 \\ \vdots \\ 0 \end{pmatrix} \equiv \mathbf{e}_i^{(n)}$$

If an arbitrary $m \times n$ matrix **A** is post-multiplied by this vector, the result will obviously be an $m \times 1$ column vector. What is likely not so obvious initially, however, is that this product column vector will be nothing more than the ith column of **A**: as each row of the product is determined, only the ith entry of **A**'s row survives the "inner product" operation used to find the corresponding row in the product vector. Symbolically, using the notation of (2.1-1),

$$\mathbf{Ae}_i^{(n)} \equiv (\mathbf{a}_1\mathbf{a}_2\ldots\mathbf{a}_n)\,\mathbf{e}_i^{(n)} = \mathbf{a}_i.$$

Similarly, if we *pre*-multiply **A** by one of the *row* n-tuple basis vectors, which can be written as just the *transpose* of one of the above **e**'s (now of length m

so that the multiplication is defined in the first place), and use (2.1-2) [with *different* "$\mathbf{a}_i$"],

$$\left(\mathbf{e}_i^{(m)}\right)^{\mathsf{T}} \mathbf{A} \equiv \left(\mathbf{e}_i^{(m)}\right)^{\mathsf{T}} \begin{pmatrix} \mathbf{a}_1^{\mathsf{T}} \\ \mathbf{a}_2^{\mathsf{T}} \\ \vdots \\ \mathbf{a}_m^{\mathsf{T}} \end{pmatrix} = \mathbf{a}_i^{\mathsf{T}},$$

utilizing block multiplication of the two quantities. Thus

- *pre*-multiplication by $\mathbf{e}_i^{\mathsf{T}}$ picks off the ith *row*,

- *post*-multiplication by $\mathbf{e}_i$ picks off the ith *column*.

of $\mathbf{A}$.

We now consider similar operations with *matrices*. In Example 1.2.8 we exhibited a basis for $m \times n$ matrices, consisting of matrices of that dimension, $\mathbf{E}_{ij}$ consisting of all 0's except for a 1 in the ith row and j^{th} column. Consider, for the time being, such a *square* basis matrix of dimension $n \times n$:

$$i \rightarrow \quad \begin{matrix} & & j \\ & & \downarrow \\ \end{matrix}$$
$$\begin{pmatrix} 0 & \cdots\cdots & 0 & 0 & 0 & \cdots & 0 \\ \vdots & & \vdots & \vdots & \vdots & & \vdots \\ \vdots & & \vdots & \vdots & \vdots & & \vdots \\ 0 & \cdots\cdots & 0 & 0 & 0 & \cdots & 0 \\ 0 & \cdots\cdots & 0 & 1 & 0 & \cdots & 0 \\ 0 & \cdots\cdots & 0 & 0 & 0 & \cdots & 0 \\ \vdots & & \vdots & \vdots & \vdots & & \vdots \\ 0 & \cdots\cdots & 0 & 0 & 0 & \cdots & 0 \end{pmatrix} \equiv \mathbf{E}_{ij}^{(n)}.$$

Let us first examine what happens if we *pre*-multiply an arbitrary $m \times n$ matrix $\mathbf{A}$ by $\mathbf{E}_{ij}^{(m)}$: We note, in the first place, that the dimensions of the two matrices are compatible with multiplication, and that this will give another product matrix of dimension $m \times n$. Clearly all rows above and below the ith row in the product $\mathbf{E}_{ij}^{(m)}\mathbf{A}$ will be zero, since each element of the *columns* of $\mathbf{A}$ will be multiplied by the 0's in those rows of $\mathbf{E}_{ij}^{(m)}$; thus it suffices to consider only the ith row of the product. When the ith row of $\mathbf{E}_{ij}^{(m)}$ multiplies the kth column of $\mathbf{A}$ (in order to find the element lying on the ith row and k^{th} column of the *product*), *only the jth entry in the kth column survives*: all other entries in that column are 0-ed out by the 0's in the ith row of $\mathbf{E}_{ij}^{(m)}$. But this puts the element (j, k) of $\mathbf{A}$ in the position (i, k) of the product. Doing this for all $k = 1, \ldots, n$, we finally see that this product has the net effect of putting the jth row of $\mathbf{A}$ in the ith of the product $\mathbf{E}_{ij}^{(m)}\mathbf{A}$; i.e., pre-multiplication by the matrix $\mathbf{E}_{ij}^{(m)}$ moves the jth row of $\mathbf{A}$ into the ith of the product.

The same analysis applies to *post*-multiplication: Now we must post-multiply by $\mathbf{E}_{ij}^{(n)}$ in order to make the two matrices amenable to multiplication, but the product will still be an $m \times n$ matrix. Now the **0** *columns* of $\mathbf{E}_{ij}^{(n)}$ will make all *columns* of the product vanish except the jth. When the kth row of $\mathbf{A}$ multiplies this column (to find the $(k,j)^{th}$ element of the product), only the ith [column] element of that row in $\mathbf{A}$ will survive, all other vanishing because of the 0's in the jth column of $\mathbf{E}_{ij}^{(n)}$; thus (for all k), the net effect is to take the ith column of $\mathbf{A}$ and put it in the jth column of the product.

Let us summarize: Given an $m \times n$ matrix $\mathbf{A}$,

- *pre*-multiplication by $\mathbf{E}_{ij}^{(m)}$ gives a product with the jth *row* of $\mathbf{A}$ in the ith of the product,
- *post*-multiplication by $\mathbf{E}_{ij}^{(n)}$ gives a product with the ith *column* of $\mathbf{A}$ in the jth of the product;

all other elements in the product vanish. Clearly, if either $\mathbf{E}_{ij}$ were multiplied by a scalar constant c, the products would yield a matrix with $c\times$ either the row or column in the final product.

Now let us introduce an $n \times n$ matrix which will become of major import when we specialize to the discussion of *square* matrices in Chapter 3 below: the nth*-order identity matrix*,[2]

$$\mathbf{1}^{(n)} \equiv \begin{pmatrix} 1 & 0 & 0 & \ldots & 0 \\ 0 & 1 & 0 & \ldots & 0 \\ 0 & 0 & 1 & \ldots & 0 \\ \vdots & \vdots & \vdots & \ddots & \vdots \\ 0 & 0 & 0 & \ldots & 1 \end{pmatrix} = \begin{pmatrix} \mathbf{e}_1 & \mathbf{e}_2 & \ldots & \mathbf{e}_n \end{pmatrix} \tag{2.1-17}$$

—the "blockwise" catenation of the $\mathbf{e}_i$, with 1's on the main diagonal and 0's elsewhere. Observe that we can write

$$\mathbf{1}^{(n)} = \mathbf{E}_{11}^{(n)} + \mathbf{E}_{22}^{(n)} + \cdots + \mathbf{E}_{nn}^{(n)},$$

The analysis above demonstrates that each of the matrices in the right-hand sum, when it pre- (or post-) multiplied an arbitrary $m \times n$ $\mathbf{A}$ [with appropriate dimension], would yield a matrix with the ith row (column) of $\mathbf{A}$. When summed, this would just return the original matrix $\mathbf{A}$:

$$\mathbf{A}\mathbf{1}^{(n)} = \mathbf{1}^{(m)}\mathbf{A} = \mathbf{A}. \tag{2.1-18}$$

With the two matrices above, now consider their sum,

$$\mathbf{E}_{ij}^{*(n)}(c) \equiv \left(\mathbf{1}^{(n)} + c\mathbf{E}_{ij}^{(n)}\right), \tag{2.1-19}$$

[2] We denote the identity matrix by "**1**" rather than the more usual "**I**" to prevent confusion with the inertia matrix, *universally* denoted **I**, in the next Part.

which can be written explicitly

$$\mathbf{E}_{ij}^{*(n)}(c) = \begin{pmatrix} 1 & 0 & \overset{\displaystyle j}{\overset{\downarrow}{0}} & \cdots & 0 \\ 0 & 1 & 0 & \cdots & 0 \\ \vdots & 0 & \ddots & \cdots & 0 \\ \vdots & \vdots & c & \ddots & 0 \\ 0 & & \cdots & & 1 \end{pmatrix} \begin{array}{l} \\ \\ \\ \leftarrow i \\ \\ \end{array} \tag{2.1-20}$$

—where the arrows point to the ith row and j^{th} column to indicate the only non-0 element aside from the 1's on the diagonal. When, say, $\mathbf{E}_{ij}^{*(m)}(c)$ *pre*-multiplies an arbitrary $m \times n$ matrix $\mathbf{A}$:

$$\mathbf{E}_{ij}^{*(m)}(c)\mathbf{A} \equiv \left(\mathbf{1}^{(m)} + c\mathbf{E}_{ij}^{(m)}\right)\mathbf{A} = \mathbf{1}^{(m)}\mathbf{A} + c\mathbf{E}_{ij}^{(m)}\mathbf{A} = \mathbf{A} + c\mathbf{E}_{ij}^{(m)}\mathbf{A}$$

—the last equation resulting from multiplication distribution over summation, (2.1-15). But the [vector] sum of the two matrices at the end is nothing more than $\mathbf{A}$ with c times the jth row in that matrix added to its ith—*i.e.*, what we referred to above as an elementary row operation! Similarly, *post*-multiplication by $\mathbf{E}_{ij}^{*(n)}(c)$ would give a matrix with c times the ith *column* of $\mathbf{A}$ added to its jth—an elementary *column* operation. (We note that this also covers the elementary operation of multiplying a row or column by a constant c, using $\mathbf{E}_{ii}^{*}(c-1)$. This also points up the fact that $\mathbf{E}_{ii}^{*}(-1)$ will make the ith row or column *vanish*!)

The above results have been obtained through a brute force examination of the matrix products involved. But there is another means of determining the [composite] $\mathbf{E}_{ij}^{*}$ matrix required to effect a given row or column operation—or even a *sequence* of them: Assume we start with an arbitrary $m \times n$ matrix $\mathbf{A}$ on which we wish to perform such operations; let us *perform the same operations on the identity matrix of appropriate order*—$\mathbf{1}^{(m)}$ if we wish to do row operations, $\mathbf{1}^{(n)}$ for column:

$$\mathbf{A} \qquad\qquad\qquad\qquad\qquad\qquad\qquad\qquad \mathbf{1}$$

Now perform an arbitrary sequence of *all* row operations:

$$\mathbf{A} \to \mathbf{E}_{i_n j_n}^{*(m)}\mathbf{E}_{i_{n-1} j_{n-1}}^{*(m)} \cdots \mathbf{A} \equiv \mathbf{E}^{*(m)}\mathbf{A} \qquad\qquad \mathbf{1}^{(m)} \to \mathbf{E}^{*(m)}\mathbf{1}^{(m)} = \mathbf{E}^{*(m)}$$

or *all* column operations:

$$\mathbf{A} \to \mathbf{A}\mathbf{E}_{i_1 j_1}^{*(n)}\mathbf{E}_{i_2 j_2}^{*(n)} \cdots \equiv \mathbf{A}\mathbf{E}^{*(n)} \qquad\qquad \mathbf{1}^{(n)} \to \mathbf{1}^{(n)}\mathbf{E}^{*(n)} = \mathbf{E}^{*(n)}.$$

We see that, by doing the same operations on $\mathbf{1}$ we *want* on $\mathbf{A}$ generates the appropriate $\mathbf{E}_{ij}^{*(n)}$ directly! Once more, however, note that we must restrict our operations to be *all* row or column ones.

Example 2.1.1. For the matrix

$$\mathsf{A} = \begin{pmatrix} 2 & 0 & 0 \\ -2 & 1 & 1 \end{pmatrix},$$

find what matrices will effect adding twice the first column to the second, then subtracting three times *that* second column from the third.

Solution: Though A is not square, since we are going to do column operations (requiring *post*-multiplication by appropriate elementary matrices) on it, we will use $\mathbf{1}^{(3)}$ to mimic these, in order to ensure dimensional compatibility:

$$\begin{pmatrix} 2 & 0 & 0 \\ -2 & 1 & 1 \end{pmatrix} \qquad \begin{pmatrix} 1 & 0 & 0 \\ 0 & 1 & 0 \\ 0 & 0 & 1 \end{pmatrix}$$

$$\begin{pmatrix} 2 & 4 & 0 \\ -2 & -3 & 1 \end{pmatrix} \begin{pmatrix} 1 & 2 & 0 \\ 0 & 1 & 0 \\ 0 & 0 & 1 \end{pmatrix}$$

$$\begin{pmatrix} 2 & 4 & -12 \\ -2 & -3 & 10 \end{pmatrix} \qquad \begin{pmatrix} 1 & 2 & -6 \\ 0 & 1 & -3 \\ 0 & 0 & 1 \end{pmatrix}.$$

The last matrix is the composite "E^*" which will effect the desired column operations on A through post-multiplication, as can be verified by direct calculation. Note the order in which operations on $\mathbf{1}$ are done: it is only after the [new] second column has been determined that this *new* column is subtracted from the third.

Though column operations were exemplified here, had row operations been specified, we would *pre*-multiply by an E^*, building this up on $\mathbf{1}^{(2)}$. |

Thus we have shown that elementary row and column operations can be obtained by matrix multiplication with one of the E^* matrices. But, by the remarks at the beginning of this topic, this means that examination of the linear independence of the rows or columns of a matrix—reduction to echelon form, recall—is tantamount to pre- or post-multiplication by a multiplicative combination of these matrices—validated by the associative property of matrix multiplication, (2.1-14). It *also* means that *multiplication by an elementary matrix leaves the rank of the matrix unchanged*—at least as long as we don't use $\mathsf{E}^*_{ii}(-1)$ (which would make a row or column disappear, artificially reducing the rank), since the latter is obtained through exactly the row/column operations these matrices effect! We shall return to these facts repeatedly in order to obtain important results below, but this observation is primarily of interest in the proofs.

Homework:

1. Consider the matrix

$$\begin{pmatrix} 1 & 2 & 3 & 4 \\ 5 & 6 & 7 & 8 \end{pmatrix}$$

(a) Pre-multiply by $\left(\mathbf{e}_2^{(n)}\right)^{\mathsf{T}}$, then

(b) post-multiply by $\mathbf{e}_2^{(n)}$. What is n in each case?

(c) What would n be if the above matrix is pre-multiplied by $\mathbf{1}^{(n)}$? Post-multiplied?

(d) Form $\mathbf{E}_{34}^{(4)}$ and thence $\mathbf{AE}_{34}^{(4)}$, by direct multiplication. Is the effect of the latter what was is expected?

(e) Form $\mathbf{E}_{21}^{(2)}$ and thence $\mathbf{E}_{21}^{(2)}\mathbf{A}$ by direct multiplication. Is the effect of the latter what was is expected?

2. Returning to the matrices of Problem 2 on page 53, we consider the matrices

$$\mathbf{A} = \begin{pmatrix} 3 & 4 \\ 1 & -1 \\ -2 & 3 \end{pmatrix} \text{ and } \mathbf{B} = \begin{pmatrix} 1 & 0 & 0 & 5 \\ 2 & 2 & 0 & 1 \end{pmatrix}.$$

Find those matrices which would be multiplied with

(a) $\mathbf{A}$ to add twice its second row to its third, and

(b) $\mathbf{B}$ to subtract the last column from 3 times its second,

verifying each by direct multiplication.

3. (*Multiplication by Diagonal Matrices*) Consider the two matrices

$$\mathbf{A} \equiv \begin{pmatrix} a_{11} & a_{12} & a_{13} \\ a_{21} & a_{22} & a_{23} \\ a_{31} & a_{32} & a_{33} \end{pmatrix} \text{ and } \mathbf{D} \equiv \begin{pmatrix} d_1 & 0 & 0 \\ 0 & d_2 & 0 \\ 0 & 0 & d_3 \end{pmatrix}.$$

(a) Show that multiplication in the order $\mathbf{AD}$ multiplies each *column* of $\mathbf{A}$ by the corresponding element of the diagonal matrix $\mathbf{D}$, while

(b) multiplication in the order $\mathbf{DA}$ multiplies each *row* by the corresponding diagonal element.

(c) Convince yourself, using block multiplication, and [independently] by expressing $\mathbf{D}$ in terms of the $E_{ii}^{*(3)}(d_i)$ matrices, that this result holds in general.

4. (*Multiplication by Matrices with Off-diagonal Terms*) Consider the three matrices

$$\mathbf{A} = \begin{pmatrix} a_{11} & a_{12} & a_{13} \\ a_{21} & a_{22} & a_{23} \\ a_{31} & a_{32} & a_{33} \end{pmatrix},$$

$$\mathbf{U} \equiv \begin{pmatrix} 0 & 1 & 0 \\ 0 & 0 & 1 \\ 0 & 0 & 0 \end{pmatrix}, \quad \mathbf{L} \equiv \begin{pmatrix} 0 & 0 & 0 \\ 1 & 0 & 0 \\ 0 & 1 & 0 \end{pmatrix}.$$

   (a) Show, perhaps using $E_{ii}^{(3)}$ matrices as in the above problem, that
      i. *post*-multiplication by the *upper* off-diagonal matrix, **AU**, yields a matrix shifting each *column* of **A** to the *right* by one, leaving the *first column* **0**, and
      ii. *post*-multiplication by the *lower* off-diagonal matrix, **AL**, yields a matrix shifting each *column* of **A** to the *left* by one, leaving the *last column* **0**.
   (b) and that
      i. *pre*-multiplication by the *upper* off-diagonal matrix, **UA**, yields a matrix shifting each *row* of **A** *up* by one, leaving the *last row* $\mathbf{0}^{\mathsf{T}}$, and
      ii. *pre*-multiplication by the *lower* off-diagonal matrix, **LA**, yields a matrix shifting each *row* of **A** *down* by one, leaving the *first row* $\mathbf{0}^{\mathsf{T}}$.

Summary. This entire section has been involved with matrices, what they are, and the operations on them, particularly important among them being *matrix multiplication.* We have defined the *transpose* of a matrix and obtained a result involving the transpose of the product of two matrices. Similarly, we have defined on column matrices (which, recall, are for us useful as the representation of vectors) the operation of the *inner product*—one reminiscent of, but distinct from, the dot product of directed line segment vectors.

Perhaps the most powerful idea introduced, however, was the concept of *partitioning* a matrix: By "partitioning" the matrix into a series of row or column matrices—"vectors" in their own right, recall—we introduced the concept of *elementary row/column operations* on matrices, which correspond to linear combinations of the resulting vectors. But this idea also carries through to *block multiplication* of matrices—a technique which allows us, without actually calculating the indicated products, to tell something about the results of matrix multiplication.

Finally, we showed that the elementary operations on matrices could themselves be represented in term of matrix multiplication. This result will be used to demonstrate many of the important properties of matrices in what follows.

2.2 Linear Transformations

We have introduced matrices and some of the operations on them. We now consider their prime utilization, linear transformations. But that means we must first know just what a "linear transformation" is! Though this has already been defined in Example 1.1.8, we repeat that definition here for sake of completeness. In essence, such a transformation is one which transforms linear combinations in one space into linear combinations in another ["over the same field"; see the discussion following the definition in that Example].

Definition 2.2.1. A *linear transformation* on a vector space $\mathcal{V}_1$ to another vector space $\mathcal{V}_2$ is a function, L, on $\mathcal{V}_1$ taking $\boldsymbol{v}_1$ and $\boldsymbol{v}_2 \in \mathcal{V}_1$:

$$L(a\boldsymbol{v}_1 + b\boldsymbol{v}_2) = aL(\boldsymbol{v}_1) + bL(\boldsymbol{v}_2) \tag{2.2-1}$$

Note that the "+" on the left-hand side applies to $\mathcal{V}_1$ while that on the right is in $\mathcal{V}_2$.

There is occasionally some confusion between the ideas of a "linear transformation" and the "equation of a line." We observe that, from the above definition, L must necessarily take $\mathbf{0} \in \mathcal{V}_1$ to its counterpart $\mathbf{0} \in \mathcal{V}_2$: for arbitrary $\boldsymbol{v} \in \mathcal{V}_1$,

$$L(\mathbf{0}) = L(\boldsymbol{v} - \boldsymbol{v}) = L(\boldsymbol{v}) - L(\boldsymbol{v}) = \mathbf{0}. \tag{2.2-2}$$

Compare this with the general equation of a line, $y = mx + b$ which, in general, takes $x = 0$ to $b \neq 0$. Note, too, that the "vector spaces" mentioned above are *arbitrary*; in particular, we can talk about linear transformations on, say, *matrices* just as well as on [n-tuple] *"vectors."*

For purposes of concreteness, as well as their intrinsic interest, we give a number of examples of such transformations. Many of these might surprise the reader!

Example 2.2.1. *Scalar Product of a Vector:* If a pair of vectors (of the same dimension, so that they can be added) is multiplied by a scalar, Properties (V2) and (V4) of Table 1.1 state that

$$c(a_1\boldsymbol{v}_1 + a_2\boldsymbol{v}_2) = a_1c\boldsymbol{v}_1 + a_2c\boldsymbol{v}_2;$$

i.e., the scalar product can be regarded as a linear transformation of a vector space onto itself. |

Example 2.2.2. *Angular Momentum of a Rigid Body:* (This is the prime example of such "transformations" in the next Part.) In a more elementary dynamics course, the angular momentum of a rigid body is expressed

$$\boldsymbol{H}_O = \mathsf{I}_O\boldsymbol{\omega},$$

where I_O is the *moment of inertia* of the body about either a fixed point about which the body is rotating, or the [*non*-fixed] center of mass of the body; in planar dynamics, this quantity is a scalar. But if there are two components to ω: $\boldsymbol{\omega} = \boldsymbol{\omega}_1 + \boldsymbol{\omega}_2$, in fact

$$\boldsymbol{H}_O = \mathbf{I}_O\boldsymbol{\omega} = \mathbf{I}_O(\boldsymbol{\omega}_1 + \boldsymbol{\omega}_2) = \mathbf{I}_O\boldsymbol{\omega}_1 + \mathbf{I}_O\boldsymbol{\omega}_2,$$

while the above example shows that the same would hold true for a linear combination of the ω's, demonstrating that the "transformation" is, indeed, linear. With reference to the discussion on page 37, this transformation could be viewed as one between two *different* spaces: the ω space, in which the basis is "measured" in terms of inverse time units $[\mathrm{T}]^{-1}$, and the $\boldsymbol{H}_O$ space, whose basis vectors have units of angular momentum, $[\mathrm{m}][\boldsymbol{r}][\boldsymbol{v}] = [\mathrm{M}][\mathrm{L}]^2[\mathrm{T}]^{-1}$; in both spaces the components would have no units. Alternatively, it could be interpreted as a transformation between vectors in the *same* space, one spanned by *dimensionless* basis vectors, with the components of each physically different type of vectors carrying the units. |

Example 2.2.3. *Inner Product:* In the vector space of n-dimensional column matrices, any such n-tuple, $\boldsymbol{v}$, can be selected to form the inner product with all other elements of the space:

$$f_{\boldsymbol{v}}(\boldsymbol{x}) \equiv \langle \boldsymbol{v}, \boldsymbol{x} \rangle.$$

By (2.1-9), this is a linear transformation (in fact, a "*bilinear*" one). It takes an element from the n-dimensional space and maps it into a *scalar*; such functions are generally referred to as "*linear functionals.*" |

Example 2.2.4. *Cross Product of Directed Line Segments:* Among the properties of the cross product (defined in 3-D space), we have that

$$(a_1\boldsymbol{a}_1 + a_2\boldsymbol{a}_2) \times \boldsymbol{b} = a_1\boldsymbol{a}_1 \times \boldsymbol{b} + a_2\boldsymbol{a}_2 \times \boldsymbol{b},$$

demonstrating that the cross product is a linear transformation on its first component as "argument." (Clearly the same could be shown for the second component vector; this is another example of a bilinear mapping, like the inner product.) |

Example 2.2.5. *"Vector" Representations under Change of Basis:* Consider a vector $\boldsymbol{v}$ which, relative to a basis $\{\boldsymbol{u}_1, \boldsymbol{u}_2, \ldots, \boldsymbol{u}_n\}$, has a representation $\boldsymbol{v} = \sum_i^n v_i\boldsymbol{u}_i$. If, now, one changes to a different basis $\{\boldsymbol{u}'_1, \boldsymbol{u}'_2, \ldots, \boldsymbol{u}'_n\}$ [note that the number of basis vectors is the same, the dimension of the vector space], $\boldsymbol{v}$ assumes a *new* representation $\boldsymbol{v} = \sum_i^n v'_i\boldsymbol{u}'_i$. Expressing the respective representations as n-tuples

$$\mathsf{v} \equiv \begin{pmatrix} v_1 \\ v_2 \\ \vdots \\ v_n \end{pmatrix}, \quad \mathsf{v}' \equiv T(\mathsf{v}) \equiv \begin{pmatrix} v'_1 \\ v'_2 \\ \vdots \\ v'_n \end{pmatrix},$$

we allege that the transformation T from v to v' is, in fact, a linear one.

Note first that, since the two bases lie in the same space, spanned by both the $\{\boldsymbol{u}_i\}$ and the $\{\boldsymbol{u'}_i\}$, we can express, in particular, the original basis vectors in terms of the new basis:

$$\boldsymbol{u}_i = \sum_j^n u_{ij}\boldsymbol{u'}_j. \tag{2.2-3}$$

But then

$$\boldsymbol{v} = \sum_i^n v_i\boldsymbol{u}_i = \sum_i^n v_i \sum_j^n u_{ij}\boldsymbol{u'}_j = \sum_j^n \left(\sum_i^n v_i u_{ij}\right)\boldsymbol{u'}_j = \sum_j^n v'_j\boldsymbol{u'}_j$$

—since the representation of a vector relative to the basis $\{\boldsymbol{u'}_i\}$ is unique (Theorem 1.3.1); *i.e.*,

$$v'_j = \sum_i^n v_i u_{ij}. \tag{2.2-4}$$

This gives us the explicit relation between the two representations.

From here, it is merely tedious to demonstrate the linearity: Given two vectors' representations v_1 and v_2, the linear combination

$$a\mathsf{v}_1 + b\mathsf{v}_2 \equiv \begin{pmatrix} av_{11} + bv_{21} \\ av_{12} + bv_{22} \\ \vdots \\ av_{1n} + bv_{2n} \end{pmatrix}$$

(by the definition in Example 1.1.3). By (2.2-4), the jth element of the new representation of this n-tuple

$$v'_j = \sum_i^n (av_{1i} + bv_{2i})u_{ij} = a\sum_i^n v_{1i}u_{ij} + b\sum_i^n v_{2i}u_{ij}$$

—which is immediately recognized as the jth component of $aT(\mathsf{v}_1) + bT(\mathsf{v}_2)$.

Note that the u_{ij} above came from (2.2-3), relating the bases; this is an important observation which will return when we consider arbitrary linear transformations in Theorem 2.2.1 below, particularly in the remarks following it. Note, however, that that equation expresses the "old" basis in terms of the "new" one; typically one expresses the "new" in terms of the "old," and it is the latter approach which will be used when we finally consider general basis transformations in Section 3.6. |

Example 2.2.6. *Derivatives of* n^{th} *Degree Polynomials:* This might be a novel perspective, but recall that the derivative of the polynomial $f(x) = a_0 + a_1x + a_2x^2 + \ldots + a_nx^n$ is just the $(n-1)^{th}$-degree polynomial $f'(x) = a_1 + 2a_2x + \cdots + na_nx^{n-1}$. But differentiation is inherently linear:

$$\frac{d}{dx}(af_1(x) + bf_2(x)) = a\frac{df_1}{dx} + b\frac{df_2}{dx}$$

Thus the operation of differentiation on such polynomials can be considered to effect a linear transformation from the space of nth-degree polynomials to that of $(n-1)^{th}$-degree ones! (Actually, one normally hears the phrase "linear *operator*" applied to differentiation; see Section 2.4.) |

Domain and Range of a [Linear] Transformation and their Dimension

Let us introduce a little more jargon used to describe transformations (linear or otherwise):

Definition 2.2.2. Given a linear transformation L from a vector space $\mathcal{V}_1$ to another, $\mathcal{V}_2$, the set of vectors in $\mathcal{V}_1$ *on which* L acts is called the *domain* of L; the set of vectors in $\mathcal{V}_2$ *to which* they transform is called its *range*.

The domain of a transformation is typically $\mathcal{V}_1$ itself; as a vector space, that domain has a dimension.[3] Assuming the domain of the transformation is a vector space, then its range is also a vector space: We have *closure*: for any $\boldsymbol{y}_1 = L(\boldsymbol{x}_1)$ and $\boldsymbol{y}_2 = L(\boldsymbol{x}_2)$ for the $\boldsymbol{x}_i$ in the domain of L, we claim that the linear combination $a\boldsymbol{y}_1 + b\boldsymbol{y}_2$ is itself in the range of L. Indeed,

$$a\boldsymbol{y}_1 + b\boldsymbol{y}_2 = aL(\boldsymbol{x}_1) + bL(\boldsymbol{x}_2) = L(a\boldsymbol{x}_1 + b\boldsymbol{x}_2),$$

where, since the domain is a vector space, $(a\boldsymbol{x}_1 + b\boldsymbol{x}_2)$ is in it, so the linear combination lies in the range of L. All other algebraic properties in Table 1.1 are inherited from the linearity as shown in Example 2.2.5, and the range of L is a vector space as alleged. Thus one commonly refers to the range of a linear transformation as the *range space.* As a vector space, it, too, has a dimension.

Though the domain be all of $\mathcal{V}_1$, the range may *not* be all of $\mathcal{V}_2$. In such cases it would be called a "proper subset" of $\mathcal{V}_2$. Consider, for example, the transformation $\boldsymbol{y} = \mathsf{A}\boldsymbol{x}$ generated by a matrix

$$\mathsf{A} = \begin{pmatrix} 1 & 3 \\ 2 & -4 \\ 0 & 0 \\ 7 & 3 \\ 0 & 0 \end{pmatrix}$$

taking elements in the space $\mathcal{V}_1$ of 2-tuples to those of the space $\mathcal{V}_2$ of 5-tuples. Every vector $\boldsymbol{y}$ in the range of this transformation will be of the form

$$\boldsymbol{y} = \begin{pmatrix} y_1 \\ y_2 \\ 0 \\ y_3 \\ 0 \end{pmatrix}$$

[3] One might artificially limit the domain of a given transformation, for example, by specifying that it operate only on vectors below a certain magnitude. In this case, linear combinations of vectors in the domain could easily fall outside it; then the domain would *not* qualify as a vector space, and the concept of dimension would be undefined. Such pathological cases are generally not encountered in practice, however!

This is clearly a proper subset of the set of *all* 5-tuples; further, it is, in fact, a *vector space*—a "*subspace*" of $\mathcal{V}_2$.

It it enticing to say this subspace has a dimension of 3—that it is spanned by the vectors, say,

$$\begin{pmatrix}1\\0\\0\\0\\0\end{pmatrix},\ \begin{pmatrix}0\\1\\0\\0\\0\end{pmatrix},\ \text{and}\ \begin{pmatrix}0\\0\\0\\1\\0\end{pmatrix};$$

but, upon closer examination, this turns out not to be the case: The three non-$\mathbf{0}$ rows of $\mathbf{A}$ are 2-tuples, elements of a space of dimension 2; thus only two of the three are linearly independent—there are c_i, not all 0, such that $c_1(1,3) + c_2(2,-4) + c_3(7,3) = 0$. In fact, $c_1 = 17$, $c_2 = 9$, and $c_3 = -5$, or $c_1 = 3.4$, $c_2 = 1.8$, and $c_3 = -1$ are possible c_i; thus

$$\mathbf{A}\begin{pmatrix}x_1\\x_2\end{pmatrix} = \begin{pmatrix}x_1+3x_2\\2x_1-4x_2\\0\\7x_1+3x_2\\0\end{pmatrix} = \begin{pmatrix}x_1+3x_2\\2x_1-4x_2\\0\\3.4(x_1+3x_2)+1.8(2x_1-4x_2)\\0\end{pmatrix}$$

$$= (x_1+3x_2)\begin{pmatrix}1\\0\\0\\3.4\\0\end{pmatrix} + (2x_1-4x_2)\begin{pmatrix}0\\1\\0\\1.8\\0\end{pmatrix},$$

and the range is only of dimension 2: regardless of the values of the x_i, all vectors in the domain are linear combinations of these two 5-tuples!

2.2.1 Linear Transformations: Basis and Representation

The careful reader might recall that in the examples in Section 1.2.3, we purposely omitted exhibiting a basis for the "vector space" of linear transformations, forestalling that until the present chapter. Having now given some concrete examples of such transformations, we shall finally get to the business of establishing a basis for them.

Recall that the basis for a vector space is comprised of elements *from* that space; in the present instance, then, our basis must, in fact, consist of certain linear transformations—*i.e.*, *functions*. And though the above examples give some evidence of these objects called "linear transformations" in particular circumstances, it is not clear that, for an arbitrary vector space, such transformations even exist! The following theorem demonstrates both the existence of such transformations for any space, and exhibits the basis for such linear transformations:

Theorem 2.2.1. Consider two vector spaces, $\mathcal{V}_1$ of dimension m with [a] basis $\{\boldsymbol{u}_1, \boldsymbol{u}_2, \ldots, \boldsymbol{u}_m\}$, and $\mathcal{V}_2$ of dimension n with [a] basis $\{\boldsymbol{u}'_1, \boldsymbol{u}'_2, \ldots, \boldsymbol{u}'_n\}$. For arbitrary $\boldsymbol{v} = v_1\boldsymbol{u}_1 + v_2\boldsymbol{u}_2 + \ldots + v_n\boldsymbol{u}_m$ in $\mathcal{V}_1$, define the functions

$$L_{ij}(\boldsymbol{v}) \equiv v_i\boldsymbol{u}'_j; \tag{2.2-5}$$

i.e., $L_{ij}(\boldsymbol{v})$ multiplies the jth basis vector in the range space by the ith component of the domain's $\boldsymbol{v}$.[4] Then

1. the L_{ij} are linear transformations, and

2. they form a basis for the set of *all* linear transformations from $\mathcal{V}_1$ to $\mathcal{V}_2$.

Proof. We begin by showing the first contention: For two arbitrary vectors $\boldsymbol{v}_1 = v_{11}\boldsymbol{u}_1 + v_{12}\boldsymbol{u}_2 + \ldots + v_{1n}\boldsymbol{u}_m$ and $\boldsymbol{v_2} = v_{21}\boldsymbol{u}_1 + v_{22}\boldsymbol{u}_2 + \ldots + v_{2n}\boldsymbol{u}_m$ in $\mathcal{V}_1$

$$L_{ij}(a\boldsymbol{v}_1 + b\boldsymbol{v}_2) \equiv av_{1i}\boldsymbol{u'}_j + bv_{2i}\boldsymbol{u'}_j \equiv aL_{ij}(\boldsymbol{v}_1) + bL_{ij}(\boldsymbol{v}_2)$$

and 1. is proved. (Thus we have at least *one* linear transformation at our disposal!)

Now address the second part: We claim that the L_{ij} form a basis for *any* linear transformation; *i.e.*, for any linear transformation L, we can write

$$L = \sum_i^m \sum_j^n \ell_{ij} L_{ij}. \tag{2.2-6}$$

We first prove the L_{ij} above *span* the "space" of linear transformations. This means that, for any $\boldsymbol{v} = \sum_k^m v_k\boldsymbol{u}_k$ in $\mathcal{V}_1$,

$$L(\boldsymbol{v}) = \sum_i^m \sum_j^n \ell_{ij}L_{ij}(\boldsymbol{v}) = \sum_i^m \sum_j^n \ell_{ij}L_{ij}\left(\sum_k^m v_k\boldsymbol{u}_k\right) = \sum_i^m \sum_j^n \sum_k^m v_k\ell_{ij}L_{ij}(\boldsymbol{u}_k),$$

by the linearity of the L_{ij}. But, by the definition of the L_{ij}, (2.2-5),

$$L_{ij}(\boldsymbol{u}_i) \equiv \boldsymbol{u}'_j, \text{ while } L_{ij}(\boldsymbol{u}_k) \equiv \boldsymbol{0} \text{ for } k \neq i; \tag{2.2-7}$$

thus the only non-$\boldsymbol{0}$ $L_{ij}(\boldsymbol{u}_k)$ terms are those for which $i = k$, and the above collapses to "merely"

$$L(\boldsymbol{v}) = \sum_j^n \sum_k^m v_k\ell_{kj}L_{kj}(\boldsymbol{u}_k), \tag{2.2-8}$$

and *this* is what we must show. But, after all these preliminaries, that is relatively easy: since

$$L(\boldsymbol{v}) = L\left(\sum_k^m v_k\boldsymbol{u}_k\right) = \sum_k^m v_kL(\boldsymbol{u}_k), \tag{2.2-9}$$

[4] We clearly are assuming that both spaces have scalar multiplication defined over the same "field"; see Definition 1.1.2 and the discussion surrounding it regarding vector spaces.

and, since $L(\boldsymbol{u}_k) \in \mathcal{V}_2$, we can express

$$L(\boldsymbol{u}_k) = \sum_j^n u'_{kj}\boldsymbol{u}'_j = \sum_j^n u'_{kj}L_{kj}(\boldsymbol{u}_k) \tag{2.2-10}$$

(again by (2.2-7)), we conclude that

$$L(\boldsymbol{v}) = \sum_k^m \sum_j^n v_k u'_{kj} L_{kj}(\boldsymbol{u}_k).$$

But this is precisely (2.2-8): the ℓ_{kj} in that equation are just the u'_{kj} in (2.2-10).

We have proven that the L_{ij} span the space, but we must still show they are linearly independent to demonstrate they form a basis for the space of linear transformations. Assume they *weren't*; *i.e.*, there are c_{ij}, not all 0, such that

$$\sum_i^m \sum_j^n c_{ij} L_{ij} = \mathbf{0}$$

—the *zero* linear transformation from $\mathcal{V}_1$ to $\mathcal{V}_2$. But then, applying this linear combination to, in particular, each of the basis vectors $\{\boldsymbol{u}_1, \boldsymbol{u}_2, \ldots, \boldsymbol{u}_m\}$ in turn, we get that

$$\mathbf{0} = \sum_i^m \sum_j^n c_{ij} L_{ij}(\boldsymbol{u}_k) = \sum_j^n c_{kj}\boldsymbol{u}'_j$$

—*again* by (2.2-7), which means that [each] $c_{kj} = 0$, by the linear independence of the $\{\boldsymbol{u}'_k\}$. □

We note several important points about this theorem—points in some ways more important than the theorem itself!—in passing: In the first place, there are mn L_{ij}'s—n corresponding to each of the $i = 1, \ldots, m$ in (2.2-7); thus the dimension of this vector space is precisely mn.

In the second place, the final remark in the proof points out that the ℓ_{ij} in (2.2-8)—and thus in (2.2-6), the equation defining the linear combination of the L_{ij}—are precisely the u'_{kj} in (2.2-10). But the last equation shows that these quantities are simply the components of the *representations* of the transformed original basis vectors in the new space! This is of fundamental importance: what effectively determines a given linear transformation between vector spaces is what that transformation does to the *basis* vectors of the original space; once that is established, the transformation of any other vector is simply the *same linear combination* of the *transformed* basis vectors as that vector's representation in terms of the *original* basis. This is precisely what (2.2-9) above demonstrates.

The ℓ_{ij}, on the other hand, are just the "components" of the "vector" L in terms of the "basis" $\{L_{ij}\}$—*i.e.*, the *representation* of the linear transformation! This observation, however, has profound implications: it indicates that, when we *change* the basis for $\mathcal{V}_1$, then we *change* the u'_{kj} in (2.2-10), and thus the ℓ_{ij} in (2.2-8); *i.e.*, the *representation changes.* Just *how* it changes is a central theme in the next chapter, but the key point for the present is that the representation of a linear transformation *does* depend on the basis of the vectors on which it operates.

But the last point is of even greater import in application: We have shown that, from the properties of matrices, matrix products are inherently linear transformations. Conversely, in the same way that *any* vector can be expressed in terms of its *representation* relative to a certain basis, these linear transformations, as "vectors," can be defined through their representations relative to a basis (such as the one in the above Theorem). But that representation requires mn ℓ_{ij}'s, the very form of which almost begs to be represented as a *matrix.* Thus we have stumbled on, and all but proved, the fundamental idea:

Theorem 2.2.2. Any linear transformation can be represented as a matrix.

In the same way that a "vector" $\boldsymbol{v}$ [in the common sense of the word] can be *represented* as an n-tuple column matrix v, any linear transformation L can be *represented* as an $m \times n$ matrix L—a matrix of *real numbers* making it easy to manipulate such linear transformations easily, regardless of the type of linear transformation (such as the L_{ij} above) involved. But, as noted above, when the basis of the vector space on which the linear transformation operates changes, so does its representation.

The study of linear transformations is tantamount to the study of matrices.

Homework:

1. Consider the two linear transformations $\mathcal{A}$ and $\mathcal{B}$:

$$\mathcal{A}: \quad \begin{aligned} y_1 &= a_{11}x_1 + a_{12}x_2 \\ y_2 &= a_{21}x_1 + a_{22}x_2 \end{aligned} \qquad \text{and} \qquad \mathcal{B}: \quad \begin{aligned} y_1 &= b_{11}x_1 + b_{12}x_2 \\ y_2 &= b_{21}x_1 + b_{22}x_2. \\ y_3 &= b_{31}x_1 + b_{32}x_2 \end{aligned}$$

   (a) By direct substitution of $\mathsf{y} \equiv (y_1 y_2)^{\mathsf{T}}$ in $\mathcal{A}$ for $\mathsf{x} \equiv (x_1 x_2)^{\mathsf{T}}$ of $\mathcal{B}$, find the *composite* transformation $\mathcal{B} \circ \mathcal{A}(\boldsymbol{x}) \equiv \mathcal{B}(\mathcal{A}(\boldsymbol{x}))$.

   (b) Representing the transformations $\mathcal{A}$ and $\mathcal{B}$ by *matrix products* $\mathsf{y} = \mathsf{Ax}$ and $\mathsf{y} = \mathsf{Bx}$, show that the *same* composite transformation is given by the *matrix product* BAx.

Dyadics

Though occasionally ventured into by scientists, the field of dyadics is one almost universally ignored by engineers. This is unfortunate, because once one gets past the somewhat dicey notation, they provide an immediate generalization from vectors to matrices and even tensors, and they explicitly demonstrate the fundamental issue above: that the representation of a linear operator on a vector space depends on the basis of *that* vector space.

Recall that we have been ubiquitous in the *representation* of vectors relative to a basis $(\boldsymbol{u}_1, \boldsymbol{u}_2, \ldots, \boldsymbol{u}_n)$: $\boldsymbol{v} = v_1\boldsymbol{u}_1 + v_2\boldsymbol{u}_2 + \ldots + v_n\boldsymbol{u}_n$; in the present context, the basis vectors would be called *first-order dyads*; scalar quantities would be referred to as *zeroth-order dyads.* Given such a basis in which an inner/dot product is defined, we define the *second-order dyad* $\boldsymbol{u}_i\boldsymbol{u}_j$—a formal juxtaposition of two basis vectors. This expression is *non-commutative*: $\boldsymbol{u}_j\boldsymbol{u}_i \neq \boldsymbol{u}_i\boldsymbol{u}_j$; furthermore, there is no multiplication implied. We can turn the set of such forms into a vector space by defining scalar multiplication and vector addition in the obvious fashion (elements $a_{ij}\boldsymbol{u}_i\boldsymbol{u}_j$ and $a_{ji}\boldsymbol{u}_j\boldsymbol{u}_i$, in different order, can be added together, they just can't be collected into a single term). The vector sum $\sum a_{ij}\boldsymbol{u}_i\boldsymbol{u}_j$ is then referred to as a *dyadic.*

We now augment these basic operations with two operations of dyadic multiplication:

- *outer multiplication*: the *outer product* of two basis vectors is simply their dyad, $\boldsymbol{u}_i\boldsymbol{u}_j$. Thus the product of two first-order dyads is a second-order one; the operation is non-commutative.
- *inner multiplication*: the *inner product* of two basis vectors is the dot/inner product of the vectors:
$$\boldsymbol{u}_i \cdot \boldsymbol{u}_j \equiv \langle \boldsymbol{u}_i, \boldsymbol{u}_j \rangle$$
—generating a scalar, a *zeroth-order dyad.* Unlike the outer product, this operation *is* commutative.

Thus the outer product *increases* the order of the result, while the inner product *reduces*—or *contracts*—it. We note that these operations enjoy no closure unless one considers the set of dyadics of *all orders.*

Now let us consider the inner product of two dyadics, one of second order and the other of first: if the basis vectors happen to be *orthogonal* (in the sense of Definition 2.1.7),

$$\left(\sum_{i,j} a_{ij}\boldsymbol{u}_i\boldsymbol{u}_j\right) \cdot \left(\sum_j v_j\boldsymbol{u}_j\right) \equiv \sum_{i,j} a_{ij}v_j(\boldsymbol{u}_i\boldsymbol{u}_j) \cdot \boldsymbol{u}_j = \sum_j a_{ij}v_j\boldsymbol{u}_i,$$

we get precisely the definition, but *along with the basis vectors*, of the matrix product, Definition 2.1-5 (which, recall, gives only the n-tuple representation of the *components* of the $\boldsymbol{u}_i$). But this is simply a representation of an arbitrary linear transformation, which we see can also be *represented* as a second-order dyadic. Even if the basis vectors don't happen to be orthogonal, however, this becomes the generalization of equation (2.1-10).

The defined operations on dyadics may seem curious, even arbitrary; but, in retrospect, they are really no more so than those for the corresponding matrix operations. And, as mentioned above, they are really just a consistent generalization of those for vectors, and they explicitly show the dependence of the linear operator on the basis of the vectors.

But they have another advantage: if one swallows the above definitions for first- and second-order dyadics, it is clear that *outer* multiplication can generate *third*-order dyadics (or those of any order); the inner product of such a dyadic with a first-order one would then result in one of second order. Such higher-order generalization of vectors, and now matrices, are referred to as *tensors*; in the same way that vectors can be *represented* as first-order dyadics, and linear operators as second-order dyadics, higher-order "tensors" can be represented as dyadics of corresponding order. In this context, *all* these quantities are really tensors, just of different order. The important point is once again *representation*: the physical quantity—the *tensor*—has an identity *independent of the coordinates used to represent it.* The intent is to keep separate that quantity and its representation.

We now demonstrate this ability to express linear transformation in terms of matrices. Of the examples given in the previous section (page 61), we shall deal with only the last three: scalar multiplication in Example 2.2.1 can formally be represented as a matrix $c\mathbf{1}$ for appropriately-dimensioned identity matrix $\mathbf{1}$, but this seems a little contrived; and the transformation from $\boldsymbol{\omega}$ to $\boldsymbol{H}_O$ in Example 2.2.2 will occupy the entire next Part!

Example 2.2.7. *Cross Product of Directed Line Segments:* We have shown previously that $\boldsymbol{a} \times \boldsymbol{b}$ is effectively a linear transformation on the first component, $\boldsymbol{a}$. From the above discussion, we should be able to *represent* this operation as some matrix multiplying the representation of $\boldsymbol{a}$. And, indeed, for $\boldsymbol{b} = (b_1, b_2, b_3)^{\mathsf{T}}$ (a column matrix, as we represent all "common vectors")

$$\boldsymbol{a} \times \boldsymbol{b} = \begin{pmatrix} 0 & b_3 & -b_2 \\ -b_3 & 0 & b_1 \\ b_2 & -b_1 & 0 \end{pmatrix} \begin{pmatrix} a_1 \\ a_2 \\ a_3 \end{pmatrix} \equiv \mathsf{ba}$$

as expected. |

Homework:

1. Show this.
2. Show that the rank of the b matrix is only 2 if all three $b_i \neq 0$.
3. Show that, rather remarkably, even if two of the three $b_i = 0$, the rank of b is *still* 2!

Example 2.2.8. *"Vector" Representations under Change of Basis:* Recall that in the presentation of Example 2.2.5, we obtained an equation explicitly giving the components of the new representation under a change of basis, (2.2-9). Though it might have been overlooked then, this is nothing more than the equation defining matrix multiplication, (2.1-5), in this case written as an *inner product* (equations (2.1-6) and (2.1-7)):

$$v'_j = \sum_i^n v_i u_{ij} = \mathsf{v}^{\mathsf{T}} \mathsf{u}_j = \mathsf{u_j}^{\mathsf{T}} \boldsymbol{v}$$

since the inner product commutes (equation (2.1-8)). Thus, expressing v' as a [column] vector,

$$\mathsf{v}' = \begin{pmatrix} \mathsf{u_1}^{\mathsf{T}}\mathsf{v} \\ \mathsf{u_2}^{\mathsf{T}}\mathsf{v} \\ \vdots \\ \mathsf{u_n}^{\mathsf{T}}\mathsf{v} \end{pmatrix} = \begin{pmatrix} \mathsf{u_1}^{\mathsf{T}} \\ \mathsf{u_2}^{\mathsf{T}} \\ \vdots \\ \mathsf{u_n}^{\mathsf{T}} \end{pmatrix} \mathsf{v}$$

(where we have made use of "block multiplication," page 52). But the second matrix above is that whose *rows* are the u_{ij}—the *representations* of the original basis vectors $\boldsymbol{u}_i$ in terms of the new basis $\boldsymbol{u'}_i$, equation (2.2-3). Once again we see the importance of the remark on page 67: what is relevant in a transformation is what happens to the basis vectors.

As noted in Example 2.2.5, this approach is not the one ultimately to be followed, where we will generally express the "new" basis vectors in terms of the "old" ones, but it does demonstrate the fact that the linear transformation from one representation to another can be expressed as a matrix product. |

Example 2.2.9. *Derivatives of nth Degree Polynomials:* Recalling that our basis for polynomials is just the set of $n+1$ powers of x, $\{x^0, x^1, \ldots, x^n\}$, differentiation effectively transforms from the representation of $f(x)$, $(a_0, a_1, \ldots, a_n)^\mathsf{T}$ to the representation of $f'(x)$ (in the space of $(n-1)^{th}$ degree polynomials), $(a_1, 2a_2, \ldots, na_n)$. But this can be represented as the product of an $(n-1) \times n$ matrix and the $f(x)$ representation:

$$\begin{pmatrix} a_1 \\ 2a_2 \\ \vdots \\ na_n \end{pmatrix} = \begin{pmatrix} 0 & 1 & 0 & \ldots & 0 \\ 0 & 0 & 2 & \ldots & 0 \\ \vdots & \vdots & \vdots & \ddots & \vdots \\ 0 & 0 & 0 & \ldots & n \end{pmatrix} \begin{pmatrix} a_0 \\ a_1 \\ \vdots \\ a_n \end{pmatrix}$$

as desired. |

We conclude these examples with one not "set up" by the previous "Examples of Linear Transformations." It is meant to drive home the fact that a linear transformation is defined in terms of what it does to the *basis* vectors, but has the practical result of, knowing explicitly what this effect on the basis vectors is, being able obtain the matrix representation of the transformation matrix itself in an almost trivial fashion! We start with a general discussion of the technique and then provide a more numerical example.

Example 2.2.10. *Explicit Determination of the Matrix Representation of a Linear Transformation:* Consider an arbitrary transformation $\mathbf{A}$ which operates on the *ordered* basis vectors $(\boldsymbol{u}_1, \boldsymbol{u}_2, \ldots, \boldsymbol{u}_m)$. In particular, if these transform to a vector space spanned by $(\boldsymbol{u}'_1, \boldsymbol{u}'_2, \ldots, \boldsymbol{u}'_n)$:

$$\boldsymbol{u}_1 \to \mathbf{A}\boldsymbol{u}_1 = u_{11}\boldsymbol{u}'_1 + u_{21}\boldsymbol{u}'_2 + \cdots + u_{n1}\boldsymbol{u}'_n$$

$$\vdots$$

$$\boldsymbol{u}_m \to \mathbf{A}\boldsymbol{u}_m = u_{1m}\boldsymbol{u}'_1 + u_{2m}\boldsymbol{u}'_2 + \cdots + u_{nm}\boldsymbol{u}'_n.$$

In terms of the n-tuple *representations* of the basis vectors $(\boldsymbol{u}'_1, \boldsymbol{u}'_2, \ldots, \boldsymbol{u}'_n)$ (*cf.*

equation (1.3-4)), these relations become:

$$\mathbf{Au}_1 = u_{11}\begin{pmatrix}1\\0\\\vdots\\0\end{pmatrix} + \cdots + u_{n1}\begin{pmatrix}0\\0\\\vdots\\1\end{pmatrix}$$

$$\mathbf{Au}_m = u_{1m}\begin{pmatrix}1\\0\\\vdots\\0\end{pmatrix} + \cdots + u_{nm}\begin{pmatrix}0\\0\\\vdots\\1\end{pmatrix}.$$

Thus, setting up the matrix of vectors of transformed basis vectors, and using block multiplication,

$$(\mathbf{Au}_1\ \mathbf{Au}_2\ \ldots \mathbf{Au}_m) = \begin{pmatrix} u_{11} & u_{12} & \cdots & u_{1m} \\ u_{21} & u_{22} & \cdots & u_{2m} \\ \vdots & \vdots & \ldots & \vdots \\ u_{n1} & u_{n2} & \cdots & u_{nm} \end{pmatrix} = \mathbf{A}(\mathbf{u}_1\ \mathbf{u}_2\ \ldots \mathbf{u}_m) = \mathbf{A1} = \mathbf{A}$$

Thus, in fact, the matrix $\mathbf{A}$ is nothing more than the matrix of the coefficients of the basis vectors' transformations! This is precisely what was shown by Theorem 2.2.1: these are the ℓ_{ij} obtained there.

As an example of *this*, consider the transformation from a 2-D space to a 5-D space such that

$$\mathbf{Au}_1 = u_1' + 2u_2' + 7u_4'$$
$$\mathbf{Au}_2 = 3u_1' - 4u_2' + 3u_4'$$

Then the matrix representation of this transformation is just

$$\mathbf{A} = \begin{pmatrix} 1 & 3 \\ 2 & -4 \\ 0 & 0 \\ 7 & 3 \\ 0 & 0 \end{pmatrix}.$$

This is, in fact, nothing more than the example on page 64, where the transformation was from the space of 2-tuples to the space of 5-tuples; but this matrix would be the representation of this transformation between *any* spaces of those dimensions! |

2.2.2 Null Space of a Linear Transformation

We observed, in equation (2.2-2), that any linear transformation takes the $\mathbf{0}$ vector from the first space into the $\mathbf{0}$ vector in the other one. We now examine the question of whether there are any *other*, *non*-**0** vectors which transform to **0**. This apparently arcane issue actually has important ramifications in application, such as the solution of linear systems, the topic of eigenvectors, and even the solution of differential equations (see Section 2.4 below).

We start with a brief example of what we are talking about. While this will be a matrix example, from what we have just discussed, it would occur for *any* linear transformation represented as a matrix.

Example 2.2.11. Consider the matrix

$$\mathsf{A} = \begin{pmatrix} 1 & -2 & -4 & 2 \\ 0 & 1 & 2 & -1 \\ -1 & 5 & 10 & -5 \end{pmatrix};$$

find those column vectors $\boldsymbol{v}$ such that $\mathsf{A}\boldsymbol{v}$ are **0**.

Since A is a 3×4 matrix, we wish to find the $\boldsymbol{v} \equiv (v_1, v_2, v_3, v_4)^{\mathsf{T}}$ such that

$$\begin{aligned} v_1 - 2v_2 - 4v_3 + 2v_4 &= 0 && \text{(a)} \\ v_2 + 2v_3 - v_4 &= 0 && \text{(b)} \\ -v_1 + 5v_2 + 10v_3 - 5v_4 &= 0 && \text{(c)} \end{aligned}$$

We solve the system:

$$\begin{aligned} &\text{(a)+2(b)}: & v_1 \qquad\qquad &= 0 && \text{(a}')\\ &\text{(c)+(a)}: & 3v_2 + 6v_3 - 3v_4 &= 0 && \text{(b}')\\ &\text{(b}')\text{−3(b)}: & 0 &= 0 & \end{aligned}$$

so that (a′) and (b′) give the answer:

$$\boldsymbol{v} = \begin{pmatrix} 0 \\ -2v_3 + v_4 \\ v_3 \\ v_4 \end{pmatrix};$$

i.e., for any values of v_3 and v_4—*even non-0 values*—$\mathsf{A}\boldsymbol{v} = \mathbf{0}$. We note in passing that the process of "solution" for this system involves adding and subtracting "rows" of equations; in the context of matrices, this is equivalent to the "elementary row operations" encountered earlier. |

Having exemplified this concept, we now introduce the term used to describe it:

Definition 2.2.3. The *null space* of a linear transformation is that set of elements which transform into the **0** vector.

The term "null *space*" arises because the elements do, in fact, define a vector space—in fact, a *sub*space of the original space on which the transformation acts. Clearly, if $\boldsymbol{v}_1$ and $\boldsymbol{v}_2$ are elements of the null space—*i.e.*, $A(\boldsymbol{v}_i) = \mathbf{0}$ under the linear transformation A, then

$$\mathsf{A}(a\boldsymbol{v}_1 + b\boldsymbol{v}_2) = aA(\boldsymbol{v}_1) + bA(\boldsymbol{v}_2) = \mathbf{0},$$

and, by Theorem 1.1.1, this forms a subspace of the original.

Dimension of the Null Space

As a [vector] *space*, it makes sense to talk about the *basis* and/or *dimension* of the null space. If we call that dimension "n" and the dimension of the domain "d," we have immediately that $n \leq d$, since the null space is, itself, a subspace of the domain. In the above example, there were the two [independent] parameters v_3 and v_4 determining the form of $\mathbf{v}$ in the null space. We can allow each to assume the value of, say, 1 while the other(s) are 0 and get two vectors

$$\boldsymbol{u}_1 \equiv \begin{pmatrix} 0 \\ -2 \\ 1 \\ 0 \end{pmatrix} \quad \boldsymbol{u}_2 \equiv \begin{pmatrix} 0 \\ 1 \\ 0 \\ 1 \end{pmatrix}$$

resulting in two linearly independent vectors. They always *will* be linearly independent, due to their positions in the final "null space vector" form. Any other vectors in the null space will be linear combinations of these; thus they comprise a basis for the null space, and we see that, in this case, the null space has a dimension of 2.

> *Important Note*: Even though the above example was cast in terms of matrices, the results carry over to *arbitrary* linear transformations on *arbitrary* vector spaces: It is necessary only to work with the matrix *representation* of the linear operator; the column matrix $\mathbf{v}$ solved for then becomes the *representation* of the element of the null space relative to the basis—of *whatever* elements comprise that vector space—in terms of which both are expressed. Thus, in the above example, the null space consists of linear combinations of $-2\boldsymbol{u}_2 + \boldsymbol{u}_3$ and $\boldsymbol{u}_2 + \boldsymbol{u}_4$ relative to the basis $\boldsymbol{u}_1, \boldsymbol{u}_2, \ldots, \boldsymbol{u}_4$.

This discussion leaves open the matter of what to call the dimension of the null space consisting only of the $\mathbf{0}$ element itself! While one might be tempted to call it "1," since there is only one element, this would clearly be at odds with null spaces which could be defined by only a single parameter, say v_4 in the above example. Little reflection makes obvious the *definition* of the null space $\{\mathbf{0}\}$ as having dimension 0; this is consistent with Theorem 1.2.1.

Relation between Dimensions of Domain, Range, and Null Space

We have mentioned above that we can assign a dimension to the domain and range of a transformation from $\mathcal{V}_1$ and $\mathcal{V}_2$; typically the domain dimension is that of $\mathcal{V}_1$, though the *range* may have one less than that of $\mathcal{V}_2$ (see the example on page 64). We now establish an important relationship between these dimensions and that of the null space (which is, recall, in $\mathcal{V}_1$):

Theorem 2.2.3. If a linear transformation L has domain of dimension d, range of dimension r, and null space of dimension n, then $r = d - n$.[5]

[5] By above discussion, r is immediately ≥ 0

Proof. Since the null space has dimension n, there is a basis $\{\mathbf{u}_1, \mathbf{u}_2, \ldots, \mathbf{u}_n\}$ in $\mathcal{V}_1$ which spans it. Since $n \leq d$, we can augment these vectors with others, $\{\mathbf{u}_{n+1}, \ldots, \mathbf{u}_d\}$, in $\mathcal{V}_1$ to form a basis for all of the domain of L.

Now consider the effect of L on each of these basis vectors, $L(\mathbf{u}_i)$; from the remark on 67, these are precisely what determine the transformation itself. But, on those vectors in the null space basis, $L(\mathbf{u}_i) \equiv \mathbf{0}$; it is only the $L(\mathbf{u}_i)$, $i = n+1, \ldots, d$—$(d-n)$ in number—which are non-$\mathbf{0}$. We claim that these latter form a *basis* in $\mathcal{V}_2$ for the *range* of L proving the assertion.

In the first place, these *span* the range space of L: For any $\mathbf{y}$ in the range, $\mathbf{y} = L(\mathbf{x})$ for some $\mathbf{x}$ in the domain of L; *i.e.*, $\mathbf{x} = \sum_1^d x_i\mathbf{u}_i$ for some $\{x_i\}$. But then

$$L(\mathbf{x}) = L\left(\sum_{i=1}^{d} x_i\mathbf{u}_i\right) = \sum_{i=1}^{d} x_i L(\mathbf{u}_i) = \sum_{i=n+1}^{d} x_i L(\mathbf{u}_i),$$

since the $\{\mathbf{u}_1, \mathbf{u}_2, \ldots, \mathbf{u}_n\}$ are in the null space. (Note that $\mathbf{0}$ in the range can be expressed in terms of $\{\mathbf{u}_{n+1}, \ldots, \mathbf{u}_d\}$ even though these vectors are *not* in the null space: simply set the $x_i \equiv 0$!) Thus the vectors $\{L(\mathbf{u}_{n+1}), \ldots, L(\mathbf{u}_d)\}$ span the range of L.

But they are also linearly independent. For if there are $\{c_i\}$ such that

$$\sum_{i=n+1}^{d} c_i L(\mathbf{u}_i) = L\left(\sum_{i=n+1}^{d} c_i\mathbf{u}_i\right) = 0,$$

$\sum_{i=n+1}^{d} c_i\mathbf{u}_i$ must itself be an element of the null space, so it can be expressed in terms of the $\{\mathbf{u}_1, \mathbf{u}_2, \ldots, \mathbf{u}_n\}$:

$$\sum_{i=n+1}^{d} c_i\mathbf{u}_i = \sum_{i=1}^{n} z_i\mathbf{u}_i \text{ or } \sum_{i=1}^{n} z_i\mathbf{u}_i - \sum_{i=n+1}^{d} c_i\mathbf{u}_i = 0$$

—a linear combination of the linearly independent $\{\mathbf{u}_1, \mathbf{u}_2, \ldots, \mathbf{u}_d\}$ which, to vanish, demands that the $\{c_i\}$ (and, for all that, the $\{z_i\}$, too) vanish.

Thus linear independence of the $\{L(\mathbf{u}_i),\ i = n+1, \ldots, d\}$ which span the range space has been established, they form a basis of that space, and are $d-n$ in number. The proof is complete. □

This theorem has been couched in terms of *general* linear transformations, but we are typically interested in the *matrix representation* of such transformations, to which we now specialize the above result. It turns out that the *rank* of the matrix once more emerges as an important measure of such transformations—that in fact the dimension of the range space, r above, is nothing more than the rank! Before establishing that, however, we must make an important observation about matrix transformations which will justify the proposition:

Consider a linear transformation generated by a matrix [representation] A, $\mathbf{y} = \mathsf{A}\mathbf{x}$. We claim that if any two rows of A are linearly dependent, then the corresponding components of $\mathbf{A}\boldsymbol{x}$ are also [for *arbitrary* $\boldsymbol{x}$]. Indeed, expressing the matrix product in terms of *inner* products, (2.1-10), we can write

$$\mathbf{A}\boldsymbol{x} = \begin{pmatrix} \mathbf{a}_1^\mathsf{T} \\ \mathbf{a}_2^\mathsf{T} \\ \vdots \\ \mathbf{a}_n^\mathsf{T} \end{pmatrix} \boldsymbol{x} = \begin{pmatrix} \langle \mathbf{a}_1^\mathsf{T}, \boldsymbol{x}\rangle \\ \langle \mathbf{a}_2^\mathsf{T}, \boldsymbol{x}\rangle \\ \vdots \\ \langle \mathbf{a}_n^\mathsf{T}, \boldsymbol{x}\rangle \end{pmatrix}.$$

If the jth row of $\mathbf{A}$ is a linear combination of the other rows:

$$\mathbf{a}_j^\mathsf{T} = \sum_{i \neq j} c_i \mathbf{a}_i^\mathsf{T}$$

then the jth component of $\mathbf{A}\boldsymbol{x}$,

$$\langle \mathbf{a}_j^\mathsf{T}, \boldsymbol{x}\rangle = \left\langle \sum_{i \neq j} c_i \mathbf{a}_i^\mathsf{T}, \boldsymbol{x} \right\rangle = \sum_{i \neq j} c_i \langle \mathbf{a}_i^\mathsf{T}, \boldsymbol{x}\rangle$$

(from the "bilinearity" of the the inner product, (2.1-9)), so that this component is not only a linear combination of the other components, it is the *same* linear combination as the corresponding row of $\mathbf{A}$! This is precisely what was observed in the example of range space on page 64.

Once this is established, we see that the number of linearly *in*dependent entries of $\mathbf{A}\boldsymbol{x}$ is exactly the same as the number of linearly independent rows of $\mathbf{A}$. But the former is precisely the dimension of the range space, while the latter is just the [row] rank of the matrix. Thus:

Theorem 2.2.4. The dimension of the range space of a matrix transformation $\mathbf{A}\boldsymbol{x}$ is the rank of $\mathbf{A}$.

With this in place, we are ready to establish the matrix analog of Theorem 2.2.3:

Theorem 2.2.5. The null space of the transformation $\boldsymbol{y} = \mathbf{A}\boldsymbol{x}$ generated by the $M \times N$ matrix $\mathbf{A}$ with rank r has dimension $n = N - r$.

Note that "r," the rank, denotes a quantity different from that in Theorem 2.2.3 (where "r" is the range dimension); don't confuse them! Further, since $r \leq N$, $n = N - r \geq 0$.

Proof. For this transformation, the range space has dimension r by Theorem 2.2.4, the domain dimension N (from the size of the matrix, operating on an N-entry column vector[6]); thus the dimension of the null space is $N - r$ by Theorem 2.2.3. □

---

[6]If the domain is artificially limited to a *subspace* of this N-dimensional domain by, say, forcing certain entries to vanish, then r would be the rank of the submatrix of $\mathbf{A}$ resulting from eliminating the associated columns.

This apparently arcane theorem actually has important application when it comes to the solution of linear systems (Section 2.3), and as a predictor for the number of eigenvectors available to a given matrix (Section 3.5), to come later. But it also appears in the theory of differential equations (Section 2.4). In all these, what becomes central is the fact that the *dimension* of the null space tells the *number of linearly independent vectors* spanning it; *i.e.*, the number of linearly independent solutions. It is clearly more than merely "arcane"!

Summary. We have defined *linear transformations* and exemplified their broad application to a number of important areas, including basis transformation and general rigid-body dynamics—both topics to which we shall return—and even differential equations. We have also discussed their *domain*, *range*, and *null space*, and obtained an important result relating the dimensions of these three. We have also gone into some detail regarding the *basis* of linear transformations (which do, after all, qualify as a vector space) and seen that these bases are inexorably intertwined with those of the vector spaces of both their domain and range. Most importantly to our purposes, however, we have seen that we can *represent* any such transformation as a *matrix*—one whose columns are just the *representations* in the range space of each basis vector in the domain. Thus the matrix representation of a given linear transformation does not exist in a vacuum; it depends on the bases used to represent the *vectors* themselves. Just *how* it depends on the latter, and particularly how it might *change* when the respective bases change, is the subject of a subsequent section.

2.3 Solution of Linear Systems

Though this is sectioned off from the previous topic, in point of fact it is just the other side of the same coin: "Linear transformations" consider the equation $\mathbf{Ax} = \mathbf{y}$ where $\mathbf{x}$ is known and we desire $\mathbf{y}$; in some sense, $\mathbf{y}$ is given *explicitly* in terms of $\mathbf{x}$. The "solution of linear systems," on the other hand, considers the same equation in which $\mathbf{y}$ is known and we wish to find $\mathbf{x}$; $\mathbf{x}$ is given *implicitly* by this equation. It is obvious that the "system" of equations

$$\begin{aligned}
a_{11}x_1 + a_{12}x_2 + \cdots + a_{1n}x_n &= b_1\\
a_{21}x_1 + a_{22}x_2 + \cdots + a_{2n}x_n &= b_2\\
&\vdots\\
a_{m1}x_1 + a_{m2}x_2 + \cdots + a_{mn}x_n &= b_m
\end{aligned}$$

can be written in the form $\mathbf{Ax} = \mathbf{b}$, where

$$\mathbf{A} \equiv \begin{pmatrix} a_{11} & a_{12} & \cdots & a_{1n}\\ a_{21} & a_{22} & \cdots & a_{2n}\\ \vdots & \vdots & & \vdots\\ a_{m1} & a_{m2} & \cdots & a_{mn} \end{pmatrix} \text{ and } \mathbf{b} \equiv \begin{pmatrix} b_1\\ b_2\\ \vdots\\ b_m \end{pmatrix}.$$

In this equation, the quantity "**x**" plays a subordinate role: the *system* is the same if we replace that variable by "**z**." Thus what really defines the system is the "coefficient matrix" **A** and the right-hand side **b**. "Solution" of the system consists in adding and subtracting equations from one another in the attempt to isolate as many of the n unknowns x_i on one side as possible. Given the fact that it is really only **A** and **b** which are important, in practice this manipulation of the two sides of the equation is often effected by operating on entire rows of the "*augmented matrix*"

$$(\mathbf{A}\ \mathbf{b}) = \begin{pmatrix} a_{11} & \cdots & a_{1n} & b_1 \\ a_{21} & \cdots & a_{2n} & b_2 \\ \vdots & \cdots & \vdots & \vdots \\ a_{m1} & \cdots & a_{mn} & b_m \end{pmatrix}. \tag{2.3-1}$$

But what exactly do we *mean* by "solution"? If, for example, $m < n$, the situation is often described by saying that there is "*no* solution"; the *best* we can hope for is to express m of the x's in terms of the remaining $n - m$:

$$\begin{aligned} x_1 &= b_1' + c_{1,m+1}x_{m+1} + \cdots + c_{1,n}x_n \\ x_2 &= b_2' + c_{2,m+1}x_{m+1} + \cdots + c_{2,n}x_n \\ &\vdots \\ x_m &= b_m' + c_{m,m+1}x_{m+1} + \cdots + c_{m,n}x_n \end{aligned}$$

while if $m = n$ we would hope for

$$\begin{aligned} x_1 &= b_1' \\ x_2 &= b_2' \\ &\vdots \\ x_m &= b_m' \end{aligned}$$

But what happens if $m > n$? And when are n equations *really* "n equations"? And if $m < n$, is it really the case that there is "no solution"? These are the questions we shall address in this section.

Before considering the problem generally, however, let us first review the types of situation that might arise in practice through a few simple examples:

Example 2.3.1. Consider the algebraic system

$$\begin{aligned} x + y &= 3 \\ x + y &= 3 \end{aligned}$$

Clearly we have not two, but only *one* equation in the two unknowns. Even if the second equation had read "$2x + 2y = 6$," we still would have had only a single equation: the last is simply $2\times$ the original one. This is obvious and means that it is simplistic merely to count the "number of equations." |

Example 2.3.2. Let us consider a slightly more optimistic set:

$$x+\ \ y+\ z=-1 \qquad \text{(a)}$$
$$x-3\ y+2z=-3 \qquad \text{(b)}$$
$$x+13y-2z=5 \qquad \text{(c)}$$

Though these "look" different, closer examination reveals that there are really only two "different" equations here: the last is just 4 times the first minus 3 times the second—*i.e.*, it is a *linear combination* of the first two! If one hadn't observed this fact and naively proceeded in "solving" the system, first by eliminating "x" from the last two:

$$\text{(b)}-\text{(a)}: \qquad -4y+\ z=-2 \qquad \text{(b}')$$
$$\text{(c)}-\text{(a)}: \qquad 12y-3z=6 \qquad \text{(c}')$$

and then "y" from the third:

$$\text{(c}')+3\text{(b}'): \qquad 0=0,$$

the last would vanish identically. This is not at all surprising: we have already seen that the third original equation is a linear combination of the first two, and *solution involves taking linear combinations of the equations!* One might, um, smell "rank" here, and that is just the point: there are only *two* linearly independent equations—*rows* of the original *augmented* matrix (2.3-1). In fact, temporarily viewing this system in its matrix representation, we could investigate the rank of that matrix by performing row operations mimicking the above solution process (giving the "*row* rank" = rank):

$$\begin{pmatrix}1&1&1&-1\\1&-3&2&-3\\1&13&-2&5\end{pmatrix}\rightarrow\begin{pmatrix}1&1&1&-1\\0&-4&1&-2\\0&12&-3&6\end{pmatrix}\rightarrow\begin{pmatrix}1&0&\frac{5}{4}&-\frac{3}{2}\\0&1&-\frac{1}{4}&\frac{1}{2}\\0&0&0&0\end{pmatrix} \qquad (2.3\text{-}2)$$

—so that the *rank* of the [augmented] matrix is, in fact, only 2. Simply dropping the last column shows that the *coefficient* matrix also has a rank of 2:

$$\begin{pmatrix}1&1&1\\1&-3&2\\1&13&-2\end{pmatrix}\rightarrow\begin{pmatrix}1&1&1\\0&-4&1\\0&12&-3\end{pmatrix}\rightarrow\begin{pmatrix}1&0&\frac{5}{4}\\0&1&-\frac{1}{4}\\0&0&0\end{pmatrix} \qquad (2.3\text{-}3)$$

This is precisely one of the "skip" cases considered in the discussion of the echelon test in Section 1.2.2: we have a 0 on the diagonal with non-0 entries above it characteristic of such skips in row n-tuples. Observe the values of those entries; they will return!

But if, undaunted, we forge ahead and try to "solve" this system, we continue to make some interesting observations: (b$'$) can be written

$$y-\frac{z}{4}=\frac{1}{2},$$

which, when substituted into (a) yields

$$x + \frac{5z}{4} = -\frac{3}{2}$$

Thus the "solution" is

$$x = -\frac{3}{2} - \frac{5z}{4}, \quad y = \frac{1}{2} + \frac{z}{4}$$

—for *arbitrary* z. Thus we can write the overall solution:

$$\begin{pmatrix} x \\ y \\ z \end{pmatrix} = \begin{pmatrix} -\frac{3}{2} \\ \frac{1}{2} \\ 0 \end{pmatrix} + \begin{pmatrix} -\frac{5z}{4} \\ \frac{z}{4} \\ z \end{pmatrix}.$$

—where we have added "z" to both sides of the "$0 = 0$" equation obtained above to amend the equation "$z = z$" to the "solution. (*Cf.* the first two coefficients of z in the last term with the "skip" entries in (2.3-3)!) We have already seen that we have only "*two* equations" in the "three unknowns," but far from having *no* solution, we have *too many*: *any* value of z can be substituted into this equation to give a different value for the solution.

But we can get even one more interesting facet of this problem: The solution consists of two parts, a numerical value and the one depending on the parameter z. Just for interest, let us substitute each of these parts into the original equation to see what it yields. In terms of the matrix product:

$$\begin{pmatrix} 1 & 1 & 1 \\ 1 & -3 & 2 \\ 1 & 13 & -2 \end{pmatrix} \begin{pmatrix} -\frac{3}{2} \\ \frac{1}{2} \\ 0 \end{pmatrix} = \begin{pmatrix} -1 \\ -3 \\ 5 \end{pmatrix}$$

while

$$\begin{pmatrix} 1 & 1 & 1 \\ 1 & -3 & 2 \\ 1 & 13 & -2 \end{pmatrix} \begin{pmatrix} -\frac{5z}{4} \\ \frac{z}{4} \\ z \end{pmatrix} = \begin{pmatrix} 0 \\ 0 \\ 0 \end{pmatrix} \tag{2.3-4}$$

—*i.e.*, it is the "particular" *numerical* part which gives the constant values on the right-hand side of the original equation, while the "general" part depending on z is in the *null space* of the coefficient matrix! This observation is of fundamental importance, as we shall see. |

While this example has been provocative, we shall undertake one more before attempting to bring all these ideas together:

Example 2.3.3. Let us consider almost the same system as above, but changing the value of the right-hand side of the last equation:

$$\begin{aligned} x+\ \ y+\ z&=-1 && \text{(a)} \\ x-3\,y+2z&=-3 && \text{(b)} \\ x+13y-2z&=8 && \text{(c*)} \end{aligned}$$

The *left*-hand side of the last equation is still the linear combination of the first two, but the *right*-hand side isn't. We concentrate on this equation in tracing the "solution" process: In eliminating x from the equation we would get, corresponding to "(c′)" in the above example,

$$12y - 3z = 9;$$

then eliminating y from this just as before, by taking this equation and adding $3\times$(b′) from above, we get

$$0 = 3$$

—a rather inauspicious result. Clearly something is wrong: there is, in fact *no* solution to this system—*no* set of x, y, and z satisfying it—due to an inconsistency on the *constant* side of the system. Having mentioned the importance of rank in the preceding example, it makes sense to investigate that issue here: The coefficient matrix is exactly the same, so its rank (see (2.3-3)) is still 2. But considering the *augmented* matrix, the calculations (2.3-2) now becomes

$$\begin{pmatrix} 1 & 1 & 1 & -1 \\ 1 & -3 & 2 & -3 \\ 1 & 13 & -2 & 8 \end{pmatrix} \to \begin{pmatrix} 1 & 1 & 1 & -1 \\ 0 & -4 & 1 & -2 \\ 0 & 12 & -3 & 9 \end{pmatrix} \to \begin{pmatrix} 1 & 0 & \frac{5}{4} & 0 \\ 0 & 1 & -\frac{1}{4} & 0 \\ 0 & 0 & 0 & 1 \end{pmatrix}$$

[again a "skip" form]—demonstrating that the augmented matrix has a rank of *3*! The additional order of rank has come from precisely the "inconsistency" cited above and clearly becomes useful as a criterion for investigating such pathological cases. |

Let us pause here briefly to review what the above examples demonstrate: In the first place, the process of "solution" of a linear system consists in performing *row* operations on the coefficient or augmented matrix. But, recall (page 54), such row operations are equivalent to *pre*-multiplying by a certain form of "elementary matrix" $\mathbf{E}^*_{ij}$; by the associative property of matrix multiplication (2.1-14), the product of such elementary matrices can be lumped together into a *single* matrix, say $\mathbf{E}^*$. But applying these row operations to the entire equation is then the same as multiplying *both* sides of the original equation by $\mathbf{E}^*$. Thus the net effect of these operations is to bring the original system $\mathbf{Ax} = \mathbf{b}$ into a form $\mathbf{E}^*\mathbf{Ax} = \mathbf{E}^*\mathbf{b}$. *The solution to the latter system is the same as the solution to the original one.*

These operations are directed toward *isolating* each unknown; equivalently, we attempt to bring the associated coefficient matrix $\mathbf{A}$ into a form reminiscent of the "identity matrix," equation (2.1-17):

$$\mathbf{E}^*\mathbf{Ax} = \begin{pmatrix} 1 & 0 & \cdots \\ 0 & 1 & \cdots \\ & & \\ 0 & 0 & \ddots \\ \vdots & \vdots & \vdots \end{pmatrix} \mathbf{x} = \begin{pmatrix} x_1 \\ x_2 \\ \vdots \end{pmatrix} = \mathbf{E}^*\mathbf{b}$$

—*precisely the echelon form of the matrix* A—providing the solution. In this regard, then, *the process of solution is tantamount to reduction of the matrix system to echelon form.* But this process also exhibits the *rank* of the matrix, which we have seen is the real measure of just "how many equations" we actually have.

Note that such operations may include [row] reordering; this doesn't change the solution, since the order of the individual equations—the *rows* of the *augmented* matrix—is irrelevant. And such reordering may be necessary if, in the process of solution, a given element on the diagonal of the coefficient matrix happens to be 0. Recall, however, that it might be the case that no subsequent row *has* a non-0 value in that column to switch with the offending row; this introduces "skips" into the final form (page 47), observed in the examples above. However annoying their presence might be, they in fact have a significant bearing on the solution, which we shall now examine.

"Skips" and the Null Space

In the discussion of Example 2.3.2, we observed that the solution to that non-homogeneous system consisted of two parts, a "particular," numerical, part giving the right-hand side, and a "general" part which, at least in that example, turned out to belong to the null space of the coefficient matrix. The latter part, in turn, was just the variable z multiplied by the *negative* of that "skip column" in the *coefficient* matrix corresponding to z—but with its signature diagonal replaced by a 1. We shall now show a generalization of this interesting result.

Though "skip" cases have thus far been modelled with a single such column, it is clear that, in general, such skips might occur more than once. Thus we must consider the possibility that there is more than one "skip column" in the reduced, echelon form of the matrix; *e.g.*, with somewhat elaborate "partitioning" of the echelon matrix (indicating the columns k_1 and k_2 in which the skips occur, and span of rows/columns by right/over-braces),

$$\mathsf{E}^*\mathsf{A} = \begin{pmatrix} \overbrace{\mathbf{1}^{(k_1-1)}}^{k_1-1} & \overset{\substack{k_1\\ \downarrow}}{\boldsymbol{\xi}_{11}} & \overbrace{\mathbf{0}}^{k_2-k_1-1} & \overset{\substack{k_2\\ \downarrow}}{\boldsymbol{\xi}_{21}} & \cdots \\ \mathbf{0} & \mathbf{0} & \mathbf{1}^{(k_2-k_1-1)} & \boldsymbol{\xi}_{22} & \cdots \\ \mathbf{0} & \mathbf{0} & \mathbf{0} & \mathbf{0} & \cdots \end{pmatrix} \begin{matrix} \}\, k_1-1 \\ \}\, k_2-k_1-1 \\ \end{matrix}.$$

We shall show that certain vectors $\boldsymbol{x}_N$, derived from the "skip columns" are, in point of fact, elements of the null space of the coefficient matrix A; *i.e.*, such that $\mathsf{A}\boldsymbol{x}_N = \mathbf{0}$. By virtue of the fact that solutions to this system are precisely the same as those to the *reduced* matrix system $\mathsf{E}^*\mathsf{A}\boldsymbol{x}_N = \mathsf{E}^*\mathbf{0} = \mathbf{0}$, we shall actually show that they are in the null space of the matrix $\mathsf{E}^*\mathsf{A}$.

Consider the first skip column; we claim that the vector

$$\boldsymbol{x}_{N_1} \equiv \begin{pmatrix} -\boldsymbol{\xi}_{11} \\ 1 \\ \mathbf{0} \end{pmatrix} \tag{2.3-5}$$

—in which the bottom of the vector is "padded" with sufficient 0's to make it an n-vector—is a member of the *null space* of $\mathbf{E}^*\mathbf{A}$; *i.e.*, $\mathbf{E}^*\mathbf{A}\boldsymbol{x}_{N_1} = \mathbf{0}$. Indeed, by direct [block] multiplication of the matrices,

$$\mathbf{E}^*\mathbf{A}\boldsymbol{x}_{N_1} = \begin{pmatrix} \mathbf{1}^{(k_1-1)}(-\boldsymbol{\xi}_{11}) + (1)\boldsymbol{\xi}_{11} + \mathbf{0} \\ 0 \\ \mathbf{0} \end{pmatrix} = \begin{pmatrix} -\boldsymbol{\xi}_{11} + \boldsymbol{\xi}_{11} + \mathbf{0} \\ 0 \\ \mathbf{0} \end{pmatrix} = \begin{pmatrix} \mathbf{0} \\ 0 \\ \mathbf{0} \end{pmatrix}.$$

Before considering the next skip column, in order to lend an air of concreteness to this proposition, let us compare this with the corresponding equation (2.3-4) in Example 2.3.2. In that example, the skip was in the third column,

$$\begin{pmatrix} \boldsymbol{\xi}_{11} \\ 0 \end{pmatrix} = \begin{pmatrix} \xi_{11_1} \\ \xi_{11_2} \\ 0 \end{pmatrix} \equiv \begin{pmatrix} \frac{5}{4} \\ -\frac{1}{4} \\ 0 \end{pmatrix};$$

$\boldsymbol{\xi}_{11}$ was a 2-vector. The element of the null space was

$$\begin{pmatrix} -\frac{5}{4} \\ \frac{1}{4} \\ 1 \end{pmatrix} z = \begin{pmatrix} -\boldsymbol{\xi}_{11} \\ 1 \end{pmatrix} z,$$

showing the "1" amended to the $-\boldsymbol{\xi}_{11}$ vector in (2.3-5); because the matrix $\mathbf{E}^*\mathbf{A}$ had as many rows as $\boldsymbol{x}$ had unknowns, it was unnecessary to further "pad" the vector with $\mathbf{0}$ in this instance. z, the variable corresponding to the third column in the coefficient matrix, was a free parameter; it is the vector *multiplying* it—an element of the null space—which made their product an element of that space.

Now consider the second skip column. We here claim that the vector

$$\boldsymbol{x}_{N_2} \equiv \begin{pmatrix} -\boldsymbol{\xi}_{21} \\ 0 \\ -\boldsymbol{\xi}_{22} \\ 1 \\ \mathbf{0} \end{pmatrix}$$

is an element of the null space of $\mathbf{E}^*\mathbf{A}$ (and so of $\mathbf{A}$). Indeed,

$$\mathbf{E}^*\mathbf{A}\boldsymbol{x}_{N_2} = \begin{pmatrix} -\boldsymbol{\xi}_{21} + \mathbf{0} - \mathbf{0} + \boldsymbol{\xi}_{21} + \mathbf{0} \\ \mathbf{0} + \mathbf{0} - \boldsymbol{\xi}_{22} + \boldsymbol{\xi}_{22} + \mathbf{0} \\ \mathbf{0} \end{pmatrix} = \begin{pmatrix} \mathbf{0} \\ \mathbf{0} \\ \mathbf{0} \end{pmatrix},$$

as expected. Having hurdled this more general instance of skip column, it is clear how to construct subsequent $\boldsymbol{x}_N$; thus the allegation is proven.

Before leaving this topic, let us make one final observation. All of this analysis has been couched in the contexts of "skips." But the above demonstration depends only the the "$\boldsymbol{\xi}$" vectors in the [reduced] matrix, not on the $\mathbf{0}$'s below them. Thus, in the case in where there *are* no such $\mathbf{0}$'s—that the matrix has the form

$$\mathsf{E}^*\mathsf{A} = \begin{pmatrix} 1 & 0 & \dots & 0 & a_{1,m+1} & \dots & a_{1n} \\ 0 & 1 & \dots & 0 & a_{2,m+1} & \dots & a_{2n} \\ \vdots & \vdots & \ddots & \vdots & \vdots & \dots & \vdots \\ 0 & 0 & \dots & 1 & a_{m,m+1} & \dots & a_{in} \end{pmatrix}. \qquad (2.3\text{-}6)$$

—exactly the same results would apply. In particular, the columns to the right would generate elements of the null space of $\mathsf{E}^*\mathsf{A}$, and thus of A itself (since, again, the solutions of $\mathsf{E}^*\mathsf{A}x = \mathbf{0}$ and $\mathsf{A}x = \mathbf{0}$ are exactly the same).

2.3.1 Theory of Linear Equations

With the background from these examples, as well as from the previous material, we can approach all these issues quite comprehensively. We have learned from the above examples that

- Solution is tantamount to reducing the coefficient and/or augmented matrix to echelon form.
- The *rank* of the appropriate matrix is a measure of the actual number of [independent] equations we have in the system.
- "*Skips*" in the matrix signal members of the null space of that matrix.

Recall the last example above, 2.3.3. We saw there that an inconsistency on the right-hand, "constant," side of the equation led to a system with no solution; had the constant side been the $\mathbf{0}$ vector, this would not have arisen, since linear combinations of 0 are still 0. This suggests that the two cases must be analyzed separately—that systems of equations $\mathsf{Ax} = \mathbf{0}$ and $\mathsf{Ax} = \mathbf{b} \neq \mathbf{0}$ are somehow fundamentally different. Thus we classify systems of linear equations according to their constant sides, as being *homogeneous* if that side is $\mathbf{0}$, and (reasonably) *non-homogeneous* if it isn't. Note that the former, homogeneous, case is precisely the problem of determining the *null space* of the *transformation* generated by the coefficient matrix!

Homogeneous Linear Equations

We have already observed that the issue of *consistency* which caused the problem in Example 2.3.3 will not arise in homogeneous equations: homogeneous equations are *always* consistent. Since we don't have to worry about the constant side of the equation at all, we shall examine merely the $m \times n$ coefficient matrix, and what the form of the *[row-] reduced* system indicates. We consider three cases, according to the rank, r:[7]

[7] In the event $r < m$, the number of equations, there would be additional rows of 0's in each example (see (2.1-4)). But these result in 0-rows in $\mathsf{A}x$, so are ignored for purposes of clarity. The issue of "skips" doesn't enter, since their effect is already accounted for in the rank; see page 24.

1. $r < n$: Here our coefficient matrix, in reduced form, becomes

$$\mathsf{E}^*\mathsf{A} = \begin{pmatrix} 1 & 0 & \dots & 0 & a'_{1,r+1} & \cdots & a'_{1n} \\ 0 & 1 & \dots & 0 & a'_{2,r+1} & \cdots & a'_{2n} \\ \vdots & \vdots & \ddots & \vdots & \vdots & \ddots & \vdots \\ 0 & 0 & \dots & 1 & a'_{r,r+1} & \cdots & a'_{rn} \end{pmatrix}$$

   (*Cf.* (2.3-6)). We obtain here solutions for only r of the n unknowns; the remaining $n - r$ variables simply become *parameters*, transposed to the "solution" side of the equation, which can be chosen freely and still satisfy the equation. In particular, they can independently be assigned *non-0* values, resulting in non-**0** solutions to the homogeneous equations. In any event, there are an infinite number of possible solutions.

2. $r = n$: Now the reduced coefficient matrix becomes simply

$$\mathsf{E}^*\mathsf{A} = \begin{pmatrix} 1 & 0 & \cdots & 0 \\ 0 & 1 & \cdots & 0 \\ \vdots & \vdots & \ddots & \vdots \\ 0 & 0 & \cdots & 1 \end{pmatrix}$$

   which results in the *single* solution $\boldsymbol{x} = \mathbf{0}$. The solution is unique.

3. $r > n$: This situation does not arise: the *maximum* rank can only be n, the number of columns in the coefficient matrix.

This is the point at which Theorem 2.2.5 becomes useful practically: *The process of determining the solution to a homogeneous equation* $\mathsf{A}\boldsymbol{x} = \mathbf{0}$ *is precisely that of finding vectors in the null space of* A. $\boldsymbol{x} = \mathbf{0}$ is always a solution; the real issue is whether there are any $\boldsymbol{x} \neq \mathbf{0}$ which are. If $\mathbf{0}$ is the only solution, the dimension of the null space is [*defined* to be] 0; this is precisely the reason for this definition. Non-**0** solutions are signalled by a null space dimension $n > 0$, giving the number of linearly independent elements in the basis spanning the null space, of which all other solutions are linear combinations. But, by Theorem 2.2.5, $n = m - r$, where m is the number of equations, and r the rank of the coefficient matrix A. If $r = m$, $n = 0$: the only solution is $\boldsymbol{x} = \mathbf{0}$. But if $r < m$, $n > 0$: there are n linearly independent solutions to the homogeneous equation.

Non-homogeneous Linear Equations

Now the issue of consistency *can* arise: it is possible that the right-hand sides of the equations result in an impossible solution. But we saw in Example 2.3.3 that this was signalled when the rank of the augmented matrix (2.3-1) was *greater* than that of the coefficient matrix. (It can't be *less*: the minimum rank of the augmented matrix is that of the coefficient matrix, which comprises part of the augmented one.) Thus we have all but proven the

Theorem 2.3.1. If the rank of the augmented matrix of a non-homogeneous system of equations is greater than that of the coefficient matrix, there is *no* solution.

Once this is recognized, the number of variables which can be solved for when there *is* a solution follow immediately in the same fashion as above, according to the rank of the coefficient matrix:

1. $r < n$: It is possible to solve for r of the unknowns in terms of the remaining $n - r$ variables. There are an infinite number of such solutions.
2. $r = n$: It is possible to solve for all n variables. The solution is unique.
3. $r > n$: Again, this situation cannot arise.

The first of these cases, in which the rank of the coefficient matrix is less than the number of unknowns, is precisely the one encountered in Example 2.3.2. We noted there that the solution consisted of two parts: one giving the constant side of the matrix, and the other being an element of the null space. This is characteristic of such systems in the sense of the following theorem:

Theorem 2.3.2. The general solution of the non-homogeneous system $\mathbf{Ax} = \mathbf{b} \neq \mathbf{0}$ is $\mathbf{x} = \mathbf{x}_P + \mathbf{x}_N$, where

$$\mathbf{A}\mathbf{x}_P = \mathbf{b}$$

and $\mathbf{x}_N$ is an element of the null space.

Proof. Assume there are two solutions, $\mathbf{x}_1$ and $\mathbf{x}_2$ to the system $\mathbf{Ax} = \mathbf{b} \neq \mathbf{0}$. Thus $\mathbf{Ax}_1 - \mathbf{Ax}_2 = \mathbf{A}(\mathbf{x}_1 - \mathbf{x}_2) = \mathbf{0}$ (by the linearity of $\mathbf{A}$); *i.e.*, $(\mathbf{x}_1 - \mathbf{x}_2)$ *is an element of the null space of* $\mathbf{A}$, say $\mathbf{x}_N$. Thus, immediately $\mathbf{x}_1 = \mathbf{x}_2 + \mathbf{x}_N$, proving the assertion. □

Now we have come full circle: Solution of linear systems is tantamount to reduction of the coefficient matrix (the augmented matrix, too, for non-homogeneous systems) to echelon form. The rank of the matrix is reduced by "skips" and can lead to systems in which the rank of the coefficient matrix is less than the number of unknowns. These result in columns generating vectors in the null space of the coefficient matrix—parts of the solution for homogeneous equations, and the $\mathbf{x}_N$ above in non-homogeneous ones.

Homework:

1. In Example 2.2.7 on page 70, we showed that cross product of two 3-D vectors $\boldsymbol{a}$ and $\boldsymbol{b}$ could be represented as the product of a *matrix* $\mathbf{b}$ with a *column matrix* $\mathbf{a}$. The Homework following demonstrated that $\mathbf{b}$ only has rank 2. Consider the non-homogeneous system of linear equations for $\boldsymbol{b}$

$$\boldsymbol{a} \times \boldsymbol{b} = \boldsymbol{c},$$

and, in view of Theorem 2.3.1, give the conditions on the components of $\boldsymbol{b}$ and $\boldsymbol{c}$ such that this system, in fact, *has* a solution. *Hint*: Clearly all three components of $\boldsymbol{b}$ cannot vanish, or $\boldsymbol{b}$ itself does! Assume, say, $b_y \neq 0$, and then find conditions under which the second row of the augmented matrix vanish. Does the result change if *both* b_x and b_z vanish identically?

2. Verify Theorem 2.3.2 with regard to solving the above system. What is the general element of the null space of the matrix **b**?

3. (*Solutions to Linear Equations*) For each pair matrix **A** and [column] vector **b** following, tell how many variables can be solved for in the system $\mathbf{Ax} = \mathbf{b}$, and how many of these solutions exist for each. It is *not* necessary to solve the systems explicitly!

a) $$\mathbf{A} = \begin{pmatrix} 1 & 2 & -5 \\ 3 & 6 & 7 \\ 2 & 1 & 1 \end{pmatrix} \quad \text{and} \quad \mathbf{b} = \begin{pmatrix} 3 \\ 8 \\ 1 \end{pmatrix}$$

Ans. single solution for 3 variables.

b) $$\mathbf{A} = \begin{pmatrix} 2 & 5 & 8 \\ 3 & 9 & -6 \\ -4 & -19 & 92 \end{pmatrix} \quad \text{and} \quad \mathbf{b} = \begin{pmatrix} 3 \\ -1 \\ 27 \end{pmatrix}$$

Ans. infinite solutions for 2 variables.

c) $$\mathbf{A} = \begin{pmatrix} 2 & 5 & 8 \\ 3 & 9 & -6 \\ -4 & -19 & 92 \end{pmatrix} \quad \text{and} \quad \mathbf{b} = \begin{pmatrix} 5 \\ 8 \\ -10 \end{pmatrix}$$

Ans. no solution.

d) $$\mathbf{A} = \begin{pmatrix} 1 & 5 & -3 \\ -2 & -10 & 6 \\ -1 & -5 & 3 \end{pmatrix} \quad \text{and} \quad \mathbf{b} = \begin{pmatrix} 3 \\ -6 \\ -3 \end{pmatrix}$$

Ans. infinite solutions for 1 variable.

2.3.2 *Solution* of Linear Systems—Gaussian Elimination

Echelon form has been the tool used for the *analysis* of systems of linear equations, but, at the end of the day, one is generally at least as interested in *solving* them! We here briefly describe possibly the most common practical method of doing so, *Gaussian Elimination.*

It is certainly quite proper to reduce an arbitrary system of linear equations to echelon form: having obtained a matrix equation of the form, say,

$$\mathbf{E}^*\mathbf{A}\mathbf{x} = \begin{pmatrix} 1 & 0 & \ldots & 0 & a'_{1,r+1} & \ldots & a'_{1n} \\ 0 & 1 & \ldots & 0 & a'_{2,r+1} & \ldots & a'_{2n} \\ \vdots & \vdots & \ddots & \vdots & \vdots & \ldots & \vdots \\ 0 & 0 & \ldots & 1 & a'_{r,r+1} & \ldots & a'_{in} \end{pmatrix} \mathbf{x} = \mathbf{b}'$$

(in which $n > r$), one has immediately that

$$\begin{pmatrix} x_1 \\ x_2 \\ \vdots \\ x_r \end{pmatrix} = \begin{pmatrix} b'_1 - \sum_{j\geq r+1}^{n} a'_{1,j}x_j \\ b'_2 - \sum_{j\geq r+1}^{n} a'_{2,j}x_j \\ \vdots \\ b'_r - \sum_{j\geq r+1}^{n} a'_{r,j}x_j \end{pmatrix},$$

giving the solution for r of the x's in term of the remaining $(n-r)$. This approach is generally called the *Gauss-Jordan Method*[8] for solving linear systems. Though straightforward, there is a great deal of effort involved, first in normalizing each coefficient on the "diagonal," and then subtracting off products of each row from the others in order to make the off-diagonal terms vanish.

As system *almost* as easy to solve, however, would be the one

$$\begin{pmatrix} a'_{11} & a'_{12} & \cdots & a'_{1r} \\ 0 & a'_{22} & \cdots & a'_{2r} \\ \vdots & \vdots & \ddots & \vdots \\ 0 & 0 & \cdots & a'_{rr} \end{pmatrix} \mathbf{x} = \mathbf{b}';$$

—so-called "*upper triangular form.*" $\mathbf{x}$ can be found first by obtaining

$$x_r = b'_r / a'_{r,r},$$

from which

$$x_{r-1} = (b'_{r-1} - a'_{r-1,r}x_r)/a'_{r-1,r-1},$$

with other components being found similarly—a process called the "*back solution.*" In comparison with the above method of explicitly putting $\mathbf{A}$ in echelon form, this approach, *Gaussian Elimination*, avoids dividing *each* coefficient in *each* row by its diagonal term (this is effected by the *single* division by that term

[8] Actually, the Gauss-Jordan method generally also involves "pivotting," described below.

in the back solution) and certainly eliminates clearing of the upper off-diagonal terms; thus it is somewhat more efficient than the Gauss-Jordan approach.

Though this describes the essence of Gaussian Elimination, any such method is clearly directed ultimately towards numerical solution, and it is here that subtleties of numerical analysis enter: It may occur that, in the process of eliminating the coefficients below a certain diagonal term, one of these coefficients is *very large* (typically orders of magnitude) compared with the diagonal term. In this case, numerical *roundoff error*, imposed either by choice or, in the [likely] case of computer implementation, by hardware limitations, can become a major problem. We demonstrate this with a simple example:

Consider the system of equations

$$10^{-5}x + y = 2$$
$$x + y = 3;$$

clearly y is approximately 2, so x will be close to 1. Now let us say that this is being run on a *very* bad computer—one with only three significant figures in the floating-point mantissa and, say, two in the exponent. When we eliminate x from the second equation, it *should* be

$$(1 - 10^5)y = 3 - 2 \times 10^5$$

But, because of the limited accuracy of our computer, it assumes the internal form

$$-10^5 y = -2 \times 10^5.$$

resulting in $y = 2.00$. Not bad so far. But, when we back-solve, the first equation now becomes

$$10^{-5}x + 2 = 2$$

yielding $x = 0$, not 1, even to the accuracy of our machine. This just isn't acceptable! If, however, we were simply to reorder the equations

$$x + y = 3$$
$$10^{-5}x + y = 2,$$

elimination of x in the [now] second equation would give

$$(1 - 10^{-5})y = 2 - 3 \times 10^{-5},$$

or, internally, "$y = 2.00$," whose back solution would give the desired $x = 1$.

This example suggests the utility of always having the *largest* term in a given column be on the diagonal, advising a *reordering* of the rows so this is assured—a device referred to a "*pivotting*." (Actually, this is called "*partial* pivotting," since we only examine elements in a given *column*; "*full* pivotting" would also examine *other* columns, but the benefits of this approach are still an issue of some debate.)

Neither this section nor this book is intended to be a discourse on numerical analysis; this topic warrants an entire course. But the above is meant at least to introduce the reader to considerations in the field.

Summary. Though the examples used to motivate the results of this section involved a great deal of detail, we have finally begun to bring together previous topics: The process of "solving" a system of linear equations is merely the "inverse" operation to a linear transformation and is tantamount to bringing the coefficient matrix to echelon form. In particular, finding a *homogeneous* solution is precisely the determination of vectors in the *null space* of the transformation, whose dimension is related to the number of linearly independent rows—*i.e.*, the *rank*—of the matrix representation of that linear transformation; this allows us to predict whether any more than the $\mathbf{0}$ vector comprises such a solution. In the case of *non*-homogeneous equations, there is the additional issue of *consistency* among the equations, where rank once again plays a role: this is assured if and only if the ranks of both the coefficient matrix and the "augmented" matrix are the same. Assuming they are, then the general solution to the non-homogeneous equations consists of two parts: a "particular" part, giving the value of the right-hand side, and a "general" part, consisting of elements of the null space multiplied by some corresponding parameters.

The end result of these considerations is given by the corresponding pair of practical criteria: Given an $m \times n$ coefficient matrix $\mathbf{A}$ for the system $\mathbf{Ax} = \mathbf{b}$,

- a solution *exists* for any homogeneous system ($\mathbf{b} = \mathbf{0}$), or for any *non*-homogeneous system ($\mathbf{b} \neq \mathbf{0}$) in which the ranks of $\mathbf{A}$ and the *augmented matrix* ($\mathbf{Ab}$) are the same; and
- if the rank, r, of $\mathbf{A}$ is less than n, there are an infinite number of solutions, while if $r = n$ the solution is unique.

2.4 Linear Operators—Differential Equations

The entire chapter thus far has been devoted to linear *transformations*—the mapping of one element in a vector space into another. We now shift the emphasis to a closely allied topic of linear *operators*, as exemplified by n^{th} order linear ordinary differential equations.

"Closely allied" is an apt description: The material here was originally going to be sorted out as examples in previous sections, illustrating how their ideas were reflected in differential equations. But, in view of the "operator" nature of the present subject, and as a means to bring together virtually *all* the concepts presented thus far, it seemed expedient pedagogically to break off this particular topic into its own section.

Solutions of Linear Differential Equations as a Vector Space. We have already observed (Example 1.1.6) that the set of *solutions* to the homogeneous

n^{th}-order linear ordinary differential equation

$$x^{[n]} + a_{n-1}(t)x^{[n-1]} + \cdots + a_o(t)x = 0 \tag{2.4-1}$$

constitute a vector space, in the sense of being closed under the customary definitions of "vector addition" and "scalar multiplication." Further, as a vector space, it has a dimension: Example 1.2.10 demonstrated that this dimension is simply n. The latter relied on associating with the *scalar* solution $x(t)$ an *n-vector* function $\boldsymbol{x} \equiv (x^{[n-1]}(t)\ x^{[n-2]}(t)\ \ldots\ x(t))^{\mathsf{T}}$, comprised of x itself and its derivatives up to order $n-1$—the so-called *phase space.* Recall (Example 1.2.10) that such differential equations are called *autonomous* (or "*homogeneous*"); in the *non*-autonomous case, one typically introduces a vector $\boldsymbol{x}' \equiv (x^{[n-1]}(t)\ x^{[n-2]}(t)\ \ldots\ x(t)\ t)^{\mathsf{T}}$ in the *extended* phase space. (Note that primes will be used to denote the addition of t as a component of the vector, not the derivative!) For purposes of generality, it is the latter we will consider here, with in-line remarks addressing the case in which the equation is homogeneous.

Linear Differential Equations as Linear Operators. We now adopt a slightly different perspective—not to be confused with the previous one—in which we view this equation as a *linear operator* on the elements of a vector space of *functions*, to *each* of which we identify a vector of the function and its derivatives. Note that, unlike the above formulation, *this space is infinite in dimension*, even though it consists of finite n-tuples: there are an infinite number of [n-differentiable] functions $x(t)$; that this is a n-tuple is merely a vestige of the number of derivatives we include. It satisfies Definition 2.2-1, being linear in its argument due to the inherently linear nature of differentiation. And it operates on one [n-tuple] function $\boldsymbol{x}$ in that space to return another; in this regard it could be viewed as a "linear transformation" of *functions.* In fact, although there is a palpably different "flavor" to operators *vis-à-vis* transformations, they are clearly fundamentally related—so much so that most texts tend to avoid a rigorous definition to differentiate one from the other! Generally, however, *operators* are distinguished from *transformations* by being applied to *functions*[9] (though the author has seen at least once the term "operator" reserved merely for "transformations" from a space into itself).

We note in passing that the above equation can be written in the form of an inner product (shown above in Example 2.2.3 to be a linear transformation) of two $(n+1)$-tuples (not to be confused with the above n-tuple functions)

$$\langle \boldsymbol{a}, \tilde{\boldsymbol{x}} \rangle = f(t), \tag{2.4-2}$$

in which $\boldsymbol{a} \equiv (1\ a_{n-1}(t)\ a_{n-2}\ \ldots\ a_o)^{\mathsf{T}}$ and $\tilde{\boldsymbol{x}} \equiv (x^{[n]}\ x^{[n-1]}\ \ldots\ x)^{\mathsf{T}}$. (Again, tildes will denote derivatives up to order n to distinguish the variable from the original $\boldsymbol{x}$.) Thus we see that the fundamental equation 2.4-1, much like *algebraic* linear systems, is a *non-homogeneous* (*homogeneous*, in the case that

[9] Kreyszig, *Advanced Engineering Mathematics*, 1982, §2.5

$f(t) \equiv 0$) "linear equation" *implicitly* defining that $\tilde{\boldsymbol{x}}$ satisfying the equation. (We begin to see the emergence of "linear systems" considered in the previous section already!)

Matrix Representation of the Linear Operator. As a linear transformation/operator, the differential equations should be expressible using matrices (Theorem 2.2.2). The governing equation 2.4-1 can easily be expressed as a matrix product with, generally, an additional additive term:

$$\frac{d\boldsymbol{x}'}{dt} \equiv \frac{d}{dt}\begin{pmatrix} x^{[n-1]} \\ x^{[n-2]} \\ \vdots \\ x \\ t \end{pmatrix}$$

$$= \begin{pmatrix} -a_{n-1}(t) & -a_{n-2}(t) & \cdots & -a_1(t) & -a_o(t) & 0 \\ 1 & 0 & \cdots & 0 & 0 & 0 \\ \vdots & \ddots & \vdots & \vdots & \vdots & \vdots \\ 0 & 0 & \cdots & 1 & 0 & 0 \\ 0 & 0 & \cdots & 0 & 0 & 0 \end{pmatrix}\begin{pmatrix} x^{[n-1]} \\ x^{[n-2]} \\ \vdots \\ x \\ t \end{pmatrix}$$

$$+ \begin{pmatrix} f(t) \\ 0 \\ \vdots \\ 0 \\ 1 \end{pmatrix}$$

$$\equiv \mathbf{A}(t)\boldsymbol{x}' + \boldsymbol{f}_t(t).$$

(The student should be careful to understand the form of the $\mathbf{A}(t)$!) Here $\mathbf{A}(t)$ is an $(n+1) \times (n+1)$ matrix, but if the system is *autonomous*—$f(t) \equiv 0$—the last row on each side reduces to "1" and becomes superfluous, and all dimensions reduce to n.

The "operator" nature of this equation can be emphasized by introducing a "matrix derivative operator" $\mathbf{D}$—note its form—and writing the system as the product of the matrix "linear operator" $(\mathbf{D} - \mathbf{A}(t))$ times $\boldsymbol{x}'$:

$$\dot{\boldsymbol{x}}' - \mathbf{A}(t)\boldsymbol{x}'$$

$$= (\mathbf{D} - \mathbf{A}(t))\boldsymbol{x}'$$

$$\equiv \left(\begin{pmatrix} \frac{d}{dt} & 0 & \cdots & 0 & 0 \\ 0 & \frac{d}{dt} & \cdots & 0 & 0 \\ \vdots & \vdots & \ddots & \vdots & \vdots \\ 0 & 0 & \cdots & \frac{d}{dt} & \vdots \\ 0 & 0 & \cdots & 0 & \frac{d}{dt} \end{pmatrix}\right.$$

$$
- \begin{pmatrix} -a_{n-1}(t) & -a_{n-2}(t) & \dots & -a_1(t) & -a_o(t) & 0 \\ 1 & 0 & \dots & 0 & 0 & 0 \\ \vdots & & \ddots & \vdots & \vdots & \vdots \\ 0 & 0 & \dots & 1 & 0 & 0 \\ 0 & 0 & \dots & 0 & 0 & 0 \end{pmatrix} \Biggr) \begin{pmatrix} x^{[n-1]} \\ x^{[n-2]} \\ \vdots \\ x \\ t \end{pmatrix}
$$

$$
= \begin{pmatrix} (\frac{d}{dt} + a_{n-1}(t)) & a_{n-2}(t) & \dots & a_1(t) & a_o(t) & 0 \\ -1 & \frac{d}{dt} & \dots & 0 & 0 & 0 \\ \vdots & & \ddots & \vdots & \vdots & \vdots \\ 0 & 0 & \dots & -1 & \frac{d}{dt} & 0 \\ 0 & 0 & \dots & 0 & 0 & \frac{d}{dt} \end{pmatrix} \begin{pmatrix} x^{[n-1]} \\ x^{[n-2]} \\ \vdots \\ x \\ t \end{pmatrix}
$$

$$
= \boldsymbol{f}_t(t);
$$

the linearity in the argument $\boldsymbol{x}'$ is then rather more explicit. But this also allows a further correspondence to be made between it and the "linear system" nature of such operators, as we shall see below. In this regard, note that if we explicitly invoke the identities that $\frac{dx^{(i)}}{dt} \equiv x^{(i+1)}$ in the above equation, it reduces to

$$
(\mathbf{D} - \mathbf{A}(t))\boldsymbol{x}' = \begin{pmatrix} \frac{d}{dt} + a_{n-1}(t) & a_{n-2}(t) & \dots & a_1(t) & a_o(t) & 0 \\ 0 & 0 & \dots & 0 & 0 & 0 \\ \vdots & & \ddots & \vdots & \vdots & \vdots \\ 0 & 0 & \dots & 0 & 0 & 0 \\ 0 & 0 & \dots & 0 & 0 & \frac{d}{dt} \end{pmatrix} \begin{pmatrix} x^{[n-1]} \\ x^{[n-2]} \\ \vdots \\ x \\ t \end{pmatrix}
$$

$$
= \boldsymbol{f}_t(t).
$$

Again (accounting for $\frac{dt}{dt} \equiv 1$ in the last row), all rows are identical on both sides except the first, which is (allowing for notation) precisely the form of equation (2.4-2) above.

In fact, the relationship between the above equations—the form usually implemented in the study of differential equations—and (2.4-2) can be made clearer by considering not $\boldsymbol{x}$ or $\boldsymbol{x}'$ (with derivatives only up to the $(n-1)^{st}$) as the unknown function, but rather, say, $\tilde{\boldsymbol{x}}$ or $\tilde{\boldsymbol{x}}'$, that vector with *all* derivatives up to the nth; the above equation then assumes the form

$$
\begin{pmatrix} 1 & a_{n-1}(t) & \dots & a_1(t) & a_o(t) & 0 \\ 0 & 0 & \dots & 0 & 0 & 0 \\ \vdots & & \ddots & \vdots & \vdots & \vdots \\ 0 & 0 & \dots & 0 & 0 & 0 \\ 0 & 0 & \dots & 0 & 0 & \frac{d}{dt} \end{pmatrix} \begin{pmatrix} x^{[n]} \\ x^{[n-1]} \\ \vdots \\ x \\ t \end{pmatrix} = \begin{pmatrix} f(t) \\ 0 \\ \vdots \\ 0 \\ 1 \end{pmatrix} \tag{2.4-3}
$$

$$
\text{or} \quad \tilde{\mathbf{A}}(t)\tilde{\boldsymbol{x}}' = \tilde{\boldsymbol{f}}_t(t),
$$

where now the [square] matrix $\tilde{\mathbf{A}}$ and vectors have dimension $n+2$; this, just as with any other differential equation, is a relation among the derivatives of the function. Like $\mathbf{A}(t)$ above, the rank of $\tilde{\mathbf{A}}(t)$ is only 2; in the event the system is autonomous, the superfluous last row would be dropped, and this matrix would then have rank 1.

Null Space of the Linear Operator. We further develop the correspondence between linear "operators" and "transformations." Given their reflexive nature, we would expect there to be a "null space" associated with this operator; it is nothing more than the set of $\tilde{x}'$ such that $\tilde{\mathbf{A}}(t)\tilde{x}' = \mathbf{0}$—*i.e.*, the *homogeneous* solution to this equation! But there's more: the fact that the rank of $\tilde{\mathbf{A}}$ is only 2 (1 if the system were autonomous) means (by Theorem 2.2.3) that the null space of this matrix has dimension n—precisely the number of linearly independent homogeneous solutions to the original differential equation!

Once again, then, the distinction between transformations and operators becomes blurred.

Linear Differential Equations as "Linear Systems". We have already characterized the original differential equation as a "system of linear equations," and have even applied the terminology "homogeneous" and "non-homogeneous" to such systems according as to whether the right-hand side vanishes or not. Continuing the connection in the language of the previous section, we note that this "linear equation" in fact *does* have a solution: due to the placement of the 0's in both $\tilde{\mathbf{A}}$ and $\tilde{f}_t$, and the fact that the last row of both $\tilde{\mathbf{A}}(t)\tilde{x}'$ and $\tilde{f}_t$ is independent of the others, the rank of the augmented matrix $(\tilde{\mathbf{A}}\ \tilde{f}_t)$ is the same as that of $\tilde{\mathbf{A}}$. Furthermore, by Theorem 2.3.2, the general solution consists of two parts: the *particular* part $\mathbf{x}_P$ satisfying this equation, plus an element $\mathbf{x}_N$ of the null space. *But the latter is precisely a solution to the* homogeneous *equation* $\tilde{\mathbf{A}}\tilde{\mathbf{x}}' = \mathbf{0}$; thus we have the familiar result that the general solution to a linear non-homogeneous differential equation consists of the "homogeneous" part plus the "particular" part. And, as we pointed out previously, the dimension of the null space of $\tilde{\mathbf{A}}$ is, by Theorem 2.2.3, just n—the number of linearly independent solutions to the homogeneous equation!

Summary. We have seen that n^{th}-order linear ordinary differential equations are an instance of "linear operators"—just "linear transformations" of functions. These operate on elements of the n-dimensional vector space of scalar functions and their derivatives to return such elements, and are, in effect, *systems* of "linear equations," homogeneous or non-homogeneous, implicitly defining the desired function. The null space of such transformations, seen to have dimension n, consists of just the "homogeneous" solutions to the differential equation; like algebraic linear systems, the most general solution to a non-homogeneous differential equation consists of this homogeneous part plus a "particular" solution to the equation.

None of this helps one actually to *solve* the equation, but it does give some indication of how pervasive the theory of linear systems actually is!

Summary

This chapter has dealt with general *linear transformations* between vectors spaces, though we have quickly established the association between such transformations and *matrices*: in the same way that a vector can be *represented* as an n-tuple, linear transformations can be *represented* as matrices. Thus we spent a great deal of time early on discussing the *partitioning* of matrices into vectors and "blocks," their *transpose*, and operations on them: matrix *addition* and *scalar multiplication*—operations which qualify matrices as a vector space—but, most importantly, *matrix multiplication*, that operation on which most applications, including the *inner product*, of matrices depend. We show how partitioning can be implemented in *block multiplication* and the transpose. Further, we demonstrate how the elementary operations on matrices (viewed as partitioned vectors) can be implemented through matrix products.

We then return to the primary topic of the chapter, linear transformations: those satisfying the property $L(a\boldsymbol{v}_1 + b\boldsymbol{v}_2) = aL(\boldsymbol{v}_1) + bL(\boldsymbol{v}_2)$. We consider a number of examples of such transformations, most importantly vector representations under a change of basis and, for purposes of the next Part, the transformation between angular velocity and angular momentum. We discuss the *domain* and *range* of such transformations, as well as their *null space*—the set of vectors in the domain which transform to the $\mathbf{0}$ vector in the range—and establish a fundamental relation between the dimensions of these three spaces. We have already seen that linear transformations themselves comprise a vector space; we now exhibit the basis and dimension of that space. We establish that what really defines the transformation is what it does to the *basis* vectors in the domain space: any other vector, a linear combination of the domain's basis vectors, is simply the *same* linear combination of the transformed basis vectors. But we also see that the the linear transformation, since it relies on the effect on its domain's basis, will *change* when that basis changes. All these observations culminate in the equivalence between linear transformations and matrices—that one can always be expressed as the other. Finally there is a brief introduction to *dyadics*—a fruitful formulation of linear transformations which exhibits the effect of basis explicitly.

If we generally view "linear transformations" as determining the element in the range space knowing that in its domain, the inverse problem of establishing which element in the domain is mapped to a known element in the range is generally described as the "solution of linear systems." We categorize such systems according to whether the range element is $\mathbf{0}$ ("homogeneous") or not (non-homogeneous). After establishing how "skips" in the matrix representation of linear transformations are reflected in the null space, we finally confront a general theory of linear equations, established by recognizing that "solution," effected through row operations—equivalent to premultiplication by elementary [non-singular][10] matrices and ultimately to reduction of the coefficient matrix to echelon form—really hinges on the *rank* of that matrix. If the system is

[10]See Definition 3.1.1.

homogeneous, solution amounts to determination of vectors in the null space of the coefficient matrix: if that rank is less than the "number of equations" (the number of *rows* in the coefficient matrix), the null space has non-0 dimension, and there are an infinite number of solutions; while if it is the same, the **0** vectors is the only solution. (The case in which the number of rows is *greater* than the rank cannot occur.) For *non*-homogeneous systems, the analysis is similar, but now we must consider the *consistency* of the system: that the *augmented matrix* of the coefficient matrix catenated with the right-hand side vector be the same. If this is satisfied, the number of solutions depends on the rank of the coefficient matrix in the same way it does for homogeneous equations; otherwise there is *no* solution. We also show that the most general form of non-homogeneous solution consists of the sum of two parts: a *particular* part satisfying the non-homogeneous equation, plus an element of the null space. This section concludes with a brief discussion of the most common practical means of solving such equations, *Gaussian elimination*, and some numerical considerations regarding its implementation.

Finally we regard an application of virtually all the foregoing theory, the solution of nth-order linear differential equations, and that operation related to linear transformations, *linear operators*—linear "transformations" of *functions*. We see that solution to such systems themselves constitute an n-dimensional vector space; that the null space of this operator is just the *homogeneous* solutions of the differential equation; that, as a linear operation, these equations can be represented as a matrix; and that the rank of the associated matrix indicates that the null space is just n. Finally, by viewing the differential equation as "transformation" on the space of vectors obtained from a scalar function and its n derivatives—the *phase space* or, in the case of non-homogeneous equations, the *extended* phase space including t as an additional component—and the resulting differential equation as a "linear system," homogeneous or not according as to whether there is a non-0 $f(t)$, we see that the previous theory can be applied here too; in particular, the general solution to a *non*-homogeneous differential equation consists of a "particular" part satisfying the equation, and the "homogeneous" part residing in the null space.

While the above traces the development of linear transformations in general, the key points to recur continually are

- matrix *partitioning*, particularly the *block multiplication* it effects,
- the fact that an arbitrary linear transformation can be *represented*, *relative to a given basis*, as a matrix, and thus that this representation will *change* when the basis changes,
- the *null space* of a transformation, particularly how it relates to the *rank* of that matrix [representing the] linear transformation, and
- ability to predict the number of solutions to a system of linear equations.

These are the concepts which will be required in the sequel.

Chapter 3

Special Case—Square Matrices

Introduction

All matrices up to this point have been arbitrary rectangular arrays; we now specialize to the case of *square* matrices. After a brief presentation of the "algebra" of such matrices, we then discuss their inverses, determinants (and a standard technique for determining the linear independence of the vectors comprising these matrices), and some terms classifying them.

This done, we are in a position to consider eigenvectors: what they are, when the they will be linearly independent and, if not, how to find "generalized eigenvectors" to supplement the standard ones. We then consider the basic theme of the chapter: how the representations of either vectors or matrices in a given vector space change explicitly with a change of basis. Since basis vectors are elements of the vector space, we can represent a new basis in terms of the original one as linear combinations of the original basis vectors. Thus it should not come as a surprise that that a basis change can itself be expressed as a linear transformation—through yet *another* matrix! How that transformation can be applied to vectors and matrices to get their new representations is then revealed. We shall also consider a particularly important application of such basis transformation: *rotation matrices.*

Finally, we will have attained sufficient background to treat the subject toward which all the above has been directed: "normal" or "canonical" forms of matrices. If the *form*—*i.e.* the *representation*—of a matrix depends on its basis, is there a "standard" form of a given matrix which most simply reflects the essence of that matrix? The answer, of course, is "yes," or the question wouldn't have been posed in the first place! It turns out that, if the eigenvectors of the matrix are linearly independent, it can be *diagonalized*; if not, it can be put in a *block*-diagonalized form, "Jordan normal form," generalizing the diagonalized one. All this depends on the basis chosen in terms of which to represent the matrix. In the case of the inertia matrix of 3-D kinetics, it can always be diagonalized, and the new basis comprises the "principle axes," which turn out to be nothing more than a set *rotated* from the original ones. Virtually everything covered up to that point converges on this topic.

In the preceding, we have been discussing *arbitrary* $m \times n$ matrices—matrices which would generate linear transformations on m-tuple [representations] to result in an n-tuple [representation]. The observant reader might have noted, particularly after a somewhat lengthy discussion of the "algebra" of vectors in the first Chapter, that little has been made of that regarding matrices. That is because one of the primary properties discussed in algebra, the *closure* of operations defined on the elements of a vector space, just isn't satisfied for the central operations unique to matrices, matrix multiplication: except in the most general sense that the product of two matrices is another matrix, the product of a general $m \times k$ matrix and a $k \times n$ matrix is an $m \times n$ matrix—one whose dimension is neither of its constituents!

If we specialize to the topic of *square* matrices—$n \times n$ matrices, as it were—closure under matrix multiplication *is* satisfied. And, regarding such matrices as the [representation of a] linear transformation, this would correspond to the transformation of an n-tuple [representation of a] vector in an n-dimensional space $\mathcal{V}$ into the *same* space. This is, in fact, the primary application of such matrices: They will ultimately be utilized to effect "basis transformations"—or more correctly, the change in the *representations* of both vectors and *other* matrices—under a change of basis, including the important special case of *rotation of axes.* In this context we will examine just how one can select the basis to force a square matrix to assume a particularly simple form—the culmination of both this chapter and this Part.

But, aside from its intrinsic importance, the *raison d'etre* for this study of linear algebra is to set the stage for the next Part on 3-D rigid body dynamics: As was mentioned in Example 2.2.2, the angular momentum of a 3-D rigid body—and thus the equations of motion for such a body—can itself be regarded as a transformation between two vectors, $\boldsymbol{\omega}$ and $\mathbf{H}$, in 3-space:

$$\mathbf{H} = \mathbf{I}\boldsymbol{\omega}.$$

There the moment of inertia will be represented as a matrix—one amenable to "simplification" resulting from a particular *choice* of axes in terms of which to represent it. This choice is precisely what is called the "principal axes" in both dynamics and solid mechanics.

The "Algebra" of Square Matrices

We have already remarked a number of times that, insofar as the operations of "vector addition" and "scalar multiplication," as defined in Example 1.1.4, *any* $m \times n$ matrices form a *vector space*; in particular, under "vector addition" for matrices, the set of square matrices—let us refer to them as $\mathcal{M}$—will form an Abelian group, properties (G1)-(G5) in Table 1.1, under this operation. But we have also observed that, under *matrix multiplication, square* matrices satisfy the following properties for $\mathbf{A}$, $\mathbf{B}$, and $\mathbf{C}$ in $\mathcal{M}$:

(RM1)	$\mathbf{AB} \in \mathcal{M}$ (closure)
(RM2)	$(\mathbf{AB})\mathbf{C} = \mathbf{A}(\mathbf{BC})$ (associativity)
(RM3)	$(\mathbf{A}+\mathbf{B})\mathbf{C} = \mathbf{AC}+\mathbf{BC}$ (distribution)
(RM4)	$\mathbf{A}(\mathbf{B}+\mathbf{C}) = \mathbf{AC}+\mathbf{AC}$ (distribution)
(RM5)	$\mathbf{1A} = \mathbf{A1} = \mathbf{A}$ (identity element)
(RMA)	$c(\mathbf{AB}) = (c\mathbf{A})\mathbf{B} = \mathbf{A}(c\mathbf{B})$

Table 3.1: Properties of Multiplication on Square Matrices

—where the first follows from the definition of matrix multiplication, the second through the fourth are just equations (2.1-14)–(2.1-16) obtained in the section on block multiplication of matrices, the fifth is (2.1-18), defined for the nth*-order identity matrix*, defined in (2.1-17)

$$\mathbf{1} \equiv \mathbf{1}^{(n)} \equiv \begin{pmatrix} 1 & 0 & 0 & \ldots & 0 \\ 0 & 1 & 0 & \ldots & 0 \\ 0 & 0 & 1 & \ldots & 0 \\ \vdots & \vdots & \vdots & \ddots & \vdots \\ 0 & 0 & 0 & \ldots & 1 \end{pmatrix},$$

and the last follows from the "scalar multiplication" part of the vector space property of [square] matrices.

But the first four properties, along with the structure of an Abelian group under *vector* addition, are precisely those defining a *ring*, Table 1.1.1! The fifth, not in that table, is an additional property defining a special ring, a *ring with identity element.* The last, "(RMA)," relating the scalar multiplication [over a "field"] to this matrix multiplication, is that which makes this into what mathematicians call an "*algebra*":

Definition 3.0.1. An associative ring over a field, scalar multiplication with which satisfies property (RMA), is called an *algebra.*

Thus square matrices form an *algebra with unit element*, (RM5)—a *linear algebra*, if we focus on the linear nature of matrix multiplication, central to the above properties.

There are a number of operations and classifications peculiar to square matrices. We now consider them:

3.1 The Inverse of a Square Matrix

The identity matrix above is significant in its form: it is precisely the target *echelon form* of an $n \times n$ square matrix. *If* the rank of the matrix $\mathbf{A}$ is n, it can be reduced to this form through either a sequence of *row* operations or *column* operations. But the former result from a series of *pre*-multiplications by the $\mathbf{E}^*_{ij}$

matrices (page 54) which can be associatively combined into a single matrix, say L, to reduce it to this form:

$$\mathsf{LA} = \mathbf{1};$$

similarly, *column* operations result from *post*-multiplication by the E^*_{ij}, which can be combined into a single matrix, say R, to reduce it to this echelon form:

$$\mathsf{AR} = \mathbf{1}.$$

The first might reasonably be called, in analogy with real numbers, a "*left* inverse," the second a "*right* inverse." While, conceptually, these might be expected to be different, in the case of *matrices* at least, they are the same:

Theorem 3.1.1. If both the left inverse, L, and right inverse, R, exist for a square matrix A, they are equal.

Proof.

$$\mathsf{L} = \mathsf{L}\mathbf{1} = \mathsf{L}(\mathsf{AR}) = (\mathsf{LA})\mathsf{R} = \mathbf{1}\mathsf{R} = \mathsf{R}.$$

(Note that the fundamental property leading to this result is the *associativity* of matrix multiplication; thus we would expect this result to hold for *any* associative operation, not just matrices.) □

In this case we refer simply to *the inverse* of A, generally written A^{-1}.

$$\mathsf{AA}^{-1} = \mathsf{A}^{-1}\mathsf{A} = \mathbf{1}. \tag{3.1-1}$$

Definition 3.1.1. A non-invertible matrix is called *singular*; one which can be inverted is called *non-singular*.

Calculation of the Inverse. The inverse of a matrix at least *seems* possibly important (it is), and it would be useful to have a means of calculating it. Certainly one way would be to find the rank of the matrix using row *or* column operations, keeping tabs of the corresponding E^*_{ij} matrices which generate such operations, then find their product directly; this is just the approach used to motivate the left and right inverses above (which, by the Theorem, are both equal to *the* inverse). There is, however, an alternative which does effectively the same thing, but which doesn't require explicit introduction of the E^*_{ij}:

Assume we have a matrix A whose inverse we are trying to find. If we perform the same matrix pre- *or* post-products—elementary row and column operations are equivalent to matrix products, recall—to *both* A *and* the identity matrix $\mathbf{1}$, then what started as the identity matrix will end as that *single* matrix which, when pre- or post-multiplying A, generates the final form of A. In essence, this says that, starting with

$$\mathsf{A} \qquad\qquad\qquad\qquad\qquad \mathbf{1},$$

then, using L, the aggregate *row* operation matrix,

$$\mathsf{A} \to \mathsf{LA} = \mathsf{1} \qquad\qquad \mathsf{1} \to \mathsf{L1} = \mathsf{L}$$

or R, the final *column* operation matrix,

$$\mathsf{A} \to \mathsf{AR} = \mathsf{1} \qquad\qquad \mathsf{1} \to \mathsf{1R} = \mathsf{R}\,.$$

This appears similar to the procedure on page 57 for finding an E^* to effect the appropriate row *or* column operations on a matrix. Like that procedure, *it only works for all left products or all right products*; it is not possible to mix them, since the result of such operations on A could not be effected through a *single* left- or right-multiplication! In application, one doesn't perform matrix products at all, but simply performs the same row *or* column operations (not *both*!) on both A and $\mathbf{1}$; the latter then becomes that *matrix* which generates the same sequence of operations through a matrix product. This can best be illustrated by an example:

Example 3.1.1. Find the inverse of the matrix

$$\mathsf{A} = \begin{pmatrix} 2 & 0 & 0 \\ -2 & 1 & 1 \\ -1 & 1 & 2 \end{pmatrix}.$$

Solution: We start with the matrices A and $\mathbf{1}$ side-by-side and then operate on both simultaneously. In this example, *row* operations are used on A.

$$\begin{pmatrix} 2 & 0 & 0 \\ -2 & 1 & 1 \\ -1 & 1 & 2 \end{pmatrix} \qquad \begin{pmatrix} 1 & 0 & 0 \\ 0 & 1 & 0 \\ 0 & 0 & 1 \end{pmatrix}$$

$$\begin{pmatrix} 1 & 0 & 0 \\ -2 & 1 & 1 \\ -1 & 1 & 2 \end{pmatrix} \qquad \begin{pmatrix} \frac{1}{2} & 0 & 0 \\ 0 & 1 & 0 \\ 0 & 0 & 1 \end{pmatrix}$$

$$\begin{pmatrix} 1 & 0 & 0 \\ 0 & 1 & 1 \\ 0 & 1 & 2 \end{pmatrix} \qquad \begin{pmatrix} \frac{1}{2} & 0 & 0 \\ 1 & 1 & 0 \\ \frac{1}{2} & 0 & 1 \end{pmatrix}$$

$$\begin{pmatrix} 1 & 0 & 0 \\ 0 & 1 & 1 \\ 0 & 0 & 1 \end{pmatrix} \qquad \begin{pmatrix} \frac{1}{2} & 0 & 0 \\ 1 & 1 & 0 \\ -\frac{1}{2} & -1 & 1 \end{pmatrix}$$

$$\begin{pmatrix} 1 & 0 & 0 \\ 0 & 1 & 0 \\ 0 & 0 & 1 \end{pmatrix} \qquad \begin{pmatrix} \frac{1}{2} & 0 & 0 \\ \frac{3}{2} & 2 & -1 \\ -\frac{1}{2} & -1 & 1 \end{pmatrix}$$

The last matrix is A^{-1}, as can be verified directly. Note that no actual matrix products using the E^*_{ij} were used; the *results* of such products, elementary row operations, are merely reflected in both matrices. Since row operations were used [exclusively], the matrix is, strictly speaking, the *left* inverse of $\mathbf{A}$; but, since echelon form could be arrived at through *column* operations—*i.e.*, through *post*-multiplication by what would turn out to be the *right* inverse—*the* inverse exists and is equal to both (Theorem 3.1.1). |

Working under the assumption that the inverse of a square matrix is important (it is), it would appear useful to have some properties of the inverse. These are contained in the following theorems:

Properties of the Inverse

Theorem 3.1.2. (*Uniqueness*): If it exists, the inverse, $\mathbf{A}^{-1}$, of a square matrix $\mathbf{A}$ is *unique.*

Proof. Assume $\mathbf{A}^{-1}\mathbf{A} = \mathbf{1} = \mathbf{C}\mathbf{A}$; we claim that $\mathbf{A}^{-1} = \mathbf{C}$. Indeed, since $\mathbf{A}^{-1}\mathbf{A} = \mathbf{1} = \mathbf{A}\mathbf{A}^{-1}$,

$$\mathbf{C} = \mathbf{C}\mathbf{1} = \mathbf{C}\mathbf{A}\mathbf{A}^{-1} = \mathbf{1}\mathbf{A}^{-1} = \mathbf{A}^{-1}$$ □

Theorem 3.1.3. (*Inverse of a Product*): $(\mathbf{A}\mathbf{B})^{-1} = \mathbf{B}^{-1}\mathbf{A}^{-1}$.

Proof.

$$(\mathbf{B}^{-1}\mathbf{A}^{-1})(\mathbf{A}\mathbf{B}) = \mathbf{B}^{-1}(\mathbf{A}^{-1}\mathbf{A})\mathbf{B} = \mathbf{B}^{-1}\mathbf{B} = \mathbf{1}$$

(*Cf.* the *transpose* of a product, Theorem 2.1.2.) □

Theorem 3.1.4. (*Inverse of an Inverse*): $(\mathbf{A}^{-1})^{-1} = \mathbf{A}$

Proof.

$$\mathbf{A} = \mathbf{A}\mathbf{1} = \mathbf{A}(\mathbf{A}^{-1}(\mathbf{A}^{-1})^{-1}) = \mathbf{1}(\mathbf{A}^{-1})^{-1}) = (\mathbf{A}^{-1})^{-1}$$ □

We conclude these results with a criterion to determine when the inverse of a square matrix actually exists in a given instance. We have observed already that determination of the inverse of an $n \times n$ matrix is equivalent to being able to reduce that matrix to an echelon form which is itself the identity matrix; but this is possible if and only if all n rows or columns are, in fact, linearly independent. Thus we have obtained the following important result:

Theorem 3.1.5. The inverse of $n \times n$ $\mathbf{A}$ exists if and only if all of its rows or columns are linearly independent; *i.e.*, if and only if the rank of $\mathbf{A}$ is n.
In terms of Definition 3.1.1, a matrix is *non-singular* iff all of its rows or columns are linearly independent.

Summary. After introducing the concept of "left-" and "right-"inverses, we show that, for *associative* operations, these two coalesce into "*the*" matrix. We show how this can be calculated in practice, and obtain its properties.

3.2 The Determinant of a Square Matrix

Associated with a square matrix **A** is a real number referred to as the *determinant* of **A**, denoted $|\mathbf{A}|$. Though it can be defined rather elegantly in terms of things like "permutation operators," the definition is curious to the point of being dicey! It is rather more straightforward (if only little less dicey) to define the determinant in a somewhat more mundane fashion due to Laplace:

Calculation of the Determinant—*Expansion in Minors*: The procedure proceeds in a series of steps, at each of which we express the determinant of a matrix in certain linear combinations of determinants of *sub*matrices of that matrix; thus, if the original matrix were of dimension $n \times n$, the linear combinations would involve determinants of $(n-1) \times (n-1)$ submatrices. The process continues iteratively until we have reduced the submatrices to dimension 2×2, whose determinant we *define* to be

$$\begin{vmatrix} a & b \\ c & d \end{vmatrix} \equiv ad - bc. \tag{3.2-1}$$

In order to explain the approach, it is necessary first to establish a couple of preliminaries:

- Given any *element* of an $n \times n$ square matrix, there is associated with it *another* determinant, called the *minor* of that element; it is simply the determinant of the *submatrix* obtained from the elements remaining after striking out the row and column of the [original] matrix containing the element in question. Because of this "striking out" operation, the minor becomes the determinant of an $(n-1) \times (n-1)$ matrix.
- Associated with the matrix is a *sign matrix*, consisting of alternating +- and −-signs, obtained by starting in the upper left-hand corner of the matrix with a +:

$$\begin{pmatrix} + & - & + & \cdots \\ - & + & - & \cdots \\ \vdots & \vdots & \vdots & \ldots \end{pmatrix}.$$

The *minor* associated with a given element a_{ij}, multiplied by the appropriate *sign* from this sign matrix, is referred to as the *cofactor* of the element, $\operatorname{cof}(a_{ij})$.

With this in hand, we now describe the process in more detail: Given an $n \times n$ matrix, we choose to "expand in minors of" an arbitrary row or column by taking each element of that row or column and multiplying

1. the *element* (with its associated *sign*), times
2. the *associated sign* from the above "sign matrix" (so that, if the associated sign is "−," the sign of the element is reversed), times
3. the *minor* of that element.

—note that the last two steps generate the cofactor of the element—and summing it with the results of similar products from the other elements of that row or column. Symbolically,

$$|\mathbf{A}| \equiv \sum_{k=1}^{n}(a_{ik}\cdot \operatorname{cof}(a_{ik})) \qquad \text{(expansion on } i^{th} \textit{ row}) \tag{3.2-2a}$$

$$\equiv \sum_{k=1}^{n}(a_{ki}\cdot \operatorname{cof}(a_{ki})) \qquad \text{(expansion on } i^{th} \textit{ column}). \tag{3.2-2b}$$

Each of the minors is the determinant of a square matrix one order less than the original matrix, giving a linear combination of n terms involving determinants of $(n-1)\times(n-1)$ matrices.

The process is now repeated on the individual determinants in *this* sum; now *each* such generates a linear combination consisting of $n-1$ determinants of order $n-2$—a total of $n(n-1)$ terms.

We now operate on each of *these* determinants, continuing at each step until we ultimately end up with $(n(n-1)\ldots 2)$ 2×2 determinants, which we finally evaluate using the definition (3.2-1).

> **Warning**: If one us accustomed to the expansion of, say, 3×3 determinants by taking products of triples first to the lower right, and then to the lower left (with the opposite sign), he should be aware that this device *will not work* for higher-order determinants!

Though the above procedure lacks any degree of elegance, it is at least straightforward and likely the best for hand calculation for reasonably small matrices. Clearly, however, the amount of effort is proportional to $n!$ in the order of the matrix, so, in general (and particularly for even semi-large matrices), one uses a dodge: rather than using this definition, the matrix is reduced to the "*upper-triangular*" form already encountered in connection with Gauss Elimination (Section 2.3.2), with zeros below the main diagonal:

$$\mathbf{A}=\begin{pmatrix} a_{11} & a_{12} & \cdots & a_{1n}\\ a_{21} & a_{22} & \cdots & a_{2n}\\ \vdots & \vdots & & \vdots\\ a_{n1} & a_{n2} & \cdots & a_{nn}\end{pmatrix} \rightarrow \mathbf{A}'\equiv\begin{pmatrix} a'_{11} & a'_{12} & \cdots & a'_{1n}\\ 0 & a'_{22} & \cdots & a'_{2n}\\ \vdots & \vdots & \ddots & \vdots\\ 0 & 0 & \cdots & a'_{nn}\end{pmatrix}. \tag{3.2-3}$$

We note explicitly that this triangular form does *not* normalize the diagonal entries; doing so would simply multiply the determinant of the resulting triangularized matrix (now with 1's along the diagonal) by the product of the normalizing factors (Theorem 3.2.3 below). Thus it will have the *same* determinant: by the above definition (expanding successively in minors of respective columns, starting with the first, say), is just

$$|\mathbf{A}'| = a'_{11}a'_{22}\ldots a'_{nn}.$$

The matrix $\mathbf{A}'$ is arrived at through elementary row/column operations, involving only n^2 multiplications. We shall show below (Theorem 3.2.5) that such operations, in fact, *do* leave the determinant unaltered.

There is still one loose end to be tied up regarding the above definition of the determinant: it was couched in terms of expansion in minors of *any* row or column, and it hardly evident that the resulting number would be *independent* of this choice! But some reflection indicates that is is:

Consider first the *form* of each term in the final summation giving the determinant—a *product* of n elements of the matrix. When each element, on the row/column in terms of which the expansion is made, multiplies its minor at each intermediate step in the reduction, *it no longer appears in any subdeterminant*; thus

- each element, in turn representing one row/column, *appears in only one term* of the linear combination of subdeterminants.

Further, however, recall that we "strike out" the row *and* column in which that element appears, in order to determine the submatrix whose determinant will be multiplied by the element; thus each term in the linear combination of subdeterminants *contains no other terms in that element's row* or *column*; *i.e.*,

- each term in the final linear combination of products of elements of the original matrix contains exactly one element from each row and column of that matrix.

Since there *is* one element from each row [*and* column] in *each* term of the final summation, it will have the form $a_{1j_1}a_{2j_2}\ldots a_{nj_n}$–*i.e.*, we can order terms in the product by at least their *row* index, even though the column indices will contain [exactly one of] the numbers $1,\ldots,n$ *in any order.* But there are exactly $n!$ possible such terms (corresponding to the choice of the j_k)—exactly the number of products in the definition as stated. Thus the *products* of elements in the final sum giving the defined value of the determinant are independent of whether one expands in minors of a row or column.

The only thing left is the *sign* assigned to this particular product. But that comes from the sign matrix, applied whether expansion is done in rows or columns. Thus, in fact, the definition for the determinant is well-defined.

Having now defined the determinant, it should come as no surprise that we will examine some of its properties. These are contained in the following theorems:

Properties of the Determinant

Theorem 3.2.1. (*Determinant of Transpose*): $|\mathbf{A}^{\mathsf{T}}| = |\mathbf{A}|$.

Proof. Since the expansion of $|\mathbf{A}|$ can be made in terms of either rows or columns, and expanding, say, $|\mathbf{A}|$ in columns corresponds to expanding $|\mathbf{A}^{\mathsf{T}}|$ in *rows*, the result will be the same in either case. □

Theorem 3.2.2. (*Interchange*): The interchange of any two rows/columns of $\mathbf{A}$ reverses the sign on $|\mathbf{A}|$.

Proof. We note first that the interchange of two *adjacent* rows/columns of $\mathbf{A}$ changes the sign of $|\mathbf{A}|$: if we expand this in terms of a given row/column, then switch and re-expand in terms of the *same* row, the signs on the second expansion will be reversed because the adjacent row's *sign* matrix reverses.

Now consider two row/columns, say i and j where, without loss of generality, we can assume $j > i$: $j = i + m, m \geq 1$. To interchange these, it takes

- $(m-1)$ interchanges below (or to the right) of the ith to bring the jth to row/column $(i+1)$,
- 1 interchange to switch *these*, and
- $m - 1$ interchanges to bring the *new* $(i+1)^{st}$ row/column to the jth,

—a total of $2m - 1$ interchanges (an odd number)—*each* of which changes the sign of $|\mathbf{A}|$ by the above Theorem 3.2.2. □

Theorem 3.2.3. (*Row/Column Multiplication by a Constant*): Multiplication of a row/column of $\mathbf{A}$ by a number c changes $|\mathbf{A}|$ to $c|\mathbf{A}|$.

Proof. If one expands $|\mathbf{A}|$ in terms of that row/column, each term in the initial reduction, using elements of the row/column, will be multiplied by c; thus so will $|\mathbf{A}|$. □

Theorem 3.2.4. (*Proportional Row/Columns*): If two row/columns of $\mathbf{A}$ are proportional, then $|\mathbf{A}| \equiv 0$.

Proof. If, say, row j is k times row i, then, by the above theorem,

$$\begin{vmatrix} \cdots & \cdots & \cdots & \cdots \\ a_{i1} & a_{i2} & \cdots & a_{in} \\ \cdots & \cdots & \cdots & \cdots \\ ka_{i1} & ka_{i2} & \cdots & ka_{in} \\ \cdots & \cdots & \cdots & \cdots \end{vmatrix} = k \begin{vmatrix} \cdots & \cdots & \cdots & \cdots \\ a_{i1} & a_{i2} & \cdots & a_{in} \\ \cdots & \cdots & \cdots & \cdots \\ a_{i1} & a_{i2} & \cdots & a_{in} \\ \cdots & \cdots & \cdots & \cdots \end{vmatrix} = 0,$$

since if we expand $|\mathbf{A}|$ in all other rows *except* i and j at each step, using these to expand the final 2×2 minors at the end of the reduction, each of the latter will be of the form

$$\begin{vmatrix} a_{ii_1} & a_{ii_2} \\ a_{ii_1} & a_{ii_2} \end{vmatrix} \equiv 0$$

so that the total determinant vanishes. □

Note that, as a special case of this, a $\mathbf{0}$ row or column—just 0 times a row or column—will generate a 0 determinant.

The next theorem appears at first to directly contradict Theorem 3.2.3, but note that we *add* a constant times the row/column to another row/column; we don't simply *multiply* it!

Theorem 3.2.5. (*Addition of [Constant times] Row/Column to another Row/ Column*): Adding a c times a row/column to another row/column of $\mathbf{A}$ leaves $|\mathbf{A}|$ unchanged.

Proof. Expanding the altered matrix' determinant first in terms of the row i_1, say, that $c\times$(row i_2) has been added to, yielding terms $a_{i_1 j} + c a_{i_2 j}$, each *cofactor* $M_{i_1 j}$ of these terms is multiplied by that element; thus the resulting determinant is

$$\sum_j^n (a_{i_1 j} + c a_{i_2 j}) M_{i_1 j} = \sum_j^n a_{i_1 j} M_{i_1 j} + c \sum_j^n a_{i_2 j} M_{i_1 j}$$

But the first of these is $|\mathbf{A}|$ itself, while the second is the determinant of a matrix whose rows i_1 and i_2 are proportional (in fact *equal*), so vanishes, by the above theorem. □

(This result justifies the procedure alluded to for the determination of determinants of large matrices on page 104, particularly equation 3.2-3: if we merely add a constant times one row/column to another row/column—effectively reducing it to echelon form but *without normalizing* [which, by Theorem 3.2.3, would *scale* the determinant], the determinant remains unchanged.)

Before continuing with these important results, let us illustrate the usefulness of the above theorem with an example, which will also be used in the following theorem:

Example 3.2.1. Find the determinant of an arbitrary elementary matrix

$$\begin{array}{rcl} & \begin{array}{ccccc} & & j & & \\ & & \downarrow & & \end{array} & \\ \mathbf{E}^*_{ij}(c) \equiv & \begin{pmatrix} 1 & 0 & 0 & \cdots & 0 \\ 0 & 1 & 0 & \cdots & 0 \\ \vdots & 0 & \ddots & \cdots & 0 \\ \vdots & \vdots & c & \ddots & 0 \\ 0 & \cdots & \cdots & \cdots & 1 \end{pmatrix} & \begin{array}{l} \\ \\ \\ \leftarrow i \\ \\ \end{array} \end{array}$$

—where the arrows point to the ith row and j^{th} column to indicate the only non-0 element aside from the 1's on the diagonal.

Solution: While the determinant is not too difficult to evaluate using its definition above, a more more elegant means is to notice that $\mathbf{E}^*_{ij}(c)$ is defined to be the identity matrix $\mathbf{1}$, c times whose j^{th} row has been added to its ith (equation (2.1-19)). From the above theorem, the determinant will be simply the same as $|\mathbf{I}|$, namely 1, for $i \neq j$; if $i = j$, adding c times the ith row to itself puts the number $(c+1)$ in the ith term on the diagonal—$(c+1)$ times the ith row of $\mathbf{1}$ which, by Theorem 3.2.3, has a determinant of just $(c+1)\cdot 1 = (c+1)$:

$$|\mathbf{E}^*_{ij}(c)| = \begin{cases} 1 & i \neq j \\ (c+1) & i = j \end{cases}. \tag{3.2-4}$$

Note that the determinant of an elementary matrix will vanish for $i = j$ if $c = -1$—the very case mentioned on page 57, amounting to multiplying the ith row or column by 0! |

Note that the operations addressed in Theorems 3.2.3 and 3.2.5 give us a means of analyzing the effect of linear combinations on the determinant. If, for example, one wishes to perform column operations on a matrix denoted as an ordered collection of column matrices—"vectors" $\mathbf{a}_i$ so that $\mathbf{A} = (\mathbf{a}_1\mathbf{a}_2\ldots\mathbf{a}_n)$—the linear combination $c_1\mathbf{a}_i + c_2\mathbf{a}_j = c_1(\mathbf{a}_i + (c_2/c_1)\mathbf{a}_j)$ can be formed in the ith column by first adding (c_2/c_1) times $\mathbf{a}_j$ to that column and then multiplying the result by c_1. Thus we see that such a linear combination in fact *changes* the value of the determinant: the first operation does nothing (by Theorem 3.2.5), but the second *multiplies* the value of the original determinant by c_1 (Theorem 3.2.3). Multiplication of a column by a constant, for example to normalize it, will similarly multiply the determinant by that constant.

Thus linear combinations *change* the determinant; what they do *not* do is to make the determinant *vanish*—at least as long as one avoids multiplication of a row/column by the constant 0. In particular, the linear combinations utilized to determine the *rank* of a matrix $\mathbf{A}$ (which also avoid 0-multiplication, page 46)—*i.e.*, to bring the matrix to echelon form—cannot zero out the determinant $|\mathbf{A}|$:

$$|\mathbf{A}_{ech}| = c|\mathbf{A}|, \quad c \neq 0. \tag{3.2-5}$$

But, if a particular row/column of the original matrix was *itself* a linear combination of *other* row/columns—*i.e.*, the rank of the matrix is less than n—the result of such operations will be an $\mathbf{A}_{ech}$ with a $\mathbf{0}$ row or column. Expansion of the determinant in terms of this will thus make $|\mathbf{A}_{ech}|$ vanish, not because of the *operations* but because of the *linear dependence*. Thus, by (3.2-5), $|\mathbf{A}| = 0$, and we have immediately a test for linear independence of the rows or columns of a square matrix:

Theorem 3.2.6. The determinant of a square matrix $\mathbf{A}$ vanishes if and only if the rows and columns of $\mathbf{A}$ are linearly dependent.

Corollary. The rank of a matrix $\mathbf{A}$ is less than n if and only if $|\mathbf{A}| = 0$.

But, then, by Theorem 3.1.5 we have immediately a corresponding test for the existence of the inverse of matrix:

Theorem 3.2.7. The inverse of a square matrix $\mathbf{A}$ exists if and only if $|\mathbf{A}| \neq 0$.

As a particular instance of this, consider the matrix $\mathbf{E}^*_{ij}(c)$ again; we have already seen that $|\mathbf{E}^*_{ij}(c)| \neq 0$ as long as $c \neq -1$ (equation (3.2-4)); thus it has an inverse. In fact, it is obvious that

$$(\mathbf{E}^*_{ij}(c))^{-1} = \mathbf{E}^*_{ij}(-c)! \tag{3.2-6}$$

In particular, we observe that the inverse of an elementary matrix is *itself* an elementary matrix.

We are now ready to prove the final property of determinants:

Theorem 3.2.8. (*Determinant of a Product*): $|\mathbf{AB}|=|\mathbf{A}||\mathbf{B}|$.

To do this, we will first prove a special case of the theorem, applicable when one of the two matrices in the product is an elementary matrix:

Lemma. For arbitrary matrix $\mathbf{B}$, $|\mathbf{E}^*_{ij}(c)\mathbf{B}| = |\mathbf{E}^*_{ij}(c)||\mathbf{B}|$.

Proof. As in Example 3.2.1, we consider two cases: If $i \neq j$, $\mathbf{E}^*_{ij}(c)\mathbf{B}$ simply adds c times the jth row of $\mathbf{B}$ to the ith; but, by Theorem 3.2.5, this leaves the determinant of the matrix unchanged. If, on the other hand, $i = j$, this product effectively multiplies the ith row by $(c+1)$, just as in the above example; thus, again by Theorem 3.2.3, the determinant itself is just multiplied by $(c+1)$:

$$|\mathbf{E}^*_{ij}(c)\mathbf{B}| = \begin{cases} |\mathbf{B}| & i \neq j \\ (c+1)|\mathbf{B}| & i = j \end{cases}.$$

This proves the lemma. □

With this in hand, we are finally ready to prove the ultimate result:

Proof of the Product Determinant Theorem. Observe that we can always bring an arbitrary square matrix $\mathbf{A}$ to its echelon form by performing, say, row operations—tantamount to pre-multiplying by appropriate $\mathbf{E}^*_{ij}(c)$ matrices (to simplify the notation, simply labelled "$\mathbf{E}^*_1$," "$\mathbf{E}^*_2$," *etc.*, in what follows):

$$\mathbf{A}_{ech} = \mathbf{E}^*_{N_A} \dots \mathbf{E}^*_1\mathbf{A} \equiv \mathbf{E}^*\mathbf{A}$$

[note the order]. But, precluding the pathological case $c = -1$, each of the $\mathbf{E}^*_i$ has an inverse (see (3.2-6)), so that we can write

$$\mathbf{A} = (\mathbf{E}^*_1)^{-1} \dots (\mathbf{E}^*_{N_A})^{-1}\mathbf{A}_{ech}$$

—where we have used Theorem 3.1.3 recursively, or, by the Lemma (also applied repeatedly),

$$|\mathbf{A}| = |(\mathbf{E}^*_1)^{-1}| \dots |(\mathbf{E}^*_{N_A})^{-1}||\mathbf{A}_{ech}|. \tag{3.2-7}$$

There are now two possible cases:

1. If the rank of $\mathbf{A}$ is n, then $\mathbf{A}_{ech} = \mathbf{1}$, and

$$\mathbf{AB} = (\mathbf{E}^*_1)^{-1} \dots (\mathbf{E}^*_{N_A})^{-1}\mathbf{A}_{ech}\mathbf{B} = (\mathbf{E}^*_1)^{-1} \dots (\mathbf{E}^*_{N_A})^{-1}\mathbf{B};$$

thus, by repeated application of the Lemma (noting that the inverses are each themselves elementary matrices),

$$|\mathbf{AB}| = |(\mathbf{E}^*_1)^{-1} \dots (\mathbf{E}^*_{N_A})^{-1}\mathbf{B}| = |(\mathbf{E}^*_1)^{-1}| \dots |(\mathbf{E}^*_{N_A})^{-1}||\mathbf{B}| = |\mathbf{A}||\mathbf{B}|$$

by (3.2-7), and the fact that $|\mathbf{A}_{ech}| = |\mathbf{I}| = 1$.

2. If, on the other hand, the rank of $\mathbf{A}$ is *less* than n, $\mathbf{A}_{ech}$ has at least one $\mathbf{0}$ row (using row operations to give the *row* rank), and thus so does $\mathbf{A}_{ech}\mathbf{B}$, and thus $|\mathbf{A}_{ech}\mathbf{B}| = 0$. But then, again by the Lemma,

$$|\mathbf{AB}| = |(\mathbf{E}_1^*)^{-1} \ldots (\mathbf{E}_{N_A}^*)^{-1}||\mathbf{A}_{ech}\mathbf{B}| = 0 = |\mathbf{A}||\mathbf{B}|.$$

since $|\mathbf{A}| = 0$ by Theorem 3.2.7.

This important theorem is proven. □

As an important consequence of this theorem, we have the following:

Theorem 3.2.9. *Determinant of an Inverse*: If the inverse, $\mathbf{A}^{-1}$, of a matrix $\mathbf{A}$ exists, then $|\mathbf{A}^{-1}| = 1/|\mathbf{A}|$.

Proof. Since $\mathbf{A}^{-1}$ exists, we can form

$$|\mathbf{A}^{-1}||\mathbf{A}| = |\mathbf{AA}^{-1}| = |\mathbf{1}| = 1.$$

But, by Theorem 3.2.7, since the inverse exists, $|\mathbf{A}| \neq 0$; thus we can divide both sides of this equation by it, to get the desired result. □

There is one last fundamental result involving determinants, not always given in their exposition:

Theorem 3.2.10. *Derivative of a Determinant*: The derivative of the determinant of a square matrix $\mathbf{A}$ is

$$\begin{aligned}|\mathbf{A}|' &\equiv |(\mathbf{a}_1\, \mathbf{a}_2\, \ldots\, \mathbf{a}_n)|' \\ &= |(\mathbf{a}_1'\, \mathbf{a}_2\, \ldots\, \mathbf{a}_n)| + |(\mathbf{a}_1\, \mathbf{a}_2'\, \ldots\, \mathbf{a}_n)| + \cdots + |(\mathbf{a}_1\, \mathbf{a}_2\, \ldots\, \mathbf{a}_n')|.\end{aligned}$$

Though this has been posed in terms of the columns of $\mathbf{A}$, clearly the same result holds for *rows*, by Theorem 3.2.1.

Proof. Though this is difficult to show if one straightforwardly expresses the determinant and then differentiates directly, it can be demonstrated fairly easily if the arguments on page 105 regarding the well-defined nature of Laplace' definition are understood: Since each term in the determinant's summation contains exactly one element from each row and column of the matrix, the derivative this summation would be a sum of terms, each of which contains only one *derivative* from each row and column. Gathering together the terms corresponding to the ith column, say, these terms are nothing more than the determinant of $|(\mathbf{a}_1\, \ldots\, \mathbf{a}_i'\, \ldots\, \mathbf{a}_n)|$, and the above result follows immediately. □

We conclude this section with an equation giving the inverse of a matrix explicitly using its determinant:

Theorem 3.2.11. Let $\mathsf{C_A} = (c_{ij})$ be the *transpose* of the matrix of *cofactors* of each element of the $n \times n$ matrix A:

$$\mathsf{C_A} = \begin{pmatrix} \mathrm{cof}(a_{11}) & \mathrm{cof}(a_{21}) & \cdots & \mathrm{cof}(a_{n1}) \\ \mathrm{cof}(a_{12}) & \mathrm{cof}(a_{22}) & \cdots & \mathrm{cof}(a_{n2}) \\ \vdots & \vdots & & \vdots \\ \mathrm{cof}(a_{1n}) & \mathrm{cof}(a_{2n}) & \cdots & \mathrm{cof}(a_{nn}) \end{pmatrix}.$$

Note the ordering of terms arising from the transpose: c_{ij} is the *signed* determinant of the submatrix formed by striking out the jth row and i^{th} column of $\mathsf{A} = (a_{ij})$. Then

$$\mathsf{A}^{-1} = \frac{1}{|\mathsf{A}|}\mathsf{C_A}.$$

Proof. If we denote the right-hand side of the above equation by V, then we wish to show that $\mathsf{AV} = \mathbf{1}$. Writing $\mathsf{AV} \equiv (b_{ij})$, the definition of matrix multiplication, Definition 2.1-5, gives us immediately that

$$b_{ij} = \frac{1}{|\mathsf{A}|}\sum_{k=1}^{n}(a_{ik}c_{kj}) \equiv \frac{1}{|\mathsf{A}|}\sum_{k=1}^{n}(a_{ik} \cdot \mathrm{cof}(a_{jk}))$$

We consider first the off-diagonal terms of this product, $i \neq j$: recalling the definition of the determinant, (3.2-2), this is the determinant of that matrix obtained by replacing the jth row of A by its ith—*i.e.*, the determinant of a matrix whose i^{th} and j^{th} rows are identical! But, by Theorem 3.2.4, the determinant of such a matrix vanishes (the constant of proportionality is 1!), so $b_{ij} = 0,\ i \neq j$.

Now consider the diagonal terms of this product, $i = j$:

$$b_{ii} = \frac{1}{|\mathsf{A}|}\sum_{k=1}^{n}(a_{ik}c_{ki}) \equiv \frac{1}{|\mathsf{A}|}\sum_{k=1}^{n}(a_{ik} \cdot \mathrm{cof}(a_{ik}))$$

But the summation part of this is nothing more than the expansion of $|\mathsf{A}|$ in cofactors of the ith row; thus, after division by $|\mathsf{A}|$, we have that $b_{ii} = 1$.

Thus we have shown that $\mathsf{AV} = (b_{ij}) = \mathbf{1}$, as desired. □

Recall that we have already given a reasonably straightforward method of calculating the inverse of a matrix (see Example 3.1.1). To be brutally frank, this theorem is of little utility: *no* one calculates the inverse of a matrix this way! The only reason it is given here is to use the result in a later section on the Cayley-Hamilton Theorem, though it is also cited in Section 3.4 in connection with Cramer's Rule.

Summary. We introduce a scalar associated with a square matrix, its *determinant*, and also enumerate its properties.

3.3 Classification of Square Matrices

We categorize square matrices according to certain properties they might enjoy:

Definition 3.3.1. A square matrix A such that $\mathsf{A}^{\mathsf{T}} = \mathsf{A}$ is called *symmetric*.

Observe that, though we can talk about the transpose of an arbitrary matrix, it is only *square* matrices that have any hope of being symmetric: the transpose of a general $m \times n$ matrix will be an $n \times m$ matrix which clearly cannot be the same as the original!

Definition 3.3.2. A square matrix $\mathbf{A}$ such that $\mathbf{A}^{\mathsf{T}} = -\mathbf{A}$ is called *skew-* [or *anti-*] *symmetric.*

Definition 3.3.3. A square matrix $\mathbf{A}$ such that $\mathbf{A}^{\mathsf{T}} = \mathbf{A}^{-1}$ is called *orthogonal.*

Orthogonal matrices enjoy a very special form:

Theorem 3.3.1. *The columns of an orthogonal matrix* $\mathbf{A} = (\boldsymbol{a}_1, \boldsymbol{a}_2, \ldots, \boldsymbol{a}_n)$ *form a set of "orthogonal" n-tuples with unit length, in the sense that their inner products satisfy:*

$$\langle \mathbf{a}_i, \mathbf{a}_j \rangle = \begin{cases} 1 & i = j \\ 0 & i \neq j \end{cases}.$$

(*Cf.* Definitions 2.1.7 and 2.1.8.)

Proof. Using the interpretation afforded by equation (2.1-10),

$$\begin{aligned} \mathbf{A}^{\mathsf{T}}\mathbf{A} &= \begin{pmatrix} \mathbf{a}_1^{\mathsf{T}} \\ \mathbf{a}_2^{\mathsf{T}} \\ \vdots \\ \mathbf{a}_n^{\mathsf{T}} \end{pmatrix} (\mathbf{a}_1, \mathbf{a}_2, \ldots \mathbf{a}_n) = (\langle \mathbf{a}_i, \mathbf{a}_j \rangle) \\ &= \mathbf{A}^{-1}\mathbf{A} = \mathbf{1} \end{aligned}$$

from which, equating the right hand sides of each line, the result follows immediately. □

In this sense, it is likely that a better name for "orthogonal" matrices might be "orthonormal" matrices! It is also the case that the product of two orthogonal matrices is itself orthogonal:

Theorem 3.3.2. *If* $\mathbf{A}$ *and* $\mathbf{B}$ *are orthogonal,* $\mathbf{AB}$ *is orthogonal.*

Proof. If $\mathbf{A}^{-1} = \mathbf{A}^{\mathsf{T}}$ and $\mathbf{B}^{-1} = \mathbf{B}^{\mathsf{T}}$, then, by Theorems 3.1.3 and 2.1.2

$$(\mathbf{AB})^{-1} = \mathbf{B}^{-1}\mathbf{A}^{-1} = \mathbf{B}^{\mathsf{T}}\mathbf{A}^{\mathsf{T}} = (\mathbf{AB})^{\mathsf{T}}$$

as desired. □

Thus orthogonal matrices are "closed" under matrix multiplication, with which they form a "group."

Homework:

1. Show that, for an arbitrary matrix $\mathbf{A}$, $\mathbf{A}^{\mathsf{T}}\mathbf{A}$ is a symmetric matrix. (*Hint*: Use Theorem 2.1.2.)

2. Show that, for a non-singular $n \times n$ matrix A, $|\mathsf{A}| = -|-\mathsf{A}|$ if and only if n is *odd*.
3. Show that the determinant of any $n \times n$ skew-symmetric matrix, *for n odd*, vanishes. [*Cf.* homework on page 70, problems 2-3.]

3.3.1 Orthogonal Matrices—Rotations

"Symmetric" square matrices seem intuitively to be a Good Thing; it is not so clear why "orthogonal" ones might be any more than a curiosity. But consider the results of operating on an arbitrary vector [representation] v by such an orthogonal matrix A. If, perchance, we should happen to consider the *inner product* of the result of this operation on two n-tuples

$$\langle \mathsf{A}\mathsf{v}_1, \mathsf{A}\mathsf{v}_2\rangle \equiv (\mathsf{A}\mathsf{v}_1)^\mathsf{T}\mathsf{A}\mathsf{v}_2 = \mathsf{v}_1^\mathsf{T}\mathsf{A}^\mathsf{T}\mathsf{A}\mathsf{v}_2 = \mathsf{v}_1^\mathsf{T}\mathbf{1}\mathsf{v}_2 \equiv \langle \mathsf{v}_1, \mathsf{v}_2\rangle, \tag{3.3-1}$$

(where we have used Theorem 2.1.2 to "unpack" $(\mathsf{A}\mathsf{v})^\mathsf{T}$), we see that *inner products are preserved under the transformation generated by an orthogonal matrix.* This is important in application when we are transforming *representations* of vectors relative to *orthonormal* bases of directed line segments:

Recall that, in general, the *inner product* and the *dot product* are different (page 50). But, when the v above corresponds to the representation of a vector in terms of an *orthonormal* basis, the two operations coalesce; this is precisely the reason that the dot product of two vectors $\boldsymbol{v}_1$ and $\boldsymbol{v}_2$ in their Cartesian representations $v_{1x}\hat{\boldsymbol{\imath}} + v_{1y}\hat{\boldsymbol{\jmath}} + v_{1z}\hat{\boldsymbol{k}}$ and $v_{2x}\hat{\boldsymbol{\imath}} + v_{2y}\hat{\boldsymbol{\jmath}} + v_{2z}\hat{\boldsymbol{k}}$ can be calculated

$$\boldsymbol{v}_1 \cdot \boldsymbol{v}_2 = v_{1x}v_{2x} + v_{1y}v_{2y} + v_{1z}v_{2z}.$$

Thus, in fact, what equation (3.3-1) shows is that, when an orthogonal matrix operates on the orthonormal representation of a [single] vector, *it preserves dot products*; thus it preserves *magnitudes*—$|\mathsf{v}|^2 = \boldsymbol{v} \cdot \boldsymbol{v}$—and the transformations represented by such matrices will change the vector, at worst, through some form of *rotation.*

Note the qualification "some form of" in the above statement: preservation of the dot product of two directed line segment vectors means that the *magnitude* of a directed line segment vector is preserved and that the *cosine* of the angle between two of them is, but it doesn't mean that the angle *itself* is: $\cos(\theta) = \cos(2\pi - \theta)$, so it is at least possible that the angle between the two vectors has been altered, from θ to $2\pi - \theta$.

This fact becomes important when one considers what happens not just to a single vector, or a pair of vectors, but to *all* the basis vectors under such an orthogonal transformation. To indicate the problem we are trying to address, consider first a standard Cartesian basis $(\hat{\boldsymbol{\imath}}, \hat{\boldsymbol{\jmath}}, \hat{\boldsymbol{k}})$—note that these are *ordered*—under a transformation [represented by a matrix] A which leaves the first two vectors unchanged, but which *reflects* the vector $\hat{\boldsymbol{k}}$:

$$\hat{\boldsymbol{\imath}} \to \hat{\boldsymbol{\imath}}' \equiv \hat{\boldsymbol{\imath}},\ \hat{\boldsymbol{\jmath}} \to \hat{\boldsymbol{\jmath}}' \equiv \hat{\boldsymbol{\jmath}},\ \hat{\boldsymbol{k}} \to \hat{\boldsymbol{k}}' \equiv -\hat{\boldsymbol{k}}.$$

In terms of the n-tuple *representation* of the basis vectors (see the discussion on page 36), this means that

$$\begin{pmatrix}1\\0\\0\end{pmatrix} \to \mathbf{A}\begin{pmatrix}1\\0\\0\end{pmatrix} = \begin{pmatrix}1\\0\\0\end{pmatrix},\ \begin{pmatrix}0\\1\\0\end{pmatrix} \to \mathbf{A}\begin{pmatrix}0\\1\\0\end{pmatrix} = \begin{pmatrix}0\\1\\0\end{pmatrix},\ \begin{pmatrix}0\\0\\1\end{pmatrix} \to \mathbf{A}\begin{pmatrix}0\\0\\1\end{pmatrix} = \begin{pmatrix}0\\0\\-1\end{pmatrix}.$$

Cartesian vectors are defined to be *ordered* (in the sense of the *ordered* set we use to express their n-tuple representation) in "*right-hand orientation*": $\hat{\boldsymbol{k}} \equiv \hat{\boldsymbol{\imath}} \times \hat{\boldsymbol{\jmath}}$. But the new set is a *left*-hand orientation: $\hat{\boldsymbol{k}}' \equiv \hat{\boldsymbol{\jmath}}' \times \hat{\boldsymbol{\imath}}'$. Thus this transformation *reverses the orientation* of the original basis. What would signal this potentially unfortunate circumstance?

A similar situation would hold if, rather than explicitly *multiplying* one of the original basis vectors by a negative, we had simply *interchanged* the order of the original [ordered] set:

$$\hat{\boldsymbol{\imath}} \to \hat{\boldsymbol{\imath}}' \equiv \hat{\boldsymbol{\imath}},\ \hat{\boldsymbol{\jmath}} \to \hat{\boldsymbol{\jmath}}' \equiv \hat{\boldsymbol{k}},\ \hat{\boldsymbol{k}} \to \hat{\boldsymbol{k}}' \equiv \hat{\boldsymbol{\jmath}}.$$

or

$$\begin{pmatrix}1\\0\\0\end{pmatrix} \to \mathbf{A}\begin{pmatrix}1\\0\\0\end{pmatrix} = \begin{pmatrix}1\\0\\0\end{pmatrix},\ \begin{pmatrix}0\\1\\0\end{pmatrix} \to \mathbf{A}\begin{pmatrix}0\\1\\0\end{pmatrix} = \begin{pmatrix}0\\0\\1\end{pmatrix},\ \begin{pmatrix}0\\0\\1\end{pmatrix} \to \mathbf{A}\begin{pmatrix}0\\0\\1\end{pmatrix} = \begin{pmatrix}0\\1\\0\end{pmatrix};$$

once again, the cross product of the first two transformed vectors gives $-\hat{\boldsymbol{k}}'$. And again we ask what criterion might be used to determine when this might occur.

Note that the above two transformations in one case multiplied a basis vector by a negative number, in the other simply interchanged the order of the original set—*precisely the operations which, when applied to determinants, change their sign*! This at least suggests the criterion which indicates a reversal of orientation: that the determinant of the matrix of the [representations of the] transformed basis vectors is negative. Indeed, applying this somewhat *ad hoc* criterion to the above two examples' transformed bases,

$$\begin{vmatrix}1 & 0 & 0\\0 & 1 & 0\\0 & 0 & -1\end{vmatrix} = -1; \qquad \begin{vmatrix}1 & 0 & 0\\0 & 0 & 1\\0 & 1 & 0\end{vmatrix} = -1.$$

While this goes a long way to showing what is happening, it would be nice to have a criterion which could be applied to the matrix $\mathbf{A}$ itself, since, after all, it is this matrix which really determines the transformations in the first place! But both of the above examples are really of the form, writing the indicated matrices as column vectors,

$$|\hat{\boldsymbol{\imath}}'\hat{\boldsymbol{\jmath}}'\hat{\boldsymbol{k}}'| = |\mathbf{A}\hat{\boldsymbol{\imath}}\ \mathbf{A}\hat{\boldsymbol{\jmath}}\ \mathbf{A}\hat{\boldsymbol{k}}| = |\mathbf{A}\ (\hat{\boldsymbol{\imath}}\hat{\boldsymbol{\jmath}}\hat{\boldsymbol{k}})| = |\mathbf{A}||(\hat{\boldsymbol{\imath}}\hat{\boldsymbol{\jmath}}\hat{\boldsymbol{k}})| \tag{3.3-2}$$

—where we have applied Theorem 3.2.8 to the block multiplication of $(\hat{\boldsymbol{\imath}}\hat{\boldsymbol{\jmath}}\hat{\boldsymbol{k}})$ by the "scalar" $\mathbf{A}$. Thus we make the important observation: *the sign on* $|\hat{\boldsymbol{\imath}}'\hat{\boldsymbol{\jmath}}'\hat{\boldsymbol{k}}'|$ *will change that on* $|\hat{\boldsymbol{\imath}}\hat{\boldsymbol{\jmath}}\hat{\boldsymbol{k}}|$ *if and only if* $|\mathbf{A}|$ *is negative.*

We can see this explicitly in the above examples by exhibiting the respective transformation matrices explicitly. In the first case

$$\mathbf{A} = \begin{pmatrix} 1 & 0 & 0 \\ 0 & 1 & 0 \\ 0 & 0 & -1 \end{pmatrix},$$

in the second

$$\mathbf{A} = \begin{pmatrix} 1 & 0 & 0 \\ 0 & 0 & 1 \\ 0 & 1 & 0 \end{pmatrix};$$

(see Example 2.2.10); in both $|\mathbf{A}| = -1$.

Actually we can say a little more about orthogonal matrices. In terms of their determinants:

Theorem 3.3.3. The determinant of an orthogonal matrix is ± 1.

Proof.

$$|\mathbf{A}|^2 = |\mathbf{A}||\mathbf{A}| = |\mathbf{A}^\mathsf{T}||\mathbf{A}| = |\mathbf{A}^{-1}||\mathbf{A}| = |\mathbf{A}^{-1}\mathbf{A}| = |\mathbf{1}| = 1;$$

—where we have used Theorem 3.2.1, the definition of orthogonal matrices, and Theorem 3.2.8 in the second through fourth equalities, respectively. □

The above examples demonstrate this.

We have seen that orthogonal matrices correspond to a "rotation" of the vectors on which it acts. From the above theorem, they will have a determinant of ± 1; further, from what we have observed in equation (3.3-2) and the examples preceding it, $|\mathbf{A}| = +1$ preserves the *orientation* of the basis when $\mathbf{A}$ is applied to it, while $|\mathbf{A}| = -1$ reverses it. Thus it appears useful to categorize such matrices:

Definition 3.3.4. When the determinant of an orthogonal matrix is $+1$, the basis rotation it generates are referred to as *proper*; when it is -1, they are called *improper*.

3.3.2 The Orientation of Non-orthonormal Bases

The above has been couched in terms of orthonormal [directed line segment] bases, both because these are the ones—specifically *Cartesian* vectors—we shall be primarily interested in in terms of applications; and they are the ones for which the *inner* product, which is always preserved under orthogonal transformations, can be interpreted as the *dot* product on which the above interpretation was based. But we can generalize the results of this discussion if we mandate the intermediate step of expressing even *non*-orthonormal basis vectors in terms of Cartesian ones: the dot product of vectors in such a basis can still be found in terms of the inner product of their representations and the results of operating on them through an orthogonal transformation will still correspond to rotations.

This is useful in interpreting the *sign* of the determinant as giving the *orientation* of a basis. Let us consider an arbitrary basis in three dimensions on which an orthogonal transformation "reflects" the basis vector u_3 across the *plane* determined by the [non-parallel] first two: if originally u_3 made an angle θ with this plane *above* it, the reflection will carry it across the plane to make the same angle with the plane, but now from *below* it. (Recall that, to preserve the *cos* of this angle, $\theta \to 2\pi - \theta$ or, in this case, $\theta \to -\theta$, which also enjoys the same value of the cos.)

Though the transformation does this reflection "instantaneously," it is useful to visualize it as doing so *continuously* from its initial angle θ to its final orientation $-\theta$. As it does so, u_3 eventually lies in the plane of the other two; but then the three vectors are *coplanar*, so that the *triple scalar product* $u_1 \cdot u_2 \times u_3 \equiv 0$. *But the triple scalar product of Cartesian vectors can be calculated by taking the determinant of their representations.* Further, in order for the sign reversal of the vectors' orientation to change under this conceptually continuous transformation, *it must go through 0* in order to pass from a positive to a negative value.

This is effectively the connection between *orientations* and *determinants* (and their signs), observed above in a somewhat *ad hoc* manner. It does show, however, how the discussion can be applied to even more general bases: orientation is a concept which can be applied to an arbitrary basis, and the determinant remains a measure of it.

Summary. We categorize square matrices as being *symmetric* and/or *orthogonal*. The latter become of particular significance, since they preserve inner products and correspond to *rotations* of basis. We show the determinant of any such orthogonal matrix is ± 1 and introduce the concept of *orientation* of such rotations—whether they preserve the "handedness" of the original basis.

3.4 Linear Systems: n Equations in n Unknowns

In Section 2.3 we discussed *arbitrary* systems of linear equations—an *arbitrary* number of equations in an *arbitrary* number of unknowns. In the same way that we are specializing to square matrices in this chapter, we will consider the case in which the number of equations and unknowns are the *same*.

In this case, solution of the system $\mathbf{Ax} = \mathbf{b}$ is tantamount to finding

$$\mathbf{x} = \mathbf{A}^{-1}\mathbf{b}. \tag{3.4-1}$$

We observe that we can only talk about $\mathbf{A}^{-1}$ (whether it exists or not) in the case that $\mathbf{A}$ is square—*i.e.*, that there are as many [formal] equations as unknowns. We can only *find* $\mathbf{x}$ if $\mathbf{A}^{-1}$ exists, *i.e.*, if and only if $|\mathbf{A}| \neq 0$ and/or $\mathbf{A}$ has rank n (Theorem 3.1.5); in this regard, the present treatment is equivalent to that in Section 2.3. But, unlike the more general case considered previously, the distinction between *homogeneous* and *non-homogeneous* equations becomes moot: *both* cases are treated by this single equation.

Theorem 3.2.11 gives us an explicit expression for $\mathbf{A}^{-1}$:

$$\mathbf{A}^{-1} = \frac{1}{|\mathbf{A}|}\mathbf{C}_{\mathbf{A}},$$

in which $\mathbf{C_A}$ is the *transpose* of the matrix of *cofactors* of each element in $\mathbf{A}$. Substituting this expression into (3.4-1), writing $\mathbf{C_A} \equiv (c_{ij})$, we get the solution as a column matrix

$$\mathbf{x} = \frac{1}{|\mathbf{A}|}\mathbf{C_A}\mathbf{b} = \frac{1}{|\mathbf{A}|}\left(\sum_{j=1}^{n} c_{kj}b_j\right),$$

—where c_{kj} is the cofactor of a_{jk} due to the transposition—using the definition of the matrix product, equation (2.1-5). We claim that the last summation, the kth component of $\mathbf{x}$, is nothing more than the *determinant* of that matrix obtained by replacing the kth *column* of $\mathbf{A}$ by $\mathbf{b}$. Indeed, consider that matrix, $\mathbf{A_b}^{(k)} \equiv (a_{ij}^{(k)})$:

$$a_{ij}^{(k)} \equiv \begin{cases} a_{ij}, & j \neq k \\ b_j, & j = k \end{cases}.$$

From the equation defining the determinant in minors of a column, (3.2-2b), if we choose to expand $|\mathbf{A_b}^{(k)}|$ in minors of the kth column,

$$\begin{aligned} |\mathbf{A_b}^{(k)}| &\equiv \sum_{j=1}^{n} (a_{jk}^{(k)} \cdot \operatorname{cof}(a_{jk}^{(k)})) \\ &\equiv \sum_{j=1}^{n} (b_j \cdot \operatorname{cof}(a_{jk}^{(k)})) \end{aligned}$$

by the definition of $\mathbf{A_b}^{(k)}$ itself, and the fact that the cofactors of the kth column are the same for that matrix and $\mathbf{A}$ (since they involve only elements of the respective matrices *off* the kth column). This is precisely what was alleged above.

Thus we have established *Cramer's Rule*: that the kth component of $\mathbf{x}$,

$$x_k = \frac{|\mathbf{A_b}^{(k)}|}{|\mathbf{A}|}.$$

Once again, in order that a solution be found, it is necessary that $|\mathbf{A}| \neq 0$, for otherwise we divide by 0 in each component.

Cramer's Rule seems to be one of those things the student remembers from high school: it is neat and tidy, resulting in a certain intellectual comfort. But it is rather de-emphasized in this book: In the first place, it *only* holds when there are as many [independent] equations as unknown, whereas we are viewing these systems somewhat more generally. But the primary reason for ignoring it is its gross inefficiency: We have noted that it requires approximately $n!$ calculations to evaluate an $n \times n$ determinant, and to find all n components of $\mathbf{x}$ would require $(n+1)$ determinants (one for the denominator plus one for the numerator of each component)—a total of $(n+1)!$ calculations! Gaussian elimination (Section 2.3.2), on the other hand, can be shown to require fewer than $n^3/3$ calculations to solve the *entire* system, and is always a more efficient method. In fact, the only reason we even mention Cramer's Rule, aside from completeness and the desire to demystify it, is in connection with the following point:

The perspicacious student might have observed that the non-vanishing of this determinant is at least *consistent* with the more general theory outlined in Section 2.3. But the latter made a great deal of the fact that we also had to be sure that the *augmented matrix*, $(\mathbf{A}\ \mathbf{b})$, not have *rank* greater than that of $\mathbf{A}$ itself. Yet nothing has been said about that here, either regarding (3.4-1) or in Cramer's Rule. Are we leaving something out?

In point of fact, we *aren't*: Assuming the determinant of the coefficient matrix doesn't vanish, that matrix must have rank n (by Theorem 3.1.5). But forming the augmented matrix cannot *increase* the rank: the rank is the number of linearly independent rows *or* columns [Theorem 2.1.1], and although the augmented matrix has $(n+1)$ columns, it still has only n *rows*; thus the *maximum* rank that matrix can have is n (which it does). Thus, in effect, examination of the coefficient matrix itself is sufficient in this case to prove sufficiency of uniqueness, and we have a special criterion for solution in this case:

Theorem 3.4.1. A unique solution to a system of n equations in n unknowns exists if and only if the *coefficient* matrix has rank n; *i.e.*, if and only if its determinant doesn't vanish.

Of course, by the general theory, if the determinant *does* vanish, we have an infinite number of solutions.

Summary. In the special case that we have the same number of equations as unknowns, it is unnecessary to examine the augmented matrix: a unique solution exists if and only if the determinant of the *coefficient* matrix is non-zero. If it vanishes, it is necessary to revert to the more general theory of Section 2.3 to determine just how many parameters the infinite number of possible solutions depend on.

3.5 Eigenvalues and Eigenvectors of a Square Matrix

Eigenvalues and eigenvectors of a matrix are oftentimes introduced in the course of presentation of other material: they are *defined* and immediately *applied*, but rarely is the luxury given to explain just what they *are* or why—or even *whether*—they are special. But they are useful, and it seems appropriate to explain briefly just why an eigenvector is a distinguished vector associated with a linear transformation.

For the time being, let us consider "vectors" as being the directed line segments with which we are all familiar. In general, if one takes an arbitrary vector v with representation $\mathbf{v}$, and multiplies that by the [matrix representation of] the linear transformation A, *the matrix product* Av *will not represent a vector parallel to* $\boldsymbol{v}$*:*

$$\mathsf{Av} \nparallel \mathsf{v}.$$

Vectors which *are*, then, are indeed distinguished.

We pause here for a brief legalism: Were we to have been somewhat precious and taken $\boldsymbol{v} = \mathbf{0}$ above, the result *would* have been satisfied (whatever "parallel" means for $\mathbf{0}$); thus we will explicitly preclude this possibility in the following definition:

Definition 3.5.1. Given a linear transformation [represented by] A, those vectors [represented by] $\boldsymbol{\xi} \neq \mathbf{0}$ on which it acts such that $\mathsf{A}\boldsymbol{\xi} = \lambda\boldsymbol{\xi}$ are called *eigenvectors*; the corresponding λ are the *eigenvalues associated with* $\boldsymbol{\xi}$.

Though we explicitly prohibit $\mathbf{0}$ from being an eigen*vector*, it may be the case that 0 is an eigen*value*: if $\boldsymbol{\xi} \neq \mathbf{0}$ happens to belong to the *null space* (Section 2.2.2) of the matrix A, then $\mathsf{A}\boldsymbol{\xi} \equiv \mathbf{0} = 0\boldsymbol{\xi}$.

Notice, too, that if $\boldsymbol{\xi}$ is an eigenvector, then so is $c\boldsymbol{\xi}$.

We observe that eigenvectors and eigenvalues are not unique: for any scalar c, if $\boldsymbol{\xi}$ is an eigenvector, so is $c\boldsymbol{\xi}$; if λ is an eigenvalue, so is $c\lambda$.

Assuming eigenvectors and eigenvalues are important (they are), is there a systematic means of finding them? The answer of course is "yes": In order to find $\boldsymbol{\xi}$ and λ satisfying the definition 3.5.1, it is necessary that $\mathsf{A}\boldsymbol{\xi} - \lambda\boldsymbol{\xi} = \mathbf{0}$; *i.e.*, that $(\mathsf{A} - \lambda\mathbf{1})\boldsymbol{\xi} = \mathbf{0}$. But this says that $\boldsymbol{\xi}$ *lies in the null space of the matrix* $(\mathsf{A} - \lambda\mathbf{1})$, and we have already seen (Section 2.3.1) that, although $\boldsymbol{\xi} = \mathbf{0}$ is *always* a solution to this homogeneous equation, the only way to get a $\boldsymbol{\xi} \neq \mathbf{0}$ is for the *rank* of $(\mathsf{A} - \lambda\mathbf{1})$ to be less than n. But, by Theorem 3.2.6, the only way *this* can happen is for $|\mathsf{A} - \lambda\mathbf{1}| = 0$. This equation depends only on λ; it will be a polynomial of degree n in the variable λ, and is called the *characteristic equation.* This determines the eigen*value*, whose value can be used to find $\boldsymbol{\xi}$ by solving the equation $(\mathsf{A} - \lambda\mathbf{1})\boldsymbol{\xi} = \mathbf{0}$ for $\boldsymbol{\xi}$.

Actually, given the theory we have obtain earlier, we can tell a little more about the eigenvectors associated with A: Theorem 2.2.3 allows us to predict the *dimension* of the null space of $|\mathsf{A} - \lambda\mathbf{1}|$—*i.e.*, the number of [linearly independent] vectors in the null space. But this is nothing more than the *number of eigenvectors* available to A corresponding to λ.

We illustrate these ideas with an example:

Example 3.5.1. Find the eigenvectors of the matrix

$$\mathsf{A} = \begin{pmatrix} 2 & 1 & 1 \\ 2 & 3 & 2 \\ 1 & 1 & 2 \end{pmatrix}$$

Solution: We must first find the eigen*values*: the characteristic equation corresponding to this matrix is

$$p(\lambda) \equiv |\mathbf{A} - \lambda\mathbf{1}| = \begin{vmatrix} 2-\lambda & 1 & 1 \\ 2 & 3-\lambda & 2 \\ 1 & 1 & 2-\lambda \end{vmatrix}$$
$$= -\lambda^3 + 7\lambda^2 - 11\lambda + 5$$

—the third-degree polynomial expected. It has three roots: $\lambda = 5$, 1, and 1; in this case, one of the roots is *repeated*—a *root of multiplicity 2.*

We now find the eigen*vectors*: writing $\boldsymbol{\xi} = (\xi_1, \xi_2, \xi_3)^\mathsf{T}$, for either of the λ,

$$(\mathbf{A} - \lambda\mathbf{1})\boldsymbol{\xi} = \begin{pmatrix} (2-\lambda) & 1 & 1 \\ 2 & (3-\lambda) & 2 \\ 1 & 1 & (2-\lambda) \end{pmatrix} \begin{pmatrix} \xi_1 \\ \xi_2 \\ \xi_3 \end{pmatrix}$$
$$= \begin{pmatrix} (2-\lambda)\xi_1 + \xi_2 + \xi_3 \\ 2\xi_1 + (3-\lambda)\xi_2 + 2\xi_3 \\ \xi_1 + \xi_2 + (2-\lambda)\xi_3 \end{pmatrix} = 0.$$

$\underline{\lambda = 5}$: Here

$$\mathbf{A} - 5\mathbf{1} = \begin{pmatrix} -3 & 1 & 1 \\ 2 & -2 & 2 \\ 1 & 1 & -3 \end{pmatrix}$$

—a matrix of rank 2 (the third row is minus the sum of the first two), so having a null space dimension of $3 - 2 = 1$, so giving only a single eigenvector; equivalently, we expect to be able to solve for only two of the ξ_i in terms of the third (recall Case 1. of the "Homogeneous Linear Equations" in Section 2.3.1). Solving the system, we get that

$$\boldsymbol{\xi}_1 = \begin{pmatrix} 1 \\ 2 \\ 1 \end{pmatrix} \xi_3$$

(where we have solved for ξ_1 and ξ_2 in terms of ξ_3). We see exhibited explicitly the "multiplicative constant" up to which $\boldsymbol{\xi}_1$ can be determined. Having done so, however, we will simply set $\xi_3 \equiv 1$.

$\underline{\lambda = 1}$ (the "repeated root"): Now

$$\mathbf{A} - (1)\mathbf{1} = \begin{pmatrix} 1 & 1 & 1 \\ 2 & 2 & 2 \\ 1 & 1 & 1 \end{pmatrix}$$

—a matrix clearly only of rank 1, with a null space of dimension 2, and a system $(\mathbf{A} - \lambda\mathbf{1})\boldsymbol{\xi} = \mathbf{0}$ which allows for only *1* unknown to be solved for in terms of the other two; in fact, the single equation embodied in the above system is just $\xi_1 = -\xi_2 - \xi_3$. Observe that, although ξ_2 and ξ_3 can be chosen independently, if we are so indiscreet as to select pairs of values which are proportional, then the resulting components will be also, leading to linearly *de*pendent eigenvectors, which we would like to avoid. Likely the easiest way to do this is to take, for example, $(\xi_2, \xi_3) = (1, 0)$ and then $(0, 1)$: these choices yield

$$\boldsymbol{\xi}_2 = \begin{pmatrix} -1 \\ 1 \\ 0 \end{pmatrix}, \text{ and } \boldsymbol{\xi}_3 = \begin{pmatrix} -1 \\ 0 \\ 1 \end{pmatrix},$$

The resulting eigenvectors are, indeed, linearly independent:

$$c_1 \begin{pmatrix} 1 \\ 2 \\ 1 \end{pmatrix} + c_2 \begin{pmatrix} -1 \\ 1 \\ 0 \end{pmatrix} + c_3 \begin{pmatrix} -1 \\ 0 \\ 1 \end{pmatrix} = \begin{pmatrix} c_1 - c_2 - c_3 \\ 2c_1 + c_2 \\ c_1 + c_3 \end{pmatrix} = 0$$

can only be true for $c_1 = c_2 = c_3 = 0$.

Let us make a couple of notes about this example:

1. Since $c\boldsymbol{\xi}$ is always an eigenvector if $\boldsymbol{\xi}$ is, we can *normalize* the above eigenvectors, by dividing each by its magnitude, so that $\|\boldsymbol{\xi}\| = 1$.
2. If the above happen to take place in a space of directed line segments (so that the *inner* product is the same as the *dot* product), the above three vectors are not orthogonal: though $\boldsymbol{\xi}_1^\mathsf{T}\boldsymbol{\xi}_3 = 0$, but $\boldsymbol{\xi}_1^\mathsf{T}\boldsymbol{\xi}_2 \neq 0$ and $\boldsymbol{\xi}_2^\mathsf{T}\boldsymbol{\xi}_3 \neq 0$. Nonetheless, we can at least find new eigenvectors $\boldsymbol{\xi}_2'$ and $\boldsymbol{\xi}_3'$ *corresponding to the repeated root* orthogonal to each other: We note first that any *linear combination* of these two eigenvectors is also an eigenvector corresponding to the root $\lambda = 1$:

$$\mathbf{A}(x_1\boldsymbol{\xi}_2 + x_2\boldsymbol{\xi}_3) = x_1\mathbf{A}\boldsymbol{\xi}_2 + x_2\mathbf{A}\boldsymbol{\xi}_3 = x_1\lambda\boldsymbol{\xi}_2 + x_2\lambda\boldsymbol{\xi}_3 = \lambda(x_1\boldsymbol{\xi}_2 + x_2\boldsymbol{\xi}_3)$$

(This would clearly hold for *any* set of eigenvectors corresponding to a repeated root of the characteristic equation.) The idea is to find an appropriate linear combination of these two eigenvectors such that $\boldsymbol{\xi}_2'^\mathsf{T}\boldsymbol{\xi}_3' = 0$. In this case, because $\boldsymbol{\xi}_3$ *is* already orthogonal to $\boldsymbol{\xi}_1$, we choose to leave it alone—$\boldsymbol{\xi}_3' \equiv \boldsymbol{\xi}_3$—and instead modify $\boldsymbol{\xi}_2$:

$$\boldsymbol{\xi}_2' \equiv \alpha\boldsymbol{\xi}_2 + \boldsymbol{\xi}_3' = \begin{pmatrix} -\alpha \\ \alpha \\ 0 \end{pmatrix} + \begin{pmatrix} -1 \\ 0 \\ 1 \end{pmatrix} = \begin{pmatrix} -(\alpha+1) \\ \alpha \\ 1 \end{pmatrix}.$$

We then select α such that the inner product of these vectors vanishes:

$$\boldsymbol{\xi}_2'^\mathsf{T}\boldsymbol{\xi}_3' = (\alpha + 1) + 1 = \alpha + 2 = 0,$$

or $\alpha = -2$. Substituting this into the above equation for $\boldsymbol{\xi}_2'$,

$$\boldsymbol{\xi}_2' = \begin{pmatrix} 1 \\ -2 \\ 1 \end{pmatrix}.$$

In the end, $\boldsymbol{\xi}_1$ and $\boldsymbol{\xi}_3'$ are orthogonal as are $\boldsymbol{\xi}_2'$ and $\boldsymbol{\xi}_3'$, but $\boldsymbol{\xi}_1$ and $\boldsymbol{\xi}_2'$ are still not.

The technique implemented in the second note is called "*Gram-Schmidt Orthogonalization.*" |

We conclude this section with a useful observation: If a matrix happens to be *upper triangular*,

$$\mathbf{A} = \begin{pmatrix} a_{11} & a_{12} & \cdots & a_{1n} \\ 0 & a_{22} & \cdots & a_{2n} \\ \vdots & \vdots & \ddots & \vdots \\ 0 & 0 & \cdots & a_{nn} \end{pmatrix}$$

(*cf.* page 104), the eigenvalues can be uncovered by inspection: Forming $(\mathbf{A}-\lambda\mathbf{1})$ would simply amend $-\lambda$ to each term on the diagonal, and the expansion of the characteristic equation, the determinant of that matrix, would be merely the product of these terms (as remarked on the above-cited page). Thus *the eigenvalues of an upper- [or lower-] triangular matrix are just the numbers on the diagonal.* This is useful to look for. But note that the technique cited on that page to find the *determinant*—performing row/column operations (without "scaling") to reduce $\mathbf{A}$ to it upper-diagonal form, $\mathbf{A}' = \mathbf{BA}$—will *not* work to find the characteristic equation: λ satisfying $|\mathbf{A} - \lambda\mathbf{1}| = 0$ are not necessarily the same as those satisfying $|\mathbf{BA} - \lambda\mathbf{1}| = 0$!

Homework:

1. Given matrix $\mathbf{A}$, pick [almost] *any* vector $\mathbf{v}$ and show that $\mathbf{Av}$ is not parallel to $\mathbf{v}$.
2. Prove that, if $\boldsymbol{\xi}$ and λ are, respectively, an eigenvector and eigenvalue of a matrix, then so are, respectively, $c\boldsymbol{\xi}$ and $c\lambda$.
3. Prove that if a given eigenvalue happens to be 0, then the matrix itself must be singular.
4. (*Linearly Independent Eigenvectors Associated with Repeated Eigenvalues*) For the matrix

$$\mathbf{A} = \begin{pmatrix} 2 & -3 & 3 \\ 0 & 5 & -3 \\ 0 & 6 & -4 \end{pmatrix}$$

   (a) find the eigenvalues and eigenvectors, and

   (b) show that these eigenvectors are linearly independent.

   [Partial] Answer: $\lambda = -1, 2, 2$.

5. (*Repeated Eigenvalues* without *Linearly Independent Eigenvectors*) Find the eigenvalues and eigenvectors of the matrix

$$\mathbf{A} = \begin{pmatrix} -6 & -7 & -13 \\ 5 & 6 & 9 \\ 2 & 2 & 5 \end{pmatrix}$$

[Partial] Answer: $\lambda = 1, 1, 3$. Note that in the above problem, there were two linearly independent eigenvectors associated with the repeated root $\lambda = 2$; here there is only *one* associated with $\lambda = 1$. This is generally the case with repeated roots; it is only if the eigenvalues are *distinct* that the associated eigenvalues are assured to be linearly independent. (See Theorem 3.5.1.)

6. (*"Generalized Eigenvectors"*) Problem 5. resulted in only a single eigenvector, say $\boldsymbol{\xi}_1$, associated with the eigenvalue $\lambda = 1$. But it is possible to find a second, "generalized," eigenvector, $\boldsymbol{\xi}'_1$, associated with this value by solving the equation

$$(\mathbf{A} - \lambda\mathbf{1})\boldsymbol{\xi}'_1 = \boldsymbol{\xi}_1.$$[1]

(See Section 3.5.3.)

(a) Solve for ξ'_1 in this case and

(b) show that this generalized eigenvector is linearly independent of the first.

(c) Attempt to find a *second* generalized eigenvector corresponding to the first:

$$(\mathbf{A} - \lambda\mathbf{1})\boldsymbol{\xi}''_1 = \boldsymbol{\xi}'_1,$$

showing that this leads to an impossible condition on $\boldsymbol{\xi}''_1$, thus terminating the procedure with one "real" eigenvector and one "generalized" eigenvector corresponding to the eigenvalue $\lambda = 1$ [of "multiplicity 2"].

Note: $\boldsymbol{\xi}'_1$ is *not* a true "eigenvector": the latter are defined to be solutions to the *homogeneous* equation $(\mathbf{A}-\lambda\mathbf{1})\boldsymbol{\xi}_1 = \mathbf{0}$, while the "generalized" eigenvector satisfies the *non*-homogeneous equation $(\mathbf{A} - \lambda\mathbf{1})\boldsymbol{\xi}'_1 = \boldsymbol{\xi}_1 \neq \mathbf{0}$!

3.5.1 Linear Independence of Eigenvectors

Given all the attention which has been paid to the linear dependence and independence of vectors, it is no surprise that we might be interested in this characteristic in reference to *eigen*vectors. We give here a couple of predictive tests allowing us to tell *a priori* whether the eigenvectors associated with a given matrix might be linearly independent, without resorting to find any linear combinations, or even the eigenvectors themselves!

[1] Note that, multiplying each wide of this equation by $(\mathbf{A} - \lambda\mathbf{1})$ gives:

$$(\mathbf{A} - \lambda\mathbf{1})^2\boldsymbol{\xi}'_1 = (\mathbf{A} - \lambda\mathbf{1})\boldsymbol{\xi}_1 \equiv \mathbf{0}$$

—the form in which this equation is normally shown.

Theorem 3.5.1. Eigenvectors corresponding to *distinct eigenvalues* of a matrix A are linearly independent.

Proof. Assume that A has eigenvalues $\{\lambda_1, \lambda_2, \ldots, \lambda_n\}$ which are *distinct—i.e.*, $\lambda_i \neq \lambda_j$ for $i \neq j$—and associated eigenvectors $\{\boldsymbol{\xi}_1, \boldsymbol{\xi}_2, \cdots, \boldsymbol{\xi}_n\}$. The claim is that the eigenvectors are linearly independent; *i.e.*,

$$\sum_{i=1}^{n} c_i \boldsymbol{\xi}_i = 0 \Rightarrow c_i = 0 \text{ for all } i.$$

Indeed, assume they are *not* linearly independent; *i.e.*, there are some $c_i \neq 0$ such that $\sum c_i \boldsymbol{\xi}_i = 0$. In particular, then, there is some i_o, $2 \leq i_o \leq n$ such that $\{\boldsymbol{\xi}_1, \boldsymbol{\xi}_2, \ldots, \boldsymbol{\xi}_{i_o-1}\}$ are linearly *in*dependent but $\{\boldsymbol{\xi}_1, \boldsymbol{\xi}_2, \ldots, \boldsymbol{\xi}_{i_o}\}$ are linearly *de*pendent; *i.e.*,

$$\sum_{i=1}^{i_o} c_i \boldsymbol{\xi}_i = 0 \text{ or } - c_{i_o} \boldsymbol{\xi}_{i_o} = \sum_{i=1}^{i_o-1} c_i \boldsymbol{\xi}_i;$$

thus

$$-\boldsymbol{\xi}_{i_o} = \sum_{i=1}^{i_o-1} \frac{c_i}{c_{i_o}} \boldsymbol{\xi}_i. \tag{3.5-1}$$

(Note that we can assume that $c_{i_o} \neq 0$, since otherwise $\boldsymbol{\xi}_{i_o}$ wouldn't enter at all!) But, then,

$$\mathsf{A}(-\boldsymbol{\xi}_{i_o}) = -\lambda_{i_o} \boldsymbol{\xi}_{i_o} = \sum_{i=1}^{i_o-1} \frac{c_i}{c_{i_o}} \mathsf{A} \boldsymbol{\xi}_i = \sum_{i=1}^{i_o-1} \frac{c_i}{c_{i_o}} \lambda_i \boldsymbol{\xi}_i \tag{3.5-2}$$

Multiplying equation 3.5-1 by λ_{i_o},

$$-\lambda_{i_o} \boldsymbol{\xi}_{i_o} = \sum_{i=1}^{i_o-1} \frac{c_i}{c_{i_o}} \lambda_{i_o} \boldsymbol{\xi}_i$$

and subtracting (3.5-2) from this,

$$0 = \sum_{i=1}^{i_o-1} \frac{c_i}{c_{i_o}} (\lambda_{i_o} - \lambda_i) \boldsymbol{\xi}_i.$$

But $\{\boldsymbol{\xi}_1, \boldsymbol{\xi}_2, \cdots, \boldsymbol{\xi}_{i_o-1}\}$ are linearly *in*dependent, so $\frac{c_i}{c_{i_o}}(\lambda_{i_o} - \lambda_i) = 0$ for all $i \leq i_o - 1$. And since $\lambda_{i_o} \neq \lambda_i$ by the distinctness hypothesis, this means that $c_i = 0$ (for all i); thus, by (3.5-1), $\boldsymbol{\xi}_{i_o} = \mathbf{0}$, violating the defining characteristic that eigenvectors *not* be $\mathbf{0}$. This contradiction proves the assertion. □

The above characteristic is precisely what was observed in Example 3.5.1: the eigenvector corresponding to $\lambda = 5$ in that example was linearly independent of those corresponding to $\lambda = 1$.

It is important to note just what this theorem *says* and what it *doesn't*: it *says* that eigenvectors corresponding to distinct eigenvalues are linearly independent; it *doesn't* say that eigenvectors associated with a *repeated* root are linearly *de*pendent. Again, this is exactly what happened in Example 3.5.1: we were able to find two linearly independent eigenvectors corresponding to $\lambda = 1$—a fact predicted by the rank of the matrix $\mathbf{A} - (1)\mathbf{1}$.

But that doesn't mean that *all* eigenvectors corresponding to a repeated root will be linearly independent:

Example 3.5.2. *(Linearly Dependent Eigenvectors of Repeated Root):* Find the number of linearly independent eigenvectors of the matrix

$$\mathbf{A} = \begin{pmatrix} 2 & 1 & 0 \\ 0 & 2 & 1 \\ 0 & 0 & 2 \end{pmatrix}.$$

Solution: The characteristic equation for this matrix is just $p(\lambda) = (2 - \lambda)^3$, so we have a repeated root, 2, "of multiplicity 3." (We would have the same result by noting that this is in "upper triangular form"—see page 122.) Forming the matrix

$$\mathbf{A} - 2\mathbf{1} = \begin{pmatrix} 0 & 1 & 0 \\ 0 & 0 & 1 \\ 0 & 0 & 0 \end{pmatrix},$$

we see that we have one of rank 2; thus there is only a *single* eigenvector.

Observe that had one of the "1" 's in $\mathbf{A}$ been a "0," $p(\lambda)$ would have been the same, but $\mathbf{A} - 2\mathbf{1}$ would only have had a rank of *1*, and there would have been *two* linearly independent eigenvectors. |

Thus, relative to the number of linearly independent eigenvectors corresponding to a *repeated* root, anything can happen; each case must be considered on its own. In order to find anything *like* linearly independent eigenvectors, we must resort to "generalized eigenvectors," considered below.

We now obtain a powerful result pertaining to a special form of square matrix, a *real, symmetric* matrix. This might seem *so* special that it would only arise under very curious circumstances. In point of fact, however, it happens to be precisely the case we are going to be using in the next Part: the "inertia tensor" is (more correctly, can be *represented* as) a real symmetric matrix.

Before doing so, however, we must take note of the character of roots of the characteristic equation. In general, the roots of this polynomial—the eigenvalues—will be complex numbers. But if the coefficients of this polynomial are *real*, precisely the case that will apply if the *matrix* is real, any complex roots will occur in complex conjugate pairs.

Now if the eigen*values* of a matrix are complex, then so generally will be the eigen*vectors*. In fact for a *real* matrix, they necessarily *must* be complex, in order that the equation $\mathbf{A}\boldsymbol{\xi} = \lambda\boldsymbol{\xi}$ be satisfied! With an eye towards applications (in, one hopes, *real* space!), this will require some adjustment in our definition of the *inner product*:

Recall that we have pointed out that, at least in the space of directed line segments with an orthonormal basis (for example the common Cartesian space spanned by the unit vectors $\hat{\imath}$, $\hat{\jmath}$, and $\hat{k}$), the inner product and dot product coalesce. But the *dot* product of a vector with itself gives the magnitude of that vector squared: $\boldsymbol{v} \cdot \boldsymbol{v} = |\boldsymbol{v}|^2$. It would clearly be an awkward development to have the magnitude of a vector be the square root of a complex number, as it would be if the dot product (*i.e.* the *inner* product in this case) of a *complex* vector were taken: for *real* $\boldsymbol{u}$ and $\boldsymbol{v}$ we can write the dot product of an arbitrary complex vector,

$$\begin{aligned}(\boldsymbol{u} + \imath\boldsymbol{v}) \cdot (\boldsymbol{u} + \imath\boldsymbol{v}) &= \langle \mathbf{u} + \imath\mathbf{v}, \mathbf{u} + \imath\mathbf{v} \rangle \\ &\equiv (\mathbf{u} + \imath\mathbf{v})^{\mathsf{T}}(\mathbf{u} + \imath\mathbf{v}) \\ &= (\mathbf{u}^{\mathsf{T}}\mathbf{u} - \mathbf{v}^{\mathsf{T}}\mathbf{v}) + \imath(\mathbf{u}^{\mathsf{T}}\mathbf{v} + \mathbf{v}^{\mathsf{T}}\mathbf{u}) \\ &= (|\mathbf{u}|^2 - |\mathbf{v}|^2) + 2\imath\mathbf{u}^{\mathsf{T}}\mathbf{v}.\end{aligned}$$

(Note that we have used the commutativity of the transpose product to combine the last two terms.) Thus in this case (and by extension to the reals, merely a special case of the complex numbers), we define the inner product for complex vectors to be

$$\langle \mathbf{v}_1, \mathbf{v}_2 \rangle \equiv \mathbf{v}_1^{\mathsf{T}}\overline{\mathbf{v}}_2, \tag{3.5-3}$$

in which $\overline{\mathbf{v}}$ denotes the *complex conjugate* of [complex] $\mathbf{v}$. Note, in this regard, properties of the complex conjugate: for two complex numbers z_1 and z_2,

$$z_1\overline{z}_2 = \overline{\overline{z}_1 z_2} \tag{3.5-4a}$$

$$\overline{z_1 z_2} = \overline{z}_1\overline{z}_2 \tag{3.5-4b}$$

$$\overline{\overline{z}} = z \tag{3.5-4c}$$

Furthermore, we observe that the only time a [complex] number can equal its conjugate, $z = \overline{z}$, is if that number is itself *real.*

In this case, the above equation now becomes

$$\begin{aligned}(\boldsymbol{u} + \imath\boldsymbol{v}) \cdot (\boldsymbol{u} + \imath\boldsymbol{v}) &= \langle \mathbf{u} + \imath\mathbf{v}, \mathbf{u} + \imath\mathbf{v} \rangle \\ &\equiv (\mathbf{u} + \imath\mathbf{v})^{\mathsf{T}}(\overline{(\mathbf{u} + \imath\mathbf{v})} \\ &= (\mathbf{u} + \imath\mathbf{v})^{\mathsf{T}}(\mathbf{u} - \imath\mathbf{v}) \\ &= (\mathbf{u}^{\mathsf{T}}\mathbf{u} + \mathbf{v}^{\mathsf{T}}\mathbf{v}) + \imath(-\mathbf{u}^{\mathsf{T}}\mathbf{v} + \mathbf{v}^{\mathsf{T}}\mathbf{u}) \\ &= (|\mathbf{u}|^2 + |\mathbf{v}|^2).\end{aligned} \tag{3.5-5}$$

This is rather more in keeping with what we expect: the magnitude squared is precisely a *real* number, equal to the hypotenuse of the triangle whose sides are the real and complex parts in the complex plane. We note that *this* inner product has a peculiar form of commutativity for *different* [complex] vectors:

$$\langle \mathbf{v}_2, \mathbf{v}_1 \rangle \equiv \mathbf{v}_2^{\mathsf{T}}\overline{\mathbf{v}}_1 = (\overline{\mathbf{v}}_1^{\mathsf{T}}\mathbf{v}_2)^{\mathsf{T}} = \overline{\mathbf{v}}_1^{\mathsf{T}}\mathbf{v}_2 = \overline{\mathbf{v}_1^{\mathsf{T}}\overline{\mathbf{v}}_2} \equiv \overline{\langle \mathbf{v}_1, \mathbf{v}_2 \rangle}, \tag{3.5-6}$$

where the transpose equals its untransposed counterpart because the inner product is a *scalar*, and the last equation results from (3.5-4a). But, as we showed above, if we take the inner product of a complex vector with *itself*, we end up with a real number.

Observe that for any complex scalars z_1 and z_2, because of the linearity of the inner product,

$$\langle z_1\mathbf{v}_1 + z_2\mathbf{v}_2, \mathbf{v}_3\rangle = z_1\langle \mathbf{v}_1, \mathbf{v}_3\rangle + z_2\langle \mathbf{v}_2, \mathbf{v}_3\rangle. \tag{3.5-7}$$

We remark, in passing, that any vector space with a "$\langle , \rangle$" satisfying the above properties (3.5-5), (3.5-6), and (3.5-7) is called an *inner product space* over the complex numbers.

We note that the properties of complex conjugates, equations (3.5-4), will apply to each term in the products comprising a *matrix* product. Thus the above definition of the inner product for *complex* numbers can be applied to the situation in which one of the vectors is *itself* the product of a matrix and another vector:

$$\begin{aligned}
\langle \mathbf{A}\mathbf{v}_1, \mathbf{v}_2\rangle &\equiv (\mathbf{A}\mathbf{v}_1)^{\mathsf{T}}\overline{\mathbf{v}_2} \\
&= \mathbf{v}_1^{\mathsf{T}}\mathbf{A}^{\mathsf{T}}\overline{\mathbf{v}_2} && \text{(by Theorem (2.1.2))} \\
&= \mathbf{v}_1^{\mathsf{T}}\overline{\overline{\mathbf{A}}}^{\mathsf{T}}\,\overline{\mathbf{v}_2} && \text{(by (3.5-4c))} \\
&= \mathbf{v}_1^{\mathsf{T}}\overline{\overline{\mathbf{A}}^{\mathsf{T}}\mathbf{v}_2} && \text{(by (3.5-4b))} \\
&\equiv \langle \mathbf{v}_1, \overline{\mathbf{A}}^{\mathsf{T}}\mathbf{v}_2\rangle
\end{aligned}$$

In particular, for matrices $\mathbf{A}$ such that

$$\mathbf{A} = \overline{\mathbf{A}}^{\mathsf{T}}$$

—so-called *Hermitian* matrices—we have that $\langle \mathbf{A}\mathbf{v}_1, \mathbf{v}_2\rangle = \langle \mathbf{v}_1, \mathbf{A}\mathbf{v}_2\rangle$. Observe that the real, symmetric matrices we are ultimately to deal with (lest we forget!) enjoy this property. Note, too, that we could repeat the above sequence of equations in the case that $\mathbf{A}$ was a "1×1 matrix"—a *scalar*, say λ; then $\overline{\lambda}^{\mathsf{T}} = \overline{\lambda}$ and

$$\langle \lambda\mathbf{v}_1, \mathbf{v}_2\rangle = \langle \mathbf{v}_1, \overline{\lambda}\mathbf{v}_2\rangle,$$

or, factoring this scalar outside the inner products, obtain

$$\lambda\langle \mathbf{v}_1, \mathbf{v}_2\rangle = \overline{\lambda}\langle \mathbf{v}_1, \mathbf{v}_2\rangle \tag{3.5-8}$$

With this background in hand, we are finally ready to obtain the desired result:

Theorem 3.5.2. If $\mathbf{A}$ is a *real, symmetric* matrix, then its eigen*values* are *real* and its eigen*vectors* are *orthogonal.*

Proof. We first show the eigenvalues are real: If $\mathbf{A}\boldsymbol{\xi} = \lambda\boldsymbol{\xi}$, then we can form the inner product

$$\langle \mathbf{A}\boldsymbol{\xi}, \boldsymbol{\xi}\rangle = \langle \lambda\boldsymbol{\xi}, \boldsymbol{\xi}\rangle = \lambda\langle \boldsymbol{\xi}, \boldsymbol{\xi}\rangle. \tag{3.5-9}$$

But, as observed above, real symmetric matrices satisfy the property that $\mathbf{A} = \overline{\mathbf{A}}^{\mathsf{T}}$; thus

$$\langle \mathbf{A}\boldsymbol{\xi}, \boldsymbol{\xi}\rangle = \langle \boldsymbol{\xi}, \mathbf{A}\boldsymbol{\xi}\rangle = \overline{\langle \mathbf{A}\boldsymbol{\xi}, \boldsymbol{\xi}\rangle}$$

(by the curious commutativity property (3.5-6) of complex inner products). But this means that $\langle \mathbf{A}\boldsymbol{\xi}, \boldsymbol{\xi}\rangle$ is *real*; further $\langle \boldsymbol{\xi}, \boldsymbol{\xi}\rangle$ is *already* real, since the inner product of a [complex] number with itself is real. Thus λ in (3.5-9) is the quotient of two real numbers, so itself real. Thus the eigenvalues are real.

We now demonstrate the orthogonality. Assume that $\boldsymbol{\xi}_1$ and $\boldsymbol{\xi}_2$ are two eigenvectors of $\mathbf{A}$ with *different* associated eigenvalues λ_1 and λ_2; *i.e.*,

$$\mathbf{A}\boldsymbol{\xi}_1 = \lambda_1\boldsymbol{\xi}_1 \text{ and } \mathbf{A}\boldsymbol{\xi}_2 = \lambda_2\boldsymbol{\xi}_2.$$

We wish to show that $\langle \boldsymbol{\xi}_1, \boldsymbol{\xi}_2\rangle = 0$ (*Cf.* Definition 2.1.7). We can once again form inner products:

$$\langle \mathbf{A}\boldsymbol{\xi}_1, \boldsymbol{\xi}_2\rangle = \langle \lambda_1\boldsymbol{\xi}_1, \boldsymbol{\xi}_2\rangle = \lambda_1\langle \boldsymbol{\xi}_1, \boldsymbol{\xi}_2\rangle, \text{ and}$$
$$\langle \boldsymbol{\xi}_1, \mathbf{A}\boldsymbol{\xi}_2\rangle = \langle \boldsymbol{\xi}_1, \lambda_2\boldsymbol{\xi}_2\rangle = \overline{\lambda_2}\langle \boldsymbol{\xi}_1, \boldsymbol{\xi}_2\rangle$$

But, from the oft-cited Hermitian property, $\langle \mathbf{A}\boldsymbol{\xi}_1, \boldsymbol{\xi}_2\rangle = \langle \boldsymbol{\xi}_1, \mathbf{A}\boldsymbol{\xi}_2\rangle$; thus

$$\lambda_1\langle \boldsymbol{\xi}_1, \boldsymbol{\xi}_2\rangle = \overline{\lambda_2}\langle \boldsymbol{\xi}_1, \boldsymbol{\xi}_2\rangle = \lambda_2\langle \boldsymbol{\xi}_1, \boldsymbol{\xi}_2\rangle$$

because the eigenvalues of a Hermitian matrix are real, or

$$(\lambda_1 - \lambda_2)\langle \boldsymbol{\xi}_1, \boldsymbol{\xi}_2\rangle = 0.$$

But λ_1 and λ_2 are different; thus $\langle \boldsymbol{\xi}_1, \boldsymbol{\xi}_2\rangle = 0$, as desired.

There is still one loose end to be tied up: the orthogonality proof depended on the two different eigenvectors having *different* eigenvalues; this was so that we knew $(\lambda_1 - \lambda_2) \neq 0$ and thus the above inner product had to vanish. But it is at least conceivable that these different eigenvectors might belong to the *same* eigenvalue; *i.e.*, that they are [linearly independent] eigenvectors corresponding to a *repeated* root. But in this case we can carry out the Gram-Schmidt Orthogonalization noted at the end of of Example 3.5.1 and *force* the eigenvectors to be orthogonal! □

3.5.2 The Cayley-Hamilton Theorem

We now obtain a curious, and even marginally interesting, result uniformly described as "fundamental" in texts on linear algebra. It deals with a *polynomial* equation a *matrix* satisfies. Thus we must briefly set the stage for such *matrix polynomials*:

The product of an $n \times n$ matrix with itself is another $n \times n$ matrix; *this* product can then be multiplied again by $\mathbf{A}$ to yield yet another $n \times n$ matrix. This suggests that we can recursively define the *power* of a matrix:

$$\mathbf{A}^n \equiv \mathbf{A}\mathbf{A}^{n-1}.$$

But we have made a great deal about the *non*-commutativity of [square] matrix multiplication, and the thoughtful reader might ask whether this definition is *well-defined* in the sense that, for example, the power $\mathbf{A}^{n+1}$ can be calculated as either

$$\mathbf{A}^{n+1} \equiv \mathbf{A}\mathbf{A}^n \quad \text{or} \quad \mathbf{A}^{n+1} = \mathbf{A}^n\mathbf{A}.$$

It can be shown inductively that this product *is* well-defined no matter which way $\mathbf{A}^{n+1}$ is found: It certainly is true for $n = 1$: $\mathbf{A}\mathbf{A} = \mathbf{A}\mathbf{A}$! And if it is true for all powers up to n, then

$$\mathbf{A}^{n+1} \equiv \mathbf{A}\mathbf{A}^n = \mathbf{A}(\mathbf{A}^{n-1}\mathbf{A}) \equiv (\mathbf{A}\mathbf{A}^{n-1})\mathbf{A} = \mathbf{A}^n\mathbf{A}$$

by the associativity of matrix multiplication. Thus we can speak simply of "$\mathbf{A}^{n+1}$."

$n \times n$ matrices are closed under the "vector addition" and "scalar multiplication" of matrices; thus one can form a "matrix polynomial" resulting in yet another $n \times n$ matrix:

$$p(\mathbf{A}) = a_0 + a_1\mathbf{A} + a_2\mathbf{A}^2 + \ldots + a_n\mathbf{A}^n.$$

On the other hand, it is at least possible to conceive of a matrix whose *elements are polynomials*:

$$\mathbf{Q}(\lambda) = (q_{ij}(\lambda));$$

such matrices could reasonably be called "*polynomial matrices*." Thus one must be careful which he is talking about, and we are going to use *both* in this section!

In essence, what the Cayley-Hamilton Theorem says is that *a matrix satisfies its characteristic equation*:

Theorem 3.5.3 (Cayley-Hamilton Theorem). If the characteristic equation of a square matrix $\mathbf{A}$ is $p(\lambda) \equiv |\mathbf{A} - \lambda\mathbf{1}|$, then $p(\mathbf{A}) = \mathbf{0}$.

One must be careful to observe that the "**0**" in this statement is the **0** *matrix*!

In order to prove this, it will be necessary to use a fact from algebra:

Lemma. $x^n - y^n$ is always divisible by $x - y$.

Proof. (*by induction*) This is clearly true for $n = 1$. Assume it is for all orders up to n; we must show it true for $x^{n+1} - y^{n+1}$. But, writing

$$\begin{aligned} x^{n+1} - y^{n+1} &= x^{n+1} + x^n y - xy^n - y^{n+1} - x^n y + xy^n \\ &= (x + y)(x^n - y^n) - xy(x^{n-1} - y^{n-1}), \end{aligned}$$

both products in the sum are divisible by $x - y$ by the induction hypothesis. □

As a consequence, we have what we *really* need:

Corollary. For an arbitrary polynomial $P(x)$, $P(x) - P(y)$ is divisible by $x - y$.

Proof. When one forms $P(x) - P(y)$, any constant in $P(x)$ subtracts out, leaving a polynomial with the *same* [remaining] coefficients a_i multiplying $(x^i - y^i)$. But each of these, by the Lemma, is divisible by $x - y$. □

This means that we can write $P(x) - P(y) = (x-y)P'(x,y)$, where P' is of degree 1 less than that of P.

With this in hand, we are finally ready to prove the Cayley-Hamilton Theorem:

Proof of Cayley-Hamilton Theorem. Consider the $n \times n$ matrix $\mathbf{A} - \lambda\mathbf{1}$, from which the characteristic equation $p(\lambda) = |\mathbf{A} - \lambda\mathbf{1}|$ is obtained. Form the matrix $\mathbf{Q}(\lambda)$ consisting of the *transpose* of the matrix of *cofactors* of each element of $\mathbf{A} - \lambda\mathbf{1}$ (*cf.* the proof of Theorem 3.2.11!); this is a "polynomial matrix" introduced above, with elements themselves polynomials in λ. By Theorem 3.2.11—this is the only reason we presented it!—

$$(\mathbf{A} - \lambda\mathbf{1})^{-1} = \frac{1}{|\mathbf{A} - \lambda\mathbf{1}|}\mathbf{Q}(\lambda) = \frac{1}{p(\lambda)}\mathbf{Q}(\lambda);$$

i.e.,

$$(\mathbf{A} - \lambda\mathbf{1})\mathbf{Q}(\lambda) = p(\lambda)\mathbf{1}. \tag{3.5-10}$$

Now, by the above Corollary, we can form $p(a) - p(\lambda)$ and write it as $p(a) - p(\lambda) = (a-\lambda)p'(a,\lambda)$, or $p(a) = p(\lambda) + (a-\lambda)p'(a,\lambda)$, where $p'(\lambda)$ is a polynomial in λ of degree $n-1$. Forming the "matrix polynomial" of *this* (where the scalar λ is replaced by $\lambda\mathbf{1}$):

$$\begin{aligned}
p(\mathbf{A}) &= p(\lambda\mathbf{1}) + (\mathbf{A} - \lambda\mathbf{1})p'(\mathbf{A}, \lambda\mathbf{1}) \\
&= p(\lambda)\mathbf{1} + (\mathbf{A} - \lambda\mathbf{1})p'(\mathbf{A}, \lambda\mathbf{1}) \\
&= (\mathbf{A} - \lambda\mathbf{1})\mathbf{Q}(\lambda) + (\mathbf{A} - \lambda\mathbf{1})p'(\mathbf{A}, \lambda\mathbf{1}) \\
&= (\mathbf{A} - \lambda\mathbf{1})\left(\mathbf{Q}(\lambda) + p'(\mathbf{A}, \lambda\mathbf{1})\right) \\
&= \mathbf{A}\left(\mathbf{Q}(\lambda) + p'(\mathbf{A}, \lambda\mathbf{1})\right) - \lambda\left(\mathbf{Q}(\lambda) + p'(\mathbf{A}, \lambda\mathbf{1})\right)
\end{aligned}$$

—where (3.5-10) has been introduced in the third equality. Examining this equation, we note that the left-hand side, $p(\mathbf{A})$, must be a constant, and thus so must the right. But, whereas the first term on that side has the quantity $(\mathbf{Q}(\lambda) + p'(\mathbf{A}, \lambda\mathbf{1}))$ multiplied by [the *constant*] $\mathbf{A}$, the second has this same quantity factored by λ, so there is no way the right-hand side can be independent of that variable—*unless* $(\mathbf{Q}(\lambda) + p'(\mathbf{A}, \lambda\mathbf{1})) \equiv 0$! But if *that* is the case, then the right-hand side vanishes altogether, and thus $p(\mathbf{A}) = 0$—precisely the assertion to be proved.[2] □

We conclude this section with a simple example of the above Theorem, also demonstrating matrix polynomials:

Example 3.5.3. *Cayley-Hamilton Theorem* We consider the matrix of Example 3.5.1:

$$\mathbf{A} = \begin{pmatrix} 2 & 1 & 1 \\ 2 & 3 & 2 \\ 1 & 1 & 2 \end{pmatrix}$$

We have already obtained the characteristic equation: $p(\lambda) = -\lambda^3 + 7\lambda^2 - 11\lambda + 5$. Forming the matrix products

$$\mathbf{A}^2 = \begin{pmatrix} 7 & 6 & 6 \\ 12 & 13 & 12 \\ 6 & 6 & 7 \end{pmatrix}, \quad \mathbf{A}^3 = \begin{pmatrix} 32 & 31 & 31 \\ 62 & 63 & 62 \\ 31 & 31 & 32 \end{pmatrix}$$

[2]The suspicious reader might note that, since $|\mathbf{A} - \lambda\mathbf{1}| = 0$ for eigenvalues, the $(\mathbf{A} - \lambda\mathbf{1})^{-1}$, corresponding to $\mathbf{Q}(\lambda)$ above, might not even exist! But we are dealing here with *matrix polynomials* here, not those matrices resulting from substituting a particular value for λ.

and substituting into the matrix polynomial [note that $5 \rightarrow 5\mathbf{1}$!]:

$$\begin{aligned} p(\mathbf{A}) &= -\mathbf{A}^3 + 7\mathbf{A}^2 - 11\mathbf{A} + 5\mathbf{1} \\ &= -\begin{pmatrix} 32 & 31 & 31 \\ 62 & 63 & 62 \\ 31 & 31 & 32 \end{pmatrix} + \begin{pmatrix} 49 & 42 & 42 \\ 84 & 91 & 84 \\ 42 & 42 & 49 \end{pmatrix} - \begin{pmatrix} 22 & 11 & 11 \\ 22 & 33 & 22 \\ 11 & 11 & 22 \end{pmatrix} + \begin{pmatrix} 5 & 0 & 0 \\ 0 & 5 & 0 \\ 0 & 0 & 5 \end{pmatrix} \\ &= \begin{pmatrix} 0 & 0 & 0 \\ 0 & 0 & 0 \\ 0 & 0 & 0 \end{pmatrix} \end{aligned}$$

—as expected!

Note that we can also write the characteristic equation in terms of its factors: $p(\lambda) = -(\lambda - 1)^2(\lambda - 5)$. This suggests another means of calculating $p(\mathbf{A})$: using the expressions for $(\mathbf{A} - 1\mathbf{1})$ and $(\mathbf{A} - 5\mathbf{1})$ obtained in Example 3.5.1,

$$(\mathbf{A} - 1\mathbf{1})^2 = \begin{pmatrix} 4 & 4 & 4 \\ 6 & 6 & 6 \\ 4 & 4 & 4 \end{pmatrix};$$

thus

$$p(\mathbf{A}) = (\mathbf{A} - 1\mathbf{1})^2(\mathbf{A} - 5\mathbf{1}) = \begin{pmatrix} 4 & 4 & 4 \\ 6 & 6 & 6 \\ 4 & 4 & 4 \end{pmatrix} \begin{pmatrix} -3 & 1 & 1 \\ 2 & -2 & 2 \\ 1 & 1 & -3 \end{pmatrix} = \begin{pmatrix} 0 & 0 & 0 \\ 0 & 0 & 0 \\ 0 & 0 & 0 \end{pmatrix}$$

Note that, since we can order the characteristic equation $p(\lambda) = -(\lambda - 5)(\lambda - 1)^2$, we could also find $p(\mathbf{A}) = (\mathbf{A} - 5\mathbf{1})(\mathbf{A} - 1\mathbf{1})^2$.

It is important to note that, in saying $p(\mathbf{A}) = \mathbf{0}$, this does *not* say that either factor vanishes; see the above calculations. |

3.5.3 Generalized Eigenvectors

We have noted above that, for *distinct* eigenvalues, we have *distinct*, *linearly independent* eigenvectors. We also observed that for *repeated* eigenvalues—roots of the characteristic equation with multiplicity greater than 1—we *might* have distinct eigenvectors, or we might just as easily *not*. In the latter case, there is clearly a deficiency in eigenvectors; the question is, what to do to come up with the balance. We now address that issue.

Consider, for example, the case that λ_o is a repeated root—one with multiplicity m_o—of the characteristic equation for an $n \times n$ matrix $\mathbf{A}$:

$$|\mathbf{A} - \lambda\mathbf{1}| \equiv p(\lambda) = (\lambda - \lambda_o)^{m_o} p'(\lambda),$$

where p' is of degree $(n - m_o)$ in λ. Further, let us say that $(\mathbf{A} - \lambda_o\mathbf{1})$ is of rank $n - 1$; *i.e.*, there is only *one* eigenvector, since the dimension of the null space is 1:

$$(\mathbf{A} - \lambda_o\mathbf{1})\boldsymbol{\xi}_1 = \mathbf{0}. \tag{3.5-11}$$

What are we to do in this case?

Well, in the first instance, *there are no more eigenvectors*: $\boldsymbol{\xi}_1$ is *it*! But, as in Problem 6 on page 123, consider the possibility that there *is* another vector, $\boldsymbol{\xi}_2$, satisfying the *non*-homogeneous equation (since, by the definition of eigenvectors, we only accept $\boldsymbol{\xi}_1 \neq \mathbf{0}$)

$$(\mathsf{A} - \lambda_o \mathbf{1})\boldsymbol{\xi}_2 = \boldsymbol{\xi}_1. \tag{3.5-12}$$

$\boldsymbol{\xi}_2$ *is not an eigenvector*: eigenvectors satisfy only the appropriate *homogeneous* equation (3.5-11). Observe, however, that it satisfies something *like* that equation:

$$(\mathsf{A} - \lambda_o \mathbf{1})^2\boldsymbol{\xi}_2 = (\mathsf{A} - \lambda_o \mathbf{1})\boldsymbol{\xi}_1 = \mathbf{0}$$

Such a vector is referred to as a *generalized* eigenvector, here of order 2. Similarly, any vector $\boldsymbol{\xi}_k$ satisfying

$$(\mathsf{A} - \lambda_o \mathbf{1})^{k-1}\boldsymbol{\xi}_k = \boldsymbol{\xi}_{k-1} \neq \mathbf{0}, \quad (\mathsf{A} - \lambda_o \mathbf{1})^{k}\boldsymbol{\xi}_k = \mathbf{0} \tag{3.5-13}$$

is called a *generalized eigenvector of order k*. In this context, normal, *proper* eigenvectors might be categorized as "generalized eigenvectors of order 1." This might all appear somewhat desperate, and, to a certain extent, it is. But, to the extent we can *find* such non-**0** vectors, they do enjoy at least one property we have come to value highly:

Theorem 3.5.4. Any set of non-**0** vectors $\boldsymbol{\xi}_k, k \leq m_o$ satisfying (3.5-13) are linearly independent.

Proof. Assume we have a set $\{\boldsymbol{\xi}_1, \boldsymbol{\xi}_2, \ldots, \boldsymbol{\xi}_k\}$ satisfying (3.5-13) for $k \leq m_o$. Were $\sum_{i=1}^{k} c_i \boldsymbol{\xi}_i = \mathbf{0}$, then

$$(\mathsf{A} - \lambda_o \mathbf{1})^{k-1} \sum_{i=1}^{k} c_i \boldsymbol{\xi}_i = \sum_{i=1}^{k-1} c_i (\mathsf{A} - \lambda_o \mathbf{1})^{k-1}\boldsymbol{\xi}_i + c_k (\mathsf{A} - \lambda_o \mathbf{1})^{k-1}\boldsymbol{\xi}_k = \mathbf{0}.$$

But the first term on the right-hand side—the summation—vanishes identically, since, for *each* $j \leq (k-1)$,

$$(\mathsf{A} - \lambda_o \mathbf{1})^{k-1}\boldsymbol{\xi}_j = (\mathsf{A} - \lambda_o \mathbf{1})^{k-1-j}(\mathsf{A} - \lambda_o \mathbf{1})^{j}\boldsymbol{\xi}_j = \mathbf{0}$$

by (3.5-13). But this means that $c_k(\mathsf{A} - \lambda_o \mathbf{1})^{k-1}\boldsymbol{\xi}_k = \mathbf{0}$ which, since $(\mathsf{A} - \lambda_o \mathbf{1})^{k-1}\boldsymbol{\xi}_k \neq \mathbf{0}$, means that c_k vanishes. With this shown, we can repeat the process with $(\mathsf{A} - \lambda_o \mathbf{1})^{k-2}$ applied to the sum $\sum_{i=1}^{k} c_i \boldsymbol{\xi}_i$, now $\sum_{i=1}^{k-1} c_i \boldsymbol{\xi}_i$, showing that $c_{k-1} = 0$, and continue back until ending with $c_1 = 0$. Thus the $\{\boldsymbol{\xi}_i\}$ *are* linearly independent. □

If we can find *any*, it is tempting to expect that we can find a infinite number; after all, we are solving a homogeneous equation $(\mathsf{A} - \lambda_o \mathbf{1})^{k}\boldsymbol{\xi}_k = \mathbf{0}$, and the rank of $(\mathsf{A} - \lambda_o \mathbf{1})^{k}$ is always less than n by the Corollary on page 108:

$$|(\mathsf{A} - \lambda_o \mathbf{1})^{k}| = |\mathsf{A} - \lambda_o \mathbf{1}|^{k} = 0$$

by Theorem 3.2.8. But, in point of fact, the process eventually terminates: while we might be able to *find* $\boldsymbol{\xi}$ satisfying $(\mathsf{A} - \lambda_o \mathbf{1})^{k}\boldsymbol{\xi}_k = \mathbf{0}$, $\boldsymbol{\xi}$ is only an n-dimensional vector, and thus there can be at most n of them which are linearly independent.

In fact, the number of linearly independent eigenvectors corresponding to an arbitrary eigenvalue λ_i is just m_i, the multiplicity of its root in the factored form of the characteristic equation

$$p(\lambda) = \prod_{i=1}^{n_\lambda} (\lambda - \lambda_i)^{m_i} = 0$$

—where n_λ is the number of distinct eigenvalues of $p(\lambda)$. This is certainly suggested by the corresponding form the the Cayley-Hamilton Theorem

$$p(\mathbf{A}) = \prod_{i=1}^{n_\lambda}(\mathbf{A} - \lambda_i \mathbf{1})^{m_i} = \mathbf{0}$$

which, in fact, is central to the proof of this assertion, which we now sketch. [For details, see A.D. Myškis, *Advanced Mathematics for Engineers*]. We warn the reader that it is a bit "technical"!

The set of all vectors $\boldsymbol{\xi}^{(i)}$—normal and generalized—such that $(\mathbf{A} - \lambda_i\mathbf{1})^k\boldsymbol{\xi}^{(i)} = \mathbf{0}$ for some $k \leq m_i$ form a *subspace*: given two such, $\boldsymbol{\xi}_1^{(i)}$ and $\boldsymbol{\xi}_2^{(i)}$, it is trivial to show that $c_1\boldsymbol{\xi}_1^{(i)} + c_2\boldsymbol{\xi}_2^{(i)}$ satisfies $(\mathbf{A} - \lambda_i\mathbf{1})^k(c_1\boldsymbol{\xi}_1^{(i)} + c_2\boldsymbol{\xi}_2^{(i)}) = 0$ for some $k \leq m_i$. It is not surprising, in view of the fact that standard eigenvectors corresponding to distinct eigenvalues are linearly independent, that these subspaces are *disjoint*: that $\boldsymbol{\xi}^{(i)}$ corresponding to λ_i will *not* satisfy $(\mathbf{A} - \lambda_j\mathbf{1})^k\boldsymbol{\xi}^{(i)} = \mathbf{0}$ for $k \leq m_j$if $i \neq j$. One can then show craftily that *any* n-tuple can be expressed as a linear combination of the vectors in these subspaces corresponding to the different eigenvalues,

$$\mathbf{v} = \sum_{j=1}^{n_\lambda} v_j \mathbf{C}_j \mathbf{v},$$

where

$$\mathbf{C}_j = c_j \prod_{k \neq j}^{n_\lambda}(\mathbf{A} - \lambda_k\mathbf{1})^{m_k}$$

so that, by the Cayley-Hamilton Theorem,

$$(\mathbf{A} - \lambda_j\mathbf{1})^{m_j}\mathbf{C}_j\mathbf{v} = c_j(\mathbf{A} - \lambda_j\mathbf{1})^{m_j}\prod_{k \neq j}^{n_\lambda}(\mathbf{A} - \lambda_k\mathbf{1})^{m_k}\mathbf{v} = \mathbf{0};$$

thus $\mathbf{C}_j\mathbf{v}$ is in the subspace generated by λ_j. In particular, then, for $\mathbf{v}$ *in* that subspace resulting from λ_i, the $v_j = 0$ for $j \neq i$, showing that $(\mathbf{A} - \lambda_i\mathbf{1})^k\mathbf{v}$ for such vectors will terminate for $k = m_i$; *i.e.*, there is no $\boldsymbol{\xi}_k$ for $k > m_i$ linearly independent of the others, since *any* vector will satisfy $(\mathbf{A} - \lambda_i\mathbf{1})^{m_i}\boldsymbol{\xi} = \mathbf{0}\boldsymbol{\xi} = \mathbf{0}$.

This is, admittedly, all a little involved. The bottom line is this: in such cases that a given eigenvalue λ_i has a multiplicity m_i greater than 1 (if it *is* 1, its associated eigenvector is already linearly independent of all others, remember), it is possible to find m_i *linearly independent, standard or generalized,* eigenvectors associated with that eigenvalue. In practice, the details take care of themselves, as the following examples show.

Example 3.5.4. *Eigenvectors of Repeated Roots* We return to the matrices of Example 3.5.2, which we have already seen to be deficient in normal eigenvectors.

$$\mathbf{A} = \begin{pmatrix} 2 & 1 & 0 \\ 0 & 2 & 1 \\ 0 & 0 & 2 \end{pmatrix}.$$

Solution: The characteristic equation for this matrix is just $p(\lambda) = (2 - \lambda)^3$, so we have a repeated root, 2, "of multiplicity 3." Forming the matrix

$$\mathbf{A} - 2\mathbf{1} = \begin{pmatrix} 0 & 1 & 0 \\ 0 & 0 & 1 \\ 0 & 0 & 0 \end{pmatrix},$$

we see that we have one of rank 2; thus there is only a *single* eigenvector. Though we did not do it in the example cited, we now proceed to find that eigenvector, say $\boldsymbol{\xi}_1$:

$$(\mathbf{A}-2\mathbf{1})\boldsymbol{\xi}_1 = \begin{pmatrix} 0 & 1 & 0 \\ 0 & 0 & 1 \\ 0 & 0 & 0 \end{pmatrix} \begin{pmatrix} \xi_1 \\ \xi_2 \\ \xi_3 \end{pmatrix} = \begin{pmatrix} \xi_2 \\ \xi_3 \\ 0 \end{pmatrix} = \mathbf{0} \Rightarrow \boldsymbol{\xi}_1 = \begin{pmatrix} 1 \\ 0 \\ 0 \end{pmatrix}$$

According to the above discussion, we expect there to be two other *generalized* eigenvectors satisfying equation (3.5-13), for $k = 2, 3$. Consider the first:

$$(\mathbf{A}-2\mathbf{1})\boldsymbol{\xi}_2 = \begin{pmatrix} 0 & 1 & 0 \\ 0 & 0 & 1 \\ 0 & 0 & 0 \end{pmatrix} \begin{pmatrix} \xi_1 \\ \xi_2 \\ \xi_3 \end{pmatrix} = \begin{pmatrix} \xi_2 \\ \xi_3 \\ 0 \end{pmatrix} = \boldsymbol{\xi}_1 = \begin{pmatrix} 1 \\ 0 \\ 0 \end{pmatrix}$$

In terms of the theory of linear equations, we see that the coefficient matrix has a rank of two, while the *augmented* matrix

$$\begin{pmatrix} 0 & 1 & 0 & 1 \\ 0 & 0 & 1 & 0 \\ 0 & 0 & 0 & 0 \end{pmatrix}$$

has the same rank; thus a solution exists. But, since the rank of both is only two, we can, at best, solve for two in terms of a third; indeed,

$$\boldsymbol{\xi}_2 = \begin{pmatrix} \xi_1 \\ 1 \\ 0 \end{pmatrix}, \text{ say } \boldsymbol{\xi}_2 = \begin{pmatrix} 0 \\ 1 \\ 0 \end{pmatrix}$$

[we could have taken a non-0 value for the first component]. As alleged by the above, this is linearly independent of $\boldsymbol{\xi}_1$. (It automatically satisfies $(\mathbf{A}-2\mathbf{1})^2\boldsymbol{\xi}_2 = \mathbf{0}$ by (3.5-12).)

Finally, we find $\boldsymbol{\xi}_3$:

$$(\mathbf{A}-2\mathbf{1})\boldsymbol{\xi}_3 = \begin{pmatrix} \xi_2 \\ \xi_3 \\ 0 \end{pmatrix} = \boldsymbol{\xi}_2 = \begin{pmatrix} 0 \\ 1 \\ 0 \end{pmatrix}$$

Again, both coefficient and augmented matrices have rank 2, and we can solve for two components in terms of the third. *Because of our selection of* 0 for the first component of $\boldsymbol{\xi}_2$, $\xi_2 = 0$, but ξ_1 can be arbitrary; we take it to be 0:

$$\boldsymbol{\xi}_3 = \begin{pmatrix} \xi_1 \\ 0 \\ 1 \end{pmatrix}, \text{ say } \boldsymbol{\xi}_3 = \begin{pmatrix} 0 \\ 0 \\ 1 \end{pmatrix},$$

which is, as expected, linearly independent of the others. Note that choices of the arbitrary component propagated forward to the later solutions at each step.

Now consider the possibility mentioned at the end of Example 3.5.2: that one of the 1's were a 0. Say

$$\mathbf{A} = \begin{pmatrix} 2 & 1 & 0 \\ 0 & 2 & 0 \\ 0 & 0 & 2 \end{pmatrix};$$

the characteristic equation is the same, but now

$$\mathbf{A} - 2\mathbf{1} = \begin{pmatrix} 0 & 1 & 0 \\ 0 & 0 & 0 \\ 0 & 0 & 0 \end{pmatrix},$$

with rank 1, suggesting the existence of two linearly independent eigenvectors. Indeed, writing

$$\mathbf{A} - 2\mathbf{1} = \begin{pmatrix} 0 & 1 & 0 \\ 0 & 0 & 0 \\ 0 & 0 & 0 \end{pmatrix} \boldsymbol{\xi} = \begin{pmatrix} 0 & 1 & 0 \\ 0 & 0 & 0 \\ 0 & 0 & 0 \end{pmatrix} \begin{pmatrix} \xi_1 \\ \xi_2 \\ \xi_3 \end{pmatrix} = \begin{pmatrix} \xi_2 \\ 0 \\ 0 \end{pmatrix} = \mathbf{0},$$

we get that

$$\boldsymbol{\xi} = \begin{pmatrix} \xi_1 \\ 0 \\ \xi_3 \end{pmatrix}, \text{ say } \boldsymbol{\xi}_1 = \begin{pmatrix} 1 \\ 0 \\ 0 \end{pmatrix} \text{ and } \boldsymbol{\xi}_2 = \begin{pmatrix} 0 \\ 0 \\ 1 \end{pmatrix}.$$

We now have a choice to make: do we find $\boldsymbol{\xi}_3$ such that $(\mathbf{A} - 2\mathbf{1})\boldsymbol{\xi}_3 = \boldsymbol{\xi}_1$ or $\boldsymbol{\xi}_2$? Here the theory we have obtained for linear equations in Section 2.3.1 comes to our aid: were we to select $\boldsymbol{\xi}_2$, *there would be no solution*, since the augmented matrix

$$(\mathbf{A} - 2\mathbf{1})\boldsymbol{\xi}_2 = \begin{pmatrix} 0 & 1 & 0 & 0 \\ 0 & 0 & 0 & 0 \\ 0 & 0 & 0 & 1 \end{pmatrix}$$

has rank 2, while, as observed, $(\mathbf{A} - 2\mathbf{1})$ only has rank 1. Thus we *must* use the first:

$$(\mathbf{A} - 2\mathbf{1})\boldsymbol{\xi}_3 = \begin{pmatrix} \xi_2 \\ 0 \\ 0 \end{pmatrix} = \boldsymbol{\xi}_1 = \begin{pmatrix} 1 \\ 0 \\ 0 \end{pmatrix} \Rightarrow \boldsymbol{\xi}_3 = \begin{pmatrix} 0 \\ 1 \\ 0 \end{pmatrix},$$

independent of the first two, as expected. Note that the above two [standard] eigenvectors have a different character: though $\boldsymbol{\xi}_1$ remains "single," $\boldsymbol{\xi}_2$ has started a "family" of generalized eigenvector offspring. |

We now consider an example on a somewhat grander scale:

Example 3.5.5. Consider the matrix

$$\mathbf{A} = \begin{pmatrix} 5 & 1 & 0 & 0 & 0 \\ 0 & 5 & 1 & 0 & 0 \\ 0 & 0 & 5 & 0 & 0 \\ 0 & 0 & 0 & 5 & 1 \\ 0 & 0 & 0 & 0 & 5 \end{pmatrix}$$

Again (*cf.* page 122), we have a repeated eigenvalue $\lambda = 5$ of multiplicity 5; thus

$$\mathbf{A} - 5\mathbf{1} = \begin{pmatrix} 0 & 1 & 0 & 0 & 0 \\ 0 & 0 & 1 & 0 & 0 \\ 0 & 0 & 0 & 0 & 0 \\ 0 & 0 & 0 & 0 & 1 \\ 0 & 0 & 0 & 0 & 0 \end{pmatrix}$$

has rank 3, predicting two linearly independent [standard] eigenvectors. Direct calculation of $(\mathbf{A}-5\mathbf{1})\boldsymbol{\xi}=0$ yields

$$\begin{pmatrix}\xi_2\\ \xi_3\\ 0\\ \xi_5\\ 0\end{pmatrix}=\mathbf{0}\text{ or, say, }\boldsymbol{\xi}_1=\begin{pmatrix}1\\0\\0\\0\\0\end{pmatrix}\text{ and }\boldsymbol{\xi}_2=\begin{pmatrix}0\\0\\0\\1\\0\end{pmatrix}$$

We note that the augmented matrices of *both* of these vectors with $(\mathbf{A}-5\mathbf{1})$ will have rank 3; thus *each* can be used to establish the other three linearly independent eigenvectors predicted by the theory. We consider these in turn; to keep the genealogies of the two families straight, we shall introduce a double subscripting for the generalized eigenvectors:

<u>$\boldsymbol{\xi}_1$:</u> Let us solve for, say, $\boldsymbol{\xi}_{11}$:

$$(\mathbf{A}-5\mathbf{1})\boldsymbol{\xi}_{11}=\begin{pmatrix}\xi_2\\ \xi_3\\ 0\\ \xi_5\\ 0\end{pmatrix}=\boldsymbol{\xi}_1=\begin{pmatrix}1\\0\\0\\0\\0\end{pmatrix}\Rightarrow\boldsymbol{\xi}_{11}=\begin{pmatrix}\xi_1\\1\\0\\ \xi_4\\0\end{pmatrix},\text{ say }\boldsymbol{\xi}_{11}=\begin{pmatrix}0\\1\\0\\0\\0\end{pmatrix}.$$

There is no reason not to continue with this; let us seek $\boldsymbol{\xi}_{12}$:

$$(\mathbf{A}-5\mathbf{1})\boldsymbol{\xi}_{12}=\begin{pmatrix}\xi_2\\ \xi_3\\ 0\\ \xi_5\\ 0\end{pmatrix}=\boldsymbol{\xi}_{11}=\begin{pmatrix}0\\1\\0\\0\\0\end{pmatrix}\Rightarrow\boldsymbol{\xi}_{12}=\begin{pmatrix}\xi_1\\0\\1\\ \xi_4\\0\end{pmatrix},\text{ say }\boldsymbol{\xi}_{12}=\begin{pmatrix}0\\0\\1\\0\\0\end{pmatrix}$$

...and $\boldsymbol{\xi}_{13}$:

$$(\mathbf{A}-5\mathbf{1})\boldsymbol{\xi}_{13}=\begin{pmatrix}\xi_2\\ \xi_3\\ 0\\ \xi_5\\ 0\end{pmatrix}=\boldsymbol{\xi}_{12}=\begin{pmatrix}0\\0\\1\\0\\0\end{pmatrix}$$

Oops! We've hit a snag: there *is* no solution to this! If we'd been fussy about keeping track of the augmented matrices, we'd have seen that the augmented matrix here would have had rank 4, even though the coefficient matrix has rank of only 3! Thus this "family" consists of only $\boldsymbol{\xi}_1$ and its two offspring $\boldsymbol{\xi}_{11}$ and $\boldsymbol{\xi}_{12}$.

<u>$\boldsymbol{\xi}_2$:</u> Now let us find $\boldsymbol{\xi}_{21}$. We will be a little more careful about whether such solutions exist! The augmented matrix $(\mathbf{A}-5\mathbf{1}\ \boldsymbol{\xi}_2)$ still has rank 3:

$$(\mathbf{A}-5\mathbf{1})\boldsymbol{\xi}_{21}=\begin{pmatrix}\xi_2\\ \xi_3\\ 0\\ \xi_5\\ 0\end{pmatrix}=\boldsymbol{\xi}_2=\begin{pmatrix}0\\0\\0\\1\\0\end{pmatrix}\Rightarrow\boldsymbol{\xi}_{21}=\begin{pmatrix}\xi_1\\0\\0\\ \xi_4\\1\end{pmatrix},\text{ say }\boldsymbol{\xi}_{21}=\begin{pmatrix}0\\0\\0\\0\\1\end{pmatrix},$$

but if we [foolishly!] look for, say, $\boldsymbol{\xi}_{22}$, the augmented matrix $((\mathbf{A}-5\mathbf{1})\ \boldsymbol{\xi}_{21})$ has rank 4 again; thus there is no solution.

Thus we have the five linearly independent vectors, in two families, $\{\boldsymbol{\xi}_1, \boldsymbol{\xi}_{11}, \boldsymbol{\xi}_{12}\}$ and $\{\boldsymbol{\xi}_2, \boldsymbol{\xi}_{21}\}$. Note that different choices of the free parameters ξ_1 and ξ_4 right at the determination of the [standard] eigenvectors would have changed the form of the generalized ones; similarly, such a choice in determining the first of these would have propagated to later progeny. But, at the end of the day, it was never necessary to muck about in all the details of the theory; with the assurance that it was worth looking for such eigenvectors in the first place, the math "bailed us out." |

Let us summarize the results of this rather extensive section: To find eigenvectors, we first solve the *characteristic equation* $|\mathbf{A}-\lambda\mathbf{1}| = 0$ for the eigenvalues λ; having those we determine the eigenvectors $\boldsymbol{\xi}$ such that $(\mathbf{A}-\lambda\mathbf{1})\boldsymbol{\xi}=0$, but the details of this solution diverge depending on the multiplicity of the eigenvalues:

- If the eigenvalues are *distinct*—each has a multiplicity of 1—then we know the $\boldsymbol{\xi}$ are *linearly independent.*
- If a given eigenvalue has a multiplicity m *greater* than 1, then
  - it *might* be the case that there is still a full set of m linearly independent [true] eigenvectors corresponding to that λ; these would be linearly independent of eigenvectors corresponding to the other eigenvalues. OR
  - there might be *fewer* than m eigenvectors, in which case there are m linearly independent standard *and generalized* eigenvectors corresponding to λ.

3.5.4 Application of Eigenvalues/Eigenvectors

Eigenvalues and eigenvectors pervade all of science and engineering, at least when dealing with *linear* systems. We briefly mention several examples:

Example 3.5.6. *Solution of Systems of Linear Ordinary Differential Equations with Constant Coefficients:* We can represent any system of this form as a vector/matrix system

$$\dot{\boldsymbol{x}} = \mathbf{A}\boldsymbol{x}$$

where $\mathbf{A}$ is the matrix of [constant] coefficients and "$\dot{}$" stands for, say, differentiation with respect to the time t. Recalling that the solution of the corresponding scalar system $\dot{x} = a\,x$ is $x = c\,e^{at}$ for some constant c, we *try* for a solution of the same form, $\boldsymbol{x} = \boldsymbol{\xi}e^{\lambda t}$ for some constant $\boldsymbol{\xi}$ and *some* value of λ; then

$$\dot{\boldsymbol{x}} = \lambda\boldsymbol{\xi}e^{\lambda t} = \mathbf{A}\boldsymbol{\xi}e^{\lambda t}$$

or, since $e^{\lambda t} \neq 0$, we can divide it out, giving $\lambda\boldsymbol{\xi} = \mathbf{A}\boldsymbol{\xi}$—precisely the defining equation for eigenvalue/eigenvectors!

For example, consider the second-order equation for *scalar* x: $\ddot{x}+a\dot{x}+bx=0$. This can be put in the form of a *first*-order equation by putting it in vector form by defining the components of $\boldsymbol{x}$ by $x_1 \equiv \dot{x}$ and $x_2 \equiv x$:

$$\dot{\boldsymbol{x}} \equiv \begin{pmatrix} \dot{x_1} \\ \dot{x_2} \end{pmatrix} = \begin{pmatrix} -ax_1 - bx_2 \\ x_1 \end{pmatrix} = \begin{pmatrix} -a & -b \\ 1 & 0 \end{pmatrix} \begin{pmatrix} x_1 \\ x_2 \end{pmatrix} \equiv \mathsf{A}\boldsymbol{x};$$

Trying $\boldsymbol{x} = \boldsymbol{\xi} e^{\lambda t}$, we get, as above, the *eigen*equation $(\mathsf{A} - \lambda \mathbf{1})\boldsymbol{\xi} = 0$, with characteristic equation

$$\begin{vmatrix} -a-\lambda & -b \\ 1 & -\lambda \end{vmatrix} = \lambda^2 + a\lambda + b = 0$$

—precisely the "characteristic equation" of the theory of such differential equations for the original system! |

Example 3.5.7. *Generalized Eigenvectors and the Solution of Linear Differential Equations with Constant Coefficients:* Again we consider the differential equation

$$\dot{\mathsf{x}} = \mathsf{A}\mathsf{x} \tag{3.5-14}$$

in which x is [the representation of] an n-vector and A a constant matrix. We know that, for *scalar* differential equations of order greater than 1, if λ is a root of the characteristic equation with multiplicity $m > 1$, then the functions $te^{\lambda t}$, $\frac{t^2}{2}e^{\lambda t}$, ... are also solutions of the original equation which are linearly independent. We now generalize that case to one in which there are eigenvalues λ_i of multiplicity $m_i > 1$ to *vector* systems of ordinary differential equations:

Assume that λ_i is an eigenvalue but that the null space of $\mathsf{A} - \lambda_i\mathbf{1}$ only has dimension 1; *i.e.*, there is only a single eigenvector $\boldsymbol{\xi}_i \neq \mathbf{0}$ associated with λ_i. Clearly, by our discussion above, $\boldsymbol{x}(t) = \boldsymbol{\xi}_i e^{\lambda_i t}$ is one solution to (3.5-14); we claim (assuming that $m_i \geq 3$) that the next two linearly independent solutions are

$$\boldsymbol{x}_1(t) = \boldsymbol{\xi}_i t e^{\lambda_i t} + \boldsymbol{\xi}_{i1} e^{\lambda_i t} = (\boldsymbol{\xi}_i t + \boldsymbol{\xi}_{i1}) e^{\lambda_i t} \tag{3.5-15a}$$

in which $(\mathsf{A} - \lambda_i\mathbf{1})\boldsymbol{\xi}_{i1} = \boldsymbol{\xi}_i$, and

$$\boldsymbol{x}_2(t) = (\boldsymbol{\xi}_i \frac{t^2}{2!} + \boldsymbol{\xi}_{i1} t + \boldsymbol{\xi}_{i2}) e^{\lambda_i t} \tag{3.5-15b}$$

in which $(\mathsf{A} - \lambda_i\mathbf{1})\boldsymbol{\xi}_{i2} = \boldsymbol{\xi}_{i1}$ (so that

$$(\mathsf{A} - \lambda_i\mathbf{1})^3 \boldsymbol{\xi}_{i2} = (\mathsf{A} - \lambda_i\mathbf{1})^2 \boldsymbol{\xi}_{i1} = (\mathsf{A} - \lambda_i\mathbf{1})\boldsymbol{\xi} \equiv \mathbf{0}$$

and $\boldsymbol{\xi}_{i1}$ and $\boldsymbol{\xi}_{i2}$ are generalized eigenvectors of orders 2 and 3, respectively).

Clearly these are linearly independent: if $c\boldsymbol{x}(t) + c_1\boldsymbol{x}_1(t) + c_2\boldsymbol{x}_2(t) = \mathbf{0}$, then, in order for the successive powers of t to vanish in the above, the c's must vanish. That these *are* additional solutions can be demonstrated directly: for example, differentiation of (3.5-15a) gives that

$$\dot{\boldsymbol{x}}_1(t) = \lambda_i(\boldsymbol{\xi}_i t + \boldsymbol{\xi}_{i1})e^{\lambda_i t} + \boldsymbol{\xi}_i e^{\lambda_i t} = (\lambda_i \boldsymbol{\xi}_{i1} + \boldsymbol{\xi}_i)e^{\lambda_i t} + t\lambda_i \boldsymbol{\xi}_i e^{\lambda_i t},$$

while

$$\mathbf{A}\boldsymbol{x}_1 = t\mathbf{A}\boldsymbol{\xi}_i e^{\lambda_i t} + \mathbf{A}\boldsymbol{\xi}_{i1} e^{\lambda_i t}.$$

But $\mathbf{A}\boldsymbol{\xi}_i e^{\lambda_i t} = \lambda_i \boldsymbol{\xi}_i e^{\lambda_i t}$ by the definition of the eigenvector $\boldsymbol{\xi}_i$, so the terms factored by t^1 satisfy equation (3.5-14); and, by the definition of the *generalized* eigenvector $\boldsymbol{\xi}_{i1}$, $\mathbf{A}\boldsymbol{\xi}_{i1} e^{\lambda_i t} = (\lambda_i \boldsymbol{\xi}_{i1} + \boldsymbol{\xi}_i)e^{\lambda_i t}$, so that factored by t^0 does. Thus $\boldsymbol{x}_1(t)$ *is* a solution, as desired.

One of the reasons for talking about *scalar* differential equations in the lead-in to this example was to emphasize the distinction between these and systems in general: Inducing from the scalar case, one might expect a second linearly independent solution associated with λ_i to be $t\boldsymbol{\xi}_i e^{\lambda_i t}$; it *isn't!*[3]

Though the above has implicitly presumed there to be on a *single* "family" of generalized eigenvectors associated with λ_i, the generalization to *multiple* families is obvious. |

Homework:

1. Show that $\boldsymbol{x}_2(t)$, equation (3.5-15a), does, indeed, satisfy (3.5-14).

Diagonalization of Matrices: This is the primary result we are aiming for in the Part; it will be treated later in its own section.

Stability Analysis: This will also be considered later.

Summary. We come finally to a central, albeit often unfortunately ill-understood, concept of square matrices, their *eigenvectors.* These are distinguished vectors $\boldsymbol{\xi}$ whose *orientation* is preserved under transformation by the original matrix: $\mathbf{A}\boldsymbol{\xi} = \lambda\boldsymbol{\xi}$, the corresponding λ being referred to as the *eigenvalue.* We note that such eigenvector/value pairs satisfy the equation $(\mathbf{A} - \lambda\mathbf{1})\boldsymbol{\xi} = \mathbf{0}$—the "eigen equation" of $\mathbf{A}$—and thus the $\boldsymbol{\xi}$ are solutions of a homogeneous equation, determined only up to a constant, and members of the null space of $(\mathbf{A} - \lambda\mathbf{1})$; the dimension of that null space thus predicts the number of linear independent eigenvectors available to the associated eigenvalue λ.

---

[3] A nice presentation of this and motivation for the *correct* form of such solutions can be found in Boyce & DiPrima, *Elementary Differential Equations and Boundary Value Problems,* in the Example in §7.8 in the Third Edition.

We investigate the linear independence of eigenvectors, showing, in particular, that eigen*vectors* corresponding to *distinct* eigen*values* are linearly independent—an only *sufficient*, not necessary, condition; thus even repeated eigenvalues *may* have associated linearly independent eigenvectors. We then obtain a result associated with the apparently very special case of *real, symmetric* matrices: that its eigenvalues are *real* and its eigenvectors are not only linearly independent, but *orthogonal*!

We also discuss the *Cayley-Hamilton Theorem*, which states that the *matrix polynomial* of a matrix satisfies its characteristic equation. This is of importance primarily in the discussion of *generalized eigenvectors*—"eigenvectors" satisfying not the homogeneous equation $(\mathsf{A} - \lambda\mathbf{1}) = \mathbf{0}$, but the *non*-homogeneous equation $(\mathsf{A} - \lambda_o\mathbf{1})\boldsymbol{\xi}_k = \boldsymbol{\xi}_{k-1} \neq \mathbf{0}$. It turns out that, even if there are not a full complement of linearly independent *true* eigenvectors, there are of true and *generalized* eigenvectors, shown with the Cayley-Hamilton Theorem.

Finally, we discuss the applications of these curious eigenvectors and their eigenvalues. These include the solution of linear systems of differential equations, the diagonalization [and "Jordan normal form"] of matrices, and stability analysis, the latter of which will be discussed presently.

3.6 Application—Basis Transformations

We now come to the one of the primary applications of linear transformations and their representations, here as a square matrix: basis transformations. In what follows it is essential that the reader make the distinction between the *vector*, $\boldsymbol{v}$, and its *representation*, v: the first may be an element of *any* vector space—of directed line segments, n-tuples, even *functions*—while the second is invariably a n-tuple of real (or complex) numbers of the *components* of $\boldsymbol{v}$ relative to a given basis.

The issue which will be addressed in this section is this: Consider an arbitrary vector $\boldsymbol{v}$ in an n-dimensional space spanned by two *different, ordered* bases, say $(\boldsymbol{u}_1, \boldsymbol{u}_2, \ldots, \boldsymbol{u}_n)$ and $(\boldsymbol{u}'_1, \boldsymbol{u}'_2, \ldots, \boldsymbol{u}'_n)$. These two bases give rise to two *different* representations, though the *original vector remains unchanged*:

$$\boldsymbol{v} = \sum_{i=1}^{n} v_i \boldsymbol{u}_i = \sum_{i=1}^{n} v'_i \boldsymbol{u}'_i,$$

whose representations can be expressed as two different n-tuples:

$$\mathsf{v} \equiv \begin{pmatrix} v_1 \\ v_2 \\ \vdots \\ v_n \end{pmatrix} \neq \mathsf{v}' \equiv \begin{pmatrix} v'_1 \\ v'_2 \\ \vdots \\ v'_n \end{pmatrix}.$$

Knowing the relation between the $(\boldsymbol{u}_i)$ and the $(\boldsymbol{u}'_i)$, the question is: What is the relation between v and v'? Furthermore, we have already pointed out (page 67) that the *representation* of an arbitrary linear transformation *also changes when we change the basis for vectors on which the transformation acts*; thus we are also interested in how, given such a transformation [representation] A applied to vectors in the $(\boldsymbol{u}_i)$, what is the *new* representation, A', relative to $(\boldsymbol{u}'_i)$?

Actually, we have already dealt with how the *vector* representations change in Example 2.2.5. But, as remarked at the end of that example, we considered there the representation of the original basis in terms of the new basis, whereas typically one represents the *new* in terms of the *old*. Nonetheless, essentially what we do here is just what we did in the original example, and, as demonstrated by that example, the effect of a change of basis on a vector's representation is, in fact, given by a linear transformation. We shall consider two instances of basis transformation: *arbitrary* basis transformations, and the special case of those resulting from a *rotation*.

3.6.1 General Basis Transformations

We pointed out on page 67 that the effect of *any* linear transformation is really defined by what it does to the *basis* vectors; thus it is natural to determine the transformation that relates the *representations* of the basis vectors $(\boldsymbol{u}_i)$ and $(\boldsymbol{u}'_i)$. Since both sets of vectors lie in the same space, we can use either set to express the other; let us express the primed basis in terms of the *un*primed one: We can write each $\boldsymbol{u}'_i$ in terms of the $(\boldsymbol{u}_i)$:

$$\boldsymbol{u}'_i = u'_{1i}\boldsymbol{u}_1 + u'_{2i}\boldsymbol{u}_2 + \cdots + u'_{ni}\boldsymbol{u}_n = \sum_{j=1}^{n} u'_{ji}\boldsymbol{u}_j; \tag{3.6-1}$$

i.e., the *representation* of the basis vector $\boldsymbol{u}'_i$ *relative to the basis* $(\boldsymbol{u}_i)$

$$\mathbf{u}'_i = \begin{pmatrix} u'_{1i} \\ u'_{2i} \\ \vdots \\ u'_{ni} \end{pmatrix} \equiv \mathbf{u}'_i(\boldsymbol{u}). \tag{3.6-2}$$

This business of the representations is important. Recall (page 36) that the representation of $\boldsymbol{u}'_i$ itself *in its own basis* is *always* given by equation (1.3-4), repeated here for reference:

$$i \rightarrow \begin{pmatrix} 0 \\ 0 \\ \vdots \\ 0 \\ 1 \\ 0 \\ \vdots \\ 0 \end{pmatrix} = \mathbf{u}'_i, \tag{3.6-3}$$

which could be written $\mathbf{u}'_i(\boldsymbol{u}')$ in the notation of (3.6-2).

We now carry out the representations (3.6-2) for *all* the $\boldsymbol{u}'_i$ and form the matrix obtained by catenating these column n-matrix representations side-by-side:

$$\mathsf{U} \equiv (\mathsf{u}'_1(\boldsymbol{u})\ \mathsf{u}'_2(\boldsymbol{u})\ \ldots\ \mathsf{u}'_n(\boldsymbol{u})) \tag{3.6-4a}$$

$$= \begin{pmatrix} u'_{11} & u'_{12} & \cdots & u'_{1n} \\ u'_{21} & u'_{22} & \cdots & u'_{2n} \\ \vdots & \vdots & \cdots & \vdots \\ u'_{n1} & u'_{n2} & \cdots & u'_{nn} \end{pmatrix} \tag{3.6-4b}$$

Note that U *is simply the matrix of* $\mathsf{u}'_i(\boldsymbol{u})$*—the representations of the* $\boldsymbol{u}'_i$ *in terms of the* $(\boldsymbol{u}_i)$. We also observe a rather remarkable relation that exists between the *representations* u'_i and the u_i *in the* $\{\boldsymbol{u}\}$ *basis*:

$$\mathsf{u}'_i(\boldsymbol{u}) = \mathsf{U}\mathsf{e}_i = \mathsf{U}\mathsf{u}_i(\boldsymbol{u}) \tag{3.6-5}$$

(where we have applied (3.6-3) to the $\mathsf{u}_i(\boldsymbol{u})$ in the latter equation); note that this relates [the representations of] two *different vectors'*, $\boldsymbol{u}'_i$ and $\boldsymbol{u}_i$, in the *same basis* of the $(\boldsymbol{u}_i)$—the very opposite of what we are trying to do, namely relate the representations of the *same vector* $\boldsymbol{v}$ in two *different bases* $(\boldsymbol{u}_i)$ and $(\boldsymbol{u}'_i)$! But it holds nonetheless.

We now effect what we *are* trying to do: For [arbitrary] vector $\boldsymbol{v}$

$$\boldsymbol{v} = \sum_{j=1}^{n} v_j \boldsymbol{u}_j = \sum_{i=1}^{n} v'_i \boldsymbol{u}'_i = \sum_{i=1}^{n} v'_i \left(\sum_{j=1}^{n} u'_{ji} \boldsymbol{u}_j \right) = \sum_{j=1}^{n} \left(\sum_{i=1}^{n} u'_{ji} v'_i \right) \boldsymbol{u}_j.$$

Setting the coefficients of $\boldsymbol{u}_j$ in the second and last expressions equal, we get immediately that $v_j = \sum_{i=1}^{n} u'_{ji} v'_i$; but, by the definition (2.1-5), this is nothing more than the jth row of the matrix product $\mathsf{U}\mathsf{v}'$. Thus we have that

$$\mathsf{v} = \mathsf{U}\mathsf{v}'.$$

We note, however, that since the $(\boldsymbol{u}'_i)$ form a basis, they are linearly independent; thus their *representations* are (by Theorem 1.3.2); thus the *rank* of U is n; thus, by Theorem 3.1.5, U has an inverse, and we can write

$$\mathsf{v}' = \mathsf{U}^{-1}\mathsf{v} \tag{3.6-6}$$

—the form in which we will normally use the *basis transformation matrix*, U. Note that by (3.6-4), U is the matrix whose columns are the *representations* of the *new* basis vectors in terms of the old, the order in which such transformations are usually specified.

At first blush, we expect that we should be able to apply this to *any* vectors, including the basis vectors themselves; in particular, then, we should be able to write $\mathsf{u}'_i = \mathsf{U}^{-1}\mathsf{u}_i$. (We *can*!) But this seems to run directly counter to (3.6-5)! (It *does*.) This is an error which can result from the thoughtless application of mere equations without consideration of what they *mean.* Once again: (3.6-5) relates the representations of two *different [basis] vectors*, $\boldsymbol{u}$ and $\boldsymbol{u}'$ in the *same basis*; (3.6-6) relates the representations of the *same vector* $\boldsymbol{v}$ in two *different bases*! Thus, for example, the [correct] equation $\mathbf{u}_i' = \mathbf{U}^{-1}\mathbf{u}_i$ would relate the *representations* $\mathbf{u}_i'$ and $\mathbf{u}_i$ of the *same* basis vector $\boldsymbol{u}_i$ in the ("new" and "old") $\{\boldsymbol{u}_i'\}$ and $\{\boldsymbol{u}_i\}$ bases respectively. (Note the distinctions implied by the notations "$\boldsymbol{v}$" and "$\mathbf{v}$" throughout this paragraph!). The student must keep this critical difference in mind.

With this in hand, we are now ready to determine a similar relation holding between the *representations* of a *[single] linear transformation* in the two different bases. Once again, however, we must specify exactly what we mean by this: If a matrix $\mathbf{A}$ [*representing* a linear transformation] is applied to a *representation* of the vector $\boldsymbol{v}$, $\mathbf{v}$, relative to the basis $(\boldsymbol{u}_i)$, we seek that *representation* $\mathbf{A}'$ such that, when it is applied to the *representation*, $\mathbf{v}'$, of [the *same*] $\boldsymbol{v}$ in the *new* basis $(\boldsymbol{u}_i')$, it will return that *representation* in the new basis corresponding to the *same* vector as $\mathbf{Av}$; in other words, if the representation $\mathbf{v} \to \mathbf{v}'$, we want the new representation $\mathbf{A} \to \mathbf{A}'$ such that $(\mathbf{Av}) \to (\mathbf{Av})' \equiv \mathbf{A}'\mathbf{v}'$. But, by (3.6-6), the last is simply $\mathbf{U}^{-1}(\mathbf{Av}) = \mathbf{U}^{-1}\mathbf{Av}$ by the associative property of matrix multiplication. Thus we require that $\mathbf{A}'\mathbf{v}' = \mathbf{A}'\mathbf{U}^{-1}\mathbf{v} = \mathbf{U}^{-1}\mathbf{Av}$ hold for *any* $\mathbf{v}$, so that

$$\mathbf{U}\mathbf{A}'\mathbf{U}^{-1} = \mathbf{A}$$

or

$$\mathbf{A}' = \mathbf{U}^{-1}\mathbf{A}\mathbf{U}. \tag{3.6-7}$$

A matrix transformation of the form (3.6-7) is called a *similarity transformation*, and we have shown that the relation between the *representations* of an arbitrary linear transformation in two different bases is given by such a transformation. Conversely, any relation between two representations of a linear transformation given by a similarity transformation can be *interpreted* as one resulting *from* a basis transformation given by the matrix $\mathbf{U}$, whose columns are nothing more than the *representations* of the *new* basis in terms of the *old.* We shall have occasion to implement just such an interpretation when we diagonalize matrices.

Homework:

1. Show that the determinant of a matrix under a similarity transformation remains unchanged: $|\mathbf{U}^{-1}\mathbf{A}\mathbf{U}| = |\mathbf{A}|$.

Let us make two general observations concerning this:

- If there is a sequence of *two* basis transformations, say $\{\boldsymbol{u}\} \to \{\boldsymbol{u}'\}$, generated by $\mathbf{U}$, and then another such that $\{\boldsymbol{u}'\} \to \{\boldsymbol{u}''\}$, generated by $\mathbf{U}'$, the first transformation would give $\mathbf{v}' = \mathbf{U}^{-1}\mathbf{v}$ while the second would give $\mathbf{v}'' = (\mathbf{U}')^{-1}\mathbf{v}' = (\mathbf{U}')^{-1}\mathbf{U}^{-1}\mathbf{v} = (\mathbf{U}\mathbf{U}')^{-1}\mathbf{v}$; thus the *composite* transformation would be generated by the matrix $\mathbf{U}\mathbf{U}'$ (which, if you think about it, seems *backwards*!).

- The above transformation formulae were predicated on having specified the *new* basis vectors in terms of the *old*; that's where the equation $\mathbf{U} \equiv (\mathbf{u}'_1(u)\ \mathbf{u}'_2(u)\ \ldots\ \mathbf{u}'_n(u))$ comes from. If we were given the transformation in the *opposite* order—*old* basis vectors in terms of the *new*—we would have the *inverse* of the above $\mathbf{U}$ matrix, $\mathbf{U}' \equiv (\mathbf{u}_1(u')\ \mathbf{u}_2(u')\ \ldots\ \mathbf{u}_n(u')) = \mathbf{U}^{-1}$, and the equations would then be $v' = \mathbf{U}'v$ and $\mathbf{A}' = \mathbf{U}'\mathbf{A}(\mathbf{U}')^{-1}$.

Successive Basis Transformations—Composition

We have already made explicit the confusion which might arise between equations (3.6-5) and (3.6-6), repeated here for reference:

$$\mathbf{v}' = \mathbf{U}\mathbf{v} \tag{3.6-5'}$$

$$\mathbf{v}' = \mathbf{U}^{-1}\mathbf{v}, \tag{3.6-6}$$

where we have intentionally written the first equation in a form closer to that of the second in order to highlight the similarities. We have also indicated that the former relates the representation of *two different vectors in the same basis*, whereas the latter relates the representations of the *same vector in two different bases.* We observe, too, that the first can be interpreted as giving a *new vector* [representation] resulting from the application of the matrix $\mathbf{U}$ to $\mathbf{v}$—an "active" transformation, carrying one vector to another; as opposed to the second, in which the vector v remains "passive" under a change of basis used to represent it. But it is beneficial, at this juncture, to examine these two apparently contradictory equations in more detail, particularly as regards the composition of such transformations:

Though the above relations have been derived through direct appeal to the matrices involved, we must not lose sight of the fact that this chapter itself is actually about "linear transformations." The matrices are merely introduced to *represent* such linear transformations, using Theorem 2.2.2; and such representations depend on the basis which they reference. From this perspective, let us now examine these equations.

In the first place, we might better emphasize the "two vector" nature of (3.6-5′) by writing it in the form

$$\mathbf{v}_2(u) = \mathbf{U}(u)\mathbf{v}_1(u), \tag{3.6-8}$$

where the arguments of the various matrix representations (as in (3.6-6)) invoke the common [incorrect![4]] notation to signify "the variables on which the function depends." Here they suggest the basis relative to which the respective representations are cast and emphasize that the underlying basis remains the same for both vectors. On the other hand, (3.6-6) retains its notation, with only the addition of the arguments:

$$\mathbf{v}'(u') = \mathbf{U}^{-1}(u)\mathbf{v}(u). \tag{3.6-9}$$

[4]It actually means the *value* of the function at that value of argument.

Observe, in both of the above expressions for a *single* transformation, that *all quantities on each side of the equals are expressed in the same basis.* Furthermore, in both cases we utilize the same transformation matrix, $\mathbf{U}$, describing the basis transformation from $(\boldsymbol{u}_i) \rightarrow (\boldsymbol{u}'_i)$, equation (3.6-4):

$$\mathbf{U}(\boldsymbol{u}) \equiv (\mathbf{u}'_1(\boldsymbol{u})\ \mathbf{u}'_2(\boldsymbol{u})\ \ldots\ \mathbf{u}'_n(\boldsymbol{u})).$$

Now let us consider the simple composition of two ("passive") basis transformations, already discussed above, but looking more closely at the bases involved in the representations: In addition to the above transformation $(\boldsymbol{u}_i) \rightarrow (\boldsymbol{u}'_i)$, we now make a second one $(\boldsymbol{u}'_i) \rightarrow (\boldsymbol{u}''_i)$ with associated matrix $\mathbf{U}'$:

$$\mathbf{U}'(\boldsymbol{u}') \equiv (\mathbf{u}''_1(\boldsymbol{u}')\ \mathbf{u}''_2(\boldsymbol{u}')\ \ldots\ \mathbf{u}''_n(\boldsymbol{u}')).$$

Here

$$\mathbf{v}''(\boldsymbol{u}'') = (\mathbf{U}'(\boldsymbol{u}'))^{-1}\mathbf{v}'(\boldsymbol{u}'),$$

—now all bases on the right-hand side involve the $(\boldsymbol{u}')$—and thus, substituting in (3.6-9) (which gives precisely the *representation* $\mathbf{v}'(\boldsymbol{u}')$)

$$\begin{aligned} &= (\mathbf{U}'(\boldsymbol{u}'))^{-1}\mathbf{U}^{-1}(\boldsymbol{u})\mathbf{v}(\boldsymbol{u}) \\ &= (\mathbf{U}\mathbf{U}')^{-1}\mathbf{v}, \end{aligned}$$

as already obtained before.[5]

But now regard a second "active" transformation resulting from the [same] $\mathbf{U}'$ above:

$$\mathbf{v}_3(\boldsymbol{u}) = \mathbf{U}'(\boldsymbol{u}')\mathbf{v}_2(\boldsymbol{u}). \tag{3.6-10}$$

It is enticing not to track the bases involved at each step, merely substituting into $\mathbf{v}_3 = \mathbf{U}'\mathbf{v}_2$ that $\mathbf{v}_2 = \mathbf{U}\mathbf{v}_1$, to get $\mathbf{v}_3 = \mathbf{U}'\mathbf{U}\mathbf{v}_1$. Unfortunately, this approach is too simplistic, because the right-hand side of (3.6-10) (representing a *single* transformation, just like (3.6-8)) does *not* involve just the $(\boldsymbol{u}_i)$ but also the $(\boldsymbol{u}'_1)$ through $\mathbf{U}'$! In order to bring these representations to the *same* basis $(\boldsymbol{u}_i)$ ($\mathbf{v}_2(\boldsymbol{u})$ already necessarily represented in that basis), we must transform $\mathbf{U}'(\boldsymbol{u}')$ *to* that basis—effectively re-representing $\mathbf{U}'$ resulting from the ["*passive*"] transformation $(\boldsymbol{u}'_i) \rightarrow (\boldsymbol{u}_i)$, the *inverse* of $(\boldsymbol{u}_i) \rightarrow (\boldsymbol{u}'_i)$: by (3.6-7),

$$\mathbf{U}'(\boldsymbol{u}) = \mathbf{U}\mathbf{U}'\mathbf{U}^{-1},$$

so that

$$\mathbf{v}_3(\boldsymbol{u}) = \mathbf{U}'(\boldsymbol{u})\mathbf{v}_2(\boldsymbol{u}) = (\mathbf{U}\mathbf{U}'\mathbf{U}^{-1})\mathbf{v}_2(\boldsymbol{u})$$

[5]Though it might appear that the first of the above equations involves a mismatch of the bases, referencing both the $(\boldsymbol{u})$ and the $(\boldsymbol{u}')$, this, unlike (3.6-8), involves *more* than a single transformation.

into which we can substitute (3.6-8):

$$= (\mathsf{UU'U^{-1}})\mathsf{U}\mathsf{v}_1(u)$$
$$= (\mathsf{UU'})\mathsf{v}_1(u).$$

We note that, in the same way that the matrices representing single "active" and "passive" transformations are inverses of one another, those representing the *compose* of such transformations, $\mathsf{UU'}$ and $(\mathsf{UU'})^{-1}$, are also.

We now give two examples of such basis transformations. The second, presenting basis *rotations*, is sufficiently important to warrant its own section. The first, though relatively more routine, deals with the vector space of n-tuples; the critical distinction between such n-tuple *vectors* and their [n-tuple] *representations* must be clearly observed, as well just *which* basis all calculations are being done in terms of, or the example will be more confusing than elucidating!

Example 3.6.1. Consider two bases in the space of triples:

$$(\boldsymbol{u}_1, \boldsymbol{u}_2, \boldsymbol{u}_3): \quad \boldsymbol{u}_1 = \begin{pmatrix}1\\1\\0\end{pmatrix}, \quad \boldsymbol{u}_2 = \begin{pmatrix}0\\1\\-1\end{pmatrix}, \quad \boldsymbol{u}_3 = \begin{pmatrix}0\\0\\1\end{pmatrix}$$

$$(\boldsymbol{u}'_1, \boldsymbol{u}'_2, \boldsymbol{u}'_3): \quad \boldsymbol{u}'_1 = \begin{pmatrix}2\\0\\1\end{pmatrix}, \quad \boldsymbol{u}'_2 = \begin{pmatrix}0\\1\\0\end{pmatrix}, \quad \boldsymbol{u}'_3 = \begin{pmatrix}0\\1\\1\end{pmatrix}$$

—each of which is expressed in terms of the *natural basis.*

We note first that each of these sets of vectors truly *is* a basis: the constituent vectors span the space and are *linearly independent* by Theorem 3.1.5:

$$|\boldsymbol{u}_1\ \boldsymbol{u}_2\ \boldsymbol{u}_3| = \begin{vmatrix}1 & 0 & 0\\1 & 1 & 0\\1 & -1 & 1\end{vmatrix} = 1 \neq 0; \quad |\boldsymbol{u}'_1\ \boldsymbol{u}'_2\ \boldsymbol{u}'_3| = \begin{vmatrix}2 & 0 & 0\\0 & 1 & 1\\1 & 0 & 1\end{vmatrix} = 2 \neq 0. \quad (3.6\text{-}11)$$

Now consider a vector in the [original] n-tuple space [using the *natural* basis],

$$\boldsymbol{v} = \begin{pmatrix}2\\3\\-1\end{pmatrix}.$$

We find its *representations*—yet other triples—in terms of the two *new* bases. Considering the first basis, we wish to find the $\{v_i\}$ such that $\boldsymbol{v} = v_1\boldsymbol{u}_1 + v_2\boldsymbol{u}_2 + v_3\boldsymbol{u}_3$. Expressing both $\boldsymbol{v}$ and the $\{\boldsymbol{u}_i\}$ *in the natural basis*, then, we wish

$$\boldsymbol{v} = \begin{pmatrix}2\\3\\-1\end{pmatrix} = v_1\begin{pmatrix}1\\1\\0\end{pmatrix} + v_2\begin{pmatrix}0\\1\\-1\end{pmatrix} + v_3\begin{pmatrix}0\\0\\1\end{pmatrix} = \begin{pmatrix}v_1\\v_1+v_2\\-v_2+v_3\end{pmatrix}$$

Note that this can be written as the linear system

$$\begin{pmatrix}1 & 0 & 0\\1 & 1 & 0\\0 & -1 & 1\end{pmatrix}\begin{pmatrix}v_1\\v_2\\v_3\end{pmatrix} = \begin{pmatrix}2\\3\\-1\end{pmatrix},$$

whose solution exists if and only if

$$\begin{vmatrix} 1 & 0 & 0 \\ 1 & 1 & 0 \\ 0 & -1 & 1 \end{vmatrix} \neq 0$$

(see Section 3.4)—the condition included in equation (3.6-11) above. Solving the separate components—three equations in the v_i—for these variables we get that $v_1 = 2$, $v_2 = 1$, and $v_3 = 0$. Similarly, for the $\{v'_i\}$,

$$\boldsymbol{v} = \begin{pmatrix} 2 \\ 3 \\ -1 \end{pmatrix} = v'_1 \begin{pmatrix} 2 \\ 0 \\ 1 \end{pmatrix} + v'_2 \begin{pmatrix} 0 \\ 1 \\ 0 \end{pmatrix} + v'_3 \begin{pmatrix} 0 \\ 1 \\ 1 \end{pmatrix} = \begin{pmatrix} 2v'_1 \\ v'_2 + v'_3 \\ v'_1 + v'_3 \end{pmatrix},$$

from which $v'_1 = 1$, $v'_2 = 5$, and $v'_3 = -2$. Thus the two representations are, respectively,

$$\mathsf{v} = \begin{pmatrix} 2 \\ 1 \\ 0 \end{pmatrix}, \quad \mathsf{v}' = \begin{pmatrix} 1 \\ 5 \\ -2 \end{pmatrix}.$$

Perhaps a little more should be said regarding the role of the natural basis in this step. In order even to be able to write the above equations for the determination of the $\{v_i\}$ and $\{v'_i\}$ in the first place, *both sides of the equation had to be expressed in terms of the same basis.* This is because what we are really doing here is to conclude, from

$$v_1\boldsymbol{u}_1 + v_2\boldsymbol{u}_2 + \ldots + v_n\boldsymbol{u}_n = a_1\boldsymbol{u}_1 + a_2\boldsymbol{u}_2 + \ldots + a_n\boldsymbol{u}_n$$

that $v_i = a_i$ for all i; this, in turn, can only be done in the assurance that both sides are expressed relative to the *same* basis. In the present case, the natural basis is that in terms of which both the vector and the two bases are expressed in the first place; it becomes the "basis of last resort" in such problems.

We now find the transformation matrix $\mathsf{U} \equiv (\mathsf{u}'_1(\boldsymbol{u})\ \mathsf{u}'_2(\boldsymbol{u})\ \ldots\ \mathsf{u}'_n(\boldsymbol{u}))$ between the two bases; thus we seek the *representations* of the $\{\boldsymbol{u}'_i\}$ in terms of the $\{\boldsymbol{u}_i\}$; *i.e.*, the $\{u'_{ij}\}$ satisfying equation (3.6-1):

$$\boldsymbol{u}'_i = \sum_{j=1}^{n} u'_{ji}\boldsymbol{u}_j = u'_{1i} \begin{pmatrix} 1 \\ 1 \\ 0 \end{pmatrix} + u'_{2i} \begin{pmatrix} 0 \\ 1 \\ -1 \end{pmatrix} + u'_{3i} \begin{pmatrix} 0 \\ 0 \\ 1 \end{pmatrix} = \begin{pmatrix} u'_{1i} \\ u'_{1i} + u'_{2i} \\ -u'_{2i} + u'_{3i} \end{pmatrix}$$

(Again, all vectors are expressed in terms of the *original*, natural basis.) For $i = 1$, this equation becomes

$$\boldsymbol{u}'_1 = \begin{pmatrix} 2 \\ 0 \\ 1 \end{pmatrix} = \begin{pmatrix} u'_{11} \\ u'_{11} + u'_{21} \\ -u'_{21} + u'_{31} \end{pmatrix},$$

from which $u'_{11} = 2$, $u'_{21} = -2$, and $u'_{31} = -1$. Proceeding similarly for $\boldsymbol{u}'_2$ and $\boldsymbol{u}'_3$ we get $u'_{12} = 0$, $u'_{22} = 1$, and $u'_{32} = 1$ and $u'_{13} = 0$, $u'_{23} = 1$, and $u'_{33} = 2$.

Thus

$$\mathsf{U} \equiv (\mathbf{u}_1'(\boldsymbol{u})\ \mathbf{u}_2'(\boldsymbol{u})\ \ldots\ \mathbf{u}_n'(\boldsymbol{u})) = \begin{pmatrix} 2 & 0 & 0 \\ -2 & 1 & 1 \\ -1 & 1 & 2 \end{pmatrix}$$

—that matrix whose columns are just the *components* of the $\boldsymbol{u}_i'$ relative to the $\boldsymbol{u}_i$.

Having determined U, we can verify equation (3.6-6), or, equivalently, the equation from which it came, $\mathsf{v} = \mathsf{U}\mathsf{v}'$:

$$\mathsf{U}\mathsf{v}' = \begin{pmatrix} 2 & 0 & 0 \\ -2 & 1 & 1 \\ -1 & 1 & 2 \end{pmatrix} \begin{pmatrix} 1 \\ 5 \\ -2 \end{pmatrix} = \begin{pmatrix} 2 \\ 1 \\ 0 \end{pmatrix} = \mathsf{v}$$

—as expected. Note here that we deal here with the *representations* of $\boldsymbol{v}$ in the respective, *different* bases. We can also verify (3.6-6) directly: Determining U^{-1} using the procedure in Example 3.1.1 (as opposed to the result in Theorem 3.2.11!),

$$\mathsf{U}^{-1} = \begin{pmatrix} \frac{1}{2} & 0 & 0 \\ \frac{3}{2} & 2 & -1 \\ -\frac{1}{2} & -1 & 1 \end{pmatrix}.$$

(It can be verified—in fact, is generally a good idea *to* verify after the calculation of the inverse!—that $\mathsf{U}\mathsf{U}^{-1} = \mathbf{1}$.) thus

$$\mathsf{U}^{-1}\mathsf{v} = \begin{pmatrix} \frac{1}{2} & 0 & 0 \\ \frac{3}{2} & 2 & -1 \\ -\frac{1}{2} & -1 & 1 \end{pmatrix} \begin{pmatrix} 2 \\ 1 \\ 0 \end{pmatrix} = \begin{pmatrix} 1 \\ 5 \\ -2 \end{pmatrix} = \mathsf{v}'$$

Finally we consider the representations of a linear transformation in the two bases. Consider that transformation, represented by a matrix *in the* $\{\boldsymbol{u}\}$ *basis*

$$\mathsf{A} = \begin{pmatrix} 2 & 8 & 2 \\ -4 & 0 & 2 \\ 0 & 6 & 4 \end{pmatrix}$$

This transforms $\boldsymbol{v}$—more correctly, its *representation* v—into a vector *in* $\{\boldsymbol{u}\}$

$$\mathsf{w} \equiv \mathsf{A}\mathsf{v} = \begin{pmatrix} 2 & 8 & 2 \\ -4 & 0 & 2 \\ 0 & 6 & 4 \end{pmatrix} \begin{pmatrix} 2 \\ 1 \\ 0 \end{pmatrix} = \begin{pmatrix} 12 \\ -8 \\ 6 \end{pmatrix}.$$

But, under the basis transformation $\{\boldsymbol{u}\} \rightarrow \{\boldsymbol{u}'\}$ generated by U, $\mathsf{A} \rightarrow \mathsf{A}'$ given by (3.6-7):

$$\mathsf{A}' = \mathsf{U}^{-1}\mathsf{A}\mathsf{U} = \begin{pmatrix} \frac{1}{2} & 0 & 0 \\ \frac{3}{2} & 2 & -1 \\ -\frac{1}{2} & -1 & 1 \end{pmatrix} \begin{pmatrix} 2 & 8 & 2 \\ -4 & 0 & 2 \\ 0 & 6 & 4 \end{pmatrix} \begin{pmatrix} 2 & 0 & 0 \\ -2 & 1 & 1 \\ -1 & 1 & 2 \end{pmatrix}$$

$$= \begin{pmatrix} -7 & 5 & 6 \\ -25 & 9 & 12 \\ 1 & 3 & 4 \end{pmatrix}$$

We claim that $\mathbf{A}'$ transforms the $\{u'\}$ *representation*, $\mathbf{v}'$, into $\mathbf{w}' = \mathbf{A}'\mathbf{v}'$ which is exactly the *same* vector represented by $\mathbf{w}$ in $\{u\}$; indeed,

$$\mathbf{w}' = \mathbf{A}'\mathbf{v}' = \begin{pmatrix} -7 & 5 & 6 \\ -25 & 9 & 12 \\ 1 & 3 & 4 \end{pmatrix} \begin{pmatrix} 1 \\ 5 \\ -2 \end{pmatrix} = \begin{pmatrix} 6 \\ -4 \\ 8 \end{pmatrix}$$

But, according to (3.6-6), $\mathbf{w}$ and $\mathbf{w}'$ should be related by

$$\mathbf{w}' = \mathbf{U}^{-1}\mathbf{w} = \begin{pmatrix} \frac{1}{2} & 0 & 0 \\ \frac{3}{2} & 2 & -1 \\ -\frac{1}{2} & -1 & 1 \end{pmatrix} \begin{pmatrix} 12 \\ -8 \\ 6 \end{pmatrix} = \begin{pmatrix} 6 \\ -4 \\ 8 \end{pmatrix}$$

—as expected.

The simultaneous solution for the u_{ij} in the determination of $\mathbf{U}$ here was necessitated by the fact that, in the interests of generality, we expressed both bases relative to yet a *third* (the natural basis); had we had the basis transformation giving the new directly in terms of the old—the form of equation 3.6-1—we could have read off the entries in $\mathbf{U}$ directly; *e.g.*, if we knew that

$$\begin{aligned} u_1' &= 3u_1 + 5u_2 + 8u_3 \\ u_2' &= -2u_1 + 5u_2 + 3u_3 \\ u_3' &= u_1 + 8u_2 - 2u_3, \end{aligned}$$

then the transformation matrix would be simply

$$\mathbf{U} = \begin{pmatrix} 3 & -2 & 1 \\ 5 & 5 & 8 \\ 8 & 3 & -2 \end{pmatrix}.$$

This is usually the form of the basis transformation given. |

Homework:

1. Consider the pairs of [column matrix] vectors (given in the natural basis)

$$u_1 = \begin{pmatrix} 1 \\ 0 \end{pmatrix}, \qquad u_2 = \begin{pmatrix} 1 \\ 1 \end{pmatrix},$$

$$u_1' = \begin{pmatrix} 2 \\ 1 \end{pmatrix}, \qquad u_2' = \begin{pmatrix} 0 \\ -1 \end{pmatrix}.$$

   (a) Show the pairs in each set are linearly independent.

   (b) Express the vector $V \equiv \begin{pmatrix} 0 \\ 1 \end{pmatrix}$ (again, given in the natural basis) in terms of both the $\{u_i\}$ and the $\{u_i'\}$.

   Ans: $\mathbf{V} = \begin{pmatrix} -1 \\ 1 \end{pmatrix}, \quad \mathbf{V}' = \begin{pmatrix} 0 \\ -1 \end{pmatrix}.$

(c) Find the matrix $\mathbf{U} = (t_{ij})$ such that

$$\boldsymbol{u}'_j = \Sigma\, t_{ij}\boldsymbol{u}_i$$

and its inverse, $\mathbf{U}^{-1}$.

Ans: $\mathbf{U} = \begin{pmatrix} 1 & 1 \\ 1 & -1 \end{pmatrix}, \quad \mathbf{U}^{-1} = \frac{1}{2}\mathbf{U}.$

(d) Show that, indeed, $\mathbf{V} = \mathbf{U}\mathbf{V}'$ and $\mathbf{V}' = \mathbf{U}^{-1}\mathbf{V}$.

(e) Consider the linear transformation on vectors in the $\{\boldsymbol{u}_i\}$ basis:

$$\mathbf{A} = \begin{pmatrix} 1 & -5 \\ -2 & 3 \end{pmatrix}.$$

Find its representation, $\mathbf{A}'$, in the $\{\boldsymbol{u}'_i\}$ basis, and verify that $\mathbf{A}'$ takes $\mathbf{V}'$ into the same vector (in the $\{\boldsymbol{u}'_i\}$ basis) that $\mathbf{A}$ takes $\mathbf{V}$ to.

(f) Now consider a *second* transformation, $\mathbf{U}'$: $\{\boldsymbol{u}'_i\} \to \{\mathbf{u}''_i\}$, generated by the matrix

$$\mathbf{U}' = \begin{pmatrix} 2 & -1 \\ 1 & 3 \end{pmatrix}.$$

(*Note:* this is represented *in* the $\{\boldsymbol{u}'_i\}$ basis!)

i. Find the vectors $\boldsymbol{u}''_1$ and $\boldsymbol{u}''_2$

A. in terms of the $\{\boldsymbol{u}'_i\}$ basis, and

B. in terms of the natural basis.

ii. Find $\mathbf{V}''$, the representation in the $\{\boldsymbol{u}''_i\}$ basis of $\boldsymbol{V}$, and check that $\mathbf{V}' = \mathbf{U}'\mathbf{V}''$ holds.

iii. Show that, under the transformation generated by $\mathbf{U}'' \equiv \mathbf{U}\mathbf{U}'$,

A. the $\{\boldsymbol{u}_i\}$ are transformed to the $\{\boldsymbol{u}''_i\}$ [*i.e.*, $\mathbf{u}_i = \mathbf{U}''\mathbf{u}''_i$—*be careful what this means!*]; thus that $\mathbf{U}''$ is precisely the transformation from the $\{\boldsymbol{u}_i\}$ to the $\{\boldsymbol{u}''_i\}$.

B. Verify that the *coordinate* transformation satisfies $\mathbf{V} = \mathbf{U}''\mathbf{V}''$ as expected. Note that this means that $\mathbf{V}'' = (\mathbf{U}\mathbf{U}')^{-1}\mathbf{V} = \mathbf{U}'^{-1}\mathbf{U}^{-1}\mathbf{V}$, which is consistent with the coordinate transformations done separately: $\mathbf{V}' = \mathbf{U}^{-1}\mathbf{V}, \mathbf{V}'' = \mathbf{U}'^{-1}\mathbf{V}'$.

2. Find that transfomation taking $\{\hat{\imath},\hat{\jmath},\hat{k}\}$ into the vectors $\{\hat{\imath}+2\hat{\jmath}$,-$\hat{\imath}+3\hat{k},\hat{\jmath}\}$.

3.6.2 Basis Rotations

As mentioned above, rotations are sufficiently important a topic that they are given their own section here. The very concept of "rotation" virtually demands that we be dealing with the vector space of directed line segments; their application in almost invariably to *Cartesian* vectors $\hat{\imath}$, $\hat{\jmath}$, and $\hat{k}$. Thus we will first develop rotations relative to such an orthonormal set.

Consider first the rotation to a new basis $(\hat{\imath}', \hat{\jmath}', \hat{k}')$ resulting from rotation through an angle $(+\alpha)$—positive in the *right-hand* sense about the [old] $\hat{\imath}$ direction—about the x-axis: From the figure—note that we are looking "down" the positive x-axis, defining the positive sense of rotation—we see that the new basis vectors are related to the old by

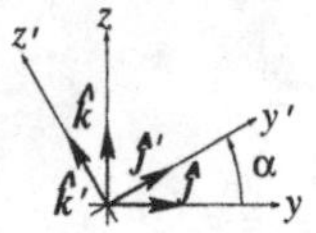

$$\hat{\imath}' = \hat{\imath}$$
$$\hat{\jmath}' = \cos\alpha\hat{\jmath} + \sin\alpha\hat{k}$$
$$\hat{k}' = -\sin\alpha\hat{\jmath} + \cos\alpha\hat{k}.$$

Thus we have immediately that the vectors transform representations

$$\mathsf{v} = \mathsf{A}(\alpha)\,\mathsf{v}' = \begin{pmatrix} 1 & 0 & 0 \\ 0 & \cos\alpha & -\sin\alpha \\ 0 & \sin\alpha & \cos\alpha \end{pmatrix} \mathsf{v}' \tag{3.6-12}$$

We claim that this matrix is, in fact, *orthogonal.* In view of our discussion of such matrices as being tantamount to rotations, this should come as no surprise, but it can be shown directly: noting that the reverse rotation through $(-\alpha)$ is just the *inverse* of this matrix:

$$\mathsf{A}^{-1}(\alpha) = \mathsf{A}(-\alpha) = \begin{pmatrix} 1 & 0 & 0 \\ 0 & \cos\alpha & \sin\alpha \\ 0 & -\sin\alpha & \cos\alpha \end{pmatrix} = \mathsf{A}^{\mathsf{T}}(\alpha),$$

we see that A satisfies the definition of such matrices. But this means that, in the form in which we normally express the new representation, $\mathsf{v}' = \mathsf{A}^{-1}(\alpha)\mathsf{v} = \mathsf{A}^{\mathsf{T}}(\alpha)\mathsf{v}$. We note too that $|\mathsf{A}(\alpha)| = +1$; *i.e.*, this corresponds to a *proper* rotation.

In a similar fashion, it is easy to show that, corresponding to a rotation about the y-axis through $(+\beta)$, we get a rotation matrix

$$\mathsf{B}(\beta) \equiv \begin{pmatrix} \cos\beta & 0 & \sin\beta \\ 0 & 1 & 0 \\ -\sin\beta & 0 & \cos\beta \end{pmatrix}, \tag{3.6-13}$$

and about the z-axis through $(+\gamma)$,

$$\mathsf{C}(\gamma) \equiv \begin{pmatrix} \cos\gamma & -\sin\gamma & 0 \\ \sin\gamma & \cos\gamma & 0 \\ 0 & 0 & 1 \end{pmatrix}. \tag{3.6-14}$$

Both are also orthogonal matrices generating proper rotations. The *new* representations will be given by the *transposes* of these matrices applied to the old representations.

Homework:

1. Derive $\mathsf{B}(\beta)$ and $\mathsf{C}(\gamma)$.
2. Show that, under a *rotation* of basis, a symmetric matrix [the representation of a linear operator] remains symmetric.
3. Express
   (a) the vector representation $\mathbf{v} = 2\hat{\imath} + 3\hat{\jmath} + 5\hat{\boldsymbol{k}}$, and
   (b) the linear operator, represented as a matrix relative to the $(\hat{\imath},\hat{\jmath},\hat{\boldsymbol{k}})$ basis

$$\mathsf{A} = \begin{pmatrix} 0 & 1 & 3 \\ 2 & 0 & 0 \\ -1 & 1 & 0 \end{pmatrix}$$

   in terms of a new basis obtained by rotating about the [old] y-axis through an angle of 20°.
   Ans:

$$\text{a)} \quad \mathbf{v}' = \begin{pmatrix} 0.16928 \\ 3 \\ 5.38250 \end{pmatrix},$$

$$\text{b)} \quad \mathsf{A}' = \begin{pmatrix} -0.64279 & 0.59767 & 2.76604 \\ 1.87939 & 0 & 0.68404 \\ -1.23396 & 1.28171 & 0.64279 \end{pmatrix}$$

   (c) Now consider *two* basis rotations, $\mathcal{R}_1$ the +20° y-axis rotation above and $\mathcal{R}_2$ one about the z-axis through an angle of -10°. Find the [*single*] matrices generating the *composite* rotations in the order $\mathcal{R}_1\mathcal{R}_2$ and then $\mathcal{R}_2\mathcal{R}_1$, showing that these are different. Of course they are expected to be: matrix multiplication doesn't, in general, commute. But this demonstrates that rotations through *finite* angles don't either.

   Note: The use of one of computer mathematics/algebra packages is strongly recommended!

In the previous section, we proposed the interpretations of "active" and "passive" transformations on a vector, the former generating a *new* vector from the old (in the *same* basis), the latter merely a new representation of the *same* vector (in a *new* basis). We also pointed out the inverse nature of these two forms of transformation relative to one another. Rotations [of magnitude/direction vectors] provide a ready example of this, particularly easy to visualize in 2-D:

We note first that the representation $\mathbf{V}$ of an arbitrary vector $\boldsymbol{V}$ ultimately depends on the angle $\boldsymbol{V}$ makes with the basis. Now let us examine the two cases of basis rotation ("passive") and vector rotation ("active"):

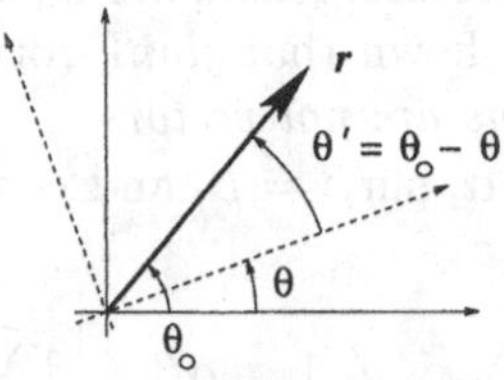

"passive" transformation

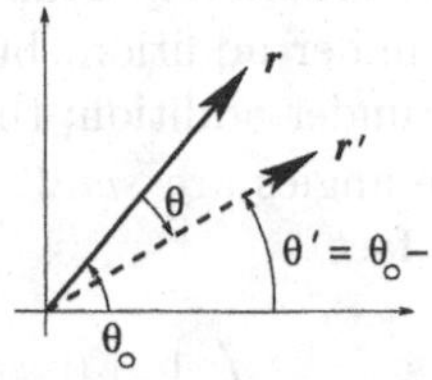

"active" transformation

- Consider, first, a vector $\boldsymbol{r}$ represented in a basis with which it makes an angle θ_o. If, now, the *same* vector is represented relative to a *new basis* rotated from the old through an angle θ as shown, the new representation $\mathbf{r}'$ will reference the angle $\theta' = (\theta_o - \theta)$ made by $\boldsymbol{r}$ with the new basis.
- Now consider the same $\boldsymbol{r}$ rotated ["actively"] through an angle $(-\theta)$ to a *new* vector $\boldsymbol{r}'$. *Its* representation $\mathbf{r}'$ reflects the angle $\theta' = (\theta_o - \theta)$ vector $\boldsymbol{r}'$ now makes with the *same* [original] basis.

Both representations $\mathbf{r}'$—of the *original* vector relative to *new* axes, and of the *new* [rotated] vector in the *original* axes—have components of a vector making the *same* angle $\theta' = (\theta_o - \theta)$ with the ultimate basis, so are the same. This can be seen in terms of the transformations themselves: the first case has $\mathbf{r}' = \mathbf{C}^{-1}(\theta)\mathbf{r}$ by (3.6-6); the second results from rotating $\boldsymbol{r}$ through $(-\theta)$ using (3.6-5): $\mathbf{r}' = \mathbf{C}(-\theta)\mathbf{r} = \mathbf{C}^{-1}(\theta)\mathbf{r}$—the same as the first case. The inverse nature of the two transformations evidences itself through the opposite signs on the rotation angles, forcing the final representations to be the same in this example.

> **Warning**: As has been the approach in this text, the above matrices were derived from the appropriate *basis* transformation. Many books jump directly to the $\mathbf{v}' = \mathbf{U}^{-1}\mathbf{v} = \mathbf{U}^{\mathsf{T}}\mathbf{v} \equiv \mathbf{R}\mathbf{v}$ representations (where we use the transpose) and call *these* $\mathbf{R}$ "rotation matrices." Others use the same term to denote the "*active*" transformation; this is precisely the problem encountered relative to equations (3.6-5) and (3.6-6). The reader must be cognizant of which is being used in a given presentation!

As was noted in the previous section, the result of two *successive* rotations about axes—a sequence of *two* basis transformations—will be just the *product* of two of the above matrices; by Theorem 3.3.2 this product will itself be orthogonal, corresponding in some sense to a "single rotation." Thus, for example, if one rotates first about the x-axis through $+\theta_x$ and then about the *new!* z-axis through $-\theta_z$, the *composite* rotation matrix would be just $(\mathbf{A}(\theta_x)\mathbf{C}(-\theta_z))$ [recall the remark on page 143], generating representations $\mathbf{v} = \mathbf{A}(\theta_x)\mathbf{C}(-\theta_z)\mathbf{v}''$ or $\mathbf{v}'' = (\mathbf{A}(\theta_x)\mathbf{C}(-\theta_z))^{-1}\mathbf{v} = \mathbf{C}^{-1}(-\theta_z)\mathbf{A}^{-1}(\theta_x)\mathbf{v} = \mathbf{C}(+\theta_z)\mathbf{A}^{\mathsf{T}}(\theta_x)\mathbf{v}$!

In this regard, recall that, even though multiplication of these square matrices is *defined* in either order, matrix multiplication is *not commutative*; *i.e.*, $\mathsf{B}(\beta)\mathsf{A}(\alpha) \neq \mathsf{A}(\alpha)\mathsf{B}(\beta)$. This means that two *finite* rotations result in different rotations. This has an important ramification: Vectors, recall, are *defined* to be commutative under addition, but we have just shown that *finite* rotations do *not* commute under addition; thus *finite rotations are not vectors.*

But, if the angles are *small*, so that $\sin\alpha \doteq \alpha$, $\sin\beta \doteq \beta$, and $\cos\alpha \doteq 1 \doteq \cos\beta$, then in fact

$$\mathsf{B}(\beta)\mathsf{A}(\alpha) \doteq \begin{pmatrix} 1 & 0 & \beta \\ 0 & 1 & 0 \\ -\beta & 0 & 1 \end{pmatrix}\begin{pmatrix} 1 & 0 & 0 \\ 0 & 1 & -\alpha \\ 0 & \alpha & 1 \end{pmatrix} = \begin{pmatrix} 1 & \alpha\beta & \beta \\ 0 & 1 & -\alpha \\ -\beta & \alpha & 1 \end{pmatrix}$$

while

$$\mathsf{A}(\alpha)\mathsf{B}(\beta) \doteq \begin{pmatrix} 1 & 0 & 0 \\ 0 & 1 & -\alpha \\ 0 & \alpha & 1 \end{pmatrix}\begin{pmatrix} 1 & 0 & \beta \\ 0 & 1 & 0 \\ -\beta & 0 & 1 \end{pmatrix} = \begin{pmatrix} 1 & 0 & \beta \\ \alpha\beta & 1 & -\alpha \\ -\beta & \alpha & 1 \end{pmatrix}.$$

In the limit of *really* small angles, $\alpha\beta \to 0$, and the above products *are* equal ("to first order"):

$$\mathsf{B}(\beta)\mathsf{A}(\alpha) \doteq \begin{pmatrix} 1 & 0 & \beta \\ 0 & 1 & -\alpha \\ -\beta & \alpha & 1 \end{pmatrix} \doteq \mathsf{A}(\alpha)\mathsf{B}(\beta),$$

and "infinitesimal rotations" *do* commute: *infinitesimal rotations are vectors.* (We recall that Theorem 3.3.2 alleges that the product of these two matrices should be orthogonal. Though this doesn't "look" orthogonal, it is *to first order*, as can be seen either by multiplying the above matrix by its transpose directly, or by examining the individual columns of the matrix and showing they are orthonormal (using Theorem 3.3.1); in either approach, second-order terms must be set to 0 to remain consistent with the first-order approach used to get the above product in the first place.)

We now illustrate the use of explicit rotation matrices with a couple of examples.

Example 3.6.2. The first example is drawn from the area of Solid Mechanics: Consider the *stress matrix*

$$\mathsf{S} \equiv \begin{pmatrix} \sigma_x & \tau_{xy} & \tau_{xz} \\ \tau_{xy} & \sigma_y & \tau_{yz} \\ \tau_{xz} & \tau_{yz} & \sigma_z \end{pmatrix}$$

—note that it is a real, symmetric matrix—in the special case of *plane stress*: $\tau_{xz} = \tau_{yz} = \sigma_z \equiv 0$. We examine how this matrix—the *representation* of a given transformation—changes under a rotation about the z-axis through an angle, say θ. From the discussion above, its representation will be transformed into the form

$$\begin{aligned}
\mathsf{S}' &= \mathsf{C}^{-1}\mathsf{S}\mathsf{C} = \mathsf{C}^{\mathsf{T}}\mathsf{S}\mathsf{C} \\
&= \begin{pmatrix} \cos\theta & \sin\theta & 0 \\ -\sin\theta & \cos\theta & 0 \\ 0 & 0 & 1 \end{pmatrix} \begin{pmatrix} \sigma_x & \tau_{xy} & 0 \\ \tau_{xy} & \sigma_y & 0 \\ 0 & 0 & 0 \end{pmatrix} \begin{pmatrix} \cos\theta & -\sin\theta & 0 \\ \sin\theta & \cos\theta & 0 \\ 0 & 0 & 1 \end{pmatrix} \\
&= \begin{pmatrix} \sigma'_x & \tau'_{xy} & 0 \\ \tau'_{xy} & \sigma'_y & 0 \\ 0 & 0 & 0 \end{pmatrix}
\end{aligned}$$

in which

$$\sigma'_x \equiv \sigma_x \cos^2\theta + \sigma_y \sin^2\theta + 2\tau_{xy}\sin\theta\cos\theta \tag{3.6-15a}$$

$$\tau'_{xy} \equiv (\sigma_x - \sigma_y)(-\sin\theta\cos\theta) + \tau_{xy}(\cos^2\theta - \sin^2\theta) \tag{3.6-15b}$$

$$\sigma'_y \equiv \sigma_x \sin^2\theta + \sigma_y \cos^2\theta - 2\tau_{xy}\sin\theta\cos\theta \tag{3.6-15c}$$

Clearly, different choices for θ will result in different forms of S', a fact returned to below in Example 3.7.2. |

Example 3.6.3. We illustrate the above ideas with an example anticipating the material of the next Part. It deals with the description of a gyroscope—an assumed axially-symmetric body moving about a fixed point at the base of that axis. Though one possible mode of motion of the gyroscope simply keeps that axis vertical, the really interesting motion occurs when that axis goes off-vertical: in that case (as anyone getting a toy gyroscope as a child knows), the top "*precesses*" about the vertical. The question is just how we might describe that motion; *i.e.*, its orientation in time.

Even *that* issue requires some judgment: Since the body (here simplified to a top) is moving about a fixed point, its position can be completely specified by three coordinates—those of some point on its rim, say. But, at the end of the day, we are not as interested in the specific orientation of the *top* about its axis of symmetry as we are in that of its *axis*. And the latter can be specified by giving two angles: the angle θ between the spin axis and the vertical; and that, ψ, between the *plane* containing both the vertical Z reference axis and the spin axis, and the XZ-plane—one which will, in fact, measure the precession.[6] In fact, it is customary to set up a set of axes spinning not with the gyroscope itself, but rather with its *spin* axis (suppressing the spin itself): the z-axis is oriented along the top's axis of symmetry, and the x-axis is taken to be perpendicular to the plane containing the moving z-axis and the fixed Z one, thus lying in the XY-plane; the y axis then follows from the right-hand rule. The result is that the angle between the fixed Z-axis and the moving z is precisely θ, and that between the fixed X-axis and the moving x is $(\psi + 90°)$.

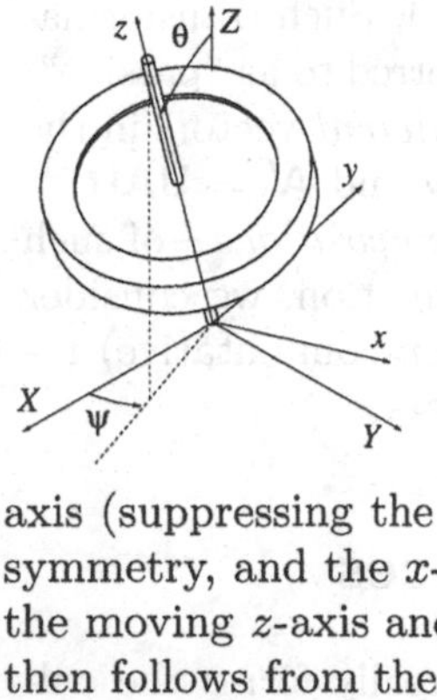

[6]These are precisely the "Euler angles" described in Section 4.2 in the next Part.

With all that in hand, we come to the real issue of this example: How can one relate quantities measured in the xyz coordinate system to the fixed XYZ? The answer is clearly through the use of rotation matrices: we go from the set of axes xyz (the "old") to the set XYZ (the "new"), where the latter are obtained from the former through a pair of rotations. We must, however, obtain those rotations in the correct order and direction. The key is to recognize that the above rotations are defined in such a fashion to bring the "old" system (really its *basis* vectors!) *coincident* with the "new" through the relevant rotations. In the present case, such coincidence can be effected through first a rotation about the ["*old*"] x-axis through an angle of $(-\theta)$—note the rotation in the *negative* sense—bringing the z and Z axes coincident, and then one about the [intermediate] z-axis through $-(\psi + 90°)$, bringing in a single operation the y and Y and x and X axes together; note that rotating first about the z-axis through $-(\psi+90°)$ would *not* bring the x- (or any *other*!) axes coincident. The composite of these *basis* transformations would thus be $\mathsf{A}(-\theta)\mathsf{C}(-(\psi+90°)) = \mathsf{A}^{-1}(\theta)\mathsf{C}^{-1}(\psi+90°) = \mathsf{A}^{\mathsf{T}}(\theta)\mathsf{C}^{\mathsf{T}}(\psi+90°) = (\mathsf{C}(\psi+90°)\mathsf{A}(\theta))^{\mathsf{T}}$. Thus the ultimate transformation giving the *representation* V in the XYZ coordinates of a vector with *representation* v in the xyz system would be given by

$$\mathsf{V} = ((\mathsf{C}(\psi+90°)\mathsf{A}(\theta))^{\mathsf{T}})^{-1}\mathsf{v} = \mathsf{C}(\psi+90°)\mathsf{A}(\theta)\mathsf{v}.$$

Facility with the relations among inverses, transposes, and negative values of transformation matrices is strongly recommended! |

Summary. Given a specific basis $\{u_1, u_2, \ldots, u_n\}$ of an n-dimensional space in terms of which vectors and linear transformations are *represented* by column vectors v and matrices A, respectively, their representations v' and A' in a different basis $\{u_1', u_2', \ldots, u_n'\}$ are given by $\mathsf{v}' = \mathsf{U}^{-1}\mathsf{v}$ and $\mathsf{A}' = \mathsf{U}^{-1}\mathsf{A}\mathsf{U}$, in which U is just the matrix whose columns are the representations of the $\{u_i'\}$ in terms of the *original* basis: $\mathsf{U} = (\mathsf{u}_1'(u), \mathsf{u}_2'(u), \ldots, \mathsf{u}_n'(u))$. Such transformations, in which *one* vector is represented in *two* bases, is referred to as "passive"; the corresponding "active" transformation, between two *different* vectors in the *same* basis, is given by the *inverse* of the above: $\mathsf{v}' = \mathsf{U}\mathsf{v}$ and $\mathsf{A}' = \mathsf{U}\mathsf{A}\mathsf{U}^{-1}$. We showed, in particular, how we could form sequences—*compositions*—of such transformations. As a specific case of such basis transformation, we consider *rotation matrices*—matrices corresponding to *finite* (so *non*-commutative) rotations around each of the original orthonormal basis vectors.

3.7 Normal Forms of Square Matrices

We finally embark on the last section of this subject. It really forms the culmination of all the previous material, which possibly seems to have gone on at greater length than necessary. But it provides the foundation for all of the next part and is presumed in the subsequent exposition.

Let us first summarize the knowledge base to date: We have seen that linear transformations (among them, the physically important case of that "transformation" which carries an angular velocity $\boldsymbol{\omega}$ to an angular momentum $\boldsymbol{H}$) can be *represented* as matrices; that, as expected, this *representation* depends on the *basis* to which it is referred; that, rather less obviously, the basis for such linear transformations ultimately refers back to the basis used to represent the *vectors* (of whatever ilk) on which such transformations operate; and that the transformation matrix to a new basis, vectors of the new basis' representation in terms of the old, predicts the representations of both vectors and transformations, the former through multiplication by, and the latter through a "similarity transformation" involving, that matrix.

But if the representation—*i.e.*, the *form*—of such a transformation depends on the basis, it is reasonable to ask whether there might be a "simplest" representation the transformation representation can assume. The answer [of course, or the question wouldn't have been posed in the first place!] is "yes," and the "simplest representation" is referred to as the *normal form* of the matrix [representation]. Since this form is dependent on the basis, determination of that form is tantamount to determination of the corresponding basis—that basis in terms of which both the transformation *and* the vectors on which it operates—which most simplifies the form of the transformation matrix.

What might be the "simplest form" of a matrix? For a *vector*, if the basis is chosen to have one axis along that vector, *its* simplest form has only a single component; but it is too much to expect matrices to be so simple. It is perhaps easy to accept, however, that there's not a much simpler matrix than one consisting *mostly* of 0's, with any non-0 entries on its diagonal. To anticipate the outcome of this discussion, let it be said that the new basis is nothing more than the set of *eigenvectors* of the matrix. And it cannot then come as too great a surprise that the corresponding *eigenvalues* of the matrix play a pivotal role, appearing as the primary entries in the simplified, normal form of the original matrix; in fact, the diagonal entries just referred to are nothing more than the eigenvalues of the matrix.

It was remarked before that this section is the "culmination" of the study thus far. A brief review of the above "set-up" indicates just why this is the case: the topics mentioned include *eigenvectors*, *eigenvalues* (both of which require the solution of *homogeneous linear equations*), *rotations*, *basis transformations*, and ultimately *representation*, *basis*, and *linear independence*—all key concepts which have been covered; in fact, the proof of the diagonalizing methods will also utilize *block multiplication* and the properties of *inverses*. The reader is expected to have reasonable facility with all these ideas, which will appear with some rapidity!

3.7.1 Linearly Independent Eigenvectors—Diagonalization

Let us assume that the $n \times n$ matrix A in question enjoys a full set of n linearly independent eigenvectors $\boldsymbol{\xi}_1, \boldsymbol{\xi}_2, \ldots, \boldsymbol{\xi}_n$. In this case, consider the matrix formed of the side-by-side catenation of these vectors,

$$\mathbf{\Xi} \equiv (\boldsymbol{\xi}_1\, \boldsymbol{\xi}_2 \ldots \boldsymbol{\xi}_n)$$

Now, more or less arbitrarily, consider the matrix product

$$\mathbf{A\Xi} \equiv \mathbf{A}(\boldsymbol{\xi}_1\, \boldsymbol{\xi}_2 \ldots \boldsymbol{\xi}_n) = (\mathbf{A}\boldsymbol{\xi}_1\, \mathbf{A}\boldsymbol{\xi}_2 \ldots \mathbf{A}\boldsymbol{\xi}_n) = (\lambda_1\boldsymbol{\xi}_1\, \lambda_2\boldsymbol{\xi}_2 \ldots \lambda_n\boldsymbol{\xi}_n)$$

—where we have utilized block multiplication [treating $\mathbf{A}$ as a "scalar"] in the second equation and the definition of eigenvector/values in the third (noting the order in which the eigenvalues appear, the same as that of the eigenvectors in $\mathbf{\Xi}$). The last term can be written

$$(\lambda_1\boldsymbol{\xi}_1\, \lambda_2\boldsymbol{\xi}_2 \ldots \lambda_n\boldsymbol{\xi}_n) = (\boldsymbol{\xi}_1\, \boldsymbol{\xi}_2 \ldots \boldsymbol{\xi}_n)\begin{pmatrix} \lambda_1 & 0 & \ldots & 0 \\ 0 & \lambda_2 & \ldots & 0 \\ \vdots & \vdots & \ddots & \vdots \\ 0 & 0 & \ldots & \lambda_n \end{pmatrix} \equiv \mathbf{\Xi\Lambda},$$

in which $\mathbf{\Lambda}$ is just that matrix consisting of the eigenvalues of $\mathbf{A}$ arrayed along its diagonal, in the same order as the $\boldsymbol{\xi}_i$ in $\mathbf{\Xi}$. (This equation can be verified either directly; by noting that each $\boldsymbol{\xi}_i = \mathbf{\Xi} e_i^{(n)}$, where $e_i^{(n)}$ is the $n \times 1$ *natural basis* vector—dropping the superscript, it being understood that we are talking of n-dimensional vectors here—so

$$(\lambda_1\boldsymbol{\xi}_1\, \lambda_2\boldsymbol{\xi}_2 \ldots \lambda_n\boldsymbol{\xi}_n) = (\lambda_1\mathbf{\Xi}e_1\, \lambda_2\mathbf{\Xi}e_2 \ldots \lambda_n\mathbf{\Xi}e_n) = \mathbf{\Xi}(\lambda_1 e_1\, \lambda_2 e_2 \ldots \lambda_n e_n),$$

from which the above relation follows; or by using block multiplication of the "$1 \times n$" row matrix $\mathbf{\Xi}$ by the "$n \times n$" matrix $\mathbf{\Lambda}$.) Collecting the above two equations into one, we conclude that

$$\mathbf{A\Xi} = \mathbf{\Xi\Lambda},$$

or, since the columns of $\mathbf{\Xi}$ are linearly independent, its inverse exists (by Theorem 3.2.7), so we can write simply

$$\mathbf{\Xi}^{-1}\mathbf{A\Xi} = \mathbf{\Lambda}. \tag{3.7-1}$$

Thus we have the rather remarkable result that the above product—a *similarity transformation* (3.6-7)—is precisely that which puts the matrix $\mathbf{A}$ in a *diagonal* form $\mathbf{\Lambda}$—one in which the entries on the diagonal are nothing more than the *eigenvalues* of the original matrix $\mathbf{A}$! But recall that, in the remarks following (3.6-7), on page 143, we observed that such a transformation can be interpreted as a *basis transformation*, with the columns of the transformation matrix (here $\mathbf{\Xi}$) being [representations of] the *new basis* in terms of the old. In the present case, those columns are just the $\boldsymbol{\xi}_i$, the *eigenvectors* of $\mathbf{A}$; thus these eigenvectors are nothing more than the new basis [representations relative to the old]. Thus we have shown that: *the eigenvectors of a matrix* $\mathbf{A}$ *are just the basis vectors in terms of which the linear transformation [represented by]* $\mathbf{A}$

is [represented by] a diagonal matrix, whose entries are the eigenvalues arrayed in the same order as the eigenvectors in the transformation matrix!

(3.7-1) is the point at which the linear independence of the eigenvectors becomes important: All the equations up to this one would have been true even if there had been some linear *de*pendence among the eigenvectors. But in order to be able to take the inverse of the matrix of eigenvectors $\mathbf{\Xi}$, they must be linear independent (Theorem 3.1.5). We recall the results of Theorem 3.5.1: that eigenvectors corresponding to *distinct* eigenvalues *are* linearly independent; on the other hand, eigenvectors associated with *multiple* eigenvalues—repeated roots of the characteristic equation—*may or may not be* linearly independent, as the examples related to this Theorem demonstrate.

Before proceeding further, let us consider a simple example of this technique:

Example 3.7.1. *Diagonalization of a Matrix.* Consider again the matrix of Example 3.5.1,

$$\mathsf{A} = \begin{pmatrix} 2 & 1 & 1 \\ 2 & 3 & 2 \\ 1 & 1 & 2 \end{pmatrix}$$

(In the absence of any other information, this matrix is implicitly assumed expressed in terms of the "natural basis"!) We have already seen in that example that the eigenvalues are $\lambda_1 = 5$ and a multiple eigenvalue $\lambda_2 = \lambda_3 = 1$. Though we cannot conclude that the eigenvectors associated with the repeated root are linearly independent, we have shown that the corresponding eigenvectors are

$$\boldsymbol{\xi}_1 = \begin{pmatrix} 1 \\ 2 \\ 1 \end{pmatrix}; \quad \boldsymbol{\xi}_2 = \begin{pmatrix} -1 \\ 1 \\ 0 \end{pmatrix}, \quad \boldsymbol{\xi}_3 = \begin{pmatrix} -1 \\ 0 \\ 1 \end{pmatrix}$$

and, indeed, linearly independent. This means that the transformation matrix from the original [natural] basis to that diagonalizing A is given by

$$\mathbf{\Xi} = \begin{pmatrix} 1 & -1 & -1 \\ 2 & 1 & 0 \\ 1 & 0 & 1 \end{pmatrix}$$

Using the techniques of Section 3.1.1 we obtain in a straightforward fashion that

$$\mathbf{\Xi}^{-1} = \begin{pmatrix} \frac{1}{4} & \frac{1}{4} & \frac{1}{4} \\ -\frac{1}{2} & \frac{1}{2} & -\frac{1}{2} \\ -\frac{1}{4} & -\frac{1}{4} & \frac{3}{4} \end{pmatrix},$$

and just as straightforward (if tedious) multiplication results in

$$\mathbf{\Xi}^{-1}\mathsf{A}\mathbf{\Xi} = \begin{pmatrix} 5 & 0 & 0 \\ 0 & 1 & 0 \\ 0 & 0 & 1 \end{pmatrix},$$

as expected: the eigenvalues are arrayed along the diagonal in the same order as their corresponding eigenvectors in $\mathbf{\Xi}$. In this regard, switching the last two columns of $\boldsymbol{\Xi}$ would have made no difference, but if we had defined that matrix instead to have the *first* two columns interchanged, the result of the final product would have been

$$\boldsymbol{\Xi}^{-1}\mathbf{A}\boldsymbol{\Xi} = \begin{pmatrix} 1 & 0 & 0 \\ 0 & 5 & 0 \\ 0 & 0 & 1 \end{pmatrix}.$$

We note in passing that these matrix multiplications *are* both "straightforward" *and* "tedious"! It is useful for the engineer/scientist to be familiar with one of the many excellent algebra/math programs available to expedite these calculations, and their use is highly recommended in working both these and other problems. |

The discussion of ordering in the above example simply dramatizes the arbitrary nature of the basis transformation. Though not as "neat and tidy" as one might like, this character is actually useful in application: We have pointed out (Section 3.3.1) that [orthogonal] rotations (and, more generally, even *non*-orthogonal transformations) can be classified as "proper" or "improper," according to the *sign of the determinant* of the transformation matrix; for example, the above *non*-orthogonal transformation has $|\boldsymbol{\Xi}| = +5$, so would be "proper." But, ever if it were not, switching the columns of this matrix would switch the sign of its determinant (Theorem 3.2.2); thus we have the freedom to *force* the above transformation to be proper simply by selecting the order of the eigenvectors correctly.

In the same vein, we recall that if $\boldsymbol{\xi}$ is an eigenvector, then so is $c\boldsymbol{\xi}$—eigenvectors are only determinable up to an arbitrary constant. This means that we also have the freedom to *normalize* the eigenvectors and make the new basis above consist of unit vectors. Both this and the ability to make the transformation *proper* will be important in the following special case, characterizing the inertia matrix in the next Part.

To motivate the importance of this case, we make an observation regarding this procedure: Recall that in the discussion of eigenvalues and eigenvectors, we pointed out that the eigenvalues, roots of a general polynomial equation—even one with real coefficients—were in general going to be complex; thus the associated eigenvectors could be—if the matrix were real, *would* be—complex. However appealing aesthetically the concept of matrix diagonalization might be, this eventuality would certainly put a damper on any enthusiasm one might generate for the technique, at least when applied to "real, physical" problems. The following addresses precisely this issue, particularly as it relates to the inertia matrix of primary interest in this book.

Homework:

1. For the matrix

$$\mathbf{A} = \begin{pmatrix} 2 & -3 & 3 \\ 0 & 5 & -3 \\ 0 & 6 & -4 \end{pmatrix}$$

we have already shown (Problem 4, page 122) that, corresponding to the eigenvalues $\lambda_1 = -1, \lambda_2 = 2, \lambda_3 = 2$ are three *linearly independent* eigenvectors; for example,

$$\boldsymbol{\xi}_1 = \begin{pmatrix} -1 \\ 1 \\ 2 \end{pmatrix}, \quad \boldsymbol{\xi}_2 = \begin{pmatrix} 1 \\ 0 \\ 0 \end{pmatrix}, \quad \boldsymbol{\xi}_3 = \begin{pmatrix} 0 \\ 1 \\ 1 \end{pmatrix},$$

respectively. Diagonalize this matrix, and show that the elements on the diagonal are, indeed, the above eigenvalues, arranged in order corresponding to that of the eigenvectors.

2. The matrix

$$\mathsf{A} = \begin{pmatrix} 2 & -3 & 3 \\ 0 & 5 & -3 \\ 0 & 6 & -4 \end{pmatrix}$$

has already been shown (Problem 5, page 122) to have eigenvalues $\lambda = 3, 1, 1$. Corresponding to the first is the eigenvector [say], $\begin{pmatrix} -3 \\ 2 \\ 1 \end{pmatrix}$, while corresponding to the eigenvalue $\lambda = 1$ one can find only the single eigenvector $\begin{pmatrix} -1 \\ 1 \\ 0 \end{pmatrix}$.

   (a) Show that the vector $\boldsymbol{\xi}' = \begin{pmatrix} 1 \\ 1 \\ -1 \end{pmatrix}$ is a *generalized eigenvector* (Section 3.5.3) for the root $\lambda = 1$, satisfying the equation $(\mathsf{A} - 1\mathbf{1})^2\boldsymbol{\xi}' = \mathbf{0}$, but that it is *not* a true eigenvector: $(\mathsf{A} - 1\mathbf{1})\boldsymbol{\xi}'! = \mathbf{0}$.

   (b) Show that the above three vectors—the two *real* eigenvectors and the [bogus] *generalized* eigenvector in a)—*are* linearly independent.

   (c) Show that the same diagonalization procedure used above, where there are three linearly independent eigenvectors, *almost* works here; in particular, show that, for P the matrix formed by concatenating the two normal eigenvectors and the *generalized* eigenvector in part a), the matrix product $\mathsf{P}^{-1}\mathsf{AP}$ still has the eigenvalues on the diagonal, but that, corresponding to the repeated root $\lambda = 1$, there is now an additional "1" on the upper off-diagonal. This [almost] diagonalized form is called *Jordan normal form* (Section 3.7.2 below), and is that resulting from repeated eigenvalues lacking the full complement of linearly independent *true* eigenvectors.

Diagonalization of Real Symmetric Matrices

Despite their apparently over-special nature, real symmetric matrices are actually quite common in physical application: as mentioned above, the stress matrix from solid mechanics is real and symmetric, and the inertia matrix in the next Part will be shown to be. We now consider the above procedure of diagonalization in the particular case that the original matrix is of this form.

According to the above exposition, a matrix can be diagonalized when it is expressed relative to a basis given by its eigenvectors. But, in the discussion of eigenvectors, we had occasion to prove Theorem 3.5.2: that the eigenvalues of a real, symmetric matrix are real, and its eigenvectors orthogonal. The former property assures us that the diagonalized matrix will always be *real.* It also, accounting for the real nature of the matrix, means the eigen*vectors*—the new basis vectors—will be real. And the latter property, along with the above observation that we can always *normalize* the eigenvectors, means that we can always force the diagonalizing basis to be *orthonormal.* If the basis in terms of which the matrix was originally expressed is also orthonormal, we would expect that this diagonalization could be carried out by means of a mere *rotation* of axes from the original to the new, diagonalizing, ones. And, except for a couple of technical details, this is precisely the case:

In the first place, we claim that the above $\boldsymbol{\Xi}$ matrix is *orthogonal* for orthonormal eigenvectors, $\boldsymbol{\xi}_i$, resulting from the diagonalization of a real symmetric matrix and normalization of its eigenvectors; indeed, for

$$\boldsymbol{\Xi} \equiv (\boldsymbol{\xi}_1\, \boldsymbol{\xi}_2\, \cdots\, \boldsymbol{\xi}_n)$$

we can directly compute

$$\boldsymbol{\Xi}^{\mathsf{T}}\boldsymbol{\Xi} = \begin{pmatrix} \boldsymbol{\xi}_1^{\mathsf{T}} \\ \boldsymbol{\xi}_2^{\mathsf{T}} \\ \vdots \\ \boldsymbol{\xi}_n^{\mathsf{T}} \end{pmatrix} (\boldsymbol{\xi}_1\, \boldsymbol{\xi}_2\, \cdots\, \boldsymbol{\xi}_n) = (\langle \boldsymbol{\xi}_i, \boldsymbol{\xi}_j \rangle) = \mathbf{1};$$

thus $\boldsymbol{\Xi}$ is, in fact, orthogonal. And (Section 3.3.1) orthogonal matrices in an orthonormal basis correspond to a *rotation.* This rotation may be proper or improper; but, by the discussion leading into this section above, we can always force the matrix—*i.e.*, the *basis transformation*—to be proper simply by ordering the eigenvectors correctly: if it turns out that the determinant of the transformation [representation] is -1, we merely switch a pair of vectors to make the determinant $+1$, and we are done!

Thus we have proven the important result:

Theorem 3.7.1. Any real, symmetric matrix represented relative to an orthonormal basis can be diagonalized to a real matrix by a [proper] *rotation* to a new set of orthonormal axes.

Example 3.7.2. *Diagonalization of Real Symmetric Matrix through Rotation.* We illustrate this concept by returning to a previous example, 3.6.2, in which we explicitly showed the form, equations (3.6-15) of the planar stress matrix—itself a real symmetric matrix, as noted there—under a rotation through arbitrary θ about the z-axis. It happened in that case that the new form of the matrix, $\mathbf{S}'$, was also symmetric; in particular, the off-diagonal term in that matrix, (3.6-15b), is reproduced here:

$$\tau'_{xy} \equiv (\sigma_x - \sigma_y)(-\sin\theta\cos\theta) + \tau_{xy}(\cos^2\theta - \sin^2\theta).$$

We observe that if we choose θ:

$$\frac{\sin\theta\cos\theta}{\cos^2\theta - \sin^2\theta} = \frac{\frac{1}{2}\sin 2\theta}{\cos 2\theta} = \frac{\tau_{xy}}{\sigma_x - \sigma_y} \quad \text{or} \quad \tan 2\theta = \frac{2\tau_{xy}}{\sigma_x - \sigma_y},$$

this off-diagonal term will vanish, and the resulting $\mathbf{S}'$ will be diagonalized by means of a simple rotation through the angle θ. |

3.7.2 Linearly *De*pendent Eigenvectors—Jordan Normal Form

Recall that the technique in the above section worked precisely because the original matrix had a full set of linearly independent eigenvectors. This meant that the Ξ matrix of these eigenvectors had an inverse, allowing the similarity transformation—*i.e.*, the *basis transformation*—to be formed. As has been pointed out repeatedly, if the eigenvalues of a matrix are distinct, then its eigenvectors *will* be linearly independent; if there is any multiplicity associated with a given eigenvalue, the associated eigenvectors may or may not be linearly independent (in the above example, they were). But in general the eigenvectors corresponding to a repeated root of the characteristic equation will be linearly *de*pendent, in which case the above approach fails. What would be the "simplest" matrix corresponding to such a case?

It turns out that, aside from a certain accommodation required, the situation in this case is remarkably similar to that for the less pathological one: the only difference is that, instead of "normal" eigenvectors, we utilize the *generalized* eigenvectors (see page 131). In order to set the scene, let us briefly review the conditions surrounding the case in which there is not a full set of linearly independent eigenvectors and we must resort to those of the generalized variety:

Consider an $n \times n$ matrix $\mathbf{A}$ for which we have $n_O < n$ distinct eigenvalues $\lambda_1, \ldots, \lambda_{n_O}$ lacking a full set of linearly independent eigenvectors (since, if there *were* a full set, the above technique is applicable). Let us denote the multiplicity of each λ_i by $m_i \geq 1$; note that $\sum m_i = n$. Then, for each m_i, one of the following two cases holds: either

1. $\underline{m_i = 1}$: $(\mathbf{A} - \lambda_i \mathbf{1})\boldsymbol{\xi}_i = 0$; *i.e.*, $\mathbf{A}\boldsymbol{\xi}_i = \lambda_i \boldsymbol{\xi}_i$ ("normal" eigenvectors)
2. $\underline{m_i > 1}$: There are $N_i \leq m_i$ linearly independent eigenvectors $\boldsymbol{\xi}_{i1}^{(1)}, \boldsymbol{\xi}_{i2}^{(1)}, \ldots, \boldsymbol{\xi}_{iN_i}^{(1)}$ associated with λ_i. Again, there are two possible cases to be considered: either
   (a) $\underline{N_i = m_i}$: We have a full set of linearly independent eigenvectors $\boldsymbol{\xi}_{ij}$: $(\mathbf{A} - \lambda_i \mathbf{1})\boldsymbol{\xi}_{ij} = 0$; *i.e.*, $\mathbf{A}\boldsymbol{\xi}_{ij} = \lambda_i \boldsymbol{\xi}_{ij}$ ("normal" eigenvectors) associated with λ_i despite the multiplicity. For all intents and purposes, this is equivalent to the above case.

(b) $\underline{N_i < m_i}$: For each $j = 1, \dots, N_i$ there is a "family" of linearly independent *generalized* eigenvectors, $\{\boldsymbol{\xi}_{ij}^{(k)}\}$:

$\boldsymbol{\xi}_{ij}^{(1)}$: ("normal" eigenvectors) For $k = 1$, $(\mathbf{A} - \lambda_i \mathbf{1})\boldsymbol{\xi}_{ij}^{(1)} = 0$; *i.e.*, $\mathbf{A}\boldsymbol{\xi}_{ij}^{(1)} = \lambda_i \boldsymbol{\xi}_{ij}^{(1)}$

$\boldsymbol{\xi}_{ij}^{(k)}$: ("generalized" eigenvectors) For $k > 1$, $(\mathbf{A} - \lambda_i \mathbf{1})\boldsymbol{\xi}_{ij}^{(k)} = \boldsymbol{\xi}_{ij}^{(k-1)}$; *i.e.*, $\mathbf{A}\boldsymbol{\xi}_{ij}^{(k)} = \lambda_i \boldsymbol{\xi}_{ij}^{(k)} + \boldsymbol{\xi}_{ij}^{(k-1)}$, for $j = 1, \dots, N_{ij}$, N_{ij} denoting the number of linearly independent generalized eigenvectors in the jth "family" of λ_i. Note that $\sum_j N_{ij} = m_i$.

Note the structure implied by the index notation: subscripts in the absence of superscripts denote "normal" eigenvectors; a double subscript indicates a repeated root of the characteristic equation. Superscripts are introduced when a repeated root is deficient in the number of linearly independent eigenvectors: the "(1)" indicates, again, the "normal" eigenvector; higher values its "offspring."

As mentioned above, we regard cases "1." and "2(a)" as equivalent; thus we consider case "2(b)." Let, then, n_d denote the number of eigenvalues which are deficient in the number of linearly independent eigenvectors: $N_i < m_i$ for $i = 1, \dots, n_d$. Having developed the process of diagonalization for a full set of ["normal"] linearly independent eigenvectors, we now try to apply the same procedure to the generalized variety: More or less arbitrarily, let us form the matrix of eigenvectors, now both "normal" *and* generalized:

$$\boldsymbol{\Xi} \equiv \Big(\overbrace{(\boldsymbol{\xi}_{11}^{(j)})}^{N_{11}} \; \overbrace{(\boldsymbol{\xi}_{12}^{(j)})}^{N_{12}} \; \cdots \; \overbrace{(\boldsymbol{\xi}_{1N_1}^{(j)})}^{N_{1N_1}} \; \overbrace{(\boldsymbol{\xi}_{21}^{(j)})}^{N_{21}} \; \overbrace{(\boldsymbol{\xi}_{22}^{(j)})}^{N_{22}} \; \cdots \; \overbrace{\boldsymbol{\xi}_{n_d+1}}^{1} \; \overbrace{\boldsymbol{\xi}_{n_d+2}}^{1} \; \cdots \; \overbrace{\boldsymbol{\xi}_{n_0}}^{1} \Big)$$

Observe that we have started with the families of generalized eigenvectors, catenated within their own submatrices; the numbered brace over each entry indicates the number of columns occupied by that family or eigenvector.

As before, we shall apply $\mathbf{A}$ to $\boldsymbol{\Xi}$:

$$\begin{aligned}\mathbf{A}\boldsymbol{\Xi} &= \mathbf{A}\Big((\boldsymbol{\xi}_{11}^{(j)}) \; (\boldsymbol{\xi}_{12}^{(j)}) \; \cdots \; (\boldsymbol{\xi}_{1N_1}^{(j)}) \; (\boldsymbol{\xi}_{21}^{(j)}) \; (\boldsymbol{\xi}_{22}^{(j)}) \; \cdots \; \boldsymbol{\xi}_{n_d+1} \; \boldsymbol{\xi}_{n_d+2} \; \cdots \; \boldsymbol{\xi}_{n_0} \Big) \\ &= \Big(\mathbf{A}(\boldsymbol{\xi}_{11}^{(j)}) \; \mathbf{A}(\boldsymbol{\xi}_{12}^{(j)}) \cdots \mathbf{A}(\boldsymbol{\xi}_{1N_1}^{(j)}) \; \mathbf{A}(\boldsymbol{\xi}_{21}^{(j)}) \; \mathbf{A}(\boldsymbol{\xi}_{22}^{(j)}) \cdots \mathbf{A}\boldsymbol{\xi}_{n_d+1} \; \mathbf{A}\boldsymbol{\xi}_{n_d+2} \cdots \mathbf{A}\boldsymbol{\xi}_{n_0} \Big).\end{aligned}$$

To see what happens in this case, consider the first N_{11} columns of the $\boldsymbol{\Xi}$ matrix, $(\boldsymbol{\xi}_{11}^{(j)})$. As before, block multiplication, use of $\boldsymbol{\xi}_i = \boldsymbol{\Xi} e_i^{(n)}$ [see page 56], and the equations satisfied by the [generalized] eigenvectors yield

$$\begin{aligned}\mathbf{A}(\boldsymbol{\xi}_{11}^{(j)}) &= \Big(\mathbf{A}\boldsymbol{\xi}_{11}^{(1)} \; \mathbf{A}\boldsymbol{\xi}_{11}^{(2)} \; \cdots \; \mathbf{A}\boldsymbol{\xi}_{11}^{(N_{11})} \Big) \\ &= \Big(\lambda_1 \boldsymbol{\xi}_{11}^{(1)} \;\; \lambda_1 \boldsymbol{\xi}_{11}^{(2)} + \boldsymbol{\xi}_{11}^{(1)} \;\; \lambda_1 \boldsymbol{\xi}_{11}^{(3)} + \boldsymbol{\xi}_{11}^{(2)} \; \cdots \Big) \\ &= (\lambda_1 \boldsymbol{\Xi} e_1 \;\; \lambda_1 \boldsymbol{\Xi} e_2 + \boldsymbol{\Xi} e_1 \;\; \lambda_1 \boldsymbol{\Xi} e_3 + \boldsymbol{\Xi} e_2 \; \dots) \\ &= \boldsymbol{\Xi} (\lambda_1 e_1 \;\; \lambda_1 e_2 + e_1 \;\; \lambda_1 e_3 + e_2 \; \dots)\end{aligned}$$

$$
= \boldsymbol{\Xi}
\begin{pmatrix}
\lambda_1 & 1 & 0 & \cdots & \cdots & 0\\
0 & \lambda_1 & 1 & \cdots & \cdots & \vdots\\
0 & 0 & \lambda_1 & \ddots & \cdots & \vdots\\
\vdots & \vdots & 0 & \ddots & \ddots & 0\\
\vdots & \vdots & \vdots & 0 & \lambda_1 & 1\\
0 & \vdots & \vdots & \vdots & 0 & \lambda_1\\
0 & \vdots & \vdots & \vdots & \vdots & 0\\
\vdots & \vdots & \vdots & \vdots & \vdots & \vdots\\
0 & \vdots & \vdots & \vdots & \vdots & 0
\end{pmatrix}
\equiv \boldsymbol{\Xi}
\begin{pmatrix}
\mathbf{J}_{11}\\
\mathbf{0}_{(n-N_{11})\times N_{11}}
\end{pmatrix},
$$

where $\mathbf{J}_{11}$ is a *square* $N_{11}\times N_{11}$ matrix with the eigenvalue λ_1 arrayed along its diagonal and 1's along the upper off-diagonal, and $\mathbf{0}_{m\times n}$ denotes the rectangular $m \times n$ matrix of 0's, acting here as a "filler" to make the matrix post-multiplying $\boldsymbol{\Xi}$ an $n \times N_{11}$ one. $\mathbf{J}_{11}$ is called a *Jordan block*; in the special case that its dimension (here N_{11}) is 1, it reduces to the 1×1 "matrix" of the eigenvalue itself.

In a similar fashion (noting that $\boldsymbol{\xi}_{12}^{(j)} = \boldsymbol{\Xi} e_{(N_{11}+j)}$),

$$
\begin{aligned}
\mathbf{A}\left(\boldsymbol{\xi}_{12}^{(j)}\right) &= \left(\mathbf{A}\boldsymbol{\xi}_{12}^{(1)}\ \mathbf{A}\boldsymbol{\xi}_{12}^{(2)}\ \ldots\ \mathbf{A}\boldsymbol{\xi}_{12}^{(N_{12})}\right)\\
&= \left(\lambda_1\boldsymbol{\xi}_{12}^{(1)}\ \ \lambda_1\boldsymbol{\xi}_{12}^{(2)}+\boldsymbol{\xi}_{12}^{(1)}\ \ \lambda_1\boldsymbol{\xi}_{12}^{(3)}+\boldsymbol{\xi}_{12}^{(2)}\ \ldots\right)\\
&= \left(\lambda_1\boldsymbol{\Xi} e_{(N_{11}+1)}\ \ \lambda_1\boldsymbol{\Xi} e_{(N_{11}+2)}+\boldsymbol{\Xi} e_{(N_{11}+1)}\ \ \lambda_1\boldsymbol{\Xi} e_{(N_{11}+3)}+\boldsymbol{\Xi} e_{(N_{11}+2)}\ \ldots\right)\\
&= \boldsymbol{\Xi}\left(\lambda_1 e_{(N_{11}+1)}\ \ \lambda_1 e_{(N_{11}+2)}+e_{(N_{11}+1)}\ \ \lambda_1 e_{(N_{11}+3)}+e_{(N_{11}+2)}\ \ \ldots\right)
\end{aligned}
$$

$$
= \boldsymbol{\Xi}
\begin{pmatrix}
0 & \vdots & \vdots & \vdots & \vdots & 0\\
\vdots & \vdots & \vdots & \vdots & \vdots & \vdots\\
0 & \vdots & \vdots & \vdots & \vdots & 0\\
\lambda_1 & 1 & 0 & \cdots & \cdots & 0\\
0 & \lambda_1 & 1 & \cdots & \cdots & \vdots\\
0 & 0 & \lambda_1 & \ddots & \cdots & \vdots\\
\vdots & \vdots & 0 & \ddots & \ddots & 0\\
\vdots & \vdots & \vdots & 0 & \lambda_1 & 1\\
0 & \vdots & \vdots & \vdots & 0 & \lambda_1\\
0 & \vdots & \vdots & \vdots & \vdots & 0\\
\vdots & \vdots & \vdots & \vdots & \vdots & \vdots\\
0 & \vdots & \vdots & \vdots & \vdots & 0
\end{pmatrix}
\equiv \boldsymbol{\Xi}
\begin{pmatrix}
\mathbf{0}_{N_{11}\times N_{12}}\\
\mathbf{J}_{12}\\
\mathbf{0}_{(n-N_{11}-N_{12})\times N_{12}}
\end{pmatrix},
$$

—$\boldsymbol{\Xi}$ post-multiplied by an $n \times N_{12}$ matrix consisting of the $N_{12} \times N_{12}$ Jordan block $\mathbf{J}_{12}$—λ_1 still goes along the diagonal, since $\boldsymbol{\xi}_{12}^{(1)}$ is still a [true] eigenvector of that eigenvalue—filled above and below by rectangular blocks of 0's.

This procedure proceeds similarly for all the remaining families ($\boldsymbol{\xi}_{ij}^{(k)}$), generating subsequent rectangular $n \times N_{ij}$ submatrices of $\mathbf{A\Xi}$. Isolated eigenvectors $\boldsymbol{\xi}_{n_d+j}$ at the end are treated exactly the same, generating only a single $n \times 1$ column instead.

Thus, finally, catenating side-by-side the above intermediate results for each submatrix of the product, we get that

$$\mathbf{A\Xi} = \mathbf{\Xi}\begin{pmatrix} \mathbf{J}_{11} & \mathbf{0} & \mathbf{0} & \cdots & & & & \cdots & \mathbf{0} \\ \mathbf{0} & \mathbf{J}_{12} & \mathbf{0} & \cdots & & & & \cdots & \vdots \\ \vdots & \mathbf{0} & \ddots & \mathbf{0} & & & & & \vdots \\ \vdots & \vdots & & \mathbf{J}_{1N_1} & & & & & \vdots \\ & & & & \mathbf{J}_{21} & & & & \vdots \\ & & & & & \ddots & & & \vdots \\ & & & & & & \lambda_{n_d+1} & & \vdots \\ & & & & & & & \ddots & \vdots \\ & & & & & & & & \lambda_{n_O} \end{pmatrix} \equiv \mathbf{\Xi J}$$

—$\mathbf{J}$ being a diagonalized matrix of Jordan *blocks* extending upward off the main diagonal, or degenerate "blocks" consisting of the single eigenvalue corresponding to the isolated, linearly independent eigenvectors of the original matrix. The structure of these blocks reflects the nature of each linearly independent eigenvector spawning its "family": the dimension of that block indicates the size of the family (including the parent linearly independent eigenvector), and again these block are arrayed in the same order as the eigenvectors, normal and generalized, in the matrix $\mathbf{\Xi}$.

Recall that the matrix $\mathbf{\Xi}$ is constructed of the linearly independent set of eigenvectors, both "normal" and "generalized"; thus it has an inverse (just as the corresponding $\mathbf{\Xi}$ for the case of a full set of "normal" eigenvectors); thus we can ultimately write the above equation in the same form:

$$\mathbf{\Xi}^{-1}\mathbf{A\Xi} = \mathbf{J}. \tag{3.7-2}$$

Again we have a similarity transformation, so again the columns of $\mathbf{\Xi}$ correspond to the [representations, relative to the original basis, of] new basis in terms of which $\mathbf{A}$ is brought to this *block*-diagonalized form $\mathbf{J}$—the *Jordan normal form* of $\mathbf{A}$. Again we observe the generalizing nature of this construction: in the case all the ["normal"] eigenvectors of $\mathbf{A}$ are linearly independent, $\mathbf{J}$ simply collapses to the truly diagonal form $\mathbf{\Lambda}$ obtained in the previous section.

Example 3.7.3. We have already laid the groundwork for this in the previous Example 3.5.5: corresponding to the repeated eigenvalue $\lambda = 5$ there were only two linearly independent eigenvectors, each of which spawned a self-terminating family of generalized eigenvectors. In fact, in that case the matrix $\mathbf{A}$ was *already* in Jordan form; the generalized eigenvectors

$$\boldsymbol{\xi}_{11} = e_1 \quad \boldsymbol{\xi}_{12} = e_2 \quad \boldsymbol{\xi}_{13} = e_3 \quad \boldsymbol{\xi}_{21} = e_4 \quad \boldsymbol{\xi}_{22} = e_5$$

would generate the matrix $\mathbf{\Xi} = \mathbf{1}$, the identity matrix, so that $\mathbf{\Xi}^{-1}\mathbf{A\Xi} = \mathbf{A}$. But the structure of $\mathbf{A}$ exhibits the structure of the families corresponding to the two [true] eigenvectors: $\mathbf{A}$ consists of two Jordan blocks, the 3×3 block corresponding to the [true] eigenvector $\boldsymbol{\xi}_{11}$ and its two offspring, and the 2×2 block associated with $\boldsymbol{\xi}_{12}$ and its single child, $\boldsymbol{\xi}_{22}$. |

Thus we have rounded out the possible normal forms of an arbitrary matrix. Either its [true] eigenvectors are linearly independent or they aren't. In the former case, the matrix can be diagonalized; in the latter, it can't, though it can be *block*-diagonalized to Jordan normal form. There is no other case.

Summary. We have investigated the "normal forms" of a matrix (the *representation* of a linear transformation, recall). In particular, if the eigenvectors of the matrix are linearly independent, its normal form is a *diagonalized* matrix, with the eigen*values* arrayed along the diagonal. [If they are not, the normal form is *block*-diagonalized Jordan normal form.] Such normal forms are arrived at through a *similarity transformation*—a *change of basis*—from that in which the linear transformation was originally represented. The columns of the similarity transformation's matrix are just the eigen*vectors* [or *generalized* eigenvectors] of the original matrix; thus they are nothing more than the new basis vectors, represented in terms of the original basis.

It is too easy to be caught up in the techniques—the *formalism*—of the above "normal forms" and miss the importance of these forms. In point of fact, however, they are actually quite significant: the normal form contains, in a very real sense, the *essence* of a given matrix and the linear transformation it represents. Yet the monicer "normal" (or "standard") form is really a misnomer: it is not so much the *form* which is standard as it is the *basis* in which the matrix representation of a linear transformation assumes that form. Recall, in this regard, the observation (on page 67) that what *defines* a linear transformation on a vector space is what it does to the *basis* vectors. The basis vectors bringing a linear transformation to normal form are just the eigenvectors ("real" and/or generalized). And, in this basis' representation, the transformation merely *stretches* these eigenvectors by a factor of the corresponding eigenvalue—$\mathbf{A}\boldsymbol{\xi}_i = \lambda_i \boldsymbol{\xi}_i$, *independently* of the other basis vectors—in the case the eigenvectors are linearly independent; or transforms them to a simple linear combination of generalized eigenvectors—$\mathbf{A}\boldsymbol{\xi}_{ij}^{(k)} = \lambda_i \boldsymbol{\xi}_{ij}^{(k)} + \boldsymbol{\xi}_{ij}^{(k-1)}$, *almost* independently of other "eigenvectors" and certainly independent of the other "families"—if they aren't.

Though the normal form is arrived at in the above approaches through a similarity transformation, recall that such transformations represent nothing more than a *change* of basis. Conversely, any other representation of this linear transformation would result from another similarity transformation—a change of basis—*of* this normal form; in this regard, then, any other matrix resulting from a similarity transformation of the normal form is truly *equivalent* to it, merely *representing* the *same* linear transformation in another, non-standard, basis.

Summary

This has been an extended chapter, because however "special" the case of square matrices might be, it comprises the one of most practical importance. We discussed the *inverse* of such a matrix, obtaining various properties and showing finally that the inverse will exist if and only if its rows/columns are linearly independent; *i.e.*, the matrix of the $n \times n$ matrix is itself n.

We then discussed *determinant* of a given square matrix, enumerating its properties formally in a series of theorems. In particular, the echelon test determining the rank of a [square] matrix was recast in terms of determinants: that the rows/columns of such a matrix are linearly independent if and only if its determinant is non-0; this, in turn, led to the fact that the *inverse* existed under the same criterion, completing a circle of equivalences connecting linear independence, inverses, and determinants.

After briefly introducing the classification of such matrices as being "symmetric" and/or "orthogonal," we then discussed the importance of the latter, showing that they were equivalent to [the *representation* of] *rotations*, briefly introducing the concepts of "proper" and "improper" rotations [and, more generally, transformations].

We applied the results of linear system solution in general to that of the case in which the number of equations and unknowns were the same. As in the general case, this was tantamount to reducing the coefficient matrix to echelon form, but in the present case, the matrix was square, and the echelon form was nothing more than the identity matrix. A brief discussion of Cramer's Rule followed, leading to the simplified criterion for the unique solution to a system of linear equations was that the determinant of the coefficient matrix [*alone*!] not vanish.

We then turned to the determination of *eigenvalues* and *eigenvectors*—a procedure involving homogeneous systems of equations, providing a useful criterion to predict when the eigenvectors of a given square matrix would be linearly independent. In particular, we showed that eigenvectors corresponding to *distinct* eigen*values* were always linearly independent, though in the case an eigenvalue had multiplicity greater than one, nothing could be said. But, since the eigenvectors of $\mathbf{A}$ corresponding to an eigenvalue λ satisfies the homogeneous equation $|\mathbf{A} - \lambda \mathbf{1}| = \mathbf{0}$, the dimension of the null space of $\mathbf{A} - \lambda \mathbf{1}$ could predict the number of linearly independent eigenvectors even in the case of a repeated root of the characteristic equation. [In the case they *weren't* linearly independent, we could still generate a full set of linearly independent *generalized eigenvectors*—"families" of vectors spawned by the [true] eigenvectors of the original matrix, related to them through the solution of a *non*-homogeneous "eigenvector-like" equation.] We also showed how, in the case $\mathbf{A}$ was *real and symmetric*, the eigenvectors were not only linearly independent, they were actually *orthogonal* (and the eigenvalues are all real). This special case of apparently only academic interest is precisely the one that holds for both the stress matrix and the inertia matrix.

We then considered a critically important application of square matrices:

just how the *representations* of both vectors and linear transformations in an n-dimensional vector space would change under a transformation of that space's basis, explicitly introducing matrices which would effect a *rotation* of orthonormal basis in three dimensions. Finally we considered "normal forms" of a square matrix—"simplified" forms of a matrix resulting from an appropriate choice of basis in terms of which to represent them. It turned out that, if the eigenvectors of the matrix were linearly independent, those eigenvectors were nothing more than the *representation* of the new basis (in terms of the original) in which the matrix representation was *diagonalized.* [In the case the matrix was deficient in linearly independent (true) eigenvectors, the same procedure gave a *block*-diagonal form, the Jordan normal form.]

The length of this chapter merely reflects the importance of the subject to the material in this book, particularly the next Part, where we consider 3-D rigid-body dynamics. There the inertia matrix will be a real, symmetric one; thus all its eigenvectors are linearly independent and, in fact, form an orthonormal set comprising a basis simply a rotation from the original set. These are the *principal axes* in which the matrix becomes diagonalized—the very same principal axes likely already encountered in Solid Mechanics. The orthogonal nature of this basis will often allow us to determine the principal axes without resorting to the diagonalization described above; the important thing is that, even though we might *not* be able to determine principal axes by inspection, they nevertheless always exist!

With these ideas in mind, those concepts to be used in the sequel include:

- *Any* matrix—the *representation* of a linear operator, recall—can be brought into a "standard form," either *diagonalized* (if the matrix' eigenvectors are linearly independent) or "Jordan normal form" (if they are not). This fact allows the analysis of *all* matrices, and thus the linear transformations they represent, in terms of just these two forms!

- As a special case of the above, *real, symmetric* matrices can *always* be diagonalized; in the event the original basis relative to which it was represented is orthonormal, this can be effected through a mere [proper] *rotation* of axes. The latter axes are referred to as *principal axes*, and become fundamental in the study of 3-D rigid body motion in the next part.

- Though the thrust of this Chapter has been on putting an arbitrary matrix in standard form, once that has been done, it can be represented relative to an *arbitrary* set of axes through a *basis transformation*, particularly those effected through *rotation matrices.* This approach will be particularly fruitful in the following Part.

Epilogue

This entire part has been concerned with *linear* systems (as in "*Linear* Algebra"). It is too easy to become comfortable with linear systems, particularly given the amount of material devoted to it thus far. Yet, at the end of the day, *physical systems are inherently nonlinear*, starting with that problem which launched the modern study of mechanics, the "two-body problem" of celestial mechanics, in which the law of gravitation varies as the *inverse square* of the distance (itself a non-linear function of the coordinates) between the two particles. There are just about as many physically linear systems as there are frictionless pins—try to get one of *those* at your local hardware store!

So why this emphasis on linear systems in the face of contrary physical evidence? The reason, quite frankly, is purely a practical one: we can *solve* linear systems of equations; we can *solve* linear differential equations of any order (including the second-order equations resulting from Newton's Second Law). In general we *can't* solve nonlinear systems (at least not analytically).

So how can we reconcile the physical reality of nonlinear systems with the practical inability to solve them? Certainly one possibility is to resort to purely *numerical* solution—numerical integration of the equations of motion for a given system. This is an effective (if somewhat inelegant) means of determining what an arbitrary system is going to do, and is limited only by the accuracies of the analytical model and computational hardware employed. But in order to ascertain just how a given system depends on its various parameters it is necessary to make multiple runs—effectively to make "numerical experiments"—and analyze the results in exactly the same way one does in the laboratory.

A second possibility—and this characterizes "classical" (read "undergraduate") physics and engineering—is to *force* the system to be linear by explicitly *linearizing* it: thus undergraduates study "Vibrations" by studying *linear* vibrations; "Solid Mechanics" relies on the *linear* Hooke's Law relating stress and strain; typical problems in "Fluid Mechanics" generally give conditions which *linearize* the resulting Navier-Stokes equations, precisely to *enable solution.* There may occasionally be nonlinear problems encountered at the lower levels, but these are *soluble* cases whose results apply only to these special cases. (We note, however, that such nonlinearities arise only on the force/moment *kinetic* side of the equations of motion; the derivatives of linear and angular momentum on the *kinematic* side of these equations *are* inherently linear.) The resulting equations at least have the virtue of being able to be solved (as mentioned above); the question remains, however, of whether the process of linearization effectively "throws the baby out with the [nonlinear] bathwater." If nonlinearities are "*small*" (whatever that means precisely), however, one implicitly assumes that the full system will behave "basically" like its linearized one—that the system primarily behaves in the way its linearized model predicts. (And then there's always the "safety factor"—factor of ignorance?—which one invokes to cover any inaccuracies the linearization might introduce.)

To the extent the above assumptions hold, however, linear systems *are* applicable, justifying the emphasis placed on such systems in this Part. In any event, the linear kinematic side of the equations of motion must be dealt with, and in three dimensions—the subject of the next Part—the [linear] "inertia matrix" will be employed extensively.

In the case the nonlinearities are "small," there is a third approach which can be employed to give some analytical handle on the system, *perturbation analysis*. This, however, makes the same implicit assumption that linearization does, albeit more subtly; namely, that the nonlinear system behaves *qualitatively* like the "unperturbed" solution does; if the latter is taken to be linear, however, this may well not be the case.

But the latter suggests a slightly more abstract approach which, nonetheless, has proven valuable in the analysis of physical systems: a *purely qualitative* analysis of the equations of motion. Here we lower our sights somewhat, becoming less interested in the exact nature of the solution than in whether it evolves to a stable final state, effectively escapes the system altogether ("blows up"), or something else. While this may initially seem less satisfying than a full-blown solution, if one is designing a piece of machinery which *will* explode, he is less concerned with the details of just *how* it does than the fact that it *does*! And this approach has a well-founded tradition: in [linear] vibrations, for example, one typically invokes the truism that friction is always present and will damp out any "transient solution" before finally settling to "the steady-state solution," on which primary attention is focussed. This "qualitative approach" has proven most fruitful and is the object of most current research in mechanics. It, too, will be the subject of a later chapter.

Part II

3-D Rigid Body Dynamics

Prologue

In a Statics course, one generally covers both 2- and 3-dimensional equilibrium of both "particles" and "rigid bodies." Two-dimensional equilibrium of particles is easy: one has merely that $\sum \boldsymbol{F}_i = \mathbf{0}$—a *pair* of equations typically involving the support forces as unknowns, whose components are commonly easy to obtain by inspection. That of rigid bodies is little harder: in addition to the above two equations, there is a moment equation, that $\sum \boldsymbol{M}_{A_i} = \mathbf{0}$ about an arbitrary point A in the body. The moments are not too difficult to calculate in this case, since they are invariably assumed to be in a direction perpendicular to the forces, and one can find the *magnitude* using the fact that $\|\boldsymbol{r} \times \boldsymbol{F}\| \equiv \|\boldsymbol{r}\| \, \|\boldsymbol{F}\| \sin\theta = \|\boldsymbol{r}_\perp\| \, \|\boldsymbol{F}\| = \|\boldsymbol{r}\| \, \|\boldsymbol{F}_\perp\|$, and the *direction* by, effectively, "hand waving"; again, these can be found by inspection.

Three-dimensional equilibrium of "particles" is also a straightforward generalization of the 2-D problem: one has merely an additional component of force to include. But that of "rigid bodies" is somewhat more involved: although the force equations are the same as for the "particle" problem, now one must, in one way or another, calculate the cross products defining the moments. At that stage in the student's career, it does pose somewhat of a challenge.

In *Dynamics*, however, the right-hand sides of the above equilibrium equations no longer vanish, although the left-hand sides are precisely the *same.* That of the force equation, giving the *translational* motion of the particle/rigid body, is just $m\boldsymbol{a}$—or $m\bar{\boldsymbol{a}}$ for a rigid body—in whatever coordinate system one chooses to express this acceleration; again, this is not too difficult in either two or three dimensions, as long as vector coordinate systems are well understood. The moment equation, giving the *rotational* motion of a rigid body, is again reasonably straightforward for *planar*, 2-D motion: the moments can be calculated in the same way they are in Statics, and the right-hand side involves a *scalar* expression for the "moment of inertia," I, calculated about an appropriate point.

But experience with *three*-dimensional motion in Statics at least suggests that the generalization of the moment equation to the higher dimension in Dynamics will be a little more involved than simply adding an additional equation, and this is, in fact, correct: In the first place, the moment of inertia is no longer a scalar, but a *matrix*—the "linear transformation" from $\boldsymbol{\omega}$ to $\boldsymbol{H} = \mathsf{I}\boldsymbol{\omega}$ alluded to in Example 2.2.2. The other difficulty is the fact that we no longer have a simple equation of the form "$\sum \boldsymbol{M} = \mathsf{I}\boldsymbol{\alpha}$" as might be expected from experience with planar motion: in the previous material, $\boldsymbol{\omega}$, so its time derivative

$\boldsymbol{\alpha}$, were always assumed to be in a *fixed direction*; in general three-dimensional motion, however, the direction of $\boldsymbol{\omega}$ is no longer fixed—a fact appreciated if one considers the motion of a gyroscope, for example, which generally "precesses."

These, then, are the primary issues to be considered in this Part: determination of the angular acceleration in the case in which the angular acceleration is no longer fixed in direction; how this changing direction of angular velocity affects the angular momentum, whose equation of motion is precisely that given by the moment equation; and the form of the "moment of inertia" in general three-dimensional motion. The matrix nature of the last, in fact, is the primary reason for the first Part of this book: as a linear transformation, its representation will change with a change of basis, and we shall see that this representation in a Cartesian basis is precisely a *real, symmetric matrix*—that special case considered in Theorem 3.5.2 and ultimately culminating in Section 3.7.1.

As is customary in the presentation of dynamics, we shall first consider "kinematics"—the *mathematical* field dealing with the *expression* of translational and rotational velocities and accelerations, and the *relations* among them; then we shall examine *kinetics*—the effect of the *physical* forces and/or couples acting on a given system. Aside from merely organizing the material, this division of labor is actually quite necessary: we can't even write down the equations of motion without a vector coordinate system in terms of which to do so, and having done so, we cannot solve the resulting equations (second-order ordinary differential equations) without knowing how these derivatives interrelate.

Chapter 4

Kinematics

Introduction

As mentioned in the Prologue to this Part, kinematics is effectively the description of motion, here the motion of three-dimensional rigid bodies. The primary distinction between that of a *rigid body* and a *particle* is *rotation*: the former rotate, while the latter do not. Thus we discuss this first, showing that it is the *rigid* nature of a "rigid" body which leads to the *single* quantities of "angular velocity" $\boldsymbol{\omega}$ and "angular acceleration" $\boldsymbol{\alpha}$ characterizing such motion. These are *vectors* so can be added vectorially, while *finite* rotations are not.

But if rotation—the time rate of change of an *angle*—is so central to the motion of rigid bodies, just how *do* we measure such angles in three dimensions? In planar motion, all such angles are confined to remain in the plane of motion, but in general, how can we measure angular orientation? The answer is *Euler angles*, which are introduced next, as contrasted with *direction angles*.

We then consider a primary theme recurring continually in this Part, *rotating coordinate systems*. These will allow us to determine the velocity and acceleration of any point in a rigid body (or *system* of rigid bodies—a "machine") given that of a single point and the rotational motion of the body itself.

The inverse problem—that of determining the angular velocity and acceleration of a body, either isolated or in a machine—generally goes under the rubric of "machine kinematics." This relies on rotating coordinate systems, already considered. One of the primary considerations in such analysis is that of *constraints* on the motion; these will occupy a great deal of attention in this Section, which concludes the discussion of the Chapter.

4.1 Motion of a Rigid Body

It has been mentioned that it is *rotation* which characterizes the motion of a "rigid body" *vis-à-vis* that of a "particle." These classifications are traditional, so observed here, but they can be somewhat misleading: one might infer that a particle is "small," and thus a rigid body "large," yet we will model a 1.5-ton car travelling on a straight road as being a "particle," and a yard stick rotating about one end as a "rigid body." In fact, it is the *rotational* motion of the latter which distinguishes it from the other.

4.1.1 General Motion of a Rigid Body

Consider the arbitrary, *infinitesimal* displacement of a rigid body, shown at left; we see that two points, A and B under such displacement move to the new positions A' and B'. Observe, however, that it is possible to conceptualize this motion by considering it to consist of two parts: the purely *parallel translation* of the points A and B to positions A' and B'' by moving *both* by an amount $d\boldsymbol{r}_A$ (where $\boldsymbol{r}_A$ is measured from some fixed reference point), followed a a *rotation* about A' (leaving that point unchanged) through a certain angle $d\theta$ (where θ is measured from some fixed reference line) to bring B'' to B'. This is the essence of what is generally called *Chasles' Theorem*: that the general motion of a rigid body consists of a *translation* plus a *rotation.*[1]

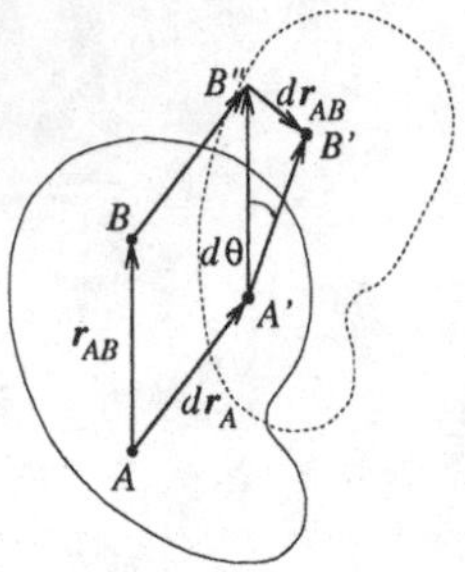

Note that the *absolute* positions of A and B, change to the new positions $\boldsymbol{r}_{A'} = \boldsymbol{r}_A + d\boldsymbol{r}_A$ and $\boldsymbol{r}_{B'} = \boldsymbol{r}_B + d\boldsymbol{r}_B$ (of which only $d\boldsymbol{r}_A$ is shown). On the other hand, if we consider a "*relative* position vector" $\boldsymbol{r}_{AB}$ (as opposed to the *absolute* ones), that vector itself changes from its initial state to the new one $\boldsymbol{r}_{A'B'}$. Thus the [infinitesimal] *vector* change in this vector, $d\boldsymbol{r}_{AB}$, will be as indicated, and we can represent $d\boldsymbol{r}_B = d\boldsymbol{r}_A + d\boldsymbol{r}_{AB}$. Observe that $d\boldsymbol{r}_{AB} \equiv 0$ during translation; it is only *rotation* which generates a change in this relative position vector.

This last characteristic can be seen more analytically by writing $\boldsymbol{r}_{AB} = r_{AB}\hat{\boldsymbol{r}}_{AB}$, where $r_{AB} \equiv \|\boldsymbol{r}_{AB}\|$. Then

$$d\boldsymbol{r}_{AB} = dr_{AB}\hat{\boldsymbol{r}}_{AB} + r_{AB}d\hat{\boldsymbol{r}}_{AB} = r_{AB}d\hat{\boldsymbol{r}}_{AB}, \tag{4.1-1}$$

since $dr_{AB} \equiv 0$ because $\|\boldsymbol{r}_{AB}\|$ is constant. As a unit vector, $\hat{\boldsymbol{r}}_{AB}$ always has a magnitude of unity by definition; thus it can only change due to a change in *direction*—*i.e.*, through *rotation* of $\boldsymbol{r}_{AB}$ due to a corresponding rotation of the rigid body in which it is embedded.

Differentials

We have rather blithely throwing around the term "differential" above. While most might be willing to accept this informally as being an "infinitesimal change," or little less so as being the "derivative times $dx \equiv (x - x_o)$"

$$df \equiv \frac{df(x)}{dx}(x - x_o), \tag{4.1-2}$$

[1] Actually, there are *two* "Chasles' Theorems," the one stated above, and a second, stronger version: that there can always be found a point in the body [extended] moving with a velocity *parallel* to the angular velocity of the body.

—at least for functions of a *scalar* variable—in point of fact there is a rather more rigorous definition with which the student should be familiar:

For a start, we note that the above traditional "definition" of the differential allows us to approximate the function $f(x)$ at a point x_o by a *linear* function:

$$f(x) \doteq f(x_o) + df \equiv f(x_o) + \frac{df(x_o)}{dx}(x - x_o)$$

in the sense that the approximation becomes *identity* in the limit as $x \to x_o$

$$\lim_{x\to x_o} f(x) = f(x_o) + \frac{df(x_o)}{dx}(x - x_o).$$

But the latter definition can immediately be generalized to vector functions of a *vector* variable:

$$\begin{aligned}\lim_{\boldsymbol{x}\to\boldsymbol{x}_o} \boldsymbol{f}(\boldsymbol{x}) &= \boldsymbol{f}(\boldsymbol{x}_o) + d\boldsymbol{f}\\ &\equiv \boldsymbol{f}(\boldsymbol{x}_o) + \frac{d\boldsymbol{f}(\boldsymbol{x}_o)}{d\boldsymbol{x}}(\boldsymbol{x} - \boldsymbol{x}_o)\\ &\equiv \boldsymbol{f}(\boldsymbol{x}_o) + \boldsymbol{D}_{\boldsymbol{f}(\boldsymbol{x}_o)}(\boldsymbol{x} - \boldsymbol{x}_o).\end{aligned}$$

It is this last which can be recast to define the differential:[2]

Definition 4.1.1. The *differential* of a vector function $\boldsymbol{f}(\boldsymbol{x})$ of a vector $\boldsymbol{x}$ is that function $d\boldsymbol{f} \equiv \boldsymbol{D}_{\boldsymbol{f}(\boldsymbol{x}_o)}(\boldsymbol{x} - \boldsymbol{x}_o)$ satisfying

$$\lim_{\boldsymbol{x}\to\boldsymbol{x}_o} \frac{\boldsymbol{f}(\boldsymbol{x}) - \boldsymbol{f}(\boldsymbol{x}_o) - \boldsymbol{D}_{\boldsymbol{f}(\boldsymbol{x}_o)}(\boldsymbol{x} - \boldsymbol{x}_o)}{\|\boldsymbol{x} - \boldsymbol{x}_o\|} = 0, \tag{4.1-3}$$

where $\boldsymbol{D}_{\boldsymbol{f}(\boldsymbol{x}_o)}$ is a linear transformation on $(\boldsymbol{x} - \boldsymbol{x}_o)$ (though $\boldsymbol{D}_{\boldsymbol{f}(\boldsymbol{x}_o)}$ is not necessarily linear in $\boldsymbol{x}$). In particular, then, if $\boldsymbol{f}$ is an m-vector function and $\boldsymbol{x}$ an n-vector variable, $\boldsymbol{D}_{\boldsymbol{f}(\boldsymbol{x}_o)}(\boldsymbol{x} - \boldsymbol{x}_o)$ can be *represented* as the product of a *constant* $m \times n$ matrix $\boldsymbol{D}_{\boldsymbol{f}(\boldsymbol{x}_o)}$ with a column n-matrix $(\boldsymbol{x} - \boldsymbol{x}_o)$.

In effect, what this somewhat austere definition says is that we can *approximate the function* $\boldsymbol{f}(\boldsymbol{x})$ *by* $(\boldsymbol{f}(\boldsymbol{x}_o) + \boldsymbol{D}_{\boldsymbol{f}(\boldsymbol{x}_o)}(\boldsymbol{x} - \boldsymbol{x}_o))$—a "straight line" *tangent* to $\boldsymbol{f}$ at $\boldsymbol{x}_o$. If we slightly rewrite the above definition, we get an equivalent formulation

$$\lim_{\boldsymbol{x}\to\boldsymbol{x}_o} \frac{\boldsymbol{f}(\boldsymbol{x}) - \boldsymbol{f}(\boldsymbol{x}_o)}{\|\boldsymbol{x} - \boldsymbol{x}_o\|} = \lim_{\boldsymbol{x}\to\boldsymbol{x}_o} \frac{\boldsymbol{D}_{\boldsymbol{f}(\boldsymbol{x}_o)}(\boldsymbol{x} - \boldsymbol{x}_o)}{\|\boldsymbol{x} - \boldsymbol{x}_o\|} = \boldsymbol{D}_{\boldsymbol{f}(\boldsymbol{x}_o)}\hat{\boldsymbol{u}}, \tag{4.1-4}$$

where $\hat{\boldsymbol{u}}$ is in the direction of [the limit as $\boldsymbol{x} \to \boldsymbol{x}_o$ of] $(\boldsymbol{x} - \boldsymbol{x}_o)$—the *directional derivative* of $\boldsymbol{f}$ in the $\hat{\boldsymbol{u}}$-direction: the term on the left is essentially the derivative of $\boldsymbol{f}$ at $\boldsymbol{x}_o$, with the denominators involving the *scalar* distance between the two *vectors* $\boldsymbol{x}$ and $\boldsymbol{y}$ to preserve the same vector direction of the approximating differential as the original function $\boldsymbol{f}$ has.

In all the discussion leading up to this, note that our "relative position" $\boldsymbol{r}$ is really a *function* of the *scalar* variable t, allowing the definition to make sense. In this case the definition (4.1-3) becomes simply

$$\lim_{\Delta t\to 0} \frac{\boldsymbol{f}(t + \Delta t) - \boldsymbol{f}(t) - d\boldsymbol{f}}{\|\Delta t\|} = \lim_{\Delta t\to 0} \frac{\boldsymbol{f}(t + \Delta t) - \boldsymbol{f}(t) - \boldsymbol{D}_{\boldsymbol{f}(t)}\Delta t}{\mathrm{sgn}(\Delta t)\,\Delta t} = 0, \tag{4.1-3$'$}$$

[2] and hence the *derivative*—that entity multiplying $(\boldsymbol{x} - \boldsymbol{x}_o)$. On the other hand, consider the difficulties trying to generalize the limit quotient definition of the derivative with respect to a *scalar* variable to a *vector* one!

in which sgn(x) is the sign of the scalar x—*note that $\Delta t \neq 0$ in the definition of the limit*, so this is well-defined—while the equivalent form (4.1-4) becomes

$$\begin{aligned}\lim_{\Delta t \to 0} \frac{\boldsymbol{f}(t+\Delta t) - \boldsymbol{f}(t)}{\Delta t} &\equiv \frac{d\boldsymbol{f}}{dt} \\ &= \lim_{\Delta t \to 0} \frac{\boldsymbol{D}_{\boldsymbol{f}(t)} \Delta t}{\Delta t} = \boldsymbol{D}_{\boldsymbol{f}(t)}\end{aligned} \tag{4.1-4$'$}$$

—demonstrating that the term multiplying Δt in the definition (4.1-3$'$), in fact, corresponds to nothing more than the *derivative* of $\boldsymbol{f}$ at t; in the general vector case (4.1-4), $\boldsymbol{D}_{\boldsymbol{f}(\boldsymbol{x}_o)}$ is just the *Jacobian matrix* of partial derivatives of the components of $\boldsymbol{f}$ with respect to the components of $\boldsymbol{x}$, evaluated at $\boldsymbol{x}_o$: $\left(\frac{\partial f_i(\boldsymbol{x}_o)}{\partial x_j}\right)$, i and j being the row and column indices, respectively. "Multiplying both sides" of (4.1-4$'$) by dt clearly reduces to the more popular definition of differential as (4.1-2): $\lim_{\Delta t \to 0} \boldsymbol{D}_{\boldsymbol{f}(t)} \Delta t \equiv \boldsymbol{D}_{\boldsymbol{f}(t)} dt \equiv d\boldsymbol{f}$, from the above definition (4.1-3$'$).

4.1.2 Rotation of a Rigid Body

Rotation, then, is characteristic of the motion of rigid bodies. But how do we *measure* such "rotation"? Rotation involves the change in an *angle*. And angles are measured between *lines*, such as $A'B''$ and $A'B'$ in the discussion introducing the previous section. Thus, in order to measure the rotation of a body, we must have a *line* fixed in the body against which to measure any possible rotation. This, in turn, requires that the body have *extent*: a "point mass"—conceptualize it as an *infinitesimal* point—might well be "rotating," *but without some line attached to this mass we cannot discern any rotation*! This is precisely the sense in which particles' motion does not involve rotation, but rigid bodies' does: without extent, we cannot draw a "relative position vector" between *two* points in the body in the first place.

The concept of "rotation" relates merely to the [angular] motion of a *line in the body*. The rectangular body to the left, for example, is moving with its center located on a circle about the fixed point O, *but it is not rotating*: a line fixed in the body—its orientation is not important—always remains at a fixed angle to, say, the horizontal. We say the body is *revolving* about O, "revolution" being a *translational* motion rather than a rotational one.

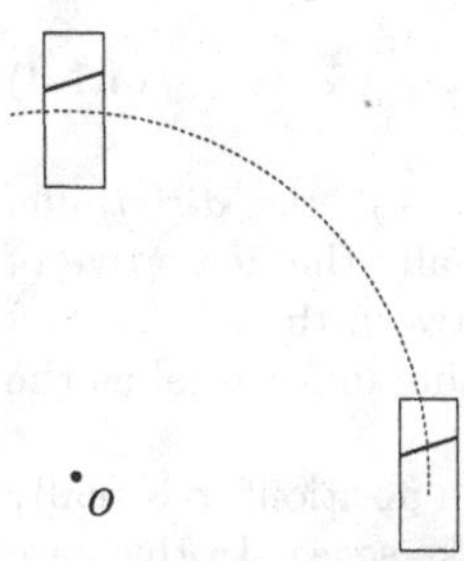

Students will often describe a rotating body as "rotating about [thus and such] a *point*." While there is, in fact, always some such point about which a body is *instantaneously* rotating, the "instantaneous center of zero velocity," one should not become too fixated on it: this "point about which the body is rotating" is fixed only *instantaneously*, changing as the body moves. Trying to ascertain the location of such a point is conceptually irrelevant and misleading—a sort of "red herring"—and altogether a waste of time. Don't bother.

Differential Rigid Body Rotation

Consider a pair of points in a rigid body; the position of one relative to the other can be specified by a vector $\boldsymbol{r}$, as was done above in Section 4.1.1. When that body moves infinitesimally, the positions of the two points change, resulting in a change, $d\boldsymbol{r}$ in this relative position vector. As mentioned above, the general [infinitesimal] motion of a body can be broken up into its translational and rotational components. *But under translation,* $d\boldsymbol{r} \equiv \mathbf{0}$: $\boldsymbol{r}$ merely moves parallel to itself. It is only *rotation* which can result in a $d\boldsymbol{r} \neq \mathbf{0}$.

We shall here obtain a fundamental result: that this *infinitesimal* change in the relative $\boldsymbol{r}$ resulting from a *differential* rotation can be characterized through the use of a cross product with another infinitesimal vector: $d\boldsymbol{r} = d\boldsymbol{\theta} \times \boldsymbol{r}$, *where* $d\boldsymbol{\theta}$ *is a single, unique vector for all points in the body.* What is notable about this result is the fact that it relies only on the stricture that *all points remain a fixed distance apart*! We first present a formal statement of this important fact:

Theorem 4.1.1. The differential of the relative position $\boldsymbol{r}_i$ between *any* points in a rigid body can be expressed $d\boldsymbol{r}_i = d\boldsymbol{\theta} \times \boldsymbol{r}_i$, where $d\boldsymbol{\theta}$ is the *same* for all points.

Note that, in introducing the cross product, we have implicitly characterized the $\boldsymbol{r}_i$ as being a *directed line segments*; thus it will be possible to assess the "parallel" and/or "perpendicular" nature of all vectors under consideration. Such geometrical characteristics will become central to the arguments of the proof:

Proof. We present a constructive proof, exhibiting explicitly the $d\boldsymbol{\theta}$ predicted by the theorem. We warn the reader that this is one of those that goes on and on, riddled with technicalities. On the other hand, remember that we're not going on very much: all we're starting with is the *scalar* fixed-distance hypothesis, and we're trying to prove a *vector* uniqueness result! Bear with us.

Copious use from here on will be made of a fundamental vector identity expressing the *triple vector product* in terms of vector dot products and scalar multiplication: Since (unlike the triple *scalar* product) the operation is defined regardless of the grouping of vectors used to find the first cross product, we must consider two possible cases of such triple vector products. The identities in each case are

$$\boldsymbol{A} \times (\boldsymbol{B} \times \boldsymbol{C}) = (\boldsymbol{A} \cdot \boldsymbol{C})\boldsymbol{B} - (\boldsymbol{A} \cdot \boldsymbol{B})\boldsymbol{C} \tag{4.1-5a}$$

$$(\boldsymbol{A} \times \boldsymbol{B}) \times \boldsymbol{C} = (\boldsymbol{A} \cdot \boldsymbol{C})\boldsymbol{B} - (\boldsymbol{B} \cdot \boldsymbol{C})\boldsymbol{A}; \tag{4.1-5b}$$

note that this means that the different groupings lead to different results. But there is also a pattern in both the above identities: the right-hand side of each starts with the *middle* term multiplied by the dot product of the other two, from which is subtracted the *other* term in the parentheses multiplied by the dot product of the two remaining; this provides a neat mnemonic to remember these. The triple vector product will return repeatedly in this Part; it is useful to remember the above identities.

We consider a rigid body, in which three arbitrary points, O, A, and B are chosen, subject only to the stricture that the relative positions $\boldsymbol{r}_1 \equiv \boldsymbol{r}_{OA}$ and $\boldsymbol{r}_2 \equiv \boldsymbol{r}_{OB}$ not be parallel, and that their *differentials* $d\boldsymbol{r}_1$ and $d\boldsymbol{r}_2$ also not be parallel. (In particular, the last hypothesis means the body must be rotating in the first place: if the relative motion differential $d\boldsymbol{r}_i \equiv 0$ for *all* pairs of points in the body, it must be purely translating.)

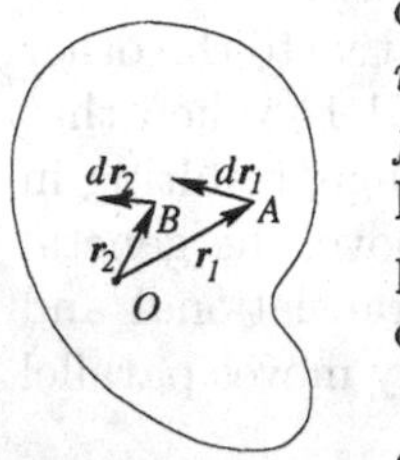

By virtue of the fact that the body is rigid, we have immediately that the $\|\boldsymbol{r}_i\|^2 = \boldsymbol{r}_i \cdot \boldsymbol{r}_i = d_i^2$ are constant. But this means that $d\|\boldsymbol{r}_i\|^2 = 2\boldsymbol{r}_i \cdot d\boldsymbol{r}_i \equiv 0$—*i.e.*, that $\boldsymbol{r}_i$ and $d\boldsymbol{r}_i$ are *perpendicular* (note again that we are dealing with "vectors" as being directed line segments here). It might be possible that *one* of the $d\boldsymbol{r}_i = \boldsymbol{0}$, but not *both* (since the $\boldsymbol{r}_i$ aren't parallel) unless there's no motion *at all*. We will assume initially that both $d\boldsymbol{r}_i \neq \boldsymbol{0}$.

The above hypotheses on the non-parallel nature of the $\boldsymbol{r}_i$ and $d\boldsymbol{r}_i$ might seem at first to be overly restrictive: is it even possible to *find* such points O, A, and B satisfying them? In point of fact, it is: If, given two non-parallel $\boldsymbol{r}_i$, it turns out that $d\boldsymbol{r}_1 \parallel d\boldsymbol{r}_2$, then we consider, rather than $\boldsymbol{r}_2$, the vector $\boldsymbol{r}_2' \equiv \boldsymbol{r}_1 \times \boldsymbol{r}_2$. Clearly this is not parallel to $\boldsymbol{r}_1$; furthermore, its differential change, $d\boldsymbol{r}_2' = d\boldsymbol{r}_1 \times \boldsymbol{r}_2 + \boldsymbol{r}_1 \times d\boldsymbol{r}_2$, is not parallel to $d\boldsymbol{r}_1$:

$$\begin{aligned} d\boldsymbol{r}_2' \times d\boldsymbol{r}_1 &= (d\boldsymbol{r}_1 \times \boldsymbol{r}_2) \times d\boldsymbol{r}_1 + (\boldsymbol{r}_1 \times d\boldsymbol{r}_2) \times d\boldsymbol{r}_1 \\ &= (d\boldsymbol{r}_1 \cdot d\boldsymbol{r}_2)\boldsymbol{r}_2 - (d\boldsymbol{r}_1 \cdot \boldsymbol{r}_2) d\boldsymbol{r}_1 \\ &\quad + (\boldsymbol{r}_1 \cdot d\boldsymbol{r}_1) d\boldsymbol{r}_2 - (d\boldsymbol{r}_1 \cdot d\boldsymbol{r}_2)\boldsymbol{r}_1 \\ &= (d\boldsymbol{r}_1 \cdot d\boldsymbol{r}_2)(\boldsymbol{r}_2 - \boldsymbol{r}_1) \end{aligned}$$

—the terms multiplying the $d\boldsymbol{r}_i$ in the second line vanishing because $d\boldsymbol{r}_1 \cdot \boldsymbol{r}_1 \equiv 0$ while, by the hypothesis on this situation, $d\boldsymbol{r}_1 \parallel d\boldsymbol{r}_2$, so its dot product with $\boldsymbol{r}_2$ also vanishes. But the last term *doesn't* vanish, again by the parallelism of the two differentials, and the fact that the $\boldsymbol{r}_i$ are presumed *not* to be parallel (the reason for this assumption). Note that it is conceivable that $\boldsymbol{r}_2'$ might lie *outside* the body if it doesn't happen to extend that far in this direction (*e.g.*, "thin plates"); in this case we consider the "body extended."

The fact that each $d\boldsymbol{r}_i$ is perpendicular to its $\boldsymbol{r}_i$ means that we can represent it as a cross product of some arbitrary vector with $\boldsymbol{r}_i$. Recall that the $d\boldsymbol{r}_i$ are differential quantities; this motivates our using a similarly differential vector, $d\boldsymbol{\theta}_i$, in the indicated cross products:

$$d\boldsymbol{r}_1 = d\boldsymbol{\theta}_1 \times \boldsymbol{r}_1 \qquad d\boldsymbol{\theta}_1 \nparallel \boldsymbol{r}_1 \tag{4.1-6a}$$

$$d\boldsymbol{r}_2 = d\boldsymbol{\theta}_2 \times \boldsymbol{r}_2 \qquad d\boldsymbol{\theta}_2 \nparallel \boldsymbol{r}_2 \tag{4.1-6b}$$

The above $d\boldsymbol{\theta}_i$ are here arbitrary, in both magnitude and direction. But we claim that we can get an explicit expression for both:

The differential $d\boldsymbol{r}_i$ have directions; in particular, we can write $d\boldsymbol{r}_i \equiv dr_i\hat{\boldsymbol{u}}_i$. We allege that we can express $d\boldsymbol{\theta}_i \equiv d\theta_i\hat{\boldsymbol{u}}_1 \times \hat{\boldsymbol{u}}_2$; indeed, substituting these into

(4.1-6) and expanding the resulting triple cross products using (4.1-5), we get a consistent set

$$d\boldsymbol{r}_1 = dr_1\hat{\boldsymbol{u}}_1 = d\theta_1[(\hat{\boldsymbol{u}}_1 \cdot \boldsymbol{r}_1)\hat{\boldsymbol{u}}_2 - (\hat{\boldsymbol{u}}_2 \cdot \boldsymbol{r}_1)\hat{\boldsymbol{u}}_1] = -d\theta_1(\hat{\boldsymbol{u}}_2 \cdot \boldsymbol{r}_1)\hat{\boldsymbol{u}}_1 \quad (4.1\text{-}7a)$$

$$d\boldsymbol{r}_2 = dr_2\hat{\boldsymbol{u}}_2 = d\theta_2[(\hat{\boldsymbol{u}}_1 \cdot \boldsymbol{r}_2)\hat{\boldsymbol{u}}_2 - (\hat{\boldsymbol{u}}_2 \cdot \boldsymbol{r}_2)\hat{\boldsymbol{u}}_1] = d\theta_2(\hat{\boldsymbol{u}}_1 \cdot \boldsymbol{r}_2)\hat{\boldsymbol{u}}_2 \quad (4.1\text{-}7b)$$

—a pair of equations for the $d\theta_i$, all other quantities being known since the $\boldsymbol{r}_i$ and $d\boldsymbol{r}_i$ are.

But more than that, we claim that $d\boldsymbol{\theta}_1 = d\boldsymbol{\theta}_2$; *i.e.*, that $d\theta_1\hat{\boldsymbol{u}}_1 \times \hat{\boldsymbol{u}}_2 = d\theta_2\hat{\boldsymbol{u}}_1 \times \hat{\boldsymbol{u}}_2$. Clearly these are parallel (or antiparallel), but it remains to show that the $d\theta_i$ are equal. But consider the vector $(\boldsymbol{r}_2 - \boldsymbol{r}_1)$ going from A to B: this has a fixed length, so, by the same reasoning that led to equations (4.1-6), we get that

$$\begin{aligned}
0 &= d(\boldsymbol{r}_2 - \boldsymbol{r}_1) \cdot (\boldsymbol{r}_2 - \boldsymbol{r}_1) = (d\boldsymbol{r}_2 - d\boldsymbol{r}_1) \cdot (\boldsymbol{r}_2 - \boldsymbol{r}_1) \\
&= (d\theta_2(\hat{\boldsymbol{u}}_1 \cdot \boldsymbol{r}_2)\hat{\boldsymbol{u}}_2 + d\theta_1(\hat{\boldsymbol{u}}_2 \cdot \boldsymbol{r}_1)\hat{\boldsymbol{u}}_1) \cdot (\boldsymbol{r}_2 - \boldsymbol{r}_1) \\
&= d\theta_2(\hat{\boldsymbol{u}}_1 \cdot \boldsymbol{r}_2)\hat{\boldsymbol{u}}_2 \cdot \boldsymbol{r}_2 - d\theta_2(\hat{\boldsymbol{u}}_1 \cdot \boldsymbol{r}_2)\hat{\boldsymbol{u}}_2 \cdot \boldsymbol{r}_1 \\
&\quad + d\theta_1(\hat{\boldsymbol{u}}_2 \cdot \boldsymbol{r}_1)\hat{\boldsymbol{u}}_1 \cdot \boldsymbol{r}_2 - d\theta_1(\hat{\boldsymbol{u}}_2 \cdot \boldsymbol{r}_1)\hat{\boldsymbol{u}}_1 \cdot \boldsymbol{r}_1 \\
&= (d\theta_1 - d\theta_2)(\hat{\boldsymbol{u}}_1 \cdot \boldsymbol{r}_2)(\hat{\boldsymbol{u}}_2 \cdot \boldsymbol{r}_1)
\end{aligned}$$

—the second line resulting from substitution of equations 4.1-7 into the [end of the] first, and the first and last terms in the penultimate line vanishing because the $\hat{\boldsymbol{u}}_i \parallel d\boldsymbol{r}_i$, so $\hat{\boldsymbol{u}}_i \cdot \boldsymbol{r}_i \equiv 0$. This implies that $d\theta_1 = d\theta_2$ as long as *neither* of the other two dot products vanishes. But the quantity $\boldsymbol{r}_1 \cdot \boldsymbol{r}_2 = \|\boldsymbol{r}_1\|\,\|\boldsymbol{r}_2\| \cos\theta_{12}$ is a constant since, again, both A and B are fixed in the body; thus

$$\begin{aligned}
d(\boldsymbol{r}_1 \cdot \boldsymbol{r}_2) = d\boldsymbol{r}_1 \cdot \boldsymbol{r}_2 + \boldsymbol{r}_1 \cdot d\boldsymbol{r}_2 = 0 \\
\therefore \quad dr_1\hat{\boldsymbol{u}}_1 \cdot \boldsymbol{r}_2 = -dr_2\hat{\boldsymbol{u}}_2 \cdot \boldsymbol{r}_1;
\end{aligned}$$

thus either *both* vanish, or *neither* vanishes. But if one (so both) vanishes, then so do both of the $d\boldsymbol{r}_i$ in (4.1-7), violating the preliminary hypothesis that both *don't.* This demonstrates the fact that the $d\theta_i$ *are*, in fact, equal, as desired.

Thus we can write that

$$d\boldsymbol{\theta}_i \equiv d\theta_i\hat{\boldsymbol{u}}_1 \times \hat{\boldsymbol{u}}_2 \equiv d\theta\hat{\boldsymbol{u}}_1 \times \hat{\boldsymbol{u}}_2 \equiv d\theta\hat{\boldsymbol{\theta}} \quad (4.1\text{-}8)$$

—an equation essentially *defining* the vector $d\boldsymbol{\theta}$ desired. We observe that this vector is perpendicular to the plane determined by the directions of the $d\boldsymbol{r}_i$.[3]

Lest we forget, recall that the above was derived under the assumption, not only that the $d\boldsymbol{r}_i$ weren't parallel, but also that neither vanished. It might be

[3]Note that, viewing equations (4.1-6) as a pair of non-homogeneous equations for the $d\boldsymbol{\theta}_i$, we have found only the *particular* solutions of each; the *general* solutions would include an element of the null space of [the *matrix*—see Homework 2.2.7 on page 70—corresponding to] the $\times\boldsymbol{r}_i$:

$$d\boldsymbol{\theta}_i = d\boldsymbol{\theta} + \lambda_i\boldsymbol{r}_i$$

(see Theorem 2.3.2). But the only solutions such that $d\boldsymbol{\theta}_1 = d\boldsymbol{\theta}_2$ require that $\lambda_1\boldsymbol{r}_1 = \lambda_2\boldsymbol{r}_2$—*i.e.*, that the $\boldsymbol{r}_i$ be parallel, thus violating the hypotheses of the Theorem.

the case that *one* of them vanishes, however. In this case we merely *define* the direction of $d\boldsymbol{\theta}$ to be parallel to the corresponding vanishing $\mathbf{r}_i$ (violating one of the conditions in equations (4.1-6), but at least giving the appropriate $d\mathbf{r}_i = \mathbf{0}$); the value of $d\theta$ then follows from the *non*-vanishing equation in (4.1-6). (It might be the case that $d\boldsymbol{\theta}$ is actually *anti*-parallel to the offending $\mathbf{r}$; in this case the *scalar* $d\theta$ obtained from the last will be negative, but the sense of the vector will be correct.) □

Recall that we have noted that the $d\mathbf{r}_i$ above will be non-$\mathbf{0}$ only as a result of *rotation.* The above proof is constructive, giving an expression, (4.1-8), for the characteristic $d\boldsymbol{\theta}$ explicitly. The direction of $d\boldsymbol{\theta}$ defines the *[instantaneous] axis of rotation.* In fact, this direction gives that of the rotation *via* the right-hand rule, just as with moments due to forces.

Equation (4.1-1) can be combined with the results of Theorem 4.1.1 to give an important result: dropping the subscripts in that equation and equating it to the expression for $d\mathbf{r}$ in the Theorem, we have

$$d\mathbf{r} = r d\hat{\mathbf{r}} = d\boldsymbol{\theta} \times \mathbf{r} = d\boldsymbol{\theta} \times r\hat{\mathbf{r}} = r\, d\boldsymbol{\theta} \times \hat{\mathbf{r}}, \tag{4.1-9}$$

from which we get immediately that $d\hat{\mathbf{r}} = d\boldsymbol{\theta} \times \hat{\mathbf{r}}$. Here $\hat{\mathbf{r}}$ is the unit vector giving the direction of a relative position $\mathbf{r}$, but clearly this result would hold for *any* unit vector undergoing a rotation. Thus we have

Corollary. The differential change of a unit vector $\hat{\mathbf{u}}$ under a rotation given by $d\boldsymbol{\theta}$ is $d\hat{\mathbf{u}} = d\boldsymbol{\theta} \times \hat{\mathbf{u}}$.

Note that, in contrast with the Theorem, in which, given the $\mathbf{r}_i$ and $d\mathbf{r}_i$, we *determined* the $d\boldsymbol{\theta}$, this Corollary presumes we *know* $d\boldsymbol{\theta}$ and determine $d\hat{\mathbf{u}}$. This result will prove fundamental in what is to follow.

The reader will note that we have been emphasizing the *differential* nature of the vector rotations $d\boldsymbol{\theta}$. This is because, as we noted in the discussion of rotation matrices on page 154, *finite* rotations are *not* vectors. The results of Theorem 4.1.1 gives another demonstration of this fact and the same result utilized previously using rotation matrices. Once again, in order to be able even to *observe* a rotation, we need some line against which to measure an angular change, even though we are interested only in the angular change itself; let such a line be given by the vector $\mathbf{r}$. Just to be contrary, let us attempt to apply the results of the Theorem to *finite* rotations, $\Delta\boldsymbol{\theta}$, whose direction could certainly be defined according to the above paragraph. *Were* we to do so—*this would be incorrect!*—the vector $\mathbf{r}$ under the sequence of rotations $\Delta\boldsymbol{\theta}_1$ and $\Delta\boldsymbol{\theta}_2$ would then become

$$\begin{aligned}
\text{under } \Delta\boldsymbol{\theta}_1:\quad & \mathbf{r} \to \mathbf{r}'_1 = \mathbf{r} + \Delta\boldsymbol{\theta}_1 \times \mathbf{r}\\
\text{under } \Delta\boldsymbol{\theta}_2:\quad & \mathbf{r}'_1 \to \mathbf{r}'_2 = \mathbf{r}'_1 + \Delta\boldsymbol{\theta}_2 \times \mathbf{r}'_1\\
& = (\mathbf{r} + \Delta\boldsymbol{\theta}_1 \times \mathbf{r}) + \Delta\boldsymbol{\theta}_2 \times (\mathbf{r} + \Delta\boldsymbol{\theta}_1 \times \mathbf{r})\\
& = \mathbf{r} + \Delta\boldsymbol{\theta}_1 \times \mathbf{r} + \Delta\boldsymbol{\theta}_2 \times \mathbf{r} + \Delta\boldsymbol{\theta}_2 \times (\Delta\boldsymbol{\theta}_1 \times \mathbf{r}).
\end{aligned}$$

On the other hand, doing the operations in the opposite order:

$$\begin{aligned}
\text{under } \Delta\boldsymbol{\theta}_2: &\quad \boldsymbol{r} \to \boldsymbol{r}_1' = \boldsymbol{r} + \Delta\boldsymbol{\theta}_2 \times \boldsymbol{r} \\
\text{under } \Delta\boldsymbol{\theta}_1: &\quad \boldsymbol{r}_1' \to \boldsymbol{r}_2' = \boldsymbol{r}_1' + \Delta\boldsymbol{\theta}_1 \times \boldsymbol{r}_1' \\
&\quad = \boldsymbol{r} + \Delta\boldsymbol{\theta}_2 \times \boldsymbol{r} + \Delta\boldsymbol{\theta}_1 \times \boldsymbol{r} + \Delta\boldsymbol{\theta}_1 \times (\Delta\boldsymbol{\theta}_2 \times \boldsymbol{r}).
\end{aligned}$$

We see that the two results differ in the last term of each: $\Delta\boldsymbol{\theta}_2 \times (\Delta\boldsymbol{\theta}_1 \times \boldsymbol{r}) \neq \Delta\boldsymbol{\theta}_1 \times (\Delta\boldsymbol{\theta}_2 \times \boldsymbol{r})$. But if we consider *differential* quantities $d\boldsymbol{\theta}_1$ and $d\boldsymbol{\theta}_2$, where we retain only *first*-order terms in such products,[4] the results would be the same: $\boldsymbol{r} \to \boldsymbol{r}_2' = \boldsymbol{r} + (d\boldsymbol{\theta}_1 + d\boldsymbol{\theta}_2) \times \boldsymbol{r}$. How does this demonstrate that finite rotations are not vectors and infinitesimal ones are? Recall that we have defined vector addition to be *commutative*—Property (G5) of Table 1.1 in Chapter 1. If we attempt to call such finite rotations vectors under the "addition" operation of rotations, the above demonstrates that we don't get the same result—that such addition is *not* commutative; the same operation on infinitesimal rotations is.

The non-vector nature of finite rotations can easily be demonstrated: Laying a book flat on a table, it can be rotated about directions through the bottom of the book (the "x-axis") and its spine (the "y-axis"). Rotating about these two axes through an angle of +90°in opposite orders will result in different orientations of the book: the finite additions of such rotations are not commutative.

Angular Velocity and Acceleration

The above section has characterized the rotational motion of a rigid body through the differential change in *relative* position vectors in the body—*lines* against which such rotation can be observed. These differential changes in both $d\boldsymbol{r}$ and $d\boldsymbol{\theta}$ will occur over a similarly differential interval in time, on which both these must depend; in particular, then, we can define the *angular velocity* $\boldsymbol{\omega}$ implicitly:

$$\boldsymbol{v} \equiv \frac{d\boldsymbol{r}}{dt} = \frac{d\boldsymbol{\theta}}{dt} \times \boldsymbol{r} \equiv \boldsymbol{\omega} \times \boldsymbol{r}. \tag{4.1-10}$$

The above informal "derivation" is pretty typical of ones involving differentials. But a more rigorous demonstration of this fact returns to our discussion of differentials above: Knowing that the differential $d\boldsymbol{r} = d\boldsymbol{\theta} \times \boldsymbol{r}$ means, in terms of Definition 4.1.1, that $d\boldsymbol{r}$, so $d\boldsymbol{\theta} \times \boldsymbol{r}$, must satisfy equation 4.1-3:

$$\begin{aligned}
\lim_{\Delta t \to 0} \frac{\boldsymbol{r}(t+\Delta t) - \boldsymbol{r}(t) - d\boldsymbol{r}\,\Delta t}{\|\Delta t\|} &= \lim_{\Delta t \to 0} \frac{\boldsymbol{r}(t+\Delta t) - \boldsymbol{r}(t) - d\boldsymbol{\theta} \times \boldsymbol{r}\,\Delta t}{\operatorname{sgn}(\Delta t)\Delta t} \\
&\equiv \lim_{\Delta t \to 0} \frac{\boldsymbol{r}(t+\Delta t) - \boldsymbol{r}(t) - \boldsymbol{D}_{\boldsymbol{\theta}(t)} \times \boldsymbol{r}\,\Delta t}{\operatorname{sgn}(\Delta t)\Delta t} = 0.
\end{aligned}$$

[4] Ignoring "second-order" terms in the differential can be justified by examining equation (4.1-4′), where such terms in Δt in the numerator will be reduced to only *first* order upon division by the denominator; thus, in the limit, these terms vanish.

But, then,

$$\lim_{\Delta t \to 0} \frac{\boldsymbol{r}(t+\Delta t) - \boldsymbol{r}(t)}{\Delta t} \equiv \frac{d\boldsymbol{r}}{dt}$$

$$= \lim_{\Delta t \to 0} \frac{\boldsymbol{D}_{\boldsymbol{\theta}(t)} \times \boldsymbol{r}\,\Delta t}{\Delta t} = \boldsymbol{D}_{\boldsymbol{\theta}(t)} \times \boldsymbol{r} = \frac{d\boldsymbol{\theta}}{dt} \times \boldsymbol{r} \equiv \boldsymbol{\omega} \times \boldsymbol{r},$$

the penultimate equality by (4.1-4′).

We observe that the direction of $\boldsymbol{\omega}$, $\hat{\boldsymbol{\omega}}$, is the same as that of $d\boldsymbol{\theta}$; this effectively *defines* the direction of the angular velocity: $\boldsymbol{\omega} \equiv \omega\hat{\boldsymbol{\omega}}$,. Note, however, that we do not attempt to define this through the normal definition of the derivative,

$$\frac{d\boldsymbol{\theta}}{dt} = \lim_{\Delta t \to 0} \frac{\Delta\boldsymbol{\theta}}{\Delta t},$$

since, as we have already noted, $\Delta\boldsymbol{\theta}$ is itself not a vector; we *start* with the derivative (the angular velocity). The situation is rather like the definition of tangential-normal coordinates, in which *position* is not defined: such coordinates start with the definition of $\boldsymbol{v} \equiv v\hat{\boldsymbol{t}} = \frac{dr}{dt}\hat{\boldsymbol{t}}$, *defining* the $\hat{\boldsymbol{t}}$-direction to be parallel to the differential change in [absolute] position, $d\boldsymbol{r}$.

Since the angular velocity is defined in terms of the *vector* $d\boldsymbol{\theta}$, $\boldsymbol{\omega}$ itself is a vector; thus angular velocities *add* vectorially. In particular, the net result of imposing two components of angular velocity, an *absolute* $\boldsymbol{\omega}_1$ and $\boldsymbol{\omega}_2$ measured *relative to* $\boldsymbol{\omega}_1$, on a rigid body is a *single, absolute* angular velocity

$$\boldsymbol{\omega} = \boldsymbol{\omega}_1 + \boldsymbol{\omega}_2. \tag{4.1-11}$$

Thus, for example, in gyroscopic motion, the *total* angular velocity of the rotor would be the [vector] sum of its spin and any precession about a vertical axis.

Having established angular velocity, we can now define the *angular acceleration:*

$$\boldsymbol{\alpha} \equiv \frac{d\boldsymbol{\omega}}{dt} = \frac{d(\omega\hat{\boldsymbol{\omega}})}{dt} = \frac{d\omega}{dt}\hat{\boldsymbol{\omega}} + \omega\frac{d\hat{\boldsymbol{\omega}}}{dt}. \tag{4.1-12}$$

The last term above, resulting from a change in the *direction* of the angular velocity, vanishes in *planar* problems: $\boldsymbol{\omega} \equiv \omega\hat{\boldsymbol{k}}$ is *defined* to be fixed in direction, so that $\dot{\hat{\boldsymbol{\omega}}} \equiv \dot{\hat{\boldsymbol{k}}} \equiv 0$. In general however, particularly in 3-D motion, it isn't. For example, the gyroscopic spin of the rotor *precesses* about a vertical axis; thus, even if both these components were constant in *magnitude*, there would be a contribution to $\boldsymbol{\alpha}$ due to the change in direction of the spin axis. As with the vector angular velocity, angular accelerations also add:

$$\boldsymbol{\alpha} = \boldsymbol{\alpha}_1 + \boldsymbol{\alpha}_2. \tag{4.1-13}$$

(though, if we start with the angular velocities, the individual accelerations will include the effects of changes in the directions of both angular velocities).

An examination of (4.1-12) is instructive: In the fist place, we see that $\frac{d\hat{\boldsymbol{\omega}}}{dt} \perp \hat{\boldsymbol{\omega}}$, since

$$2\hat{\boldsymbol{\omega}} \cdot \frac{d\hat{\boldsymbol{\omega}}}{dt} = \frac{d}{dt}(\hat{\boldsymbol{\omega}} \cdot \hat{\boldsymbol{\omega}}) = \frac{d(1)}{dt} \equiv 0;$$

thus we see that $\boldsymbol{\alpha}$ in general has two components: The first, $\frac{d\omega}{dt}\hat{\boldsymbol{\omega}}$, is parallel to $\boldsymbol{\omega}$ and corresponds to a change in $\omega \equiv \|\boldsymbol{\omega}\|$; the other, $\omega\frac{d\hat{\boldsymbol{\omega}}}{dt}$, will be perpendicular to the first, and arises from a change in *direction* of $\boldsymbol{\omega}$. Clearly, a *known* $\boldsymbol{\alpha}$ can always be split up into components parallel and perpendicular to $\boldsymbol{\omega}$:

$$\boldsymbol{\alpha} = (\boldsymbol{\alpha}\cdot\hat{\boldsymbol{\omega}})\,\hat{\boldsymbol{\omega}} + (\boldsymbol{\alpha} - (\boldsymbol{\alpha}\cdot\hat{\boldsymbol{\omega}})\hat{\boldsymbol{\omega}}),$$

with the first term corresponding to $\frac{d\omega}{dt}\hat{\boldsymbol{\omega}}$ in (4.1-12).

Homework:

1. Given $\boldsymbol{\alpha}$ and $\boldsymbol{\omega}$, find $\frac{d\omega}{dt}$.

Time Derivative of a Unit Vector with respect to Rotation

We have already seen in (4.1-12) that the change in direction of a vector will *contribute* to its time derivative; what has yet to be shown is how to *evaluate* this contribution. But we already have the mechanism to do this: We have seen, in the Corollary, on page 184, to Theorem 4.1.1, that the differential change, $d\hat{\boldsymbol{u}}$ of a unit vector $\hat{\boldsymbol{u}}$ under a differential rotation $d\boldsymbol{\theta}$ is just $d\hat{\boldsymbol{u}} = d\boldsymbol{\theta}\times\hat{\boldsymbol{u}}$. "Dividing by dt"[5] we have immediately that

Theorem 4.1.2. The time rate of change of a unit vector $\hat{\boldsymbol{u}}$ due to an angular velocity $\boldsymbol{\omega}$ is $\dot{\hat{\boldsymbol{u}}} = \boldsymbol{\omega}\times\hat{\boldsymbol{u}}$.

Homework:

1. Assume axes are rotating with [absolute] angular velocity $\boldsymbol{\omega} = \boldsymbol{\omega}_1 + \boldsymbol{\omega}_2$, where $\boldsymbol{\omega}_2$ is measured *relative to* $\boldsymbol{\omega}_1$. *Denoting* the time derivative of vector $\boldsymbol{V}$ observed with respect to a coordinate system rotating with $\boldsymbol{\omega}_i$ by "$\left.\frac{d\boldsymbol{V}}{dt}\right|_{\boldsymbol{\omega}_i}$," show that these time derivatives are related:

$$\left.\frac{d\boldsymbol{V}}{dt}\right|_{\boldsymbol{\omega}_1+\boldsymbol{\omega}_2} = \left.\frac{d\boldsymbol{V}}{dt}\right|_{\boldsymbol{\omega}_1} - \boldsymbol{\omega}_2\times\boldsymbol{V}.$$

With all this in hand, we are now ready to apply it to a couple of "practical" problems:

Example 4.1.1 *A Simple Rotating System.* We consider a rotor attached perpendicularly to the end of an arm. The arm rotates about the fixed origin O in the XY-plane with at a constant rate of $\dot{\psi}$, while the rotor rotates *relative to*

[5] This can be done rigorously in precisely the same way as showing that $\boldsymbol{v} = \boldsymbol{\omega}\times\boldsymbol{r}$ on page 185: simply replace "$\boldsymbol{r}$" in those equations by "$\hat{\boldsymbol{u}}$."

the arm at a constant rate p. We wish to determine the total angular velocity of the rotor, as well as its angular acceleration.

We note that, from the definition of angular velocity, the directions of the two rotations are given by vectors along the axes of rotation *via* the right-hand rule. (In the diagram, both the "rotation" and the defined angular velocity are indicated; in general, however, we will only diagram the latter in the balance of this exposition.) Since the arm is constrained to move in the XY-plane, its angular velocity is [always] perpendicular to that plane, so $\dot{\boldsymbol{\psi}} = \dot{\psi}\,\hat{\boldsymbol{K}}$; while $\boldsymbol{p}$ is in the direction of the arm itself, here [only instantaneously] in the $\hat{\boldsymbol{I}}$-direction: $\boldsymbol{p} = p\hat{\boldsymbol{I}}$.

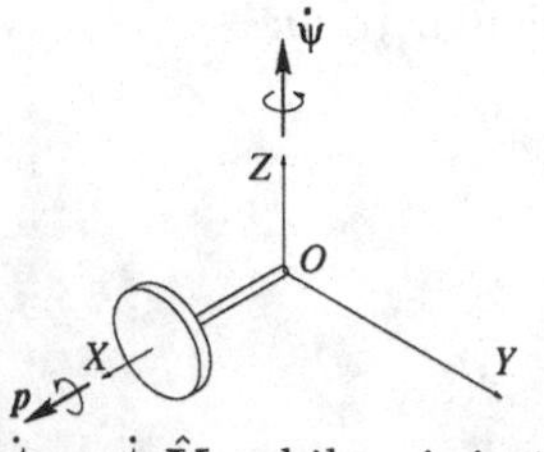

The angular velocity is easy: since we "start" at this level in defining the angular motion, we have immediately that, *at this instant*

$$\boldsymbol{\omega} = \dot{\boldsymbol{\psi}} + \boldsymbol{p} = \dot{\psi}\,\hat{\boldsymbol{K}} + p\hat{\boldsymbol{I}}.$$

To determine the angular acceleration, however, we must now account for any change in the *direction* of these vectors. To keep track of these, it is useful to write the above equation in a form amenable to the use of equation (4.1-12):

$$\boldsymbol{\omega} = \dot{\psi}\hat{\dot{\boldsymbol{\psi}}} + p\hat{\boldsymbol{p}};$$

(note that "$\hat{\dot{\boldsymbol{\psi}}}$" stands for the *direction* of $\dot{\boldsymbol{\psi}}$—here $\hat{\boldsymbol{K}}$) so that

$$\boldsymbol{\alpha} \equiv \frac{d\boldsymbol{\omega}}{dt} = \frac{d\dot{\psi}}{dt}\hat{\dot{\boldsymbol{\psi}}} + \dot{\psi}\frac{d\hat{\dot{\boldsymbol{\psi}}}}{dt} + \frac{dp}{dt}\hat{\boldsymbol{p}} + p\frac{d\hat{\boldsymbol{p}}}{dt},$$

in which $\ddot{\psi} \equiv 0 \equiv \dot{p}$ by the presumed constancy of the two angular rates, so the first and third terms above vanish. But we must evaluate the directions' time derivatives. This is where Theorem 4.1.2 comes in: Each of the derivatives arises as a result of *rotation*; in particular, we note that

- $\hat{\dot{\boldsymbol{\psi}}} = \hat{\boldsymbol{K}}$ is *constant* in direction, while
- $\hat{\boldsymbol{p}}$ is carried about by $\dot{\boldsymbol{\psi}}$.

This means that $\frac{d\hat{\dot{\boldsymbol{\psi}}}}{dt} \equiv 0$ (so the second term vanishes also), while, applying the Theorem,

$$\frac{d\hat{\boldsymbol{p}}}{dt} = \dot{\boldsymbol{\psi}} \times \hat{\boldsymbol{p}} = \dot{\psi}\hat{\boldsymbol{K}} \times \hat{\boldsymbol{I}} = \dot{\psi}\hat{\boldsymbol{J}}.$$

Note that we are not saying that $\hat{\boldsymbol{I}}$ is rotating with $\dot{\boldsymbol{\psi}}$: $\hat{\boldsymbol{p}}$ at this *instant* is in the $\hat{\boldsymbol{I}}$-direction. Thus, finally,

$$\boldsymbol{\alpha} = p\frac{d\hat{\boldsymbol{p}}}{dt} = p\dot{\psi}\hat{\boldsymbol{J}}.$$

We observe that $\boldsymbol{\alpha} \perp \boldsymbol{\omega}$; in light of the discussion on page 187, this is the result completely of a change in the direction of $\boldsymbol{\omega}$, as hypothesized in the example itself.

Again, it must be emphasized that all these results hold *only at this instant*, since the rotor arm will continue to rotate from its present position, changing the direction of $\mathbf{p}$ (if not $\dot{\boldsymbol{\psi}}$). |

The above example was intentionally a simple "starter." The next is rather more impressive, but no less "practical":

Example 4.1.2 *Gyroscopic Motion.* Diagrammed here is the gyroscope first encountered in Example 3.6.3 in the section on rotation matrices. But now we have included the possibility that the body is rotating: It will have a certain *spin*, $\mathbf{p}$, along its own axis; note that this axis always lies in a plane passing through the Z-axis: since one end lies on the origin, its other will determine a unique plane through that point and the coordinate axis. But in addition, there can be a *precession* of this axis about the $\hat{\mathbf{K}}$ axis (assumed fixed), as well as a *nutation* or "nodding" of that axis relative to the Z-axis. From the definition of angular velocity, each of these will be a vector parallel to its axis of rotation. If the angle between the spin and Z axes is θ, and that between the plane containing the spin axis and the XZ-plane is ψ, we can express these angular velocities relative to the basis vectors $\hat{\mathbf{I}}$, $\hat{\mathbf{J}}$, and $\hat{\mathbf{K}}$. These will be $\mathbf{p} \equiv p\hat{\mathbf{p}} = p\,(\sin\theta\cos\psi\hat{\mathbf{I}} + \sin\theta\sin\psi\hat{\mathbf{J}} + \cos\theta\hat{\mathbf{K}})$, $\dot{\boldsymbol{\psi}} = \dot{\psi}\,\hat{\mathbf{K}}$, and $\dot{\boldsymbol{\theta}} = \dot{\theta}\,(-\sin\psi\hat{\mathbf{I}} + \cos\psi\hat{\mathbf{J}})$, the direction of the latter following from the fact that it is perpendicular to the plane containing both $\hat{\mathbf{K}}$ and the rotor, so lies in the XY-plane.

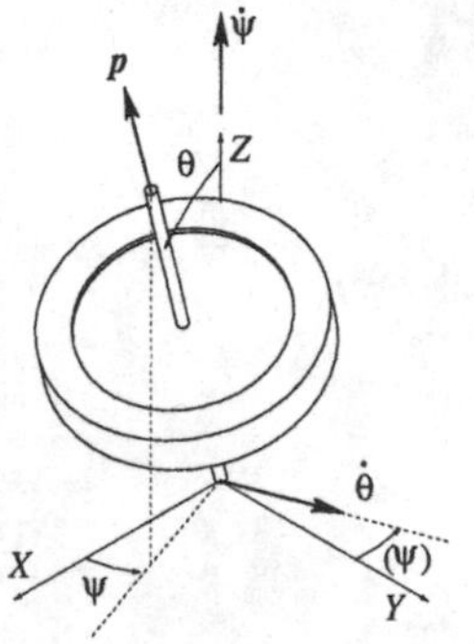

Since these angular velocities *are* vectors, they can be added together immediately to obtain the *total* angular velocity of the gyroscope:

$$\begin{aligned}\boldsymbol{\omega} &= \mathbf{p} + \dot{\boldsymbol{\psi}} + \dot{\boldsymbol{\theta}} \\ &= (p\sin\theta\cos\psi - \dot{\theta}\sin\psi)\,\hat{\mathbf{I}} + (p\sin\theta\sin\psi + \dot{\theta}\cos\psi)\,\hat{\mathbf{J}} + (p\cos\theta + \dot{\psi})\,\hat{\mathbf{K}};\end{aligned}$$

again, the relatively straightforward nature of this addition results from the fact that our description of angular motion *starts* from the "velocity level." But when it comes to determination of the angular *acceleration*, we must now account for the changes in *direction* of these vectors: Writing the angular velocity in terms of each contribution's component and its direction,

$$\boldsymbol{\omega} = p\hat{\mathbf{p}} + \dot{\psi}\hat{\dot{\boldsymbol{\psi}}} + \dot{\theta}\hat{\dot{\boldsymbol{\theta}}},$$

the angular acceleration will be the total time derivative of this quantity:

$$\begin{aligned}\boldsymbol{\alpha} \equiv \frac{d\boldsymbol{\omega}}{dt} &= \frac{dp}{dt}\hat{\boldsymbol{p}} + p\frac{d\hat{\boldsymbol{p}}}{dt} + \frac{d\dot{\psi}}{dt}\hat{\dot{\boldsymbol{\psi}}} + \dot{\psi}\frac{d\hat{\dot{\boldsymbol{\psi}}}}{dt} + \frac{d\dot{\theta}}{dt}\hat{\dot{\boldsymbol{\theta}}} + \dot{\theta}\frac{d\hat{\dot{\boldsymbol{\theta}}}}{dt} \\ &= \dot{p}\hat{\boldsymbol{p}} + p\frac{d\hat{\boldsymbol{p}}}{dt} + \ddot{\psi}\hat{\dot{\boldsymbol{\psi}}} + \dot{\psi}\frac{d\hat{\dot{\boldsymbol{\psi}}}}{dt} + \ddot{\theta}\hat{\dot{\boldsymbol{\theta}}} + \dot{\theta}\frac{d\hat{\dot{\boldsymbol{\theta}}}}{dt}.\end{aligned} \tag{4.1-14}$$

The time derivatives of p, $\dot{\psi}$, and $\dot{\theta}$, measuring the rates of change of the various *components* of $\boldsymbol{\omega}$ *along their directions*, are presumably known, but it is necessary to evaluate these directions' time derivatives. Again, we note that

- $\hat{\dot{\boldsymbol{\psi}}}$ is constant,
- $\hat{\dot{\boldsymbol{\theta}}}$ is carried about by $\dot{\boldsymbol{\psi}}$, while
- $\hat{\boldsymbol{p}}$ is carried about by *both* $\dot{\boldsymbol{\psi}}$ *and* $\dot{\boldsymbol{\theta}}$;

thus, applying Theorem 4.1.2,

$$\begin{aligned}\frac{d\hat{\dot{\boldsymbol{\theta}}}}{dt} &= \dot{\boldsymbol{\psi}} \times \hat{\dot{\boldsymbol{\theta}}} \\ &= (\dot{\psi}\hat{\boldsymbol{K}}) \times (-\sin\psi\hat{\boldsymbol{I}} + \cos\psi\hat{\boldsymbol{J}}) \\ &= -\sin\psi\dot{\psi}\hat{\boldsymbol{J}} - \cos\psi\dot{\psi}\hat{\boldsymbol{I}}\end{aligned}$$

while, since the angular velocity of $\hat{\boldsymbol{p}}$ is the vector sum of its constituent parts,

$$\begin{aligned}\frac{d\hat{\boldsymbol{p}}}{dt} &= (\dot{\boldsymbol{\psi}} + \dot{\boldsymbol{\theta}}) \times \hat{\boldsymbol{p}} \\ &= (\dot{\psi}\hat{\boldsymbol{K}} + \dot{\theta}(-\sin\psi\hat{\boldsymbol{I}} + \cos\psi\hat{\boldsymbol{J}})) \times (\sin\theta\cos\psi\hat{\boldsymbol{I}} + \sin\theta\sin\psi\hat{\boldsymbol{J}} + \cos\theta\hat{\boldsymbol{K}}) \\ &= \sin\theta\cos\psi\dot{\psi}\hat{\boldsymbol{J}} - \sin\theta\sin\psi\dot{\psi}\hat{\boldsymbol{I}} - \sin\theta\sin^2\psi\dot{\theta}\hat{\boldsymbol{K}} + \cos\theta\sin\psi\dot{\theta}\hat{\boldsymbol{J}} \\ &\qquad - \sin\theta\cos^2\psi\dot{\theta}\hat{\boldsymbol{K}} + \cos\theta\cos\psi\dot{\theta}\hat{\boldsymbol{I}}.\end{aligned}$$

These, along with the values for $\dot{p}$, $\ddot{\psi}$, and $\ddot{\theta}$, are substituted into equation (4.1-14) to get the final answer. (This will not be done here: the point of this example is merely to demonstrate the technique for a somewhat general problem.) The procedure, though tedious, is reasonably straightforward, at least once the application of (4.1-14), and the reasoning bulleted above, is understood. Note that, in contrast with the previous example, in which the spin direction was specified only *instantaneously*, here the directions of the various rotations are given in terms of arbitrary values of the angles; thus the result would hold *generally*. |

Summary. In general, the displacement of a rigid body consists of two components: *translation* and *rotation*; it is the latter which distinguishes its motion from that of "particles." The *infinitesimal* change in a *relative* position vector

$\boldsymbol{r}$ in a rigid body, due only to the constraint that this position have a constant length, is given by $d\boldsymbol{r} = d\boldsymbol{\theta} \times \boldsymbol{r}$, for some characteristic "rotation" $d\boldsymbol{\theta}$ which is the *same* for all pairs of points in the body. This *infinitesimal* rotation is a vector, though *finite* rotations generally aren't. As a result of this fact, the *angular velocity* $\boldsymbol{\omega} \equiv \frac{d\boldsymbol{\theta}}{dt}$ and *angular acceleration* $\boldsymbol{\alpha} \equiv \frac{d\boldsymbol{\omega}}{dt}$ are also vectors, so add vectorially; furthermore, the time derivative of any unit vector $\hat{\boldsymbol{u}}$ (which is itself fixed in magnitude) can only arise from rotation $\boldsymbol{\omega}$, and is given by $\dot{\hat{\boldsymbol{u}}} = \boldsymbol{\omega} \times \hat{\boldsymbol{u}}$. In particular, given the angular velocity itself, expressed as $\boldsymbol{\omega} = \omega\hat{\boldsymbol{\omega}}$, the corresponding angular acceleration will generally have two terms, including one due to the change in direction, due to some rotation $\boldsymbol{\Omega}$, of $\boldsymbol{\omega}$:

$$\boldsymbol{\alpha} = \dot{\omega}\hat{\boldsymbol{\omega}} + \omega\dot{\hat{\boldsymbol{\omega}}} = \dot{\omega}\hat{\boldsymbol{\omega}} + \omega\boldsymbol{\Omega} \times \hat{\boldsymbol{\omega}} = \dot{\omega}\hat{\boldsymbol{\omega}} + \boldsymbol{\Omega} \times \boldsymbol{\omega}.$$

Since $\boldsymbol{\alpha}$ can always be split up into components parallel and perpendicular to $\boldsymbol{\omega}$, this provides us with a means of determining the separate contributions due to $\dot{\omega}$ and $\boldsymbol{\Omega}$.

4.2 Euler Angles

As mentioned in the Prologue to this Part, there are a number of non-trivial generalizations required to go from planar, two-dimensional motion to general three-dimensional motion of rigid bodies. One of the most notable is merely how to *describe* the motion in 3-D: while the higher-dimensional formulations of the standard vector coordinate systems are easy to apply to the description of [3-D] *points*, it is rotation—the *angular* motion—which primarily distinguishes "rigid bodies" from "particles," as discussed in the previous section. While that section focussed on the time derivatives of such angles, we never quite came to grips with just what angles might be *used* to describe the orientation of a rigid body. That is the topic of the present section.

But what ultimately do we have to describe? It is the *position* and *orientation* of a rigid body which are required to specify its configuration; these, in turn, could be given by the position of a point in the body, and the rotation of the body relative to some axes. Both of these, in turn, can be expressed using a *coordinate system* fixed in the body: the origin of that system gives a reference point for position, while the directions of its axes determine its rotation. More importantly, we are ultimately interested in describing the positions of various *other* points in the body, since, as we shall see, the distribution of mass in the system affects the motion through the "inertia tensor" measuring this distribution; again, this requires a coordinate system. Thus, in the end, what we are interested in specifying is such coordinate systems.

In order to elucidate this topic, we first consider the angular description of mere *vectors*, utilizing ideas from this simple case to motivate the specification of coordinate systems. In both cases, we will consider two alternatives: *direction angles* and their cosines (the "direction cosines," appropriately enough), and then "*Euler angles*" more generally used for this purpose. As we shall see, the reason for preferring the latter is the fact that direction angles actually require one more parameter to define direction for vectors, *six* for coordinate systems, than absolutely necessary. The angles are, in fact, a *de*pendent set of variables, as opposed to the Euler angles which overcome this redundancy; it is this contrast providing the pedagogical rationale for resurrecting them here at all. It is also here that material from Part I first starts to return: we shall see that the *rotation matrices* (Section 3.6.2) can be implemented directly using the Euler angles, another advantage of their use.

4.2.1 Direction Angles and Cosines

It is common to represent vectors in Cartesian coordinates using *direction cosines*—the cosines of angles between the axes and the vector. As mentioned above, we present these first in the context of vectors (in terms of which, as we shall see, they are primarily defined), then in terms of the coordinate systems which form our central concern.

Vector Description

Given an arbitrary Cartesian vector $\boldsymbol{V} = V_X\hat{\boldsymbol{I}} + V_Y\hat{\boldsymbol{J}} + V_Z\hat{\boldsymbol{K}}$, let us, more or less arbitrarily, consider the dot products of that vector with the respective unit vectors along the axes: Taking first $\hat{\boldsymbol{I}}$, $\boldsymbol{V}\cdot\hat{\boldsymbol{I}} \equiv \|\boldsymbol{V}\|\,\|\hat{\boldsymbol{I}}\|\cos\theta_X = \|\boldsymbol{V}\|\cos\theta_X$, where θ_X is defined to be *that angle between* $\hat{\boldsymbol{I}}$ *and* $\boldsymbol{V}$—*i.e.*, between $\boldsymbol{V}$ and the *positive* X-axis, *measured in their common* $\boldsymbol{V}X$*-plane.* Taking the same dot product with the Cartesian representation, we have immediately that $\boldsymbol{V}\cdot\hat{\boldsymbol{I}} \equiv (V_X\hat{\boldsymbol{I}} + V_Y\hat{\boldsymbol{J}} + V_Z\hat{\boldsymbol{K}})\cdot\hat{\boldsymbol{I}} = V_X$. Proceeding similarly with the other two unit vectors, we have ultimately that

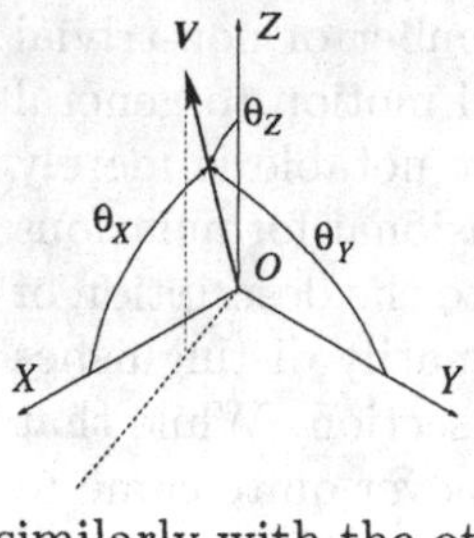

$$\begin{aligned}\boldsymbol{V} &= \|\boldsymbol{V}\|\,(\cos\theta_X\hat{\boldsymbol{I}} + \cos\theta_Y\hat{\boldsymbol{J}} + \cos\theta_Z\hat{\boldsymbol{K}})\\ &= \|\boldsymbol{V}\|\hat{\boldsymbol{V}}\\ &= V_X\hat{\boldsymbol{I}} + V_Y\hat{\boldsymbol{J}} + V_Z\hat{\boldsymbol{K}}.\end{aligned}$$

Thus we obtain immediately two results:

$$V_X = \|\boldsymbol{V}\|\cos\theta_X, \qquad V_Y = \|\boldsymbol{V}\|\cos\theta_Y, \qquad V_Z = \|\boldsymbol{V}\|\cos\theta_Z,$$

giving a relation between the Cartesian components and $\|\boldsymbol{V}\|$ and the angles θ_X, θ_Y, and θ_Z—the "magnitude-direction" formulation of $\boldsymbol{V}$—and

$$\hat{\boldsymbol{V}} = \cos\theta_X\hat{\boldsymbol{I}} + \cos\theta_Y\hat{\boldsymbol{J}} + \cos\theta_Z\hat{\boldsymbol{K}},$$

showing how an arbitrary unit vector can be expressed in terms of the direction cosines. It is common to denote the direction cosines by

$$l \equiv \cos\theta_X, \qquad m \equiv \cos\theta_Y, \qquad n \equiv \cos\theta_Z. \tag{4.2-1}$$

We note immediately that

$$l^2 + m^2 + n^2 \equiv \cos^2\theta_X + \cos^2\theta_Y + \cos^2\theta_Z = \|\hat{\boldsymbol{V}}\|^2 = 1 \tag{4.2-2}$$

is an identity which must hold among these direction cosines; thus they are, in fact, *not independent*: given two of these, as well as the *sense* of the third (or whether its angle is greater or less than 90°), we have the third.

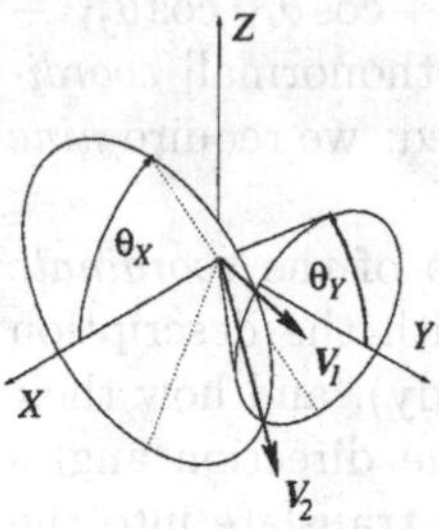

That only two direction angles are independent is reasonable: If we were given two angles, say θ_X and θ_Y, the vector would have to lie simultaneously on one cone making an angle of θ_X with the X-axis and another making one of θ_Y with the Y-axis—*i.e.*, on the *intersection* of these two cones. But two cones will intersect most generally along two lines, as illustrated in the diagram at left. Note that the two vectors $\boldsymbol{V}_1$ and $\boldsymbol{V}_2$ make angles θ_Z with the Z-axis satisfying $\cos^2\theta_Z = 1 - \cos^2\theta_X - \cos^2\theta_Y$; the two cosines differ only in sign, showing that one angle is merely the supplement of the other. (Don't forget that θ_Z is measured from the *positive* Z-axis.) Were the two cones tangent, there would be only one intersection, and it is even possible that, if the two given angles were too small, the cones wouldn't intersect at all! (This would be signalled by $\cos^2\theta_Z$ becoming negative, due to the subtraction of two cosines with squares close to 1.) Another way of seeing that only two angles need—*can*—be specified independently is to note that the *direction* $\hat{\boldsymbol{V}}$ must lie on the surface of a sphere of radius 1, and points on this sphere can be specified with only two angles, say "latitude" and "longitude."

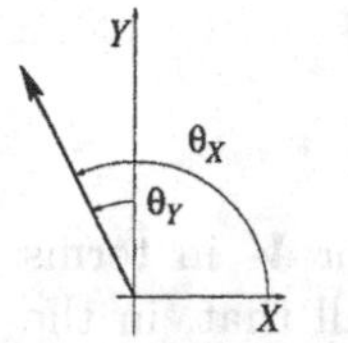

This result is neither that abstruse nor surprising: it is merely a generalization of the same situation in *planar* problems, in which, again taking the angles between the *positive* X- and Y-axes and a vector $\boldsymbol{V}$ to be "θ_X" and "θ_Y," we have also that $V_X = \|\boldsymbol{V}\|\cos\theta_X$ and $V_Y = \|\boldsymbol{V}\|\cos\theta_Y$, and $\cos^2\theta_X + \cos^2\theta_Y = 1$. Once again, knowing either angle and how the other compares with 90°, the latter is determined; this makes sense, since we clearly can specify the direction $\hat{\boldsymbol{V}}$ with only the *single* angle θ_X, say. In either event, it takes two or three angles to specify the vector's direction, only one or two of which are actually independent.

Coordinate System Description

Direction angles are defined in terms of the orientation of a *vector*, but we are interested in describing the orientation of a *coordinate system*—a set of *three* [unit] vectors. To do this with direction angles would thus require *nine* direction angles, θ_{ij} say, three for each axis, of which, from the above discussion, only six are expected to be independent: $\cos^2\theta_{iX} + \cos^2\theta_{iY} + \cos^2\theta_{iZ} \equiv 1$ for each axis $i = 1, 2, 3$ by (4.2-2).

For three *arbitrary* vectors, only six *would* be independent, but we generally demand that these vectors, in addition, be *orthonormal* to one another;

$\cos\theta_{iX}\cos\theta_{jX} + \cos\theta_{iY}\cos\theta_{jY} + \cos\theta_{iZ}\cos\theta_{jZ} = \delta_{ij}$, where δ_{ij} is the "*Kronecker delta*":

$$\delta_{ij} \equiv \begin{cases} 1 & \text{for } i = j \\ 0 & \text{for } i \neq j \end{cases}. \tag{4.2-3}$$

While, at first blush, this appears to be an additional six relations, three of them (for $i = j$) are already covered by (4.2-2); thus only three of these are "new." [Even if these orthonormal basis vectors were *not* unit vectors, but each had a magnitude d_i, a similar condition, that $\cos\theta_{iX}\cos\theta_{jX} + \cos\theta_{iY}\cos\theta_{jY} + \cos\theta_{iZ}\cos\theta_{jZ} = d_i d_j \delta_{ij}$, would hold.] So, in the case of [orthonormal] *coordinate systems*, the situation observed for vectors is exacerbated: we require *nine* angles, only *three* of which are independent.

More importantly, the above deals only with description of the *coordinate system itself*; we haven't even started to come to grips with the description of various points *in* that coordinate system (in the rigid body), and how their representations could be related to the original system: the direction angles describing the orientation of the various axes do not easily translate into the representation of, say, Cartesian vectors in the new axes to those in the old. Clearly there must be a better way; that better way is provided through the use of Euler Angles:

4.2.2 Euler Angles

Direction angles and cosines have been resurrected in order to motivate the alternative means of specifying unit vectors using only *independent* angles. As before, we will consider first the description of vectors, then of coordinate systems.

Vector Description

We have really already cleverly shown how to describe a vector $\boldsymbol{V}$ in terms of what we know to require only two independent angles: Recall that, in the gyroscope Example 4.1.2 above, we indicated that the axis of rotation always

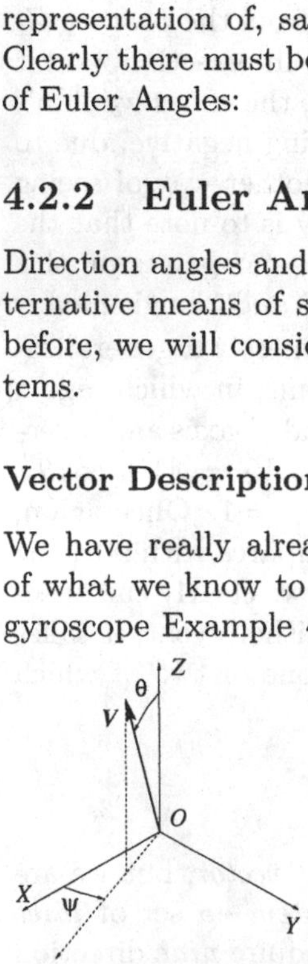

lies in a plane passing through the vertical axis. Furthermore, we specified the orientation of that axis using only *two* angles: ψ, between the XZ-plane and the axis plane, as well as θ, *in* that plane, made by the vector and the Z-axis. These are, in fact, precisely the Euler angles for that vector. This, of course, doesn't *define* what we mean by the Euler angles; in fact, unlike direction angles, which are defined for individual vectors, Euler angles' principal definition results from a definition for *coordinate systems* themselves:

We will observe, even for a single vector, that the angles defined above are (unlike direction angles) amenable to simple rotations: This vector would have a direction $\hat{\boldsymbol{V}}$ assumed by the original $\hat{\boldsymbol{K}}$ if the latter were to undergo two rotations: first one about the [original] Z-axis through an angle $+\psi$ [note the sign!]—bringing the original X- and Y-axes to the indicated intermediate X^*- and Y^*-axes indicated (though leaving the "Z^*" axis unchanged)—followed by another about the [*new*] "y" axis, Y^*, through an angle $+\theta$ bringing it to the (x, y, z) system. Recall, in this regard, equation (3.6-5): $\mathbf{u}'_i = \mathbf{U}\mathbf{e}_i = \mathbf{U}\mathbf{u}_i$—one relating the representations of *two different basis vectors* in the *same basis* of the $\boldsymbol{u}_i$, under the transformation given by $\mathbf{U} \equiv (\mathbf{u}'_1(\boldsymbol{u})\ \mathbf{u}'_2(\boldsymbol{u})\ \ldots\ \mathbf{u}'_n(\boldsymbol{u}))$. In the discussion surrounding these transformations, the above equation was potentially the source of some confusion between itself and its counterpart (3.6-6), $\mathbf{v}' = \mathbf{U}^{-1}\mathbf{v}$, relating the representations of the *same vector* in *two different bases.* Here, however, the first becomes a means of specifying the vector $\hat{\boldsymbol{k}}$:

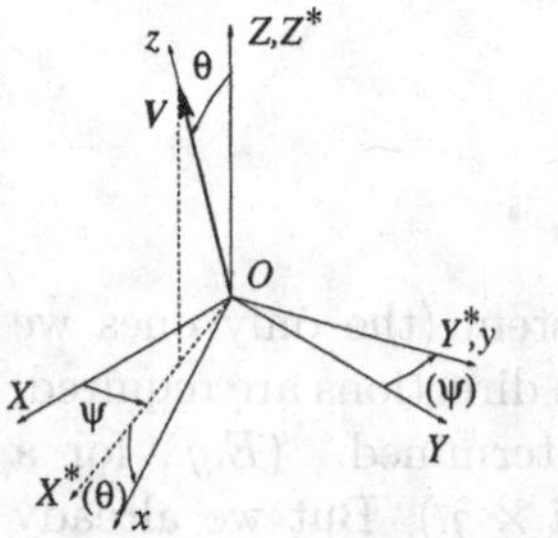

The *representations* resulting from the above two operations are precisely those given by the transformation matrix describing them; this, in turn, is nothing more than the *product* of the corresponding *rotation matrices*, Section 3.6.2:

$$\hat{\mathbf{k}} = \mathbf{U}\hat{\mathbf{K}} = \mathbf{C}(\psi)\mathbf{B}(\theta)\hat{\mathbf{K}}$$

(note the order of the matrix product here, an "active" transformation in the jargon on page 144, corresponding to (3.6-5) [and recalling the first "observation" on page 143]), where $\mathbf{B}$ and $\mathbf{C}$ are given by equations (3.6-13) and (3.6-14), respectively. But, again, the representation of $\hat{\boldsymbol{K}}$ *in its own basis* is just $\hat{\mathbf{K}} \equiv (0, 0, 1)^{\mathsf{T}}$ (see (3.6-3)); thus the only terms of $\mathbf{C}(\psi)\mathbf{B}(\theta)$ which will survive the complete product will be those of the third column. (This is a useful observation, and such foresight can save a great deal of calculation with these matrix products.) Noting that, we explicitly find only that column by multiplying each row of $\mathbf{C}(\psi)$ with just the last column (which is all that's needed) of $\mathbf{B}(\theta)$, in

$$\mathbf{C}(\psi)\mathbf{B}(\theta) \equiv \begin{pmatrix} \cos\psi & -\sin\psi & 0 \\ \sin\psi & \cos\psi & 0 \\ 0 & 0 & 1 \end{pmatrix} \begin{pmatrix} \ldots & \ldots & \sin\theta \\ \ldots & \ldots & 0 \\ \ldots & \ldots & \cos\theta \end{pmatrix}$$

$$= \begin{pmatrix} \ldots & \ldots & \cos\psi\sin\theta \\ \ldots & \ldots & \sin\psi\sin\theta \\ \ldots & \ldots & \cos\theta, \end{pmatrix}$$

whose last column—that vector resulting from the product with $\hat{\mathbf{K}}$—has precisely the components of $\hat{\boldsymbol{p}}$ in the gyroscope, Example 4.1.2.

Homework:

1. Show that the direction cosines l, m, and n are related to the Euler angles for a vector $\boldsymbol{V}$ through

$$l = \sin\theta\cos\psi, \qquad m = sin\theta\sin\psi, \qquad n = \cos\theta;$$

and, conversely,

$$\theta = \arccos n, \qquad \psi = \arctan \frac{m}{l}.$$

Coordinate System Description

Given the orientation of an orthogonal coordinate system (the only ones we will consider), only two basis directions are required; from these, the third is determined. (*E.g.*, for a right-handed system, $\hat{\boldsymbol{k}} \equiv \hat{\boldsymbol{\imath}} \times \hat{\boldsymbol{\jmath}}$.) But we already have shown that two Euler angles are necessary to specify the direction of one direction; the second, already known to be perpendicular to the first, is thus defined by the [single] angle ϕ it makes rotated about $\boldsymbol{V}$ away from, for example, the plane containing $\hat{\boldsymbol{k}}$ and $\hat{\boldsymbol{K}}$. In particular, if $\boldsymbol{V}$ is again taken to be the $\hat{\boldsymbol{k}}$-vector of the new system and $\hat{\boldsymbol{\imath}}$is specified by ϕ, then $\hat{\boldsymbol{\jmath}} = \hat{\boldsymbol{k}} \times \hat{\boldsymbol{\imath}}$. The three angles ψ, θ, and ϕ then define the three Euler angles describing the orientation of the coordinate system, and thus of *all* points in the rigid body: one rotation ψ about the $\hat{\boldsymbol{K}}$-axis—this brings the (X, Y, Z) coordinate system at left coincident with the intermediate $(X^*, Y^*, Z^* = Z)$ system [the Y^*-axis lying in the XY-plane]; the second θ about the [*new*] $\hat{\boldsymbol{J}}$-axis—bringing (X^*, Y^*, Z^*) coincident with $(X^{**}, Y^{**} = Y^*, Z^{**})$; followed by a final one ϕ about the [still *newer*!] $\hat{\boldsymbol{K}}$-axis—making (X^{**}, Y^{**}, Z^{**}) coincident with the final (x, y, z) system. We see how, in some sense, Euler angles are *defined* to be amenable to coordinate system rotations, since each corresponds to a rotation about an axis.

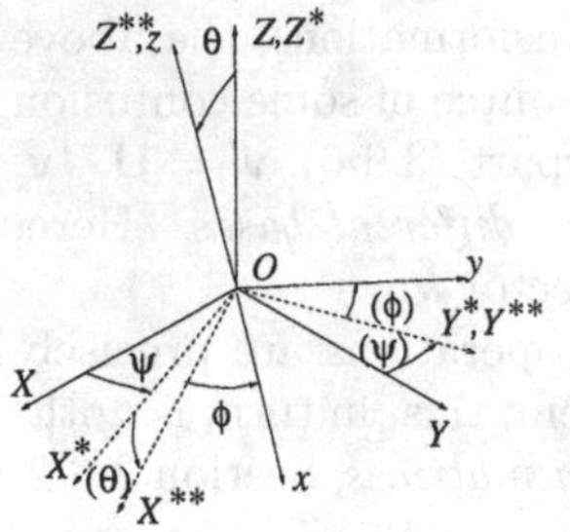

The advantage of this particular set of axes is immediately evident: the [representation of] point $\mathsf{v} = (x, y, z)$ in the final system is immediately related to that of the same point $\mathsf{V} = (X, Y, Z)$ in the original system through the use of rotation matrices (Section 3.6.2): since the final system results from the above three rotations, generated by the matrix product $\mathsf{C}(\psi)\mathsf{B}(\theta)\mathsf{C}(\phi)$, we have that

$$\mathsf{V} = \mathsf{C}(\psi)\mathsf{B}(\theta)\mathsf{C}(\phi)\,\mathsf{v};$$

interpreting the rotated system as one residing in the rigid body, this would give the value of points in that body relative to coordinates in the "fixed" original system. Conversely, points in the original system could be related to the body system by

$$\mathsf{v} = (\mathsf{C}(\psi)\mathsf{B}(\theta)\mathsf{C}(\phi))^{-1}\,\mathsf{V} = \mathsf{C}^{-1}(\phi)\mathsf{B}^{-1}(\theta)\mathsf{C}^{-1}(\psi)\,\mathsf{V} = \mathsf{C}^{\mathsf{T}}(\phi)\mathsf{B}^{\mathsf{T}}(\theta)\mathsf{C}^{\mathsf{T}}(\psi)\,\mathsf{V}.$$

Summary. Euler angles are a set of angles aptly suited to the description of rigid bodies. Unlike their counterpart direction angles (which suffer from an interdependence), precisely the "right number" of such angles are required to describe the orientation of vectors (2) and coordinate systems (3). Furthermore, since these can be interpreted as simple *rotations*, they are amenable to the use of rotation matrices to relate coordinates in the rigid body to a fixed system.

4.3 Moving Coordinate Systems

Time derivatives pervade all of mechanics: time is the ultimate variable on which all others depend, and Newton's Second Law involves the acceleration of the particle—a second *time* derivative. More importantly, this time derivative is defined to be the *absolute* time derivative, measured relative to a *fixed*[6] frame: clearly, chaotic non-uniqueness would result if we could measure such time derivatives relative to *moving* systems! We have already obtained a rather general expression for such time derivatives: equation (4.1-12), for the angular acceleration, involves a term accounting for the change in direction of $\boldsymbol{\omega}$. There is nothing special about the form of this equation pertaining to the angular velocity; *any* vector can be written in the form $\boldsymbol{V} = V\hat{\boldsymbol{V}}$, for which the time derivative would then be

$$\frac{d\boldsymbol{V}}{dt} = \frac{d(V\hat{\boldsymbol{V}})}{dt} = \frac{dV}{dt}\hat{\boldsymbol{V}} + V\frac{d\hat{\boldsymbol{V}}}{dt}, \tag{4.3-1}$$

in which the last term, arising from a rotation, say $\boldsymbol{\Omega}$, would involve the use of Theorem 4.1.2:

$$\frac{d\hat{\boldsymbol{V}}}{dt} = \boldsymbol{\Omega} \times \hat{\boldsymbol{V}}.$$

This equation is, in some sense, *generic*: nothing is said regarding how $\boldsymbol{V}$ is expressed, and the result is independent of this. But, as a practical matter, we are almost invariably forced to utilize not the vectors themselves, but their *representations* in terms of some basis; this is precisely what was done in Examples 4.1.1 and 4.1.2.

But there is more than a basis involved in these examples: their very statement is predicated on *knowing* the [absolute] rates at which the various angles are changing; this, in turn, requires the existence of some *fixed* lines or planes relative to which such changes can be measured in the first place. In a similar fashion, the measurement of accelerations, velocities, and ultimately position itself, demand some *fixed* point from which these quantities can be measured. Once again, the concept of *absolute time derivative* emerges. Implicit in all of these, then, is some form of *coordinate system*. It is important to note that such a coordinate system is not required for the vectors intrinsically; these require only a *basis* in terms of which to represent them. Rather, the coordinate system is implicit in the *definition of the quantities themselves*—positions measured relative to some point, or [differential] angles from some line—which, in turn, happen to be vectors.

[6]Actually, Newton's Second Law, giving the *acceleration*, demands only that the frame be "*inertial*"—non-*accelerating*. But we are interested in this section in *velocities*, also, so will require the frame to be fixed.

The previous section motivated its discussion of coordinate systems by pointing out that one is generally required to measure the distribution of mass in a rigid body. But these bodies are *moving*—this is *Dynamics*, after all!—and thus our coordinate systems themselves will be moving. The central theme of this section, then, will be the examination of such moving coordinate systems, with an eye to their ultimate application to the description of rigid body motion. More particularly, we shall examine how the *representations* of various vectors in such moving coordinate systems change, especially their time derivatives which will, after all, depend on this movement. Before embarking on this, however, we will indulge in one last foray into the motion of *points*—*particle* motion, as it were. This provides both a foundation for, and perspective on, the moving coordinate systems so crucial to the description of rigid bodies.

4.3.1 Relative Motion: Points

Diagrammed here are two points, A and B, whose positions, measured relative to a *fixed*, "*absolute*," coordinate system, denoted $O(X, Y, Z)$ to specify both the origin and coordinates, are given by $\boldsymbol{r}_A$ and $\boldsymbol{r}_B$, respectively. Though O is the origin of the system, we can at least imagine measuring the *relative* position $\boldsymbol{r}_{A/B}$ (read "$\boldsymbol{r}$ of A *relative to* B") of one point to the other.

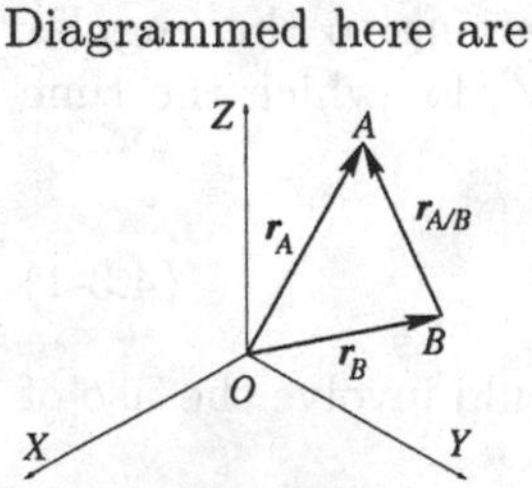

Without regard to representation, from the definition of the "head to tail" vector addition of the directed line segments at issue here, we can say that $\boldsymbol{r}_A = \boldsymbol{r}_B + \boldsymbol{r}_{A/B}$. As a general expression, this can be differentiated once, then once again, to give that $\boldsymbol{v}_A = \boldsymbol{v}_B + \boldsymbol{v}_{A/B}$ and $\boldsymbol{a}_A = \boldsymbol{a}_B + \boldsymbol{a}_{A/B}$. These expressions can be rewritten to express these "relative" position, velocity, and acceleration of A relative to B:

$$\begin{aligned} \boldsymbol{r}_{A/B} &= \boldsymbol{r}_A - \boldsymbol{r}_B \\ \boldsymbol{v}_{A/B} &= \boldsymbol{v}_A - \boldsymbol{v}_B \\ \boldsymbol{a}_{A/B} &= \boldsymbol{a}_A - \boldsymbol{a}_B. \end{aligned} \tag{4.3-2}$$

If, on the other hand, one wishes to *represent* these vectors in terms of the *coordinate system* $O(X, Y, Z)$ (and its implicitly-associated basis $\hat{\boldsymbol{I}}$, $\hat{\boldsymbol{J}}$, and $\hat{\boldsymbol{K}}$), we can express the above quantities as

$$\begin{aligned} \boldsymbol{r}_{A/B} &= X_A\hat{\boldsymbol{I}} + Y_A\hat{\boldsymbol{J}} + Z_A\hat{\boldsymbol{K}} - (X_B\hat{\boldsymbol{I}} + Y_B\hat{\boldsymbol{J}} + Z_B\hat{\boldsymbol{K}}) \\ &\equiv X_{A/B}\hat{\boldsymbol{I}} + Y_{A/B}\hat{\boldsymbol{J}} + Z_{A/B}\hat{\boldsymbol{K}} \\ \boldsymbol{v}_{A/B} &= \dot{X}_A\hat{\boldsymbol{I}} + \dot{Y}_A\hat{\boldsymbol{J}} + \dot{Z}_A\hat{\boldsymbol{K}} - (\dot{X}_B\hat{\boldsymbol{I}} + \dot{Y}_B\hat{\boldsymbol{J}} + \dot{Z}_B\hat{\boldsymbol{K}}) \\ &\equiv \dot{X}_{A/B}\hat{\boldsymbol{I}} + \dot{Y}_{A/B}\hat{\boldsymbol{J}} + \dot{Z}_{A/B}\hat{\boldsymbol{K}} \\ \boldsymbol{a}_{A/B} &= \ddot{X}_A\hat{\boldsymbol{I}} + \ddot{Y}_A\hat{\boldsymbol{J}} + \ddot{Z}_A\hat{\boldsymbol{K}} - (\ddot{X}_B\hat{\boldsymbol{I}} + \ddot{Y}_B\hat{\boldsymbol{J}} + \ddot{Z}_B\hat{\boldsymbol{K}}) \\ &\equiv \ddot{X}_{A/B}\hat{\boldsymbol{I}} + \ddot{Y}_{A/B}\hat{\boldsymbol{J}} + \ddot{Z}_{A/B}\hat{\boldsymbol{K}}. \end{aligned}$$

Note that, in general, these expressions would involve time derivatives of the basis vectors $\hat{\boldsymbol{I}}$, $\hat{\boldsymbol{J}}$, and $\hat{\boldsymbol{K}}$; but, because the system is *fixed* (*i.e.*, not rotating), such terms vanish.

It is critically important to keep in mind what these "relative" quantities *mean*: "*point-to-point*" relative position, velocity, and acceleration, without regard to any coordinate system other than the $O(X,Y,Z)$ in terms of which they are *represented*. This is important because we are shortly going to introduce *another* set of "relative" position, velocity, and acceleration with which the above can be confused!

4.3.2 Relative Motion: Coordinate Systems

Not content with the above relatively simple system, now let us regard B as the origin of a *coordinate system*—not merely a coordinate system, but one which is *translating*, with the point B having *absolute* velocity and acceleration $\boldsymbol{v}_B$ and $\boldsymbol{a}_B$, and *rotating* with *absolute* angular velocity $\boldsymbol{\omega}$ and angular acceleration $\boldsymbol{\alpha}$. Note that the latter have directions along the axis of rotation, given by the right-hand rule. The "point-to-point" relations above still hold; in particular, one can write $\boldsymbol{r}_{A/B}$ in terms of *either* coordinate system's basis, the original $\{\hat{\boldsymbol{I}},\hat{\boldsymbol{J}},\hat{\boldsymbol{K}}\}$ or the new $\{\hat{\boldsymbol{\imath}},\hat{\boldsymbol{\jmath}},\hat{\boldsymbol{k}}\}$:

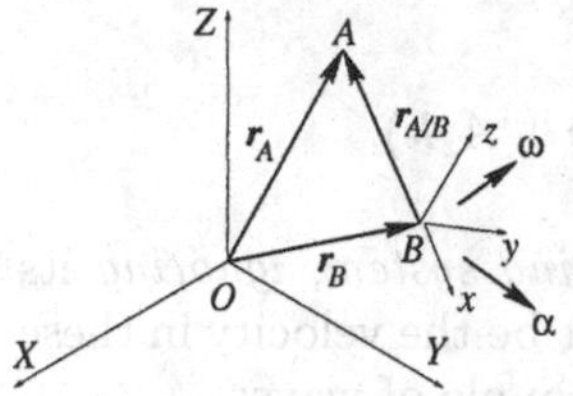

$$\begin{aligned} \boldsymbol{r}_{A/B} &\equiv x_{A/B}\hat{\boldsymbol{\imath}} + y_{A/B}\hat{\boldsymbol{\jmath}} + z_{A/B}\hat{\boldsymbol{k}} & \text{(in } B(x,y,z)) \\ &= X_{A/B}\hat{\boldsymbol{I}} + Y_{A/B}\hat{\boldsymbol{J}} + Z_{A/B}\hat{\boldsymbol{K}} & \text{(in } O(X,Y,Z)) \end{aligned}$$

(the second equation included to emphasize the fact that the *vector* $\boldsymbol{r}_{A/B}$ is the *same*, it is only its *representation* that is different). But now, for example, in terms of $B(x,y,z)$, the *same* $\boldsymbol{v}_{A/B}$ can be represented

$$\begin{aligned} \boldsymbol{v}_{A/B} &= \dot{x}_{A/B}\hat{\boldsymbol{\imath}} + \dot{y}_{A/B}\hat{\boldsymbol{\jmath}} + \dot{z}_{A/B}\hat{\boldsymbol{k}} + x_{A/B}\dot{\hat{\boldsymbol{\imath}}} + y_{A/B}\dot{\hat{\boldsymbol{\jmath}}} + z_{A/B}\dot{\hat{\boldsymbol{k}}} & \text{(in } B(x,y,z)) \\ &\equiv \dot{X}_{A/B}\hat{\boldsymbol{I}} + \dot{Y}_{A/B}\hat{\boldsymbol{J}} + \dot{Z}_{A/B}\hat{\boldsymbol{K}} & \text{(in } O(X,Y,Z)) \end{aligned}$$

—the last terms in the first equation arising because of the rotation of the system. Now there is a *second coordinate system* against which one might measure velocity and acceleration (and, for all that, even *angular* velocity and acceleration). These quantities are different conceptually from their "absolute" counterparts in the fixed $O(X,Y,Z)$ system, and should be different notationally; we will denote them $\boldsymbol{v}_{rel}$ and $\boldsymbol{a}_{rel}$ relative to the *coordinate system* $B(x,y,z)$, as distinct from $\boldsymbol{v}_{A/B}$ and $\boldsymbol{a}_{A/B}$ relative to merely the *points* A and B. These new "relative" quantities are those observed *in the coordinate system*—effectively *ignoring* its rotation, even though they generally arise from this; they only have meaning when there *is* such a second coordinate system, unlike $\boldsymbol{v}_{A/B}$ and $\boldsymbol{a}_{A/B}$, which depend only on the points themselves.

The issue of this section is to determine, knowing the motion of the system $B(x,y,z)$—*i.e.*, knowing the *absolute* $\boldsymbol{v}_B$, $\boldsymbol{a}_B$, $\boldsymbol{\omega}$, and $\boldsymbol{\alpha}$—just what v_{rel} and

a_{rel} *are.* We shall return to that presently, but in order to obtain the desired result, we need another of arguably even more fundamental import: the relation generally between *absolute* time derivatives and those observed from a rotating system. Though presented here as a means of dealing with the *kinematics* of arbitrary motion of coordinate systems, this result will have profound implications for the *kinetics* in the next Chapter.

Time Derivatives in Rotating Coordinate Systems

Rather than dealing with the above question of velocities and accelerations in coordinate systems, we will treat the more general question of the time derivative of an *arbitrary* vector quantity $\boldsymbol{A} = A_x\hat{\boldsymbol{\imath}} + A_y\hat{\boldsymbol{\jmath}} + A_z\hat{\boldsymbol{k}}$ *represented* relative to a coordinate system rotating with *absolute* angular velocity and acceleration $\boldsymbol{\omega}$ and $\boldsymbol{\alpha}$. The *absolute* time derivative

$$\frac{d\boldsymbol{A}}{dt} = (\dot{A}_x\hat{\boldsymbol{\imath}} + \dot{A}_y\hat{\boldsymbol{\jmath}} + \dot{A}_z\hat{\boldsymbol{k}}) + (A_x\dot{\hat{\boldsymbol{\imath}}} + A_y\dot{\hat{\boldsymbol{\jmath}}} + A_z\dot{\hat{\boldsymbol{k}}}).$$

The first set of terms is that *observed in the rotating system, ignoring* its rotation: were $\boldsymbol{A}$ a position, for example, this would just be the velocity in these Cartesian coordinates. We denote this derivative in a couple of ways:

$$\dot{\boldsymbol{A}}_{rel} \equiv \left.\frac{d\boldsymbol{A}}{dt}\right|_{rel} \equiv \dot{A}_x\hat{\boldsymbol{\imath}} + \dot{A}_y\hat{\boldsymbol{\jmath}} + \dot{A}_z\hat{\boldsymbol{k}}. \tag{4.3-3}$$

Note that, operationally, $\dot{\boldsymbol{A}}_{rel}$ merely differentiates its *components* with respect to time, leaving the basis vectors "fixed."

The second set of terms arises from time derivatives of the basis vectors; *i.e.*, because of *rotation*—the only way, recall, unit vectors *can* change. But this is precisely the case treated by Theorem 4.1.2: from that theorem $\dot{\hat{\boldsymbol{\imath}}} = \boldsymbol{\omega} \times \hat{\boldsymbol{\imath}}$, with similar results for $\hat{\boldsymbol{\jmath}}$ and $\hat{\boldsymbol{k}}$. Thus

$$\begin{aligned} A_x\dot{\hat{\boldsymbol{\imath}}} + A_y\dot{\hat{\boldsymbol{\jmath}}} + A_z\dot{\hat{\boldsymbol{k}}} &= A_x\,\boldsymbol{\omega} \times \hat{\boldsymbol{\imath}} + A_y\,\boldsymbol{\omega} \times \hat{\boldsymbol{\jmath}} + A_z\,\boldsymbol{\omega} \times \hat{\boldsymbol{k}} \\ &= \boldsymbol{\omega} \times (A_x\hat{\boldsymbol{\imath}} + A_y\hat{\boldsymbol{\jmath}} + A_z\hat{\boldsymbol{k}}) \\ &= \boldsymbol{\omega} \times \boldsymbol{A}. \end{aligned}$$

Thus we have the fundamental result:

Theorem 4.3.1. The *absolute* time derivative $\dot{\boldsymbol{A}}$ of a vector $\boldsymbol{A}$ is related to the *relative* time derivative $\dot{\boldsymbol{A}}_{rel}$ observed in a coordinate system rotating with *absolute* angular velocity $\boldsymbol{\omega}$ through $\dot{\boldsymbol{A}} = \boldsymbol{\omega} \times \boldsymbol{A} + \dot{\boldsymbol{A}}_{rel}$.

This is likely one of the most important propositions in this entire Part: it will be used immediately and will return in the next Chapter, and should be thoroughly understood.

We now apply this theorem to a number of examples, ultimately answering the question of how the "A/B" and "*rel*" quantities are related:

Applications of Theorem 4.3.1

Example 4.3.1. *Absolute and Relative Angular Acceleration.* Letting "$\boldsymbol{A}$" above correspond to the [absolute] angular velocity $\boldsymbol{\omega}$, we get immediately that

$$\dot{\boldsymbol{\omega}} = \boldsymbol{\omega} \times \boldsymbol{\omega} + \dot{\boldsymbol{\omega}}_{rel}$$

or, since the cross product vanishes identically,

$$\boldsymbol{\alpha} = \boldsymbol{\alpha}_{rel}.$$

This rather remarkable result says that the angular acceleration observed from the rotating system is unaffected by the rotation—either angular velocity *or* acceleration—itself! Clearly, the same result would hold for *any* vector parallel to $\boldsymbol{\omega}$; it does not depend on the special character of $\boldsymbol{\omega}$. |

Example 4.3.2. $\boldsymbol{v}_{A/B}$. Noting that $\boldsymbol{r}_{A/B} \equiv x_{A/B}\hat{\boldsymbol{\imath}} + y_{A/B}\hat{\boldsymbol{\jmath}} + z_{A/B}\hat{\boldsymbol{k}}$ is, in fact, a quantity measured relative to the rotating system $B(x, y, z)$, we simply substitute this in for "$\boldsymbol{A}$" in the general Theorem:

$$\boldsymbol{v}_{A/B} \equiv \frac{d\boldsymbol{r}_{A/B}}{dt} = \boldsymbol{\omega} \times \boldsymbol{r}_{A/B} + \left.\frac{d}{dt}\right|_{rel} \boldsymbol{r}_{A/B}.$$

But the last term, by the definition of the "*rel*" derivative (note the alternate notation), is precisely *the derivative of* $\boldsymbol{r}_{A/B}$ *as observed from* $B(x, y, z)$*—i.e.*, the elusive $\boldsymbol{v}_{rel}$ we have been seeking! Thus

$$\boldsymbol{v}_{A/B} = \boldsymbol{\omega} \times \boldsymbol{r}_{A/B} + \boldsymbol{v}_{rel},$$

relating the *point-to-point* $\boldsymbol{v}_{A/B}$ and the *coordinate system* $\boldsymbol{v}_{rel}$. |

Example 4.3.3. $\boldsymbol{a}_{A/B}$. We play much the same game we have before, but now we substitute the above result in for $\boldsymbol{v}_{A/B}$:

$$\begin{aligned}
\boldsymbol{a}_{A/B} &\equiv \frac{d\boldsymbol{v}_{A/B}}{dt} = \boldsymbol{\omega} \times \boldsymbol{v}_{A/B} + \left.\frac{d}{dt}\right|_{rel} \boldsymbol{v}_{A/B} \\
&= \boldsymbol{\omega} \times (\boldsymbol{\omega} \times \boldsymbol{r}_{A/B} + \boldsymbol{v}_{rel}) + \left.\frac{d}{dt}\right|_{rel} (\boldsymbol{\omega} \times \boldsymbol{r}_{A/B} + \boldsymbol{v}_{rel}) \\
&= [\boldsymbol{\omega} \times (\boldsymbol{\omega} \times \boldsymbol{r}_{A/B}) + \boldsymbol{\omega} \times \boldsymbol{v}_{rel}] \\
&\quad + \left[\left.\frac{d}{dt}\right|_{rel} \boldsymbol{\omega} \times \boldsymbol{r}_{A/B} + \boldsymbol{\omega} \times \left.\frac{d}{dt}\right|_{rel} \boldsymbol{r}_{A/B} + \left.\frac{d}{dt}\right|_{rel} \boldsymbol{v}_{rel}\right]
\end{aligned}$$

—where the results of each term in the second line is bracketed in the third; note the application of the derivative product rule to the *rel* differentiation. But

$$\left.\frac{d}{dt}\right|_{rel} \boldsymbol{\omega} \equiv \boldsymbol{\alpha}_{rel} = \boldsymbol{\alpha}, \quad \left.\frac{d}{dt}\right|_{rel} \boldsymbol{r}_{A/B} \equiv \boldsymbol{v}_{rel}, \quad \text{and} \quad \left.\frac{d}{dt}\right|_{rel} \boldsymbol{v}_{rel} \equiv \boldsymbol{a}_{rel},$$

the first by Example 4.3.1, and the second and third by the "*rel*" definitions. Thus, finally,

$$\boldsymbol{a}_{A/B} = \boldsymbol{\omega} \times (\boldsymbol{\omega} \times \boldsymbol{r}_{A/B}) + 2\boldsymbol{\omega} \times \boldsymbol{v}_{rel} + \boldsymbol{\alpha} \times \boldsymbol{r}_{A/B} + \boldsymbol{a}_{rel}$$

—an equation corresponding to the previous example's, but now relating the two types of acceleration. |

Rotating Coordinate System Equations

We now collect the last two examples' results with those of equations (4.3-2) for the system at the beginning of this section, to relate the *absolute* velocity and acceleration of the point A, to those measured *relative to a coordinate system* whose origin B moves with *absolute* velocity and acceleration $\boldsymbol{v}_B$ and $\boldsymbol{a}_B$, and which rotates with *absolute* angular velocity and acceleration $\boldsymbol{\omega}$ and $\boldsymbol{\alpha}$:

Velocity:

$$\boldsymbol{v}_A = \boldsymbol{v}_B + \boldsymbol{v}_{A/B} \tag{4.3-4a}$$
$$= \boldsymbol{v}_B + \boldsymbol{\omega} \times \boldsymbol{r} + \boldsymbol{v}_{rel} \tag{4.3-4b}$$

Acceleration:

$$\boldsymbol{a}_A = \boldsymbol{a}_B + \boldsymbol{a}_{A/B} \tag{4.3-5a}$$
$$= \boldsymbol{a}_B + \boldsymbol{\alpha} \times \boldsymbol{r} + \boldsymbol{\omega} \times (\boldsymbol{\omega} \times \boldsymbol{r}) + 2\boldsymbol{\omega} \times \boldsymbol{v}_{rel} + \boldsymbol{a}_{rel} \tag{4.3-5b}$$

Let us make several observations regarding this important set of equations:

- Note in these relations that the second and following terms on the right-hand side of each "b" equation correspond to the "A/B" term in the "a" line. In each case, the latter quantity is the *same*; it is only *represented* in terms of quantities related to the *absolute* and *rotating* systems, respectively. Further, all quantities—the $\boldsymbol{v}$'s, $\boldsymbol{a}$', $\boldsymbol{\omega}$, and $\boldsymbol{\alpha}$—are *absolute*, themselves implicitly measured relative to a *fixed* system $O(X, Y, Z)$ *except* for the two terms $\boldsymbol{v}_{rel}$ and $\boldsymbol{a}_{rel}$, which are measured *relative to the coordinate system* $B(x, y, z)$ rotating with [absolute] $\boldsymbol{\omega}$ and $\boldsymbol{\alpha}$.
- We have dropped the somewhat cumbersome A/B subscript on $\boldsymbol{r}$ for reasons of economy: it is measured *from the origin of the rotating system*, whose velocity and acceleration start the "b" equations. This bears on a point important in application: any coordinate system is defined given two parameters—its origin and how it's rotating—both of which are given relative, implicitly, to some *fixed* system. Students will often start problems dealing with rotating coordinate systems by going to great lengths to *define* a fixed system. But it's already there: without it, there would be no velocities and accelerations of the various points to begin with!

- The above note deals with definition of the [rotating] coordinate system. The other point that can cause confusion is the *directions* of the axes in a given problem. Recall that one must generally express the various vectors relative to some *basis*, and the question is just which basis one must use. With all the emphasis on *absolute* velocity and acceleration, one might be inclined to refer all these to the *fixed* coordinate system's basis. But there is no need to conform doggedly to this: it is always possible to use the *rotating* coordinate system's basis since, even though it is rotating conceptually, it still has a set of directions *at that instant* relative to which all other vectors can be represented. One can choose *either* fixed or rotating basis in good conscience.

Distinction between the "A/B" and "*rel*" Quantities

There is almost invariably some student confusion between "$\mathbf{v}_{A/B}$" and "$\mathbf{a}_{A/B}$," and "$\mathbf{v}_{rel}$" and "$\mathbf{a}_{rel}$": one often rather simplistically refers to just the "relative velocity" and "relative acceleration," without keeping in mind just what each is relative *to*. Once again: The "A/B" quantities are *point-to-point*, referring to the motion of the *point* A relative to the *point* B. The "*rel*" quantities are those measured relative to a *coordinate system*; they have no *meaning* unless a coordinate system has been specified against which to measure them. In the attempt to clarify this issue, we include a rather "hokey" example, for which the author apologizes in advance:

Example 4.3.4. *Distinction between* $\mathbf{v}_{A/B}$ *and* $\mathbf{v}_{\rm rel}$. Out in the hinterlands, one often see travelling "carney" shows which set up for a week during late spring and summer in a local mall's parking lot. They generally feature *small* rides, *small* merry-go-rounds, and the like, in order to be able to travel the back roads necessary to reach their clientele. As might be expected, these tend to be somewhat low-budget affairs designed mainly to appeal to small children who are supposed to drag their parents to the mall in the hopes of generating revenue for the surrounding merchants.

Diagrammed here is one such amusement ride, a *small* merry-go-round with radius r. It is understandably somewhat shop-worn, with only half the fluorescent lights adorning the canopy still working. The owner, A (call him Alphonse), stands at the middle of the apparatus, which is turning in the direction indicated with angular velocity ω. Just at the edge is his loyal worker, B (say Beauregard). It is night. But both A and B are wearing miner's caps, similarly shop-worn with the lights bent upwards. You are supposed to be hovering over the scene in a helicopter. (I *warned* you about this example!)

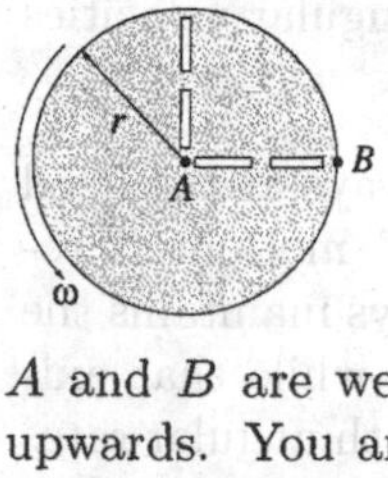

The first observation to be made is that *without the lights one cannot discern any rotation*: the positions of A and B can be seen, but without some reference lines—here the "coordinate system" generated by the lights—there is no line to measure any angles whether the merry-go-round is turning or not! Furthermore, assuming the truck on which the ride is situated is not moving, the *absolute* velocity $\boldsymbol{v}_A$ of A is zero. With that in hand, we now consider two scenarios:

1. *B stands off the merry-go-round.* Now the *absolute* $\boldsymbol{v}_B \equiv 0$. But, due to the rotation, there is a *non*-zero velocity of B *relative to the coordinate system*, $\boldsymbol{v}_{rel}$, which A observes of B: it has a magnitude of $r\omega$, and points down in the diagram.

   Now consider this situation in the light of equations (4.3-4): The first of these states that $\boldsymbol{v}_{B/A}$ [note the notation] is $\boldsymbol{v}_B - \boldsymbol{v}_A \equiv \mathbf{0}$. But this is allegedly supposed to equal the last two terms of (4.3-4b):

   $$\boldsymbol{\omega} \times \boldsymbol{r} + \boldsymbol{v}_{rel} = r\omega\!\uparrow + r\omega\!\downarrow \equiv \mathbf{0}$$

   and we see that, indeed, $\boldsymbol{v}_{B/A} = \boldsymbol{\omega} \times \boldsymbol{r} + \boldsymbol{v}_{rel}$, as expected.

2. *B stands* on *the edge of the merry-go-round.* Now $\boldsymbol{v}_A$ is still $\mathbf{0}$, but $\boldsymbol{v}_B$, the *absolute* velocity of B, has a magnitude of $r\omega$ and points *up*, while the velocity of B *relative to the merry-go-round*—the *coordinate system* in this case—vanishes. Thus, once again

   $$\begin{aligned} \boldsymbol{v}_{B/A} &= \boldsymbol{v}_B - \boldsymbol{v}_A = r\omega\!\uparrow - \mathbf{0} \\ &= \boldsymbol{\omega} \times \boldsymbol{r} + \boldsymbol{v}_{rel} = r\omega\!\uparrow + \mathbf{0}, \end{aligned}$$

   and the relation holds.

As embarrassing as this example might be for a college text, it is hoped that it points up somewhat more precisely what the terms in equations (4.3-4) and (4.3-5) actually *mean.*

The next example actually applies virtually all the kinematics we have learned up to this point to a "real-life" problem. This is a situation in which we know, or can find, all angular motion and determine the absolute velocity and acceleration of a point utilizing the rotating coordinate systems—something which might be called the "forward" problem. This point is made to contrast the example with those in the next section, where we know [something about] the velocities and accelerations of the points, and determine the angular velocities and accelerations—the "inverse" problem.

Example 4.3.5 We consider a $b \times a$ plate with one corner always at the fixed point O undergoing a sort of "tumbling" motion: it rotates about its diagonal OA—which always maintains the angle θ with the Z-axis—at a rate of p, while that axis rotates about the vertical at a rate $\dot{\psi}$—both angular rates considered constant. Sliding in a track along one edge, AB, of the plate is a small particle B, which moves with constant speed $v_B = v$ in the track. We are asked to determine the *absolute* acceleration, $\boldsymbol{a}_B$, of B.

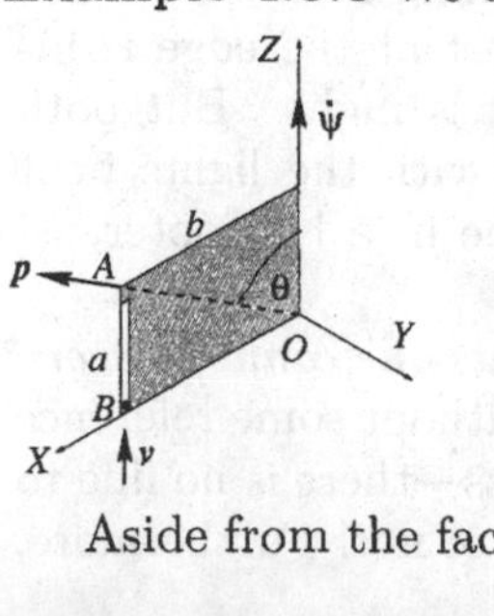

Aside from the fact that this is allegedly supposed to exemplify rotating coordinate systems, why is that technique implemented in this problem? Observe that we are given v *relative to* the plate; this almost begs to be identified with a "$\boldsymbol{v}_{rel}$"-type velocity—*i.e.*, one measured relative to a *coordinate system*. The only issue remaining is to choose just what that coordinate system is; aside from the fixed $O(XYZ)$-system, no other is indicated. This is done intentionally: someone once said that "Coordinate systems are a device of Man, not of God," meaning that, aside from the *fixed*, *absolute* coordinate system against which all other quantities are measured, any other systems are introduced merely at convenience. Just to emphasize that point, this problem is going to worked utilizing *three different* rotating systems; this is done to show that the final answer is the same for all three—it is, after all, the *absolute* acceleration we are trying to find—as well as to drive home the *meaning* of the various terms in the general rotating coordinate system equations. Recall, in this regard, just what is needed to *define* a rotating coordinate system: the motion of its *origin*, and just how it is *rotating*. Thus each choice carefully starts by declaring just what the coordinate system *is*. Such a declaration, conceptualized explicitly from the start, is useful when problems arise regarding the identity of the various terms in the rotating coordinate system equations.

Coordinate System with Origin at A, Rotating with the Plate. We consider a system $A(x, y, z)$ *fixed in the plate* with the stated origin, and whose basis vectors $\hat{\boldsymbol{\imath}}$, $\hat{\boldsymbol{\jmath}}$, and $\hat{\boldsymbol{k}}$ are *instantaneously* parallel to the $\hat{\boldsymbol{I}}$, $\hat{\boldsymbol{J}}$, and $\hat{\boldsymbol{K}}$ vectors in the fixed $O(X, Y, Z)$ system. From this declaration of coordinate system, in particular its origin, the rotating coordinate system equation becomes

$$\boldsymbol{a}_B = \boldsymbol{a}_A + \boldsymbol{\alpha} \times \boldsymbol{r} + \boldsymbol{\omega} \times (\boldsymbol{\omega} \times \boldsymbol{r}) + 2\boldsymbol{\omega} \times \boldsymbol{v}_{rel} + \boldsymbol{a}_{rel}, \tag{4.3-6}$$

where the angular velocity is

$$\boldsymbol{\omega} = \dot{\psi}\hat{\boldsymbol{K}} + p(\sin\theta\hat{\boldsymbol{\imath}}_\psi + \cos\theta\hat{\boldsymbol{K}}),$$

in which the changing direction of $\boldsymbol{p}$ is emphasized by expressing it in terms of $\hat{\boldsymbol{\imath}}_\psi$, a horizontal vector in the ZOA-plane rotating with $\dot{\boldsymbol{\psi}}$, while its $\hat{\boldsymbol{K}}$-component is fixed. Note, however, that *at the instant considered,* $\hat{\boldsymbol{\imath}}_\psi$ (if not $\dot{\hat{\boldsymbol{\imath}}}_\psi$) can be replaced by $\hat{\boldsymbol{I}}$. All that remains is to evaluate the various terms in (4.3-6):

<u>$\boldsymbol{a}_A$</u>: We actually have a couple of options here: One is to utilize cylindrical coordinates centered, say, at the fixed point O, $O(r, \theta, z)$, with the [fixed] z-axis parallel to the Z-axis; then the acceleration of A can be found using these coordinates' acceleration:

$$\begin{aligned}\boldsymbol{a}_A &= (\ddot{r} - r\dot{\theta}^2)\hat{\boldsymbol{r}} + (2\dot{r}\dot{\theta} + r\ddot{\theta})\hat{\boldsymbol{\theta}} + \ddot{z}\hat{\boldsymbol{k}} \\ &= -b\dot{\psi}^2\hat{\boldsymbol{I}}\end{aligned}$$

—since $r = b$ and $\dot{\theta} = -\dot{\psi}$ are constant, while $\hat{\boldsymbol{r}} = \hat{\boldsymbol{I}}$. The latter is, in effect, a basis transformation from the vectors $\hat{\boldsymbol{r}}$, $\hat{\boldsymbol{\theta}}$, and $\hat{\boldsymbol{k}}$ to their equivalents $\hat{\boldsymbol{I}}$, $\hat{\boldsymbol{J}}$, and

$\hat{\boldsymbol{K}}$, respectively; in particular, choosing $\hat{\boldsymbol{k}} \equiv \hat{\boldsymbol{K}}$, and the cylindrical coordinate direction $\hat{\boldsymbol{r}} \equiv \hat{\boldsymbol{I}}$ by definition, makes $\hat{\boldsymbol{\theta}} \equiv \hat{\boldsymbol{k}} \times \hat{\boldsymbol{r}} = \hat{\boldsymbol{J}}$ (imposed to keep the cylindrical coordinates right-handed), producing the minus sign in $\dot{\theta} = -\dot{\psi}$.

Alternatively, we can introduce an intermediate rotating coordinate system with origin at O and rotating with $\boldsymbol{\omega}' \equiv \dot{\psi}\hat{\boldsymbol{K}}$; then

$$\begin{aligned}\boldsymbol{a}_A &= \boldsymbol{a}_O + \boldsymbol{\alpha}' \times \boldsymbol{r}' + \boldsymbol{\omega}' \times (\boldsymbol{\omega}' \times \boldsymbol{r}') + 2\boldsymbol{\omega}' \times \boldsymbol{v}'_{rel} + \boldsymbol{a}'_{rel}\\ &= \mathbf{0} + \mathbf{0} + \dot{\psi}\hat{\boldsymbol{K}} \times (\dot{\psi}\hat{\boldsymbol{K}} \times (b\hat{\boldsymbol{I}} + a\hat{\boldsymbol{K}})) + \mathbf{0} + \mathbf{0} = -b\dot{\psi}^2\hat{\boldsymbol{I}},\end{aligned}$$

since $\boldsymbol{\omega}'$ is constant in both magnitude *and* direction, while $\boldsymbol{v}'_{rel} \equiv \mathbf{0} \equiv \boldsymbol{a}'_{rel}$ because A is fixed in this [intermediate] system.

$\underline{\boldsymbol{\alpha} \times \boldsymbol{r}}$: Recalling that $\boldsymbol{\alpha}$ is the *absolute* angular acceleration of our coordinate system, we must calculate it in the same fashion as we have before:

$$\begin{aligned}\boldsymbol{\alpha} \equiv \frac{d\boldsymbol{\omega}}{dt} &= \frac{d}{dt}\left(\dot{\psi}\hat{\boldsymbol{K}} + p(\sin\theta\hat{\boldsymbol{\imath}}_\psi + \cos\theta\hat{\boldsymbol{K}})\right)\\ &= p\sin\theta\dot{\hat{\boldsymbol{\imath}}}_\psi = p\sin\theta(\dot{\psi}\hat{\boldsymbol{K}} \times \hat{\boldsymbol{I}})\\ &= p\dot{\psi}\sin\theta\hat{\boldsymbol{J}}\end{aligned}$$

—since the angular rates are constant, and only $\hat{\boldsymbol{\imath}}$ changes its direction, rotating with $\dot{\psi}\hat{\boldsymbol{K}}$; and, *at that instant*, $\hat{\boldsymbol{\imath}}_\psi = \hat{\boldsymbol{I}}$—leading to the final result in terms of the fixed system's basis.

$\boldsymbol{r}$ is measured from the origin of the rotating coordinate system, here A; thus $\boldsymbol{r} = -a\hat{\boldsymbol{k}}$ which equals $-a\hat{\boldsymbol{K}}$ at this instant, and so, finally,

$$\boldsymbol{\alpha} \times \boldsymbol{r} = p\dot{\psi}\sin\theta\hat{\boldsymbol{J}} \times (-a\hat{\boldsymbol{K}}) = -ap\dot{\psi}\sin\theta\hat{\boldsymbol{I}}.$$

$\underline{\boldsymbol{\omega} \times (\boldsymbol{\omega} \times \boldsymbol{r})}$: This is relatively straightforward, if a little tedious:

$$\begin{aligned}\boldsymbol{\omega} \times (\boldsymbol{\omega} \times \boldsymbol{r}) &= \left(\dot{\psi}\hat{\boldsymbol{K}} + p(\sin\theta\hat{\boldsymbol{\imath}}_\psi + \cos\theta\hat{\boldsymbol{K}})\right)\\ &\quad \times \left((\dot{\psi}\hat{\boldsymbol{K}} + p(\sin\theta\hat{\boldsymbol{\imath}}_\psi + \cos\theta\hat{\boldsymbol{K}}) \times (-a\hat{\boldsymbol{K}})\right)\\ &= \left(\dot{\psi}\hat{\boldsymbol{K}} + p(\sin\theta\hat{\boldsymbol{I}} + \cos\theta\hat{\boldsymbol{K}})\right) \times ap\sin\theta\hat{\boldsymbol{J}}\\ &= ap^2\sin^2\theta\hat{\boldsymbol{K}} - ap(\dot{\psi} + p\cos\theta)\hat{\boldsymbol{I}}.\end{aligned}$$

$\underline{2\boldsymbol{\omega} \times \boldsymbol{v}_{rel}}$: $\boldsymbol{v}_{rel}$ is, recall, defined to be measured relative to the *coordinate system*, here the plate; thus, in fact, $\boldsymbol{v}_{rel} \equiv v\hat{\boldsymbol{K}}$ for our choice of coordinate system:

$$\begin{aligned}2\boldsymbol{\omega} \times \boldsymbol{v}_{rel} &= 2\left(\dot{\psi}\hat{\boldsymbol{K}} + p(\sin\theta\hat{\boldsymbol{\imath}}_\psi + \cos\theta\hat{\boldsymbol{K}})\right) \times v\hat{\boldsymbol{K}}\\ &= -2pv\sin\theta\hat{\boldsymbol{J}}.\end{aligned}$$

$\underline{\boldsymbol{a}_{rel}}$: Since $v\hat{\boldsymbol{K}} \equiv \boldsymbol{v}_{rel}$ is constant, we have immediately that $\boldsymbol{a}_{rel} \equiv \mathbf{0}$.

Finally, collecting all the results of the above calculations, we get that

$$\boldsymbol{a}_B = -(b\dot{\psi}^2 + 2ap\dot{\psi}\sin\theta + ap^2\sin\theta\cos\theta)\hat{\boldsymbol{I}} \\ - 2pv\sin\theta\hat{\boldsymbol{J}} + ap^2\sin^2\theta\hat{\boldsymbol{K}}. \tag{4.3-7}$$

Coordinate System with Origin at O, Rotating with the Plate. Now the expression for acceleration becomes, with the new origin,

$$\boldsymbol{a}_B = \boldsymbol{a}_O + \boldsymbol{\alpha}\times\boldsymbol{r} + \boldsymbol{\omega}\times(\boldsymbol{\omega}\times\boldsymbol{r}) + 2\boldsymbol{\omega}\times\boldsymbol{v}_{rel} + \boldsymbol{a}_{rel}, \tag{4.3-8}$$

in which the angular velocity, and acceleration, will be precisely those obtained in the previous coordinate system, since this one has the same rotation. This allows us to condense the calculations, with reference back to the preceding incantation of this problem:

$\underline{\boldsymbol{a}_O}\equiv \boldsymbol{0}$.

$\underline{\boldsymbol{\alpha}\times\boldsymbol{r}}$: Although $\boldsymbol{\alpha}$ is the same, now $\boldsymbol{r}$ is measured from the new origin, O. Utilizing the previous calculation for $\boldsymbol{\alpha}$,

$$\boldsymbol{\alpha}\times\boldsymbol{r} = p\dot{\psi}\sin\theta\hat{\boldsymbol{J}}\times b\hat{\boldsymbol{I}} = -bp\dot{\psi}\sin\theta\hat{\boldsymbol{K}}.$$

$\underline{\boldsymbol{\omega}\times(\boldsymbol{\omega}\times\boldsymbol{r})}$: Again, this is almost the same as above, but with a new $\boldsymbol{r}$:

$$\begin{aligned}\boldsymbol{\omega}\times(\boldsymbol{\omega}\times\boldsymbol{r}) &= \left(\dot{\psi}\hat{\boldsymbol{K}} + p(\sin\theta\hat{\boldsymbol{\imath}}_{\psi} + \cos\theta\hat{\boldsymbol{K}})\right) \\ &\quad\times\left((\dot{\psi}\hat{\boldsymbol{K}} + p(\sin\theta\hat{\boldsymbol{\imath}}_{\psi} + \cos\theta\hat{\boldsymbol{K}})\times(b\hat{\boldsymbol{I}})\right) \\ &= \left(\dot{\psi}\hat{\boldsymbol{K}} + p(\sin\theta\hat{\boldsymbol{I}} + \cos\theta\hat{\boldsymbol{K}})\right)\times b(\dot{\psi} + p\cos\theta)\hat{\boldsymbol{J}} \\ &= pb\sin\theta(\dot{\psi} + p\cos\theta)\hat{\boldsymbol{K}} - b(\dot{\psi} + p\cos\theta)^2\hat{\boldsymbol{I}}.\end{aligned}$$

$\underline{2\boldsymbol{\omega}\times\boldsymbol{v}_{rel}} = -2pv\sin\theta\hat{\boldsymbol{J}}$ as before.

$\underline{\boldsymbol{a}_{rel}}\equiv \boldsymbol{0}$ as before.

Thus, collecting *these* results (expanding the square in the $\hat{\boldsymbol{I}}$-component of $\boldsymbol{\omega}\times(\boldsymbol{\omega}\times\boldsymbol{r})$), we get that

$$\boldsymbol{a}_B = -b(\dot{\psi}^2 + 2p\dot{\psi}\cos\theta + p^2\cos^2\theta)\hat{\boldsymbol{I}} - 2pv\sin\theta\hat{\boldsymbol{J}} + bp^2\sin\theta\cos\theta\hat{\boldsymbol{K}}. \tag{4.3-9}$$

A comparison of this result with that obtained from the previous choice of coordinate system, equation (4.3-7), exhibits a disturbing inconsistency: though the $\hat{\boldsymbol{J}}$ term is the same, the $\hat{\boldsymbol{K}}$-term, and the second and third terms in the $\hat{\boldsymbol{I}}$-component aren't! This unsettling development can be reconciled if we recognize that the angle θ appearing above satisfies the equations

$$\sin\theta = \frac{b}{\sqrt{a^2+b^2}}, \quad \cos\theta = \frac{a}{\sqrt{a^2+b^2}};$$

substituting these relations into, say, the $\hat{\boldsymbol{K}}$-component of equations (4.3-7) and (4.3-9), respectively, we get that

$$ap^2 \sin^2\theta = \frac{ab^2}{a^2+b^2}p^2 = bp^2 \sin\theta\cos\theta,$$

with similar results for the other two offending terms in the $\hat{\boldsymbol{I}}$-component. Thus, in fact, the results *are* the same.

Coordinate System with Origin at O, Rotating with $\dot{\psi}\hat{\boldsymbol{K}}$. Now the expression for acceleration is still

$$\boldsymbol{a}_B = \boldsymbol{a}_O + \boldsymbol{\alpha}\times\boldsymbol{r} + \boldsymbol{\omega}\times(\boldsymbol{\omega}\times\boldsymbol{r}) + 2\boldsymbol{\omega}\times\boldsymbol{v}_{rel} + \boldsymbol{a}_{rel}, \tag{4.3-10}$$

but with $\boldsymbol{\omega} \equiv \dot{\psi}\hat{\boldsymbol{K}}$ and $\boldsymbol{\alpha} \equiv \boldsymbol{0}$. Calculations start off the same:

$\underline{\boldsymbol{a}_O}\equiv \boldsymbol{0}$.

$\underline{\boldsymbol{\alpha}\times\boldsymbol{r}}\equiv 0$, since $\boldsymbol{\alpha}\equiv\boldsymbol{0}$.

$\underline{\boldsymbol{\omega}\times(\boldsymbol{\omega}\times\boldsymbol{r})}$: $\boldsymbol{r}$ is the same as in the second try, but now $\boldsymbol{\omega}$ is different:

$$\boldsymbol{\omega}\times(\boldsymbol{\omega}\times\boldsymbol{r}) = \dot{\psi}\hat{\boldsymbol{K}}\times(\dot{\psi}\hat{\boldsymbol{K}}\times b\hat{\boldsymbol{I}}) = -b\dot{\psi}^2\hat{\boldsymbol{I}}$$

(*cf.* the expression for $\boldsymbol{a}_A$ in the first coordinate system).

Up to this point, calculations have been, if anything, easier than either of the previous choices. But here is where things begin to get a little more "interesting":

$\underline{2\boldsymbol{\omega}\times\boldsymbol{v}_{rel}}$: $\boldsymbol{\omega}\equiv\dot{\psi}\hat{\boldsymbol{K}}$ is no problem. But what is $\boldsymbol{v}_{rel}$? Recall that this is defined to be the velocity of B *relative to the coordinate system*; this, in turn, arises due to rotation of the plate about the OA-axis *which is fixed in the coordinate system we have chosen*, along with the motion of B along the slot AB. To measure this, it is useful to introduce, again, an intermediate coordinate system, this time rotating with the plate *in the original system*, and with *its* origin at O:

$$\boldsymbol{v}_{rel} = \boldsymbol{v}'_O + \boldsymbol{\omega}'\times\boldsymbol{r}' + \boldsymbol{v}'_{rel},$$

in which now all quantities are *relative to the $\dot{\psi}\hat{\boldsymbol{K}}$-rotating coordinate system*; in particular, $\boldsymbol{r}' = \boldsymbol{r} = b\hat{\boldsymbol{I}}$, and we have angular velocity and acceleration

$$\boldsymbol{\omega}' \equiv p(\sin\theta\hat{\boldsymbol{\imath}}_\psi + \cos\theta\hat{\boldsymbol{K}})$$
$$\boldsymbol{\alpha}' \equiv \boldsymbol{0},$$

because $\boldsymbol{\omega}'$ is constant in both magnitude and direction *relative to the original rotating coordinate system.* Since $\boldsymbol{v}'_O\equiv\boldsymbol{0}$ in this [original] system,

$$\begin{aligned}\boldsymbol{v}_{rel} &= \boldsymbol{v}'_O + \boldsymbol{\omega}'\times\boldsymbol{r}' + \boldsymbol{v}'_{rel}\\ &= \boldsymbol{0} + p(\sin\theta\hat{\boldsymbol{\imath}}_\psi + \cos\theta\hat{\boldsymbol{K}})\times b\hat{\boldsymbol{I}} + v\hat{\boldsymbol{K}}\\ &= pb\cos\theta\hat{\boldsymbol{J}} + v\hat{\boldsymbol{K}}\end{aligned}$$

so that, finally,

$$2\boldsymbol{\omega}\times\boldsymbol{v}_{rel} = 2\dot{\psi}\hat{\boldsymbol{K}}\times(pb\cos\theta\hat{\boldsymbol{J}} + v\hat{\boldsymbol{K}}) = -2bp\dot{\psi}\cos\theta\hat{\boldsymbol{I}}.$$

$\underline{\boldsymbol{a}_{rel}}$: This is calculated similarly to the $\boldsymbol{v}_{rel}$, using the same intermediate coordinate system:

$$\begin{aligned}\boldsymbol{a}_{rel} &= \boldsymbol{a}'_O + \boldsymbol{\alpha}'\times\boldsymbol{r}' + \boldsymbol{\omega}'\times(\boldsymbol{\omega}'\times\boldsymbol{r}') + 2\boldsymbol{\omega}'\times\boldsymbol{v}'_{rel} + \boldsymbol{a}'_{rel}\\ &= \boldsymbol{0} + \boldsymbol{0} + p(\sin\theta\hat{\boldsymbol{i}}_\psi + \cos\theta\hat{\boldsymbol{K}})\times\left(p(\sin\theta\hat{\boldsymbol{i}}_\psi + \cos\theta\hat{\boldsymbol{K}})\times b\hat{\boldsymbol{I}}\right)\\ &\quad + 2p(\sin\theta\hat{\boldsymbol{i}}_\psi + \cos\theta\hat{\boldsymbol{K}})\times v\hat{\boldsymbol{K}} + \boldsymbol{0}\\ &= p(\sin\theta\hat{\boldsymbol{i}}_\psi + \cos\theta\hat{\boldsymbol{K}})\times pb\cos\theta\hat{\boldsymbol{J}} - 2pv\sin\theta\hat{\boldsymbol{J}}\\ &= bp^2\sin\theta\cos\theta\hat{\boldsymbol{K}} - bp^2\cos^2\theta\hat{\boldsymbol{I}} - 2pv\sin\theta\hat{\boldsymbol{J}}.\end{aligned}$$

Again, collecting results, we get finally that

$$\boldsymbol{a}_B = -(b\dot{\psi}^2 + 2bp\dot{\psi}\cos\theta + bp^2\cos^2\theta)\hat{\boldsymbol{I}} - 2pv\sin\theta\hat{\boldsymbol{J}} + bp^2\sin\theta\cos\theta\hat{\boldsymbol{K}}$$

—matching exactly the second coordinate system's result, equation (4.3-9).

The above presentation of alternate choices of coordinate systems may seem at first to be "overkill"; but, once again, it is presented to demonstrate that the correct result will be obtained no matter *what* rotating coordinate system is selected: *there is no "correct" coordinate system.* It gives some idea, however, of the complexity of various choices, and also attempts to reinforce the meaning of the various terms in the rotating coordinate system equations. Finally, the freedom of selection of basis—fixed or rotating—in terms of which to express the various vector quantities is demonstrated. |

The major problem students have with these rotating systems is a tendency to wander off the particular one they have chosen to use. There is, however, a certain commitment—a pledge of *faithfulness*, in some sense—one implicitly makes when he chooses to refer quantities to a given system. This, perhaps human, inclination towards unfaithfulness should be discouraged in Dynamics, as well as in Life.

The Need for Rotating Coordinate Systems

Given their somewhat daunting form, the student might ask whether rotating coordinate systems are actually *necessary.* In point of fact, they are: Recall that we have stated that Newton's Second Law prescribes that the acceleration it embodies be the *absolute* one; in fact, one can interpret $\sum \boldsymbol{F} = m\boldsymbol{a}$ as demonstrating that we *observe* the existence of [kinetic] forces through the [kinematic] *acceleration* they produce. This makes sense when one realizes that many of the most important physical forces—gravity, for example—aren't even *visible*! Thus, if we observe an [absolute] acceleration, we infer the existence of a force generating it. But we are always observing various phenomena from a rotating, accelerating coordinate system: we are on a *rotating* earth which revolves about the Sun in a non-straight-line path, so is *accelerating.*[7]

One can certainly argue that, over the time scales of most observations, the accelerations resulting from this motion has little practical effect, and it is, for all but the most precise experiment, admissible to ignore them. But rotating coordinate systems furnish a tool for the analysis of systems accelerating relative to the earth itself, exemplified below. And they will provide the basis for the next section on "*machine* kinematics," which *does* have practical importance. They are more than merely academic.

For the time being, however, we shall apply these rotating coordinate system equations (4.3-4) and (4.3-5) in the context of the way that they were derived: that we *know* (or can find simply) the motion of the "rotating system" and desire to find the absolute velocity and/or acceleration of some other point. This is, in some sense, the "forward" problem of rotating coordinates.

Summary. Translational velocity and acceleration are time rates of change of a vector measured from some point; rotational velocity and acceleration are those measured from a line. Such points and lines implicitly define *coordinate systems.* If this system is *fixed*, the time rates are *absolute*, but if it is moving, one measures quantities "$\boldsymbol{v}_{rel}$" and "$\boldsymbol{a}_{rel}$," or "$\boldsymbol{\omega}_{rel}$" and "$\boldsymbol{\alpha}_{rel}$." The theme of this section has been to relate these two types of quantities through general *rotating coordinate systems.*

In order to develop these relations, we considered first only *points*, A and B, say, in terms of which which we measured the "point-to-point" velocity $\boldsymbol{v}_{A/B}$ and acceleration $\boldsymbol{a}_{A/B}$; such quantities required no coordinate system other than the absolute one, in order to be able to measure the respective absolute motions. We then allowed one of these itself to be the origin of a coordinate system—one which itself was *rotating* with a given angular velocity and acceleration. We obtained a fundamental relation between the *time derivatives* of quantities in the absolute and rotating systems, Theorem 4.3.1. Though this will return in the next Chapter, in the present context its primary significance was to obtain equations relating the absolute velocity and acceleration of a point and that *observed* in the rotating coordinate system, $\boldsymbol{v}_{rel}$ and $\boldsymbol{a}_{rel}$. Such relations are useful when one knows only these observed quantities and wants the absolute ones—particularly the absolute acceleration, required by Newton's Laws.

4.4 Machine Kinematics

The previous section dealt with "moving coordinate systems"—coordinate systems whose motion (more precisely, whose origin's *translation* and axes' *rota-*

[7] This doesn't even begin to scratch the surface regarding the motion of our reference frame: The Sun revolves about the center of our Milky Way galaxy. But "our" galaxy congregates with *other* galaxies constituting our "Local Group," which all move, more or less together, in a "cluster" of galaxies. These clusters, in turn, have been observed to form *super*clusters similarly moving together, and there is no reason to expect this clustering to stop at that level. It's all "fleas on fleas" on a literally cosmic scale!

tion) are known, allowing us to determine the absolute velocities and accelerations and/or those observed in the system. Though the present section goes under the rubric of "machine kinematics," it, too, will involve such coordinate systems; the difference is that now $\boldsymbol{\omega}$ and $\boldsymbol{\alpha}$ will be considered as *unknowns* to be determined from the velocities and accelerations. If the approach of the previous section was described at its end as the "forward" problem, that of the current section might similarly be characterized as the "inverse" problem.

This distinction is significant because of the manner in which $\boldsymbol{\omega}$ and $\boldsymbol{\alpha}$ actually enter the rotating coordinate system equations; namely, through cross products. If these quantities are known, it is straightforward merely to evaluate the indicated products and include them with the balance of the terms in the equations. But if they are *unknown*, it is necessary somehow to "undo" these cross products to recover the desired quantities.

The fundamental approach taken here is to fix a coordinate system in a given body; the motion (particularly the rotational motion) of the coordinate system then mimics that of the body itself. In particular, if some point is known to move with a certain velocity, or at least in a certain direction (things which typically *are* known about a machine, which is designed to move in a certain way), that point makes a good candidate for the origin of the coordinate system: since its [*translational*] motion is known, there remains only the *rotation* of the coordinate system—*i.e.*, of the body—to determine. This procedure applies to a *single* body; if this is part of a proper "machine"—a *system* of interconnected rigid bodies, then the process is merely applied recursively. This brief outline of the methodology is mirrored in the current presentation: we shall consider first *single* rigid bodies to ensure the procedure is thoroughly understood; then "*machines*" will be dealt with. Along the way we shall consider precisely what *kinematical* constraints are imposed on the motion of various bodies by the connections between them.

There remains, however, the question if just why one even *wants* to determine the angular velocity and acceleration of various members in a machine. The answer is reasonably straightforward: To design a machine, the first criterion for successful operation is that it stay together! This, in turn, demands an analysis of the forces and moments acting on the various members; and *that* requires knowledge of rotational and center-of-mass translational motion—i.e., *kinetics*—which we shall pursue in the next Chapter.

4.4.1 Motion of a *Single* Body

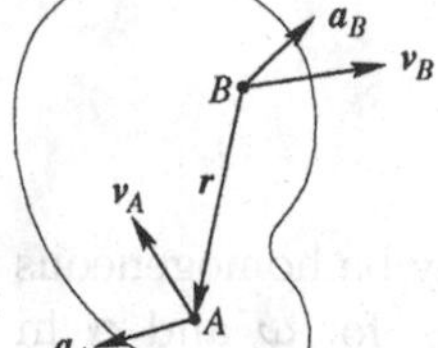

Consider an arbitrary body, diagrammed at left, in which two points, A and B, have velocities $\boldsymbol{v}_A$ and $\boldsymbol{v}_B$ and accelerations $\boldsymbol{a}_A$ and $\boldsymbol{a}_B$, respectively. If we choose to fix a coordinate system in the body at, say, the point B, then *the angular velocity and angular acceleration of the coordinate system are precisely those of the body itself.* In particular, it is possible to describe the motion of point A relative to that coordinate system through the rotating coordinate system equations obtained in the previous section, equations (4.3-4b) and (4.3-5b):

$$\boldsymbol{v}_A = \boldsymbol{v}_B + \boldsymbol{\omega} \times \boldsymbol{r} + \boldsymbol{v}_{rel}$$
$$\boldsymbol{a}_A = \boldsymbol{a}_B + \boldsymbol{\alpha} \times \boldsymbol{r} + \boldsymbol{\omega} \times (\boldsymbol{\omega} \times \boldsymbol{r}) + 2\boldsymbol{\omega} \times \boldsymbol{v}_{rel} + \boldsymbol{a}_{rel}$$

—in which $\boldsymbol{\omega}$ and $\boldsymbol{\alpha}$ are the angular velocity and acceleration *of the body.* Note that if the body is *rigid*, then *viewed from the coordinate system* there is no motion of A relative to B—*i.e.*, $\boldsymbol{v}_{rel} \equiv 0 \equiv \boldsymbol{a}_{rel}$. These terms can be non-0 only if the body itself admits of a changing distance between the two points, *e.g.*, through a "telescoping" construction.

It might be thought that the motions of A and B can be arbitrary, but this is generally not the case: If the body is rigid (as it often is), then the above equations can be written

$$\boldsymbol{v}_A - \boldsymbol{v}_B = \boldsymbol{\omega} \times \boldsymbol{r}$$
$$\boldsymbol{a}_A - \boldsymbol{a}_B - \boldsymbol{\omega} \times (\boldsymbol{\omega} \times \boldsymbol{r}) = \boldsymbol{\alpha} \times \boldsymbol{r}.$$

The first of these says that the vector $\boldsymbol{v}_A - \boldsymbol{v}_B$ must be perpendicular to $\boldsymbol{r}$, since it is expressed as a cross product with $\boldsymbol{r}$ as one component; the second says the same thing with regard to $\boldsymbol{a}_A - \boldsymbol{a}_B - \boldsymbol{\omega} \times (\boldsymbol{\omega} \times \boldsymbol{r})$. Since $\boldsymbol{r}$ is completely determined by the positions of A and B, these conditions will only hold true under special circumstances; the only way that the motions of A and B *can* be arbitrary is for there to be an additional *non*-0 $\boldsymbol{v}_{rel}$ and $\boldsymbol{a}_{rel}$ to "take up the slack." Using exactly the same arguments demonstrating that both left-hand sides must be perpendicular to $\boldsymbol{r}$, they must also be perpendicular to $\boldsymbol{\omega}$ and $\boldsymbol{\alpha}$, respectively.

Recall that we are primarily interested in determination of $\boldsymbol{\omega}$ and $\boldsymbol{\alpha}$. These can be isolated by rewriting the rotating coordinate system equations in the form

$$\boldsymbol{\omega} \times \boldsymbol{r} = \boldsymbol{v}_A - \boldsymbol{v}_B - \boldsymbol{v}_{rel} \equiv \boldsymbol{V} \tag{4.4-1a}$$
$$\boldsymbol{\alpha} \times \boldsymbol{r} = \boldsymbol{a}_A - \boldsymbol{a}_B - \boldsymbol{\omega} \times (\boldsymbol{\omega} \times \boldsymbol{r}) - 2\boldsymbol{\omega} \times \boldsymbol{v}_{rel} - \boldsymbol{a}_{rel} \equiv \boldsymbol{A}; \tag{4.4-1b}$$

just as above, $\boldsymbol{V}$ must simultaneously be perpendicular to $\boldsymbol{r}$ and $\boldsymbol{\omega}$, and $\boldsymbol{A}$ perpendicular to both $\boldsymbol{r}$ and $\boldsymbol{\alpha}$. At first blush, it might be expected that these equations would completely determine the three components of $\boldsymbol{\omega}$ and $\boldsymbol{\alpha}$, but this, too, is not correct: Using Example 2.2.7, we can *represent* the two cross products as products of a *matrix* r with the column matrices ω and α and write the above equations

$$\mathsf{r}\omega = \mathsf{V}$$
$$\mathsf{r}\alpha = \mathsf{A}.$$

These can be viewed as systems of linear equations—they may be homogeneous or non-homogeneous, depending on the values of $\boldsymbol{V}$ and $\boldsymbol{A}$—for $\boldsymbol{\omega}$ and $\boldsymbol{\alpha}$ in which each is multiplied by a "coefficient matrix" r. But, from the homework problem 3 on page 70, this matrix *always* has rank only two 2 for $\mathbf{r} \neq \mathbf{0}$;[8] thus, by the theory of systems in Section 2.3, there are an *infinite* number of possible values of $\boldsymbol{\omega}$ and $\boldsymbol{\alpha}$: the solution is not unique unless there is some additional constraint equation on these unknowns! There is a geometrical interpretation of this malady: Dotting the two vector equations with $\boldsymbol{\omega}$ and $\boldsymbol{\alpha}$, respectively,

$$\boldsymbol{\omega} \cdot \boldsymbol{\omega} \times \boldsymbol{r} \equiv 0 = \boldsymbol{\omega} \cdot \boldsymbol{V} \qquad (4.4\text{-}2a)$$

$$\boldsymbol{\alpha} \cdot \boldsymbol{\alpha} \times \boldsymbol{r} \equiv 0 = \boldsymbol{\alpha} \cdot \boldsymbol{A}, \qquad (4.4\text{-}2b)$$

showing that the two angular variables lie in planes perpendicular to $\boldsymbol{V}$ and $\boldsymbol{A}$; but there are an infinite number of such vectors, and these equations alone do not specify *which* direction the unknowns have. Once these directions $\hat{\boldsymbol{\omega}}$ and $\hat{\boldsymbol{\alpha}}$ were established, however, any component of the equations would determine the scalar ω and α to fix $\boldsymbol{\omega} = \omega\hat{\boldsymbol{\omega}}$ and $\boldsymbol{\alpha} = \alpha\hat{\boldsymbol{\alpha}}$. However it is viewed, this means that we must have a minimum of 4 equations: the three components of the above rotating coordinate system equations, plus the fourth additional condition on the unknowns $\boldsymbol{\omega}$ and $\boldsymbol{\alpha}$.

The student with experience in planar problems might ask why the necessity of such an "additional condition" never arose there. In point of fact, there *was* such an additional condition implicit in the planar nature of the problem itself: since $\boldsymbol{\omega}$ is always assumed to be perpendicular to the plane of motion—in the $\hat{\boldsymbol{k}}$-direction, say—there were *two* additional conditions assumed: that $\omega_x \equiv 0 \equiv \omega_y$. These, along with the fact that there were only two components of the rotating coordinate system equations surviving the calculations, gave unique results for $\boldsymbol{\omega}$: we had *two* fewer components to find, but only *one* less equation.

Three of the unknowns which can be found are the components of angular velocity and acceleration; what is the fourth? In the case the body is rigid—$\boldsymbol{v}_{rel} \equiv 0 \equiv \boldsymbol{a}_{rel}$—we noted above that the motion of both A and B cannot be arbitrary; in fact, if, say, $\boldsymbol{v}_B$ and $\boldsymbol{a}_B$ are given completely, only the *directions* of $\boldsymbol{v}_A$ and $\boldsymbol{a}_A$ can be specified; the scalar *components* v_A and a_A along these directions then typically comprise the last unknowns. On the other hand, if the body is *not* rigid, $\boldsymbol{v}_A$ and $\boldsymbol{a}_A$ can be specified completely and the directions of $\boldsymbol{v}_{rel}$ and $\boldsymbol{a}_{rel}$ specified, leaving the components v_{rel} and a_{rel} to be found. These are the typical cases, and they comprise the elusive fourth unknown in the respective circumstances. But it must be kept in mind that one generally cannot find all three components of the angular velocity and acceleration without a fourth equation for each, and that there are generally four unknowns, of whatever ilk, to be found from these.

There is another important observation to be made with regard to actually using these rotating coordinate system equations to determine the angular motion of a body: Note that the equation for $\boldsymbol{\alpha}$ itself involves $\boldsymbol{\omega}$ through the triple

[8]This, of course, assumes the equation *has* a solution in the first place! By the theory in Section 2.3.1, it always will if the equation is homogeneous, but if $\boldsymbol{V} \neq \mathbf{0}$ and/or $\boldsymbol{A} \neq \mathbf{0}$, the *rank* of the augmented matrix, $(\mathbf{r}\ \mathbf{V})$ or $(\mathbf{r}\ \mathbf{A})$, must also be less than 2. See Homework Problem 2 on page 218.

cross product $\boldsymbol{\omega} \times (\boldsymbol{\omega} \times \boldsymbol{r})$; thus $\boldsymbol{\alpha}$ cannot be determined without knowing $\boldsymbol{\omega}$! Even though one might be interested only in the value of the angular acceleration, *it is necessary to solve for* $\boldsymbol{\omega}$ *from the velocity equation in order to find* $\boldsymbol{\alpha}$. Similar considerations hold with regard to $\boldsymbol{v}_{rel}$ and $\boldsymbol{a}_{rel}$ in the case the body is not rigid.

The mathematics notwithstanding, there remains the practical question of just which points "A" and "B" to consider in these equations. Clearly if one knows completely the velocity and acceleration of a point, there is a certain psychological security generated by picking that as the origin of the rotating system, though this is not a necessary practice: if one interchanges the identities of "A" and "B," $\boldsymbol{r} \equiv \boldsymbol{r}_{A/B} \to \boldsymbol{r}_{B/A}$, effectively reversing the sign on the cross products in the original assignment and leaving the final $\boldsymbol{\omega}$ and $\boldsymbol{\alpha}$ the same. But what would be the other point? At the end of the day, the motion of certain points—generally only two such—ultimately determine the motion of the body as a whole; the other points in the body simply "come along for the ride." It is more than merely expedient to determine such points from the start—this requires only a minimal sense of physical intuition—use *them* to determine the angular motion of the body, and then determine the motion of other points from the known motion of one and the angular motion of the body, using now the "forward" rotating coordinate system equations. This is exemplified in the following:

Example 4.4.1. *Ball-and-socket Connection.* We treat here a fairly general example, one in which the body is not rigid. As noted above, this allows us to specify completely the motion of the two end points, in the assurance that the "*rel*" quantities will adjust for it.

We consider a piston AB, connected by balls-and-sockets at each end to collars which move along two rods, one of which is aligned with the X-axis and the other of which moves along one parallel to the Y-axis and 2 m above the first as shown. If end A moves with a constant speed $v_A = 2\,\text{m/s}$ while B moves with constant $v_B = 1\,\text{m/s}$, find the acceleration, $\overline{\boldsymbol{a}}$, of the center of mass G of the piston, assumed always to lie 0.47 of the length of AB from A, at the instant A and B are 1 m and 2 m, respectively, away from the Z-axis. Assume there to be no rotation of the piston about its axis.

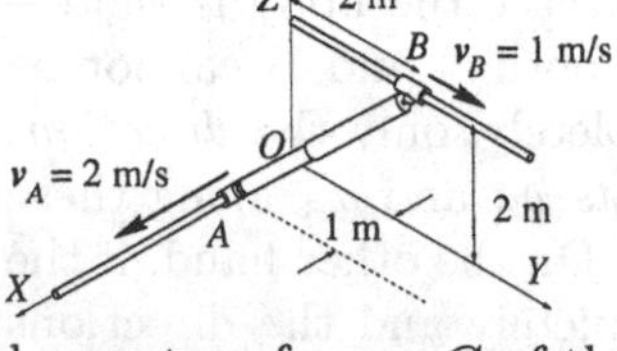

The reader will recall our discussion of the "additional condition" required to determine angular velocity and acceleration. The final sentence above, regarding the rotational motion (or *lack* of motion) about the piston's axis is precisely such a statement. This particular choice may appear to be somewhat *ad hoc* at present; we shall shortly see the rationale for this. For the present we merely accept it as a given condition and solve the problem.

In view of the fact that the rotating coordinate equations are the approach to determine these quantities, we will, more or less arbitrarily, consider a coordinate system with origin at B and rotating with AB; then $\boldsymbol{\omega}$ and $\boldsymbol{\alpha}$ correspond to $\boldsymbol{\omega}_{AB}$ and $\boldsymbol{\alpha}_{AB}$, respectively, while $\boldsymbol{r}$ will be $\boldsymbol{r}_{A/B}$. Though this defines our coordinate system, we have yet to declare the *directions* of the basis vectors in that system: for convenience, we will assume the rotating system has directions *instantaneously* the same as the fixed system's, expressing all quantities in terms of $\hat{\boldsymbol{I}}$, $\hat{\boldsymbol{J}}$, and $\hat{\boldsymbol{K}}$.

Since we know the accelerations of both A and B, we can express

$$\overline{\boldsymbol{a}} = \boldsymbol{a}_B + \boldsymbol{\alpha} \times \overline{\boldsymbol{r}} + \boldsymbol{\omega} \times (\boldsymbol{\omega} \times \overline{\boldsymbol{r}}) + 2\boldsymbol{\omega} \times \overline{\boldsymbol{v}}_{rel} + \overline{\boldsymbol{a}}_{rel}, \tag{4.4-3}$$

where $\overline{\boldsymbol{r}}$ measures the position of G from the rotating system's origin, here B, and $\overline{\boldsymbol{v}}_{rel}$ and $\overline{\boldsymbol{a}}_{rel}$ account for any motion of G *in* the rotating system. This requires that we know both $\boldsymbol{\omega}$ and $\boldsymbol{\alpha}$, the former demanding a previous velocity analysis. Both angular quantities, in turn, are determined by the motion of A and B, so we shall initially focus attention on these points rather than G. Furthermore, even though we are not interested in them, it will be necessary to include the *rel* quantities resulting from the expandable nature of the piston just to find the angular quantities. In summary, even though we wish to find the *acceleration* of G, we must start with the *velocities* of A and B, contending along the way with apparently "extraneous" $\boldsymbol{v}_{rel}$ and $\boldsymbol{a}_{rel}$. [Actually, we shall need even these, as we shall see.]

We consider the velocity first: From our choice of coordinate system, we have immediately that

$$\boldsymbol{v}_A = \boldsymbol{v}_B + \boldsymbol{\omega} \times \boldsymbol{r} + \boldsymbol{v}_{rel}. \tag{4.4-4a}$$

Unlike the above equation, $\boldsymbol{v}_{rel}$ in this one stands for the motion of A measured relative to our coordinate system. Since in this coordinate system $\boldsymbol{v}_{rel}$ is directed along the line from A to B, and writing $\boldsymbol{\omega} = \omega_X\hat{\boldsymbol{I}} + \omega_Y\hat{\boldsymbol{J}} + \omega_Z\hat{\boldsymbol{K}}$,

$$\begin{aligned} 2\hat{\boldsymbol{I}} = 1\hat{\boldsymbol{J}} &+ (\omega_X\hat{\boldsymbol{I}} + \omega_Y\hat{\boldsymbol{J}} + \omega_Z\hat{\boldsymbol{K}}) \times (1\hat{\boldsymbol{I}} - 2\hat{\boldsymbol{J}} - 2\hat{\boldsymbol{K}}) \\ &+ v_{rel}\frac{1\hat{\boldsymbol{I}} - 2\hat{\boldsymbol{J}} - 2\hat{\boldsymbol{K}}}{3} \end{aligned}$$

(the sign on v_{rel} will determine the *actual* vector direction of the last term). Carrying out the indicated cross products and collecting terms, we get three component equations

$$\begin{aligned} 2 &= \qquad -2\omega_Y + 2\omega_Z + \frac{1}{3}v_{rel} \\ -1 &= \quad 2\omega_X \qquad + \omega_Z - \frac{2}{3}v_{rel} \\ 0 &= -2\omega_X - \omega_Y \qquad - \frac{2}{3}v_{rel}. \end{aligned} \tag{4.4-4b}$$

[The v_{rel} terms are necessary: if they were absent, the coefficient matrix determining the components of $\boldsymbol{\omega}$ would have rank of only 2, while that of the augmented matrix, including the non-homogeneous column, would have rank 3; thus the equations would be inconsistent, and there would be no solution. See Section 2.3.] This is a set of three equations in the three components of $\boldsymbol{\omega}$ as well as v_{rel}; clearly there is yet another equation required. This is the condition that there be no component of $\boldsymbol{\omega}$ along AB:

$$0 \equiv \boldsymbol{\omega} \cdot \boldsymbol{r} \tag{4.4-5a}$$

or

$$0 \equiv \omega_X - 2\omega_Y - 2\omega_Z. \tag{4.4-5b}$$

The "b" equations of (4.4-4) and (4.4-5) constitute a set of four equations in four unknowns. Solving these—note that we will need $\boldsymbol{v}_{rel}$ as well as $\boldsymbol{\omega}$ in the acceleration equation—we get that

$$\omega_X = -\frac{2}{9}, \quad \omega_Y = -\frac{4}{9}, \quad \omega_Z = \frac{1}{3}, \quad v_{rel} = \frac{4}{3};$$

thus

$$\boldsymbol{\omega} = -\frac{2}{9}\hat{\boldsymbol{I}} - \frac{4}{9}\hat{\boldsymbol{J}} + \frac{1}{3}\hat{\boldsymbol{K}}, \quad \boldsymbol{v}_{rel} = \frac{4}{9}\hat{\boldsymbol{I}} - \frac{8}{9}\hat{\boldsymbol{J}} - \frac{8}{9}\hat{\boldsymbol{K}}$$

—quantities which are needed following.

We briefly interject the fact that, knowing now $\boldsymbol{\omega}$ and $\boldsymbol{v}_A$ (or $\boldsymbol{v}_B$), we could find the velocity of the center of mass:

$$\begin{aligned}\overline{\boldsymbol{v}} &= \boldsymbol{v}_A + \boldsymbol{\omega} \times \boldsymbol{r}_{G/A} + \boldsymbol{v}_{rel}\\ &= \boldsymbol{v}_A + \boldsymbol{\omega} \times 0.47\boldsymbol{r}_{B/A} + 0.47\boldsymbol{v}_{rel_{B/A}}\end{aligned}$$

since G is located 0.47 the distance along AB from A, and thus its *rel* velocity is the same fraction of $\boldsymbol{v}_{B/A}$.

We now consider the acceleration: Once again taking B to be the origin of the system rotating with the piston, we use

$$\boldsymbol{a}_A = \boldsymbol{a}_B + \boldsymbol{\alpha} \times \boldsymbol{r} + \boldsymbol{\omega} \times (\boldsymbol{\omega} \times \boldsymbol{r}) + 2\boldsymbol{\omega} \times \boldsymbol{v}_{rel} + \boldsymbol{a}_{rel}. \tag{4.4-6a}$$

Since both A and B move with constant speeds, so [because their *directions* don't change] velocities, their accelerations vanish; taking $\boldsymbol{\alpha} \equiv \alpha_x\hat{\boldsymbol{I}} + \alpha_y\hat{\boldsymbol{J}} + \alpha_z\hat{\boldsymbol{K}}$ and noting that, once again, $\boldsymbol{a}_{rel}$ is along $\boldsymbol{r}_{AB}$, this equation becomes

$$\begin{aligned}\boldsymbol{0} = \boldsymbol{0} &+ (\alpha_X\hat{\boldsymbol{I}} + \alpha_Y\hat{\boldsymbol{J}} + \alpha_Z\hat{\boldsymbol{K}}) \times (1\hat{\boldsymbol{I}} - 2\hat{\boldsymbol{J}} - 2\hat{\boldsymbol{K}})\\ &+ (-\frac{2}{9}\hat{\boldsymbol{I}} - \frac{4}{9}\hat{\boldsymbol{J}} + \frac{1}{3}\hat{\boldsymbol{K}}) \times \left((-\frac{2}{9}\hat{\boldsymbol{I}} - \frac{4}{9}\hat{\boldsymbol{J}} + \frac{1}{3}\hat{\boldsymbol{K}}) \times (1\hat{\boldsymbol{I}} - 2\hat{\boldsymbol{J}} - 2\hat{\boldsymbol{K}})\right)\\ &+ 2(-\frac{2}{9}\hat{\boldsymbol{I}} - \frac{4}{9}\hat{\boldsymbol{J}} + \frac{1}{3}\hat{\boldsymbol{K}}) \times (\frac{4}{9}\hat{\boldsymbol{I}} - \frac{8}{9}\hat{\boldsymbol{J}} - \frac{8}{9}\hat{\boldsymbol{K}}) + a_{rel}\frac{1\hat{\boldsymbol{I}} - 2\hat{\boldsymbol{J}} - 2\hat{\boldsymbol{K}}}{3}.\end{aligned}$$

Again carrying out the indicated calculations, we get that

$$\begin{aligned}-\frac{83}{81} &= \qquad -2\alpha_Y + 2\alpha_Z + \frac{1}{3}a_{rel}\\ \frac{50}{81} &= \quad 2\alpha_X \qquad + \alpha_Z - \frac{2}{3}a_{rel}\\ -\frac{122}{81} &= -2\alpha_X - \alpha_Y \qquad - \frac{2}{3}a_{rel}.\end{aligned} \tag{4.4-6b}$$

Again, the fact that there is supposed no motion along the piston means that

$$\boldsymbol{\alpha} \cdot \boldsymbol{r} = \alpha_X - 2\alpha_Y - 2\alpha_Z \equiv 0; \tag{4.4-7}$$

this, along with (4.4-6b) constitute the four equations in the variables α_X, α_Y, α_Z, and a_{rel}. The solution is

$$\alpha_X = \frac{16}{81}, \quad \alpha_Y = \frac{32}{81}, \quad \alpha_Z = -\frac{8}{27}, \quad a_{rel} = \frac{29}{27};$$

so that

$$\boldsymbol{\alpha} = \frac{16}{81}\hat{\boldsymbol{I}} + \frac{32}{81}\hat{\boldsymbol{J}} - \frac{8}{27}\hat{\boldsymbol{K}}, \quad \boldsymbol{a}_{rel} = \frac{29}{81}\hat{\boldsymbol{I}} - \frac{58}{81}\hat{\boldsymbol{J}} - \frac{58}{81}\hat{\boldsymbol{K}}.$$

Again, it was necessary to include the *rel* terms in order to find $\boldsymbol{\alpha}$.

The above would be the procedure required to determine the angular motion of the rod alone. But, lest we forget, the example asked us to determine the acceleration of the center of mass! This is now relatively straightforward, implementing rotating coordinate systems in the "forward" direction using equation (4.4-3) in which $\boldsymbol{a}_B$ is known [to vanish], and we have determined $\boldsymbol{\omega}$ and $\boldsymbol{\alpha}$ above. From the assumed placement of the center of mass 0.47 of the distance from A—0.53 that from B—we get that $\overline{\boldsymbol{r}} \equiv 0.53\hat{\boldsymbol{I}} - 1.06\hat{\boldsymbol{J}} - 1.06\hat{\boldsymbol{K}}$. But what will we use for $\boldsymbol{v}_{rel}$ and $\boldsymbol{a}_{rel}$? Again, the presumed placement of G along $\boldsymbol{r}_{AB}$ allows us to conclude that these quantities for that point are just 0.53 the values for A obtained above. Substituting these into (4.4-3), we get that

$$\overline{\boldsymbol{a}} = -2.80\hat{\boldsymbol{I}} + 1.20\hat{\boldsymbol{J}} - 1.40\hat{\boldsymbol{K}} \times 10^{-10} \text{ m/s}^2.$$

It is not so surprising that this so small (the accelerations of the ends of the piston are both zero, recall), but that there is one at all (rather like Johnson's dog walking on its hind legs)!

This might seem an awfully lot of work to get such a small number, but in actual problems it would be necessary to find that acceleration in order to determine the forces acting at ends A and B; this will be the subject of the next Chapter. But however small this acceleration might be, the forces acting on the piston will sum to the *mass* of the piston times this value; and the piston *is* [at least] 3 m long, recall!

Before we leave this example, let us briefly consider how it would change if AB were *not* telescoping—*i.e.*, it were a *rigid* body. In the first place, $\boldsymbol{v}_{rel} \equiv 0 \equiv \boldsymbol{a}_{rel}$. But, more importantly, complete specification of *both* $\boldsymbol{v}_A$ and $\boldsymbol{v}_B$, and/or *both* $\boldsymbol{a}_A$ and $\boldsymbol{a}_B$ would be impossible: the system would then become overdetermined. Thus let us assume that only the motion of A is known; the fact that B moves along the bar means that $\boldsymbol{v}_B = v_B\hat{\boldsymbol{J}}$ and $\boldsymbol{a}_B = a_B\hat{\boldsymbol{J}}$, and the scalars v_B and a_B replace v_{rel} and a_{rel} above, all other quantities remaining the same.

Corresponding now to Equations (4.4-4), we have

$$\boldsymbol{v}_A = \boldsymbol{v}_B + \boldsymbol{\omega} \times \boldsymbol{r}$$

(since $v_{rel} \equiv 0$), or

$$\begin{aligned} 2 &= \qquad\quad -2\omega_Y + 2\omega_Z \\ 0 &= v_B + 2\omega_X \qquad + \omega_Z \\ 0 &= \quad -2\omega_X - \omega_Y \quad ; \end{aligned}$$

while (4.4-5) still holds:

$$0 \equiv \omega_X - 2\omega_Y - 2\omega_Z.$$

Solving these four [component] equations for $\boldsymbol{\omega}$ and v_B, we get, finally, that

$$\boldsymbol{\omega} = \frac{2}{9}\hat{\boldsymbol{I}} - \frac{4}{9}\hat{\boldsymbol{J}} + \frac{5}{9}\hat{\boldsymbol{K}}, \quad \boldsymbol{v}_B = -\hat{\boldsymbol{J}}.$$

Compare this with the previous answer. The acceleration is similar and will not be worked.

Homework:

1. Show that, indeed, were the v_{rel} terms absent from (4.4-4b), the coefficient matrix has the expected rank of 2, while that of the augmented matrix is 3.
2. In equations 4.4-2 we showed that $\boldsymbol{\omega}$ and $\boldsymbol{\alpha}$ must each be perpendicular to the vectors $\boldsymbol{V} \neq \boldsymbol{0}$ and $\boldsymbol{A} \neq \boldsymbol{0}$, respectively, in equations 4.4-1. A similar condition regarding $\boldsymbol{V}$ and $\boldsymbol{A}$ can be obtained involving the vector $\boldsymbol{r}$ in these [latter] equations. a) Obtain that condition, and b) reconcile it with the results of Homework 1 on page 86. Again, note that at least one component of $\boldsymbol{r}$ must be non-0 or $\boldsymbol{r} \equiv \boldsymbol{0}$!

A Useful Trick

As we have seen above, one typically has four unknown to solve: the three components of the angular variable under consideration, as well as that of either an absolute or "*rel*" quantity in known direction. Most generally one is interested in the angular motion, but it might be the case that only the single latter type of variable is desired. Though one might expect it still to be necessary to solve a system of four equations just to find that value, it turns out that that can be determined from a single equation independent of the angular variable:

Consider first the velocity equation, (4.3-4b):

$$\boldsymbol{v}_A = \boldsymbol{v}_B + \boldsymbol{\omega} \times \boldsymbol{r} + \boldsymbol{v}_{rel}.$$

If we dot this equation with $\boldsymbol{r}$,

$$\begin{aligned} \boldsymbol{v}_A \cdot \boldsymbol{r} &= \boldsymbol{v}_B \cdot \boldsymbol{r} + \boldsymbol{\omega} \times \boldsymbol{r} \cdot \boldsymbol{r} + \boldsymbol{v}_{rel} \cdot \boldsymbol{r} \\ &= \boldsymbol{v}_B \cdot \boldsymbol{r} + \boldsymbol{v}_{rel} \cdot \boldsymbol{r} \end{aligned}$$

since the triple scalar product of three vectors, two of which are parallel, vanishes. This equation *eliminates* the variable $\boldsymbol{\omega}$ from the equation, leaving just the fourth unknown, be it v_A, v_B, or v_{rel} (in known direction). This can be exemplified in the above Example, where, dotting the velocity equation (4.4-4a) with $\boldsymbol{r}$ we get that $2 = -2 + 9v_{rel}/3$, yielding $v_{rel} = 4/3$ as before.

In a similar fashion, dotting the acceleration equation (4.3-5b) with $\boldsymbol{r}$,

$$\boldsymbol{a}_A \cdot \boldsymbol{r} = \boldsymbol{a}_B \cdot \boldsymbol{r} + \boldsymbol{\alpha} \times \boldsymbol{r} \cdot \boldsymbol{r} + [\boldsymbol{\omega} \times (\boldsymbol{\omega} \times \boldsymbol{r})] \cdot \boldsymbol{r} + 2\boldsymbol{\omega} \times \boldsymbol{v}_{rel} \cdot \boldsymbol{r} + \boldsymbol{a}_{rel} \cdot \boldsymbol{r}.$$

The triple scalar product in $\boldsymbol{\alpha}$ vanishes in the same way the corresponding term in $\boldsymbol{\omega}$ did before, but what about the third and fourth terms, which seem to depend on $\boldsymbol{\omega}$?

Let us consider first the third term: It is useful for this analysis to break up $\boldsymbol{\omega}$ into two components:

$$\boldsymbol{\omega} \equiv \boldsymbol{\omega}_{\parallel} + \boldsymbol{\omega}_{\perp}, \tag{4.4-8a}$$

in which $\boldsymbol{\omega}_{\parallel}$ is parallel to $\boldsymbol{r}$ and $\boldsymbol{\omega}_{\perp}$ [in a plane] perpendicular to $\boldsymbol{r}$. This can always be done; indeed,

$$\boldsymbol{\omega}_{\parallel} = (\boldsymbol{\omega} \cdot \hat{\boldsymbol{r}})\hat{\boldsymbol{r}}, \quad \boldsymbol{\omega}_{\perp} = \boldsymbol{\omega} - \boldsymbol{\omega}_{\parallel}. \tag{4.4-8b}$$

We note from the start that

$$\boldsymbol{\omega} \times \boldsymbol{r} = (\boldsymbol{\omega}_{\parallel} + \boldsymbol{\omega}_{\perp}) \times \boldsymbol{r} = \boldsymbol{\omega}_{\perp} \times \boldsymbol{r} \tag{4.4-9}$$

by virtue of the fact that the parallel component vanishes in the cross product. This shows that it is only $\boldsymbol{\omega}_{\perp}$ which *enters* the velocity equation, and only that which can be *determined* from the equation; it is just another way of looking at the fact that the rank of the matrix representation of the cross product is only 2, and that it does not determine a *unique* value of $\boldsymbol{\omega}$. (Still *another* way is to note that the *general* solution to equation (4.4-1a), $\boldsymbol{\omega} \times \boldsymbol{r} = \boldsymbol{V}$, involves an element of the null space of $\boldsymbol{r}$ [Theorem 2.3.2 and the Homework on page 87)]—here called $\boldsymbol{\omega}_{\parallel}$.)

Expanding now the triple vector product using (4.1-5)

$$\begin{aligned}\boldsymbol{\omega} \times (\boldsymbol{\omega} \times \boldsymbol{r}) &= (\boldsymbol{\omega}_{\parallel} + \boldsymbol{\omega}_{\perp}) \times (\boldsymbol{\omega}_{\perp} \times \boldsymbol{r}) \\ &= \left((\boldsymbol{\omega}_{\parallel} + \boldsymbol{\omega}_{\perp}) \cdot \boldsymbol{r}\right) \boldsymbol{\omega}_{\perp} - \left((\boldsymbol{\omega}_{\parallel} + \boldsymbol{\omega}_{\perp}) \cdot \boldsymbol{\omega}_{\perp}\right) \boldsymbol{r} \\ &= (\boldsymbol{\omega}_{\parallel} \cdot \boldsymbol{r})\boldsymbol{\omega}_{\perp} - \omega_{\perp}^2 \boldsymbol{r}\end{aligned} \tag{4.4-10}$$

since the dot product retains only components parallel to its constituents. Now let us consider what happens to this term when we dot it with $\boldsymbol{r}$:

$$[\boldsymbol{\omega} \times (\boldsymbol{\omega} \times \boldsymbol{r})] \cdot \boldsymbol{r} = [(\boldsymbol{\omega}_{\parallel} \cdot \boldsymbol{r})\boldsymbol{\omega}_{\perp} - \omega_{\perp}^2 \boldsymbol{r}] \cdot \boldsymbol{r} = -\omega_{\perp}^2 r^2;$$

thus we see that *the only part of* $\boldsymbol{\omega}$ *which survives the dot product is* $\boldsymbol{\omega}_{\perp}$—precisely, incidentally, that part remaining from the cross product in the velocity equation! [Though the triple cross product has been expanded to show the need for $\boldsymbol{\omega}$, it is usually most straightforward to calculate it directly in a given problem.]

Now consider the fourth term, $2\boldsymbol{\omega}\times\boldsymbol{v}_{rel}$: In general, this will vary depending on the nature of $\boldsymbol{v}_{rel}$, and impossible to predict which parts of $\boldsymbol{\omega}$ will be required to evaluate it; thus we cannot make any sweeping statements regarding what happens when this term is dotted with $\boldsymbol{r}$. However, it is common for $\boldsymbol{v}_{rel}$ to be parallel to $\boldsymbol{r}$ itself, as we saw in the previous example. In *this* case, then, the term $2\boldsymbol{\omega}\times\boldsymbol{v}_{rel}\cdot\boldsymbol{r}$ vanishes once again because two components of the triple scalar product are parallel. Even in the full equation, however, only $\boldsymbol{\omega}_\perp$ will survive the cross product $2\boldsymbol{\omega}\times\boldsymbol{v}_{rel}$!

We conclude that, in general,

$$\boldsymbol{a}_A\cdot\boldsymbol{r} = \boldsymbol{a}_B\cdot\boldsymbol{r} - \omega_\perp^2 r^2 + 2\boldsymbol{\omega}\times\boldsymbol{v}_{rel}\cdot\boldsymbol{r} + \boldsymbol{a}_{rel}\cdot\boldsymbol{r}, \tag{4.4-11}$$

requiring that we know $\boldsymbol{\omega}$. But we note also that third term on the right-hand side drops out if $\boldsymbol{v}_{rel}\parallel\boldsymbol{r}$. As noted above, this is precisely what happens in the previous example, where, substituting the value of $\boldsymbol{\omega} = -\frac{2}{9}\hat{\boldsymbol{I}} - \frac{4}{9}\hat{\boldsymbol{J}} + \frac{1}{3}\hat{\boldsymbol{K}}$ and $\boldsymbol{v}_{rel} = \frac{4}{9}\hat{\boldsymbol{I}} - \frac{8}{9}\hat{\boldsymbol{J}} - \frac{8}{9}\hat{\boldsymbol{K}}$ into (4.4-6a), we get that $0 = -\frac{87}{27} + 3a_{rel}$, yielding $a_{rel} = \frac{29}{27}$, again as before.

Despite the fact that we must still solve the full velocity equation set, this does provide a useful means of finding the fourth acceleration unknown: instead of solving six equations in six unknowns, we need obtain the solution to only four equations in four. It is at least an efficient means of finding the fourth *velocity* unknown in a single equation.

The Non-slip Condition

We now give another example of single-body "machine" kinematics, this time one whose results are of intrinsic importance: the "non-slip condition." This ultimately will serve to relate the velocities and accelerations of points in a rigid body to the angular velocity and acceleration of the body itself (and, if the surface on which the body is moving, the velocity and acceleration of the surface itself).

In order to develop this subject, we will first treat an absurdly simple example: the motion of a sphere. But before we can even do this, we shall state formally exactly what the "non-slip condition" is; this, in effect, becomes the *definition* of that condition. That done, we shall obtain the *ramifications* of this condition: relations among the velocities and accelerations of the *center* of the sphere, those of the surface on which the body is moving without slipping, and the angular velocity and acceleration of the sphere—relations which, unfortunately are *also* informally referred to in practice as "non-slip conditions."

Definition 4.4.1 (Non-slip Condition). Consider an arbitrary body contacting a surface at a single point C. Associated with the point of contact on the surface, there is a point C' in the body itself. The *non-slip condition* says that the *relative* velocity $\boldsymbol{v}_{C'/C}$ and acceleration $\boldsymbol{a}_{C'/C}$ *tangent to the surface*,

$\boldsymbol{v}^t_{C'/C}$ and $\boldsymbol{t}^t_{C'/C}$, both vanish:

$$\boldsymbol{v}^t_{C'/C} \equiv \mathbf{0} \tag{4.4-12a}$$

$$\boldsymbol{a}^t_{C'/C} \equiv \mathbf{0} \tag{4.4-12b}$$

Note that nothing is said regarding the component of these quantities *normal* to the surface.

Example: Motion of a Sphere under the Non-slip Condition. Now consider a sphere of radius r moving without slipping on a surface. The velocity of the center O of the sphere can be expressed in terms of the motion of the point of contact C' in that sphere through rotating coordinates:

$$\boldsymbol{v}_O = \boldsymbol{v}_{C'} + \boldsymbol{\omega} \times \boldsymbol{r}_{O/C'} + \boldsymbol{v}_{rel} = (\boldsymbol{v}_C + \boldsymbol{v}_{C'/C}) + \boldsymbol{\omega} \times \boldsymbol{r}_{O/C'}, \tag{4.4-13}$$

since $\boldsymbol{v}_{rel} \equiv \mathbf{0}$. Clearly $\boldsymbol{v}_C$ is tangent to the surface at that point, but so is $\boldsymbol{v}_O$. Both will be in a plane tangent to the surface (in tangential-normal coordinates at C); and it is possible to break $\boldsymbol{v}_{C'/C}$ up into two components, one [in a *plane*] tangent to the surface and one normal to it:

$$\boldsymbol{v}_{C'/C} = \boldsymbol{v}^t_{C'/C}\hat{\boldsymbol{t}} + \boldsymbol{v}^n_{C'/C}\hat{\boldsymbol{n}} = \boldsymbol{v}^n_{C'/C}\hat{\boldsymbol{n}}$$

by the non-slip definition (4.4-12a). Furthermore, $\boldsymbol{\omega} \times \boldsymbol{r}_{O/C'}$ is perpendicular to $\boldsymbol{r}_{O/C'}$, so perpendicular to $\hat{\boldsymbol{n}}$. In particular, then, the component of equation (4.4-13) in the tangent plane reads

$$\boldsymbol{v}_O = \boldsymbol{v}_C + \boldsymbol{\omega} \times \boldsymbol{r}_{O/C'}; \tag{4.4-14a}$$

while the $\hat{\boldsymbol{n}}$-component gives that $v^n_{C'/C} = 0$ so that $\boldsymbol{v}_{C'/C} \equiv \mathbf{0}$, and we get an equivalent formulation for the non-slip condition on $\boldsymbol{v}_{C'} = \boldsymbol{v}_C + \boldsymbol{v}_{C'/C}$:

$$\boldsymbol{v}_{C'} = \boldsymbol{v}_C; \tag{4.4-14b}$$

i.e., the point of contact in the sphere moves with the same [absolute] velocity as the surface. This latter has a rather remarkable consequence: it says that, as one is travelling down the highway,[9] four points in the car—those in contact with the [stationary] pavement—aren't moving!

We now obtain an analogous result for acceleration: The acceleration of O can be referred to coordinates in the sphere centered at C':

$$\begin{aligned}\boldsymbol{a}_O &= \boldsymbol{a}_{C'} + \boldsymbol{\alpha} \times \boldsymbol{r}_{O/C'} + \boldsymbol{\omega} \times (\boldsymbol{\omega} \times \boldsymbol{r}_{O/C'})\\ &= (\boldsymbol{a}_C + \boldsymbol{a}_{C'/C}) + \boldsymbol{\alpha} \times \boldsymbol{r}_{O/C'} + \boldsymbol{\omega} \times (\boldsymbol{\omega} \times \boldsymbol{r}_{O/C'}).\end{aligned}$$

Now, however, a little more is involved in sorting out the directions of the various components: As above, $\boldsymbol{\alpha} \times \boldsymbol{r}_{O/C'}$ will lie in a plane parallel to the tangent plane

[9] at *precisely* the speed limit!

at C. But now $\boldsymbol{a}_C$ no longer lies completely in a plane tangent to the surface; rather it has a component normal to that plane, as does $\boldsymbol{a}_{C'/C}$:

$$\boldsymbol{a}_C = \boldsymbol{a}_C^t + a_{C_n}\hat{\boldsymbol{n}}$$
$$\boldsymbol{a}_{C'/C} = \boldsymbol{a}_{C'/C}^t + a_{C'/C_n}\hat{\boldsymbol{n}} = a_{C'/C_n}\hat{\boldsymbol{n}}.$$

—again, by the non-slip definition (4.4-12b). Furthermore, $\boldsymbol{a}_O$ also has two components: using tangential-normal coordinates at O,

$$\boldsymbol{a}_O = \dot{v}_O\hat{\boldsymbol{t}}^* + \frac{v_O^2}{\rho_C + r}\hat{\boldsymbol{n}},$$

where $\hat{\boldsymbol{t}}^*$ is in the plane parallel to the tangent plane at C, and ρ_C is the radius of the "osculating sphere" approximating the surface at C, so $\rho_C + r$ is the radius of the path of point O.

The fact that the radius of curvature is $\rho_C + r$ is initially "obvious," but more reflection can generate some doubts. Consider, however, a pair of geometrically similar curves at left. Measuring the two vectors $\boldsymbol{r}$ and $\boldsymbol{r}'$ from some fixed point O, we can clearly express $\boldsymbol{r}' = \boldsymbol{r} - r\hat{\boldsymbol{n}}$ (note that, in tangential-normal coordinates, $\hat{\boldsymbol{n}}$ is the *inward* normal to the curve; due to the similarity of the two curves, both $\hat{\boldsymbol{t}}$ and $\hat{\boldsymbol{n}}$ are the same for both at the points indicated).

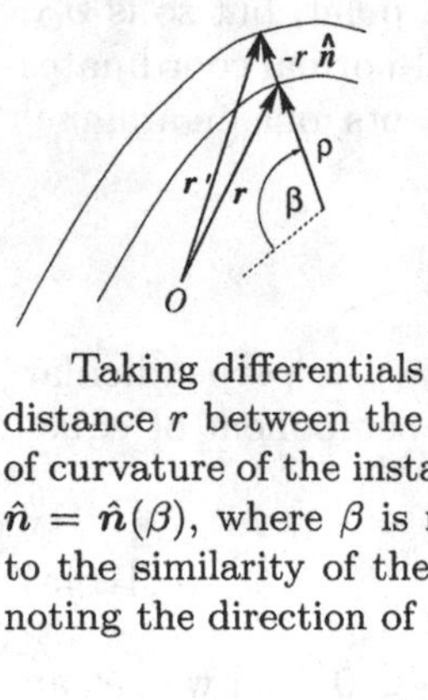

Taking differentials of this general relation we get that $d\boldsymbol{r}' = d\boldsymbol{r} - r d\hat{\boldsymbol{n}}$, since the distance r between the two curves is constant. But $d\boldsymbol{r} = \rho\, d\beta\hat{\boldsymbol{t}}$, where ρ is the radius of curvature of the instantaneous osculating circle to the inner curve as indicated. and $\hat{\boldsymbol{n}} = \hat{\boldsymbol{n}}(\beta)$, where β is measured from the center of the osculating circle, which, due to the similarity of the two curves, is the same for both. Using Corollary 4.1.2 and noting the direction of positive β, we get that

$$d\hat{\boldsymbol{n}} = \frac{d\hat{\boldsymbol{n}}}{d\beta}d\beta = -\hat{\boldsymbol{t}}d\beta,$$

so that

$$d\boldsymbol{r}' = \rho\, d\beta\hat{\boldsymbol{t}} + r\, d\beta\hat{\boldsymbol{t}} = (\rho + r)\, d\beta\hat{\boldsymbol{t}}$$

as desired.

The term $\boldsymbol{\omega} \times (\boldsymbol{\omega} \times \boldsymbol{r}_{O/C'})$ presents a special problem. Expanding this using (4.1-5),

$$\boldsymbol{\omega} \times (\boldsymbol{\omega} \times \boldsymbol{r}_{O/C'}) = (\boldsymbol{\omega} \cdot \boldsymbol{r}_{O/C'})\boldsymbol{\omega} - \omega^2\boldsymbol{r}_{O/C'}.$$

While the last term on the right-hand side is in the $\hat{\boldsymbol{n}}$-direction, the first is in the direction of $\boldsymbol{\omega}$ and generally has both a tangential and a normal component [unlike the planar case, in which it is always perpendicular to the plane of motion]. Perhaps the best that can be done to separate out its contributions in the two directions is to say

$$\begin{aligned}\boldsymbol{\omega} \times (\boldsymbol{\omega} \times \boldsymbol{r}_{O/C'}) &= \left(\boldsymbol{\omega} \times (\boldsymbol{\omega} \times \boldsymbol{r}_{O/C'}) - (\boldsymbol{\omega} \times (\boldsymbol{\omega} \times \boldsymbol{r}_{O/C'}) \cdot \hat{\boldsymbol{n}})\,\hat{\boldsymbol{n}}\right)\\ &\quad + (\boldsymbol{\omega} \times (\boldsymbol{\omega} \times \boldsymbol{r}_{O/C'}) \cdot \hat{\boldsymbol{n}})\,\hat{\boldsymbol{n}};\end{aligned}$$

the first is in a plane tangent to the surface, while the second *is* in the $\hat{\boldsymbol{n}}$-direction.

Again taking the components of the acceleration equation in the tangent plane,

$$\begin{aligned}\dot{v}_O\hat{\boldsymbol{t}}^* &= \boldsymbol{a}_C^t + \boldsymbol{\alpha}\times\boldsymbol{r}_{O/C'}\\ &\quad+\left(\boldsymbol{\omega}\times(\boldsymbol{\omega}\times\boldsymbol{r}_{O/C'}) - (\boldsymbol{\omega}\times(\boldsymbol{\omega}\times\boldsymbol{r}_{O/C'})\cdot\hat{\boldsymbol{n}})\,\hat{\boldsymbol{n}}\right),\end{aligned} \tag{4.4-15a}$$

while the $\hat{\boldsymbol{n}}$-component of the equation gives that

$$a_{C'/C_n}\hat{\boldsymbol{n}} = \frac{v_O^2}{\rho_C + r}\hat{\boldsymbol{n}} - a_{C_n}\hat{\boldsymbol{n}} - (\boldsymbol{\omega}\times(\boldsymbol{\omega}\times\boldsymbol{r}_{O/C'})\cdot\hat{\boldsymbol{n}})\,\hat{\boldsymbol{n}}$$

so that

$$\boldsymbol{a}_{C'} = \boldsymbol{a}_C^t + \frac{v_O^2}{\rho_C + r}\hat{\boldsymbol{n}} - (\boldsymbol{\omega}\times(\boldsymbol{\omega}\times\boldsymbol{r}_{O/C'})\cdot\hat{\boldsymbol{n}})\,\hat{\boldsymbol{n}} \tag{4.4-15b}$$

—the pair of equations (4.4-15) corresponding respectively to the above equations (4.4-14) for the velocities. Unlike velocity, the acceleration of C' does *not* match that of C, but has an additional component in the normal direction.

Such generality is all well and good, but let us consider a common case in which these simplify: the situation in which the "surface" is *flat*: There $\rho_C \to \infty$ so that $\boldsymbol{a}_O = \dot{v}_O\hat{\boldsymbol{t}}^*$ and $\boldsymbol{a}_C^t = \boldsymbol{a}_C$; thus equations (4.4-15) become "simply"

$$\begin{aligned}\boldsymbol{a}_O &= \boldsymbol{a}_C + \boldsymbol{\alpha}\times\boldsymbol{r}_{O/C'}\\ &\quad+\left(\boldsymbol{\omega}\times(\boldsymbol{\omega}\times\boldsymbol{r}_{O/C'}) - (\boldsymbol{\omega}\times(\boldsymbol{\omega}\times\boldsymbol{r}_{O/C'})\cdot\hat{\boldsymbol{n}})\,\hat{\boldsymbol{n}}\right),\end{aligned} \tag{4.4-16a}$$

$$\boldsymbol{a}_{C'} = \boldsymbol{a}_C - (\boldsymbol{\omega}\times(\boldsymbol{\omega}\times\boldsymbol{r}_{O/C'})\cdot\hat{\boldsymbol{n}})\,\hat{\boldsymbol{n}} \tag{4.4-16b}$$

equations bearing so little resemblance to (4.4-14) to be of little practical use, even though similar simplification for $\boldsymbol{a}_C \equiv \boldsymbol{0}$ also holds. Their primary utility is to demonstrate the presence of a normal component of acceleration for C'—and to use them to obtain the equations of motion for a rigid body in the non-slip case in the next Chapter.

But if the above demonstrates how *not* to find $\boldsymbol{a}_O$, how *do* we impose the non-slip condition? Certainly we can deal with the velocity using equations (4.4-14) above, but that still leaves $\boldsymbol{a}_O$ to determine. Likely the best way to deal with this quandary is, rather than implementing the rotating coordinate system equations we did above to find accelerations, to go one step back and use the foundation on which these equations rest in the first place, Theorem 4.3.1: that $\dot{\boldsymbol{A}} = \boldsymbol{\omega}\times\boldsymbol{A} + \dot{\boldsymbol{A}}_{rel}$. In particular, then, we can obtain $\boldsymbol{v}_O$ from (4.4-14), and then obtain the acceleration by considering a system moving with the body. We illustrate this in the following example:

Example 4.4.2 *Non-slip Motion* A thin disk of radius r rolls without slipping over a fixed horizontal plane. If the orientation of the disk is described by the angles θ it makes with the vertical and ψ its path makes with the X-axis, obtain expressions for the velocity and acceleration of the center O of the disk, $\boldsymbol{v}_O$ and $\boldsymbol{a}_O$.

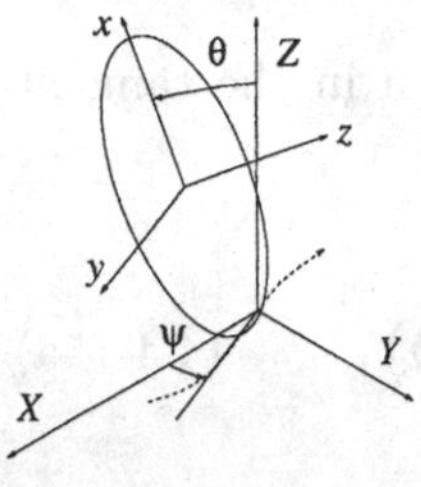

With an eye towards the determination of the acceleration, we consider a coordinate system (x, y, z) fixed in and rotating *with the disk.* In particular, we take O as its center, with $\hat{\boldsymbol{k}}$ perpendicular to the disk, $\hat{\boldsymbol{\imath}}$ in the direction of the vector from the point of contact C to O, and then define $\hat{\boldsymbol{\jmath}} \equiv \hat{\boldsymbol{k}} \times \hat{\boldsymbol{\imath}}$ which, we observe, will thus be parallel to the XY-plane. Note that these directions are all defined *at the instant of interest*: all will change as the disk moves.

In these directions we can represent $\boldsymbol{\omega} = \omega_x \hat{\boldsymbol{\imath}} + \omega_y \hat{\boldsymbol{\jmath}} + \omega_z \hat{\boldsymbol{k}}$, where, in particular, $\omega_y = \dot{\theta}$ and ω_z is the *spin* of the disk about its axis. Since the surface is not moving, (4.4-14) reduces to just

$$\boldsymbol{v}_O = \boldsymbol{\omega} \times \boldsymbol{r}_{O/C} = (\omega_x \hat{\boldsymbol{\imath}} + \omega_y \hat{\boldsymbol{\jmath}} + \omega_z \hat{\boldsymbol{k}}) \times r\hat{\boldsymbol{\imath}} = r\omega_z \hat{\boldsymbol{\jmath}} - r\omega_y \hat{\boldsymbol{k}}.$$

That's easy. Now, to find $\boldsymbol{a}_O$, we return to Theorem 4.3.1: the desired acceleration $\boldsymbol{a}_O = \dot{\boldsymbol{v}}_O = (\dot{\boldsymbol{v}}_O)_{rel} + \boldsymbol{\omega} \times \boldsymbol{v}_O.$ in which, recall, the "*rel*" time derivative *only differentiates the components* of the vector. Thus, since $\dot{r}_{rel} \equiv 0$,

$$\begin{aligned}
\boldsymbol{a}_O &= (r\dot{\omega}_z \hat{\boldsymbol{\jmath}} - r\dot{\omega}_y \hat{\boldsymbol{k}}) + \begin{vmatrix} \hat{\boldsymbol{\imath}} & \hat{\boldsymbol{\jmath}} & \hat{\boldsymbol{k}} \\ \omega_x & \omega_y & \omega_z \\ 0 & r\omega_z & -r\omega_y \end{vmatrix} \\
&= (r\dot{\omega}_z \hat{\boldsymbol{\jmath}} - r\dot{\omega}_y \hat{\boldsymbol{k}}) + \Big(-r(\omega_y^2 + \omega_z^2)\hat{\boldsymbol{\imath}} - (-r\omega_x\omega_y)\hat{\boldsymbol{\jmath}} + r\omega_x\omega_z \hat{\boldsymbol{k}}\Big) \\
&= -r(\omega_y^2 + \omega_z^2)\hat{\boldsymbol{\imath}} + (r\omega_x\omega_y + r\dot{\omega}_z)\hat{\boldsymbol{\jmath}} + (r\omega_x\omega_z - r\dot{\omega}_y)\hat{\boldsymbol{k}}.
\end{aligned}$$

At this stage, values of the components of $\boldsymbol{\omega}$ could be substituted in; similarly, this representation could be transformed to one in $\hat{\boldsymbol{I}}$, $\hat{\boldsymbol{J}}$, and $\hat{\boldsymbol{K}}$. Neither will be done here, but the advantage of this approach relative to the one above is obvious! |

Motion of an *Arbitrary* Body under the Non-slip Condition. While the above study of non-slip motion of a sphere was clearly a special case, its results can be applied immediately to the motion of an arbitrary body contacting a surface at a single point C. Obviously, we must know the geometry of the body; in particular, we have the radius of the osculating sphere approximating the body at the point of contact C. *This* radius then replaces "r" in the above equations, and the rest of the analysis goes through directly.

Why is all this useful? Recall that, once we know the velocity and acceleration of *one* point in the body, then the velocity and acceleration of any *other* point can be found from $\boldsymbol{\omega}$ and $\boldsymbol{\alpha}$ using rotating coordinates centered at the point whose motion is known. In particular, then, motion of the center of mass of a rigid body—a quantity central to the analysis of the dynamics of such bodies—can be determined.

The Instantaneous Center of Zero Velocity

An idea which gets a great deal of attention in *planar*, *rigid* body kinematics is the "instantaneous center of zero velocity"—a point[10] about which the body actually *is* rotating at a particular instant. If, in planar systems, this provides a useful alternative to the more direct vector approach espoused here, it does not generally in 3-D motion.

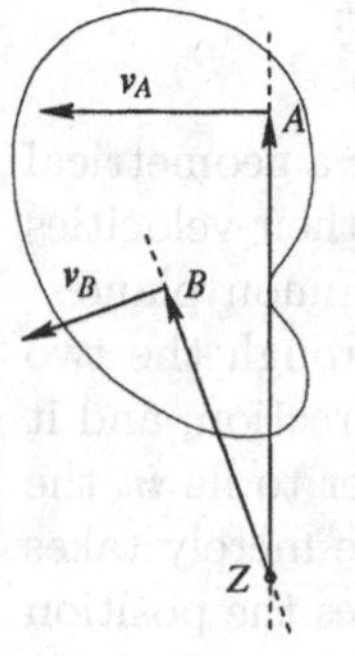

Let us first briefly review the concept itself: Consider an arbitrary rigid body diagrammed at left. If two points A and B have velocities $\boldsymbol{v}_A$ and $\boldsymbol{v}_B$ respectively, we seek a point Z, in the body or "body extended," with zero velocity such that the body will appear *instantaneously* to rotate about that point. What does that mean?

By the hypotheses on this point, we can express the velocity of each point A and B in the form $\boldsymbol{v}_i = \boldsymbol{v}_Z + \boldsymbol{\omega} \times \boldsymbol{r}_i + \boldsymbol{v}_{rel} = \boldsymbol{\omega} \times \boldsymbol{r}_i$ (in which the $\boldsymbol{r}_i$ are measured *from* Z), since $\boldsymbol{v}_Z \equiv \boldsymbol{0} \equiv \boldsymbol{v}_{rel}$:

$$\boldsymbol{v}_A = \boldsymbol{\omega} \times \boldsymbol{r}_{A/Z} \tag{4.4-17a}$$
$$\boldsymbol{v}_B = \boldsymbol{\omega} \times \boldsymbol{r}_{B/Z}. \tag{4.4-17b}$$

But, as cross products, the $\boldsymbol{v}_i$ must be simultaneously perpendicular to their respective $\boldsymbol{r}_i$ and the *common* $\boldsymbol{\omega}$; equivalently, the $\boldsymbol{r}_i$ and $\boldsymbol{\omega}$ must be perpendicular to the $\boldsymbol{v}_i$—they must lie in planes perpendicular to the velocities. Though the $\boldsymbol{r}_i$ are different, $\boldsymbol{\omega}$ is *common* and thus must lie in *both* planes; *i.e.*, $\boldsymbol{\omega}$ lies on the [common] intersection of these two planes, and this intersection forms the desired axis. Any point on this axis then serves as the "instantaneous center." Clearly the planes must intersect in the first place; thus the $\boldsymbol{v}_i$ must be non-parallel—unless both points share a common [perpendicular] plane, in which case the intersection is not unique. Assuming it is, however, *the* $\boldsymbol{\omega}$ *thus found will necessarily be perpendicular to the* $\boldsymbol{v}_i$.

Since the bodies are rigid, one typically knows one velocity completely and has $\boldsymbol{\omega}$ and the other velocity in known direction to find (see the modification of Example 4.4.1 at the end). Having located the instantaneous center and knowing the direction $\hat{\boldsymbol{\omega}}$, it then becomes an easy matter to determine the *magnitude* ω from the known velocity, and, from that, the second velocity:

$$\frac{\|\boldsymbol{v}_B\|}{\|\boldsymbol{v}_A\|} = \frac{\|\boldsymbol{\omega} \times \boldsymbol{r}_B\|}{\|\boldsymbol{\omega} \times \boldsymbol{r}_A\|} \tag{4.4-18}$$

The above is a more complete development than typically given in an introductory Dynamics course. It shows an important point universally ignored in such presentations: *instantaneous centers rely on the rotating coordinate equations alone!* The latter, in turn, utilize only that component of the angular

[10] Actually, it should likely be called an "instantaneous *axis* of zero velocity: *any* point on such an axis would serve as an "instantaneous center."

velocity *perpendicular* to the $\boldsymbol{r}$ vector; they neither reference any component *along* the rod, nor can they *find* it.

We now consider both Planar and 3-D formulations of the concept, in order to show why it may be useful in the former, but not in the latter:

Planar Motion. Equations (4.4-17) provide the foundation for a geometrical approach useful in 2-D problems: Since the points A and B, and their velocities (assumed non-parallel for the nonce), are known to lie in a common plane—say the xy-plane, those planes perpendicular to the $\boldsymbol{v}$'s and through the two points will intersect in the $\hat{\boldsymbol{k}}$-direction; $\boldsymbol{\omega}$ is then also in the $\hat{\boldsymbol{k}}$-direction, and it suffices (though is not necessary) to take the instantaneous center to lie *in* the xy-plane. Knowing the *directions* of the two velocities, then, one merely takes lines perpendicular to the $\boldsymbol{v}$'s, and where those lines intersect gives the position of Z. (Refer again to the diagram on page 225.)

We observe that the $\boldsymbol{v}_i = \boldsymbol{\omega} \times \boldsymbol{r}_i$, coupled with the fact that $\boldsymbol{\omega}$ is known to be perpendicular to the plane of the motion (and so to the $\boldsymbol{r}_i$), means that equation (4.4-18) becomes

$$\frac{\|\boldsymbol{v}_B\|}{\|\boldsymbol{v}_A\|} = \frac{\|\boldsymbol{\omega} \times \boldsymbol{r}_B\|}{\|\boldsymbol{\omega} \times \boldsymbol{r}_A\|} = \frac{\|\boldsymbol{\omega}\|\,\|\boldsymbol{r}_B\|}{\|\boldsymbol{\omega}\|\,\|\boldsymbol{r}_A\|} = \frac{\|\boldsymbol{r}_B\|}{\|\boldsymbol{r}_A\|}.$$

Knowing either of the $\boldsymbol{v}$'s, along with knowledge of the corresponding $\boldsymbol{r}$ from the determined point Z, enables us to find $\|\boldsymbol{\omega}\|$; the *direction* of $\boldsymbol{\omega}$ can then be determined by inspection, and the velocity of the other point calculated. But all this is predicated on being able to locate Z in the first place, and in the pathological event the $\boldsymbol{v}$'s happen to be *parallel*, this is not generally possible. If they are *equal but not sharing a common perpendicular*, this corresponds to pure *translation*:

$$\mathbf{0} = \boldsymbol{v}_B - \boldsymbol{v}_A = \boldsymbol{\omega} \times (\boldsymbol{r}_B - \boldsymbol{r}_A),$$

$\boldsymbol{\omega} = \mathbf{0}$ because $(\boldsymbol{r}_B - \boldsymbol{r}_A)$ neither vanishes nor is parallel to $\boldsymbol{\omega}$; otherwise the method fails.[11]

Note the central role the assumption that $\boldsymbol{\omega}$ is perpendicular to the $\boldsymbol{r}$'s plays here: it means that "$\boldsymbol{\omega}$" determined by the procedure is *always* the "$\boldsymbol{\omega}_\perp$" utilized by the rotating coordinate equations.

3-D Motion. As with the planar case, equations (4.4-17) provide a useful tool in simple cases: Assuming the velocities to be non-parallel, the intersection of planes, perpendicular to them and through the two points, will give both the desired axis and the direction of $\boldsymbol{\omega}$. (If the $\boldsymbol{v}$'s are parallel, on the other hand—*assuming both points lie on the same plane perpendicular to these*—the problem is reduced to a 2-D case in that plane containing the velocities themselves.)

The angular velocity so obtained will generally have a component along $\boldsymbol{r}$ (see Example below). But, as we have observed several times, such a component

[11] The position of Z *can* be found for parallel $\boldsymbol{v}$'s *if both are known*, simply by using the above ratio and similarity of the $\boldsymbol{r}/\boldsymbol{v}$ triangles.

is meaningless in terms of the motion of [the endpoints of] $\boldsymbol{r}$, which, as noted above, depends only on that component of $\boldsymbol{\omega}$ *perpendicular* to $\boldsymbol{r}$. Furthermore, such an analysis examines only the motions of the *points* in a body; it does not make any accommodation regarding the *constraints* imposed on its motion by connections. In the next section we examine these conditions in more detail; for present purposes, however, it suffices to note that there are two possibilities: either $\boldsymbol{\omega}$ is *completely* determined (as exemplified by the "clevis" in the next section), or (as with the ball-and-socket joints in Example 4.4.1) we can only find that component, $\boldsymbol{\omega}_\perp$, perpendicular to $\boldsymbol{r}_{A/B}$ such that $\boldsymbol{\omega}_\perp \cdot \boldsymbol{r}_{A/B} = 0$. In the former case, it is quite possible that the $\boldsymbol{\omega}$ determined by the clevis is *not* perpendicular to the $\boldsymbol{r}$'s in (4.4-17); in such case there *is* no instantaneous center of zero velocity.

In sum, then, we see that the instantaneous axis of zero velocity can only be found if $\boldsymbol{\omega}$ of the body is perpendicular to the known $\boldsymbol{v}$'s; this is not known *a priori*, of course, and generally is not even the case! Generally we are limited, at best, to finding only $\boldsymbol{\omega}_\perp$—that component of $\boldsymbol{\omega}$ perpendicular to the $\boldsymbol{r}$'s.

Example 4.4.3. *Instantaneous Center of Zero Velocity.* We briefly return to Example 4.4.1, particularly the rigid-body modification at the end, where $\boldsymbol{v}_B$ is regarded as unknown. The planes perpendicular to $\boldsymbol{v}_A = 2\hat{\boldsymbol{I}}\,\text{m/s}$ and $\boldsymbol{v}_B$ (whose *direction* is known) obviously intersect along a vertical line through $\boldsymbol{r}_Z = \hat{\boldsymbol{I}} + 2\hat{\boldsymbol{J}}$, say, making $\boldsymbol{\omega} = 1\hat{\boldsymbol{K}}\,\text{s}^{-1}$ to produce the given $\boldsymbol{v}_A$. This turns out to give the correct $\boldsymbol{v}_B = \boldsymbol{\omega} \times \boldsymbol{r}_{B/Z} = -1\hat{\boldsymbol{J}}\,\text{m/s}$, yet it is *not* the $\boldsymbol{\omega}_\perp = \frac{2}{9}\hat{\boldsymbol{I}} - \frac{4}{9}\hat{\boldsymbol{J}} + \frac{5}{9}\hat{\boldsymbol{K}}\,\text{s}^{-1}$ found previously. But, decomposing $\boldsymbol{\omega}$ into components parallel and perpendicular to $\boldsymbol{r}_{A/B}$:

$$\begin{aligned}\boldsymbol{\omega}_\parallel &= (\boldsymbol{\omega} \cdot \hat{\boldsymbol{r}}_{A/B})\hat{\boldsymbol{r}}_{A/B} = \left(\hat{\boldsymbol{K}} \cdot \frac{-\hat{\boldsymbol{I}} + 2\hat{\boldsymbol{J}} + 2\hat{\boldsymbol{K}}}{3}\right)\frac{-\hat{\boldsymbol{I}} + 2\hat{\boldsymbol{J}} + 2\hat{\boldsymbol{K}}}{3} \\ &= \frac{-2\hat{\boldsymbol{I}} + 4\hat{\boldsymbol{J}} + 4\hat{\boldsymbol{K}}}{9} \\ \therefore \quad \boldsymbol{\omega}_\perp &= \boldsymbol{\omega} - \boldsymbol{\omega}_\parallel = \frac{2}{9}\hat{\boldsymbol{I}} - \frac{4}{9}\hat{\boldsymbol{J}} + \frac{5}{9}\hat{\boldsymbol{K}}\end{aligned}$$

—precisely the value found before. |

It is not surprising that we cannot utilize such a methodology alone: as the above example shows, to be able to do so would allow determination of the velocities *without regard to the kinematic constraints imposed by the linkages*! This clearly flies in the face of the previously-presented theory.

The upshot of all this is the fact that instantaneous centers of zero velocity are essentially useless in general motion. And they don't apply to *accelerations* even in planar problems. And they only apply to *rigid* bodies. For present purposes, it's likely best simply to ignore them altogether!

4.4.2 Kinematic Constraints Imposed by Linkages

We have seen that the solution for angular velocity and acceleration requires an "additional condition"—a form of *constraint* on $\boldsymbol{\omega}$ and $\boldsymbol{\alpha}$—because their appearance in the cross product generates only two independent equations. Example 4.4.1 showed how such a condition could be used to obtain the solution, but it did nothing to elucidate the origin of such conditions. Before embarking on the discussion of multibody "machines," then, it is appropriate that we first clarify this issue.

We shall see that the physical nature of the connection between two bodies imposes a kinematic constraint on their motions. Though there are many types of such linkages, two of the most common are the ball-and-socket (exemplified above), and the yoke or "clevis." Though this hardly seems an exhaustive enumeration of the possibilities, they do characterize the problem in the following sense: We get conditions on $\boldsymbol{\omega} \cdot \boldsymbol{r}$ and $\boldsymbol{\alpha} \cdot \boldsymbol{r}$ to either vanish or not. Since we seek a scalar condition on these vector angular quantities, such a condition is essentially forced to involve the dot product; and since this product must satisfy either a homogeneous or a non-homogeneous equation, there is little option left!

Though the clevis connection is the more involved, unlike the ball-and-socket, it is completely determinate. For this reason we consider the clevis first.

Clevis Connections

Illustrated here is a close-up of a yoke rigidly connected to a rod whose axis is given by the vector $\boldsymbol{r}$ from some other point who motion is known; the clevis connection is comprised of this yoke passing over two pins fixed perpendicular to a collar which is moving along a guide, which will be assumed here to be straight.[12] The [instantaneous] directions of the pin and guide axes are given by $\hat{\boldsymbol{u}}_p$ and $\hat{\boldsymbol{u}}_c$, respectively; observe that these are also perpendicular due to the orientation of the pins. We note that the orientation of the clevis and collar are *completely constrained* once $\boldsymbol{r}$ is specified relative to the guide: given $\boldsymbol{r}$, the collar must rotate about the guide, and the clevis about the pins, to adjust to this orientation; this fixes the angles the collar makes with the Z-axis (as measured with a pin's axis), and the yoke makes with a plane perpendicular to the collar.

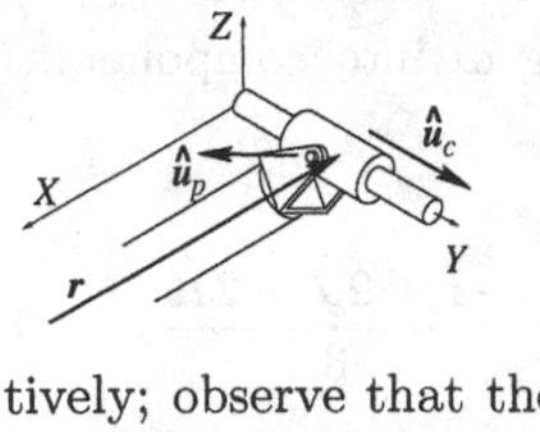

Velocity. But if these angles are determined, their time rates of change—the angular velocity $\boldsymbol{\omega}$ of the clevis, and thus of the body to which it is attached—are also. This shows immediately that we can completely express

$$\boldsymbol{\omega} = \omega_c \hat{\boldsymbol{u}}_c + \omega_p \hat{\boldsymbol{u}}_p \tag{4.4-19}$$

—at least once we know what the two directions actually are! The first of these is easy: $\hat{\boldsymbol{u}}_c$ is presumably known from the given geometry of the guide along which the collar moves. But what is the other?

[12] We briefly generalize this at the end of this subsection.

$\hat{\boldsymbol{u}}_p$ is already known to be perpendicular to the collar, and thus to $\hat{\boldsymbol{u}}_c$. But, since the yoke swings in a plane perpendicular to the pins, $\hat{\boldsymbol{u}}_p$ *must also be perpendicular to* $\boldsymbol{r}$! Since $\hat{\boldsymbol{u}}_p$ is simultaneously perpendicular to $\hat{\boldsymbol{u}}_c$ and $\boldsymbol{r}$, we can write that, up to a sign, $\hat{\boldsymbol{u}}_p \parallel \hat{\boldsymbol{u}}_c \times \boldsymbol{r}$ so that

$$\hat{\boldsymbol{u}}_p = \frac{\hat{\boldsymbol{u}}_c \times \boldsymbol{r}}{\|\hat{\boldsymbol{u}}_c \times \boldsymbol{r}\|}. \tag{4.4-20}$$

If $\hat{\boldsymbol{u}}_c$ and $\boldsymbol{r}$ are parallel, $\hat{\boldsymbol{u}}_p$ vanishes making the pin direction indeterminate—a circumstance making sense physically. As long as they aren't, this equation determines $\hat{\boldsymbol{u}}_p$; it along with (4.4-19) fixes the direction of $\boldsymbol{\omega}$.

That's all well and good, but how are we to cajole a *scalar* equation out of this to give the additional condition on $\boldsymbol{\omega}$? Equation (4.4-19) shows that $\boldsymbol{\omega}$ must lie in the plane determined by $\hat{\boldsymbol{u}}_c$ and $\hat{\boldsymbol{u}}_p$; thus a vector *perpendicular* to this plane must be *perpendicular* to $\boldsymbol{\omega}$. And such a vector perpendicular to the plane containing $\hat{\boldsymbol{u}}_c$ and $\hat{\boldsymbol{u}}_p$ is just their cross product, $\hat{\boldsymbol{u}}_c \times \hat{\boldsymbol{u}}_p$:

$$\boldsymbol{\omega} \cdot \hat{\boldsymbol{u}}_c \times \hat{\boldsymbol{u}}_p \equiv 0. \tag{4.4-21}$$

[This same equation can be obtained directly by dotting $\boldsymbol{\omega}$ expressed in equation (4.4-19) with $\hat{\boldsymbol{u}}_c \times \hat{\boldsymbol{u}}_p$: the resulting triple scalar products will vanish identically because of two identical constituent vectors.] We note that this equation says that $\boldsymbol{\omega}$ lies in a plane perpendicular to the vector $\hat{\boldsymbol{u}}_c \times \hat{\boldsymbol{u}}_p$; geometrically, then, $\boldsymbol{\omega}$ lies simultaneously on this plane and that perpendicular to the right-hand side "$\boldsymbol{V}$" of the velocity equation—*i.e.*, on their intersection—fixing the *actual* direction $\hat{\boldsymbol{\omega}}$.

We can investigate the pathology that there *be* no such intersection—that these two planes be *parallel*: This will occur only if the *normals* to the two planes, $\boldsymbol{V}$ and $\hat{\boldsymbol{u}}_c \times \hat{\boldsymbol{u}}_p$, are parallel; *i.e.*, that $\boldsymbol{V} \times (\hat{\boldsymbol{u}}_c \times \hat{\boldsymbol{u}}_p) \equiv \boldsymbol{0}$. But

$$\boldsymbol{V} = \boldsymbol{\omega} \times \boldsymbol{r} = (\omega_c \hat{\boldsymbol{u}}_c + \omega_p \hat{\boldsymbol{u}}_p) \times \boldsymbol{r},$$

so

$$\begin{aligned}
\boldsymbol{V} \times (\hat{\boldsymbol{u}}_c \times \hat{\boldsymbol{u}}_p) &= (\omega_c \hat{\boldsymbol{u}}_c \times \boldsymbol{r} + \omega_p \hat{\boldsymbol{u}}_p \times \boldsymbol{r}) \times (\hat{\boldsymbol{u}}_c \times \hat{\boldsymbol{u}}_p) \\
&= \omega_c[(\hat{\boldsymbol{u}}_c \times \boldsymbol{r}) \times (\hat{\boldsymbol{u}}_c \times \hat{\boldsymbol{u}}_p)] + \omega_p[(\hat{\boldsymbol{u}}_p \times \boldsymbol{r}) \times (\hat{\boldsymbol{u}}_c \times \hat{\boldsymbol{u}}_p)] \\
&= \omega_c[((\hat{\boldsymbol{u}}_c \times \boldsymbol{r}) \cdot \hat{\boldsymbol{u}}_p)\hat{\boldsymbol{u}}_c - ((\hat{\boldsymbol{u}}_c \times \boldsymbol{r}) \cdot \hat{\boldsymbol{u}}_c)\hat{\boldsymbol{u}}_p] \\
&\quad + \omega_p[((\hat{\boldsymbol{u}}_p \times \boldsymbol{r}) \cdot \hat{\boldsymbol{u}}_p)\hat{\boldsymbol{u}}_c - ((\hat{\boldsymbol{u}}_p \times \boldsymbol{r}) \cdot \hat{\boldsymbol{u}}_c)\hat{\boldsymbol{u}}_p]
\end{aligned}$$

by the use of identities (4.1-5),

$$= \omega_c[((\hat{\boldsymbol{u}}_c \times \boldsymbol{r}) \cdot \hat{\boldsymbol{u}}_p)\hat{\boldsymbol{u}}_c - 0] + \omega_p[0 - ((\hat{\boldsymbol{u}}_p \times \boldsymbol{r}) \cdot \hat{\boldsymbol{u}}_c)\hat{\boldsymbol{u}}_p],$$

the vanishing terms being triple scalar products with two vectors parallel. Now consider the bracketed expressions in these two terms: by (4.4-20),

$$\hat{\boldsymbol{u}}_c \times \boldsymbol{r} \cdot \hat{\boldsymbol{u}}_p \equiv \|\hat{\boldsymbol{u}}_c \times \boldsymbol{r}\| \hat{\boldsymbol{u}}_p \cdot \hat{\boldsymbol{u}}_p = \|\hat{\boldsymbol{u}}_c \times \boldsymbol{r}\|,$$

while, interchanging the $\times$ and $\cdot$ in the triple scalar product in the second,

$$\hat{\mathbf{u}}_p \times \mathbf{r} \cdot \hat{\mathbf{u}}_c = \hat{\mathbf{u}}_p \cdot \mathbf{r} \times \hat{\mathbf{u}}_c = \hat{\mathbf{u}}_p \cdot (-\|\hat{\mathbf{u}}_c \times \mathbf{r}\|\hat{\mathbf{u}}_p) = -\|\hat{\mathbf{u}}_c \times \mathbf{r}\|;$$

thus finally

$$\mathbf{V} \times (\hat{\mathbf{u}}_c \times \hat{\mathbf{u}}_p) = \omega_c\|\hat{\mathbf{u}}_c \times \mathbf{r}\|\hat{\mathbf{u}}_c + \omega_p\|\hat{\mathbf{u}}_c \times \mathbf{r}\|\hat{\mathbf{u}}_p = \|\hat{\mathbf{u}}_c \times \mathbf{r}\|\boldsymbol{\omega}!$$

This vanishes—*i.e.*, the two planes will be parallel—if and only if either $\|\hat{\mathbf{u}}_c \times \mathbf{r}\| \equiv 0$ (the situation in which $\hat{\mathbf{u}}_p$ is indeterminate because $\hat{\mathbf{u}}_c \parallel \mathbf{r}$) or $\boldsymbol{\omega}$ itself vanishes (in which case $\mathbf{V} = \boldsymbol{\omega} \times \mathbf{r} \equiv \mathbf{0}$ already).

With the direction $\hat{\boldsymbol{\omega}}$ established, then, any component of the velocity equation is sufficient to determine ω, so $\boldsymbol{\omega}$, uniquely. In application, one simply solves the four equations simultaneously, first finding $\hat{\mathbf{u}}_p$ using (4.4-20), calculating $\hat{\mathbf{u}}_c \times \hat{\mathbf{u}}_p$, and then implementing (4.4-21) as the fourth, condition, equation for $\boldsymbol{\omega}$. (Since the last is a homogeneous equation, it is not really necessary to normalize $\hat{\mathbf{u}}_p$ as in (4.4-20) for the velocity analysis; we shall see, however, that this *is* necessary for the acceleration.)

Acceleration. In general, the angular acceleration of the clevis is found from the time derivative of (4.4-19):

$$\boldsymbol{\alpha} = \dot{\omega}_c\hat{\mathbf{u}}_c + \dot{\omega}_p\hat{\mathbf{u}}_p + \omega_c\dot{\hat{\mathbf{u}}}_c + \omega_p\dot{\hat{\mathbf{u}}}_p.$$

For the straight guide presumed in this discussion, however, $\dot{\hat{\mathbf{u}}}_c \equiv \mathbf{0}$; while, since the pin is rotating with the collar's angular velocity $\omega_c\hat{\mathbf{u}}_c$, $\dot{\hat{\mathbf{u}}}_p = \omega_c\hat{\mathbf{u}}_c \times \hat{\mathbf{u}}_p$, so $\omega_p\dot{\hat{\mathbf{u}}}_p = \omega_p\omega_c\hat{\mathbf{u}}_c \times \hat{\mathbf{u}}_p$. Thus

$$\boldsymbol{\alpha} = \dot{\omega}_c\hat{\mathbf{u}}_c + \dot{\omega}_p\hat{\mathbf{u}}_p + \omega_c\omega_p\hat{\mathbf{u}}_c \times \hat{\mathbf{u}}_p.$$

We note that $\hat{\mathbf{u}}_p$, by (4.4-20), must be perpendicular to $\hat{\mathbf{u}}_c$; by this, $\hat{\mathbf{u}}_c \times \hat{\mathbf{u}}_p$ is itself a unit vector perpendicular to them both; thus the above expression for $\boldsymbol{\alpha}$ expresses it relative to an orthonormal set of vectors. This observation allows us to get a scalar expression giving a condition on $\boldsymbol{\alpha}$ rather similar to that on $\boldsymbol{\omega}$, equation (4.4-21): taking a cue from the form of that equation and dotting $\boldsymbol{\alpha}$ with $\hat{\mathbf{u}}_c \times \hat{\mathbf{u}}_p$,

$$\boldsymbol{\alpha} \cdot \hat{\mathbf{u}}_c \times \hat{\mathbf{u}}_p = \omega_c\omega_p, \tag{4.4-22}$$

each of the first two terms vanishing as a triple scalar product with two parallel vectors. Geometrically, this says that $\boldsymbol{\alpha}$ makes an angle of $\theta_a = \arccos(\omega_c\omega_p/\|\boldsymbol{\alpha}\|)$ with the vector $\hat{\mathbf{u}}_c \times \hat{\mathbf{u}}_p$; thus $\hat{\boldsymbol{\alpha}}$ lies simultaneously on this *cone* with vertex angle θ_a about $\hat{\mathbf{u}}_c \times \hat{\mathbf{u}}_p$ and the plane perpendicular to the right-hand side "$\mathbf{A}$" of the acceleration equation, generally giving *two* possible lines of intersection. This means that we generally need *both* the independent components of $\boldsymbol{\alpha} \times \mathbf{r} = \mathbf{A}$ (expressing the components of $\boldsymbol{\alpha}$ in terms of $\|\boldsymbol{\alpha}\|$). Again, in practice, one merely solves the four equations—three components of the acceleration rotating coordinates plus the additional condition (4.4-22)—simultaneously. $\boldsymbol{\omega}$ at this point has been determined from the velocity analysis, but it remains to find the separate components ω_c and ω_p in order to implement the condition equation. Because of the perpendicularity of $\hat{\mathbf{u}}_c$ and $\hat{\mathbf{u}}_p$, this is basically trivial:

$$\omega_c = \boldsymbol{\omega} \cdot \hat{\mathbf{u}}_c \quad \text{and} \quad \omega_p = \boldsymbol{\omega} \cdot \hat{\mathbf{u}}_p.$$

Though the above is a "direct" approach to obtaining equation (4.4-22) through the straightforward differentiation of $\boldsymbol{\omega}$ and evaluation of the various time derivatives arising due to rotation, there is another, more "elegant," so more austerely abstruse, means of deriving it—one which we employ extensively in the sequel: Since the velocity constraint (4.4-21) is a *general* condition, we can differentiate it to obtain

$$\boldsymbol{\alpha} \cdot [\hat{\mathbf{u}}_c \times \hat{\mathbf{u}}_p] + \boldsymbol{\omega} \cdot [\dot{\hat{\mathbf{u}}}_c \times \hat{\mathbf{u}}_p + \hat{\mathbf{u}}_c \times \dot{\hat{\mathbf{u}}}_p] \equiv 0.$$

But the bracketed term

$$\dot{\hat{\mathbf{u}}}_c \times \hat{\mathbf{u}}_p + \hat{\mathbf{u}}_c \times \dot{\hat{\mathbf{u}}}_p = \mathbf{0} + \hat{\mathbf{u}}_c \times \omega_c(\hat{\mathbf{u}}_c \times \hat{\mathbf{u}}_p),$$

since $\hat{\mathbf{u}}_c$ is fixed and $\hat{\mathbf{u}}_p$ rotates with $\omega_c\hat{\mathbf{u}}_c$; expanding this using (4.1-5),

$$= \omega_c[(\hat{\mathbf{u}}_c \cdot \hat{\mathbf{u}}_p)\hat{\mathbf{u}}_c - (\hat{\mathbf{u}}_c \cdot \hat{\mathbf{u}}_c)\hat{\mathbf{u}}_p] = -\omega_c\hat{\mathbf{u}}_p$$

because $\hat{\mathbf{u}}_c$ and $\hat{\mathbf{u}}_p$ are perpendicular by (4.4-20). Thus, finally,

$$\boldsymbol{\omega} \cdot [\dot{\hat{\mathbf{u}}}_c \times \hat{\mathbf{u}}_p + \hat{\mathbf{u}}_c \times \dot{\hat{\mathbf{u}}}_p] = (\omega_c\hat{\mathbf{u}}_c + \omega_p\hat{\mathbf{u}}_p) \cdot (-\omega_c\hat{\mathbf{u}}_p) = -\omega_p\omega_c$$

so

$$\boldsymbol{\alpha} \cdot [\hat{\mathbf{u}}_c \times \hat{\mathbf{u}}_p] - \omega_p\omega_c \equiv 0,$$

precisely the condition (4.4-22).

The two conditions on $\boldsymbol{\omega}$ and $\boldsymbol{\alpha}$, equations (4.4-21) and (4.4-22), are remarkably similar in form: that for $\boldsymbol{\omega}$ is homogeneous while that for $\boldsymbol{\alpha}$ is non-homogeneous. We note that is *is* necessary to normalize $\hat{\mathbf{u}}_c \times \hat{\mathbf{u}}_p$, however.

What to do if the guide is curved. If the collar is not moving along a straight guide, we can no longer say that its angular velocity is just $\boldsymbol{\omega}_c \equiv \omega_c\hat{\boldsymbol{\omega}}_c$ for fixed $\hat{\boldsymbol{\omega}}_c$: there will be an additional component due to its changing orientation along the track. How will this complicate the analysis?

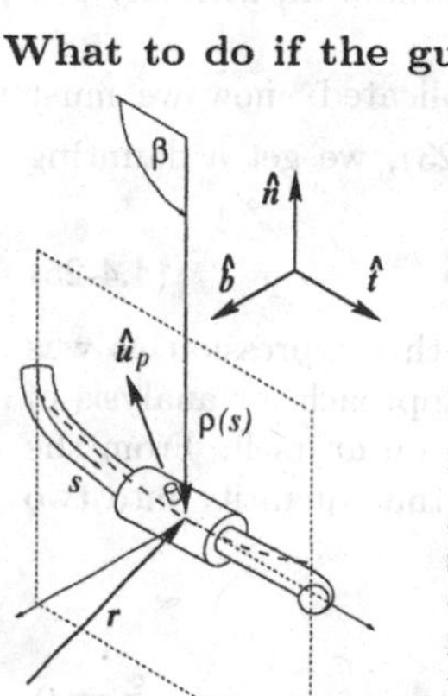

Clearly, we must know the geometry of the guide; otherwise the problem is indeterminate. In particular, then, we must know the *shape* of the curve; this means we can determine its radius of curvature ρ at each point along its arc length s: $\rho = \rho(s)$. We will also know the *tangent* to the track $\hat{\mathbf{t}}$, corresponding to $\hat{\mathbf{u}}_c$ above, as well as the *normal* $\hat{\mathbf{n}}$ to the curve (in the "osculating plane"), and its radius of curvature ρ, at the particular instant considered; these quantities are indicated in the diagram at left, with the instantaneous osculating plane and approximating circle of radius ρ dotted. From the two directions, we can calculate the *binormal* vector $\hat{\boldsymbol{b}} \equiv \hat{\boldsymbol{t}} \times \hat{\boldsymbol{n}}$. These are just the basis vectors for three-dimensional tangential-normal coordinates.

We assume the collar is *small*—it almost *has* to be to follow a curved guide!. This means that any *rotational* motion of the collar will be mimicked by its *translational* motion in the following sense: any component of angular velocity of the collar in the $\hat{\boldsymbol{b}}$-direction must be accompanied by motion around the path's instantaneous circle of curvature, and any rotation in the $\hat{\boldsymbol{n}}$-direction must correspond to translation out of the osculating plane. (Rotation in the $\hat{\boldsymbol{t}}$-direction is precisely what has been dealt with above.) *But there is no binormal component of translational motion*, in either velocity or acceleration; thus there will be *no* $\hat{\boldsymbol{n}}$*-component of rotation.* This leave only the additional binormal component of angular motion to be accounted for.

For *velocity*, $\hat{\boldsymbol{b}}$ gives the *direction* of this added rotation of $\boldsymbol{\omega}_c$:

$$\boldsymbol{\omega}_c = \omega_t \hat{\boldsymbol{t}} + \omega_b \hat{\boldsymbol{b}}$$

in the total angular velocity

$$\boldsymbol{\omega} = \boldsymbol{\omega}_c + \omega_p \hat{\boldsymbol{u}}_p, \qquad \hat{\boldsymbol{u}}_p = \frac{\hat{\boldsymbol{t}} \times \boldsymbol{r}}{\|\hat{\boldsymbol{t}} \times \boldsymbol{r}\|},$$

the last supplanting equation (4.4-20). ω_b is then related to the velocity through tangential-normal coordinates expressed in terms of coordinates ρ and β, where β measures the rate at which the radius-of-curvature vector rotates about the instantaneous center of curvature: $\omega_b = \dot{\beta}$, where $\boldsymbol{v} = v\hat{\boldsymbol{t}} = \rho\dot{\beta}\hat{\boldsymbol{t}}$. In terms of the previous discussion, one would likely write

$$\boldsymbol{\omega} = \omega_t \hat{\boldsymbol{t}} + \dot{\beta}\hat{\boldsymbol{b}} + \omega_p \hat{\boldsymbol{u}}_p \equiv \boldsymbol{\omega}' + \dot{\beta}\hat{\boldsymbol{b}} = \boldsymbol{\omega}' + \frac{v}{\rho}\hat{\boldsymbol{b}} \tag{4.4-23}$$

with $\boldsymbol{\omega}'$ taking the place of the unknown $\boldsymbol{\omega}$ above: since it has only $\hat{\boldsymbol{t}}$- and $\hat{\boldsymbol{u}}_p$-components, we have

$$\boldsymbol{\omega}' \cdot \hat{\boldsymbol{t}} \times \hat{\boldsymbol{u}}_p \equiv 0, \tag{4.4-24}$$

corresponding to (4.4-21) above. If v were *known* in the problem, $\frac{v}{\rho}$ would be also, leaving one more quantity, typically a v_{rel}, to find. If $\boldsymbol{v}$ were *unknown*, it would enter ω through only *linearly*, through $\boldsymbol{\omega} \times \boldsymbol{r}$; v would be the fourth unknown, with any $\boldsymbol{v}_{rel}$ necessarily known.

For *acceleration*, the situation becomes much more complicated: now we must account for $\ddot{\beta}$ or $\dot{v}$, as well as $\dot{\hat{\boldsymbol{b}}}$ above. Differentiating (4.4-23), we get a daunting expression:

$$\boldsymbol{\alpha} = \dot{\omega}_t \hat{\boldsymbol{t}} + \omega_t \dot{\hat{\boldsymbol{t}}} + \dot{\omega}_p \hat{\boldsymbol{u}}_p + \omega_p \dot{\hat{\boldsymbol{u}}}_p + \ddot{\beta}\hat{\boldsymbol{b}} + \dot{\beta}\dot{\hat{\boldsymbol{b}}}. \tag{4.4-25}$$

While it would be possible to evaluate the various terms in this expression as was done above initially for the clevis, we will take the alternate approach for analysis of the angular acceleration there, differentiation of the *condition* on ω itself. From the general expression for ω, equation (4.4-23), we can separate that quantity into two parts,

$$\boldsymbol{\alpha} = \boldsymbol{\alpha}' + \ddot{\beta}\hat{\boldsymbol{b}} + \dot{\beta}\dot{\hat{\boldsymbol{b}}}$$

Working backwards, we see immediately that the last term drops out, since $\dot{\hat{\boldsymbol{b}}} \equiv 0$. Furthermore, we claim that we can express $\ddot{\beta}$:

If $\dot{\beta}$ above was related to v, it is not surprising that $\ddot{\beta}$ must be determined in conjunction with the acceleration of the collar, $\boldsymbol{a}_C$. In the tangential-normal components used,

$$\boldsymbol{a}_C = \dot{v}\hat{\boldsymbol{t}} + \frac{v^2}{\rho}\hat{\boldsymbol{n}},$$

in which v is known either *á priori* or from the velocity analysis; this leaves $\dot{v}$, in the known $\hat{\boldsymbol{t}}$-direction as either already known (if $\boldsymbol{a}_C$ is given), or to determine from the acceleration equation. This is related to $\dot{\beta}$ through

$$\dot{v} = \frac{d}{dt}(\rho\dot{\beta}) = \dot{\rho}\dot{\beta} + \rho\ddot{\beta}.$$

ρ is known from the geometry, and $\dot{\beta}$ either known from known v or determined from the velocity analysis, yet to find $\ddot{\beta}$, we must know $\dot{\rho}$. But, recall, the geometry is known well enough to know $\rho(s)$, where $s = s(t)$; thus

$$\dot{\rho} = \frac{d\rho}{ds}\frac{ds}{dt} = v\frac{d\rho}{ds}$$

in which v is known, so that, *finally*,

$$\dot{v} = v\frac{d\rho}{ds}\dot{\beta} + \rho\ddot{\beta}.$$

Again, $\dot{v}$ is either known from $\boldsymbol{a}_C$ (allowing us to determine $\ddot{\beta}$) or not. In the former case, so is $\ddot{\beta}$, and there is a fourth variable, typically an a_{rel}, to find. In the latter,

$$\ddot{\beta} = \frac{\dot{v} - v\frac{d\rho}{ds}\dot{\beta}}{\rho}$$

can be substituted into $\boldsymbol{\alpha}$; $\dot{v}$ enters linearly through the term $\boldsymbol{\alpha} \times \boldsymbol{r}$, as well as in its $\boldsymbol{a}_C = \dot{v}\hat{\boldsymbol{t}} + \frac{v^2}{\rho}\hat{\boldsymbol{n}}$ expression. It is thus *totally* a linear unknown which "can" be solved for.

This leaves us with $\boldsymbol{\alpha}'$ to determine. It will enter the acceleration equation through $\boldsymbol{\alpha} \times \boldsymbol{r}$, resulting in its appearance as components of $\boldsymbol{\alpha}' \times \boldsymbol{r}$. As with any cross product, it is necessary to have an additional condition to uniquely determine the unknown. This condition can be found by differentiation of (4.4-24)

$$\boldsymbol{\alpha}' \cdot \hat{\boldsymbol{t}} \times \hat{\boldsymbol{u}}_p + \boldsymbol{\omega}' \cdot [\dot{\hat{\boldsymbol{t}}} \times \hat{\boldsymbol{u}}_p + \hat{\boldsymbol{t}} \times \dot{\hat{\boldsymbol{u}}}_p] \equiv 0.$$

But the first term in brackets can be found immediately: since $\hat{\boldsymbol{t}}$ rotates with $\dot{\beta}\hat{\boldsymbol{b}}$, we have that $\dot{\hat{\boldsymbol{t}}} = \dot{\beta}\hat{\boldsymbol{b}} \times \hat{\boldsymbol{t}} = \dot{\beta}\hat{\boldsymbol{n}}$, so $\dot{\hat{\boldsymbol{t}}} \times \hat{\boldsymbol{u}}_p = \dot{\beta}\hat{\boldsymbol{n}} \times \hat{\boldsymbol{u}}_p$. And since the pin rotates with the collar's angular velocity so that $\dot{\hat{\boldsymbol{u}}}_p = (\omega_t\hat{\boldsymbol{t}} + \omega_b\hat{\boldsymbol{b}}) \times \hat{\boldsymbol{u}}_p$,

$$\begin{aligned}
\hat{\boldsymbol{t}} \times \dot{\hat{\boldsymbol{u}}}_p &= \hat{\boldsymbol{t}} \times [\omega_t\hat{\boldsymbol{t}} \times \hat{\boldsymbol{u}}_p + \omega_b\hat{\boldsymbol{b}} \times \hat{\boldsymbol{u}}_p] \\
&= \omega_t[(\hat{\boldsymbol{t}} \cdot \hat{\boldsymbol{u}}_p)\hat{\boldsymbol{t}} - (\hat{\boldsymbol{t}} \cdot \hat{\boldsymbol{t}})\hat{\boldsymbol{u}}_p] + \omega_b[(\hat{\boldsymbol{t}} \cdot \hat{\boldsymbol{u}}_p)\hat{\boldsymbol{b}} - (\hat{\boldsymbol{t}} \cdot \hat{\boldsymbol{b}})\hat{\boldsymbol{u}}_p] \\
&= -\omega_t\hat{\boldsymbol{u}}_p
\end{aligned}$$

since $\hat{\boldsymbol{t}}$ is perpendicular to both $\hat{\boldsymbol{u}}_p$ and $\hat{\boldsymbol{b}}$. Thus, finally,

$$\boldsymbol{\alpha}' \cdot \hat{\boldsymbol{t}} \times \hat{\boldsymbol{u}}_p + \boldsymbol{\omega}' \cdot (-\omega_t\hat{\boldsymbol{u}}_p) \equiv 0, \tag{4.4-26}$$

in which $\boldsymbol{\omega}'$ has been determined from the velocity analysis above, becomes the "additional condition" on $\boldsymbol{\alpha}'$.

It's not pretty, but at least it can be done!

We note in passing that the "right" way to do this—one not requiring the assumption that the collar is "small" (an assumption implicit in even the normal analysis)—is to consider motion of the contact points of the collar with the guide, relating these to the motion of, say, the center of the collar and its angular motion.

Example 4.4.4 *Clevis Connection:* We conclude this section with an example of this type of connection—merely a restatement of Example 4.4.1 with the ball-and-socket end B replaced by a clevis, though we shall be content with finding only $\boldsymbol{\omega}$ and $\boldsymbol{\alpha}$ here. This reprise is done for pedagogical reasons: In the first place, we shall see the similarities with the previous example, as well as the differences between the two. Just as importantly, however, it gives us the opportunity to discuss the correspondence between the two [different] answers obtained, at the end of the example.

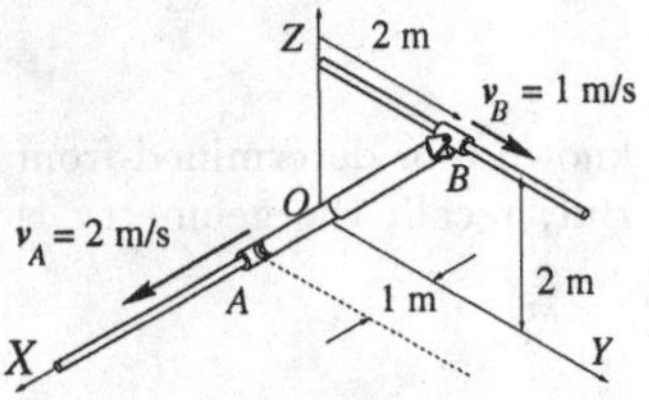

We note that the *velocity* rotating coordinate system equations for this problem, equations (4.4-4), remain unchanged, since it is only the *constraint* equation which depends on the additional condition. Just for easy reference, the calculated "b" form of the original incantation of the system's velocity equation is reproduced here:

$$\begin{aligned}
2 &= \quad -2\omega_Y + 2\omega_Z + \frac{1}{3}v_{rel}\\
-1 &= 2\omega_X \qquad + \omega_Z - \frac{2}{3}v_{rel}\\
0 &= -2\omega_X - \omega_Y \qquad - \frac{2}{3}v_{rel}.
\end{aligned}$$

The constraint equation 4.4-21, $\boldsymbol{\omega}\cdot\hat{\boldsymbol{u}}_c\times\hat{\boldsymbol{u}}_p\equiv 0$, on $\boldsymbol{\omega}$ requires that we know $\hat{\boldsymbol{u}}_c$ of the collar's rotational axis, and $\hat{\boldsymbol{u}}_p$ of the pin's. The former is immediately recognized to be $\hat{\boldsymbol{J}}$; the latter is perpendicular to the plane containing $\hat{\boldsymbol{u}}_c$ and $\boldsymbol{r}$ and must be normalized:

$$\begin{aligned}
\hat{\boldsymbol{u}}_p &\equiv \frac{\hat{\boldsymbol{u}}_c\times\boldsymbol{r}}{\|\hat{\boldsymbol{u}}_c\times\boldsymbol{r}\|}\\
&= \frac{\hat{\boldsymbol{J}}\times(1\hat{\boldsymbol{I}}-2\hat{\boldsymbol{J}}-2\hat{\boldsymbol{K}})}{\|\hat{\boldsymbol{J}}\times(1\hat{\boldsymbol{I}}-2\hat{\boldsymbol{J}}-2\hat{\boldsymbol{K}})\|}\\
&= \frac{-2\hat{\boldsymbol{I}}-\hat{\boldsymbol{K}}}{\sqrt{5}} = -\frac{2\sqrt{5}}{5}\hat{\boldsymbol{I}} - \frac{\sqrt{5}}{5}\hat{\boldsymbol{K}};
\end{aligned}$$

thus

$$\hat{\boldsymbol{u}}_c\times\hat{\boldsymbol{u}}_p = -\frac{\sqrt{5}}{5}\hat{\boldsymbol{I}} + \frac{2\sqrt{5}}{5}\hat{\boldsymbol{K}},$$

so that finally the condition on $\boldsymbol{\omega}$ is

$$\boldsymbol{\omega}\cdot\hat{\boldsymbol{u}}_c\times\hat{\boldsymbol{u}}_p=-\frac{\sqrt{5}}{5}\omega_X+\frac{2\sqrt{5}}{5}\omega_Z\equiv 0.$$

Amending this to the three component equations for the velocity above, we get finally that

$$\omega_X=-\frac{2}{45},\quad \omega_Y=-\frac{4}{5},\quad \omega_Z=-\frac{1}{45},\quad v_{rel}=\frac{4}{3}.$$

Comparing these with the answers to Example 4.4.1, we are not surprised to see that $\boldsymbol{\omega}$ is different: the above additional condition on this variable is, after all, different. But $v_{\rm rel}$ *is the same!* In some ways, this is to be expected physically: if the velocities of A and B are the same, regardless of how the piston is connected, it must be expanding at the same rate in order to "keep up" with its endpoints. Yet $\boldsymbol{\omega}$ is radically different, and that, too, goes into the $\boldsymbol{v}$'s. On the other hand, we know that v_{rel} could be found using the "useful trick" on page 218, and that gives a result independent of $\boldsymbol{\omega}$ because it *eliminates* that unknown from the equation. So it isn't really so surprising after all!

A similar situation holds for the accelerations, though now, due to the different value for $\boldsymbol{\omega}$, the terms $\boldsymbol{\omega}\times(\boldsymbol{\omega}\times\boldsymbol{r})$ and $2\boldsymbol{\omega}\times\boldsymbol{v}_{rel}$ change the constant side:

$$\begin{aligned}-\frac{271}{405}&=-2\alpha_Y+2\alpha_Z+\frac{1}{3}a_{rel}\\ \frac{38}{405}&=2\alpha_X\qquad+\ \alpha_Z-\frac{2}{3}a_{rel}\\ -\frac{826}{405}&=-2\alpha_X-\ \alpha_Y\qquad-\frac{2}{3}a_{rel}.\end{aligned}$$

Now we must obtain the additional condition on $\boldsymbol{\alpha}$, equation (4.4-22), that $\boldsymbol{\alpha}\cdot\hat{\boldsymbol{u}}_c\times\hat{\boldsymbol{u}}_p=\omega_c\omega_p$. The left-hand side will be identical in form to that for $\boldsymbol{\omega}$ above, but we need the two components ω_c and ω_p of (4.4-19): $\boldsymbol{\omega}=\omega_c\hat{\boldsymbol{u}}_c+\omega_p\hat{\boldsymbol{u}}_p$. But, since $\hat{\boldsymbol{u}}_c$ and $\hat{\boldsymbol{u}}_p$ are perpendicular, we have immediately that $\omega_c=\boldsymbol{\omega}\cdot\hat{\boldsymbol{u}}_c=-\frac{4}{5}$ and $\omega_p=\boldsymbol{\omega}\cdot\hat{\boldsymbol{u}}_p=\frac{\sqrt{5}}{45}$. Substituting these values into the condition on $\boldsymbol{\alpha}$, we get

$$\boldsymbol{\alpha}\cdot\hat{\boldsymbol{u}}_c\times\hat{\boldsymbol{u}}_p=-\frac{\sqrt{5}}{5}\alpha_X+\frac{2\sqrt{5}}{5}\alpha_Z=\omega_c\omega_p=-\frac{4\sqrt{5}}{225}.$$

Amending this equation to the three above yields a solution

$$\alpha_X=\frac{692}{2025},\quad \alpha_Y=\frac{16}{25},\quad \alpha_Z=\frac{256}{2025},\quad a_{rel}=\frac{29}{27}.$$

Once again $\boldsymbol{\alpha}$ is drastically different, while a_{rel} is precisely the same. While a similar result was able to be rationalized with regard to the velocity equation above, and it again certainly makes sense physically that the expansion rate along the piston should match what's happening at the ends, our discussion of the "useful trick" points out that the [different] angular velocity *does* enter the acceleration equation through the triple cross product $\boldsymbol{\omega}\times(\boldsymbol{\omega}\times\boldsymbol{r})$ as well as through the term $2\boldsymbol{\omega}\times\boldsymbol{v}_{rel}$, and these terms survive the dot product with $\boldsymbol{r}$ there. So what's going on?

Again we appeal to the device of dotting the acceleration equation with $\boldsymbol{r}$, as expanded in equation (4.4-11). In the summary following that expression, we observe that if $\boldsymbol{v}_{rel}\parallel\boldsymbol{r}$, the term $2\boldsymbol{\omega}\times\boldsymbol{v}_{rel}\cdot\boldsymbol{r}\equiv 0$; this is precisely the situation in the present example. And if we explicitly determine the value of $\boldsymbol{\omega}_{\perp}$ in this problem using (4.4-8b),

$$\begin{aligned}
\boldsymbol{\omega}_{\parallel} &= (\boldsymbol{\omega}\cdot\hat{\boldsymbol{r}})\hat{\boldsymbol{r}}\\
&= \left(\left(-\frac{2}{45}\hat{\boldsymbol{I}}-\frac{4}{5}\hat{\boldsymbol{J}}-\frac{1}{45}\hat{\boldsymbol{K}}\right)\cdot\left(\frac{\hat{\boldsymbol{I}}-2\hat{\boldsymbol{J}}-2\hat{\boldsymbol{K}}}{3}\right)\right)\frac{\hat{\boldsymbol{I}}-2\hat{\boldsymbol{J}}-2\hat{\boldsymbol{K}}}{3}\\
&= \frac{8}{45}\hat{\boldsymbol{I}}-\frac{16}{45}\hat{\boldsymbol{J}}-\frac{16}{45}\hat{\boldsymbol{K}}\\
\therefore\quad \boldsymbol{\omega}_{\perp} &= \boldsymbol{\omega}-\boldsymbol{\omega}_{\parallel}\\
&= \left(-\frac{2}{45}\hat{\boldsymbol{I}}-\frac{4}{45}\hat{\boldsymbol{J}}-\frac{1}{45}\hat{\boldsymbol{K}}\right)-\left(\frac{8}{45}\hat{\boldsymbol{I}}-\frac{16}{45}\hat{\boldsymbol{J}}-\frac{16}{45}\hat{\boldsymbol{K}}\right)\\
&= -\frac{2}{9}\hat{\boldsymbol{I}}-\frac{4}{9}\hat{\boldsymbol{J}}+\frac{1}{3}\hat{\boldsymbol{K}},
\end{aligned}$$

$\boldsymbol{\omega}_{\perp}$ is exactly the same value as obtained for "$\boldsymbol{\omega}$" in Example 4.4.1. $\boldsymbol{a}_A$, $\boldsymbol{a}_B$, and $\boldsymbol{r}$ are the same in both examples, $\boldsymbol{\omega}_{\perp}$ is the same, and $2\boldsymbol{\omega}\times\boldsymbol{v}_{rel}\cdot\boldsymbol{r}\equiv 0$ for both; thus the right-hand side of (4.4-11) is the same, and the a_{rel} obtained from each will be identical, just as observed above.

We make one more remark regarding this: We explicitly calculated $\boldsymbol{\omega}_{\perp}$ here in order to exemplify the use of equations (4.4-8). But we should have known from the start that $\boldsymbol{\omega}_{\perp}$ would be the same for both problems: the value of $\boldsymbol{\omega}$ in each satisfied the same velocity equation and, as pointed out in the discussion regarding the "useful trick," this leads to both having the same $\boldsymbol{\omega}_{\perp}$. Everything hangs together.

On the other hand, if we calculate the component of $\boldsymbol{\alpha}$ *perpendicular* to $\boldsymbol{r}$ in the two problems, we get that

$$\begin{aligned}
\text{ball-and-socket:}\qquad \boldsymbol{\alpha}_{\perp} &= \frac{16}{81}\hat{\boldsymbol{I}}+\frac{32}{81}\hat{\boldsymbol{J}}-\frac{8}{27}\hat{\boldsymbol{K}}\\
\text{clevis:}\qquad \boldsymbol{\alpha}_{\perp} &= \frac{1496}{2025}\hat{\boldsymbol{I}}-\frac{104}{675}\hat{\boldsymbol{J}}-\frac{1352}{2025}\hat{\boldsymbol{K}}.
\end{aligned}$$

These are *not* the same because $\boldsymbol{\alpha}$ is determined from the *full* acceleration equation, not just that component remaining when that equation is dotted with $\boldsymbol{r}$. And the *total* value of $\boldsymbol{\omega}$ is folded into this equation through both the terms $\boldsymbol{\omega}\times(\boldsymbol{\omega}\times\boldsymbol{r})$ and $2\boldsymbol{\omega}\times\boldsymbol{v}_{rel}$ (the different constant side of the acceleration equation). Although we pointed out in the "trick" discussion that only $\boldsymbol{\omega}_{\perp}$ affects the first (so that term is the same for both cases), equation (4.4-10) demonstrates the $\boldsymbol{\omega}_{\parallel}$ enters the second. For the ball-and-socket, $\boldsymbol{\omega}_{\parallel}\equiv 0$, since we enforced the condition that $\boldsymbol{\omega}\cdot\mathbf{r}\equiv 0$; but for the clevis, we have explicitly calculated $\boldsymbol{\omega}_\parallel \neq 0$ above. |

Ball-and-socket Connections

It was mentioned that this analysis is rather more simple than that of the clevis preceding. We shall see, however, that what becomes a practical simplification also generates a conceptual conundrum.

We pointed out, in the development of the kinematic constraint imposed on the motion of the body due to the clevis, that the orientation of the body was completely specified physically knowing $\mathbf{r}$ and the tangential direction of the guide. This led to a constraint completely specifying the direction of $\boldsymbol{\omega}$, and this, in turn, resulted in a condition on $\boldsymbol{\alpha}$. But the ball-and-socket enjoys no such constraint: physically it is possible to rotate the body freely about the $\mathbf{r}$ axis—this corresponds to what we have been calling "$\boldsymbol{\omega}_\parallel$"—without violating any conditions on the motion of the end points of the body itself. This observation is completely consistent with the mathematics involved: the velocity analysis invoking the rotating coordinate systems determines $\boldsymbol{\omega}$ by solving for it from its appearance in $\boldsymbol{\omega}\times\mathbf{r}=\boldsymbol{\omega}_\perp\times\mathbf{r}$—*i.e.*, it is only the *perpendicular* component of $\boldsymbol{\omega}$ which can be found from this condition, and any further component of that vector is determined by the additional [scalar] condition imposed. The analysis for $\boldsymbol{\alpha}$ is similar, in that any angular acceleration along $\mathbf{r}$ will have no effect on the accelerations of the endpoints.

As a result of these observations, it is often the practice to admit defeat regarding determination of the components of $\boldsymbol{\omega}$ and $\boldsymbol{\alpha}$ parallel to the rod, $\boldsymbol{\omega}_\parallel$ and $\boldsymbol{\alpha}_\parallel$, and determine only the perpendicular components $\boldsymbol{\omega}_\perp$ and $\boldsymbol{\alpha}_\perp$. These conditions can be stated simply as

$$\boldsymbol{\omega}\cdot\mathbf{r}=\boldsymbol{\omega}_\perp\cdot\mathbf{r}\equiv 0 \quad\text{and}\quad \boldsymbol{\alpha}\cdot\mathbf{r}=\boldsymbol{\alpha}_\perp\cdot\mathbf{r}\equiv 0 \tag{4.4-27}$$

—precisely those used in Example 4.4.1—effectively *forcing* $\boldsymbol{\omega}$ and $\boldsymbol{\alpha}$ to be perpendicular to $\mathbf{r}$. Geometrically, these conditions describe a plane perpendicular to $\mathbf{r}$; $\boldsymbol{\omega}$ and $\boldsymbol{\alpha}$ lie on the intersections of this plane with those perpendicular to the right-hand sides "$\mathbf{V}$" and "$\mathbf{A}$" of the velocity and acceleration equations (4.4-1), just as with the angular velocity in the clevis. The same condition for indeterminacy—that the normals to these planes, $\mathbf{r}$, and $\mathbf{V}$ or $\mathbf{A}$ be parallel—applies. But now this simplifies somewhat: $\mathbf{V}=\boldsymbol{\omega}\times\mathbf{r}$ and $\mathbf{A}=\boldsymbol{\alpha}\times\mathbf{r}$ are *always* perpendicular to $\mathbf{r}$, never parallel, so the only way these planes are "parallel" is if $\boldsymbol{\omega}$ or $\boldsymbol{\alpha}$ vanishes.

There is, however, a subtle problem, rarely stated explicitly, regarding this practice: equation (4.4-10)

$$\boldsymbol{\omega}\times(\boldsymbol{\omega}\times\mathbf{r})=(\boldsymbol{\omega}_\parallel\cdot\mathbf{r})\boldsymbol{\omega}_\perp-\omega_\perp^2\mathbf{r}$$

shows that the value of $\boldsymbol{\omega}_\parallel$ enters the acceleration equation in the determination of $\boldsymbol{\alpha}$ (if not of the acceleration of an endpoint or the a_{rel}); the direction of that term involving $\boldsymbol{\omega}_{\|}$ is that of $\boldsymbol{\omega}_{\perp}$—*i.e.*, perpendicular to $\boldsymbol{r}$ itself, and thus in the direction of the equation determining even $\boldsymbol{\alpha}_{\perp}$ from $\boldsymbol{\alpha}\times\boldsymbol{r}$. We saw this, in fact, at the end of Example 4.4.4, where, although the $\boldsymbol{\omega}_{\perp}$ were the same for both that example and Example 4.4.1, the non-0 $\boldsymbol{\omega}_{\|}$ in the latter altered even the value of $\boldsymbol{\alpha}_{\perp}$. Thus the ignorance made explicit in the above conditions (4.4-27) comes at a cost: not only can we determine only $\boldsymbol{\omega}_{\perp}$, we can't even obtain all of $\boldsymbol{\alpha}_{\perp}$! This rather remarkable circumstance means that it is necessary to have determined $\boldsymbol{\omega}_{\|}$ by observation or some independent means; it *cannot* be determined from a kinematic analysis alone. This stricture does *not* hold in planar problems: there $\boldsymbol{\omega}\equiv\omega\hat{\boldsymbol{k}}\perp\boldsymbol{r}$ [by hypothesis], so $\boldsymbol{\omega}_{\|}\equiv\mathbf{0}$ from the start!

4.4.3 Motion of *Multiple* Rigid Bodies ("Machines")

With the nature of constraints imposed by connections between various bodies [one hopes!] well in hand, we can embark on the study systems of interconnected rigid bodies—*i.e.*, "*machines.*" As mentioned in the prologue to this section, the basic idea to the analysis of such a system is to "chain" together the velocity and/or acceleration analyses of those for a single rigid body by fixing a rotating coordinate system in each. Just as with the analysis of single bodies, the points chosen for this description will be those determining the motion of each body—points invariably including the points of connection between pairs of bodies. This technique will typically involve expressing the motion of points in one body in terms of those in another and equating the components of the vector velocity and acceleration equations; to expedite this equating, it is useful—and this is really the only "new" idea in this application—to utilize the same basis for all bodies. We exemplify this in the following:

Example 4.4.5 *"Chainable" Machine.* Diagrammed here is a 4-ft rod OA, making a fixed 30° angle with the vertical, rotating around that axis at a constant rate $\boldsymbol{\omega}=\omega\hat{\boldsymbol{K}}$. It is attached by means of a ball-and-socket to a second 4-ft rod AB, whose other end B is attached to a collar sliding along the X-axis by means of a second ball-and-socket joint. It is desired to examine the acceleration of point B as it varies with the angle θ made by the vertical plane through A with the XZ-plane.

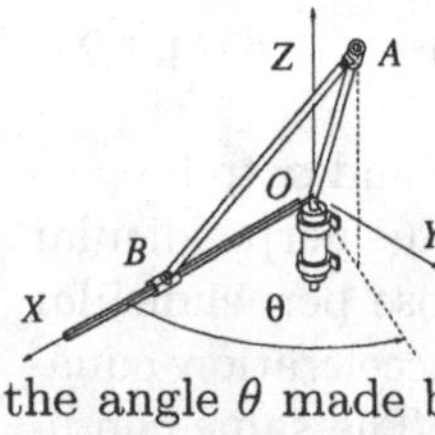

This becomes a relatively straightforward application of the single-body approach discussed previously: In effect we "know everything" about the motion of OA; from that we can determine, in particular, the motion of point A—that point connecting the two bodies and common to both—in that member, which is precisely the same as the absolute velocity of the corresponding point in AB. Between that and the constraint imposed on the motion of B due to the collar, we can solve for $\boldsymbol{a}_B$. The situation is slightly muddied due to this business regarding determination of the angle θ, but, if we express all quantities in terms of *general* θ, we will then have $\boldsymbol{a}_B(\theta)$.

That is precisely the strategy employed: We want $\boldsymbol{a}_B$, so will need the $\boldsymbol{\omega}$'s, requiring a velocity analysis. Looking first to OA, we see immediately that the velocity of A satisfies

$$\begin{aligned}
\boldsymbol{v}_A &= \boldsymbol{v}_O + \boldsymbol{\omega} \times \boldsymbol{r}_{A/O} + \boldsymbol{v}_{rel} \\
&= \mathbf{0} + \omega \hat{\boldsymbol{K}} \times (2\cos\theta \hat{\boldsymbol{I}} + 2\sin\theta \hat{\boldsymbol{J}} + 2\sqrt{3}\hat{\boldsymbol{K}}) + \mathbf{0} \\
&= -2\omega \sin\theta \hat{\boldsymbol{I}} + 2\omega\cos\theta \hat{\boldsymbol{J}} \qquad \text{(relative to } OA) \\
&= \boldsymbol{v}_B + \boldsymbol{\omega}_{AB} \times \boldsymbol{r}_{A/B} + \boldsymbol{v}_{rel} \\
&= v_B \hat{\boldsymbol{I}} \\
&\quad + (\omega_{AB_X} \hat{\boldsymbol{I}} + \omega_{AB_Y} \hat{\boldsymbol{J}} + \omega_{AB_Z} \hat{\boldsymbol{K}}) \times \left(-2\cos\theta \hat{\boldsymbol{I}} + 2\sin\theta \hat{\boldsymbol{J}} + 2\sqrt{3}\hat{\boldsymbol{K}}\right) + \mathbf{0} \\
&= (v_B + 2\sqrt{3}\omega_{AB_Y} - 2\omega_{AB_Z}\sin\theta)\hat{\boldsymbol{I}} - (2\omega_{AB_Z}\cos\theta + 2\sqrt{3}\omega_{AB_X})\hat{\boldsymbol{J}} \\
&\quad + (2\omega_{AB_X}\sin\theta + 2\omega_{AB_Y}\cos\theta)\hat{\boldsymbol{K}} \qquad \text{(relative to } AB),
\end{aligned}$$

r_{A/B_X} being found by solving $(2\cos\theta - X_B)^2 + (2\sin\theta)^2 + (2\sqrt{3})^2 = 4^2$ for *either* $X_B = 4\cos\theta$—so that $r_{A/B_X} = 2\cos\theta - 4\cos\theta$, *or* $X_B = 0$—in which case $r_{A/B_X} = +2\cos\theta$.

This "machine" is going to arise again in a couple of guises, so it is beneficial to examine the actual motion of the system in some more detail. Assuming that AB has end B pulled in the positive X-direction—by a spring, say—then, for $\theta \in [0, \pi/4)$, X_B is, indeed, $4\cos\theta$; it starts at $4a$ and is drawn toward 0 as $\theta \to \pi/4$. When it reaches that limit, however, AB lies along OA and cannot pass "through" it; thus AB becomes carried coincident with the latter arm until $\theta \to 3\pi/4$, when the spring gradually pulls end B away from O until its $X_B = 4a$ again at $\theta = 2\pi$. While end A moves in the circle described by that end of OA, end B traces out the motion given by X_B at right. For that reason, we will implicitly be assuming that $\theta \in (-\pi/4, \pi/4)$ in the subsequent analysis. At the same time, it should be recognized that, although X_B is *continuous*, it is *not differentiable*: there is a discontinuity in slope–the velocity—at $\pi/2$ and $3\pi/2$ which carries through to the acceleration.

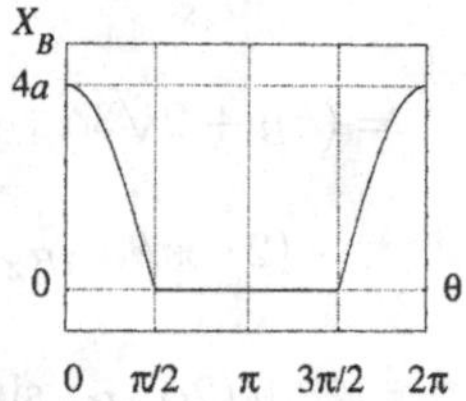

In addition, we have the constraint on AB that we can find only the component of $\boldsymbol{\omega}_{AB}$ perpendicular to $\boldsymbol{r}_{A/B}$:

$$\boldsymbol{\omega}_{AB} \cdot \boldsymbol{r}_{AB} = -2\omega_{AB_X}\cos\theta + 2\omega_{AB_Y}\sin\theta + 2\sqrt{3}\omega_{AB_Z} \equiv 0.$$

Note that we have expressed a pair of vector equations *represented in terms of the* same *basis vectors* $\hat{\boldsymbol{I}}$, $\hat{\boldsymbol{J}}$, *and* $\hat{\boldsymbol{K}}$. This allows us to utilize either or both sides of *both* equations (relative to OA and AB) with impunity. The above is a set of seven scalar equations in the unknown components of $\boldsymbol{v}_A$, $\boldsymbol{\omega}_{AB}$, and the scalar v_B. But we don't really *care* about $\boldsymbol{v}_A$ (though in this case its "solution" is just the OA equation!); because both sets are relative to the same basis, we can simply set the *right*-hand sides of the equations equal and use the resulting set of *four* equations[13] (the first three, along with the above equation $\boldsymbol{\omega}_{AB} \cdot \boldsymbol{r}_{AB} \equiv 0$) to find $\boldsymbol{\omega}_{AB}$ necessary for the acceleration analysis:

$$v_B = -4\omega \sin\theta; \qquad \boldsymbol{\omega}_{AB} = -\frac{\sqrt{3}}{4}\omega\cos\theta\hat{\boldsymbol{I}} + \frac{\sqrt{3}}{4}\omega\sin\theta\hat{\boldsymbol{J}} - \frac{1}{4}\omega\hat{\boldsymbol{K}},$$

completing the velocity analysis.

We are now ready to tackle the acceleration analysis. In precisely the same way as above, we get that

$$\begin{aligned}
\boldsymbol{a}_A &= \boldsymbol{a}_O + \boldsymbol{\alpha}_{OA} \times \boldsymbol{r}_{A/O} + \boldsymbol{\omega} \times (\boldsymbol{\omega} \times \boldsymbol{r}_{A/O}) + 2\boldsymbol{\omega} \times \boldsymbol{v}_{rel} + \boldsymbol{a}_{rel} \\
&= \boldsymbol{0} + \boldsymbol{0} + \omega\hat{\boldsymbol{K}} \times (\omega\hat{\boldsymbol{K}} \times (2\cos\theta\hat{\boldsymbol{I}} + 2\sin\theta\hat{\boldsymbol{J}} + 2\sqrt{3}\hat{\boldsymbol{K}}) + \boldsymbol{0} + \boldsymbol{0} \\
&= 2\omega^2\cos\theta\hat{\boldsymbol{I}} + 2\omega^2\sin\theta\hat{\boldsymbol{J}} \qquad \text{(relative to OA)} \\
&= \boldsymbol{a}_B + \boldsymbol{\alpha}_{AB} \times \boldsymbol{r}_{A/B} + \boldsymbol{\omega}_{AB} \times (\boldsymbol{\omega}_{AB} \times \boldsymbol{r}_{A/B}) + 2\boldsymbol{\omega}_{AB} \times \boldsymbol{v}_{rel} + \boldsymbol{a}_{rel} \\
&= a_B\hat{\boldsymbol{I}} \\
&\quad + (\alpha_{AB_X}\hat{\boldsymbol{I}} + \alpha_{AB_Y}\hat{\boldsymbol{J}} + \alpha_{AB_Z}\hat{\boldsymbol{K}}) \times (-2\cos\theta\hat{\boldsymbol{I}} + 2\sin\theta\hat{\boldsymbol{J}} + 2\sqrt{3}\hat{\boldsymbol{K}}) \\
&\quad + \left(-\frac{\sqrt{3}}{4}\omega\cos\theta\hat{\boldsymbol{I}} + \frac{\sqrt{3}}{4}\omega\sin\theta\hat{\boldsymbol{J}} - \frac{1}{4}\omega\hat{\boldsymbol{K}}\right) \\
&\qquad \times \left(\left(-\frac{\sqrt{3}}{4}\omega\cos\theta\hat{\boldsymbol{I}} + \frac{\sqrt{3}}{4}\omega\sin\theta\hat{\boldsymbol{J}} - \frac{1}{4}\omega\hat{\boldsymbol{K}}\right)\right. \\
&\qquad\qquad \left.\times (-2\cos\theta\hat{\boldsymbol{I}} + 2\sin\theta\hat{\boldsymbol{J}} + 2\sqrt{3}\hat{\boldsymbol{K}})\right) + \boldsymbol{0} + \boldsymbol{0} \\
&= (a_B + 2\sqrt{3}\alpha_{AB_Y} - 2\alpha_{AB_Z}\sin\theta + \frac{1}{2}\omega^2\cos\theta)\hat{\boldsymbol{I}} \\
&\quad - (2\cos\theta\alpha_{AB_Z} + 2\alpha_{AB_X} - \frac{1}{2}\omega^2\sin\theta)\hat{\boldsymbol{J}} \\
&\quad + (2\alpha_{AB_X}\sin\theta + 2\alpha_{AB_Y}\cos\theta - \frac{\sqrt{3}}{2}\omega^2\sin^2\theta)\hat{\boldsymbol{K}} \qquad \text{(relative to } AB\text{);}
\end{aligned}$$

once more we have to make $\boldsymbol{\alpha}_{AB}$ be perpendicular to $\boldsymbol{r}_{AB}$:

$$\boldsymbol{\alpha}_{AB} \cdot \boldsymbol{r}_{AB} = -2\alpha_{AB_X}\cos\theta + 2\alpha_{AB_Y}\sin\theta + 2\sqrt{3}\alpha_{AB_Z} \equiv 0.$$

Again, we can obviate solution for $\boldsymbol{a}_A$ by simply setting the right-hand sides of the above expressions for that variable equal and including the ball-and-socket condition on $\boldsymbol{\alpha}_{AB}$. The solution for the desired variables is

$$a_B = -4\omega^2\frac{\sin^2\theta}{\cos\theta}; \qquad \boldsymbol{\alpha}_{AB} = -\frac{\sqrt{3}}{4}\omega^2\sin\theta\hat{\boldsymbol{I}} + \frac{\sqrt{3}}{4}\omega^2\frac{1+\sin^2\theta}{\cos\theta}\hat{\boldsymbol{J}} - \frac{1}{2}\omega^2\tan\theta\hat{\boldsymbol{K}}.$$

We observe that, although $\boldsymbol{a}_B$ vanishes at $\theta = 0$ and $\theta = \pi$, it becomes *singular* at $\theta = \pm\frac{\pi}{2}$. This is a rather remarkable result, almost counterintuitive, and

[13] This is rather similar to the device in planar machine kinematics, where it typically results in a *pair* of scalar components with the angular velocity or acceleration as unknowns. But in planar problems, there is only one component of such angular quantities to find; here there are *three*.

would not have been uncovered without the above analysis. We also note, in passing, similar behavior for $\boldsymbol{\alpha}$ through its Y- and Z-components. |

If the above example was marginally interesting, we now discuss a system both larger and of intrinsic practical import:

Example 4.4.6 *Universal Joint.* The universal joint, schematized here, is really itself another form of linkage between rigid bodies—one designed to change the direction of rotational motion. It consists, however, of a system of *three* rigid bodies, two forked-shaped assemblies, enclosing a "spider" suspended between the arms of each fork. Here the width of the forks on both AO and OB is $2a$; the shaft of each is held in place by pillow blocks fixed, respectively, parallel to the XY- and YZ-planes, with the latter arm making an angle of θ with the Y-axis, If AO is rotating with an angular velocity of $\boldsymbol{\omega}$ about its axis, we wish to determine the angular velocity of the second arm, $\boldsymbol{\omega}_{OB}$.

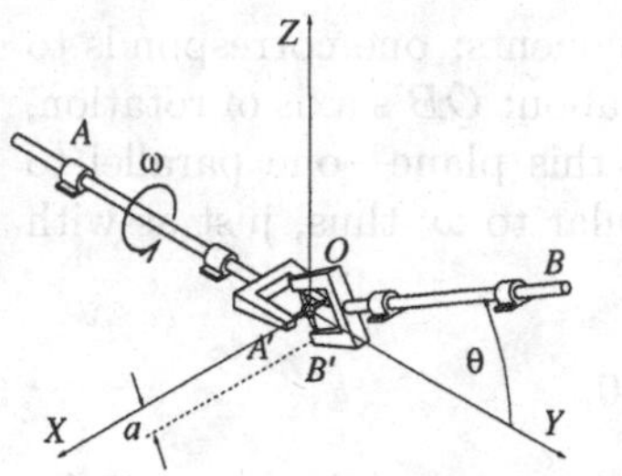

As has been mentioned before, it is useful to assess exactly what is actually driving the motion of the system: in this case arm AO drives the spider $OA'B'$ which, in turn, drives OB. Thus it makes intuitive sense to analyze the system in this order, and this is, indeed, a judicious approach. In particular, we shall first determine the angular motion of the spider by implementing the fact that the point A' on the spider has the same [absolute] velocity as that point on AO; the angular velocity of the spider will then give the velocity of point B' on that member, which in turn will be used to determine the final angular velocity of OB.

We know that A' is *one* point driving the spider, but what *other* point do we use for the latter? The key to this body's motion is the observation that, since the center O of the spider always lies on the axes of the two shafts, and since those are fixed, $\boldsymbol{v}_O \equiv \mathbf{0}$. In this case, that point can serve as the origin of rotating coordinates for both AO and the spider. Fixing such systems in both $OA'B'$ and AO, we have immediately that

$$\begin{aligned}\boldsymbol{v}_{A'} &= \boldsymbol{v}_O + \boldsymbol{\omega}_{AO} \times \boldsymbol{r}_{A'/O} + \boldsymbol{v}_{rel_{AO}} && \text{(relative to } AO\text{)}\\ &= \boldsymbol{v}_O + \boldsymbol{\omega}_{OA'B'} \times \boldsymbol{r}_{A'/O} + \boldsymbol{v}_{rel_{OA'B'}}. && \text{(relative to } OA'B'\text{)}\end{aligned}$$

As noted, $\boldsymbol{v}_O \equiv \mathbf{0}$, and, in this case, both $\boldsymbol{v}_{rel}$'s vanish because the A' does not move *relative to the respective coordinate systems*; thus all that remains to determine is $\boldsymbol{\omega}_{OA'B'}$. Representing these in terms of the fixed basis,

$$\begin{aligned}\boldsymbol{v}_{A'} &= \mathbf{0} + \omega\hat{\boldsymbol{J}} \times a\hat{\boldsymbol{I}} + \mathbf{0} = -a\omega\hat{\boldsymbol{K}}\\ &= \mathbf{0} + (\omega_X\hat{\boldsymbol{I}} + \omega_Y\hat{\boldsymbol{J}} + \omega_Z\hat{\boldsymbol{K}}) \times a\hat{\boldsymbol{I}} + \mathbf{0} = a\omega_Z\hat{\boldsymbol{J}} - a\omega_Y\hat{\boldsymbol{K}}\end{aligned}$$

[The fact that the first equation has only a $\hat{\boldsymbol{K}}$-component results from the special position of point A': in general there would be components in both the

$\hat{\boldsymbol{I}}$- and $\hat{\boldsymbol{K}}$-directions.] The first equation will completely determine $\boldsymbol{v}_{A'}$, since $\boldsymbol{\omega}$ is known, but the second only allows determination of ω_Y and the fact that $\omega_Z = 0$; there is nothing said about ω_X! Looking at this equation carefully, we note that ω_X doesn't even appear, due to the cross product with $a\hat{\boldsymbol{I}}$ in the first equation! But the cross product $\boldsymbol{\omega} \times \boldsymbol{r}$ has, recall, only two independent components, demanding some additional scalar to find $\boldsymbol{\omega}$; how are we do get an additional condition here?

Once again, we note that $\boldsymbol{\omega}_{OA'B'}$ has two components: one corresponds to motion about the $\hat{\boldsymbol{J}}$-direction due to AO, the other about OB's axis of rotation, $\cos\theta\hat{\boldsymbol{J}} + \sin\theta\hat{\boldsymbol{K}}$. Thus a vector perpendicular to this plane—one parallel to $\hat{\boldsymbol{J}} \times (\cos\theta\hat{\boldsymbol{J}} + \sin\theta\hat{\boldsymbol{K}}) = \sin\theta\hat{\boldsymbol{I}}$—will be perpendicular to $\boldsymbol{\omega}$; thus, just as with the clevis (to which the universal joint is related),

$$\boldsymbol{\omega} \cdot \sin\theta\hat{\boldsymbol{I}} = -\omega_X \sin\theta \equiv 0,$$

becomes the determining condition on $\boldsymbol{\omega}$. But this just means that $\omega_X \equiv 0$ for $\theta \neq 0$. (Observe that if $\theta = 0$, then all of $\boldsymbol{\omega}$ is in the Y-direction, and ω_X *still* vanishes!) Thus ω_X *is* determined and, equating the remaining components in the velocity equation, we have immediately that $\boldsymbol{\omega}_{OA'B'} = \omega\hat{\boldsymbol{J}}$—a rather remarkable result, if you think about it! Note that this equating of components assumed the representations on both sides were relative to the same basis—the reason this was mentioned in the remarks beginning the section.

We are now prepared to determine $\boldsymbol{\omega}_{OB}$: The motion of OB is a consequence of the fixed nature of *its* shaft supports and the motion of B'; the latter, in turn, results from the spider's rotation, now known, about the fixed point O. Again fixing coordinates systems in these two bodies,

$$\begin{aligned} \boldsymbol{v}_{B'} &= \boldsymbol{v}_O + \boldsymbol{\omega}_{OA'B'} \times \boldsymbol{r}_{B'/O} + \boldsymbol{v}_{rel_{OA'B'}} && \text{(relative to } OA'B'\text{)} \\ &= \boldsymbol{v}_O + \boldsymbol{\omega}_{OB} \times \boldsymbol{r}_{B'/O} + \boldsymbol{v}_{rel_{AO}}. && \text{(relative to } OB\text{)} \end{aligned}$$

Once again, O doesn't move, and the $\boldsymbol{v}_{rel}$'s both vanish. Substituting in the above-determined value for $\boldsymbol{\omega}_{OA'B'}$, and noting that the *direction* of $\boldsymbol{\omega}_{OB}$ is known: $\boldsymbol{\omega}_{OB} = \omega_{OB}(\cos\theta\hat{\boldsymbol{J}} + \sin\theta\hat{\boldsymbol{K}})$, ω_{OB} becomes the *only* [scalar] unknown, despite the alleged three-dimensional nature of the problem. But when the calculations are actually carried out,

$$\begin{aligned} \boldsymbol{v}_{B'} &= \boldsymbol{0} + \omega\hat{\boldsymbol{J}} \times a(\sin\theta\hat{\boldsymbol{J}} + \cos\theta\hat{\boldsymbol{K}}) + \boldsymbol{0} = a\omega\cos\theta\hat{\boldsymbol{I}} \\ &= \boldsymbol{0} + \omega_{OB}(\cos\theta\hat{\boldsymbol{J}} + \sin\theta\hat{\boldsymbol{K}}) \times a(-\sin\theta\hat{\boldsymbol{J}} + \cos\theta\hat{\boldsymbol{K}}) + \boldsymbol{0} \\ &= a\omega_{OB}(\cos^2\theta + \sin^2\theta)\hat{\boldsymbol{I}}, \end{aligned}$$

so that only a single component remains, and $\omega_{OB} = \omega\cos\theta$.

Note that there is a singularity in this solution when $\theta = 90°$; what does this tell us? This occurs when the two shafts are *perpendicular*, but at that point, the system would "lock up," and there would be *no* rotation! The student should always be sensitive to such pathological circumstances, since they invariably indicate something potentially terribly, terribly wrong. |

Their results notwithstanding, these examples have been "comfortable" in the sense that we were able completely to determine the motion of each body in the chain: at each, we were able to obtain a simple set of four equations which could be solved to describe both the translational and rotation motion of that body. More generally, however, we can end up with rather more impressive sets of equations which must be solved simultaneously. This occurs typically in systems in which a pair of bodies are linked together, but *no* body's motion is completely known; in some ways, it is the general situation and is the focus of the balance of the examples:

Example 4.4.7 *"Non-chainable" Machine:* Diagrammed here is a pair of rods, AC, 7 m long, and BC, 3 m long, connected by ball-and-sockets at their ends to the collar A and a piston B, respectively. The assembly leans against the YZ-plane as shown so that C just rests against that plane. If A and B are moving at a constant rate of 1 m/s along the X-axis and vertically as indicated, we are asked to find the acceleration $\boldsymbol{a}_C$ of point C in that plane, and the angular accelerations $\boldsymbol{\alpha}_{AC}$ and $\boldsymbol{\alpha}_{BC}$ of the two rods when the collar is 3 m from the origin of coordinates, B 4 m above the XY-plane, and C located at the point $(0, 2, 6)$. We explicitly assume that there is "no rotation" of the two rods along their respective axes.

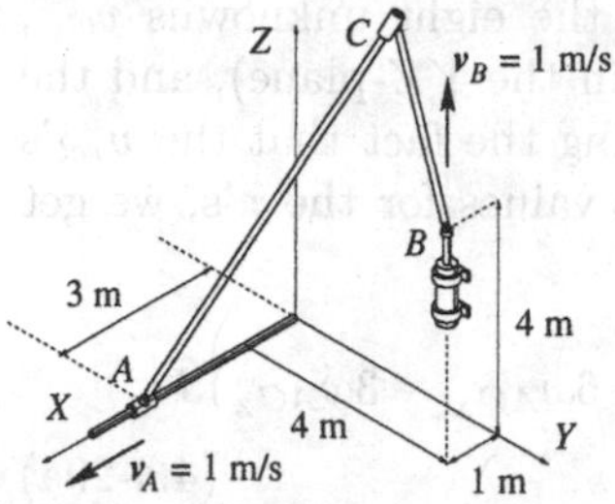

Before continuing with the problem, let us observe precisely how the position of C itself was determined: it is located at a position $(0, Y, Z)$ satisfying the conditions that this point is, respectively, 7 m and 3 m from A and B:

$$7^2 = 3^2 + Y^2 + Z^2$$
$$3^2 = 1^2 + (4 - Y)^2 + (4 - Z)^2.$$

Expanding these, it turns out that terms quadratic in Y and Z both appear with unit coefficient; thus the two equations can be subtracted to yield a *linear* equation $Y + Z = 8$. This can be used to eliminate one of the two variables, say Z, from either of the above two quadratic equations, yielding two solutions: $(Y, Z) = (2, 6)$ or $(Y, Z) = (6, 2)$; it is the former chosen for this problem.

Unlike the previous examples, we cannot simply "chain" together the velocity and accelerations equations for the respective bodies: though we know the motion of the ends A and B, we don't have the angular velocities and/or accelerations of the two rods themselves from which to determine the motion of C. Nonetheless, we can at least write the equations describing the motion of C relative to coordinate systems fixed in these bodies, with origins at the points whose motion are known. To emphasize the overall structure of this, we shall not be quite so detailed as usual, giving only the final equations obtained. The reader is invited to verify these calculations!

Starting, as usual, with the velocities (since we need the angular velocities in order to determine the accelerations):

$$\boldsymbol{v}_C = \boldsymbol{v}_A + \boldsymbol{\omega}_{AC} \times \boldsymbol{r}_{C/A} + \boldsymbol{v}_{rel\,AC} \qquad \text{(relative to } AC) \qquad (4.4\text{-}28a)$$

$$= \boldsymbol{v}_B + \boldsymbol{\omega}_{BC} \times \boldsymbol{r}_{C/B} + \boldsymbol{v}_{rel\,BC} \qquad \text{(relative to } BC) \qquad (4.4\text{-}28b)$$

augmented by the fact that the respective angular velocities are purely normal to the two rods, due to assumption that there be no rotation along the rods:

$$\boldsymbol{\omega}_{AC} \cdot \boldsymbol{r}_{AC} \equiv 0 \qquad (4.4\text{-}28c)$$

$$\boldsymbol{\omega}_{BC} \cdot \boldsymbol{r}_{BC} \equiv 0 \qquad (4.4\text{-}28d)$$

These constitute a set of eight linear equations in the eight unknowns v_{C_Y}, v_{C_Z} ($v_{C_X} \equiv 0$ because motion is constrained to be in the YZ-plane), and the three components of each $\boldsymbol{\omega}_{AC}$ and $\boldsymbol{\omega}_{BC}$. Recognizing the fact that the $\boldsymbol{v}_{rel}$'s both vanish in this problem and substituting in the values for the $\boldsymbol{r}$'s, we get evaluated equations corresponding to the above:

$$v_{C_Y}\hat{\boldsymbol{J}} + v_{C_Z}\hat{\boldsymbol{K}} = (1 + 6\omega_{AC_Y} - 2\omega_{AC_Z})\hat{\boldsymbol{I}} + (-6\omega_{AC_X} - 3\omega_{AC_Z})\hat{\boldsymbol{J}} + (2\omega_{AC_X} + 3\omega_{AC_Y})\hat{\boldsymbol{K}} \qquad (4.4\text{-}29a)$$

$$= (2\omega_{BC_Y} + 2\omega_{BC_Z})\hat{\boldsymbol{I}} + (-2\omega_{BC_X} - \omega_{BC_Z})\hat{\boldsymbol{J}} + (1 - 2\omega_{BC_X} + \omega_{BC_Y})\hat{\boldsymbol{K}} \qquad (4.4\text{-}29b)$$

$$0 = -3\omega_{AC_X} + 2\omega_{AC_Y} + 6\omega_{AC_Z} \qquad (4.4\text{-}29c)$$

$$0 = -\omega_{BC_X} - 2\omega_{BC_Y} + 2\omega_{BC_Z}. \qquad (4.4\text{-}29d)$$

In the first example, we set the right-hand sides of the equations giving the motion of the interconnecting point equal to one another; the resulting equation, along with the perpendicularity condition on $\boldsymbol{\omega}_{AB}$, allowed us to solve directly for that angular variable. But if we try that trick here, we see that we will have three component equations from the right-hand side equality, along with *two* perpendicularity conditions—a total of only *five* equations for the *six* unknown components of the angular velocities $\boldsymbol{\omega}_{AC}$ and $\boldsymbol{\omega}_{BC}$! Thus we are forced to introduce the additional equation on $\boldsymbol{v}_C$, adding three more equations (including the fact that $v_{C_X} \equiv 0$) with only two accompanying unknowns, the non-0 components of $\boldsymbol{v}_C$. That done, the solutions are "immediate":

$$\boldsymbol{v}_C = -\frac{9}{8}\hat{\boldsymbol{J}} - \frac{1}{8}\hat{\boldsymbol{K}},$$

$$\boldsymbol{\omega}_{AC} = \frac{13}{98}\hat{\boldsymbol{I}} - \frac{51}{392}\hat{\boldsymbol{J}} + \frac{43}{392}\hat{\boldsymbol{K}}, \quad \boldsymbol{\omega}_{BC} = \frac{1}{2}\hat{\boldsymbol{I}} - \frac{1}{8}\hat{\boldsymbol{J}} + \frac{1}{8}\hat{\boldsymbol{K}}.$$

Were there to have been any components of the $\boldsymbol{\omega}$'s along their respective rods—these would have to have been determined by some analysis independent of this purely kinematical one, recall—they would be added in at this point.

The angular velocities, in particular, are needed in the acceleration equations; relative to the two bodies AC and BC, respectively, these are:

$$\begin{aligned}\boldsymbol{a}_C &= \boldsymbol{a}_A + \boldsymbol{\alpha} \times \boldsymbol{r}_{C/A} + \boldsymbol{\omega}_{AC} \times (\boldsymbol{\omega}_{AC} \times \boldsymbol{r}_{C/A}) \\ &\quad + 2\boldsymbol{\omega}_{AC} \times \boldsymbol{v}_{rel\,AC} + \boldsymbol{a}_{rel\,AC} \qquad (4.4\text{-}30a) \\ &= \boldsymbol{a}_A + \boldsymbol{\alpha} \times \boldsymbol{r}_{C/B} + \boldsymbol{\omega}_{BC} \times (\boldsymbol{\omega}_{BC} \times \boldsymbol{r}_{C/B}) \\ &\quad + 2\boldsymbol{\omega}_{BC} \times \boldsymbol{v}_{rel\,BC} + \boldsymbol{a}_{rel\,BC} \qquad (4.4\text{-}30b)\end{aligned}$$

and the associated constraints

$$\boldsymbol{\alpha}_{AC} \cdot \boldsymbol{r}_{AC} \equiv 0 \qquad (4.4\text{-}30c)$$

$$\boldsymbol{\alpha}_{BC} \cdot \boldsymbol{r}_{BC} \equiv 0. \qquad (4.4\text{-}30d)$$

The *rel* quantities, of course all vanish; substituting in the above values of the $\boldsymbol{\omega}$'s resulting from the velocity analysis, we get the corresponding evaluated set

$$\begin{aligned}a_{C_Y}\hat{\boldsymbol{J}} + a_{C_Z}\hat{\boldsymbol{K}} &= (6\alpha_{AC_Y} - 2\alpha_{AC_Z} + \frac{219}{1568})\hat{\boldsymbol{I}} + (-6\alpha_{AC_X} - 3\alpha_{AC_Z} - \frac{73}{784})\hat{\boldsymbol{J}} \\ &\quad + (2\alpha_{AC_X} + 3\alpha_{AC_Y} - \frac{219}{784})\hat{\boldsymbol{I}} \qquad (4.4\text{-}31a) \\ &= (2\alpha_{BC_Y} + 2\alpha_{BC_Z} + \frac{9}{32})\hat{\boldsymbol{I}} + (-2\alpha_{BC_X} - \alpha_{BC_Z} + \frac{9}{16})\hat{\boldsymbol{J}} \\ &\quad + (-2\alpha_{BC_X} + \alpha_{BC_Y} - \frac{9}{16})\hat{\boldsymbol{K}} \qquad (4.4\text{-}31b) \\ 0 &= -3\alpha_{AC_X} + 2\alpha_{AC_Y} + 6\alpha_{AC_Z} \qquad (4.4\text{-}31c) \\ 0 &= -\alpha_{BC_X} - 2\alpha_{BC_Y} + 2\alpha_{BC_Z}. \qquad (4.4\text{-}31d)\end{aligned}$$

Once more, we must solve the full system of equations simultaneously for the angular accelerations and $\boldsymbol{a}_C$:

$$\boldsymbol{a}_C = \frac{85}{128}\hat{\boldsymbol{J}} - \frac{77}{128}\hat{\boldsymbol{K}},$$

$$\boldsymbol{\alpha}_{AC} = -\frac{83}{784}\hat{\boldsymbol{I}} - \frac{33}{896}\hat{\boldsymbol{J}} - \frac{255}{6272}\hat{\boldsymbol{K}}, \quad \boldsymbol{\alpha}_{BC} = -\frac{1}{72}\hat{\boldsymbol{I}} - \frac{77}{1152}\hat{\boldsymbol{J}} - \frac{85}{1152}\hat{\boldsymbol{K}}.$$

Again, recall that the angular accelerations could not have been determined without knowledge of those components of the $\boldsymbol{\omega}$'s parallel to the rods (see the discussion on page 238). |

Example 4.4.8 *Non-vanishing "rel" Terms* A 6-ft rod AB is supported at A by a ball-and-socket, while its other end leans against the YZ-"wall." Sliding along this rod at a speed of 3 ft/s is a collar C, attached by a ball-and-socket to a 3-ft rod CD whose other end is also free to rotate in the fixed ball-and-socket joint at D. At the instant the collar is in the middle of AB, we are asked to determine the velocity $\boldsymbol{v}_B$ of end B, under the explicit assumption that there is no rotation of either body along its axis.

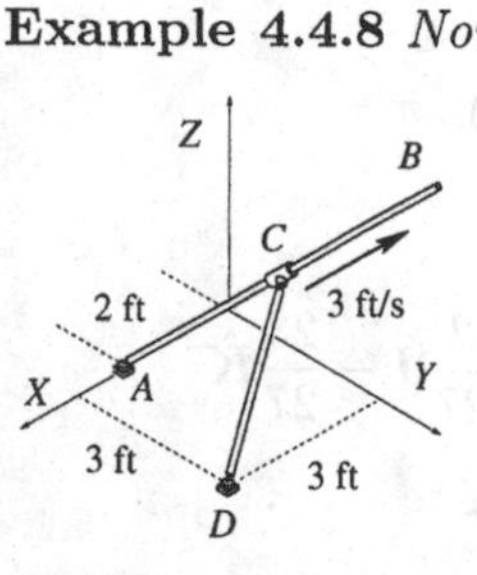

Similarly to Example 4.4.7, we can find the position of collar C in this one; in particular, we select the solution $\boldsymbol{r}_C = \hat{\boldsymbol{I}} + 2\hat{\boldsymbol{J}} + 2\hat{\boldsymbol{K}}$ ft here.

Body CD is simple, with velocity equation

$$\begin{aligned}\boldsymbol{v}_C &= \boldsymbol{v}_D + \boldsymbol{\omega}_{CD} \times \boldsymbol{r}_{C/D} + \boldsymbol{v}_{rel}\\ &= \mathbf{0} + (\omega_{CD_X}\hat{\boldsymbol{I}} + \omega_{CD_Y}\hat{\boldsymbol{J}} + \omega_{CD_Z}\hat{\boldsymbol{K}}) \times (-2\hat{\boldsymbol{I}} - \hat{\boldsymbol{J}} + 2\hat{\boldsymbol{K}}) + \mathbf{0}\\ &= (2\omega_{CD_Y} + 2\omega_{CD_Z})\hat{\boldsymbol{I}} - (2\omega_{CD_Z} + 2\omega_{CD_X})\hat{\boldsymbol{J}} - (\omega_{CD_X} - 2\omega_{CD_Y})\hat{\boldsymbol{K}}.\end{aligned}$$

The same point can be referred to AB: since $\boldsymbol{v}_{rel} = v_{rel}\hat{\boldsymbol{r}}_{rel}$ where the latter vector is along AB,

$$\begin{aligned}&= \boldsymbol{v}_A + \boldsymbol{\omega}_{AB} \times \boldsymbol{r}_{C/A} + \boldsymbol{v}_{rel}\\ &= \mathbf{0} + (\omega_{AB_X}\hat{\boldsymbol{I}} + \omega_{AB_Y}\hat{\boldsymbol{J}} + \omega_{AB_Z}\hat{\boldsymbol{K}}) \times (-\hat{\boldsymbol{I}} + 2\hat{\boldsymbol{J}} + 2\hat{\boldsymbol{K}})\\ &\quad + 3\frac{-\hat{\boldsymbol{I}} + 2\hat{\boldsymbol{J}} + 2\hat{\boldsymbol{K}}}{3}\\ &= (2\omega_{AB_Y} - 2\omega_{AB_Z} - 1)\hat{\boldsymbol{I}} - (\omega_{AB_Z} + 2\omega_{AB_X} - 2)\hat{\boldsymbol{J}}\\ &\quad + (2\omega_{AB_X} + \omega_{AB_Y} + 2)\hat{\boldsymbol{K}}.\end{aligned}$$

As in the previous example, we see that there is little benefit in trying to set the right-hand sides of these equations equal: even with the perpendicularity conditions on the angular velocities, we have only five equations to find the $\boldsymbol{\omega}$'s. But we know a further condition on the motion of the system: since end B is constrained to move along the "wall,"

$$\begin{aligned}\boldsymbol{v}_B &= v_{B_Y}\hat{\boldsymbol{J}} + v_{B_Z}\hat{\boldsymbol{K}}\\ &= \boldsymbol{v}_A + \boldsymbol{\omega}_{AB} \times \boldsymbol{r}_{B/A} + \boldsymbol{v}_{rel}\\ &= \mathbf{0} + (\omega_{AB_X}\hat{\boldsymbol{I}} + \omega_{AB_Y}\hat{\boldsymbol{J}} + \omega_{AB_Z}\hat{\boldsymbol{K}}) \times (-2\hat{\boldsymbol{I}} + 4\hat{\boldsymbol{J}} + 4\hat{\boldsymbol{K}}) + \mathbf{0}\\ &= (4\omega_{AB_Y} - 4\omega_{AB_Z})\hat{\boldsymbol{I}} - (2\omega_{AB_Z} + 4\omega_{AB_X})\hat{\boldsymbol{J}}\\ &\quad + (4\omega_{AB_X} + 2\omega_{AB_Y})\hat{\boldsymbol{K}}.\end{aligned}$$

This introduces another three equations with only two additional unknowns, the components of $\boldsymbol{v}_B$. Amending to the above the perpendicularity condition on the angular velocities:

$$\begin{aligned}\boldsymbol{\omega}_{AB} \cdot \boldsymbol{r}_{AB} &= -2\omega_{AB_X} + 4\omega_{AB_Y} + 4\omega_{AB_Z} \equiv 0\\ \boldsymbol{\omega}_{CD} \cdot \boldsymbol{r}_{CD} &= 2\omega_{CD_X} + \omega_{CD_Y} - 2\omega_{CD_Z} \equiv 0,\end{aligned}$$

we finally get the solution

$$\boldsymbol{\omega}_{AB} = -\frac{16}{27}\hat{\boldsymbol{I}} - \frac{4}{27}\hat{\boldsymbol{J}} - \frac{4}{27}\hat{\boldsymbol{K}}, \qquad \boldsymbol{\omega}_{CD} = -\frac{22}{27}\hat{\boldsymbol{I}} - \frac{2}{27}\hat{\boldsymbol{J}} - \frac{23}{27}\hat{\boldsymbol{K}}$$

$$\boldsymbol{v}_C = -\hat{\boldsymbol{I}} + \frac{10}{3}\hat{\boldsymbol{J}} + \frac{2}{3}\hat{\boldsymbol{K}}, \quad \boldsymbol{v}_B = \frac{8}{3}\hat{\boldsymbol{J}} - \frac{8}{3}\hat{\boldsymbol{K}}.$$

We were asked to find the velocity of B here, so it was appropriate to reference that point in the above equations. But if all we were interested in was the angular velocities, there is an alternative approach which could be used: An equation implementing the planar motion of B could be transferred to the point C' *on the rod*:

$$\boldsymbol{v}_C = \boldsymbol{v}_{C'} + \boldsymbol{v}_{rel}$$

in which we could implement the constraint on B as reflected in the motion of C' by observing that

$$\boldsymbol{r}_B = \boldsymbol{r}_A + \boldsymbol{r}_{B/A} = \boldsymbol{r}_{C'} + \boldsymbol{r}_{B/C'}$$

$$\therefore \boldsymbol{r}_{C'} = \boldsymbol{r}_A + \boldsymbol{r}_{B/A} - \boldsymbol{r}_{B/C'} = \boldsymbol{r}_A + \frac{\boldsymbol{r}_{B/A}}{2}$$

$$\therefore \boldsymbol{v}_{C'} = \boldsymbol{v}_A + \frac{\boldsymbol{v}_{B/A}}{2} = \frac{\boldsymbol{v}_{B/A}}{2} = \frac{1}{2}(v_{B_Y}\hat{\boldsymbol{J}} + v_{B_Z}\hat{\boldsymbol{K}})$$

since $\boldsymbol{v}_A \equiv \boldsymbol{0}$ here. |

Curved Interconnections

Though we were interested only in the velocity analysis in the previous example, a similar approach could be used to find accelerations: now $\boldsymbol{a}_{rel}$ would be along AB *because the rod is straight.* And although $\boldsymbol{v}_{rel}$ (and, if it had been given, $\boldsymbol{a}_{rel}$) was known in the example, such quantities could also be scalar *unknowns*, as in Examples 4.4.1 and 4.4.4. But what if the rod *weren't* straight? How would the curved nature of such interconnections affect the motion and resulting equations? We now briefly indicate how the analysis would change.

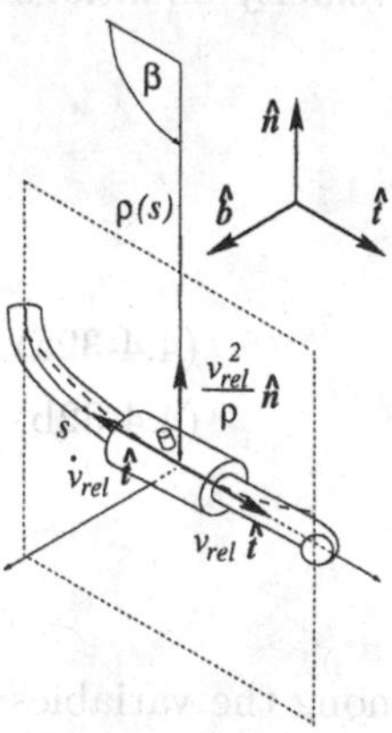

It is convenient once again to introduce Tangential-normal coordinates here. Diagrammed at left is a typical collar moving along a curved guide. We must know the geometry of that guide; in particular, we know $\rho(s)$, the radius of curvature of the guide whose instantaneous circle in the indicated osculating plane is dashed, $\hat{\boldsymbol{t}}$, the [instantaneous] tangent vector, and $\hat{\boldsymbol{n}}$, the [instantaneous] normal to the path at this point.

We know that $\boldsymbol{v}_{rel} = v_{rel}\hat{\boldsymbol{t}}$, $\hat{\boldsymbol{t}}$ replacing the direction of the straight rod in the above example; there's really nothing startling about that, and, if that quantity were unknown, it could be determined just as in the initial Examples 4.4.1 and 4.4.4 from a velocity analysis. The difference arises in the $\boldsymbol{a}_{rel}$ term: It has, in general, *two* components, $\dot{v}_{rel}\hat{\boldsymbol{t}}$ and $\frac{v_{rel}^2}{\rho}\hat{\boldsymbol{n}}$. *But the latter of these is now known*, from the previous velocity analysis. Thus the single scalar unknown a_{rel}, in a known direction along the straight rod above, is merely replaced by $\dot{v}_{rel}$ in the known $\hat{\boldsymbol{t}}$-direction. Everything else is the same!

General Analysis of Universal Joints

In the previous section, we first worked an example of the ball-and-socket connection to gain some familiarity with the issues related to it, then later returned to it to give a more general discussion of the analysis of this linkage. We do the same here, returning to a rather more exhaustive analysis of the universal joint, first encountered in Example 4.4.6. We shall view this in the context introduced in the above examples characterized by the *simultaneous* analysis of multiple rigid bodies, and see that approach to be endemic to this form of linkage.

Diagrammed again here is the universal joint: the motion of the input arm, say AO, is known, and it is desired to determine the angular velocity of the output arm OB making an angle of θ with the horizontal plane as shown. Unlike the previous incarnation of this system, however, we shall impose no conditions regarding the motion of any of the three components (such as point O being fixed, or, despite the diagram, even that AO is).

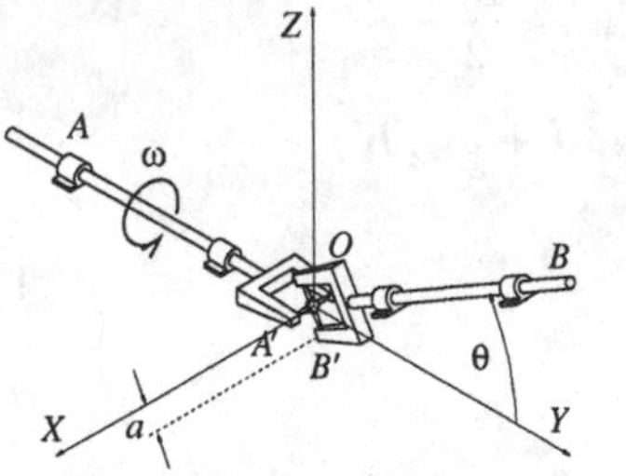

The points determining the motion of the three constituent parts are A, O, and A' on the fork AO, O, A', and B' on the spider, and B, O, and B' on the fork OB. Though we worked Example 4.4.6 sequentially piece by piece, certainly we can write down the *entire* system of velocity equations describing the motion of these various points:

$$
\begin{aligned}
AO: \quad & \boldsymbol{v}_O = \boldsymbol{v}_A + \boldsymbol{\omega}_{AO} \times \boldsymbol{r}_{O/A} \\
& \boldsymbol{v}_{A'} = \boldsymbol{v}_A + \boldsymbol{\omega}_{AO} \times \boldsymbol{r}_{A'/A} \\
\text{spider}: \quad & \boldsymbol{v}_{A'} = \boldsymbol{v}_O + \boldsymbol{\omega}_O \times \boldsymbol{r}_{A'/O} && \text{(4.4-32a)} \\
& \boldsymbol{v}_{B'} = \boldsymbol{v}_O + \boldsymbol{\omega}_O \times \boldsymbol{r}_{B'/O} && \text{(4.4-32b)} \\
OB: \quad & \boldsymbol{v}_{B'} = \boldsymbol{v}_O + \boldsymbol{\omega}_{OB} \times \boldsymbol{r}_{B'/O} \\
& \boldsymbol{v}_B = \boldsymbol{v}_O + \boldsymbol{\omega}_{OB} \times \boldsymbol{r}_{B/O}
\end{aligned}
$$

The above constitute a set of *eighteen* (!) scalar equations among the variables $\boldsymbol{v}_O$, $\boldsymbol{v}_A$, $\boldsymbol{v}_{A'}$, $\boldsymbol{v}_{B'}$, $\boldsymbol{v}_B$, $\boldsymbol{\omega}_{AO}$, $\boldsymbol{\omega}_O$, and $\boldsymbol{\omega}_{OB}$. The most typical situation is one in which we know everything about the motion of AO; the first two equations then allow $\boldsymbol{v}_O$ and $\boldsymbol{v}_{A'}$ to be found and used in the later equations. But this then leaves $\boldsymbol{v}_{B'}$, $\boldsymbol{v}_B$, $\boldsymbol{\omega}_O$, and $\boldsymbol{\omega}_{OB}$ to be determined—twelve scalar components in what appear to be twelve equations. But recall that the angular velocities appear in equations involving cross products; since these products can be represented as the product of the associated unknowns by a matrix of rank only 2, this means that only ten of these equations are actually linearly independent of one another, and we need two more conditions in order to obtain unique answers.

The diagram at left is a close-up of the spider and one of the attached arms. We note that the spider axis connected to AO rotates *with* that arm;

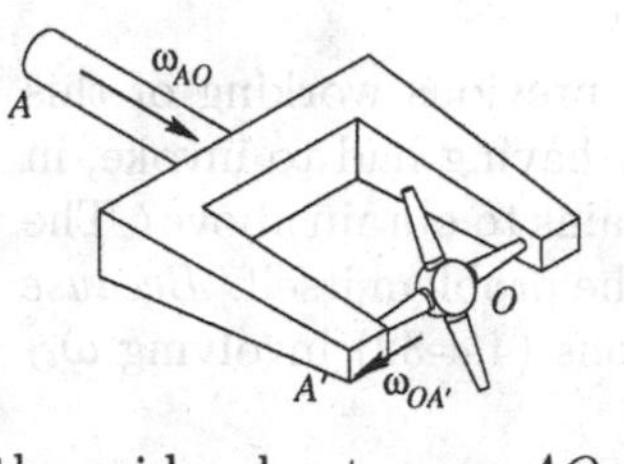

similarly, the other spider axis rotates with OB. By virtue of this observation relative to AO, we see that part of the spider's rotation is precisely the angular velocity of the connected arm; any other rotational component can only arise as a result of the spider's rotation about its OA' axis. These are the physical constraints imposed on the motion of the spider due to arm AO.

Thus we have immediately that

$$\boldsymbol{\omega}_O = \boldsymbol{\omega}_{AO} + \boldsymbol{\omega}_{O/AO} = \boldsymbol{\omega}_{AO} + \omega_{OA'}\hat{\mathbf{u}}_{OA'}.$$

This bears a striking similarity to the constraint imposed by the clevis connection, equation (4.4-19) in Section 4.4.2; like the treatment in that section, we can conclude immediately that

$$\boldsymbol{\omega}_O \cdot \boldsymbol{\omega}_{AO} \times \hat{\mathbf{u}}_{OA'} \equiv 0. \tag{4.4-33a}$$

This, then, becomes a scalar condition on the angular velocity of the spider. A similar analysis would hold for the other arm:

$$\boldsymbol{\omega}_O \cdot \boldsymbol{\omega}_{BO} \times \hat{\mathbf{u}}_{OB'} \equiv 0; \tag{4.4-33b}$$

these two arms' constraints thus provide the two scalar constraints required.

Since (4.4-33a) holds in general, we can find a similar condition on $\boldsymbol{\alpha}_O$: differentiating this equation,

$$\boldsymbol{\alpha}_O \cdot \boldsymbol{\omega}_{AO} \times \hat{\mathbf{u}}_{OA'} + \boldsymbol{\omega}_O \cdot \boldsymbol{\alpha}_{AO} \times \hat{\mathbf{u}}_{OA'} + \boldsymbol{\omega}_O \cdot \boldsymbol{\omega}_{AO} \times \dot{\hat{\mathbf{u}}}_{OA'} \equiv 0, \tag{4.4-34a}$$

We observe, however, that by Theorem 4.1.2,

$$\dot{\hat{\mathbf{u}}}_{OA'} = \boldsymbol{\omega}_{AO} \times \hat{\mathbf{u}}_{OA'}$$

since OA' is rotating with $\boldsymbol{\omega}_{AO}$, so that the last term

$$\begin{aligned} \boldsymbol{\omega}_O \cdot \boldsymbol{\omega}_{AO} \times \dot{\hat{\mathbf{u}}}_{OA'} &= \boldsymbol{\omega}_O \cdot \boldsymbol{\omega}_{AO} \times (\boldsymbol{\omega}_{AO} \times \hat{\mathbf{u}}_{OA'}) \\ &= \boldsymbol{\omega}_O \cdot \left((\boldsymbol{\omega}_{AO} \cdot \hat{\mathbf{u}}_{OA'})\boldsymbol{\omega}_{AO} - \omega_{AO}^2 \hat{\mathbf{u}}_{OA'}\right). \end{aligned} \tag{4.4-34b}$$

(This term does not fold in neatly into (4.4-34a) so is presented separately.) This equation, and the similar one for OB arising from (4.4-33b), are the two scalar constraints on $\boldsymbol{\alpha}_O$. Again we note the similarity to the analysis for the clevis: these condition on $\boldsymbol{\alpha}_O$ are generally *non*-homogeneous equations.

Before leaving this section, we note that the above conditions formulated in terms of $\hat{\mathbf{u}}_{OA'}$ can be recast in terms of $\mathbf{r}_{OA'} \parallel \hat{\mathbf{u}}_{OA'}$: equation (4.4-33a) becomes

$$\boldsymbol{\omega}_O \cdot \boldsymbol{\omega}_{AO} \times \mathbf{r}_{OA'} \equiv 0, \tag{4.4-35a}$$

while equations (4.4-34) can be combined into the single

$$\boldsymbol{\alpha}_O \cdot \boldsymbol{\omega}_{AO} \times \mathbf{r}_{OA'} + \boldsymbol{\omega}_O \cdot \boldsymbol{\alpha}_{AO} \times \mathbf{r}_{OA'} + \boldsymbol{\omega}_O \cdot \boldsymbol{\omega}_{AO} \times \mathbf{v}_{OA'} \equiv 0, \tag{4.4-35b}$$

in which $\boldsymbol{v}_{OA'}$ is the *relative* $\boldsymbol{v}_{A'} - \boldsymbol{v}_O$.

How does the present analysis compare with the previous working of this problem, Example 4.4.6, particularly in light of never having had to invoke, in the latter, the additional constraints we gone to such pains to obtain above? The fact is that these constraints were already *implicit* in the problem itself: *Because* $\boldsymbol{v}_O \equiv \mathbf{0}$ in the statement of the problem, the equations (4.4-32) involving $\boldsymbol{\omega}_O$ for the spider above, become simply

$$\boldsymbol{v}_{A'} = \boldsymbol{\omega}_O \times \boldsymbol{r}_{A'/O}$$
$$\boldsymbol{v}_{B'} = \boldsymbol{\omega}_O \times \boldsymbol{r}_{B'/O};$$

but this means that $\boldsymbol{\omega}_O \cdot \boldsymbol{v}_{A'} \equiv 0 \equiv \boldsymbol{\omega}_O \cdot \boldsymbol{v}_{B'}$—precisely the condition on $\boldsymbol{\omega}_O$, equation (4.4-35a). In this case, the two conditions reduce to $\boldsymbol{\omega}_O \cdot (-\hat{\boldsymbol{K}}) \equiv \mathbf{0} \equiv \boldsymbol{\omega}_O \cdot (-\hat{\boldsymbol{I}})$, showing that $\boldsymbol{\omega}_O = \omega_O \hat{\boldsymbol{J}}$, just as observed in the Example. And, although we did not do it there, had we chosen to determine the accelerations, the terms in equation (4.4-35b)

$$\boldsymbol{\omega}_O \cdot \boldsymbol{\alpha}_{AO} \times \boldsymbol{r}_{OA'} = \omega_O \hat{\boldsymbol{J}} \cdot \alpha_{AO} \hat{\boldsymbol{J}} \times \boldsymbol{r}_{OA'} \equiv 0$$
$$\boldsymbol{\omega}_O \cdot \boldsymbol{\omega}_{AO} \times \boldsymbol{v}_{OA'} = \omega_O \hat{\boldsymbol{J}} \cdot \omega_{AO} \hat{\boldsymbol{J}} \times \boldsymbol{v}_{OA'} \equiv 0,$$

since both are triple scalar products with two vectors parallel; thus the acceleration condition would be simply

$$\boldsymbol{\alpha}_O \cdot \boldsymbol{\omega}_{AO} \times \boldsymbol{r}_{OA'} \equiv 0$$

showing $\boldsymbol{\alpha}_O$ to be perpendicular to $\hat{\boldsymbol{K}}$. On the other hand, applying the condition analogous to (4.4-35b) to B', only

$$\boldsymbol{\omega}_O \cdot \boldsymbol{\alpha}_{OB} \times \boldsymbol{r}_{OB'} = \omega_O \hat{\boldsymbol{J}} \cdot \alpha_{OB} a \hat{\boldsymbol{J}} \equiv 0;$$

the other two terms

$$\begin{aligned}\boldsymbol{\alpha}_O \cdot \boldsymbol{\omega}_{OB} \times \boldsymbol{r}_{OB'} &+ \boldsymbol{\omega}_O \cdot \boldsymbol{\omega}_{OB} \times \boldsymbol{v}_{OB'} \\ &= \boldsymbol{\alpha}_O \cdot a\omega \cos\theta(-\hat{\boldsymbol{I}}) + \omega\hat{\boldsymbol{J}} \cdot a\omega^2 \cos\theta(-\sin\theta\hat{\boldsymbol{J}} + \cos\theta\hat{\boldsymbol{K}}) \\ &= \boldsymbol{\alpha}_O \cdot a\omega\cos\theta(-\hat{\boldsymbol{I}}) - a\omega^3\cos\theta\sin\theta \equiv 0,\end{aligned}$$

showing that $\boldsymbol{\alpha}_O$ does, in fact, have an $\hat{\boldsymbol{I}}$-component! (In fact, solving the full system, $\boldsymbol{\alpha}_O$ is *entirely* in the X-direction!)

If these conditions were redundant in Example 4.4.6, one can reasonably question the need for their development here. But $\boldsymbol{v}_O$ *vanished* in that example, leading to the equivalent conditions above; furthermore, the rotational directions of AO and OB were also *fixed*. More generally—*e.g.*, in the analysis of "trailing arms" so popular in British sports cars—OB might itself be rotating, and $\boldsymbol{\alpha}_{OB}$ would have an additional component due to the rotation of $\boldsymbol{\omega}_{OB}$.

Note that, in any event, there are still required two additional scalar equations on the bodies even to carry that off! These can assume any form: for example, knowing the direction of one of the $\boldsymbol{\omega}$'s reduces the originally three unknown variables to only one; *i.e.*, it imposes two implicit constraints on the motion. The need for such a constraint makes sense physically: the above set of kinematic equations hold for *any* system, and we must ultimately tie down the *specific* system with which we are working.

Summary. This section has been long, primarily because of the extensive number of examples (and the length of each of these). Yet, overall, the basic idea is quite simple: a rotating coordinate system fixed in each body reflects the angular motion of that body through the coordinate system's $\boldsymbol{\omega}$ and $\boldsymbol{\alpha}$. The determination of these variables—the "undoing" of the cross products through which these angular quantities enter—is effected through the simultaneous solution of these rotating coordinate system equations, written for those points which *determine* the motion of each body. Though the resulting set may grow impressively, at the end of the day, the technique reduces "simply" to the solution of a system of *linear* equations.

Summary

We started this chapter by discussing the distinction between "particles" and "rigid bodies": that while the former can only *translate*, the latter can both translate and *rotate*. Since the first is presumably (certainly *hopefully*!) well understood, the emphasis here has been on the latter. After pointing out that rotation involves merely the change in orientation of a line fixed in an [extended] rigid body, we then showed that the *differential* change in the relative position of any point in that body could be expressed as $d\boldsymbol{r} = d\boldsymbol{\theta}\times\boldsymbol{r}$, where the *infinitesimal* $d\boldsymbol{\theta}$ was the *same* vector for *all* points in the body; thus it characterizes the rotational motion of the body as a whole. We then introduced *angular velocity* $\boldsymbol{\omega} \equiv \frac{d\boldsymbol{\theta}}{dt}$ and *acceleration* $\boldsymbol{\alpha} \equiv \frac{d\boldsymbol{\omega}}{dt}$, the latter, in particular, generally involving the change in direction of $\boldsymbol{\omega}$—something which could be found by utilizing the results to that point: $\dot{\hat{\boldsymbol{\omega}}} = \boldsymbol{\Omega}\times\hat{\boldsymbol{\omega}}$, where $\boldsymbol{\Omega}$ measured the rate at which $\boldsymbol{\omega}$ rotates.

We then introduced the *Euler Angles.* We first compared them with a parallel development of the more familiar direction angles (which suffer from a somewhat extravagant redundancy). Then we showed how the Euler angles were so much better suited to the description of vector coordinate systems and were amenable to the direct application of rotation matrices discussed in the previous Part.

We then moved to the discussion of moving coordinate systems, first presenting how velocity and acceleration could be described relative either to *points*—we denote such relative velocity and acceleration $\boldsymbol{v}_{A/B}$ and $\boldsymbol{a}_{A/B}$—or *coordinate systems*—denoted $\boldsymbol{v}_{rel}$ and $\boldsymbol{a}_{rel}$—the latter, in particular, involving *rotation* once more. We obtained a fundamental result relating the *absolute* time derivative $\dot{\boldsymbol{V}}$ of a vector $\boldsymbol{V}$ to that *observed* in a coordinate system rotating with $\boldsymbol{\Omega}$, $\dot{\boldsymbol{V}}_{rel}$: $\dot{\boldsymbol{V}} = \dot{\boldsymbol{V}}_{rel} + \boldsymbol{\Omega}\times\boldsymbol{V}$, pointing out that that meant, in particular, that the angular acceleration of a coordinate system as observed *from* that coordinate system, $\boldsymbol{\alpha}_{rel}$, was precisely the same as its absolute angular acceleration $\boldsymbol{\alpha}$. Using this fact, we finally obtained a pair of equations for each velocity and acceleration, relating both the "point-to-point" "A/B" quantities and [coordinate system] "relative" "*rel*" quantities:

$$\begin{aligned}\boldsymbol{v}_A &= \boldsymbol{v}_B + \boldsymbol{v}_{A/B}\\ &= \boldsymbol{v}_B + \boldsymbol{\omega}\times\boldsymbol{r} + \boldsymbol{v}_{rel}\\ \boldsymbol{a}_A &= \boldsymbol{a}_B + \boldsymbol{a}_{A/B}\\ &= \boldsymbol{a}_B + \boldsymbol{\alpha}\times\boldsymbol{r} + \boldsymbol{\omega}\times(\boldsymbol{\omega}\times\boldsymbol{r}) + 2\boldsymbol{\omega}\times\boldsymbol{v}_{rel} + \boldsymbol{a}_{rel}\end{aligned}$$

—a set of equation which would become central in what followed. We concluded with a brief discussion of just why anyone really *needs* such relations in the first place!

The first reason they are important surfaced immediately: If these equations can be used to find, say, $\boldsymbol{v}_A$ and $\boldsymbol{a}_A$ knowing the balance of the quantities—particularly $\boldsymbol{\omega}$ and $\boldsymbol{\alpha}$—they can also be used to determine the angular quantities knowing something about the $\boldsymbol{v}$'s and $\boldsymbol{a}$'s. This "inverse" problem, involving somehow "undoing" the cross products through which the angular variables enter the equations, generally goes under the rubric of "machine kinematics." We developed this by first discussing in some detail the motion of a *single* body, showing how it might be possible to find the velocity and acceleration of points in a rigid body without resorting to finding the angular velocity and acceleration explicitly and ultimately developing the *non-slip condition* relating the motion of the point of contact, or of the center of the osculating sphere approximating the geometry of the body at an instant, to that of the surface on which the body moves. We then presented "kinematic conditions" resulting from the connections of the body with other moving points, particularly the *clevis* (whose rotational motion is completely determinate), and the *ball-and-socket* (only whose component *perpendicular* to the line between the points determining the motion can be determined). We concluded with a discussion of proper "machines"—systems of interconnected rigid bodies which can be treated by sequential application of the approach of a single body. This generally resulted in impressive, albeit normally *linear* systems of equations to solve simultaneously. As a particular application of this technique, we looked at an important application, the *universal joint*, obtaining kinematic constraints on the angular velocity and acceleration of the connecting "spider."

But just what is the place of rigid body kinematics in the scheme of things—why is it Good For You? The analysis of this section demonstrates that, once the rotational motion of the body and the translational motion of *any* point in the body are known, then the [translational] motion of *all* points are (see, *e.g.*, Example 4.4.1, where the center-of-mass acceleration of the piston was obtained from the known motion of end A and the angular acceleration obtained from the kinematic analysis). But this is precisely what *kinetics*, introducing *forces* for the first time, does: it gives the angular acceleration of the body and the acceleration of the mass center! This is the subject of the next Chapter.

Chapter 5

Kinetics

Introduction

After finally having completed the description of the general 3-D motion of rigid bodies in the previous Chapter, we now discuss the issue of how forces and moments actually affect this motion—that study which generally goes under the rubric of "kinetics." If Newton's Second Law is the determining equation relating the *physical* forces and *kinematical* acceleration for the motion of a single particle, it is not too surprising that it is for the motion of a rigid body, too. But a rigid body is *not* a "particle" at all; rather it is a *collection* of such particles—oftentimes a *continuous* collection of ["infinitesimal"] particles—moving together as a "rigid" unit, with a single angular velocity and/or angular acceleration describing its rotational motion, as described in Section 4.1.2. So how does Newton's Law actually come to bear on the motion of such a body?

The answer to this comes from the analysis of *systems* of particles—a "gas" of particles each moving *independently* of one another, and not a rigid body at all. Though this topic is invariably covered in a course of planar rigid body dynamics, it appears there as almost a curiosity—a brief interlude between the more extensive "particle dynamics" and the "good stuff" of [planar] rigid-body dynamics. But, in fact, it plays a central role in the development of rigid body dynamics, providing a touchstone between the dynamics of a single particle and the motion of rigid bodies. Given its importance—one which should be appreciated for any students having gotten to this level in the field of dynamics—this subject is presented again here, both as a review and to make the exposition self-contained. In particular, we shall see: how the fundamental "kinetic quantities" of linear momentum, angular momentum, and energy, first developed for [single] particles, are generalized into corresponding quantities for a system; that the "equations of motion" for the first two are precisely the same as for a single particle (with energy *virtually* the same); and that, in fact, the last two are themselves a consequence of the first.

That done, we are ready to consider, in particular, the motion of *rigid bodies.*

If the equations of motion are the foundation on which all of kinetics is based, it is reasonable that we first consider the equations of motion for such a body. Recall that it has two modes of motion: translation and rotation. It is not too surprising that the first is measured by linear momentum, and little more so that the second is by angular momentum. If translation becomes a direct carryover from particle motion, however, the second, requiring the preceding *system of particle* analysis, requires rather more care: in fact, the angular momentum of a rigid body turns out generally to be expressed as the product of a *tensor* with the angular velocity—one which reduces to a *scalar* only in the special case of planar motion. This characteristic of angular momentum is precisely the rationale for the development of linear algebra, of which tensors are the most general component, in the first Part of this treatise. But, as we showed there, it is a matter of practical necessity to *represent* both vectors and tensors in order to be able to work with them: vectors can be represented as column n-tuples of numbers, and [second-order] tensors as *matrices*; for three-dimensional motion, the inertia tensor will be [represented as] a 3×3 matrix, generally having nine elements. Just what these are and how to calculate them is clearly important and will be discussed. But we also noted in the first Part that the *form* of such representations changes with respect to the basis used *to* represent them; thus, in particular, the form of the inertia matrix/tensor will be changed for different choices of axes. In fact, that matrix happens to be *real* and *symmetric*, and thus there is always a basis in which it can be *diagonalized*; this basis, which can be obtained from whichever one used initially to represent the inertia tensor by a mere *rotation*, is called the *principal axes*, and is clearly of interest in the practical dealings with rigid body motion. In fact, such axes can be the starting point for the determination of the inertia *matrix*—the *representation* of the inertia tensor, in exactly the same way columns vectors *represent* vectors—in *arbitrary* axes, arrived at through translation and/or rotation from the principal ones.

With the inertia tensor firmly in hand, we will finally be in the position to obtain the *equations of motion for a rigid body.* These equations are obtained from the *system of particles* analysis and are, just as for particles, *instantaneous* second-order ordinary differential equations; in the case of three-dimensional motion, then, we have a *system* of such differential equations, generally *coupled* among the coordinates. In order to get some idea of how the body or bodies actually behave in time, however, it is necessary to *integrate* these equations over a finite time interval. As a special application of the equations of motion, we will consider the case of a symmetric body's motion—the *gyroscope*, where we will see how the forces determine its rotational motion, particularly "precession" and "nutation."

While the equations of motion are of intrinsic interest, they also have application in another important topic: *stability.* Let us say that we have somehow managed, by hook or by crook, actually to integrate the above equations. While one justifiably gets a certain sense of self-satisfaction from having done so, there is always a nagging questions: how *valid* is this solution? In theory, for example, a pencil should stand on its tip, since this solution "satisfies" the equilibrium equations; but it never happens, because there is always some form of disturbance to knock it over. The issue, then, is how such disturbances—"*perturbations*"—affect the solution, and the answer lies in finding how these perturbations themselves develop in time. But these means we need the equations *they* satisfy, and these are found using the equations of motion.

Is there a means of obtaining information about the system without resorting to explicit integration of these equations? The answer is *yes* (otherwise, the question would likely not have been posed in the first place): through the use of *conservation* of just the kinetic quantities developed, again, in *systems of particles* and specialized tot he case of rigid-body motion. These actually *do* involve integration, but it's done once and for all, *generally*, rather than for each specific problem. What information these integrated equations give us, and how we can determine when a certain quantity *is* conserved at all, are the subject of the last section.

5.1 Particles and Systems of Particles

We develop here the equations governing "systems of particles," on which those for rigid bodies are based. This is allegedly a review and will be done as briefly as possible. But, to establish a perspective on that fundamental law on which *all* of dynamics is based, Newton's Second for *particles*, we shall present them *de nuvo*.

5.1.1 Particle Kinetics

This section reviews in some detail the first appearance of linear momentum, angular momentum, and energy for particles. It is done to lead into the next section on systems of particles, but also to show the entire development of dynamics in a relatively short space—most importantly, to demonstrate how *all* of kinetics rests on Newton's Second Law! Certain parts, particularly the general discussion of Conservation of Energy or the "Conservation *Caveat*," might be new, yet are presumed in the sequel. But, with either an already firm grasp of Particle Dynamics, or a lack of interest in a presentation aimed as much at aesthetic appreciation as Dynamics, it could likely be skipped!

Linear Momentum and its Equation of Motion

One generally states Newton's Second Law as "$\boldsymbol{F}=m\boldsymbol{a}$," but this is not actually the form in which Newton himself presented it. In fact, he starts defining the "*quantity of motion*" to be the product of "mass" (which he merely described as the "quantity of matter") and "motion" (which corresponds to our *velocity*); thus, in effect, Newton's "quantity of motion is nothing more than our *linear momentum*:

$$\boldsymbol{G} \equiv m\boldsymbol{v}. \tag{5.1-1}$$

(It is important to note that the "$\boldsymbol{v}$" here is the *absolute* velocity, measured relative to a conceptually *fixed* system.) He then states that the "fluxion"—the term Newton uses to describe the time derivative[1]—of "motion" equals the net force[2] acting on a particle; in modern parlance,

$$\frac{d\boldsymbol{G}}{dt} = \sum_i \boldsymbol{F}_i. \tag{5.1-2}$$

This means that

$$\sum_i \boldsymbol{F}_i = \frac{d(m\boldsymbol{v})}{dt} = \dot{m}\boldsymbol{v} + m\dot{\boldsymbol{v}} = m\boldsymbol{a}$$

only when $\dot{m} \equiv 0$, and it *isn't* when one must invoke relativity. The distinction is unimportant for most Engineering work, but the startling thing is that Newton got it right, over two centuries before the Theory of Relativity was even a gleam in Einstein's eye! In any event, we see that $\frac{d\boldsymbol{G}}{dt}$ introduces $\boldsymbol{a}$ (and $\boldsymbol{v}$ in the relativistic case)—*kinematic* quantities.

Again, $\boldsymbol{a}$ is the *absolute* acceleration, measured relative to an "inertial"—*i.e.*, *non-accelerating*—system. In an *accelerating* system, we observe not $\boldsymbol{a}$ but what was called in the previous Chapter "$\boldsymbol{a}_{rel}$." If we measure the latter, we might ascribe to it a net force

$$m\boldsymbol{a}_{rel} = \sum_i \boldsymbol{F}_i - m(\boldsymbol{\alpha} \times \boldsymbol{r} + \boldsymbol{\omega} \times (\boldsymbol{\omega} \times \boldsymbol{r}) + 2\boldsymbol{\omega} \times \boldsymbol{v}_{rel}) \equiv \sum \boldsymbol{F}_{eff}$$

—an "effective" force summation, consisting of the *actual*, *physically-applied* forces, along with "fictitious forces" due not to any *actual* forces, but the non-inertial quality of the observing system. It is better not to invoke such "fictitious" forces at all, but rather to stick to inertial systems from the start.

It is common to describe the time derivatives of "kinetic quantities" such as $\boldsymbol{G}$ as "equations of motion." Thus Newton's Second Law is the "equation of motion for linear momentum." Though philosophers of science occasionally describe this as a "definition of mass," it is likely most fruitful to view this as a relation between [primitives of] the [*many*] *forces* and [resulting, *single*] *acceleration*—between the ***kinetics*** and the ***kinematics***, as it were, with the kinetics on one side of (5.1-2) and the kinematics on the other. This pattern will re-emerge continually.

It is important to note that Newton's Second Law is an *instantaneous* formulation: it holds at an *instant* in time. But it is, at least conceptually, possible

[1]Newton actually never *uses* calculus in the *Principia*: all his Propositions make use of purely *geometrical* arguments—some of them rather arcane! It is the author's understanding that Newton regarded the calculus as a secret he was unwilling to divulge, and he was clever enough to reformulate all his fundamental ideas in terms of geometry. When the contemporary Leibniz actually *did* publish his work on calculus, Newton accused him of stealing his proprietary technique, and the two were on the outs until Leibniz died in 1716 (Newton surviving until 1727). In fact, vestiges of both authors' work remain to this day: the symbols "$\dot{x}$," "$\ddot{x}$," (and x' and x'') *etc.* are Newton's; the notation "$\frac{dx}{dt}$" and "$\frac{d^2x}{dt^2}$" are Leibniz'.

[2]Note that *only* forces, not couple moments, act on a particle.

to integrate (5.1-2) through a simple separation of variables:

$$\int_{t_1}^{t_2} \sum_i \boldsymbol{F}_i\, dt = \boldsymbol{G}(t_2) - \boldsymbol{G}(t_1) \equiv \boldsymbol{G}_2 - \boldsymbol{G}_1. \tag{5.1-3}$$

The integral on the left is generally referred to as the *linear impulse.* We note that this *Principle of Impulse and Linear Momentum,* unlike the Newton Law from which it is derived, holds over an *interval.* But, as aesthetically pleasing as this might appear, it involves no more than a [first] time integration of the equation of motion for $\boldsymbol{G}$, including all the forces acting on the particle over the interval $[t_1, t_2]$; there's really nothing new here.

Nonetheless, we can get an interesting perspective from this equation: If the linear impulse—the integral of the net force over time—happens to *vanish,* then we see immediately that $\boldsymbol{G}_2 = \boldsymbol{G}_1$—the *Principle of Conservation of Linear Momentum.* To show the impulse vanishes would normally demand explicit integration of the net force, which is tantamount to partial solution of the differential equations afforded us by Newton's Second. But if the integrand $\sum_i \boldsymbol{F}_i \equiv \boldsymbol{0}$ *over the entire time interval,* we know *á priori* that the impulse vanishes so that linear momentum must be conserved. (Identical vanishing of the net force, of course, is precisely the condition for *Statics*; this is a course in *Dynamics,* so such an eventuality is regarded as "uninteresting.") Since $\boldsymbol{G}$ is a vector, this can also be applied component-wise in the event forces vanish in only certain directions (though note the "*caveat*" on page 262).

Angular Momentum and its Equation of Motion

Given a body of mass m moving with [*absolute*] velocity $\boldsymbol{v}$, we define the *angular momentum* relative to a *fixed*[3] point O:

$$\boldsymbol{H}_O \equiv \boldsymbol{r} \times \boldsymbol{G} = m\boldsymbol{r} \times \boldsymbol{v}, \tag{5.1-4}$$

where $\boldsymbol{r}$ is measured from O to the position of the particle. Though many students seem to identify this with motion in a *curved* path around O, in fact angular momentum is defined even for particles moving in a *straight* one!

We can obtain an equation of motion for this quantity too: by taking its time derivative to establish such an equation, we get that

$$\frac{d\boldsymbol{H}_O}{dt} \equiv \frac{d(\boldsymbol{r} \times \boldsymbol{G})}{dt} = \frac{d\boldsymbol{r}}{dt} \times \boldsymbol{G} + \boldsymbol{r} \times \frac{d\boldsymbol{G}}{dt}. \tag{5.1-5}$$

But $\dot{\boldsymbol{r}} \equiv \boldsymbol{v} \parallel \boldsymbol{G}$, so the first term vanishes identically; while, by Newton's Second,

$$\boldsymbol{r} \times \frac{d\boldsymbol{G}}{dt} = \boldsymbol{r} \times \sum_i \boldsymbol{F}_i = \sum_i \boldsymbol{r} \times \boldsymbol{F}_i$$

[3]While it is possible to define angular momentum relative to *moving* points, and many authors do so, one can get into arguments involving just what such a quantity *means.* Such a concept is really unnecessary to develop the equations of motion anyway, and the author prefers to avoid contentiousness!

because, since it is a *particle*, $\boldsymbol{r}$ is the same for each $\boldsymbol{F}$ in the summation. Thus, most simply,

$$\frac{d\boldsymbol{H}_O}{dt} = \sum_i \boldsymbol{M}_{O_i}, \tag{5.1-6}$$

where $\boldsymbol{M}_{O_i}$ is the moment about O due to $\boldsymbol{F}_i$—the "equation of motion" for angular momentum. Once again, $\dot{\boldsymbol{H}}_O$ will involve the kinematics:

$$\frac{d\boldsymbol{H}_O}{dt} \equiv \frac{d(m\boldsymbol{r}\times\boldsymbol{v})}{dt} = m(\boldsymbol{v}\times\boldsymbol{v}+\boldsymbol{r}\times\boldsymbol{a}) = m\boldsymbol{r}\times\boldsymbol{a},$$

and we have the *kinetic* moments on one side of (5.1-6) and the resulting *kinematics* on the other.[4]

As with linear momentum, the equation of motion for angular momentum is an *instantaneous* principle. As with linear momentum, however, we can separate variables and integrate (5.1-6) to obtain

$$\int_{t_1}^{t_2} \sum_i \boldsymbol{M}_{O_i}\,dt = \boldsymbol{H}_O(t_2) - \boldsymbol{H}_O(t_1) \equiv \boldsymbol{H}_{O_2} - \boldsymbol{H}_{O_1}, \tag{5.1-7}$$

to obtain the *Principle of Angular Impulse and Momentum.* The quantity on the left is, appropriately enough, the *angular impulse*, and this is once again simply a first time integration of all the moments acting on the particle over the *interval* $[t_1, t_2]$.

Just as with linear momentum, if the angular impulse vanishes, then we have the *Conservation of Angular Momentum*: $\boldsymbol{H}_{O_2} = \boldsymbol{H}_{O_1}$. Though this certainly holds if $\sum \boldsymbol{F} \equiv \boldsymbol{0}$ (the "uninteresting" Statics case), it also holds under less trivial conditions. In particular, if there is some [*fixed*] point O through which all the forces happen to pass, then $\sum \boldsymbol{M}_O \equiv \boldsymbol{0}$, the *angular* impulse vanishes, and angular momentum is conserved even if the linear impulse doesn't.

Energy

Work and Kinetic Energy. Before talking about energy, we must first talk about *work*, U, which is actually defined in terms of a *differential*: given a particle of mass m, to which there is applied a net force $\sum \boldsymbol{F}$, we define the differential of work

$$dU \equiv \sum_i \boldsymbol{F}_i \cdot d\boldsymbol{r}, \tag{5.1-8}$$

where $d\boldsymbol{r}$ is the *absolute*[5] [differential] displacement of the mass. This might be expected to be integrable, too, but to do so will involve the evaluation of a *line integral*:

$$\int dU \equiv U_{1\to 2} = \int_{\boldsymbol{r}_1}^{\boldsymbol{r}_2} \sum_i \boldsymbol{F}_i \cdot d\boldsymbol{r}.$$

---

[4]It is interesting to note that (5.1-5), and thus its consequent (5.1-6), hold even if m is not constant: any $\dot{m}$ term is accounted for in $\dot{\boldsymbol{G}}$ in the first equation. $\dot{\boldsymbol{H}}_O = m\boldsymbol{r}\times\boldsymbol{a}$, on the other hand, presumes m once again to be constant, just as does (5.1-2), that $\dot{\boldsymbol{G}} = m\boldsymbol{a}$, does.

[5]See footnote on page 257.

To do *this*, in turn, demands parameterization of the path over which the particle happens to move with respect to some scalar parameter. Fortunately, we have one immediately available: since $\boldsymbol{r} = \boldsymbol{r}(t)$, t itself will serve, and the differential $d\boldsymbol{r} = \frac{d\boldsymbol{r}}{dt}\, dt$. Thus

$$\begin{aligned} U_{1\to 2} &= \int_{\boldsymbol{r}_1}^{\boldsymbol{r}_2} \sum_i \boldsymbol{F}_i \cdot d\boldsymbol{r} \\ &= \int_{t_1}^{t_2} \sum_i \boldsymbol{F}_i \cdot \frac{d\boldsymbol{r}}{dt}\, dt \end{aligned}$$

—note that we have changed the variable of integration to t, where $\boldsymbol{r}(t_i) \equiv \boldsymbol{r}_i$—

$$\begin{aligned} &= \int_{t_1}^{t_2} m\frac{d^2\boldsymbol{r}}{dt^2} \cdot \frac{d\boldsymbol{r}}{dt}\, dt \\ &= \int_{t_1}^{t_2} m\frac{1}{2}\frac{d}{dt}\left(\frac{d\boldsymbol{r}}{dt} \cdot \frac{d\boldsymbol{r}}{dt}\right) dt \\ &= \int_{t_1}^{t_2} \frac{1}{2}\frac{d}{dt}\left(m\frac{d\boldsymbol{r}}{dt} \cdot \frac{d\boldsymbol{r}}{dt}\right) dt \qquad \textit{if m is constant!} \\ &= \int_{t_1}^{t_2} \frac{d}{dt}\left(\frac{1}{2}mv^2\right) dt = \frac{1}{2}mv^2(t_2) - \frac{1}{2}mv^2(t_1), \end{aligned}$$

or simply

$$U_{1\to 2} = \int_{\boldsymbol{r}_1}^{\boldsymbol{r}_2} \sum_i \boldsymbol{F}_i \cdot d\boldsymbol{r} = T_2 - T_1, \qquad \text{for } T_i \equiv \frac{1}{2}mv^2(t_i), \tag{5.1-9}$$

where T is called the *kinetic energy*, and this equation named the *Principle of Work and [Kinetic] Energy.* It is important to note that we explicitly presume m to be constant; this principle would not apply to relativistic cases without modification. Once again, we have the kinetics entering on one side of the equation through the work, and the resulting kinematics on the other, through the *speed* in T. Just as important to the present development, however, we note that it is *inherently* yet another integrated form—this time a *line* integral—of Newton's Second. *Everything* we have obtained to date *starts* with Newton's Second Law. That's all one needs to know about Dynamics!

Conservation of Energy. The nature of line integrals is a major inconvenience: in order to evaluate one in practice, it is necessary to express the $\boldsymbol{F}_i$ and $\boldsymbol{r}$ in some vector coordinate system, select a particular path between the points, parameterize that path with respect to t, and then, taking the indicated dot product with the parameterized form of $d\boldsymbol{r}_i$, evaluate the resulting scalar integral. The $\boldsymbol{F}_i$ generally depend on position, the $d\boldsymbol{r}_i$ definitely depend on the path chosen, so the work done is generally dependent on the path—different path, different work.

But it is at least possible to conceive of forces whose work might be *independent of the path*—that such work, if different for different endpoints, doesn't depend on how one *gets* from $\boldsymbol{r}_1$ to $\boldsymbol{r}_2$. This would be the case if the integrand of the line integral were an *exact differential*: $\sum \boldsymbol{F}_i \cdot d\boldsymbol{r}_i = -dV(\boldsymbol{r})$, say, for then

$$\begin{aligned} U_{1\to 2} &\equiv \int_{\boldsymbol{r}_1}^{a\boldsymbol{r}_2} \sum_i \boldsymbol{F}_i \cdot d\boldsymbol{r} \\ &= \int_{\boldsymbol{r}_1}^{\boldsymbol{r}_2} (-dV(\boldsymbol{r})) = -V(\boldsymbol{r}_2) + V(\boldsymbol{r}_1) = T_2 - T_1, \end{aligned} \tag{5.1-10}$$

or, for $V_i \equiv V(\boldsymbol{r}_i)$, simply

$$-V_2 + V_1 = T_2 - T_1.$$

But this says that $T_2 + V_2 = T_1 + V_1$, or defining the *[total] energy* $E \equiv T + V$, that

$$E_2 = E_1. \tag{5.1-11}$$

(The minus sign in writing the integrand of equation (5.1-10) as "$-dV$" is simply to allow writing $E \equiv T + V$ rather than $E \equiv T - V$!) This is referred to as the *conservation of energy*;[6] $V(\boldsymbol{r})$ is referred to in this context as the *potential energy*.

Although, as the above sketch of the derivation demonstrates, "conservation of energy" is merely a special case of the more general "work-energy" relation, there is a major advantage to use of the potential energy: rather than having to calculate the work through a line integral (with all its parameterizations, *etc.*), the work is determined by *evaluating a simple scalar function of position.* Little reflection (and little more experience with line integrals!) verifies this assessment.

Equation (5.1-11) hinges on work due to the force being path-independent; those forces enjoying this property warrant their own classification, *conservative forces.* Clearly it would be nice to recognize *a priori* whether a given force is conservative or not; in point of fact, there is a chain of equivalences associated with such forces:

$$\boldsymbol{F} \text{ conservative} \quad \textit{iff} \quad \boldsymbol{F} = -\nabla V \quad \textit{iff} \quad \nabla \times \boldsymbol{F} \equiv 0. \tag{5.1-12}$$

∇ is the "del" operator, which depends on the coordinates chosen to describe a given force, and $V = V(\boldsymbol{r})$ is precisely the potential energy. The second expression in the chain, $\boldsymbol{F} = -\nabla V$, essentially allows us to write $\boldsymbol{F} \cdot d\boldsymbol{r} = -\nabla V \cdot d\boldsymbol{r} \equiv -dV$ as in the above equations. The last expression is particularly useful, since it gives a criterion to establish whether or not a given force *is* conservative; it is a vector form of the test for exact differentials.

[6] Mechanicians tend to take a somewhat parochial viewpoint with regard to "energy," regarding only the *mechanical* energy; when this form is converted to others, particularly heat, they will accuse energy of having been lost. Thermodynamicists, on the other hand, adopt a much more catholic perspective!

There is an important observation, rarely made, regarding equation (5.1-12): The criterion that $\boldsymbol{F}$ be conservative, $\nabla \times \boldsymbol{F} \equiv 0$, takes derivative *only with respect to position*, not time. In particular, it is quite conceivable that a *time-dependent* force, $\boldsymbol{F}(\boldsymbol{r};t)$, might satisfy this equation; *e.g.*, $\boldsymbol{F}(\boldsymbol{r};t) = \boldsymbol{R}(\boldsymbol{r})T(t)$ in which $\nabla \times \boldsymbol{R} \equiv 0$ would, under this criterion, be regarded as "conservative," though with some abuse of terminology! Thus, by the equivalence of the conditions in this equation, there would be a V such that $\boldsymbol{F} = -\nabla V$, where now $V = V(\boldsymbol{r};t)$. We could *still* form an "energy" (more abuse!) $E \equiv T + V$, but now, through V, $E = E(\boldsymbol{r}, \boldsymbol{v}; t)$; in particular, E is no longer constant. But the time-dependent force can still be put into this "potential."

Although we have already obtained the "conservation of energy," both to reinforce the idea in the above paragraph, and to stress what we mean by "conservation," it is useful briefly to reexamine energy itself. Assuming there is a potential energy V—we will write it as $V(\boldsymbol{r};t)$ for the time being to cover the possibility of a time-dependent potential—the energy for a particle can be written in the form

$$E = T + V = \frac{1}{2}(m\boldsymbol{v}) \cdot \boldsymbol{v} + V(\boldsymbol{r};t).$$

To say that energy is "conserved," what we really mean is that it is *constant in time*; *i.e.*, $\dot{E} \equiv 0$. Differentiating the above expression with respect to time, we see that

$$\frac{dE}{dt} = m\boldsymbol{a} \cdot \boldsymbol{v} + \left(\nabla V \cdot \boldsymbol{v} + \frac{\partial V}{\partial t}\right) = (m\boldsymbol{a} + \nabla V) \cdot \boldsymbol{v} + \frac{\partial V}{\partial t}. \tag{5.1-13}$$

But the first term in the last expression vanishes by virtue of the fact that $\nabla V = -\sum \boldsymbol{F}$, and the second will vanish *as long as t does not appear in* V; on the other hand, if t *does* appear in V, energy is *not* conserved. Many *define* V to be independent of time, and they might be uncomfortable with the generalization presented here; but it is useful, and will be implemented in several examples to follow.

The above conservation of energy, and the advantages of scalar functions *vs.* line integrals, relies on *all* forces being conservative. In practice, one is rarely so lucky. But it is possible to generalize the technique used to obtain energy conservation to allow scalar functions to be used for conservative forces, while admitting the flexibility to use line integrals for the balance of non-conservative forces: if we divide the forces in equation (5.1-10) into the conservative $\boldsymbol{F}_C$ and non-conservative $\boldsymbol{F}_N$,

$$\begin{aligned} U_{1\to 2} &\equiv \int_{\boldsymbol{r}_1}^{\boldsymbol{r}_2} \sum_i \boldsymbol{F}_{i_N} \cdot d\boldsymbol{r}_i + \int_{\boldsymbol{r}_1}^{\boldsymbol{r}_2} \sum_i \boldsymbol{F}_{i_C} \cdot d\boldsymbol{r}_i \\ &= \int_{\boldsymbol{r}_1}^{\boldsymbol{r}_2} \sum_i \boldsymbol{F}_{i_N} \cdot d\boldsymbol{r}_i + (-V(\boldsymbol{r}_2) + V(\boldsymbol{r}_1)) \\ &\equiv (U_N)_{1\to 2} + (-V(\boldsymbol{r}_2) + V(\boldsymbol{r}_1)) = T_2 - T_1, \end{aligned}$$

—where we have used (5.1-10) to express the integral of the $\boldsymbol{F}_{i_C}$ in the second equation—and then bring the V's over to the right-hand side in the last equation to form $E_i \equiv T_i + V_i$, we get

$$(U_N)_{1\to 2} = E_2 - E_1, \tag{5.1-14}$$

—where E includes *conservative* forces (the only ones for which there is a V in the first place), and $(U_N)_{1\to 2}$ involves *all non*-conservative forces and any *conservative* forces that one, by choice or ignorance, doesn't put into a potential energy. This hybrid is the most general form of energy: if all forces are conservative (and one chooses to put them into potential energy), then the left-hand side vanishes and we have the conservation of energy (5.1-11); if $V \equiv 0$, *all* forces' work is included in the work term, and we have the work-energy principle (5.1-10), applicable to all forces, conservative or not.

Thus we have developed the three "kinetic quantities" of $\boldsymbol{G}$, $\boldsymbol{H}_O$, and T (or E), the *instantaneous* "equations of motion" for the first two, *integrated* forms for all three, holding over an *interval* of time (for the impulses) or space (for energy), and criteria to know when $\boldsymbol{G}$, $\boldsymbol{H}_O$, or E might be conserved.

A *Caveat* regarding Conservation

Conservation is an important tool in the repertoire of techniques for analyzing dynamics. That energy or a momentum—or even a *component* of momentum—is conserved amounts to having effected a partial integration of the system *without actually having to carry out the integration*! With enough of these "integrals," the system can be solved completely, albeit only implicitly.

There is, however, a subtle point regarding the application of conservation, particularly of [components of] momentum: In one problem, it might be the case that the sum of the forces, chosen to be represented in Cartesian coordinates, vanishes in the vertical $\hat{\boldsymbol{K}}$-direction because, say, a normal force exactly balances the weight force. One interprets equation (5.1-3) to imply that $\boldsymbol{G}$ is conserved in that direction; *i.e.*, that $(mv_Z)_2 = (mv_Z)_1$—and gets the problem correct. In another problem, represented in polar coordinates, it might be the case that the sum of the forces vanishes in the $\hat{\boldsymbol{\theta}}$-direction; one concludes that G_θ is conserved—$(mv_\theta)_2 = (mv_\theta)_1$—and get the problem *wrong*. What's going on here, and "why does it work in one case and not the other?"

Let's analyze exactly what we're saying here: $G_Z = \boldsymbol{G} \cdot \hat{\boldsymbol{K}}$, so to argue that "the vanishing of forces in the $\hat{\boldsymbol{K}}$-direction means this component of linear momentum is conserved" is actually to say that

$$\sum F_Z = \sum \boldsymbol{F} \cdot \hat{\boldsymbol{K}} = \frac{dG_Z}{dt} = \frac{d(\boldsymbol{G} \cdot \hat{\boldsymbol{K}})}{dt}.$$

Now, in point of fact,

$$\frac{d(\boldsymbol{G} \cdot \hat{\boldsymbol{K}})}{dt} = \frac{d\boldsymbol{G}}{dt} \cdot \hat{\boldsymbol{K}} + \boldsymbol{G} \cdot \frac{d\hat{\boldsymbol{K}}}{dt} = \frac{d\boldsymbol{G}}{dt} \cdot \hat{\boldsymbol{K}}$$

because $\hat{\boldsymbol{K}}$ *is fixed in direction.* And, indeed, simply dotting Newton's Second Law by $\hat{\boldsymbol{K}}$ shows that this is true:

$$\sum \boldsymbol{F} \cdot \hat{\boldsymbol{K}} = \frac{d\boldsymbol{G}}{dt} \cdot \hat{\boldsymbol{K}},$$

as alleged.

On the other hand, attempting to apply the same reasoning to the polar example is to say that

$$\sum F_\theta = \sum \boldsymbol{F} \cdot \hat{\boldsymbol{\theta}} = \frac{dG_\theta}{dt} = \frac{d(\boldsymbol{G} \cdot \hat{\boldsymbol{\theta}})}{dt},$$

whereas

$$\frac{d(\boldsymbol{G} \cdot \hat{\boldsymbol{\theta}})}{dt} = \frac{d\boldsymbol{G}}{dt} \cdot \hat{\boldsymbol{\theta}} + \boldsymbol{G} \cdot \frac{d\hat{\boldsymbol{\theta}}}{dt};$$

so, dotting Newton's Second with $\hat{\boldsymbol{\theta}}$, we get that

$$\sum \boldsymbol{F} \cdot \hat{\boldsymbol{\theta}} = \frac{d\boldsymbol{G}}{dt} \cdot \hat{\boldsymbol{\theta}} = \frac{d(\boldsymbol{G} \cdot \hat{\boldsymbol{\theta}})}{dt} - \boldsymbol{G} \cdot \frac{d\hat{\boldsymbol{\theta}}}{dt} \neq \frac{d(\boldsymbol{G} \cdot \hat{\boldsymbol{\theta}})}{dt} \qquad (5.1\text{-}15)$$

because $\dot{\hat{\boldsymbol{\theta}}} \neq \mathbf{0}$. In fact, what we *do* know is that

$$\sum F_\theta = ma_\theta = m(2\dot{r}\dot{\theta} + r\ddot{\theta}) = \frac{1}{r}\frac{d}{dt}(mr^2\dot{\theta}) = \frac{1}{r}\frac{dH_{Oz}}{dt};$$

i.e., the sum of forces vanishing in the $\hat{\boldsymbol{\theta}}$-direction doesn't mean that G_θ is conserved, but rather that H_{Oz} is! This example demonstrates that such "peeling off" of components of a momentum can be done only in a *fixed* direction [or, more generally, one whose time derivative is perpendicular to the kinetic quantity whose conservation is being alleged, as shown by (5.1-15)].

5.1.2 Particle System Kinetics

Having introduced the momenta and energy of a *single particle*, we now formulate exactly the same three quantities for a "*system* of particles." What is such a system? Fundamentally, it is anything we *want* it to be: we *identify* an arbitrary group of independently-moving particles—a "gas," as it were—as being a "system" simply by conceptually enveloping them and maintaining the identity of this particular set of particles as being part of the system, regardless of how they move.

Having done that, however, has important ramifications on the kinetics of the system: it simultaneously divides the forces acting on the various particles into two categories, those arising from outside the system—the *external* forces, and those from within—the *internal* ones.

The system and resultant force categorization is a fundamental concept transcending just Dynamics: it is exactly what the free body diagram in Statics effects, and is essentially the "control volume" in fluid mechanics. We shall return to the former construct shortly.

Having defined the system, we will now return immediately to the fundamental kinetic quantities defined above for a single particle, defining a *single* linear momentum, and *single* angular momentum, and a *single* energy—*kinetic* energy, really—characterizing the system as a whole. Clearly some information regarding the individual motion of the particles will be lost, but that is of no consequence. More remarkably, we will obtain equations of motion and integrated forms almost exactly the same for the entire system as we did above for a single particle; it is the latter which we need ultimately to study the kinetics of rigid bodies. Recall that the above development emphasized the *absolute* nature of the $\boldsymbol{v}$'s and $\boldsymbol{a}$'s that entered it; thus we first consider such an *inertial* reference.

Kinetics relative to a *Fixed* System

Shown here is just a system of particles as modelled above, enclosed by an envelope—a "baggie" as it were—identifying just which particles constitute the system. We see a typical particle, m_i, whose position relative to the *fixed* point O is given by $\boldsymbol{r}_i$. That particle is moving with [*absolute*] velocity $\boldsymbol{v}_i$ and is acted upon by two forces: $\boldsymbol{F}_i$, the *net* external force arising from outside the system, and the internal force $\boldsymbol{f}_{ij}$, resulting from the effect on m_i due to m_j, another particle in the system. By Newton's Third Law $\boldsymbol{f}_{ji} = -\boldsymbol{f}_{ij}$; further it is obviously the case that $\boldsymbol{f}_{ii} \equiv \boldsymbol{0}$. We also explicitly assume of these internal forces that they are directed along the line joining the two particles. Note that this last assumption is independent of Newton's Third, which says merely that the two forces are *parallel*, not that they are along the defined line. Gravitational forces satisfy this stricture; electromagnetic ones do not.

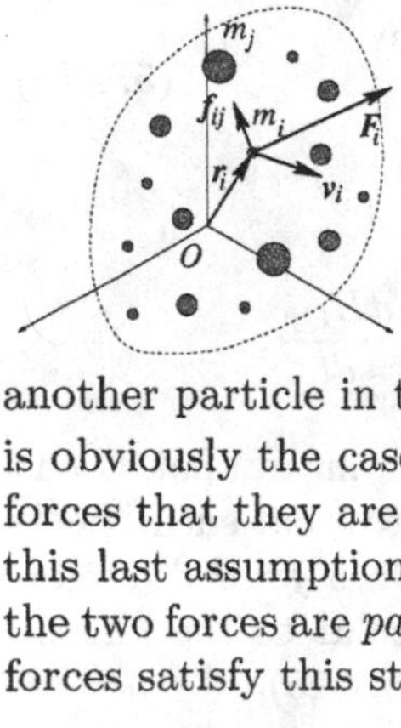

The Kinetic Quantities. We start by defining linear and angular momentum, and *kinetic* energy, for this system as a whole. The most naive such definition would be that these quantities are nothing more than the *sum* of the corresponding quantities for all the particles, and this is precisely the tack followed:

$$\boldsymbol{G} \equiv \sum_i \boldsymbol{G}_i = \sum_i m_i \boldsymbol{v}_i \tag{5.1-16a}$$

$$\boldsymbol{H}_O \equiv \sum_i \boldsymbol{H}_{O_i} = \sum_i m_i \boldsymbol{r}_i \times \boldsymbol{v}_i \tag{5.1-16b}$$

$$T \equiv \sum_i T_i = \sum_i \frac{1}{2} m_i v_i^2 \tag{5.1-16c}$$

Equations of Motion and Energy. We can define anything we want, but do these definitions make any sense? In point of fact, they do, as we will see when we develop the equations of motion and work-energy relation:

We start with the **linear momentum**. Its equation of motion is simply the time derivative of the above-defined $\boldsymbol{G}$:

$$\frac{d\boldsymbol{G}}{dt} = \frac{d}{dt}\sum_i \boldsymbol{G}_i = \sum_i \frac{d\boldsymbol{G}_i}{dt}.$$

But the equation of motion for the ith particle is just

$$\frac{d\boldsymbol{G}_i}{dt} = \boldsymbol{F}_i + \sum_j \boldsymbol{f}_{ij},$$

so

$$\frac{d\boldsymbol{G}}{dt} = \sum_i \left(\boldsymbol{F}_i + \sum_j \boldsymbol{f}_{ij}\right) = \sum_i \boldsymbol{F}_i + \sum_i \sum_j \boldsymbol{f}_{ij}.$$

But we claim that *the last, double, summation vanishes*: this can be seen most simply by arraying each $\sum_j \boldsymbol{f}_{ij}$ for fixed i in separate rows:

$$\begin{array}{rcccccccl} \sum_i \sum_j \boldsymbol{f}_{ij} & = & \boldsymbol{f}_{11} & + & \boldsymbol{f}_{12} & + & \boldsymbol{f}_{13} & + & \ldots \quad \leftarrow i=1 \\ & & \boldsymbol{f}_{21} & + & \boldsymbol{f}_{22} & + & \boldsymbol{f}_{23} & + & \ldots \quad \leftarrow i=2 \\ & & \boldsymbol{f}_{31} & + & \boldsymbol{f}_{32} & + & \boldsymbol{f}_{33} & + & \ldots \quad \leftarrow i=3 \\ & & \vdots & & \vdots & & \vdots & & \ldots \quad . \end{array}$$

All terms on the diagonal are $\boldsymbol{f}_{ii} \equiv \boldsymbol{0}$, while off the diagonal, each $\boldsymbol{f}_{ij}$ is matched by a $\boldsymbol{f}_{ji} = -\boldsymbol{f}_{ij}$ by Newton's Third Law. Thus, in the end, we have that

$$\begin{aligned} \frac{d\boldsymbol{G}}{dt} &= \sum_i \boldsymbol{F}_i \\ &= \sum_i m_i \boldsymbol{a}_i \end{aligned} \tag{5.1-17a}$$

(where we have explicitly differentiated (5.1-16a) to get the second line)—*i.e.*, the time derivative of the *system* linear momentum is just the sum of the *external* forces. Aside from the restriction to external forces, the first equation is precisely (5.1-2), that for a particle!

This done, we can obtain precisely the same linear impulse-momentum equation (5.1-3) derived before for a single particle, for the *system* $\boldsymbol{G}$, but now including only the *external* forces in the integrand. In particular, we see that linear momentum is conserved as long as the sum of the *external* forces vanishes.

The equation of motion for **angular momentum** is developed similarly: Differentiating the definition (5.1-16b), we get that

$$\frac{d\boldsymbol{H}_O}{dt} = \sum_i \frac{d\boldsymbol{H}_{O_i}}{dt} = \sum_i \boldsymbol{r}_i \times \boldsymbol{F}_i + \sum_i \sum_j \boldsymbol{r}_i \times \boldsymbol{f}_{ij}.$$

From our experience with linear momentum, we expect the double summation to vanish again, and it does. But the reasoning is a little more subtle this time: From the arraying of this sum we see that the diagonal terms will be of the form $\boldsymbol{r}_i \times \boldsymbol{f}_{ii} \equiv \boldsymbol{0}$. But now the off-diagonal terms pair up

$$\boldsymbol{r}_i \times \boldsymbol{f}_{ij} + \boldsymbol{r}_j \times \boldsymbol{f}_{ji} = (\boldsymbol{r}_i - \boldsymbol{r}_j) \times \boldsymbol{f}_{ij}$$

by Newton's Third, which is not sufficient by itself to make these terms vanish. But because we have *assumed the* $\boldsymbol{f}_{ij}$ *to act along the line joining the particles*, the two parallel vectors in the cross product literally "cross out." Thus we have that

$$\begin{aligned}\frac{d\boldsymbol{H}_O}{dt} &= \sum_i \boldsymbol{r}_i \times \boldsymbol{F}_i = \sum_i \boldsymbol{M}_{O_i} \\ &= \sum_i m_i \boldsymbol{r}_i \times \boldsymbol{a}_i\end{aligned} \tag{5.1-17b}$$

(differentiating (5.1-16b) in the second line)—that the time derivative of the angular momentum is the sum of the moments, due to *external* forces once again. And, just as for the system linear momentum above, we can obtain an angular impulse-momentum equation and conservation principle, both examining only the moments due to *external* forces.

We remarked on how free-body diagrams are really just an implementation of the concept of systems. This fact is demonstrated by the above equations of motion: since only the *external* forces and moments change the linear and angular momentum of the system, it is precisely external agents which enter the equations; and it is also those included *on the free-body diagram.* In fact, in order to find any *internal* forces or moments, it is necessary to break up the free body in order to *externalize* these forces.

Finally we come to **energy**. The most general formulation of energy is the general work-energy relation (5.1-9), precisely the reason we defined an expression for the kinetic energy of the system, rather than a "total" energy. We get that

$$T_2 - T_1 \equiv \sum_i (T_{i_2} - T_{i_1}) = \sum_i \int_{\boldsymbol{r}_{i_1}}^{\boldsymbol{r}_{i_2}} (\boldsymbol{F}_i + \sum_j \boldsymbol{f}_{ij}) \cdot d\boldsymbol{r}_i. \tag{5.1-17c}$$

We expect initially that the last terms' double summation on the internal forces will vanish. But it *doesn't*: The terms for $i = j$ certainly do because $\boldsymbol{f}_{ii} \equiv \boldsymbol{0}$; but the pairing of terms for $i \neq j$, $\boldsymbol{f}_{ij} \cdot d\boldsymbol{r}_i + \boldsymbol{f}_{ji} \cdot d\boldsymbol{r}_j$, don't because, although $\boldsymbol{f}_{ji} = -\boldsymbol{f}_{ij}$, $d\boldsymbol{r}_i$ needn't be the same as $d\boldsymbol{r}_j$ due to the independent motion of these particles. Thus, unlike linear and angular momentum, *both external* and *internal forces do work on a system.* Thus in the two-body problem of Newton, in which the *only* force in the problem is the internal gravitational one between the two bodies (so linear and angular momentum of this system of two bodies are conserved), *the gravity does work* on the bodies, changing the kinetic energy of that system.

As for a single particle, the forces doing work, be they external or internal, may themselves be conservative. It is then possible to define a potential energy for such forces and, if *all* the forces, external *and* internal, doing work are conservative, the total energy of the system is conserved.

Thus, aside from the delicate negotiation between external and internal forces when it comes to energy, the fundamental kinetic principles for systems are little different from those for single particles!

Kinetics relative to the Center of Mass

While we should perhaps be self-satisfied and leave well enough alone, we are

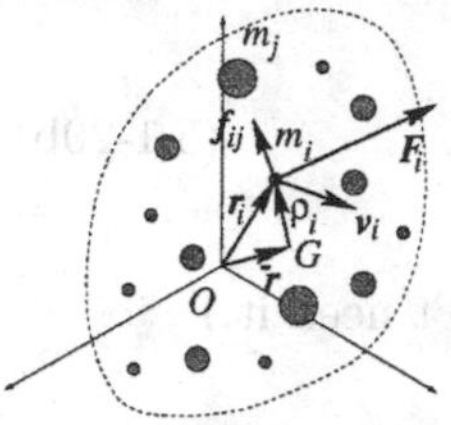

now going to do something potentially very perilous: reformulate all the above results in terms of the center of mass of the system. The reason this is risky is because the center of mass is generally an *accelerating* point, and thus a patently *non*-inertial system, not the stable *fixed, inertial* point relative to which all the above relations were defined. But it will turn out that it is these latter relations which will be the primary governing equations for rigid bodies.

We diagram here once again the system motivating the previous section, but now indicate the position, G, of its center of mass. Note that the position $\overline{\boldsymbol{r}}$ of this point, measured relative to the inertial system, satisfies the equation

$$m\overline{\boldsymbol{r}} = \sum_i m_i \boldsymbol{r}_i, \tag{5.1-18a}$$

$$m \equiv \sum_i m_i$$

—an equation holding generally and thus able to be differentiated[7] to yield an expression for $\overline{\boldsymbol{v}}$ and $\overline{\boldsymbol{a}}$, the velocity and acceleration of the center of mass:

$$m\overline{\boldsymbol{v}} = \sum_i m_i \boldsymbol{v}_i, \tag{5.1-18b}$$

$$m\overline{\boldsymbol{a}} = \sum_i m_i \boldsymbol{a}_i. \tag{5.1-18c}$$

We certainly can measure positions relative to this new reference point, even though it is moving; the position of m_i relative to the center of mass will be uniformly denoted as $\boldsymbol{\rho}_i$. These are related to the positions, $\boldsymbol{r}_i$, relative to O through the equation

$$\boldsymbol{r}_i = \overline{\boldsymbol{r}} + \boldsymbol{\rho}_i, \tag{5.1-19a}$$

which can also be differentiated to yield

$$\boldsymbol{v}_i = \overline{\boldsymbol{v}} + \dot{\boldsymbol{\rho}}_i. \tag{5.1-19b}$$

[7]Note that we once again assume the m_i, so m, to be *constant.*

(Note that $\dot{\boldsymbol{\rho}}_i = \boldsymbol{v}_i - \overline{\boldsymbol{v}}$ can be written "$\boldsymbol{v}_{i/G}$," the velocity of m_i relative to [the *point*] G.) Substituting this into equation (5.1-18a), we get that

$$m\overline{\boldsymbol{r}} = \sum_i m_i(\overline{\boldsymbol{r}} + \boldsymbol{\rho}_i) \equiv m\overline{\boldsymbol{r}} + \sum_i m_i\boldsymbol{\rho}_i,$$

from which, immediately,

$$\sum_i m_i\boldsymbol{\rho}_i \equiv \mathbf{0} \tag{5.1-20a}$$

—a relation making perfectly good sense, saying merely that the center of mass, measured in center-of-mass coordinates, is just at the origin! This, too, is general and can be differentiated to yield

$$\sum_i m_i\dot{\boldsymbol{\rho}}_i \equiv \mathbf{0}. \tag{5.1-20b}$$

(A similar relation could be obtained for the $\ddot{\boldsymbol{\rho}}_i$, but we don't need it.)

The Kinetic Quantities. We now express the *same* $\boldsymbol{G}$, $\boldsymbol{H}_O$ and T defined by relations (5.1-16) in terms of the new center-of-mass variables:

The system **linear momentum**

$$\boldsymbol{G} \equiv \sum_i m_i\boldsymbol{v}_i = m\overline{\boldsymbol{v}}, \tag{5.1-21a}$$

by (5.1-18b). This says that the linear momentum is the same as that of a *single* particle with mass m (the total mass) moving with the center-of-mass velocity—the "linear momentum of the center-of-mass particle," as it were.

The **angular momentum** of the system

$$\begin{aligned}\boldsymbol{H}_O &\equiv \sum_i m_i\boldsymbol{r}_i \times \boldsymbol{v}_i = \sum_i m_i(\overline{\boldsymbol{r}} + \boldsymbol{\rho}_i) \times \boldsymbol{v}_i \qquad \text{(by (5.1-19a))}\\ &= \overline{\boldsymbol{r}} \times \sum_i m_i\boldsymbol{v}_i + \sum_i m_i\boldsymbol{\rho}_i \times \boldsymbol{v}_i \equiv \overline{\boldsymbol{r}} \times m\overline{\boldsymbol{v}} + \overline{\boldsymbol{H}},\end{aligned}$$

where the first term in the final expression is just the "*angular momentum of the center of mass particle*," and the second, $\overline{\boldsymbol{H}} \equiv \sum_i m_i\boldsymbol{\rho}_i \times \boldsymbol{v}_i$, can be *interpreted* as the "angular momentum *about* the center of mass," even though we are mixing the $\boldsymbol{\rho}_i$ with the *absolute* $\boldsymbol{v}_i$. But it turns out that this apparently bogus mismatch can be corrected:

$$\begin{aligned}\overline{\boldsymbol{H}} &\equiv \sum_i m_i\boldsymbol{\rho}_i \times \boldsymbol{v}_i = \sum_i m_i\boldsymbol{\rho}_i \times (\overline{\boldsymbol{v}} + \dot{\boldsymbol{\rho}}_i) \qquad \text{(by (5.1-19b))}\\ &= \sum_i m_i\boldsymbol{\rho}_i \times \overline{\boldsymbol{v}} + \sum_i m_i\boldsymbol{\rho}_i \times \dot{\boldsymbol{\rho}}_i = \sum_i m_i\boldsymbol{\rho}_i \times \dot{\boldsymbol{\rho}}_i\end{aligned}$$

by (5.1-20a). So $\overline{\boldsymbol{H}}$ has exactly the *same form* as any angular momentum; it merely uses the $\boldsymbol{\rho}$'s and $\dot{\boldsymbol{\rho}}$'s in place of the $\boldsymbol{r}$'s and $\boldsymbol{v}$'s. Thus we have, finally, that

$$\boldsymbol{H}_O = \overline{\boldsymbol{r}} \times m\overline{\boldsymbol{v}} + \overline{\boldsymbol{H}} \tag{5.1-21b}$$

$$\overline{\boldsymbol{H}} \equiv \sum_i m_i \boldsymbol{\rho}_i \times \dot{\boldsymbol{\rho}}_i. \tag{5.1-21c}$$

Unlike the *linear* momentum expressed in terms of center-of-mass quantities, the *angular* momentum has *two* parts: one, like the linear momentum, consisting of angular momentum of the center-of-mass particle, and the other, new, one, consisting of motion *about* the center of mass. [Actually, we *could* effect a similar division for *linear* momentum—writing $\boldsymbol{G} = m\overline{\boldsymbol{v}} + \overline{\boldsymbol{G}}$—by *defining* a "linear momentum about the center of mass" $\overline{\boldsymbol{G}} \equiv \sum_i m_i \dot{\boldsymbol{\rho}}_i$; it's just that the latter vanishes identically, by (5.1-20b).]

Finally we come to **kinetic energy**: Again by the definition (5.1-16c),

$$\begin{aligned} T &\equiv \sum_i \frac{1}{2} m_i v_i^2 = \sum_i \frac{1}{2} m_i (\overline{\boldsymbol{v}} + \dot{\boldsymbol{\rho}}_i) \cdot (\overline{\boldsymbol{v}} + \dot{\boldsymbol{\rho}}_i) \\ &= \frac{1}{2} \sum_i m_i \overline{\boldsymbol{v}} \cdot \overline{\boldsymbol{v}} + \overline{\boldsymbol{v}} \cdot \sum_i m_i \dot{\boldsymbol{\rho}}_i + \sum_i \frac{1}{2} m_i \dot{\boldsymbol{\rho}}_i \cdot \dot{\boldsymbol{\rho}}_i \\ &= \frac{1}{2} m\overline{v}^2 + 0 + \sum_i \frac{1}{2} m_i \dot{\rho}_i^2, \end{aligned}$$

the second term vanishing, again by virtue of (5.1-20b). But, in the spirit of the $\overline{\boldsymbol{H}}$ we defined above, we can call the last term the "kinetic energy about the center of mass," and write this equation as

$$T = \frac{1}{2} m\overline{v}^2 + \overline{T} \tag{5.1-21d}$$

$$\overline{T} \equiv \sum_i \frac{1}{2} m_i \dot{\rho}_i^2. \tag{5.1-21e}$$

Like the angular momentum (but not the *linear*), kinetic energy consists of two parts: the "center of mass particle" part, and one "*about* the center of mass."

Thus we have, once again, obtained expressions for the linear and angular momentum, and energy, but now in terms of center-of-mass quantities; their reasonably simple form is a result of the special character of the $\boldsymbol{\rho}_i$ and the $\dot{\boldsymbol{\rho}}_i$ in equations (5.1-20). It must be emphasized that these are the *same* quantities defined in equations (5.1-16), merely written differently. Thus we can use the same kinetic equations (5.1-17) to describe them, precisely what we do now:

Equations of Motion and Energy. We start with the **linear momentum**. Simply substituting in (5.1-21a) into (5.1-17a), we get immediately that

$$\sum_i \boldsymbol{F}_i = \frac{d(m\overline{\boldsymbol{v}})}{dt} = m\overline{\boldsymbol{a}} \tag{5.1-22a}$$

—an equation saying that the "center-of-mass" particle moves as if all the forces were acting there. This can actually be observed on the Fourth of July, when one can visually follow the center of mass of the individual particles after a skyrocket has exploded, seeing that it merely continues the arc of the rocket on its ascent.

Once again, we can, just as for a single particle obeying an equation of motion of the same form, obtain a linear impulse-momentum equation, leading to similar considerations regarding the conservation of this quantity. But we consider only the impulse due to *external* forces.

Now consider **angular momentum**. Recall that we have seen that, when expressed in terms of center-of-mass quantities, this kinetic quantity has *two* components, one due to the center-of-mass particle, the other due to motion of the individual particles *about* the center of mass. When we examine the equations of motion for angular momentum, (5.1-17b), a rather remarkable thing happens: substituting (5.1-21b) into (5.1-17b), we get

$$\begin{aligned}\sum_i \boldsymbol{M}_{O_i} &= \sum_i \boldsymbol{r}_i \times \boldsymbol{F}_i = \sum_i (\overline{\boldsymbol{r}} + \boldsymbol{\rho}_i) \times \boldsymbol{F}_i \qquad \text{(by (5.1-19a))}\\ &= \overline{\boldsymbol{r}} \times \sum_i \boldsymbol{F}_i + \sum_i \boldsymbol{\rho}_i \times \boldsymbol{F}_i\\ &= \frac{d\boldsymbol{H}_O}{dt} = \frac{d(\overline{\boldsymbol{r}} \times m\overline{\boldsymbol{v}})}{dt} + \frac{d\overline{\boldsymbol{H}}}{dt}\end{aligned}$$

But the first term in the second line,

$$\begin{aligned}\overline{\boldsymbol{r}} \times \sum_i \boldsymbol{F}_i &= \overline{\boldsymbol{r}} \times m\overline{\boldsymbol{a}} \qquad \text{(by (5.1-22a))}\\ &= \frac{d(\overline{\boldsymbol{r}} \times m\overline{\boldsymbol{v}})}{dt},\end{aligned} \qquad (5.1\text{-}22b)$$

so all that's left of the above equation

$$\sum_i \boldsymbol{\rho}_i \times \boldsymbol{F}_i \equiv \sum_i \overline{\boldsymbol{M}}_i = \frac{d\overline{\boldsymbol{H}}}{dt}. \qquad (5.1\text{-}22c)$$

Thus, in the same way that the angular momentum itself separates into two parts, *the equation of motion separates into two corresponding parts*, one giving the center-of-mass particle's motion, the other motion *about* the center of mass! Now we can formulate *two* angular impulse-momentum principles, one for the center-of-mass "particle," the other for angular momentum *about* the center of mass; these will be generally independent of one another. And the resulting conservation principles will hold, again independently.

Finally we examine **energy**. Again substituting (5.1-21d-e) into (5.1-17c), we get

$$\sum_i \int_{\boldsymbol{r}_{i_1}}^{\boldsymbol{r}_{i_2}} (\boldsymbol{F}_i + \sum_j \boldsymbol{f}_{ij}) \cdot d\boldsymbol{r}_i = \sum_i \int_{\boldsymbol{r}_{i_1}}^{\boldsymbol{r}_{i_2}} (\boldsymbol{F}_i + \sum_j \boldsymbol{f}_{ij}) \cdot (d\overline{\boldsymbol{r}} + d\boldsymbol{\rho}_i)$$

(taking the differential of (5.1-19a))

$$\begin{aligned}
&= \int_{\overline{r}_1}^{\overline{r}_2} \sum_i (\boldsymbol{F}_i + \sum_j \boldsymbol{f}_{ij}) \cdot d\overline{\boldsymbol{r}} \\
&\quad + \sum_i \int_{\boldsymbol{\rho}_{i_1}}^{\boldsymbol{\rho}_{i_2}} (\boldsymbol{F}_i + \sum_j \boldsymbol{f}_{ij}) \cdot d\boldsymbol{\rho}_i \\
&= \left(\frac{1}{2} m\overline{v}_2^2 + \overline{T}_2\right) - \left(\frac{1}{2} m\overline{v}_1^2 + \overline{T}_1\right).
\end{aligned}$$

But

$$\sum_i (\boldsymbol{F}_i + \sum_j \boldsymbol{f}_{ij}) \cdot d\overline{\boldsymbol{r}} = \sum_i \boldsymbol{F}_i \cdot d\overline{\boldsymbol{r}} + \sum_i \sum_j \boldsymbol{f}_{ij} \cdot d\overline{\boldsymbol{r}} = \sum_i \boldsymbol{F}_i \cdot d\overline{\boldsymbol{r}},$$

since we have already shown that the double summation $\sum_i \sum_j \boldsymbol{f}_{ij}$ vanishes due to Newton's Third Law—this is the argument leading up to (5.1-17a). Then, in exactly the same way we derived the work-energy relation (5.1-9),

$$\begin{aligned}
\int_{\overline{r}_1}^{\overline{r}_2} \sum_i \boldsymbol{F}_i \cdot d\overline{\boldsymbol{r}} &= \int_{t_1}^{t_2} m \frac{d^2\overline{\boldsymbol{r}}}{dt^2} \cdot \frac{d\overline{\boldsymbol{r}}}{dt}\, dt = \int_{t_1}^{t_2} \frac{1}{2} \frac{d}{dt} \left(m \frac{d\overline{\boldsymbol{r}}}{dt} \cdot \frac{d\overline{\boldsymbol{r}}}{dt} \right) dt \\
&= \frac{1}{2} m\overline{v}_2^2 - \frac{1}{2} m\overline{v}_1^2, \qquad (5.1\text{-}22d)
\end{aligned}$$

and, once again, all that's left of the original equation is

$$\sum_i \int_{\boldsymbol{\rho}_{i_1}}^{\boldsymbol{\rho}_{i_2}} (\boldsymbol{F}_i + \sum_j \boldsymbol{f}_{ij}) \cdot d\boldsymbol{\rho}_i = \overline{T}_2 - \overline{T}_1. \qquad (5.1\text{-}22e)$$

Thus, just as with angular momentum, the work-energy relation splits up into two parts, one involving the work done on the center-of-mass particle, done only by the *external* forces (the only ones affecting this "particle's" motion, recall), and those due to *both* external *and* internal forces, changing the kinetic energy *about* the center of mass. (Again, Newton's Third Law does not eliminate the internal forces, since the independent motions of the m_i allow the $d\boldsymbol{\rho}_i$ to be "different.") And, in the same way we have done for particles, then systems of particles, we can talk about energy conservation; but now there are *two* such formulations, one for the center-of-mass particle, the other for motion *about* the center of mass.

Thus we have once again formulated the "kinetic quantities" of linear momentum, angular momentum, and [kinetic] energy—the *same* quantities expressed in terms of center-of-mass coordinates: we see that the *linear* momentum consists only of a "center-of-mass particle" part, while both *angular* momentum and kinetic energy consist of both a center-of-mass particle part and one *about* the center of mass. The equations of motion for the two momenta split up the same way: the *external* forces act as if they were all acting on the center of mass particle for *linear* momentum, while the equation of motion for *angular* momentum splits up into two parts: the moments due to the external forces on the center-of-mass particle change *that* "particle's" angular momentum, while the moments *about* the center of mass due to the external forces change the angular momentum *about* the center of mass. And the work-energy relation also divides up along the same lines: the work, due to only the *external* forces acting at the center of mass, change the kinetic energy of that "particle," while *both* external *and* internal forces, doing work *relative* to the center of mass, change the "*about* center of mass" kinetic energy.

Summary. This summary will be brief, since the two subsections have each ended with their own summaries. But we have, starting only with Newton's Second Law, obtained "kinetic quantities" of *linear momentum, angular momentum*, and [*kinetic*] *energy.* This has been done first for a *single* particle, and then for *systems* of particles—particles merely *identified* as a group, with a concomitant categorization of the forces acting on the individual particles as being "*external*" and "*internal*"; this is precisely what the free-body diagram does. Only the former forces change the two momenta, while the latter also change the kinetic energy of the system. Along the way, we have introduced *impulse*—the time integral of the forces—for the two momenta, leading to *conservation* principles for both these quantities over the interval of integration; this is analogous to energy conservation, resulting from a special case of the work-energy principle, which inherently is applied over an interval [in space]. For systems, we have also expressed the two momenta and energy both in terms of an *absolute, inertial* reference, and relative to the *non*-inertial center of mass. The latter will be the fundamental reference for rigid bodies in the next section.

5.2 Equations of Motion for Rigid Bodies

One of the primary ideas developed in the previous Chapter was that fact that the motion of rigid bodies *vis-à-vis* particles included not only translation, but also rotation. At the same time, given the motion of *any* point in the body, and its rotation, it is possible to find the motion of any *other* point in the body through the implementation of rotating coordinate systems. Thus what is needed now is to determine how the physical agents—forces and couple moments—acting on a given body can determine the translational motion of some point and the rotational motion of the body as a whole. Having been quite careful to point out where the mass of the particles was assumed constant above, we shall here make the sweeping assumption that we are talking about rigid bodies with constant mass.

Consider first the translation. What point should we reference, and just how do the kinetic forces affect it? The results of the previous Section provide the answer immediately: The equation of motion for *linear momentum* $\boldsymbol{G} \equiv$

$\sum_i m_i \boldsymbol{v}_i = m\overline{\boldsymbol{v}}$ of a system of particles, equation (5.1-22a)

$$\sum_i \boldsymbol{F}_i = m\overline{\boldsymbol{a}}$$

states that the "center-of-mass particle" acts as if all the forces were acting on it; thus, with no further development, we *have* the motion of one point in the body, the center of mass.

But that leaves the *rotational* motion of the body to be determined. This will require a little more consideration, since the motion of particles, to which the development has been limited to date, is inherently translational! The key to this is the observation that, in general, rotation generates a *transverse* motion of one point relative to another: in the particular case of a rigid body considered in Section 4.1.2, this was demonstrated explicitly in Theorem 4.1.1, where it was shown that the relative motion of one point relative to another could be expressed as $d\boldsymbol{r}_i = d\boldsymbol{\theta} \times \boldsymbol{r}_i$. As a cross product, the $d\boldsymbol{r}_i$ will necessarily be perpendicular to the particular $\boldsymbol{r}_i$ measuring the relative position of one point to the other. That vector operation which picks off the transverse component of a vector is the cross product, and the cross product is the essence of *angular momentum*: $\boldsymbol{H} \equiv \sum_i \boldsymbol{r}_i \times m\boldsymbol{v}_i$ [in a generic form]. Thus it is not too surprising that, in the same way *linear* momentum's equation of motion provides the *translational* component of motion of a rigid body [point], *angular* momentum, and *its* equation of motion, will provide the basis for analysis of the *rotational* motion of the body.

Linear momentum is relatively straightforward, and carries over without modification from systems of particles to rigid bodies. But just what *is* the angular momentum in the particular case of a rigid body? That is the subject of the following section:

5.2.1 Angular Momentum of a Rigid Body—the Inertia Tensor

We have been rather cavalier in our discussion of angular momentum leading into this section, carefully avoiding being too specific about details of the angular momentum of a rigid body. This is because there are, in fact, *two* possible formulations of angular momenta and their equations of motion for a system of particles:[8] relative to a *fixed* point O, equations (5.1-17b) and (5.1-16b)

$$\sum_i \boldsymbol{M}_{O_i} = \frac{d\boldsymbol{H}_O}{dt}, \qquad \boldsymbol{H}_O \equiv \sum_i m_i \boldsymbol{r}_i \times \boldsymbol{v}_i, \tag{5.2-1a}$$

[8] Actually, Theorem 5.2.5 allows the equation of motion for $\boldsymbol{H}_O$ to be obtained from that for $\overline{\boldsymbol{H}}$, and this is the tack taken subsequently with regard to expressions for T and $\boldsymbol{H}_O$, derived from $\overline{T}$ and $\overline{\boldsymbol{H}}$, respectively. ("$\overline{\boldsymbol{G}}$" vanishes identically.) But, in view of the insistence that angular momentum be defined relative to a *fixed* point—$\overline{\boldsymbol{H}}$ just "happens" to be of the same form as $\boldsymbol{H}_O$—$\boldsymbol{H}_O$ is retained and the two developed in parallel for the nonce.

or equations (5.1-22c) and (5.1-21c), about the [generally *accelerating*] center of mass

$$\sum_i \overline{\boldsymbol{M}}_i = \frac{d\overline{\boldsymbol{H}}}{dt}, \qquad \overline{\boldsymbol{H}} \equiv \sum_i m_i \boldsymbol{\rho}_i \times \dot{\boldsymbol{\rho}}_i. \tag{5.2-1b}$$

These two equation are of precisely the same "generic" form:

$$\sum_i \boldsymbol{M}_i = \frac{d\boldsymbol{H}}{dt}, \qquad \boldsymbol{H} \equiv \sum_i m_i \boldsymbol{r}_i \times \boldsymbol{v}_i, \tag{5.2-1c}$$

where "$\boldsymbol{M}_i$" corresponds to either $\boldsymbol{M}_{O_i}$ or $\overline{\boldsymbol{M}}_i$, and "$\boldsymbol{H}$" to either $\boldsymbol{H}_O$ or $\overline{\boldsymbol{H}}$, depending on which case we are considering; furthermore, the $\boldsymbol{H}$'s themselves are of the same form, "$\boldsymbol{r}_i$" and "$\boldsymbol{v}_i$" in $\boldsymbol{H}_O$ merely being replaced by $\boldsymbol{\rho}_i$ and $\dot{\boldsymbol{\rho}}_i$ to get $\overline{\boldsymbol{H}}$. Thus it suffices to consider this general form in the sequel, keeping in mind the replacements to be made for each case. But it turns out that this analogy is maintained in the special case of present interest, the *rigid body*:

In the first case, dealing with motion about fixed O, *if the body is rotating about that point*, we can express the $\boldsymbol{v}_i$ using a rotating coordinate system rotating with the body's $\boldsymbol{\omega}$ about O: since $\boldsymbol{r}_i$ is measured from that point,

$$\boldsymbol{v}_i = \boldsymbol{v}_O + \boldsymbol{\omega} \times \boldsymbol{r}_i + \boldsymbol{v}_{rel} = \boldsymbol{0} + \boldsymbol{\omega} \times \boldsymbol{r}_i + \boldsymbol{0}.$$

Similarly, in the evaluation of the $\dot{\boldsymbol{\rho}}_i$, a system rotating with $\boldsymbol{\omega}$ can be fixed at the center of mass G, so that

$$\dot{\boldsymbol{\rho}}_i \equiv \boldsymbol{v}_{i/G} = \boldsymbol{\omega} \times \boldsymbol{r}_{i/G} + \boldsymbol{v}_{rel} \equiv \boldsymbol{\omega} \times \boldsymbol{\rho}_i + \boldsymbol{0}. \tag{5.2-2}$$

Thus both "$\boldsymbol{v}_i$" are of exactly the same form, and we can write the generic $\boldsymbol{H}$, in the special case of rigid body motion, as

$$\boldsymbol{H} = \sum_i m_i \boldsymbol{r}_i \times (\boldsymbol{\omega} \times \boldsymbol{r}_i); \tag{5.2-3}$$

this, then, is the generic form we will use to analyze the angular momentum of a rigid body.

We see that $\boldsymbol{H}$ is sum of triple vector products (multiplied by the m_i). Though these are generally rather obnoxious to deal with, we have already introduced a vector identity, equations (4.1-5), which allow us to express such a product as the difference of two vectors; the applicable one here is (4.1-5a), $\boldsymbol{A} \times (\boldsymbol{B} \times \boldsymbol{C}) = (\boldsymbol{A} \cdot \boldsymbol{C})\boldsymbol{B} - (\boldsymbol{A} \cdot \boldsymbol{B})\boldsymbol{C}$, so that equation (5.2-3) becomes

$$\boldsymbol{H} = \sum_i m_i \left((\boldsymbol{r}_i \cdot \boldsymbol{r}_i)\boldsymbol{\omega} - (\boldsymbol{r}_i \cdot \boldsymbol{\omega})\boldsymbol{r}_i\right) = \sum_i m_i \left(\|\boldsymbol{r}_i\|^2 \boldsymbol{\omega} - (\boldsymbol{r}_i \cdot \boldsymbol{\omega})\boldsymbol{r}_i\right).$$

In order to evaluate this quantity, let us set up a Cartesian coordinate system at the appropriate origin (O or G), so that the vectors $\boldsymbol{r}_i$ and $\boldsymbol{\omega}$ will be represented relative to this basis

$$\boldsymbol{r}_i = x_i \hat{\boldsymbol{\imath}} + y_i \hat{\boldsymbol{\jmath}} + z_i \hat{\boldsymbol{k}}, \quad \boldsymbol{\omega} = \omega_x \hat{\boldsymbol{\imath}} + \omega_y \hat{\boldsymbol{\jmath}} + \omega_z \hat{\boldsymbol{k}},$$

where $\boldsymbol{r}_i$ is, recall, measured from that origin. Then the above $\boldsymbol{H}$ becomes

$$\begin{aligned}\boldsymbol{H} &= \sum_i m_i[(x_i^2 + y_i^2 + z_i^2)(\omega_x\hat{\boldsymbol{\imath}} + \omega_y\hat{\boldsymbol{\jmath}} + \omega_z\hat{\boldsymbol{k}}) \\ &\qquad - (x_i\omega_x + y_i\omega_y + z_i\omega_z)(x_i\hat{\boldsymbol{\imath}} + y_i\hat{\boldsymbol{\jmath}} + z_i\hat{\boldsymbol{k}})] \\ &= \sum_i m_i[(y_i^2 + z_i^2)\omega_x - x_iy_i\omega_y - x_iz_i\omega_z]\hat{\boldsymbol{\imath}} \\ &\quad + \sum_i m_i[-x_iy_i\omega_x + (x_i^2 + z_i^2)\omega_y - y_iz_i\omega_z]\hat{\boldsymbol{\jmath}} \\ &\quad + \sum_i m_i[-x_iz_i\omega_x - y_iz_i\omega_y + (x_i^2 + y_i^2)\omega_z]\hat{\boldsymbol{k}}.\end{aligned}$$

This is a somewhat awkward form, but it is here that the material from Part I begins to come into play: In section 2.2.1, we introduced curious quantities called *dyadics*—linear combinations of juxtaposed basis vectors $\sum a_{12\ldots n}\boldsymbol{u}_1\boldsymbol{u}_2\ldots\boldsymbol{u}_n$, along with the operations of "*inner*" and "*outer*" *multiplication.* Such objects were a means of representing *tensors* of the appropriate "*order*," the value of n. We claim that the above expression for $\boldsymbol{H}$ can itself be *represented* as the inner product of a dyadic of second order with one of first order: $\mathsf{H} = \mathsf{I}\boldsymbol{\omega}$, where

$$\begin{aligned}\mathsf{I} &\equiv \sum_i m_i[(y_i^2 + z_i^2)\hat{\boldsymbol{\imath}}\hat{\boldsymbol{\imath}} - x_iy_i\hat{\boldsymbol{\imath}}\hat{\boldsymbol{\jmath}} - x_iz_i\hat{\boldsymbol{\imath}}\hat{\boldsymbol{k}}] \\ &\quad + \sum_i m_i[-x_iy_i\hat{\boldsymbol{\jmath}}\hat{\boldsymbol{\imath}} + (x_i^2 + z_i^2)\hat{\boldsymbol{\jmath}}\hat{\boldsymbol{\jmath}} - y_iz_i\hat{\boldsymbol{\jmath}}\hat{\boldsymbol{k}}] \\ &\quad + \sum_i m_i[-x_iz_i\hat{\boldsymbol{k}}\hat{\boldsymbol{\imath}} - y_iz_i\hat{\boldsymbol{k}}\hat{\boldsymbol{\jmath}} + (x_i^2 + y_i^2)\hat{\boldsymbol{k}}\hat{\boldsymbol{k}}]\end{aligned}$$

and

$$\boldsymbol{\omega} \equiv \omega_x\hat{\boldsymbol{\imath}} + \omega_y\hat{\boldsymbol{\jmath}} + \omega_z\hat{\boldsymbol{k}}.$$

(Recall that these formal vector juxtapositions are not commutative: $\hat{\boldsymbol{\imath}}\hat{\boldsymbol{\jmath}} \neq \hat{\boldsymbol{\jmath}}\hat{\boldsymbol{\imath}}$.) This shows that I is a second-order *tensor*—the "*inertia tensor*"—and $\boldsymbol{\omega}$ one of first order. As such, $\boldsymbol{H}$ can be *represented* as the product of a *matrix* I with a *column matrix* representing $\boldsymbol{\omega}$: $\mathsf{H} = \mathsf{I}\boldsymbol{\omega}$ for

$$\mathsf{I} \equiv \begin{pmatrix} \sum_i m_i(y_i^2 + z_i^2) & -\sum_i m_ix_iy_i & -\sum_i m_ix_iz_i \\ -\sum_i m_ix_iy_i & \sum_i m_i(x_i^2 + z_i^2) & -\sum_i m_iy_iz_i \\ -\sum_i m_ix_iz_i & -\sum_i m_iy_iz_i & \sum_i m_i(x_i^2 + y_i^2) \end{pmatrix} \tag{5.2-4a}$$

$$\equiv \begin{pmatrix} I_{xx} & -I_{xy} & -I_{xz} \\ -I_{xy} & I_{yy} & -I_{yz} \\ -I_{xz} & -I_{yz} & I_{zz} \end{pmatrix} \tag{5.2-4b}$$

—the "*inertia matrix*"—and

$$\boldsymbol{\omega} \equiv \begin{pmatrix} \omega_x \\ \omega_y \\ \omega_z \end{pmatrix}.$$

Note that the inertia matrix is a *real, symmetric* one—this justifies the notation in (5.2-4b)—a fact that has important ramifications. In particular, there are only six quantities which actually determine I: the quantities I_{xx}, I_{yy} and I_{zz} are called "*moments of inertia*"; I_{xy}, I_{xz}, and I_{yz} are termed "*products of inertia*" [*note the signs with which these enter* I!]

All of the above development, resting on the previous presentation of systems of particles, has been treating individual, *discrete* particles. If the majority of "rigid bodies" treated are *continuous* in their mass distribution, however, the extension of the above to such cases is immediate by taking the limit of the above definitions for infinitesimal quantities: thus, for example,

$$I_{xx} \equiv \int_m (y^2 + z^2)\, dm, \quad \text{and} \quad I_{xy} \equiv \int_m xy\, dm,$$

in which the integration "$\int_m dm$" schematically does so over the entire [mass of the] body. In a similar fashion, equations (5.1-20) [page 268] become

$$\int_m \boldsymbol{\rho}\, dm = \mathbf{0}, \quad \int_m \dot{\boldsymbol{\rho}}\, dm = \mathbf{0}. \tag{5.2-5}$$

Since this is the more common case, most of the sequel will focus on it. But all the properties of mass moments of inertia and centers are the same, whether the mass distribution is continuous or discrete; in particular, moments of inertia always have units of $[M][L]^2$.

Before proceeding with the 3-D inertia tensor, it is useful to compare this with the planar case; why does one deal there only with the scalar "moment of inertia" I_{zz} rather than the tensor/matrix formulation presented above? After all, even if $\boldsymbol{\omega} = \omega\hat{\boldsymbol{k}}$, say (so $\omega_x \equiv 0 \equiv \omega_y$), $\mathsf{I}\boldsymbol{\omega}$ would still give terms $I_{xz}\omega_z\hat{\boldsymbol{k}}$ and $I_{yz}\omega_z\hat{\boldsymbol{k}}$ in $\boldsymbol{H}$. In fact, however, the general development *is* implicit in the planar one, which invariably assumes one of two conditions on the rigid body under consideration, forcing I_{xz} and I_{yz} to vanish: it either

- is a *thin plate*—$z \equiv 0$ (so the products/integrands vanish identically), or
- enjoys *mass symmetry about the z-axis* (so to each m_i/dm at (x, y, z) there is one at $(x, y, -z)$).

Thus, in both cases, $\boldsymbol{H}$ reduces to $I_{zz}\omega_z\hat{\boldsymbol{k}}$, and it is only I_{zz} one need consider.

Example 5.2.1 *The Inertia Matrix by Direct Integration.* Diagrammed at left is a uniform "slender rod" of mass M and length L lying in an arbitrary direction given by the direction cosines l, m, and n. We desire to ascertain the inertia matrix I for this body relative to the given axes.

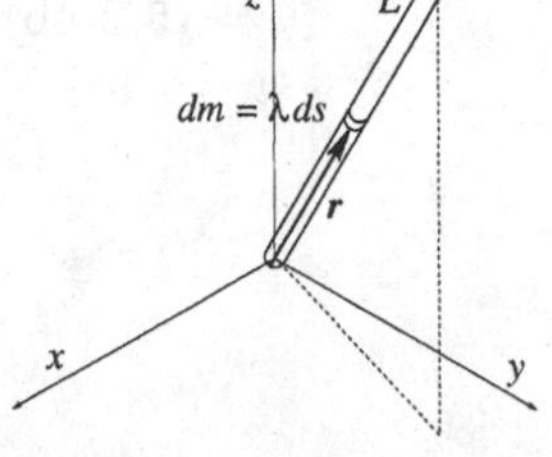

Since this is a continuous, rather than discrete, system, we shall do so by direct integration, using the above definitions of the various entries in the matrix (5.2-4b). But this means that we must have the x-,

y-, and z-components of the mass element $dm = \lambda\, ds$, where λ is the [uniform] "linear density" of the rod *per unit length* (whatever that is), and s is that variable measuring the distance along the rod from the origin, ranging from 0 to L. In particular, because we know the direction cosines of the rod's direction, it is an easy matter to obtain $\boldsymbol{r} = s(l\hat{\imath}+m\hat{\jmath}+n\hat{k})$, from which we get immediately the Cartesian components we require.

That done, it is a straightforward matter to obtain the various moments and product of inertia in the inertia matrix:

$$\begin{aligned} I_{xx} &\equiv \int_m (y^2+z^2)\, dm = \int_0^L (m^2+n^2)s^2(\lambda\, ds) \\ &= (m^2+n^2)\lambda\frac{L^3}{3} = (m^2+n^2)\frac{ML^2}{3}, \end{aligned}$$

in which we have craftily multiplied the penultimate result by 1 in the clever guise $1 = M/\lambda L$ in order to introduce the mass, M. I_{yy} and I_{zz} follow the same pattern. Similarly,

$$I_{xy} \equiv \int_m xy\, dm = \int_0^L lms^2(\lambda\, ds) = lm\frac{ML^2}{3},$$

with I_{xy} and I_{yz} following suit. Thus, finally,

$$\mathsf{I} = \begin{pmatrix} (m^2+n^2) & -lm & -ln \\ -lm & (l^2+n^2) & -mn \\ -ln & -mn & (l^2+m^2) \end{pmatrix}\frac{ML^2}{3}.$$

Note the signs on the products of inertia; this is likely the most common mistake to make! If the rod had happened to lie along an axis, say the x-axis, then the direction cosines would be $l = 1$ and $m = n = 0$; in that case *all off-diagonal terms vanish*, as does I_{xx}.

Were the rod to have been *non*-uniform, λ would not have been a constant but rather a function of s: $\lambda = \lambda(s)$ would have to be a known function in the integrand, but it would be impossible to introduce $M = \int \lambda(s)\, ds$. |

Homework:

1. Oftentimes, a mass moment or product of inertia might not be in tables, but one of *area* is. Show that if the body is *uniform* and has a density per unit area of σ, then the *mass* moment/product is merely σ times the *areal* moment/product.

2. Show that, for a *thin* triangular plate of mass m, altitude h, base b, lying in the xy-plane with its vertex a distance a from the y-axis as shown, the inertia matrix

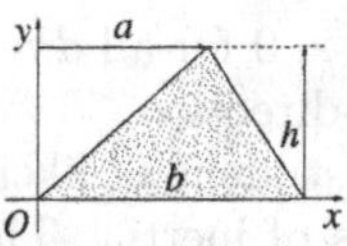

$$I_O = \begin{pmatrix} \frac{h^2}{6} & -\frac{(b+2a)h}{12} & 0 \\ -\frac{(b+2a)h}{12} & \frac{a^2+ab+b^2}{6} & 0 \\ 0 & 0 & \frac{a^2+ab+b^2+h^2}{6} \end{pmatrix} m.$$

We will now discuss briefly some of the properties of both the moments and products of inertia as individual quantities. Then we shall enumerate the important properties of the inertia matrix/tensor—despite the technical distinction between the *representation* of this object and the object itself, we tend to use the two terms interchangeably—as a whole.

Moments of Inertia. Moments of inertia are precisely the same "moments of inertia" discussed in Statics, though there they are moments of inertia of *area*, while here they are moments of inertia of *mass*. Properties, however, are precisely the same for the two cases, since the only difference [for continuous systems] is the variable over which the integration is done, and the properties of these quantities depend only on the properties of the integrals themselves, not the variable with respect to which the integration is carried out. We shall demonstrate the various properties of I_{zz}; clearly, the same ones will hold for I_{xx} and I_{yy}.

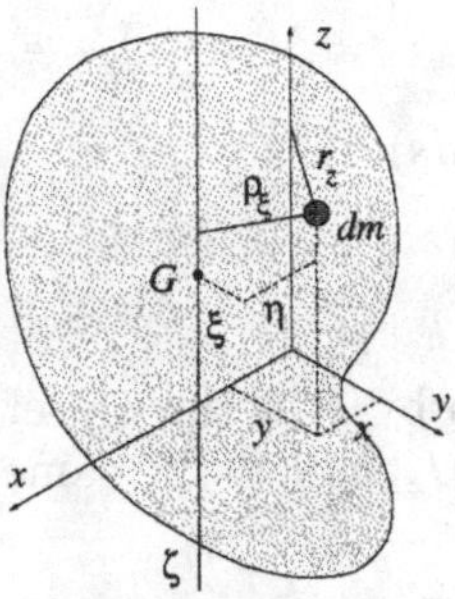

We can reformulate the mass moment of inertia integral

$$I_{zz} \equiv \int_m (x^2 + y^2)\, dm = \int_m r_z^2\, dm, \tag{5.2-6}$$

where r_z is just the *distance* from the z-axis. We see immediately one property of moments of inertia: as integrals with positive—or at least non-negative—integrands, they can *never be negative*:

$$I_{zz} \geq 0. \tag{5.2-7}$$

In fact, they only way they can *vanish* is for $x \equiv 0 \equiv y$—an infinitesimally thin *rod* lying along the z-axis; this is precisely what happened in the above example, there relative to the x-axis.

In *planar* moments of inertia, of both area and mass, one often will use the identity that "$I_{zz} = I_{xx} + I_{yy}$." The student should be warned that this is generally *not* the case:

$$\begin{aligned} I_{zz} &\equiv \int_m (x^2 + y^2)\, dm \\ &\neq \int_m (y^2 + z^2)\, dm + \int_m (x^2 + z^2)\, dm \equiv I_{xx} + I_{yy} \end{aligned}$$

unless $z \equiv 0$ for all dm—*i.e.*, the body is a "*plate*" with infinitesimal thickness in the z-direction.

The student is likely familiar with the *radius of gyration* associated with moments of inertia. This, in some ways, an analog to the equation for centers of mass, here also generalized to continuous systems from equation (5.1-18a) for discrete systems:

$$m\overline{\mathbf{r}} = \int_m \mathbf{r}\, dm,$$

where $\overline{\boldsymbol{r}}$ can be interpreted as that point at which "all the mass"—just the center-of-mass particle—could be placed to generate the same value as the integral. In the same fashion, the radius of gyration k_{zz} can be interpreted as that distance from the z-axis at which the center-of-mass particle could lie to generate the same integral, here just I_{zz}:

$$mk_{zz}^2 \equiv \int_m r_z^2\, dm \equiv \mathsf{I}_{zz}, \tag{5.2-8}$$

with similar definitions for k_{xx} and k_{yy}. There is really nothing terribly mysterious about this quantity: it is merely an alternative means of specifying the moment of inertia!

Moment of Inertia about an *arbitrary* Axis. The above discussion has centered on the moments of inertia about the x-, y- and z-axes—those axes in terms of which all the position vectors $\boldsymbol{r}$ of the dm are specified. But, at times, it is useful to be able to find the moment of inertia I_λ about an *arbitrary* axis, one whose direction can be specified by means of a unit vector $\hat{\boldsymbol{\lambda}} = l\hat{\imath} + m\hat{\jmath} + n\hat{k}$ using the direction cosines (Section 4.2.1): this is a simple *vector*, not a coordinate system, whose orientation would better be specified using Euler angles. Thus, by (5.2-6), $I_\lambda = \int_m r_\lambda^2\, dm$, where now r_λ measures the distance of dm from the $\hat{\boldsymbol{\lambda}}$-axis.

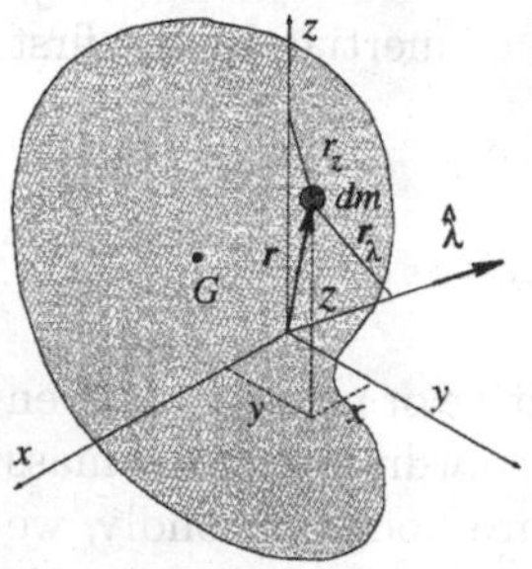

Now that $r_\lambda = r \sin\theta = \|\hat{\boldsymbol{\lambda}} \times \boldsymbol{r}\|$, where $\boldsymbol{r}$ [and $r \equiv \|\boldsymbol{r}\|$] is the position of dm measured relative to the origin, whichever (fixed point or center of mass) that might be. Thus

$$\begin{aligned}
r_\lambda^2 &= (\hat{\boldsymbol{\lambda}} \times \boldsymbol{r}) \cdot (\hat{\boldsymbol{\lambda}} \times \boldsymbol{r}) \\
&= \hat{\boldsymbol{\lambda}} \times \boldsymbol{r} \cdot (\hat{\boldsymbol{\lambda}} \times \boldsymbol{r}) && \text{(writing this as a triple scalar product)} \\
&= \hat{\boldsymbol{\lambda}} \cdot \boldsymbol{r} \times (\hat{\boldsymbol{\lambda}} \times \boldsymbol{r}) && \text{(interchanging the } \cdot \text{ and } \times\text{)} \\
&= \hat{\boldsymbol{\lambda}} \cdot [r^2 \hat{\boldsymbol{\lambda}} - (\hat{\boldsymbol{\lambda}} \cdot \boldsymbol{r})\boldsymbol{r}] && \text{(by (4.1-5))} \\
&= r^2 - (\hat{\boldsymbol{\lambda}} \cdot \boldsymbol{r})^2;
\end{aligned}$$

so, representing these quantities in terms of the $(\hat{\imath}, \hat{\jmath}, \hat{k})$ basis,

$$\begin{aligned}
r_\lambda^2 &= r^2 - (\hat{\boldsymbol{\lambda}} \cdot \boldsymbol{r})^2 \\
&= (x^2 + y^2 + z^2) - (lx + my + nz)^2 \\
&= (l^2 + m^2 + n^2)(x^2 + y^2 + z^2) \\
&\qquad - (l^2x^2 + m^2y^2 + n^2z^2 + 2lmxy + 2lnxz + 2mnyz) \\
&= l^2(y^2 + z^2) + m^2(x^2 + z^2) + n^2(x^2 + y^2) - 2lmxy - 2lnxz - 2mnyz
\end{aligned}$$

—where we have used (4.2-2) in the third line. Thus, integrating this over all of m,

$$I_\lambda = \int_m r_\lambda^2\,dm = l^2 I_{xx} + m^2 I_{yy} + n^2 I_{zz} - 2lmI_{xy} - 2lnI_{xz} - 2mnI_{yz}. \quad (5.2\text{-}9)$$

We see that not only the *moments* of inertia enter this equation, but also the *products.* The clever reader might have seen the connection here between this development and *rotations*: Euler angles could have been used to specify $\hat{\boldsymbol{\lambda}}$, and, as pointed out in Section 4.2, these not only can be used to specify vectors; they also are amenable to the implementation of *rotation matrices.* We shall shortly see an alternative means of obtaining this same result using such an approach.

Products of Inertia Though most students have likely had *moments* of inertia in Statics and Strength of Materials, *products* of inertia may be new to many. They are fundamentally different from moments of inertia: In the first place, the definition [for continuous systems] of, say

$$I_{xy} \equiv \int_m xy\,dm,$$

shows that products of inertia can assume *any* value, positive or negative or even zero, depending on how the dm and the *signed* $x-$ and y-coordinates of the mass element conspire together when integrated over the entire body. Secondly, we point out that even the *notation* of products of inertia, *vis-à-vis* moments, is somewhat inconsistent; *e.g.*,

$$I_{zz} \neq \int_m zz\,dm!$$

In the previous section we obtained equation (5.2-9) giving the moment of inertia about an arbitrary axis $\hat{\boldsymbol{u}}$, specified there in terms of its direction cosines. One might reasonably expect a parallel development here, but a little reflection shows the difficulty in as straightforward an approach as used above: the products of inertia reference *two* axes, and it would be necessary to specify *both* in terms of the $(\hat{\boldsymbol{\imath}}, \hat{\boldsymbol{\jmath}}, \hat{\boldsymbol{k}})$ basis, requiring *two* sets of direction cosines! But this is precisely the place where Euler angles—inherently amenable to *rotations*—became so useful, and thus we will forestall discussion of the products of inertia under such rotations until the end of the next section.

Properties of the Inertia Tensor

Certainly the inertia tensor is made up of its individual components, the moments and products of inertia discussed separately above. But ultimately we are interested in the matrix itself, rather the individual entries, and this is the perspective adopted here, even though we often return to the separate elements of that matrix. We will discuss two important properties of the matrix, both of which ultimately enable the use of tables to evaluate the inertia tensor of an arbitrary body.

Such tables will typically include centroids and moments of inertia, though occasionally one finds tables including products, too. Such tables are limited to common, regular figures such as triangular blocks, rectangular ones, circles and spheres, and the like. But few bodies of physical interest are so simple, generally being *composed* of parts which themselves might be more elementary in nature. The first Theorem deals with how the inertia tensor of such a *composite body* could be found:

Theorem 5.2.1 (The Composition Theorem). Assume the body of interest is made up—*composed*—of a series of smaller bodies: $m = m_1 + m_2 + \ldots$. Then, relative to a given set of axes,

$$\mathsf{I} = \mathsf{I}_1 + \mathsf{I}_2 + \ldots$$

[The student should not confuse this with the corresponding Composition Theorem for *centers of mass*, which is a *weighted* sum of the separate centers of mass:

$$m\overline{\mathbf{r}} = m_1\overline{\mathbf{r}}_1 + m_2\overline{\mathbf{r}}_2 + \ldots$$
$$m = m_1 + m_2 + \ldots]$$

Proof. Consider, for example, the moment of inertia I_{zz}; the other two would be similar:

$$\begin{aligned} I_{zz} &\equiv \int_m (x^2 + y^2)\, dm = \int_{m_1} (x^2 + y^2)\, dm + \int_{m_2} (x^2 + y^2)\, dm + \ldots \\ &\equiv I_{zz_1} + I_{zz_2} + \ldots \end{aligned}$$

since integrals sum over the domains of integration; *i.e., the moment of inertia of a body about a given axis is just the sum of the moments of inertia of its component parts.* Since the proof here hinges on the property of integrals themselves, rather than on the particular integrand, the same holds for *products* of inertia; *e.g.,*

$$I_{xy} = I_{xy_1} + I_{xy_2} + \ldots,$$

with similar results for the other two products. Assembly of the various terms into the bodies' matrices leads to the above result. □

The next results treat how to deal with changes in the *reference axes.* There are two ways in which these axes may change: either through a *translation* of axes or through a *rotation.* In the first case, the axis directions, *i.e.*, the *basis*, remain unchanged, and only the origin changes; in the second, the *basis* changes, but the origin remains fixed. Most generally, of course, *both* change, but such general alteration can be broken up into the two constituent moves: rotation, followed by translation. Thus it suffices to consider the two separately.

Translation. In this form of axis change, the basis remains the same; and, in view of our discussion in Part I, where we discussed only how a matrix changes [the *representation* of] a linear operator under a change of basis, one might reasonably question why a mere translation would have *any* effect on a matrix. The reason in the present case is the fact that *the inertia matrix is itself* defined *in terms of the coordinates*, and changing the origin changes the coordinates. Thus translation *will* alter the form of the matrix.

The first result, the *Parallel Axis Theorem*, is oftentimes somewhat mysterious to students. At its foundation, it relates the moment of inertia of a body about an axis to that about a *parallel* axis *through the center of mass* of the body; this is the *translational* change in axis:

Theorem 5.2.2 (The Parallel Axis Theorem). If the center of mass of a body with mass m is located at a position $\overline{\boldsymbol{r}} = \overline{x}\hat{\boldsymbol{\imath}} + \overline{y}\hat{\boldsymbol{\jmath}} + \overline{z}\hat{\boldsymbol{k}}$ relative to arbitrary axes, the moment of inertia I relative to these axes is related to $\overline{\mathsf{I}}$, the moment of inertia relative to *parallel* axes with origin *at the center of mass*, through

$$\mathsf{I} = \overline{\mathsf{I}} + m \begin{pmatrix} d_x^2 & -\overline{x}\,\overline{y} & -\overline{x}\,\overline{z} \\ -\overline{x}\,\overline{y} & d_y^2 & -\overline{y}\,\overline{z} \\ -\overline{x}\,\overline{z} & -\overline{y}\,\overline{z} & d_z^2 \end{pmatrix}, \tag{5.2-10a}$$

where d_x, d_y and d_z are the *distances* between the respective pairs of [parallel] x-, y-, and z-axes.

Though this result, expressed as a matrix sum in the spirit of the presentation emphasizing the matrix itself, may at first appear daunting, one should not be put off by this! In practice, the elements of I are found individually, and this is precisely how the proof is made. In any event, we note that the added matrix is itself, like the inertia matrix, real and symmetric, assuring that the final I will be.

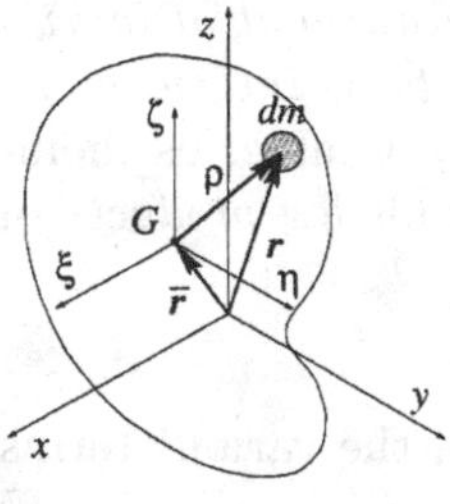

Proof. Consider first the *moment* of inertia components of I. For example, we can again find I_{zz} in terms of the original coordinates measuring the position $\boldsymbol{r} = x\hat{\boldsymbol{\imath}} + y\hat{\boldsymbol{\jmath}} + z\hat{\boldsymbol{k}}$ of dm from O. But now measure the position of dm relative to *parallel* axes through G, the center of mass at $\overline{\boldsymbol{r}} = \overline{x}\hat{\boldsymbol{\imath}} + \overline{y}\hat{\boldsymbol{\jmath}} + \overline{z}\hat{\boldsymbol{k}}$ [relative to O]: $\boldsymbol{\rho} = \xi\hat{\boldsymbol{\imath}} + \eta\hat{\boldsymbol{\jmath}} + \zeta\hat{\boldsymbol{k}}$. Since $\boldsymbol{r} = \overline{\boldsymbol{r}} + \boldsymbol{\rho}$, these two sets of coordinates are related: $x = \overline{x} + \xi$, $y = \overline{y} + \eta$, and $z = \overline{z} + \zeta$. In particular,

$$\begin{aligned} I_{zz} &\equiv \int_m (x^2 + y^2)\,dm = \int_m \left((\overline{x} + \xi)^2 + (\overline{y} + \eta)^2\right) dm \\ &= \overline{x}^2 \int_m dm + 2\overline{x} \int_m \xi\,dm + \int_m \xi^2\,dm \\ &\quad + \overline{y}^2 \int_m dm + 2\overline{y} \int_m \eta\,dm + \int_m \eta^2\,dm. \end{aligned}$$

But $\int_m \xi\, dm \equiv 0 \equiv \int_m \eta\, dm$, since these integrals are just, respectively, $m\overline{\xi}$ and $m\overline{\eta}$ for a continuous system, giving the ξ- and η-coordinates of the center of mass with respect *to* the center of mass [see equations (5.2-5)]. Thus, since $\int_m dm = m$,

$$= \overline{x}^2 m + \int_m (\xi^2 + \eta^2)\, dm + \overline{y}^2 m \equiv \overline{I}_{zz} + m d_z^2,$$

where $\overline{I}_{zz} \equiv \int_m (\xi^2 + \eta^2)\, dm$ is the moment of inertia about the parallel center-of-mass ζ-axis, and $d_z^2 = \overline{x}^2 + \overline{y}^2$ is the square of the distance between the z– and ζ-axes. Thus *the moment of inertia about an axis is that about a* parallel *axis* through the center of mass *plus the mass times the square of the distance between the two [parallel] axes.*

As a byproduct of this part of the proof, we also have a

Corollary. The minimum value of the moment of inertia about an axis in a given direction is that about a parallel axis through the center of mass [where each $d \equiv 0$].

Now consider a typical *product* of inertia, say

$$\begin{aligned} I_{xy} &\equiv \int_m xy\, dm = \int_m (\overline{x} + \xi)(\overline{y} + \eta)\, dm \\ &= \overline{x}\,\overline{y} \int_m dm + \overline{x} \int_m \eta\, dm + \overline{y} \int_m \xi\, dm + \int_m \xi\eta\, dm \\ &= m\overline{x}\,\overline{y} + \overline{\mathsf{I}}_{xy} \end{aligned}$$

in the same way we obtained the result for moments of inertia.[9] The negative signs in the added matrix in (5.2-10) then result from the fact that the *products* of inertia enter the inertia matrix with that sign.

Note the importance of having the parallel axes pass through the centroid: this was required to make the two integrals which *did* vanish in each case do so! □

Although equation (5.2-10) is the normal statement of this principle, it is instructive to write the result of the above Theorem in a slightly different form by substituting the definitions of d_x, d_y, and d_z:

$$\mathsf{I} = \overline{\mathsf{I}} + m \begin{pmatrix} \overline{y}^2 + \overline{z}^2 & -\overline{x}\,\overline{y} & -\overline{x}\,\overline{z} \\ -\overline{x}\,\overline{y} & \overline{x}^2 + \overline{z}^2 & -\overline{y}\,\overline{z} \\ -\overline{x}\,\overline{z} & -\overline{y}\,\overline{z} & \overline{x}^2 + \overline{y}^2 \end{pmatrix} \qquad (5.2\text{-}10b)$$

—one which echoes the form of the sums/integrands in the definition (5.2-4a).

[9]Though we have been strictly observant of the definitions of $\overline{x}$, $\overline{y}$, and $\overline{z}$ as being the coordinates of the center of mass *relative to the original axes*, in point of fact, product of inertia translation is rather forgiving of the distinction: had we used the coordinates of the *original* origin relative to the center of mass, this would simply change the signs of *each* in the integrand, leading to the same net result!

In planar problems (as noted on page 276), the *matrix* I in $\mathsf{I}\omega$ reduces to a *scalar* I_{zz}. Any application of the Parallel Axis Theorem there involves only distances d_z between perpendicular axes in the direction of ω and α measured in the xy-plane; as a practical issue, neither the vector nature of $\overline{r}$, nor the tensor nature of I—let alone the fact that the representation of each (as a column n-tuple or matrix) depends on the basis chosen—really matters. But in 3-D problems, it *does*. In particular, one must carefully observe that the components of $\overline{r} = (\overline{x}, \overline{y}, \overline{z})^{\mathsf{T}}$ and the values of the variables ξ, η, and ζ in the above sums/integrands are measured along the same directions; *i.e.*, *the bases for* I *and* $\overline{r}$ *must be the same*; this fact will become important in Examples 5.2.6 and 5.2.7 below.

There is a Composition Theorem for centers of mass; one might well wonder why there is not a "parallel axis theorem" for these. In point of fact, there "is," but it's so trivial that it's never stated explicitly: the center of mass $\overline{r}$ of a body in arbitrary axes is the origin of a set of parallel axes through its center of mass!

Homework:

1. Show that, writing the Parallel Axis Theorem in the form $\mathsf{I} = \overline{\mathsf{I}} + m\mathsf{D}$, the matrix D can be written

$$\mathsf{D} = \overline{r}^2\mathsf{I} - \overline{\mathsf{r}}\overline{\mathsf{r}}^{\mathsf{T}},$$

in which $\overline{\mathsf{r}}$ is the column matrix [representation] of $\overline{r}$ from the origin to the center of mass. Note that the last product, $\overline{\mathsf{r}}\overline{\mathsf{r}}^{\mathsf{T}}$, will thus be a *matrix*!

Rotation. The matrix nature of the inertia matrix—its *representation* relative to a given coordinate system—means that all the formalism we have developed for matrices in Part I can be brought to bear on I. In particular, if the vector *basis* changes, meaning that the new basis is *rotated* from the original in a known way, we can determine a *rotation matrix* R, using the methodology of Section 3.6.2. Then, in fact, the *representation* I' of the moment of inertia in the new basis will simply be given by equation (3.6-7):

$$\mathsf{I}' = \mathsf{R}^{-1}\mathsf{I}\mathsf{R}.$$

Recall that I is a [real] symmetric matrix; thus (Homework 2 on page 152) I' will be also, just as it should. This observation is useful as either a means of expediting the calculations, or of checking them.

We illustrate this with a couple of examples. The first of these is likely more typical of the application of this technique, where we first find the inertia matrix relative to a special set of axes in which that matrix is simplified, then rotate to the desired set. The second illustrates how these rotation matrices can be used to obtain a result determined previously; though one would probably never use the approach, it is perhaps interesting nonetheless.

Example 5.2.2 *A Simple Example.* Diagrammed here is a "thin" uniform semi-circular plate of mass m and radius a, inclined by an angle θ with the yz-plane and having its diameter along the y-axis. We are asked to determine its inertia matrix relative to the given axes.

In the spirit of the practical calculation of the various elements in the matrix,

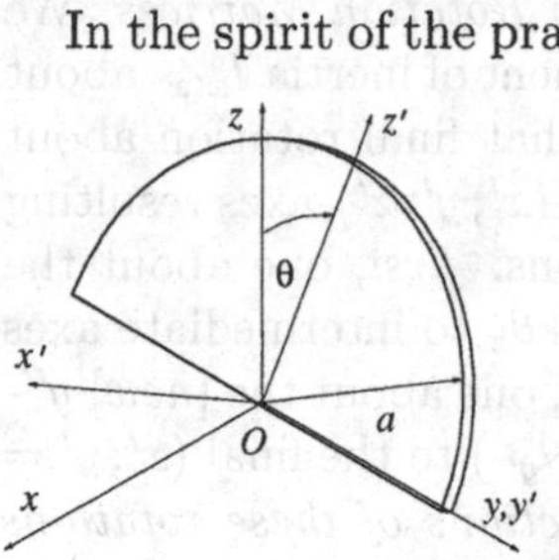

we note that tables give immediately the *moments* of inertia about both the y-axis and one perpendicular to y in the plate—call these the "y'" and "z'" respectively—are $\frac{1}{4}ma^2$, while that about one perpendicular to the plate through O—the "x'" axis—is $\frac{1}{2}ma^2$. This observation suggests using such axes in which to determine the inertia matrix, then simply *rotate* from these axes to the desired ones. Yet nothing is given regarding the *products* of inertia of such a body, even relative to this set referenced to the plane; thus it is necessary to find these directly.

But that is not too difficult: $I_{x'y'} \equiv \int_m x'y'\,dm$ has $x' = 0$ for all dm due to the "thin" nature of the plate, so this product vanishes; similarly, $I_{x'z'} \equiv 0$. This leaves only $I_{y'z'}$ to find. But, due to the symmetry about the z'-axis, to each dm at $(0, y', z')$ there is the *same* dm at $(0, -y', z')$ [we have to introduce the uniform nature of the plate explicitly here]; thus, although the integrand of $I_{y'z'}$ doesn't vanish identically, its integral over the entire mass will. The products of inertia for this particular set of axes vanish!

Thus we have the inertia matrix relative to these axes:

$$\mathsf{I}' = \begin{pmatrix} 2 & 0 & 0 \\ 0 & 1 & 0 \\ 0 & 0 & 1 \end{pmatrix} \frac{ma^2}{4},$$

where we have factored out the fraction to make the subsequent calculation easier. We observe that we can go from the (x', y', z') axes to the desired (x, y, z) through a simple rotation through *positive* θ about the ["old"] y'-axis: using equation (3.6-13) from Part I in the basis transformation (3.6-7),

$$\begin{aligned}
\mathsf{I} &= \mathsf{B}^{-1}(\theta)\mathsf{I}'\mathsf{B}(\theta) = \mathsf{B}^{\mathsf{T}}(\theta)\mathsf{I}'\mathsf{B}(\theta) \\
&= \begin{pmatrix} \cos\theta & 0 & -\sin\theta \\ 0 & 1 & 0 \\ \sin\theta & 0 & \cos\theta \end{pmatrix} \begin{pmatrix} 2 & 0 & 0 \\ 0 & 1 & 0 \\ 0 & 0 & 1 \end{pmatrix} \begin{pmatrix} \cos\theta & 0 & \sin\theta \\ 0 & 1 & 0 \\ -\sin\theta & 0 & \cos\theta \end{pmatrix} \frac{ma^2}{4} \\
&= \begin{pmatrix} 1+\cos^2\theta & 0 & \sin\theta\cos\theta \\ 0 & 1 & 0 \\ \sin\theta\cos\theta & 0 & 1+\sin^2\theta \end{pmatrix} \frac{ma^2}{4}.
\end{aligned}$$

As expected, I is, indeed, symmetric; this acts as a check on our calculations.

The technique implemented here—finding a special set of axes in which the inertia matrix is especially simple, then transforming back to the desired axes—is precisely the general approach to determining that matrix. The intermediate

(x', y', z') axes were indeed special (though their choice here was "obvious"): they are an example of *principal axes* to be discussed presently. |

As a second example, we provide an alternative derivation of the equation giving the moment of inertia about an arbitrary axis, equation (5.2-9). Here, however, we use rotation matrices:

Example 5.2.3 *Moment of Inertia about an Axis from Rotation Matrices.* We will regard I_λ to be the moment of inertia $I_{x'x'}$ about the x'-axis resulting from that final rotation about the original (x, y, z)-axes to (x', y', z')-axes resulting from two successive rotations: *first*, one about the z-axis through an angle of $+\theta_z$ to intermediate axes $(x'', y'', z'' = z)$; and *second*, one about the *[new]* y''-axis through an angle of $(-\theta_{y''})$ to the final $(x', y' = y'', z')$-axes. (*Note the directions of these rotations indicated on the diagram, as well as the signs of the rotation angles, relative to the indicated axes!*)

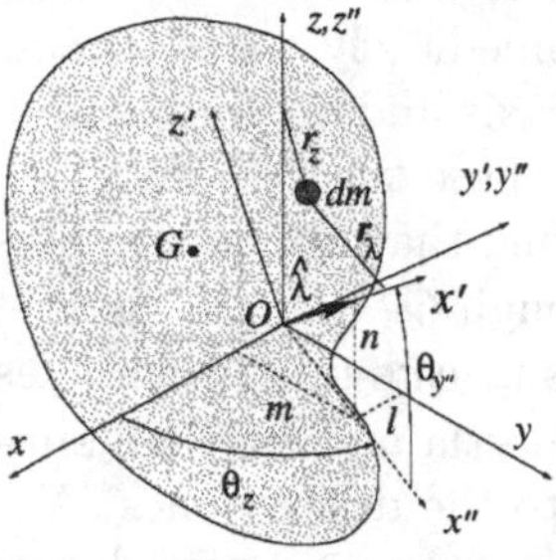

In terms of the rotation matrices given on page 154, we are going from the "old" (x, y, z) system/basis to the "new" (x', y', z'); the net effect of these two rotations is a single rotation matrix $\mathsf{R} = \mathsf{C}(+\theta_z)\mathsf{B}(-\theta_{y''})$, so that the final inertia matrix will be given by

$$\mathsf{I}' = \mathsf{R}^{-1}\mathsf{I}\mathsf{R} = \mathsf{R}^{\mathsf{T}}\mathsf{I}\mathsf{R},$$

since rotation matrices are orthogonal. But now it is necessary to calculate the individual entries in the two matrices. Noting the x- and y-components of $\hat{\boldsymbol{\lambda}}$ are l and m on the diagram,

$$\sin\theta_z = \frac{m}{b}, \qquad \cos\theta_z = \frac{l}{b},$$
$$\sin\theta_{y''} = n, \qquad \cos\theta_{y''} = b,$$

in which $b \equiv \sqrt{l^2 + m^2}$ (and recalling that $\|\hat{\boldsymbol{\lambda}}\| \equiv 1$!); substituting these into the rotation matrices, we get immediately that

$$\mathsf{C}(\theta_z) = \begin{pmatrix} \frac{l}{b} & -\frac{m}{b} & 0 \\ \frac{m}{b} & \frac{l}{b} & 0 \\ 0 & 0 & 1 \end{pmatrix}, \qquad \text{and} \qquad \mathsf{B}(-\theta_{y''}) = \begin{pmatrix} b & 0 & -n \\ 0 & 1 & 0 \\ n & 0 & b \end{pmatrix}$$

so that

$$\mathsf{R} = \mathsf{C}(\theta_z)\mathsf{B}(-\theta_{y''}) = \begin{pmatrix} l & -\frac{m}{b} & -\frac{nl}{b} \\ m & \frac{l}{b} & -\frac{mn}{b} \\ n & 0 & b \end{pmatrix}.$$

(We note that the first column of this matrix, the components of $\hat{\imath}'$ relative to the original basis, is just $(l, m, n)^{\mathsf{T}}$ as expected.)

Though we doggedly calculated all nine elements of R for pedagogical purposes, a little foresight (and planning) would have shown that the first column of that matrix was really all that's required: the desired $I_{x'x'}$, that in the first row and column of the transformed I', will result from the product of the first *row* of R^T, I itself, and the first *column* of R:

$$\mathsf{I}' = \mathsf{R}^\mathsf{T}\mathsf{I}\mathsf{R} = \begin{pmatrix} l & m & n \\ \cdots & \cdots & \cdots \\ \cdots & \cdots & \cdots \end{pmatrix} \begin{pmatrix} I_{xx} & -I_{xy} & -I_{xz} \\ -I_{xy} & I_{yy} & -I_{xz} \\ -I_{xz} & -I_{yz} & I_{zz} \end{pmatrix} \begin{pmatrix} l & \cdots & \cdots \\ m & \cdots & \cdots \\ n & \cdots & \cdots \end{pmatrix}$$

$$= \begin{pmatrix} l & m & n \\ \cdots & \cdots & \cdots \\ \cdots & \cdots & \cdots \end{pmatrix} \begin{pmatrix} lI_{xx} - mI_{xy} - nI_{xz} & \cdots & \cdots \\ -lI_{xy} + mI_{yy} - nI_{yz} & \cdots & \cdots \\ -lI_{xz} - mI_{yz} + nI_{zz} & \cdots & \cdots \end{pmatrix},$$

resulting in precisely the same expression we got for I_λ before. |

Translation *and* Rotation. As mentioned in setting up this topic, we can consider an arbitrary change of axes to be made up of a translation and a rotation. We have seen above that a translation schematically makes $\mathsf{I} \to \mathsf{I} + \mathsf{P}$ (where P is that matrix added onto the original I by the Parallel Axis Theorem), while a rotation makes $\mathsf{I} \to \mathsf{R}^{-1}\mathsf{I}\mathsf{R}$. This means that *the order in which these two operations are applied is important*—the two are not "commutative":

$$\begin{aligned} &\text{translation, then rotation:} && \mathsf{I} \to \mathsf{I} + \mathsf{P} \to \mathsf{R}^{-1}(\mathsf{I} + \mathsf{P})\mathsf{R} \\ &\text{rotation, then translation:} && \mathsf{I} \to \mathsf{R}^{-1}\mathsf{I}\mathsf{R} \to \mathsf{R}^{-1}\mathsf{I}\mathsf{R} + \mathsf{P}, \end{aligned}$$

and the end results are not the same! What is the reason for this, and which of the two orders is correct?

The key to this is realization of the fact that, in the above translations, given by the Parallel Axis Theorem, *$\bar{r}$ and I must be in the same basis* (page 284). If they are not, it is generally necessary to rotate the basis of one or the other to match them *before* the translation is done. On the other hand, it may be that the bases match, but that these are not parallel to the finally-desired ones; in that case, one can translate first, then rotate back to the ultimate axes. Examples of such considerations follow (5.2.6 and 5.2.7).

Principal Axes

In Example 5.2.2 we first obtained the inertia matrix relative to a particular set of axes in which that matrix assumed an especially simple form—that of a *diagonalized* matrix. This is precisely the subject of Section 3.7.1, in which we showed that, if the eigen*vectors* of a matrix were linearly independent, there was a particular choice of axes—a particular *basis*, in fact just the *eigenvectors* of the matrix—in which that matrix could be diagonalized. This culminated in Theorem 3.7.1, where we showed that any matrix, real and symmetric in an orthonormal basis, could be diagonalized through a simple [proper] *rotation* to a new orthonormal basis.

We have repeatedly stressed the real, symmetric nature of the inertia matrix, and the results of this Theorem can be brought to bear directly on such a matrix.

In the present context, this means that there is *always* a new set of Cartesian axes in which the [originally Cartesian] inertia matrix is diagonalized; *i.e.*, in which *the products of inertia vanish.* This basis, and the axes whose directions are given by that basis, are called the *principal axes* of the inertia matrix. Before proceeding further, let us give an example to review just how such axes could be found:

Example 5.2.4. *Principal Axes of Example 5.2.2.* Recall that in the cited example, we ultimately obtain the final matrix

$$\mathsf{I} = \begin{pmatrix} 1+\cos^2\theta & 0 & \sin\theta\cos\theta \\ 0 & 1 & 0 \\ \sin\theta\cos\theta & 0 & 1+\sin^2\theta \end{pmatrix} \frac{ma^2}{4}$$

by transforming from what we now know to be the *principal* axes using rotation matrices. We now work that problem in the opposite order, *starting* with the above I and obtaining the principal axes in which this matrix would be diagonalized.

Recall (Section 3.7.1) that the directions of these axes—the *basis* vectors—are precisely the eigenvectors of the matrix desired to be diagonalized *if the eigenvectors are linearly independent*; thus we must find these, which means finding first the eigen*values* of the matrix I. Despite all the trig functions, it is a surprisingly easy matter to find the characteristic equation for this matrix:

$$\begin{aligned} |\mathsf{I} - \lambda \mathbf{1}| &= [\lambda^3 - (3+\sin^2\theta+\cos^2\theta)\lambda^2 \\ &\qquad + (3+2\sin^2\theta+2\cos^2\theta)\lambda - (1+\sin^2\theta+\cos^2\theta)]\frac{ma^2}{4} \\ &= [\lambda^3 - 4\lambda^2 + 5\lambda - 2]\frac{ma^2}{4} = 0 \end{aligned}$$

with eigenvalues $\lambda_1 = 2$ and a *degenerate* value $\lambda_2 = \lambda_3 = 1$; thus we cannot immediately conclude that the eigenvectors *will* be linearly independent, though the fundamental Theorem 3.7.1 on which all this is based assures us that they will be!

Plugging the value for $\lambda_1 = 2$ into the above I—note the fortuitous "1" along the diagonal—we get that $\boldsymbol{\xi}_1 = (\cos\theta, 0, \sin\theta)^\mathsf{T}$. On the other hand, for $\lambda_2 = \lambda_3 = 1$, we get that

$$\mathsf{I} - \lambda_i \mathbf{1} = \frac{ma^2}{4} \begin{pmatrix} \cos^2\theta & 0 & \sin\theta\cos\theta \\ 0 & 0 & 0 \\ \sin\theta\cos\theta & 0 & \sin^2\theta \end{pmatrix} \begin{pmatrix} \xi_1 \\ \xi_2 \\ \xi_3 \end{pmatrix} = \mathbf{0},$$

—a linear homogeneous equation with rank 1, so having two linearly independent solutions (elements of the null space—see Theorem 2.2.5). From the component equations we conclude that ξ_2 is arbitrary and that $\cos\theta\xi_1 + \sin\theta\xi_3 = 0$, with two linearly independent solutions $\boldsymbol{\xi}_2 = (\sin\theta, 1, -\cos\theta)^\mathsf{T}$ and $\boldsymbol{\xi}_3 = (0,1,0)^\mathsf{T}$, the middle component chosen to ensure linear independence of $\boldsymbol{\xi}_1$.

These eigenvectors, the *representations* relative to the (x, y, z) axes of the *directions* of the principal axes guaranteed by the Theorem [they haven't been normalized!], show that the y-axis ($\boldsymbol{\xi}_3$) is already a principal axis, as is $\boldsymbol{\xi}_1$, along the x'-axis in Example 5.2.2. Furthermore, the fact that the inner products of $\boldsymbol{\xi}_1$ with the other two vanish shows that these are orthogonal so linearly independent, just as they should be, since each pair consists of eigenvectors of distinct eigenvalues (Theorem 3.5.1). And all three eigenvectors are linearly independent: using Theorem 3.2.6, the determinant of that matrix composed of the three eigenvectors

$$|\boldsymbol{\xi}_1 \quad \boldsymbol{\xi}_2 \quad \boldsymbol{\xi}_3| = \begin{vmatrix} \cos\theta & \sin\theta & 0 \\ 0 & 1 & 1 \\ \sin\theta & -\cos\theta & 0 \end{vmatrix} = +1.$$

That's the good news. But recall that Theorem 3.7.1 guaranteed that the diagonalizing axes would *themselves* be orthonormal, and, although $\boldsymbol{\xi}_1$ is perpendicular to the other two, $\boldsymbol{\xi}_2$ and $\boldsymbol{\xi}_3$ *aren't*: $\boldsymbol{\xi}_2 \cdot \boldsymbol{\xi}_3 = 1 \neq 0$; this is precisely the possibility pointed out at the end of the proof of that Theorem.

But it was also pointed out there that the Gram-Schmidt procedure, exemplified at the end of Example 3.5.1, ensured that we could *force* these two eigenvectors, corresponding to their common eigenvalue, to be orthogonal: recognizing that a linear combination of eigenvectors corresponding to the same eigenvalue are also an eigenvector, we choose to retain $\boldsymbol{\xi}_3$ with its appealing form and replace $\boldsymbol{\xi}_2$ by the linear combination $\boldsymbol{\xi}_2' \equiv a\boldsymbol{\xi}_2 + \boldsymbol{\xi}_3$ (which will thus also be an eigenvector for $\lambda_2 = 1$) such that it is perpendicular to $\boldsymbol{\xi}_3$:

$$\boldsymbol{\xi}_2' \cdot \boldsymbol{\xi}_3 = (a\boldsymbol{\xi}_2 + \boldsymbol{\xi}_3) \cdot \boldsymbol{\xi}_3 = a(1) + 1 \equiv 0.$$

Thus $a = -1$ so that the desired $\boldsymbol{\xi}_2' = (-\sin\theta, 0, \cos\theta)^{\mathsf{T}}$. This vector is still perpendicular to $\boldsymbol{\xi}_1$ ($\boldsymbol{\xi}_1 \cdot \boldsymbol{\xi}_2' \equiv 0$), so the three vectors form an orthogonal—now even ortho*normal*—set. In fact, the matrix comprised of the three vectors,

$$\boldsymbol{\Xi} \equiv (\boldsymbol{\xi}_1 \quad \boldsymbol{\xi}_3 \quad \boldsymbol{\xi}_2') = \begin{pmatrix} \cos\theta & 0 & -\sin\theta \\ 0 & 1 & 0 \\ \sin\theta & 0 & \cos\theta \end{pmatrix},$$

the new basis represented in terms of the old, is *orthogonal*—$\boldsymbol{\Xi}^{\mathsf{T}}\boldsymbol{\Xi} = \mathbf{I}$, so corresponds to a *rotation* (Section 3.3.1). And $|\boldsymbol{\Xi}| = +1$ (this is the reason the vectors were slightly reordered in $\boldsymbol{\Xi}$), so that this ordering of the three eigenvectors describes a *proper* rotation (Definition 3.3.4).[10]

We have already pointed out that $\boldsymbol{\xi}_1 = \hat{\imath}'$ and $\boldsymbol{\xi}_3 = \hat{\jmath} = \hat{\jmath}'$ in Example 5.2.2; we now observe that $\boldsymbol{\xi}_2'$ is just $\hat{k}'$ in the Example. Thus these three vectors comprise precisely those basis directions diagonalizing the $\mathbf{I}$—the principal axes predicted. Though we could continue and show explicitly that the product

[10] In fact, $\boldsymbol{\Xi}$ can be recognized to be nothing more than the rotation matrix $\mathsf{B}(-\theta)$—that matrix bringing the (x, y, z) axes to the principal (x', y', z').

$\boldsymbol{\Xi}^{-1}\mathbf{I}\boldsymbol{\Xi} = \boldsymbol{\Xi}^{\mathsf{T}}\mathbf{I}\boldsymbol{\Xi}$ diagonalizes the matrix, with the three eigenvalues 2, 1, and 1 along the diagonal in that order, this will not be done here.

Before leaving this example entirely, we point out that the same procedure could be applied to Example 5.2.1. But now the characteristic polynomial, after simplification using $l^2 + m^2 + n^2 = 1$, becomes

$$|\mathbf{I} - \lambda\mathbf{1}| = \lambda^3 - 2\lambda^2 + \lambda = 0,$$

an equation giving $\lambda = 0$ as an eigenvalue—precisely the possibility mentioned at the end of Definition 3.5.1! The corresponding eigenvalue $\boldsymbol{\xi} \neq \mathbf{0}$ must therefore be an element of the null space of $\mathbf{I} - 0\mathbf{1} = \mathbf{I}$, and, in fact, turns out to be $\boldsymbol{\xi} = (l, m, n)^{\mathsf{T}} = \hat{\boldsymbol{\lambda}}$, the direction of the rod. This corresponds to the principal "$I_{xx} = 0$" in the remark regarding this Example on page 277. |

Homework:

1. Show that $\boldsymbol{\xi} = (l, m, n)^{\mathsf{T}} = \hat{\boldsymbol{\lambda}}$ is, indeed, an element of the null space of the $\mathbf{I}$ in Example 5.2.1.

A point rarely emphasized strongly enough is the fact that principal axes for a given body are more than merely a matter of the *directions* of the axes, even though they are ultimately determined by a simple *rotation* of axes from those in which $\mathbf{I}$ is originally expressed. This is because $\mathbf{I}$ fundamentally depends on the *coordinates* used to express it—the *values* of the x, y, and z in the various integrals making up the elements of the matrix, and not only the *directions* in which they are measured—and thus the *origin* of such coordinates (see the remark on page 282). This fact is illustrated in the following:

Example 5.2.5 *Principal Axes dependent on Origin.* We have at left a simple thin, uniform, rectangular plate of mass m, lying in the xy-plane, whose center is located at the origin of coordinates. We claim that the given axes are in fact principal: by virtue of the fact that $z \equiv 0$ to all dm in the plate, we have immediately that the products of inertia $I_{xz} = I_{yz} \equiv 0$; further, $I_{xy} \equiv 0$ because, to each dm at $(x, y, 0)$ there is one at $(-x, y, 0)$, so the net integral $\int_m xy\,dm \equiv 0$. This means that there remain only the *moments* of inertia of the plate to determine. These can be found easily in the tables, leading to the matrix $\bar{\mathbf{I}}$ (to distinguish its being through the center of mass)

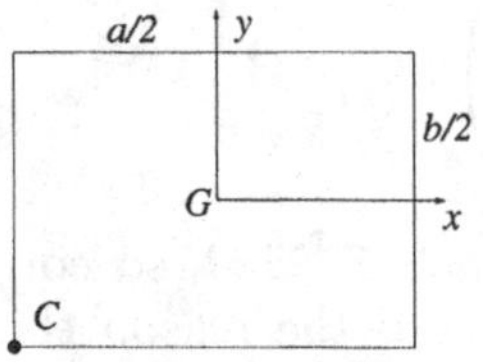

$$\bar{\mathbf{I}} = \begin{pmatrix} \frac{1}{12}mb^2 & 0 & 0 \\ 0 & \frac{1}{12}ma^2 & 0 \\ 0 & 0 & \frac{1}{12}m(a^2 + b^2) \end{pmatrix}.$$

Now let us consider the inertia matrix I_C relative to the parallel axes at the point C. A straightforward application of the Parallel Axis Theorem gives

$$I_{xx} = \overline{I}_{xx} + md_x^2 = \frac{1}{12}mb^2 + m\left(\frac{b}{2}\right)^2 = \frac{mb^2}{3}$$
$$I_{yy} = \overline{I}_{yy} + md_y^2 = \frac{1}{12}ma^2 + m\left(\frac{a}{2}\right)^2 = \frac{ma^2}{3}$$
$$I_{zz} = \overline{I}_{zz} + md_z^2 = \frac{1}{12}(a^2+b^2) + m\left(\left(\frac{a}{2}\right)^2 + \left(\frac{b}{2}\right)^2\right) = \frac{m(a^2+b^2)}{3}$$

In the same way,

$$I_{xz} = \overline{I}_{xz} + m\overline{xz} = 0 + m\left(\frac{b}{2}\right)0 = 0,$$

with I_{yz} vanishing similarly. But when it comes to the last product of inertia,

$$I_{xy} = \overline{I}_{xy} + m\overline{xy} = 0 + m\left(\frac{b}{2}\right)\left(\frac{a}{2}\right) = \frac{mab}{4},$$

which *doesn't* vanish! Thus the parallel axes at C are *not* principal. While it would be possible to determine what the principal axes at C *are*, this will not be done here; the point is that even simple translations can destroy the principal nature of the axes' directions. |

Homework:

1. Find directions of the principal axes at C in the above example, comparing these with those at O.

While these examples verify the existence of principal axes (and also bring together a great deal of the material of Part I!), the practical importance of such axes is demonstrated by Examples 5.2.2 and 5.2.5: if we can determine such principal axes *by inspection*, it becomes a simple matter to find only the *moments* of inertia in such axes, then translate and/or rotate from these axes to those in which the inertia matrix is ultimately desired, giving the complete inertia matrix in the desired coordinates. Given the advantages of such an approach, it is useful to be able to recognize in a given problem just what the principal axes might be. There are two useful criteria to apply:

Principal axes are generally regarded as the entire *set* of axes in which the inertia matrix is diagonalized. But we can also consider a *single* "principal axis" to be one whose two products with the other coordinates vanish; for example, if I_{xy} and I_{xz} both vanish, then we can regard x as *a* principal axis. This is the thrust of the first criterion:

Theorem 5.2.3 (Symmetry.). A principal axis of a rigid body is perpendicular to a plane of symmetry.

Proof. Assume that a body is symmetric with respect to, say, the xy-plane. We claim that the z-axis is principal; *i.e.*, that $I_{xz} \equiv 0 \equiv I_{yz}$. This follows immediately from the definitions of the latter products. For example, $I_{xz} \equiv \int_m xz\,dm \equiv 0$ because to each dm at (x, y, z) there is an identical one at $(x, y, -z)$. □

We remark that this phenomenon was demonstrated in the above Example 5.2.5: even though the x- and y-axes were not principal at C, the z-axis still was, because it was perpendicular to a plane of symmetry. Furthermore, in the event that there are *two* such axes of symmetry—they themselves will have to be perpendicular, by Theorem 3.7.1—the third, also perpendicular to the first two, is then completely determined.

The second criterion deals with translation along a principal axis, particularly one of the set *through the center of mass*:

Theorem 5.2.4 (Translation along Center of Mass Principal Axes). A parallel translation to a new origin along a principal axis *through the center of mass* leaves the axes principal.

Proof. Consider, for example, translation along the principal x-axis. The products of inertia calculated about such a new origin at $(\overline{x}, 0, 0)$ will be

$$\begin{aligned}
I_{x'y'} &= \overline{I}_{x'y'} + m\overline{x}'\overline{y}' = 0 + m\overline{x}'\,0 = 0 \\
I_{x'z'} &= \overline{I}_{x'z'} + m\overline{x}'\overline{z}' = 0 + m\overline{x}'\,0 = 0 \\
I_{y'z'} &= \overline{I}_{y'z'} + m\overline{y}'\overline{z}' = 0 + m\,0\,0 = 0,
\end{aligned}$$

since the original center-of-mass axes were principal themselves. □

In the event we move off the principal axes altogether, however, the original axes relative to the new point will, in general, *not* be principal, since non-vanishing products will be introduced by the Parallel Axis Theorem for these products; this is just what happened in Example 5.2.5. And, even though new axes remain principal in the above cases so the *products* of inertia are unchanged, in general the *moments* of inertia will have to be recalculated using the Parallel Axis Theorem.

We are now ready for a couple of examples showing how the inertia matrix would generally be determined in practice. We recall that this matrix will usually have both moments and products of inertia, yet few tables give the latter! The reason for this is the fact that the moments of inertia given in these tables are almost invariably along *principal* axes, and we have shown on page 284 how we can *rotate* from such [principal] axes in which there are *no* products to an arbitrary set of axes in which there *are*. This is the procedure we exemplify in the following:

Example 5.2.6 *Translation plus Rotation on Single Body.* We return to the

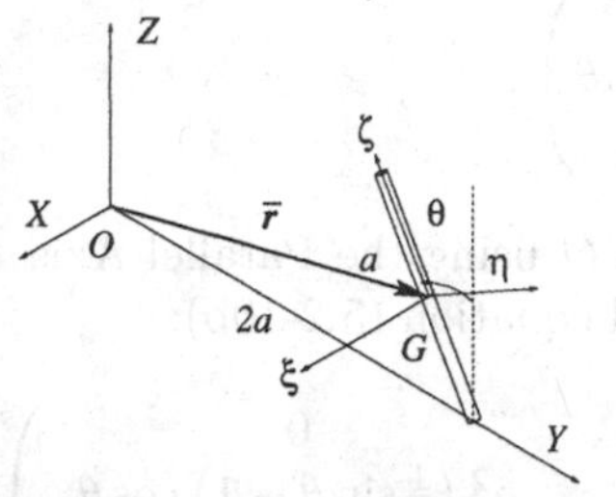

uniform slender rod of mass m and length a. It is located in the YZ-plane, with one end located $2a$ along the Y-axis from O, and makes an angle of θ with the Z-axis as shown. We desire to find the inertia matrix of this rod relative to the point O in the given axes.

Clearly we are going to have to undergo a translation to get the matrix relative to the origin at O; this means we will have to know the matrix [representation] relative to the center of mass of the rod G, lying at its middle. Tables give the *moments* of inertia about axes *perpendicular* to the rod through its center as being $\frac{1}{12}ma^2$, while, from the definition of the moment of inertia, that about the rod's axis vanishes ("$r = 0$"). From our discussion of principal axes, it is clear that one of these is the axis of the rod itself: the body is symmetric about a plane through G (remember that the identity of principal axes depends on their origin). But the rod is similarly symmetric to *any* plane through that axis, making the other two principal axes any pair of mutually perpendicular axes also perpendicular to the rod's axis (remember that principal axes always form an *orthogonal* set, by Theorem 3.7.1); for convenience we take one of these axes parallel to the X-axis, with the other perpendicular to the first two selected—the centroidal (ξ, η, ζ) on the diagram. These axes are precisely the ones cited in the tables, so we have immediately the inertia matrix relative to these center-of-mass *principal* axes:

$$\bar{\mathbf{I}}^{(p)} = \begin{pmatrix} \frac{1}{12}ma^2 & 0 & 0 \\ 0 & \frac{1}{12}ma^2 & 0 \\ 0 & 0 & 0 \end{pmatrix}.$$

The center of mass is located at the point $\bar{\mathbf{r}} = \left(2a - \frac{a}{2}\sin\theta\right)\hat{\boldsymbol{J}} + \frac{a}{2}\cos\theta\hat{\boldsymbol{K}}$ relative to the origin. *But this representation is in terms of the original* $(\hat{\boldsymbol{I}}, \hat{\boldsymbol{J}}, \hat{\boldsymbol{K}})$ *basis*, whereas $\bar{\mathbf{I}}^{(p)}$ is expressed in the *principal axis* $(\hat{\mathbf{u}}_\xi, \hat{\mathbf{u}}_\eta, \hat{\mathbf{u}}_\zeta)$ basis. According to the remark on page 284, however, these must be in the *same* basis in order to invoke the Parallel Axis Theorem.

One means of overcoming this basis mismatch is to transform the principal axis representation $\bar{\mathbf{I}}^{(p)}$ to its representation $\bar{\mathbf{I}}$ parallel to the original axes; then we can apply the Parallel Axis Theorem to the resulting $\bar{\mathbf{I}}$. With our knowledge of how matrix representations transform, this is easy: to bring the principal axes parallel to the original ones, we merely rotate about the x'-axis through an angle of $-\theta$ [note the sign!] using the matrix $\mathsf{A}(-\theta)$, equation (3.6-12), and the fact that $\mathsf{A}^{-1}(\theta) = \mathsf{A}(-\theta) = \mathsf{A}^{\mathsf{T}}(\theta)$:

$$\begin{aligned}\bar{\mathbf{I}} &= \mathsf{A}^{-1}(-\theta)\bar{\mathbf{I}}^{(p)}\mathsf{A}(-\theta) = \mathsf{A}(\theta)\bar{\mathbf{I}}^{(p)}\mathsf{A}^{\mathsf{T}}(\theta) \\ &= \begin{pmatrix} 1 & 0 & 0 \\ 0 & \cos\theta & -\sin\theta \\ 0 & \sin\theta & \cos\theta \end{pmatrix} \begin{pmatrix} \frac{1}{12}ma^2 & 0 & 0 \\ 0 & \frac{1}{12}ma^2 & 0 \\ 0 & 0 & 0 \end{pmatrix} \begin{pmatrix} 1 & 0 & 0 \\ 0 & \cos\theta & \sin\theta \\ 0 & -\sin\theta & \cos\theta \end{pmatrix}\end{aligned}$$

$$= \begin{pmatrix} \frac{1}{12}ma^2 & 0 & 0 \\ 0 & \frac{1}{12}ma^2\cos^2\theta & \frac{1}{12}ma^2\sin\theta\cos\theta \\ 0 & \frac{1}{12}ma^2\sin\theta\cos\theta & \frac{1}{12}ma^2\sin^2\theta \end{pmatrix}.$$

That done, we translate along these parallel axes to O using the Parallel Axis Theorem, using the components of $\bar{\mathbf{r}}$ cited above and equation (5.2-10b):

$$\mathsf{I}_O = \bar{\mathsf{I}} + m \begin{pmatrix} a^2\left(\frac{17}{4} - 2\sin\theta\right) & 0 & 0 \\ 0 & a^2\frac{1}{4}\cos^2\theta & a^2\left(\frac{1}{4}\sin\theta - 1\right)\cos\theta \\ 0 & a^2\left(\frac{1}{4}\sin\theta - 1\right)\cos\theta & a^2\left(4 - 2\sin\theta + \frac{\sin^2\theta}{4}\right) \end{pmatrix}$$
$$= \begin{pmatrix} \frac{13}{3} - 2\sin\theta & 0 & 0 \\ 0 & \frac{1}{3}\cos^2\theta & \left(\frac{1}{3}\sin\theta - 1\right)\cos\theta \\ 0 & \left(\frac{1}{3}\sin\theta - 1\right)\cos\theta & 4 - 2\sin\theta + \frac{1}{3}\sin^2\theta \end{pmatrix} ma^2. \quad (5.2\text{-}11)$$

But there is another way of attacking this problem: Rather than expressing $\bar{\mathsf{I}}^{(p)}$ in the *original* axes, we can instead express $\bar{\mathbf{r}}$ in the *principal* ones; this means we transform the representation $\bar{\mathbf{r}}$ in the original axes to its representation $\bar{\mathbf{r}}^{(p)}$ in the principal ones by means of a basis rotation about the x-axis through an angle of $+\theta$:

$$\bar{\mathbf{r}}^{(p)} = \mathsf{A}^{-1}(\theta)\bar{\mathbf{r}} = \mathsf{A}^{\mathsf{T}}(\theta)\bar{\mathbf{r}}$$
$$= \begin{pmatrix} 1 & 0 & 0 \\ 0 & \cos\theta & \sin\theta \\ 0 & -\sin\theta & \cos\theta \end{pmatrix} \begin{pmatrix} 0 \\ 2a - \frac{1}{2}\sin\theta \\ \frac{1}{2}a\sin\theta \end{pmatrix} = \begin{pmatrix} 0 \\ 2a\cos\theta \\ \frac{1}{2}a - 2a\sin\theta \end{pmatrix}.$$

Now we apply the Parallel Axis Theorem along *these* directions, using the components of $\bar{\mathbf{r}}^{(p)}$:

$$\mathsf{I}_O^{(p)} = \bar{\mathsf{I}}^{(p)} +$$
$$m \begin{pmatrix} a^2\left(\frac{17}{4} - 2\sin\theta\right) & 0 & 0 \\ 0 & a^2\left(\frac{1}{4} - 2\sin\theta + 4\sin^2\theta\right) & a^2(4\sin\theta - 1)\cos\theta \\ 0 & a^2(4\sin\theta - 1)\cos\theta & 4a^2\cos^2\theta \end{pmatrix}$$
$$= \begin{pmatrix} a^2\left(\frac{13}{3} - 2\sin\theta\right) & 0 & 0 \\ 0 & a^2\left(\frac{1}{3} - 2\sin\theta + 4\sin^2\theta\right) & a^2(4\sin\theta - 1)\cos\theta \\ 0 & a^2(4\sin\theta - 1)\cos\theta & 4a^2\cos^2\theta \end{pmatrix}.$$

But this is I_O represented in *principal* coordinates; to represent it in terms of the *original* basis, we merely rotate it back through an angle of $-\theta$ to bring the principal axes coincident with the original:

$$\mathsf{I}_O = \mathsf{A}^{-1}(-\theta)\mathsf{I}_O^{(p)}\mathsf{A}(-\theta) = \mathsf{A}(\theta)\mathsf{I}_O^{(p)}\mathsf{A}^{\mathsf{T}}(\theta),$$

obtaining precisely the result (5.2-11) above. (Yes, it actually *does* work!)

The primary reason for the second approach was more pedagogical than practical: the intent was to show how these representations interact and to reinforce the directions in which the various rotations and translations should be implemented. Though the reader is, perhaps, invited to verify the above calculations, the important thing is less the details and actual results than gaining some facility with the approach and, again, the various relations between rotation matrices, their inverses and transposes, and the signs of the angular arguments. |

Example 5.2.7 *Inertia Matrix for a* Composite *Body.* The last example in this section deals with the inertia matrix of a simple composite body: The body ABC diagrammed at left consists of two [again!] "uniform slender rods," AB and BC, welded together at a right angle at the point B. The mass of AB is $2m$ and it has a length of 1200 mm; BC, on the other hand, has mass and length of m and 600 mm, respectively.[11] The body is positioned with A at a point 1200 mm along the Y-axis, with BC parallel to the X-axis and AB making an angle of 30° with the vertical. We are again to determine the inertia matrix relative to the origin of coordinates, O.

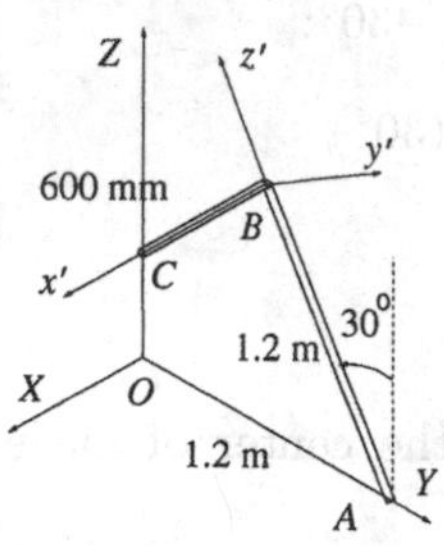

The statement of the problem is obviously meant to illustrate the Composition Theorem 5.2.1. But, as with a number of other examples presented, we shall work this problem two ways: first obtaining the desired inertia matrix of the entire body by finding each of the two parts relative to O *separately*, then by finding that for the body as a whole and translating the *single* inertia matrix to O using the Parallel Axis Theorem.

The first approach is relatively simple: We shall invoke the Composition Theorem in the form

$$\mathbf{I}_O = \mathbf{I}_{O_{AB}} + \mathbf{I}_{O_{BC}}.$$

We observe that the original basis directions are center-of-mass principal axes for BC: the body is symmetric about a plane through its center and parallel to the YZ-plane, so the $\hat{\imath}$-direction is principal; and planes through the rod's axis parallel to the other two [original] coordinate planes are also ones of symmetry. Thus we can invoke the Parallel Axis Theorem for this body directly using $\bar{r}_{BC} = .3\hat{I} + .6\hat{J} + .6\sqrt{3}\hat{K}$ m:

$$\begin{aligned}\mathbf{I}_{O_{BC}} &= \begin{pmatrix} 0 & 0 & 0 \\ 0 & \frac{1}{12}ma^2 & 0 \\ 0 & 0 & \frac{1}{12}ma^2 \end{pmatrix} + m\begin{pmatrix} d_x^2 & -\overline{xy} & -\overline{xz} \\ -\overline{xy} & d_y^2 & -\overline{yz} \\ -\overline{xz} & -\overline{yz} & d_z^2 \end{pmatrix} \\ &= \begin{pmatrix} 0 & 0 & 0 \\ 0 & 0.03m & 0 \\ 0 & 0 & .03m \end{pmatrix} + m\begin{pmatrix} 1.44 & -.18 & -.18\sqrt{3} \\ -.18 & 1.17 & -.36\sqrt{3} \\ -.18\sqrt{3} & -.36\sqrt{3} & .45 \end{pmatrix} \\ &= \begin{pmatrix} 1.44 & -.18 & -.18\sqrt{3} \\ -.18 & 1.2 & -.36\sqrt{3} \\ -.18\sqrt{3} & -.36\sqrt{3} & .48 \end{pmatrix} m\end{aligned}$$

[11] "m" is left in literal form to separate its effect from that of the various dimensions.

The inertia matrix due to AB is found similarly to the previous example: The directions for the (x', y', z') axes through the center of mass are principal, so that, since the moments of inertia $I_{x'x'} = I_{y'y'} = \frac{1}{12}(2m)(2a)^2 = \frac{2}{3}ma^2$ and $I_{z'z'} \equiv 0$,

$$\bar{\mathbf{I}}_{AB}^{(p)} = \begin{pmatrix} \frac{2}{3}ma^2 & 0 & 0 \\ 0 & \frac{2}{3}ma^2 & 0 \\ 0 & 0 & 0 \end{pmatrix} = \begin{pmatrix} .24m & 0 & 0 \\ 0 & .24m & 0 \\ 0 & 0 & 0 \end{pmatrix};$$

this representation in *principal* axes can be expressed relative to the *original* axes through a rotation about the x'-axis through the angle $-30°$:

$$\bar{\mathbf{I}}_{AB} = \mathbf{A}^{-1}(-30°)\bar{\mathbf{I}}_{AB}^{(p)}\mathbf{A}(-30°) = \mathbf{A}(30°)\bar{\mathbf{I}}_{AB}^{(p)}\mathbf{A}^{\mathsf{T}}(30°)$$
$$= \begin{pmatrix} .24m & 0 & 0 \\ 0 & .18m & .06\sqrt{3}m \\ 0 & .06\sqrt{3}m & .06m \end{pmatrix};$$

and finally we get $\mathbf{I}_{O_{AB}}$ through a *parallel* translation to the center of mass $\bar{\boldsymbol{r}}_{AB} = .9\hat{\boldsymbol{J}} + .3\sqrt{3}\hat{\boldsymbol{K}}$ [*note the "2m"*]:

$$\mathbf{I}_{O_{AB}} = \bar{\mathbf{I}}_{AB} + 2m\begin{pmatrix} 1.08 & 0 & 0 \\ 0 & .27 & -.27\sqrt{3} \\ 0 & -.27\sqrt{3} & .81 \end{pmatrix} = \begin{pmatrix} 2.4 & 0 & 0 \\ 0 & .72 & -.48\sqrt{3} \\ 0 & -.48\sqrt{3} & 1.68 \end{pmatrix} m.$$

Thus, finally, the *total* inertia matrix about O is, by the Composition Theorem,

$$\mathbf{I}_O = \mathbf{I}_{O_{AB}} + \mathbf{I}_{O_{BC}} = \begin{pmatrix} 3.84 & -.18 & -.18\sqrt{3} \\ -.18 & 1.92 & -.84\sqrt{3} \\ -.18\sqrt{3} & -.84\sqrt{3} & 2.16 \end{pmatrix} m. \qquad (5.2\text{-}12)$$

The above approach required two Parallel Axis translations, one for *each* part of the body *as well as a rotation of axes* for one of them, whose principal axes were not parallel to the primary reference axes. This is fairly typical of the general application of the Composition Theorem: sub-bodies of a composite will rarely be assembled with their individual principal axes parallel to one another; thus an axis rotation is almost unavoidable, somewhere along the way, to find the inertia matrix for the whole.

An alternative means would be to find the inertia matrix for the *entire* body about some point (and relative to some basis), followed by a *single* translation; that is what will be done next:

The (x', y', z') principal axes at point B are *also* principal *for each body*: this is a point lying on a principal axis of each through its center of mass (Theorem 5.2.4). Thus, again by the Composition Theorem,

$$\mathbf{I}_B^{(p)} = \mathbf{I}_{B_{AB}}^{(p)} + \mathbf{I}_{B_{BC}}^{(p)} = \begin{pmatrix} \frac{8}{3}ma^2 & 0 & 0 \\ 0 & \frac{8}{3}ma^2 & 0 \\ 0 & 0 & 0 \end{pmatrix} + \begin{pmatrix} 0 & 0 & 0 \\ 0 & \frac{1}{3}ma^2 & 0 \\ 0 & 0 & \frac{1}{3}ma^2 \end{pmatrix}$$
$$= \begin{pmatrix} .96 & 0 & 0 \\ 0 & 1.08 & 0 \\ 0 & 0 & .12 \end{pmatrix} m.$$

It is tempting to rotate from these principal axes to the original ones and translate back immediately using the Parallel Axis Theorem. But this would be *wrong*: *the Parallel Axis Theorem only applies to translations from the center of mass!* Thus it is necessary first to find its representation $\bar{\mathbf{I}}^{(p)}$ relative to this point: Since, relative to B, the center of mass $\bar{\mathbf{r}}^{(p)}$ in these principal coordinates satisfies

$$(m_{AB} + m_{BC})\bar{\mathbf{r}}^{(p)} = 3m\bar{\mathbf{r}}^{(p)} = m_{AB}\bar{\mathbf{r}}_{AB} + m_{BC}\bar{\mathbf{r}}_{BC} = 2m(-.6\hat{k}') + m(.3\hat{\imath}'),$$

we get that $\bar{\mathbf{r}}^{(p)} = .1\hat{\imath}' - .4\hat{k}'$. Thus, applying the Parallel Axis Theorem "backwards,"

$$\bar{\mathbf{I}}^{(p)} = \mathbf{I}_B^{(p)} - 3m\begin{pmatrix} .16 & 0 & .04 \\ 0 & .17 & 0 \\ .04 & 0 & .01 \end{pmatrix} = \begin{pmatrix} .48 & 0 & -.12 \\ 0 & -.51 & 0 \\ -.12 & 0 & .09 \end{pmatrix} m.$$

Again, however, this representation is relative to the *principal* axis directions; thus it is necessary to rotate back to the *original* ones:

$$\begin{aligned}\bar{\mathbf{I}} &= \mathbf{A}^{-1}(-30°)\bar{\mathbf{I}}^{(p)}\mathbf{A}(-30°) = \mathbf{A}(30°)\bar{\mathbf{I}}^{(p)}\mathbf{A}^{\mathsf{T}}(30°) \\ &= \begin{pmatrix} .48 & .06 & -.06\sqrt{3} \\ .06 & .45 & .12\sqrt{3} \\ -.06\sqrt{3} & .12\sqrt{3} & .21 \end{pmatrix} m.\end{aligned}$$

Finally, we are ready to translate back to O using the fact that the center of mass of the *entire* body is located at $\bar{\mathbf{r}} = .1\hat{I} + .8\hat{J} + .8\sqrt{3}\hat{K}$:

$$\mathbf{I}_O = \bar{\mathbf{I}} + 3m\begin{pmatrix} 1.12 & -.08 & -.04\sqrt{3} \\ -.08 & .49 & -.32\sqrt{3} \\ -.04\sqrt{3} & -.32\sqrt{3} & .65 \end{pmatrix},$$

which *does*, in fact, give the same result, (5.2-12)!

Again, less important than the actual results above, one should focus on the sequence of steps required to get to each solution. It is also important to note just which angles and rotation matrices were used, as well as the values of the masses multiplying each parallel translation matrix. |

The above examples have explicitly utilized rotation matrices to effect the transformation from principal axes; this is done to drive home the fact that it is simple rotations which take one from non-principal to principal axes—and the other way around. Though the last two examples required only a single rotation, in general several such, about different axes, are required, necessitating the use of matrix products; see, for example, Example 5.2.3. In that case, we were given not the angles explicitly, but rather *coordinate* information regarding the unit vector $\hat{\boldsymbol{\lambda}}$, and it was necessary to go to some lengths to obtain the requisite angles. (It is also necessary to find the inverse of such products in order to transform the inertia tensor, of course, by equation (3.6-7). But, for the rotation matrices guaranteed by the real, symmetric nature of that tensor [representation] [Theorem 3.7.1], the inverse is itself just the transpose [Theorem 3.3.2]).

Yet, as neatly as such rotations fit into the scheme expounded thus far to diagonalize—*i.e.*, find the principal axes of—the inertia matrix, it does seem superfluous to have to find rotation angles when the very same information would appear to be contained in specification of the coordinates given. We thus present an alternative approach obviating the need for the matrix products required for the composition of such rotations—at least when the appropriate information is given:

Inertial Matrices from Dimensional/Coordinate Data. In the previous Example 5.2.3, we knew I in the original *reference* axes and found I' in the newly rotated ones; this entailed a rotation from the (x, y, z) to (x', y', z') system. Now we will implement exactly the same philosophy of Section 3.6—finding the basis transformation matrix—but doing so *directly*, rather than finding them through rotation matrices (Section 3.6.2).

Recall that we argued in that Example that, in order to obtain $I_{x'x'}$ of—*there*—$\mathsf{I}' = \mathsf{R}^{\mathsf{T}}\mathsf{I}\mathsf{R}$, it was necessary only to have the product of the first row of R^{T} and the first column of R. A great deal of effort was still required, however, to determine those angles necessary to effect even this partial rotation.

But we already *have* the direction cosines necessary to describe the direction $\hat{\boldsymbol{\lambda}}$ there; indeed, that unit vector is nothing more than the *representation* of the new x'-axis, which will be the first column of a transformation matrix going from the $(\hat{\boldsymbol{\imath}}, \hat{\boldsymbol{\jmath}}, \hat{\boldsymbol{k}})$ basis to $(\hat{\boldsymbol{\imath}}', \hat{\boldsymbol{\jmath}}', \hat{\boldsymbol{k}}')$:

$$\mathsf{U} \equiv \begin{pmatrix} l & \cdots & \cdots \\ m & \cdots & \cdots \\ n & \cdots & \cdots \end{pmatrix}.$$

This will then determine the desired

$$\mathsf{I}' = \mathsf{U}^{-1}\mathsf{I}\mathsf{U} = \mathsf{U}^{\mathsf{T}}\mathsf{I}\mathsf{U},$$

where we know $\mathsf{U}^{-1} = \mathsf{U}^{\mathsf{T}}$ precisely *because rotations are generated by orthogonal matrices.* Thus we get exactly the same result we had in the previous incantation of that Example, but without the need to find $\mathsf{R} = \mathsf{C}(\theta_z)\mathsf{B}(-\theta_{y''})$ and their requisite $\sin\theta_z$, $\cos\theta_z$, $\sin\theta_{y''}$, and $\cos\theta_{y''}$.

But, more generally, we have the *principal* axes' representation $\mathsf{I}^{(p)}$ (or $\bar{\mathsf{I}}^{(p)}$) and must rotate *back* to the reference axes. In the first place, it is clearly necessary to know the orientation of the body. Knowing that and the body's principal axes, then, the orientation of the principal axes is also known; from this information we have the *representations* of the principal axes, $\hat{\boldsymbol{\imath}}$, $\hat{\boldsymbol{\jmath}}$, $\hat{\boldsymbol{k}}$ in terms of the fundamental $\hat{\boldsymbol{I}}$, $\hat{\boldsymbol{J}}$, and $\hat{\boldsymbol{K}}$:

$$\hat{\mathbf{i}}(\hat{\boldsymbol{I}},\hat{\boldsymbol{J}},\hat{\boldsymbol{K}})$$
$$\hat{\mathbf{j}}(\hat{\boldsymbol{I}},\hat{\boldsymbol{J}},\hat{\boldsymbol{K}})$$
$$\hat{\mathbf{k}}(\hat{\boldsymbol{I}},\hat{\boldsymbol{J}},\hat{\boldsymbol{K}})$$

If, now, we form the matrix

$$\mathsf{U} \equiv (\hat{\mathbf{i}}(\hat{\boldsymbol{I}},\hat{\boldsymbol{J}},\hat{\boldsymbol{K}}),\hat{\mathbf{j}}(\hat{\boldsymbol{I}},\hat{\boldsymbol{J}},\hat{\boldsymbol{K}}),\hat{\mathbf{k}}(\hat{\boldsymbol{I}},\hat{\boldsymbol{J}},\hat{\boldsymbol{K}})), \tag{5.2-13}$$

this is precisely the matrix U defined in equation (3.6-4)—that matrix generating the transformation from the $(\hat{\boldsymbol{I}},\hat{\boldsymbol{J}},\hat{\boldsymbol{K}})$ basis to the $(\hat{\boldsymbol{\imath}},\hat{\boldsymbol{\jmath}},\hat{\boldsymbol{k}})$ one. Thus to go from $\mathsf{I}^{(p)}$ in principal axes to the base set—*from the latter basis to the former*—it is necessary to apply the *inverses* of those in (3.6-7):

$$\begin{aligned}\mathsf{I} &= (\mathsf{U}^{-1})^{-1}\mathsf{I}^{(p)}(\mathsf{U})^{-1}\\ &= \mathsf{U}\mathsf{I}^{(p)}\mathsf{U}^{-1},\end{aligned}$$

in which, recall, the transformation itself is merely a *rotation*, thus orthogonal, and having an easy inverse:

$$= \mathsf{U}\mathsf{I}^{(p)}\mathsf{U}^{\mathsf{T}}$$

Thus, rather than using the rotation angles to generate the rotation matrices and their [product] composite transformation, we jump immediately to the matrix composite by having the new basis vectors' representations directly—information available from their coordinate representations.

Now let us look at another example of both approaches:

Example 5.2.8 *Determination of* I *Given only Dimensions.* We consider a slight generalization of Example 4.4.5—one which will return in Example 5.2.11: A "uniform slender rod" AC [we ignore the mass of the clevis connection at end C] of length $2a$ has end A located on a circle of radius a, parallel to the XY-plane and oriented parallel to that plane $\sqrt{3}a$ above it. Its other end C is constrained to move along the X-axis. We are asked to determine the centroidal inertia matrix $\bar{\mathsf{I}}$ relative to the reference XYZ-axes when the angle in the XY-plane between the X-axis and the line from O to the projection of A in that plane is θ.[12] We shall work this problem two ways: first using the explicit rotation matrices used heretofore, then by directly determining the transformation matrix from [known] directions of the principal axes. The thrust will be on the rotation transformations; we shall not find $\mathsf{I}^{(p)}$

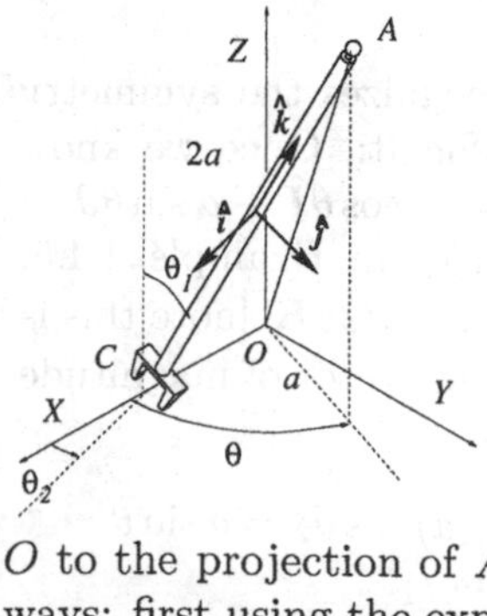

itself explicitly.

Clearly the information given fixes the orientation of AC. But no angles are given relative to which to perform the appropriate rotations to the reference axes from the principal ones. In fact, the axial symmetry of the rod makes such principal axes non-unique—a fact which we will use to our advantage. We will take the principal z-axis to be along the rod; then *any* pair of orthogonal axes perpendicular to this will also be principal, due to its very symmetry. For convenience we take the centroidal x-axis [instantaneously] to be along the intersection of two planes through the centroid of the member: one perpendicular to the rod and the other perpendicular to $\hat{\boldsymbol{K}}$:

$$\hat{\boldsymbol{\imath}} \parallel \hat{\boldsymbol{K}} \times \boldsymbol{r}_{AC};$$

thus it is parallel to the XY-plane, and perpendicular to the plane determined by these two directions. In particular, then, there is an angle in this latter plane between the Z-axis and the rod, denoted by θ_1; there is also an angle [in a plane parallel to the XY-plane] between the x- and X-axes, which we call θ_2. With $\hat{\boldsymbol{k}}$ and $\hat{\boldsymbol{\imath}}$ determined, $\hat{\boldsymbol{\jmath}}$ is also (though we won't need this one).

This choice of xyz-axes can easily be brought coincident (or, more correctly, *parallel*) with the XYZ reference axes through a pair of rotations:

1. a rotation about the x-axis through $+\theta_1$, bringing the z- and Z-axes coincident, followed by
2. a second rotation about the [new] z-axis through $-\theta_2$, bringing the x- and X-axes—and thus y and Y—coincident.

Thus the rotation matrix bringing the centroidal principal inertia tensor $\bar{\mathsf{I}}^{(p)}$ to its representation in the reference axes is just $\mathsf{A}(\theta_1)\mathsf{C}(-\theta_2) = \mathsf{A}(\theta_1)\mathsf{C}^{\mathsf{T}}(\theta_2)$ (recall the remark regarding the composition of two rotations on page 143).

It remains "merely" to determine θ_1 and θ_2. The first is easy: since AC has length $2a$ and $Z_A = \sqrt{3}a$ while $Z_B \equiv 0$,

$$\cos\theta_1 = \frac{\sqrt{3}a}{2a} = \frac{\sqrt{3}}{2};$$

θ_1 is at a constant 30°!—rather surprising until one recognizes the symmetry inherent in the problem. The second is little more difficult: Once we know the actual direction of $\hat{\boldsymbol{\imath}}$, then $\hat{\boldsymbol{\imath}} \cdot \hat{\boldsymbol{I}} = \cos\theta_2$. But $\boldsymbol{r}_A = a\cos\theta\hat{\boldsymbol{I}} + a\sin\theta\hat{\boldsymbol{J}} + \sqrt{3}a\hat{\boldsymbol{K}}$; while, in precisely the same way we found r_{AB_X} in Example 4.4.5, $\boldsymbol{r}_C = 2a\cos\theta\hat{\boldsymbol{I}}$. Thus $\boldsymbol{r}_{AC} = \boldsymbol{r}_C - \boldsymbol{r}_A = a\cos\theta\hat{\boldsymbol{I}} - a\sin\theta\hat{\boldsymbol{J}} - \sqrt{3}a\hat{\boldsymbol{K}}$ [note this is from A to C], so that $\hat{\boldsymbol{\imath}} \equiv \hat{\boldsymbol{K}} \times \boldsymbol{r}_{AC} = a\sin\theta\hat{\boldsymbol{I}} + a\cos\theta\hat{\boldsymbol{J}}$—a vector of magnitude a, and

$$\hat{\boldsymbol{\imath}} \cdot \hat{\boldsymbol{I}} \equiv (\hat{\boldsymbol{K}} \times \boldsymbol{r}_{AC}) \cdot \hat{\boldsymbol{I}} = \|\hat{\boldsymbol{I}}\|\,\|\hat{\boldsymbol{K}} \times \boldsymbol{r}_{AC}\|\cos\theta_2 = (1)(a)\cos\theta_2 = a\sin\theta$$

$$\therefore \cos\theta_2 = \sin\theta, \qquad \sin\theta_2 = \cos\theta.^{13}$$

[12] Relative to the remark concerning the motion of Example 4.4.5, we implicitly assume that $\theta \in (-3\pi/2, \pi/2)$.

This done, we are now prepared to determine the composite rotation matrix, R:

$$\begin{aligned}\mathsf{R} &= \mathsf{A}(\theta_1)\mathsf{C}^{\mathsf{T}}(\theta_2) = \mathsf{A}(30^\circ)\mathsf{C}^{\mathsf{T}}(\theta_2)\\ &= \begin{pmatrix}1 & 0 & 0\\ 0 & \cos 30^\circ & -\sin 30^\circ\\ 0 & \sin 30^\circ & \cos 30^\circ\end{pmatrix}\begin{pmatrix}\cos\theta_2 & \sin\theta_2 & 0\\ -\sin\theta_2 & \cos\theta_2 & 0\\ 0 & 0 & 1\end{pmatrix}\\ &= \begin{pmatrix}1 & 0 & 0\\ 0 & \frac{\sqrt{3}}{2} & -\frac{1}{2}\\ 0 & \frac{1}{2} & \frac{\sqrt{3}}{2}\end{pmatrix}\begin{pmatrix}\sin\theta & \cos\theta & 0\\ -\cos\theta & \sin\theta & 0\\ 0 & 0 & 1\end{pmatrix}\\ &= \begin{pmatrix}\sin\theta & \cos\theta & 0\\ -\frac{\sqrt{3}}{2}\cos\theta & \frac{\sqrt{3}}{2}\sin\theta & -\frac{1}{2}\\ -\frac{1}{2}\cos\theta & \frac{1}{2}\sin\theta & \frac{\sqrt{3}}{2}\end{pmatrix},\end{aligned} \tag{5.2-14}$$

from which we calculate $\mathsf{I} = \mathsf{R}^{\mathsf{T}}\mathsf{I}^{(p)}\mathsf{R}$.

Now consider the same problem, but determining the transformation matrix $\mathsf{U} \equiv (\hat{\mathsf{i}}(\hat{\boldsymbol{I}},\hat{\boldsymbol{J}},\hat{\boldsymbol{K}}), \hat{\mathsf{j}}(\hat{\boldsymbol{I}},\hat{\boldsymbol{J}},\hat{\boldsymbol{K}}), \hat{\mathsf{k}}(\hat{\boldsymbol{I}},\hat{\boldsymbol{J}},\hat{\boldsymbol{K}}))$ directly; this means we must calculate the respective representations explicitly:

From the diagram,

$$\begin{aligned}\hat{\boldsymbol{k}} \equiv \frac{\boldsymbol{r}_{CA}}{\|\boldsymbol{r}_{CA}\|} &= \frac{-a\cos\theta\hat{\boldsymbol{I}} + a\sin\theta\hat{\boldsymbol{J}} + \sqrt{3}a\hat{\boldsymbol{K}}}{2a}\\ &= -\frac{\cos\theta}{2}\hat{\boldsymbol{I}} + \frac{\sin\theta}{2}\hat{\boldsymbol{J}} + \frac{\sqrt{3}}{2}\hat{\boldsymbol{K}}.\end{aligned}$$

since $\|\boldsymbol{r}_{CA}\| = 2a$. Thus we get

$$\begin{aligned}\hat{\boldsymbol{\imath}} &\equiv \frac{\hat{\boldsymbol{K}}\times\boldsymbol{r}_{AC}}{\|\hat{\boldsymbol{K}}\times\boldsymbol{r}_{AC}\|}\\ &= \sin\theta\hat{\boldsymbol{I}} + \cos\theta\hat{\boldsymbol{J}},\end{aligned}$$

since, as already observed, $\|\hat{\boldsymbol{K}}\times\boldsymbol{r}_{AC}\| = a$. Thus, finally,

$$\begin{aligned}\hat{\boldsymbol{\jmath}} \equiv \hat{\boldsymbol{k}}\times\hat{\boldsymbol{\imath}} &= \begin{vmatrix}\hat{\boldsymbol{I}} & \hat{\boldsymbol{J}} & \hat{\boldsymbol{K}}\\ -\frac{\cos\theta}{2} & \frac{\sin\theta}{2}\hat{\boldsymbol{J}} & +\frac{\sqrt{3}}{2}\\ \sin\theta & \sin\theta & 0\end{vmatrix}\\ &= -\frac{\sqrt{3}}{2}\cos\theta\hat{\boldsymbol{I}} + \frac{\sqrt{3}}{2}\sin\theta\hat{\boldsymbol{J}} - \frac{1}{2}\hat{\boldsymbol{K}}\end{aligned}$$

[13] Just to be fussy, we note that, since $\theta \in (-3\pi/2, \pi/2)$, $\sin\theta_2$ and $\cos\theta_2$ are both positive for $\theta \in [0, \pi/2)$, while $\sin\theta_2 \geq 0$ and $\cos\theta_2 \leq 0$ on $\theta \in (-3\pi/2, 0]$; the first case give $\theta_2 \leq \pi/2$ for $\theta \in [0, \pi/2)$, while $\theta_2 \in [\pi/2, \pi)$ for $\theta \in (-3\pi/2, 0]$. Thus $\theta_2 \in (0, \pi)$ over the entire region of interest.

Collecting these vectors as n-tuple column vectors in the transformation matrix,

$$\mathsf{U} = \begin{pmatrix} \sin\theta & -\frac{\sqrt{3}}{2}\cos\theta & -\frac{1}{2}\cos\theta \\ \cos\theta & \frac{\sqrt{3}}{2}\sin\theta & \frac{1}{2}\sin\theta \\ 0 & -\frac{1}{2} & \frac{\sqrt{3}}{2} \end{pmatrix}. \tag{5.2-15}$$

Note that this matrix U *gives the* "new" *[*principal*] vectors' representations* $(\hat{\mathbf{i}}, \hat{\mathbf{j}}, \hat{\mathbf{k}})$ *in terms of the* "old" *[*reference*]* $(\hat{\boldsymbol{I}}, \hat{\boldsymbol{J}}, \hat{\boldsymbol{K}})$. But we wish to go *from* the "new" *to* the "old," whose transformation matrix is given by the inverse—*i.e.*, *transform*—matrix U^T:

$$\mathsf{I} = \mathsf{U}\mathsf{I}^{(p)}\mathsf{U}^\mathsf{T}.$$

rather than $\mathsf{I} = \mathsf{R}^\mathsf{T}\mathsf{I}^{(p)}\mathsf{R}$. Thus $\mathsf{U} = \mathsf{R}^\mathsf{T}$—as examination of equations (5.2-14) and (5.2-15) verifies—as expected.

Comparing the two approaches, we see that the first demanded finding θ_1 and θ_2, followed by explicit calculation of the product $\mathsf{A}(\theta_1)\mathsf{C}^\mathsf{T}(\theta_2)$, while the second required only determination of $\hat{\boldsymbol{\jmath}} \equiv \hat{\boldsymbol{k}} \times \hat{\boldsymbol{\imath}}$. |

Observe that much of the internal calculation above was necessitated by not having explicit expressions for the principal axes' directions. For a body without symmetry, on the other hand, such directions would already be determined.

As this example demonstrates, direct calculation of the transformation matrix is arguably more straightforward than going back through rotation matrices—at least, if the appropriate data are provided—for composite rotations. On the other hand, for *single* rotations—*e.g.*, Examples 5.2.2, 5.2.6, and 5.2.7—it's at least a toss-up. Use your judgment.

The primary thrust of this section has been determination of the matrix representation of the inertia tensor, which appears continually throughout the study of 3-D rigid body dynamics. But, lest we forget, the whole point of these inertia matrices is to calculate the angular momentum [in generic form] $\mathsf{H} = \mathsf{I}\boldsymbol{\omega}$! Let us make a few remarks regarding this:

- In the event that we undergo a rotation to bring I to diagonalized form $\mathsf{I}' = \mathsf{R}^{-1}\mathsf{I}\mathsf{R} = \mathsf{R}^\mathsf{T}\mathsf{I}\mathsf{R}$, *we have to express all other vectors* v *in terms of the new basis*: $\mathsf{v}' = \mathsf{R}^{-1}\mathsf{v} = \mathsf{R}^\mathsf{T}\mathsf{v}$; this is just what was done above in Examples 5.2.6 and 5.2.7 in determining $\overline{\boldsymbol{r}}$. In particular, we would also have to transform the angular velocity to $\boldsymbol{\omega}' = \mathsf{R}^\mathsf{T}\boldsymbol{\omega}$. We observe, in passing, that this makes $\mathsf{H}' = \mathsf{I}'\boldsymbol{\omega}' = \mathsf{R}^\mathsf{T}\mathsf{I}\mathsf{R}\mathsf{R}^\mathsf{T}\boldsymbol{\omega} = \mathsf{R}^\mathsf{T}\mathsf{I}\boldsymbol{\omega}$—or simply $\mathsf{H}' = \mathsf{R}^\mathsf{T}\mathsf{H} = \mathsf{R}^{-1}\mathsf{H}$, the transformed *vector* H!

- Recall that, in general the product $\mathsf{I}\boldsymbol{\omega}$ *is not parallel to* $\boldsymbol{\omega}$: only if $\boldsymbol{\omega}$ is an *eigenvector* of I—*i.e.*, $\boldsymbol{\omega}$ is itself parallel to a principal axis of I—will this occur. The last is precisely what happens in *planar* rigid-body motion, where I reduces to the *scalar* I_{zz}, so that $\mathsf{I}\boldsymbol{\omega} = I_{zz}\boldsymbol{\omega} \parallel \boldsymbol{\omega}$.

- All the above has focussed on a "generic" $\mathsf{H} = \mathsf{I}\boldsymbol{\omega}$, where "$\boldsymbol{H}$" is either $\overline{\boldsymbol{H}}$ or $\boldsymbol{H}_O$ and "I" $\bar{\mathsf{I}}$ or I_O, respectively. *But the latter holds only if the body is* rotating *about the fixed point O*; more generally, the angular momentum about an arbitrary [fixed] point O is given by equation (5.1-21b):

$$\boldsymbol{H}_O = m\overline{\boldsymbol{r}} \times \overline{\boldsymbol{v}} + \overline{\boldsymbol{H}} = m\overline{\boldsymbol{r}} \times \overline{\boldsymbol{v}} + \bar{\mathbf{I}}\boldsymbol{\omega}. \tag{5.2-16}$$

Homework:

1. Show that *if O is fixed and the body is rotating about it*, the term $m\overline{\boldsymbol{r}} \times \overline{\boldsymbol{v}}$ in equation (5.2-16) can be written

$$m\overline{\boldsymbol{r}} \times \overline{\boldsymbol{v}} = m[\overline{\boldsymbol{r}}^2\boldsymbol{\omega} - (\overline{\boldsymbol{r}} \cdot \boldsymbol{\omega})\overline{\boldsymbol{r}}].$$

2. In a manner similar to that leading to equations (5.2-4), show that *when the body is rotating about the fixed point O*,

$$m[\overline{\boldsymbol{r}}^2\boldsymbol{\omega} - (\overline{\boldsymbol{r}} \cdot \boldsymbol{\omega})\overline{\boldsymbol{r}}] = m\mathbf{D}\boldsymbol{\omega},$$

in which $\mathbf{D}$ is that same matrix in Problem 1 on page 284 using which the Parallel Axis Theorem can be written in the form $\mathbf{I}_O = \bar{\mathbf{I}} + m\mathbf{D}$. This demonstrates that, in the special case that the body is rotating about the fixed point O, the general expression (5.2-16) for $\mathbf{H}_O$ reduces to just $\mathbf{I}_O\boldsymbol{\omega}$.

3. Show that

$$m[\overline{\boldsymbol{r}}^2\boldsymbol{\alpha} - (\overline{\boldsymbol{r}} \cdot \boldsymbol{\alpha})\overline{\boldsymbol{r}}] + \bar{\mathbf{I}}\boldsymbol{\alpha} = \mathbf{I}_O\boldsymbol{\alpha}.$$

4. What are the eigenvalues of *any* inertia matrix?

Summary. This section has been rather extensive, and there has been a slow shift in emphasis toward the end. The primary conclusion, however, is that, given merely the *moments of inertia* in *principal axes* (in which the *products of inertia* are defined to vanish), it is possible to obtain the inertia matrix relative to coordinates with an arbitrary origin and orientation: from the inertia matrix represented in center-of-mass principal coordinates, one simply invokes translation (using the *Parallel Axis Theorem*) and rotation (using the *rotation matrices* on page 154). The approach enables the use of tables of such moments of inertia; a composite body made up of shapes included in such tables can be similarly composed using the *Composition Theorem* for the inertia matrix. In the event a body doesn't happen to be in the tables, it is necessary to resort to direct integration (as in, for example, Example 5.2.1).

5.2.2 Equations of Motion

The previous section is so long that it is easy to lose sight of its place in the development. But we are attempting to obtain the equations of motion for a rigid body. And Systems of Particles (Section 5.1.2) form the touchstone—the bridge—between single particles covered by Newton's Second Law and the present rigid bodies (which are just a special case of Systems). As was pointed out on the lead-in to these two sections, the center-of-mass formulation of the equation of motion for linear momentum of systems, equation (5.1-22a), carries over immediately to rigid bodies:

$$\sum_i \boldsymbol{F}_i = \frac{d\boldsymbol{G}}{dt}, \qquad \boldsymbol{G} \equiv m\overline{\boldsymbol{v}};$$

thus

$$\sum_i \boldsymbol{F}_i = m\overline{\boldsymbol{a}},$$

giving the translational motion of the [center of mass of the] body. This suggests that we *need* such center of mass; it can be determined from equation (5.1-18a) for a system of particles, or, for a *composite* rigid body made up of constituent sub-bodies, using the analogous

$$m\overline{\boldsymbol{r}} = \sum_i m_i \overline{\boldsymbol{r}}_i, \qquad (5.2\text{-}17)$$

$$m \equiv \sum_i m_i,$$

when each of the individual centers of mass $\overline{\boldsymbol{r}}_i$ is known.

Occasionally for a composite body we might happen to know the accelerations of its various parts' centers of mass; *e.g.*, when the body is known to rotate about a fixed axis, such accelerations are in circles of known radius. In this case we can bypass calculation of the *single* composite center of mass and use the analog to equation (5.1-18c) to utilize this information directly:

$$\sum_i \boldsymbol{F}_i = m\overline{\boldsymbol{a}} = \sum_i m_i \overline{\boldsymbol{a}}_i. \qquad (5.2\text{-}18)$$

We would expect that, in the same way that *linear* momentum determines the *translational* motion of the body, *angular* momentum would govern the *rotational* motion, and this is correct: Retaining, for the nonce, the "generic" form of equation (5.2-1c) $\mathsf{H} = \mathsf{I}\boldsymbol{\omega}$,

$$\sum_i \boldsymbol{M}_i = \frac{d\boldsymbol{H}}{dt}, \qquad \boldsymbol{H} \equiv \sum_i m_i \boldsymbol{r}_i \times \boldsymbol{v}_i = \mathsf{I}\boldsymbol{\omega},$$

we see that, in general, the right-hand side will yield the derivative

$$\frac{d(\mathsf{I}\boldsymbol{\omega})}{dt} = \dot{\mathsf{I}}\boldsymbol{\omega} + \mathsf{I}\dot{\boldsymbol{\omega}}.$$

These are *absolute* time derivatives—derivatives taken with respect to a *fixed* [inertial] system; in particular, then, $\dot{\boldsymbol{\omega}} \equiv \boldsymbol{\alpha}$, but $\dot{\mathbf{I}} \neq \mathbf{0}$ *because the body is rotating, so that, relative to the fixed system,* $\mathbf{I}$ *is continuously changing!* This is a remarkably awkward development, forcing us to evaluate this derivative.

If, rather than considering a *fixed* coordinate system to reference the body's motion, we introduced a *rotating* system, particularly one rotating *with the body*, the time derivative of $\mathbf{I}$ as observed *relative to the rotating system*, $\dot{\mathbf{I}}_{rel}$, would vanish: the coordinate system would rotate *with* the body, and there would be no change in $\mathbf{I}$ relative to that system. But the time derivative relative to a rotating coordinate system is precisely the quantity studied on page 200, and the derivative of $\boldsymbol{H}$ observed *in* that system is precisely what we called "$\dot{\boldsymbol{H}}_{rel}$," vanishing in the present case. And we can relate the time derivative of $\boldsymbol{H}$ in *that* system to the *absolute* time derivative required in the above equation of motion through the use of Theorem 4.3.1: since the current rotating system moves with the angular velocity, $\boldsymbol{\omega}$, of the *body*, we have immediately from that Theorem that

$$\frac{d\boldsymbol{H}}{dt} = \frac{d(\mathbf{I}\boldsymbol{\omega})}{dt} = \boldsymbol{\omega} \times \mathbf{I}\boldsymbol{\omega} + \left.\frac{d}{dt}\right|_{rel} (\mathbf{I}\boldsymbol{\omega}).$$

The second term

$$\left.\frac{d}{dt}\right|_{rel} (\mathbf{I}\boldsymbol{\omega}) = \dot{\mathbf{I}}_{rel}\boldsymbol{\omega} + \mathbf{I}\dot{\boldsymbol{\omega}}_{rel}.$$

But, since the system rotates with the body, $\dot{\mathbf{I}}_{rel} \equiv \mathbf{0}$, while, from Example 4.3.1, $\dot{\boldsymbol{\omega}}_{rel} \equiv \boldsymbol{\alpha}_{rel} = \boldsymbol{\alpha}$; thus, in the end,

$$\frac{d\boldsymbol{H}}{dt} = \boldsymbol{\omega} \times \mathbf{I}\boldsymbol{\omega} + \mathbf{I}\boldsymbol{\alpha},$$

so

$$\sum_i \boldsymbol{M}_i = \frac{d\boldsymbol{H}}{dt} = \boldsymbol{\omega} \times \mathbf{I}\boldsymbol{\omega} + \mathbf{I}\boldsymbol{\alpha}; \tag{5.2-19}$$

this is referred to as *Euler's equation.*[14]

Homework:

1. Assume that a rigid body rotates with [absolute] angular velocity $\boldsymbol{\omega}$ with two components: $\boldsymbol{\omega} = \boldsymbol{\Omega} + \boldsymbol{p}$, where $\boldsymbol{p}$ is measured *relative to* $\boldsymbol{\Omega}$. Show that the equation of motion for angular momentum satisfies the equation

$$\frac{d\boldsymbol{H}}{dt} = \left.\frac{d}{dt}\right|_{\boldsymbol{\Omega}} \mathbf{I}\boldsymbol{\omega} + \boldsymbol{\Omega} \times \mathbf{I}\boldsymbol{\omega}, \tag{5.2-20}$$

[14]This is the name given by Goldstein [*Classical Mechanics, Second Edition*, p. 204], and he should know. But the reader is warned that is it common, particularly in introductory Engineering texts, to apply the term only to these equations represented in *principal* coordinates. Such a distinction seems over-nice to the author, and runs counter to the coordinate-free thrust of this exposition, so is not observed in this text.

It is most not common not to expand $\frac{d\boldsymbol{H}}{dt}$ as is done here. The present practice is chosen for two reasons: it is closer in form to the equation "$\sum \boldsymbol{M} = \mathbf{I}\boldsymbol{\alpha}$" familiar from planar dynamics, and it stresses the fact that the moments determine the *angular* acceleration.

where "$\left.\frac{d}{dt}\right|_{\Omega}$" denotes the derivative observed in the system rotating with $\mathbf{\Omega}$ as in Homework 1 on page 187.

Students coming to this subject after a course in planar rigid body dynamics will recognize the second term on the right-hand side, but the first term comes as a bit of a shock. But, actually, it has been there all along, even in the special case of planar rigid body motion: We remarked on page 302 that when I is a tensor, $\mathsf{I}\boldsymbol{\omega} \not\parallel \boldsymbol{\omega}$. But in *scalar* problems, I becomes the *scalar* I_{zz}, so that $\mathsf{I}\boldsymbol{\omega} \parallel \boldsymbol{\omega}$, and the cross product $\boldsymbol{\omega} \times \mathsf{I}\boldsymbol{\omega} \equiv \mathbf{0}$.

If, to this point, we have been focussing on the "generic" form $\mathsf{H} = \mathsf{I}\boldsymbol{\omega}$, in which "$\mathsf{H}$" corresponds either to $\overline{\mathsf{H}}$ about the center of mass or "H_O" about a fixed point O about which the body is rotating—cases (5.2-1a-b), we shall now begin to be more specific about the equations of motion. The first case translates to

$$\sum_i \boldsymbol{M}_{O_i} = \frac{d\boldsymbol{H}_O}{dt} = \boldsymbol{\omega} \times \mathsf{I}_O\boldsymbol{\omega} + \mathsf{I}_O\boldsymbol{\alpha};$$

again let us emphasize that this is only a special case, corresponding to that in which *the body itself is rotating about the fixed point O!* The second case, however, is more general, holding for any type of motion:

$$\sum_i \overline{\boldsymbol{M}}_i = \frac{d\overline{\boldsymbol{H}}}{dt} = \boldsymbol{\omega} \times \bar{\mathsf{I}}\boldsymbol{\omega} + \bar{\mathsf{I}}\boldsymbol{\alpha};$$

as a general equation, it can be used in any situation. Given the similarity between the above equations, it is not uncommon for students to conclude, inductively,[15] that summing moments about an arbitrary point A means

$$\sum_i \boldsymbol{M}_{A_i} = \frac{d\boldsymbol{H}_A}{dt} = \boldsymbol{\omega} \times \mathsf{I}_A\boldsymbol{\omega} + \mathsf{I}_A\boldsymbol{\alpha};$$

but this is wrong. We now prove the *correct* means of summing moments about an arbitrary point:

Theorem 5.2.5 (Moment Equation about an *Arbitrary* Point). The summation of moments about an arbitrary point A satisfies the equation of motion

$$\sum_i \boldsymbol{M}_{A_i} = \boldsymbol{r}_{G/A} \times m\overline{\boldsymbol{a}} + \dot{\overline{\boldsymbol{H}}} = \boldsymbol{r}_{G/A} \times m\overline{\boldsymbol{a}} + \boldsymbol{\omega} \times \bar{\mathsf{I}}\boldsymbol{\omega} + \bar{\mathsf{I}}\boldsymbol{\alpha}.$$

Proof. We now prove the above assertion by considering a system of forces and [couple] moments acting on an arbitrary body. Shown here is such a body acted upon by a typical indexed force $\boldsymbol{F}_i$ and couple $\boldsymbol{M}_i$. The center of mass is located

[15] Induction is one of the classic examples of erroneous reasoning in Logic.

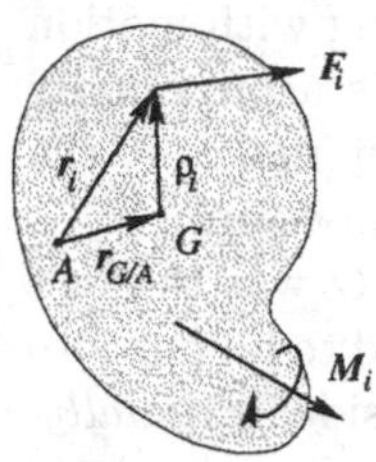

at G; as we have tried to do consistently, positions relative to this point are denoted as $\boldsymbol{\rho}_i$. Positions to the points of application of the $\boldsymbol{F}_i$ from the point A are denoted as $\boldsymbol{r}_i$; in particular, the position of G relative to that point is just $\boldsymbol{r}_{G/A}$. Though the couple moment has the same effect on all points in the body, the moment generated by $\boldsymbol{F}_i$ depends on the point about which it is calculated:

$$\begin{aligned}\sum_i \boldsymbol{M}_{A_i} &= \sum_i \boldsymbol{r}_i \times \boldsymbol{F}_i + \sum_i \boldsymbol{M}_i = \sum_i (\boldsymbol{r}_{G/A} + \boldsymbol{\rho}_i) \times \boldsymbol{F}_i + \sum_i \boldsymbol{M}_i \\ &= \boldsymbol{r}_{G/A} \times \sum_i \boldsymbol{F}_i + \sum_i \boldsymbol{\rho}_i \times \boldsymbol{F}_i + \sum_i \boldsymbol{M}_i.\end{aligned}$$

But, from the force equation, $\sum_i \boldsymbol{F}_i = m\overline{\boldsymbol{a}}$, while

$$\sum_i \boldsymbol{\rho}_i \times \boldsymbol{F}_i + \sum_i \boldsymbol{M}_i \equiv \sum_i \overline{\boldsymbol{M}}_i = \dot{\overline{\boldsymbol{H}}} = \boldsymbol{\omega} \times \bar{\mathsf{I}}\boldsymbol{\omega} + \bar{\mathsf{I}}\boldsymbol{\alpha},$$

from which the Theorem is proven. □

This is *the* most general form of the moment equation: If A is the center of mass G itself, then $\boldsymbol{r}_{G/A} \equiv \boldsymbol{0}$, and this reduces to just $\sum_i \overline{\boldsymbol{M}}_i = \boldsymbol{\omega} \times \bar{\mathsf{I}}\boldsymbol{\omega} + \bar{\mathsf{I}}\boldsymbol{\alpha}$. On the other hand, if A happens to be a *fixed* point O about which the body is rotating, then if $\overline{\boldsymbol{r}}$ is the position of the center of mass relative to O,

$$\overline{\boldsymbol{a}} = \boldsymbol{a}_O + \boldsymbol{\alpha} \times \overline{\boldsymbol{r}} + \boldsymbol{\omega} \times (\boldsymbol{\omega} \times \overline{\boldsymbol{r}}) = \boldsymbol{0} + \boldsymbol{\alpha} \times \overline{\boldsymbol{r}} + (\boldsymbol{\omega} \cdot \overline{\boldsymbol{r}})\boldsymbol{\omega} - \omega^2 \overline{\boldsymbol{r}}$$

—where we have once again utilized the triple vector product expansion, 4.1-5; thus the term $\boldsymbol{r}_{G/O} \times m\overline{\boldsymbol{a}}$ in the general moment equation becomes

$$\begin{aligned}\overline{\boldsymbol{r}} \times m\overline{\boldsymbol{a}} &= m[\overline{\boldsymbol{r}} \times (\boldsymbol{\alpha} \times \overline{\boldsymbol{r}}) + (\boldsymbol{\omega} \cdot \overline{\boldsymbol{r}})\overline{\boldsymbol{r}} \times \boldsymbol{\omega} - \omega^2 \overline{\boldsymbol{r}} \times \overline{\boldsymbol{r}}] \\ &= m[\overline{r}^2 \boldsymbol{\alpha} - (\boldsymbol{\alpha} \cdot \overline{\boldsymbol{r}})\overline{\boldsymbol{r}} - \boldsymbol{\omega} \times (\boldsymbol{\omega} \cdot \overline{\boldsymbol{r}})\overline{\boldsymbol{r}} - \boldsymbol{0}] \qquad (5.2\text{-}21\text{a}) \\ &= m\overline{r}^2 \boldsymbol{\alpha} - m(\boldsymbol{\alpha} \cdot \overline{\boldsymbol{r}})\overline{\boldsymbol{r}} + \boldsymbol{\omega} \times m\overline{r}^2 \boldsymbol{\omega} - \boldsymbol{\omega} \times m(\boldsymbol{\omega} \cdot \overline{\boldsymbol{r}})\overline{\boldsymbol{r}}, \qquad (5.2\text{-}21\text{b})\end{aligned}$$

where the third term in the last line is a thinly-disguised $\boldsymbol{0}$! Thus the general moment equation above becomes

$$\begin{aligned}\sum_i \boldsymbol{M}_{O_i} &= \overline{\boldsymbol{r}} \times m\overline{\boldsymbol{a}} + \boldsymbol{\omega} \times \bar{\mathsf{I}}\boldsymbol{\omega} + \bar{\mathsf{I}}\boldsymbol{\alpha} \\ &= \boldsymbol{\omega} \times (\bar{\mathsf{I}}\boldsymbol{\omega} + m\overline{r}^2\boldsymbol{\omega} - m(\boldsymbol{\omega} \cdot \overline{\boldsymbol{r}})\overline{\boldsymbol{r}}) + (\bar{\mathsf{I}}\boldsymbol{\alpha} + m\overline{r}^2\boldsymbol{\alpha} - m(\boldsymbol{\alpha} \cdot \overline{\boldsymbol{r}})\overline{\boldsymbol{r}}).\end{aligned}$$

But, by Homework Problem 2 on page 303, the first is nothing more than $\boldsymbol{\omega} \times \mathsf{I}_O\boldsymbol{\omega}$, while, from Problem 3 on the same page, the second is $\mathsf{I}_O\boldsymbol{\alpha}$. Thus the general moment equation reduces to the special case of rotation about a fixed point. While these facts regarding the generality of Theorem 5.2.5 would appear to render the other two equations obsolete, the machinations required to get them makes these special cases, particularly the one dealing with motion about a fixed point, useful to remember. But the general one is a profoundly useful form, and the freedom to take moments about *any* point makes it the most powerful. We shall make more and more use of it as time goes on.

What really made the general case "collapse" to that about O was the fact that $\overline{\boldsymbol{r}} \times m\overline{\boldsymbol{a}}$ reduced to equation (5.2-21a), which then was cajoled to be put into form (5.2-21b), ultimately leading to the desired result. Now consider a *radially-symmetric* body rolling *without slipping* on a *flat, horizontal surface*: we have obtained, on page 223, an expression for the acceleration of the [geometrical] *center* O of a sphere but with a surface acceleration of $\boldsymbol{a}_C$:

$$\boldsymbol{a}_O = \boldsymbol{a}_C + \boldsymbol{\alpha} \times \boldsymbol{r}_{O/C'} + \left(\boldsymbol{\omega} \times (\boldsymbol{\omega} \times \boldsymbol{r}_{O/C'}) - (\boldsymbol{\omega} \times (\boldsymbol{\omega} \times \boldsymbol{r}_{O/C'}) \cdot \hat{\boldsymbol{n}})\, \hat{\boldsymbol{n}}\right).$$

From the symmetry, $\overline{\boldsymbol{a}} = \boldsymbol{a}_O$ in the *kinetic* equations, suggesting setting $\boldsymbol{r}_{O/C'} = \overline{\boldsymbol{r}}$; and, since the surface is *fixed*, $\boldsymbol{a}_C \equiv \mathbf{0}$; thus the "correction" term in the general moment equation becomes

$$\begin{aligned}\overline{\boldsymbol{r}} \times m\overline{\boldsymbol{a}} = \overline{\boldsymbol{r}} \times m\boldsymbol{a}_O &= \overline{\boldsymbol{r}} \times m\left[\boldsymbol{\alpha} \times \overline{\boldsymbol{r}} + (\boldsymbol{\omega} \times (\boldsymbol{\omega} \times \overline{\boldsymbol{r}}) - (\boldsymbol{\omega} \times (\boldsymbol{\omega} \times \overline{\boldsymbol{r}}) \cdot \hat{\boldsymbol{n}})\, \hat{\boldsymbol{n}})\right] \\ &= m[\overline{r}^2\boldsymbol{\alpha} - (\boldsymbol{\alpha} \cdot \overline{\boldsymbol{r}})\overline{\boldsymbol{r}} - \boldsymbol{\omega} \times (\boldsymbol{\omega} \cdot \overline{\boldsymbol{r}})\overline{\boldsymbol{r}} - \mathbf{0} - \mathbf{0}]\end{aligned}$$

—the last term vanishing because $\hat{\boldsymbol{n}} \parallel \overline{\boldsymbol{r}}$. From here on out, the derivation proceeds in exactly the same way as above, demonstrating that

$$\sum_i \boldsymbol{M}_{C_i} = \boldsymbol{\omega} \times \mathsf{I}_C\boldsymbol{\omega} + \mathsf{I}_C\boldsymbol{\alpha}$$

even though C is not fixed, and the body isn't "rotating about" it! This equation is used often in such special situations, particularly in elementary courses, yet there's virtually no justification for its introduction. As helpful as it is, however, one must remember the conditions limiting its application.

We now bring all the basic equations of motion together in one place:

Translation:

$$\sum_i \boldsymbol{F}_i = m\overline{\boldsymbol{a}} \tag{5.2-22}$$

Rotation:

$$\sum_i \overline{\boldsymbol{M}}_i = \boldsymbol{\omega} \times \overline{\mathsf{I}}\boldsymbol{\omega} + \overline{\mathsf{I}}\boldsymbol{\alpha} \quad \text{(general)} \tag{5.2-23a}$$

$$\sum_i \boldsymbol{M}_{O_i} = \boldsymbol{\omega} \times \mathsf{I}_O\boldsymbol{\omega} + \mathsf{I}_O\boldsymbol{\alpha} \quad (\textit{only} \text{ if rotating about } \textit{fixed } O) \tag{5.2-23b}$$

$$\sum_i \boldsymbol{M}_{A_i} = \boldsymbol{r}_{G/A} \times m\overline{\boldsymbol{a}} + \boldsymbol{\omega} \times \overline{\mathsf{I}}\boldsymbol{\omega} + \overline{\mathsf{I}}\boldsymbol{\alpha} \quad (\textit{completely} \text{ general}) \tag{5.2-23c}$$

As usual, we make several remarks regarding these equations:

- Each equation is of precisely the same form: as written above, the left-hand side of each contains the [*several*] applied forces and/or moments—the *kinetics*—while the right-hand side has the [*single*] translational acceleration and/or [*single*] rotational acceleration—the *kinematics*—resulting from these applied agents. This is precisely what occurred in the corresponding equations in Section 5.1.

- The equations are inherently *instantaneous*, relating the forces and/or moments at each *instant* to the resulting acceleration. In particular, if we desire to know the motion of the body/system as a function of time, it is necessary that these be cast in *general* form, resulting in a generally coupled set of second-order ordinary differential equations which must be integrated.

- The set of equations for a rigid body will constitute a set of at most six scalar equations, which can thus be used to solve for [at most] six unknowns.

- As pointed out in Section 5.1.2, the forces and moments entering the equations of motion are those *external* to the "system," however that's chosen. In particular, given a system of *several* rigid bodies—a "machine," it is necessary to "break up" the system to consider the individual bodies separately, since each has its individual center-of-mass $\overline{\boldsymbol{a}}$ and rotational $\boldsymbol{\alpha}$; then included on each will be the forces and/or moments resulting from the interconnections between the body and its neighbors—a situation precisely the same as in the Static analysis of "frames and machines." Like that analysis, the assumed directions of such forces or moments on one body due to another will have to be reversed on the second body in order to account for Newton's Third Law. And like that analysis, one generally will have to deal with a set of $6N$ equations in $6N$ unknowns *coupled* through these interconnecting forces or moments.

- In the desire to generate additional equations, students will occasionally take moments about an additional point. *But there is no new information to be derived from such a practice:* applying the general moment equation (5.2-23c) to two such points A and B, we see that their difference (any common couple moments cancelling out) is

$$\begin{aligned}\sum_i \boldsymbol{M}_{A_i} - \sum_i \boldsymbol{M}_{B_i} &= \sum_i \boldsymbol{r}_{A_i} \times \boldsymbol{F}_i - \sum_i \boldsymbol{r}_{B_i} \times \boldsymbol{F}_i \\ &= \sum_i (\boldsymbol{r}_{A_i} - \boldsymbol{r}_{B_i}) \times \boldsymbol{F}_i \\ &= \sum_i ((\boldsymbol{r}_{B_i} + \boldsymbol{r}_{A/B}) - \boldsymbol{r}_{B_i}) \times \boldsymbol{F}_i \\ &= \boldsymbol{r}_{A/B} \times \sum_i \boldsymbol{F}_i = \boldsymbol{r}_{A/B} \times m\overline{\boldsymbol{a}};\end{aligned}$$

thus, in fact, the difference between any two such moment equations is nothing more than the force equation (5.2-22) cross-multiplied by $\boldsymbol{r}_{B/A}$—a linear combination of the force equation's components. Thus the new equation components are linearly *de*pendent on the original basic six equations.

- These equations involve *forces/moments* and translational/rotational *accelerations.* However trivial that observation might seem, in the vast majority of cases, if one wants one or the other of these types of quantities, the equations of motion are the way to go!

- In line with the above observation, we note that *it is necessary to know the [instantaneous] angular velocity* $\boldsymbol{\omega}$ of the body in order to find the forces (moments) and accelerations.

- For a rigid body, $\overline{\boldsymbol{a}}$ *and* $\boldsymbol{\alpha}$ *are related:* from the rotating coordinate system equations involving the acceleration of an arbitrary point A,

$$\overline{\boldsymbol{a}} = \boldsymbol{a}_A + \boldsymbol{\alpha} \times \boldsymbol{r}_{G/A} + \boldsymbol{\omega} \times (\boldsymbol{\omega} \times \boldsymbol{r}) + 2\boldsymbol{\omega} \times \boldsymbol{v}_{rel} + \boldsymbol{a}_{rel}.$$

  While the *rel* terms will vanish in most cases, Example 4.4.1 demonstrates that this is not always the case![16]

The last point, in particular, shows that generally we have not six, but *nine* equations associated with each rigid body: the six equations of motion, plus the three components relating $\overline{\boldsymbol{a}}$ and $\boldsymbol{\alpha}$. If we *know* $\boldsymbol{\alpha}$ and $\boldsymbol{a}_A$ (and, of course, $\boldsymbol{\omega}$), $\overline{\boldsymbol{a}}$ is determined, and the equations of motion allow determination of [up to] six forces and/or couple moments entering linearly on the left-hand sides of equations (5.2-22) and (5.2-23); this, in effect, would determine the forces and moments required to enforce a given motion.

A slightly more quirky case would arise with knowledge not of $\boldsymbol{a}_A$ and $\boldsymbol{\alpha}$, but, say, $\boldsymbol{a}_A$ and $\overline{\boldsymbol{a}}$: Since the angular acceleration enters the kinematic equations for a rigid body through a cross product, there result only two independent equations for its three components (see Section 4.4.1); consequently, there is some additional condition required to find $\boldsymbol{\alpha}$ in this case. Alternatively, equations (5.2-22) and (5.2-23) allow determination of only *five* scalar unknowns.

On the other hand, it might be the case that we know the forces/moments, and desire to determine the resulting accelerations; this situation is a little more complicated: $\overline{\boldsymbol{a}}$ can be found immediately from (5.2-22)—*as long as* $m \neq 0$, *i.e.*, the body is not "light." In the latter case, the appropriate I also vanishes, and *neither acceleration* $\overline{\boldsymbol{a}}$ *nor* $\boldsymbol{\alpha}$ *can be determined*! In fact, *the forces and moments acting on the body satisfy equilibrium equations*, not because there is no acceleration, but because the right-hand sides vanish by virtue of the fact terms multiplying these accelerations vanish!

[16] Actually, the kinematics in this case would be applied not to a single rigid body, but the *system* of rigid bodies consisting of the two sleeves of the piston. The forces of each sleeve on the other would be "internalized," preventing their determination, but the external forces at the ends of the complete piston *could* be found.

But, even absent that "pathological" case, there can be some subtle complications involving determination of the *angular* acceleration: Most generally, $\boldsymbol{\alpha}$ satisfies equation (5.2-23c), which can be written

$$\bar{\mathsf{I}}\boldsymbol{\alpha} = \sum_i \boldsymbol{M}_{A_i} - \boldsymbol{r}_{G/A} \times m\overline{\boldsymbol{a}} - \boldsymbol{\omega} \times \bar{\mathsf{I}}\boldsymbol{\omega}.$$

This is a non-homogeneous system of three linear equations in the three components of $\boldsymbol{\alpha}$; we know from Theorem 3.4.1 that a unique solution exits if and only if the determinant of the coefficient matrix, here $\bar{\mathsf{I}}$, doesn't vanish. But certain inertia matrices, notably the "slender rod" in its principal axes [see the remark on page 277], have a vanishing moment of inertia; thus $|\bar{\mathsf{I}}| \equiv 0$, so there is *no* unique solution in this case. And this circumstance is not merely because of the use of principal axes: any other axes' representation of $\bar{\mathsf{I}}$ would be obtained from the principal axes' by a coordinate transformation—*i.e.*, a *similarity transformation*, equation (3.6-7)—and by Homework 1 on page 143, $|\bar{\mathsf{I}}|$ remains unchanged under such a transformation. Thus the situation is endemic to the body itself.

But this development isn't as bad as it might appear: A vanishing moment of inertia will, at least for a body with mass, occur only if there is no mass distribution about the corresponding axis—the "slender rod" case above. But this means that any forces acting on this body must themselves be applied *along that axis*; *i.e.*, the *moment about this axis due to forces vanishes identically, too*. If, however, there happens to be a *couple* applied about this axis, $\boldsymbol{\alpha}$ will be indeterminate.

Machines have even more problems: Again, it is not too bad *if we know the accelerations*: The requisite forces and moments on each individual body will satisfy the equations of motion for that body, with Newton's Third applied at each interconnections; once more, impressive systems of simultaneous equations generally result. A machine consisting of n bodies will thus generate a system of $6n$ scalar equations—three components of both force and moment equations for each body. A sometimes useful form of *force* equation isolates those forces *external* to the machine as a whole: since all the interconnecting forces are purely *internal* to the machine, they will not appear in a machine analog to 5.2-18

$$\sum_i \boldsymbol{F}_{\text{ext}\,i} = \sum_i m_i \overline{\boldsymbol{a}}_i,$$

in which the right-hand side is known since the bodies' [center of mass] accelerations are. Note that this equation is *not independent* of the individual bodies' force equations: it is, after all, merely their sum! But, just as with frames and machines in statics, it can occasionally be a useful addition to the more general equations.

But *if the accelerations are unknown*, such systems must include the additional *kinematical* constraints imposed on the system relating the individual $\boldsymbol{\alpha}_i$ and $\overline{\boldsymbol{a}}_i$ resulting from the known motion of various points in the machine. Now the resulting systems of simultaneous equations transcend being merely "impressive," they become downright tedious! But this development is more a practical one than conceptual; at the end of the day, we are still fundamentally utilizing only the standard equations of motion, along with the kinematical constraints discussed in the previous Chapter. And, with the exception of $\boldsymbol{\omega}$ (which enters the kinematical equations through the triple cross product $\boldsymbol{\omega} \times (\boldsymbol{\omega} \times \boldsymbol{r})$—and, as remarked above, this value must be known anyway), all variables enter the equations of motion *linearly*, so it is always a *linear* system of equations which must be solved, however large it may be.

A final note is in order: It is common, in elementary texts, to formulate the equations of motion in *principal* axes exclusively; certainly this leads to great simplification in form. But [typically unknown] reactions may be more amenable to *non*-principal axes (see, *e.g.*, Example 5.2.9); or one might simply *prefer* them. Most generally, however, one deals with multi-body *machines*, and axes principal for one body are almost certainly not principal for the others (*e.g.*, Example 5.2.11). It is precisely the flexibility afforded by representing the inertia matrix in *any* choice of axes that justifies the present approach.

Forces/Moments at Interconnections

While on the subject of "machines," there is a point given unfortunately little attention in Mechanics texts: what to do if there is a force applied *at* a *ball-and-socket* joint connecting two components of a machine. This is the norm for "trusses" in a Statics course, and it is dealt with there using either the Method of Joints (in which the connecting joint is itself the free body) or the Method of Sections (in which the joint is included in the subsection used to find the member forces). But when "frames and machines" are covered in that course, the question is either touched only only briefly [the author recalls one text recommending merely to "include the force on one of the members"], or ignored altogether. The present exposition seems an appropriate time to discuss this important topic.

To start with a simple case, we consider two bodies, "1" and "2," connected by a ball-and-socket at their common end O. Diagrammed at top left are the free-body diagrams for each, along with that for the joint *considered as a separate body*; we include the forces $\boldsymbol{F}$ applied at the joint, as well as those forces "internal" when the machine is assembled, $\boldsymbol{f}_{O1}$ and $\boldsymbol{f}_{O2}$ on O due to the respective members, as well as their opposites $\boldsymbol{f}_{1O}$ and $\boldsymbol{f}_{2O}$ acting on the individual members, opposite by Newton's Third.

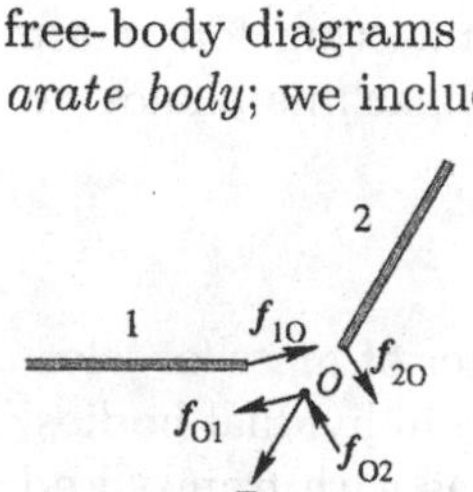

Below is an alternative free body, where now we have included the joint in member 1. This operation internalizes the forces $\boldsymbol{f}_{O1}$ and $\boldsymbol{f}_{1O}$, and there remain only $\boldsymbol{f}_{O2}$ and $\boldsymbol{f}_{2O}$—forces which might be designated "$\boldsymbol{f}_{21}$" and "$\boldsymbol{f}_{12}$" if we were to *start* with this diagram. As the maxim to "include the force in one of the members" suggests, $\boldsymbol{F}$ is applied to the chosen member, and we have precisely the forces expected.

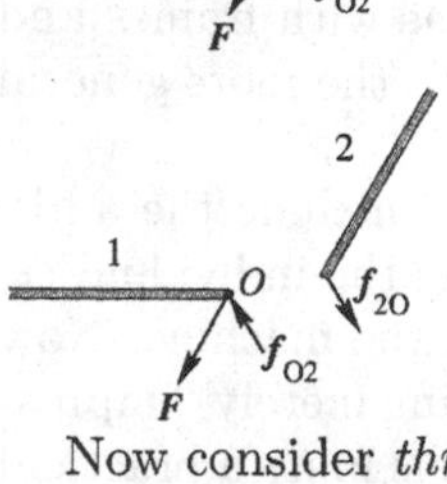

Now consider *three* bodies: As before, we start with the free-body diagrams of the three individual bodies, along with the *separate* free body of the joint O. The diagram is the same, with the exception of the additional pair of opposite forces $\boldsymbol{f}_{O3}$ and $\boldsymbol{f}_{3O}$ due to the third member. There seems to be nothing new here.

But when we once again consider the joint to be made part of member 1, effectively "applying the force $\boldsymbol{F}$ to a member," something rather unexpected happens: while it is not surprising that we have the forces $\boldsymbol{f}_{O2}$ and $\boldsymbol{f}_{O3}$ acting on the chosen member, and their opposites $\boldsymbol{f}_{2O}$ and $\boldsymbol{f}_{3O}$ on the others—"$\boldsymbol{f}_{21}$" and "$\boldsymbol{f}_{31}$" if one prefers—*there are no forces "$\boldsymbol{f}_{23}$" and "$\boldsymbol{f}_{32}$"* which might be expected from a thoughtlessly naive application of Newton's Third!

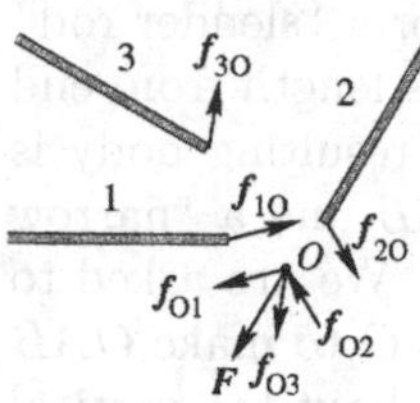

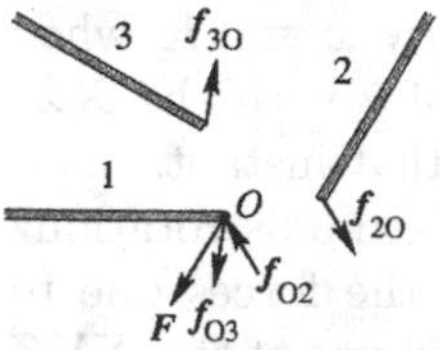

All of the above has been motivated by the application of an external force at the interconnecting joint. But say there *were* no such force—*i.e.*, that $\boldsymbol{F} \equiv 0$. *Nothing would change: there would* still *not be forces "$\boldsymbol{f}_{23}$" and "$\boldsymbol{f}_{32}$"!* The easiest way to avoid such an error is to do precisely what was done above: consider the ball-and-socket joint to be its own free body and apply the internalization *explicitly*, or, perhaps better still, to avoid the internalization altogether.

Though the latter practice introduces additional free bodies to consider, it preserves a certain symmetry in the equations of motion: We invariably consider these pins/ball-and-sockets to be *light*, of *negligible* mass; as remarked on page 310, such joints will satisfy the *equilibrium* equations. In particular, then, $\sum_i \boldsymbol{F}_i \equiv 0$, and, in the second example, we would have, in addition to the $\boldsymbol{f}_{ij}$ acting on members i, a force equation for the joint:

$$\boldsymbol{F} + \boldsymbol{f}_{O1} + \boldsymbol{f}_{O2} + \boldsymbol{f}_{O3} \equiv 0.$$

But *there is no [non-trivial] moment equation*: $\sum_i \boldsymbol{M}_{O_i} \equiv 0$ because the forces all pass through the [infinitesimal] joint. Thus, accepting the suggested approach of considering the connecting joint to be its own rigid body, we garner an additional *three* "equations of motion" (*equilibrium*, really) at each.

A similar consideration applies to *clevis* connections: For one attached to a collar free to slide along a track, for example, there would be exerted forces perpendicular to the track at that interconnection. But, recall, the angular acceleration of the body is *completely determined* by a clevis; in particular, the kinematics can determine [a component of] angular acceleration about the axis between connecting points in the body. But what physically produces such an acceleration? The very nature of a clevis generates a *couple moment* due to the fork-type connection characterizing it. This, then, must be included in the free-body diagram of bodies acting under such interconnections and will be encountered in Example 5.2.11.

Examples

With all this background in hand, we are now ready to give a number of examples. Most will be worked entirely analytically, not because this is a Good Thing, but because it makes it easier to follow the roles played by the individual parameters and to keep them separate.

Example 5.2.9 *Simple Single Rigid Body.* We start with the "simple" case of a single rigid body OAB, consisting of a uniform "slender rod" of mass m and length $3a$, bent one-third of its length from end B to make a 30° angle at O as shown. The resulting body is held with BO vertical by a "shallow cup" at B and a "narrow strap" at O, both assumed to be frictionless. We are asked to determine the couple moment $\boldsymbol{M}_O$ applied at O to make OAB rotate with an angular acceleration $\boldsymbol{\alpha} \equiv \alpha\hat{\boldsymbol{K}}$ about the vertical Z-axis with an [instantaneous] angular velocity $\boldsymbol{\omega} \equiv \omega\hat{\boldsymbol{K}}$ when the plane containing OAB makes an angle of θ with the XZ-plane as shown, as well as the forces applied at O and B at that instant.

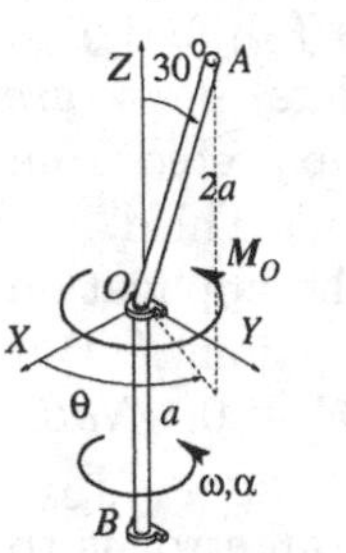

This is an instance of what is generally referred to as "fixed-axis rotation" about the fixed $\hat{\boldsymbol{K}}$-direction. The fact that the forces due to the band at O lie in the XY-plane recommend use of the XYZ coordinate system shown. We know $\boldsymbol{\alpha}$, and thus $\overline{\boldsymbol{a}}$ and desire the forces and moments acting on the body—an obvious application of the equations of motion for the body. $\sum_i \boldsymbol{F}_i = m\overline{\boldsymbol{a}}$ is always with us; since O (or B) is a fixed point about which the body is rotating, we shall use the moment equation in the form (5.2-23b):

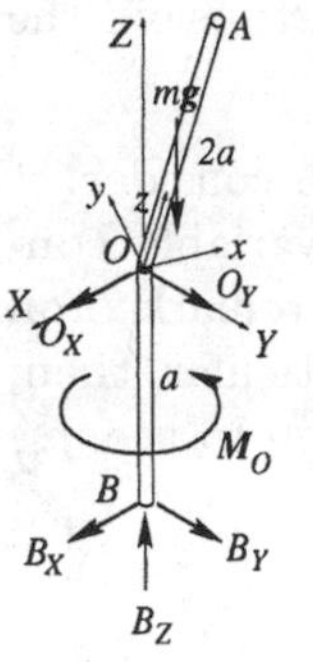

$$\sum_i \boldsymbol{M}_{O_i} = \boldsymbol{\omega} \times \mathsf{I}_O\boldsymbol{\omega} + \mathsf{I}_O\boldsymbol{\alpha}.$$

We note that the six [component] equations of motion can be used to determine the six [scalar] unknowns $\boldsymbol{F}_O$ (2 unknowns), $\boldsymbol{F}_B$ (3 unknowns) and $\boldsymbol{M}_O$ (1 unknown, in the $\hat{\boldsymbol{K}}$-direction). To implement these equations, however, we need $\overline{\boldsymbol{a}}$ and I_O.

Knowing $\boldsymbol{\alpha}$, the first of these is easy: Using rotating coordinates,

$$\overline{\boldsymbol{a}} = \boldsymbol{a}_O + \boldsymbol{\alpha} \times \overline{\boldsymbol{r}} + \boldsymbol{\omega} \times (\boldsymbol{\omega} \times \overline{\boldsymbol{r}}) + 2\boldsymbol{\omega} \times \boldsymbol{v}_{rel} + \boldsymbol{a}_{rel}, \tag{5.2-24}$$

in which $\boldsymbol{a}_O \equiv 0$, $\boldsymbol{\alpha}$ and $\boldsymbol{\omega}$ are given, and, relative to the [rotating coordinate system] origin O, $\overline{\boldsymbol{r}}$ satisfies the equation

$$m\overline{\boldsymbol{r}} = \frac{m}{3}(-\frac{a}{2})\hat{\boldsymbol{K}} + \frac{2m}{3}(a\cos\theta\hat{\boldsymbol{I}} + a\sin\theta\hat{\boldsymbol{J}} + a\sqrt{3}\hat{\boldsymbol{K}})$$

or

$$\overline{\boldsymbol{r}} = \frac{2}{3}a\cos\theta\hat{\boldsymbol{I}} + \frac{2}{3}a\sin\theta\hat{\boldsymbol{J}} + \frac{4\sqrt{3}-1}{6}a\hat{\boldsymbol{K}}.$$

Substituting these values into (5.2-24), we get finally that

$$\overline{\boldsymbol{a}} = -\frac{2}{3}a(\alpha\sin\theta + \omega^2\cos\theta)\hat{\boldsymbol{I}} + \frac{2}{3}a(\alpha\cos\theta - \omega^2\sin\theta)\hat{\boldsymbol{J}}.$$

But now we must find $\mathbf{I}_O$: The given axes are principal for part BO, and, in fact,

$$\mathbf{I}_{O_{BO}} = \begin{pmatrix} \frac{1}{9} & 0 & 0 \\ 0 & \frac{1}{9} & 0 \\ 0 & 0 & 0 \end{pmatrix} ma^2.$$

But these axes are *not* principal for OA, so a little more is involved. Introducing the indicated *principal* (x, y, z) axes for this part, we have that

$$\mathbf{I}_{OA}^{(p)} = \begin{pmatrix} \frac{8}{9} & 0 & 0 \\ 0 & \frac{8}{9} & 0 \\ 0 & 0 & 0 \end{pmatrix} ma^2.$$

From this it is straightforward to rotate back to the (X, Y, Z) axes, first through a rotation about the [original] x-axis through (-30°) and then about the [*new*] z-axis through $(-(\theta + 90°))$:

$$\begin{aligned}
\mathbf{I}_{O_{OA}} &= \mathbf{C}^{-1}(-(\theta+90°))\mathbf{A}^{-1}(-30°)\mathbf{I}_{OA}^{(p)}\mathbf{A}(-30°)\mathbf{C}(-(\theta+90°)) \\
&= \mathbf{C}(\theta+90°)\mathbf{A}(30°)\mathbf{I}_{OA}^{(p)}\mathbf{A}^{\mathsf{T}}(30°)\mathbf{C}^{\mathsf{T}}(\theta+90°) \\
&= \begin{pmatrix} \frac{8}{9} - \frac{2}{9}\cos^2\theta & -\frac{2}{9}\sin\theta\cos\theta & -\frac{2}{9}\sqrt{3}\cos\theta \\ -\frac{2}{9}\sin\theta\cos\theta & \frac{2}{3} + \frac{2}{9}\cos^2\theta & -\frac{2}{9}\sqrt{3}\sin\theta \\ -\frac{2}{9}\sqrt{3}\cos\theta & -\frac{2}{9}\sqrt{3}\sin\theta & \frac{2}{9} \end{pmatrix} ma^2,
\end{aligned}$$

so that, finally,

$$\mathbf{I}_O = \mathbf{I}_{O_{BO}} + \mathbf{I}_{O_{OA}} = \begin{pmatrix} 1 - \frac{2}{9}\cos^2\theta & -\frac{2}{9}\sin\theta\cos\theta & -\frac{2}{9}\sqrt{3}\cos\theta \\ -\frac{2}{9}\sin\theta\cos\theta & \frac{7}{9} + \frac{2}{9}\cos^2\theta & -\frac{2}{9}\sqrt{3}\sin\theta \\ -\frac{2}{9}\sqrt{3}\cos\theta & -\frac{2}{9}\sqrt{3}\sin\theta & \frac{2}{9} \end{pmatrix} ma^2.$$

Determination of the six unknowns is now straightforward: The force equations become

$$\sum F_X = O_X + B_X = m\overline{a}_X = -\frac{2}{3}ma(\alpha\sin\theta + \omega^2\cos\theta) \tag{5.2-25a}$$

$$\sum F_Y = O_Y + B_Y = m\overline{a}_Y = \frac{2}{3}ma(\alpha\cos\theta - \omega^2\sin\theta) \tag{5.2-25b}$$

$$\sum F_Z = B_Z - mg = m\overline{a}_Z \equiv 0, \tag{5.2-25c}$$

while the moment equations about the fixed point O involve only

$$\begin{aligned}\boldsymbol{M}_O(\boldsymbol{B}) &= \boldsymbol{r}_{B/O} \times \boldsymbol{B} = (-a\hat{\boldsymbol{K}}) \times (B_X\hat{\boldsymbol{I}} + B_Y\hat{\boldsymbol{J}} + B_Z\hat{\boldsymbol{K}}) \\ &= aB_Y\hat{\boldsymbol{I}} - aB_X\hat{\boldsymbol{J}} \end{aligned} \tag{5.2-26a}$$

$$\begin{aligned}\boldsymbol{M}_O(m\boldsymbol{g}) &= \left(\frac{2}{3}a\cos\theta\hat{\boldsymbol{I}} + \frac{2}{3}a\sin\theta\hat{\boldsymbol{J}} + \frac{4\sqrt{3}-1}{6}a\hat{\boldsymbol{K}}\right) \times (-mg\hat{\boldsymbol{K}}) \\ &= -\frac{2}{3}mga\sin\theta\hat{\boldsymbol{I}} + \frac{2}{3}mga\cos\theta\hat{\boldsymbol{J}} \end{aligned} \tag{5.2-26b}$$

and

$$\boldsymbol{M}_O(\boldsymbol{M}_O) = M_O\hat{\boldsymbol{K}}; \tag{5.2-26c}$$

this is the left-hand side of (5.2-23b). The right-hand side is easy to calculate:

$$\begin{aligned}\boldsymbol{\omega} \times \mathsf{I}_O\boldsymbol{\omega} &= \omega\hat{\boldsymbol{K}} \times ma^2 \begin{pmatrix} 1 - \frac{2}{9}\cos^2\theta & -\frac{2}{9}\sin\theta\cos\theta & -\frac{2}{9}\sqrt{3}\cos\theta \\ -\frac{2}{9}\sin\theta\cos\theta & \frac{7}{9} + \frac{2}{9}\cos^2\theta & -\frac{2}{9}\sqrt{3}\sin\theta \\ -\frac{2}{9}\sqrt{3}\cos\theta & -\frac{2}{9}\sqrt{3}\sin\theta & \frac{2}{9} \end{pmatrix} \begin{pmatrix} 0 \\ 0 \\ \omega \end{pmatrix} \\ &= \omega\hat{\boldsymbol{K}} \times \left(-\frac{2}{9}\sqrt{3}ma^2\omega\cos\theta\hat{\boldsymbol{I}} - \frac{2}{9}\sqrt{3}ma^2\omega\sin\theta\hat{\boldsymbol{J}} + \frac{2}{9}ma^2\omega\hat{\boldsymbol{K}}\right) \\ &= \frac{2}{9}\sqrt{3}ma^2\omega^2\sin\theta\hat{\boldsymbol{I}} - \frac{2}{9}\sqrt{3}ma^2\omega^2\cos\theta\hat{\boldsymbol{J}} \end{aligned} \tag{5.2-27a}$$

(note that we have used matrix/column vector representations for the product $\mathsf{I}_O\boldsymbol{\omega}$), while

$$\mathsf{I}_O\boldsymbol{\alpha} = -\frac{2}{9}\sqrt{3}ma^2\alpha\cos\theta\hat{\boldsymbol{I}} - \frac{2}{9}\sqrt{3}ma^2\alpha\sin\theta\hat{\boldsymbol{J}} + \frac{2}{9}ma^2\alpha\hat{\boldsymbol{K}}. \tag{5.2-27b}$$

Summing the three components of the sum of equations (5.2-26) with those of the sum of equations (5.2-27), we get the final set of three moment equations:

$$aB_Y - \frac{2}{3}mga\sin\theta = \frac{2}{9}\sqrt{3}ma^2\omega^2\sin\theta - \frac{2}{9}\sqrt{3}ma^2\alpha\cos\theta \tag{5.2-28a}$$

$$-aB_X + \frac{2}{3}mga\cos\theta = -\frac{2}{9}\sqrt{3}ma^2\omega^2\cos\theta - \frac{2}{9}\sqrt{3}ma^2\alpha\sin\theta \tag{5.2-28b}$$

$$M_O = \frac{2}{9}ma^2\alpha \tag{5.2-28c}$$

With the six equations (5.2-25) and (5.2-28) we get the answers immediately:

$$\begin{aligned}\boldsymbol{B} = &\left(\frac{2}{9}\sqrt{3}ma\omega^2\cos\theta + \frac{2}{9}\sqrt{3}ma\alpha\sin\theta + \frac{2}{3}mg\cos\theta\right)\hat{\boldsymbol{I}} \\ &+ \left(\frac{2}{9}\sqrt{3}ma\omega^2\sin\theta - \frac{2}{9}\sqrt{3}ma\alpha\cos\theta + \frac{2}{3}mg\sin\theta\right)\hat{\boldsymbol{J}} + mg\hat{\boldsymbol{K}}\end{aligned}$$

$$
\begin{aligned}
\boldsymbol{O} = &-\left(\frac{2}{9}(3+\sqrt{3})ma\left(\omega^2\cos\theta + \alpha\sin\theta\right) + \frac{2}{3}mg\cos\theta\right)\hat{\boldsymbol{I}} \\
&+\left(\frac{2}{9}(3+\sqrt{3})ma\left(\omega^2\sin\theta + \alpha\cos\theta\right) - \frac{2}{3}mg\sin\theta\right)\hat{\boldsymbol{J}}
\end{aligned}
$$

$$
\boldsymbol{M}_O = \frac{2}{9}ma^2\alpha\hat{\boldsymbol{K}}
$$

It is useful to step back briefly and take an overview of what has been done here: although, knowing $\boldsymbol{\alpha}$, we explicitly solved for $\overline{\boldsymbol{a}}$ in equation (5.2-24) and substituted the result into the $\sum \boldsymbol{F} = m\overline{\boldsymbol{a}}$ equation to get equations (5.2-25), in essence, we actually solved *nine* equations: the kinematic (5.2-25)

$$
\overline{\boldsymbol{a}} = \boldsymbol{a}_O + \boldsymbol{\alpha}\times\overline{\boldsymbol{r}} + \boldsymbol{\omega}\times(\boldsymbol{\omega}\times\overline{\boldsymbol{r}}) + 2\boldsymbol{\omega}\times\boldsymbol{v}_{rel} + \boldsymbol{a}_{rel},
$$

the force equation (5.2-22)

$$
\sum_i \boldsymbol{F}_i = m\overline{\boldsymbol{a}},
$$

and the moment equation (5.2-23b)

$$
\sum_i \boldsymbol{M}_{O_i} = \boldsymbol{\omega}\times\mathsf{I}_O\boldsymbol{\omega} + \mathsf{I}_O\boldsymbol{\alpha}
$$

in *nine* unknowns, consisting of the components of $\overline{\boldsymbol{a}}$, $\boldsymbol{F}_B$, $\boldsymbol{F}_O$, and $\boldsymbol{M}_O$ (the last two having only two and one non-zero components, respectively). This perspective is useful if we consider the problem in which $\boldsymbol{M}_O$ was given but *not* $\boldsymbol{\alpha}$: Now $\overline{\boldsymbol{a}}$, $\boldsymbol{F}_B$, and $\boldsymbol{F}_O$ remain unknown, but we have an additional *three* unknown components of the angular acceleration—a total of *11* [scalar] unknowns, yet only nine equations. [Admittedly, we know $\boldsymbol{\alpha}$ has only a $\hat{\boldsymbol{K}}$-component in this problem, but in general we would not have such knowledge.] Where would we get an additional set of equations?

In fact, we *have* such a set, one not utilized explicitly above—the *kinematical* condition that the point B is also fixed:

$$
\boldsymbol{a}_B = \boldsymbol{a}_O + \boldsymbol{\alpha}\times\boldsymbol{r}_{B/O} + \boldsymbol{\omega}\times(\boldsymbol{\omega}\times\boldsymbol{r}_{B/O}) + 2\boldsymbol{\omega}\times\boldsymbol{v}_{rel} + \boldsymbol{a}_{rel},
$$

in which we know $\boldsymbol{a}_B \equiv \boldsymbol{0}$ as are $\boldsymbol{v}_{rel}$ and $\boldsymbol{a}_{rel}$. Although this appears initially to have three components, in fact only *two* of them are non-trivial: since $\boldsymbol{r}_{B/O}$, appearing in both of the only non-vanishing terms, has only a Z-component which doesn't vanish, the right-hand side of this equation has a vanishing $\hat{\boldsymbol{K}}$-component, matching the corresponding one in $\boldsymbol{a}_B$. Thus the above four *vector* equations have only 11 non-zero scalar components, and the system is not over-determined.

Before leaving this, we make one final observation: Though $\boldsymbol{\alpha}$ above was considered known and $\boldsymbol{M}_O$ unknown, the same relation $M_O = \frac{2}{9}ma^2\alpha$ would hold if their roles were reversed. In particular, knowing that M_O was *constant*, α would be, too. |

Example 5.2.10. *A Not-so-simple Single Rigid Body.* The next example is superficially like the previous one, but there are a number of additional points involved. The basic idea is the same as before: simply write down both kinematic and kinetic equations and solve the resulting system of linear equations. It's just that there are several new features here.

We consider a uniform "thin" rectangular $a \times b$ rectangular plate OAB of mass m supported by ball-and-sockets at its *diagonal* corners A and B as shown. The latter, in turn, are attached to one arm, OA, of a Y-shaped frame whose other arm, OB is always vertical. At the instant represented, OB rotates with an angular velocity $\boldsymbol{\Omega} = \Omega\hat{\boldsymbol{K}}$ and angular acceleration $\dot{\boldsymbol{\Omega}} = \dot{\Omega}\hat{\boldsymbol{K}}$; the plate rotates about its diagonal axis OA with angular velocity $\boldsymbol{\omega}_1 = \omega_1\hat{\boldsymbol{d}}$, where $\hat{\boldsymbol{d}}$ is directed along the diagonal. (What's *making* it move like this is not indicated; it just *is*! Though we might expect pins at the points A and O to enforce this, we would be wrong, as we shall see shortly.) We are asked to determine the angular acceleration of the plate *about the OA axis.*

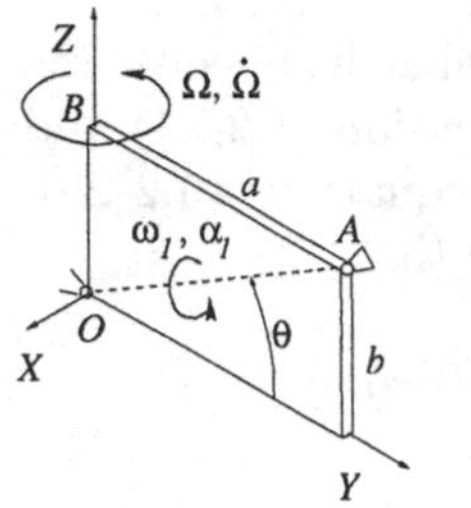

That's the first point: the rotational motion of the plate has two components, the absolute $\boldsymbol{\Omega}$ and $\boldsymbol{\omega}_1$ measured *relative to* $\boldsymbol{\Omega}$. This is precisely the situation presented in Section 4.1.2; in particular, equation (4.1-11) in that section shows that the *total, absolute* angular velocity of the plate will thus be $\boldsymbol{\omega} = \boldsymbol{\Omega} + \boldsymbol{\omega}_1$. More importantly for present purposes, however, that Section also discussed the fact that the [*absolute*] angular acceleration $\boldsymbol{\alpha}$ would have components due not only to $\dot{\boldsymbol{\Omega}}$ and any acceleration, say $\boldsymbol{\alpha}_1$, about OA, but *also one due to the change in direction of* $\boldsymbol{\omega}$; this was the thrust of Examples 4.1.1 and 4.1.2. In the present situation, that means that

$$\boldsymbol{\alpha} = \frac{d}{dt}(\Omega\hat{\boldsymbol{K}} + \omega_1\hat{\boldsymbol{d}}) = \dot{\Omega}\hat{\boldsymbol{K}} + \alpha_1\hat{\boldsymbol{d}} + \omega_1\dot{\hat{\boldsymbol{d}}} = \dot{\Omega}\hat{\boldsymbol{K}} + \alpha_1\hat{\boldsymbol{d}} + \omega_1\boldsymbol{\Omega} \times \hat{\boldsymbol{d}}.$$

If we take our coordinate system (x, y, z)—one rotating with the body, recall—for convenience to be aligned with the *fixed* (X, Y, Z) at the instant shown, $\hat{\boldsymbol{d}} = \cos\theta\hat{\boldsymbol{J}} + \sin\theta\hat{\boldsymbol{K}}$, and the angular quantities expressed in terms of this basis then become

$$\boldsymbol{\omega} = \omega_1\cos\theta\hat{\boldsymbol{J}} + (\Omega + \omega_1\sin\theta)\hat{\boldsymbol{K}}$$
$$\boldsymbol{\alpha} = -\Omega\omega_1\cos\theta\hat{\boldsymbol{I}} + \alpha_1\cos\theta\hat{\boldsymbol{J}} + (\dot{\Omega} + \alpha_1\sin\theta)\hat{\boldsymbol{K}},$$

in which

$$\cos\theta = \frac{a}{\sqrt{a^2 + b^2}}, \qquad \sin\theta = \frac{b}{\sqrt{a^2 + b^2}}. \tag{5.2-29}$$

α_1 then assumes the role of an unknown needed to determine the total $\boldsymbol{\alpha}$; note that this cannot be found from a kinematic analysis of O and B, since it measures the motion of a *rigid body*, not just of *points*.

We know everything about the motion of the line OB so, in particular, can find the center-of-mass acceleration, $\overline{\boldsymbol{a}}$, of the plate, necessary for the kinetic analysis:

$$
\begin{aligned}
\overline{\boldsymbol{a}} &= \boldsymbol{a}_O + \boldsymbol{\alpha}\times\overline{\boldsymbol{r}} + \boldsymbol{\omega}\times(\boldsymbol{\omega}\times\overline{\boldsymbol{r}}) + 2\boldsymbol{\omega}\times\boldsymbol{v}_{rel} + \boldsymbol{a}_{rel}\\
&= \mathbf{0} + \left(-\Omega\omega_1\cos\theta\hat{\boldsymbol{I}} + \alpha_1\cos\theta\hat{\boldsymbol{J}} + (\dot{\Omega}+\alpha_1\sin\theta)\hat{\boldsymbol{K}}\right)\times\left(\frac{a}{2}\hat{\boldsymbol{J}} + \frac{b}{2}\hat{\boldsymbol{K}}\right)\\
&\quad + \left(\omega_1\cos\theta\hat{\boldsymbol{J}} + (\Omega+\omega_1\sin\theta)\hat{\boldsymbol{K}}\right)\times\\
&\qquad\left(\left(\omega_1\cos\theta\hat{\boldsymbol{J}} + (\Omega+\omega_1\sin\theta)\hat{\boldsymbol{K}}\right)\times\left(\frac{a}{2}\hat{\boldsymbol{J}} + \frac{b}{2}\hat{\boldsymbol{K}}\right)\right) + \mathbf{0} + \mathbf{0}\\
&= -\frac{1}{2}a\dot{\Omega}\hat{\boldsymbol{I}} - \frac{1}{2}a\Omega^2\hat{\boldsymbol{J}} \qquad (5.2\text{-}30)
\end{aligned}
$$

Thus we now focus on the kinetics.

This is where the second point comes in: The equations of motion will comprise six scalar components of the force and moment equations—the latter surely to be taken about the fixed point O. A naive application of the ball-and-sockets at each O and A would thus lead to three unknown components at each which, along with α_1, would comprise *seven* unknowns, with only the *six* equations to determine them! So what's wrong here?

Nothing, actually, though it does introduce the concept of "*dynamic indeterminacy*," analogous to *static* indeterminacy familiar from Statics. The fact is that we don't *need* three constraints at both O and B: the Y-component of force, required to generate the centripetal acceleration to keep the center of mass moving in a circle, could be provided by a *single* force at *either* O *or* A. Again, similar situations arose in Statics, in which, for example, a shaft supported by pillow blocks at its ends would be modelled as having an axial force at only *one* of them; not to do so resulted in an indeterminate force in the axial direction. And, like classical Statics, classical Dynamics determines the *minimum* number of constraints required to enforce a given motion. While such dynamic indeterminacy can certainly be treated through the introduction of Solid Mechanics as is done in static cases, we shall be satisfied to take the more simplistic approach of merely eliminating the superfluous force F_{O_Y}.[17] But, clearly, this must be considered before drawing the free-body diagram, given at left.

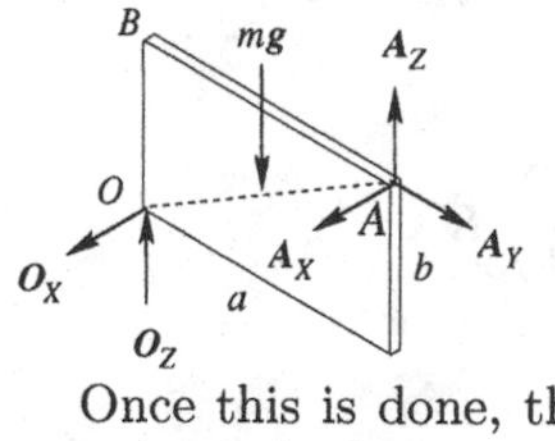

Once this is done, the balance of the problem becomes routine. The components of the force equation, into which (5.2-30) has been substituted, become simply

$$
\begin{aligned}
O_X + A_X &= m\overline{a}_X = -m\frac{1}{2}a\dot{\Omega}\\
A_Y &= m\overline{a}_Y = -m\frac{1}{2}a\Omega^2\\
O_Z + A_Z - mg &= m\overline{a}_Z = m\cdot 0.
\end{aligned}
$$

[17] Physically, such reaction force could be generated, *e.g.*, by having a pin at O connected to a rod freely sliding in the Y-direction. The choice of this particular variable is somewhat arbitrary: we could also have picked F_{O_Z} or one of the corresponding components at A.

The moment equation is little harder. As suggested above, we will take moments about the fixed point O: $\sum \boldsymbol{M}_O = \boldsymbol{\omega} \times \mathsf{I}_O\boldsymbol{\omega} + \mathsf{I}\boldsymbol{\alpha}$. But this means we need I_O, which is relatively easy because the basis directions are principal relative to the center of mass; thus we can apply the Parallel Axis Theorem, which gives

$$\mathsf{I}_O = \begin{pmatrix} \frac{1}{12}m(a^2+b^2) & 0 & 0 \\ 0 & \frac{1}{12}mb^2 & 0 \\ 0 & 0 & \frac{1}{12}ma^2 \end{pmatrix} + \begin{pmatrix} \frac{1}{4}m(a^2+b^2) & 0 & 0 \\ 0 & \frac{1}{4}mb^2 & -\frac{1}{4}mab \\ 0 & -\frac{1}{4}mab & \frac{1}{4}ma^2 \end{pmatrix}$$
$$= \begin{pmatrix} \frac{1}{3}m(a^2+b^2) & 0 & 0 \\ 0 & \frac{1}{3}mb^2 & -\frac{1}{4}mab \\ 0 & -\frac{1}{4}mab & \frac{1}{3}ma^2 \end{pmatrix}.$$

(Note that, as in Example 5.2.5 [which is the same body], axes at O are *not* principal, even though parallel axes through the center of mass *are*.) Components of the moment equation become

$$aA_Z - bA_Y - \frac{1}{2}amg = \frac{\left(\omega_1^2(a^2-b^2) + 3\Omega_1^2(a^2+b^2)\right)mab}{12\,(a^2+b^2)} - \frac{\Omega\omega_1 mab^2}{6\sqrt{a^2+b^2}}$$
$$bA_X = \frac{mab^2\alpha_1}{12\sqrt{a^2+b^2}} - \frac{mab\dot{\Omega}}{4}$$
$$-aA_X = \frac{ma^2b\alpha_1}{12\sqrt{a^2+b^2}} + \frac{ma^2\dot{\Omega}}{3}.$$

In all the above equations, θ has been eliminated in favor of a and b using (5.2-29). As expected, they are linear in the desired unknowns. The solutions are

$$\boldsymbol{O} = -\frac{5}{24}ma\dot{\Omega}\hat{\boldsymbol{I}} + m\left(\frac{2g - b\Omega^2}{4} + \frac{b\omega_1^2(a^2-b^2)}{12(a^2+b^2)} - \frac{b^2\Omega\omega_1^2}{6\sqrt{a^2+b^2}}\right)\hat{\boldsymbol{K}}$$
$$\boldsymbol{A} = -\frac{7}{24}ma\dot{\Omega}\hat{\boldsymbol{I}} - \frac{1}{2}ma\Omega^2\hat{\boldsymbol{J}}$$
$$+ m\left(\frac{2g + b\Omega^2}{4} - \frac{b\omega_1^2(a^2-b^2)}{12(a^2+b^2)} + \frac{b^2\Omega\omega_1^2}{6\sqrt{a^2+b^2}}\right)\hat{\boldsymbol{K}}$$
$$\alpha_1 = -\frac{\sqrt{a^2+b^2}}{2b}\dot{\Omega},$$

from the last of which

$$\boldsymbol{\alpha} = -\frac{a}{\sqrt{a^2+b^2}}\Omega\omega_1\hat{\boldsymbol{I}} - \frac{a}{2b}\dot{\Omega}\hat{\boldsymbol{J}} + \frac{1}{2}\dot{\Omega}\hat{\boldsymbol{K}}.$$

That $\alpha_1 \neq 0$ is at first surprising, since there is no moment being applied about the OA-axis. But its generation can be argued heuristically as follows: If we regard the two separate triangular portions of the plate above and below the diagonal, the respective centers of mass are at different distances, $\frac{a}{3}$ and $\frac{2a}{3}$, from the OB-axis. These distances, multiplied by the $\dot{\Omega}$-terms, in effect result in "forces" which are themselves different, giving a net moment about the OA-axis observed.

It is instructive to view this example from the perspective of kinematics, particularly the discussion of the ball-and-socket joints on page 237, bringing us to the final point demonstrated by this example: Despite the nature of the connections at O and A, we have never introduced the kinematical constraints, equations (4.4-27) $\boldsymbol{\omega} \cdot \boldsymbol{r} = \boldsymbol{\omega}_{\perp} \cdot \boldsymbol{r} \equiv 0$ and $\boldsymbol{\alpha} \cdot \boldsymbol{r} = \boldsymbol{\alpha}_{\perp} \cdot \boldsymbol{r} \equiv 0$, imposed on the angular motion by such constraints when such motion is *unknown*. The first is unnecessary: $\boldsymbol{\omega} = \boldsymbol{\Omega} + \boldsymbol{\omega}_1$, where both components are *known*. But, even though $\boldsymbol{\alpha}$ is not [entirely] known, attempting to implement the second would be *wrong*: both of these constraints merely make explicit our inability to determine components of the desired angular motion along the axis of rotation purely kinematically; the desired $\boldsymbol{\alpha}_1$ is in just that direction, and any attempt to introduce this constraint would result in a set of inconsistent equations! Ending the above-cited discussion, on page 238, we pointed out that such components "*cannot* be determined from a kinematic analysis." But in the present case, we *were* able to determine [all of] $\boldsymbol{\alpha}$, not from a kinematical analysis, but from a *kinetic* one.

This example was presented by first determining $\overline{\boldsymbol{a}}$ from the kinematic equation (5.2-30), then solving the force and moment equations separately. But we note again that this could be regarded as a complete system of nine equations (three each for $\overline{\boldsymbol{a}}$, $\sum \boldsymbol{F}$, and $\sum \boldsymbol{M}_O$) in the nine unknowns $\overline{\boldsymbol{a}}$ (3), $\boldsymbol{F}_O$ (2), $\boldsymbol{F}_A$ (3), and α_1 (1); the system could be written down *in toto* from the beginning, then solved simultaneously. The latter perspective is useful in unifying the approach to *all* problems, regardless of the details in each. |

Example 5.2.11. *A "Simple" Machine.* We have been conscientious about giving all the details of most problems, showing all intermediate calculations and getting final answers. This will *not* be done in the present case, because the final system is impressive and, although the answers have been obtained, the tens of terms which most contain add little to the ultimate intent of the problem—can, in fact, *obscure* the points the problem is attempting to illustrate. Thus we shall be content merely to indicate *which* equations would be used to solve it.

This problem differs from the previous ones in that here it is the applied moment which is known, with the resulting angular motion being *totally* unknown (and, of course, it is a "machine"). We consider two "slender" rods composed of the same stock: OAB, of length $3a$ and mass m with a 30°-bend one-third of its length from B—the same body considered in Example 5.2.9—connected to AC of length $2a$ and thus of mass $\frac{2m}{3}$ by a ball-and-socket. The other end of AC is attached to a collar free to slide along the X-axis by a clevis. OB is held vertical by a "narrow strap" at O and a "cup" at B—both frictionless, and has applied to it a couple moment M_O at O. We are asked to determine the angular

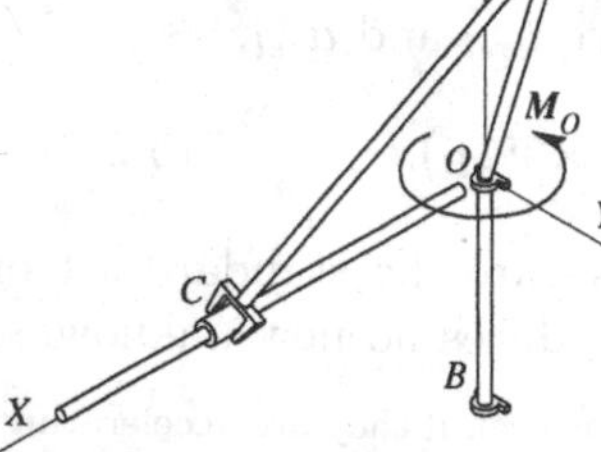

acceleration of OAB, $\boldsymbol{\alpha}_{OA}$, assuming that we know the angular velocities $\boldsymbol{\omega}_{AB}$ and $\boldsymbol{\omega}_{AC}$, as a function of the angle θ from the X-axis to the line projected from OA into the XY-plane.

As suggested in the discussion of interconnected bodies on page 312, we find it safest to consider the "ball" of a ball-and-socket joint as a separate free body; thus we consider *three* free-body diagrams and the kinematic/kinetic equations on each: OAB, AC, and the ball A. To emphasize the parallels with the cited discussion, upper case $\boldsymbol{F}$ and lower case $\boldsymbol{f}$ will denote the external and internal forces, respectively.

We first consider the free-body diagram of the bent rod OAB. We show the weight $m\boldsymbol{g}$—note that it acts at the center of mass of the *entire* body—and three components of the force $\boldsymbol{F}_B$ at the "small cup" at B. Also shown are the two components of the force at O, $\boldsymbol{F}_O$, and the applied couple moment $\boldsymbol{M}_O$. None of this is anything new. What *is* new, however, is the set of forces at A: f_{A_X}, f_{A_Y}, and f_{A_Z} are the components of the force $\boldsymbol{f}_{A_{OA}}$ at the ball-and-socket joint *generated by the ball itself*; these are not the "force of AC on OAB," and, in any event, are not directed along AC.[18]

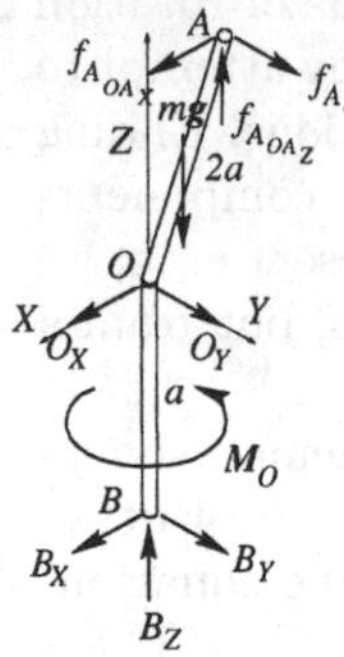

These cataloged, the equations of motion are immediate: The force equation is always the same:

$$\boldsymbol{F}_O + \boldsymbol{F}_B + \boldsymbol{f}_{A_{AB}} + m\boldsymbol{g} = m\overline{\boldsymbol{a}}_{AB}. \qquad (5.2\text{-}31a)$$

Since the body rotates about the fixed point O—B could have been chosen as well—we will sum moments about that point: denoting the position of the center of mass of this body relative to O by $\overline{\boldsymbol{r}}_{AB}$,

$$\boldsymbol{r}_{B/O} \times \boldsymbol{F}_B + \boldsymbol{r}_{A/O} \times \boldsymbol{f}_{A_{AB}} + \overline{\boldsymbol{r}}_{AB} \times m\boldsymbol{g} = \boldsymbol{\omega}_{AB} \times \mathbf{I}_{O_{AB}}\boldsymbol{\omega}_{AB} + \mathbf{I}_{O_{AB}}\boldsymbol{\alpha}_{AB}. \qquad (5.2\text{-}31b)$$

(This equation demands determination of $\mathbf{I}_{O_{AC}}$, but we have already done this as part of Example 5.2.9.) The above pair comprises a set of six scalar equations in the unknowns $\boldsymbol{F}_O$ (2 components), $\boldsymbol{F}_B$ (3), $\boldsymbol{f}_{A_{AB}}$ (3), $\overline{\boldsymbol{a}}_{AB}$ (3) and $\boldsymbol{\alpha}_{AB}$ (1)—a total of 12 unknown scalar components. Clearly we need more equations!

One set of these is immediately suggested by those above: the kinematical relations between $\overline{\boldsymbol{a}}_{AB}$ and $\boldsymbol{\alpha}_{AB}$: since $\boldsymbol{a}_O \equiv \mathbf{0}$, as are $\boldsymbol{v}_{rel}$ and $\boldsymbol{a}_{rel}$,

$$\overline{\boldsymbol{a}}_{AB} = \boldsymbol{\alpha}_{AB} \times \overline{\boldsymbol{r}}_{AB} + \boldsymbol{\omega}_{AB} \times (\boldsymbol{\omega}_{AB} \times \overline{\boldsymbol{r}}_{AB}). \qquad (5.2\text{-}31c)$$

These three vector equations—nine scalar equations—are the standard set of equations for a single rigid body. While the last introduces no new unknowns,

[18] *Light* rigid bodies formally satisfy the equilibrium equations even if they are accelerating: the mass, so both m and $\mathbf{I}$, and thus $m\overline{\boldsymbol{a}}$ and $\mathbf{I}\boldsymbol{\alpha}$, will vanish. Were AC "light," then, it would be in "equilibrium" under the two forces at A and C, and thus a "two force member" under either tension or compression. But here $m_{AC} \neq 0$.

it only furnishes us with three more scalar equations; we're still short. This is precisely the situation with frames and machines in Statics: each separate body's equations of equilibrium are always deficient in the number of equations available.

While on body ABB, there is one more equation which will turn out to be necessary for another part of the machine: the motion of A, used to determine the motion of AC. That is also routine:

$$\boldsymbol{a}_A = \boldsymbol{\alpha}_{AB} \times \boldsymbol{r}_{A/O} + \boldsymbol{\omega}_{AB} \times (\boldsymbol{\omega}_{AB} \times \boldsymbol{r}_{A/O}). \tag{5.2-31d}$$

We note that, while this gives us three more equations, it also introduces another three unknown components of $\boldsymbol{a}_A$. Thus far, then, we have 12 equations in 15 unknowns.

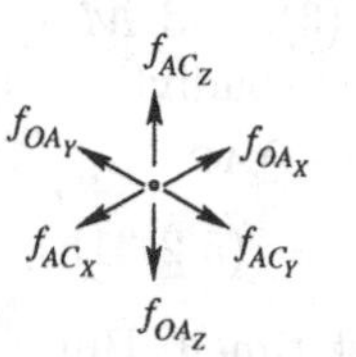

Now consider the ball at A: Along with the three components of $\boldsymbol{f}_{A_{AB}}$ already accounted for—but now with the opposite sign by Newton's Third, we have additionally the three components due to body AC, $\boldsymbol{f}_{A_{AC}}$. Thus the force equations comprise the three scalar components of

$$-\boldsymbol{f}_{A_{AB}} + \boldsymbol{f}_{A_{AC}} \equiv \mathbf{0} \tag{5.2-31e}$$

—the right-hand side vanishing because the ball is "light." There is no moment equation: it vanishes identically since all forces pass through A and $\mathsf{I}_A \equiv 0$. Thus we have picked up three new equations, but at the expense of three new unknowns, the components of $\boldsymbol{f}_{A_{AC}}$. The totals: 15 equations in 18 unknowns.

Finally we consider the clevis-ended rod AC: Once again the forces of the ball on end A, $-\boldsymbol{f}_{A_{AC}}$ is included, reversed from its direction on A above. Because the rod along which the collar at C slides is smooth, there is no reaction force in the X-direction. And there is the ubiquitous weight, here only $\frac{2}{3}mg$ because of the shorter length of this member.

But there is additionally a *couple moment due to the clevis* at end C: what direction is it in? At first [see the remark on page 313] one might expect it to be *along* AC, but this is wrong, as a more careful analysis indicates: In the first place, the collar is free to rotate about the rod—denoted "$\hat{\boldsymbol{u}}_c$" in the kinematic analysis of these connections on page 228; thus any couple moment must be *perpendicular* to $\hat{\boldsymbol{u}}_c$: $\boldsymbol{M}_C \perp \hat{\boldsymbol{u}}_c$. Similarly, $\boldsymbol{M}_C \perp \hat{\boldsymbol{u}}_p$, the direction of the pin connecting the fork to the collar. Thus $\boldsymbol{M}_C$ is mutually perpendicular to *both* these vectors: $\boldsymbol{M}_C \parallel \hat{\boldsymbol{u}}_c \times \hat{\boldsymbol{u}}_p$, or $\boldsymbol{M}_C = M_C \frac{\hat{\boldsymbol{u}}_c \times \hat{\boldsymbol{u}}_p}{\|\hat{\boldsymbol{u}}_c \times \hat{\boldsymbol{u}}_p\|}$; this is the direction indicated in the free-body diagram at left.[19] Note, however, that $\boldsymbol{M}_C$ corresponds to only a single unknown in known direction.

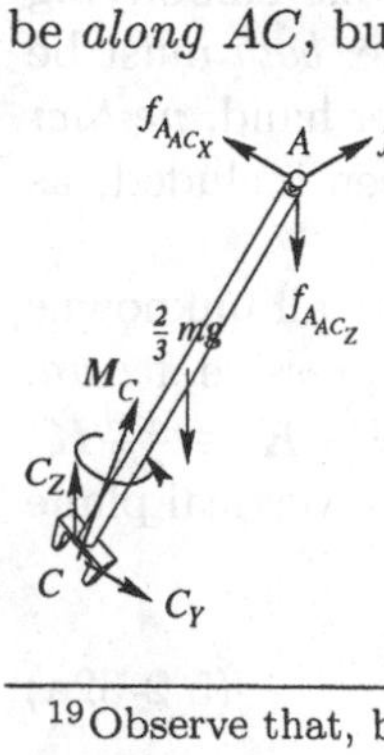

[19]Observe that, by equation (4.4-21), this makes $\boldsymbol{M}_C \perp \boldsymbol{\omega}_{AC}$!

The kinetic equations are little less immediate than they were before: The force equation is

$$\boldsymbol{F}_C - \boldsymbol{f}_{A_{AC}} - \frac{2}{3}m\boldsymbol{g} = \frac{2}{3}m\overline{\boldsymbol{a}}_{AC}, \tag{5.2-31f}$$

where the mass of this body is only two-thirds that of AB, due to its shorter length. The moment equation is different from what it was before, however: since there is no fixed point about which the body rotates, we effectively default to taking moments about the center of mass: conventionally denoting vectors measured from that point with $\boldsymbol{\rho}$'s, we get that

$$\boldsymbol{\rho}_A \times (-\boldsymbol{f}_{A_{AC}}) + \boldsymbol{\rho}_C \times \boldsymbol{F}_C + \boldsymbol{M}_C = \boldsymbol{\omega}_C \times \overline{\mathsf{I}}_{AC}\boldsymbol{\omega}_C + \overline{\mathsf{I}}_{AC}\boldsymbol{\alpha}_{AC}. \tag{5.2-31g}$$

(Once again, we need $\overline{\mathsf{I}}_{AC}$, but this is precisely the inertia matrix determined in Example 5.2.8. Note that we ignore the pathological geometry which could be introduced by the clevis at C, engaging in the standard Engineering approximation that this is "small.") The three scalar components of this equation introduce the new unknowns $\boldsymbol{F}_C$ (2 components), $\overline{\boldsymbol{a}}_{AC}$ (3), $\boldsymbol{\alpha}_{AC}$ (3), and $\boldsymbol{M}_C$ (1); now 21 equations, but 27 unknowns. We need more equations—badly!

Again, the kinematics comes to our aid: we can relate $\overline{\boldsymbol{a}}_{AC}$ and $\boldsymbol{\alpha}_{AC}$,

$$\overline{\boldsymbol{a}}_{AC} = \boldsymbol{a}_A + \boldsymbol{\alpha}_{AC} \times \overline{\boldsymbol{r}}_{AC} + \boldsymbol{\omega}_{AC} \times (\boldsymbol{\omega}_{AC} \times \overline{\boldsymbol{r}}_{AC}); \tag{5.2-31h}$$

Three more equations to give 24, *no* new unknowns [for the first time]! But, unlike AB, the center-of-mass motion is itself determined by the kinematical constraints on end A and C:

$$\boldsymbol{a}_C = \boldsymbol{a}_A + \boldsymbol{\alpha}_{AC} \times \boldsymbol{r}_{C/A} + \boldsymbol{\omega}_{AC} \times (\boldsymbol{\omega}_{AC} \times \boldsymbol{r}_{C/A}). \tag{5.2-31i}$$

But this introduces a *28*th unknown, a_C, since $\boldsymbol{a}_C$ is known to be only in the X-direction: 27 equations, 28 unknowns. Where to get that last equation?

End C is a *clevis*, and we know that the rotation about the connection axis, here AC, is kinematically determined by the scalar equation (4.4-22):

$$\boldsymbol{\alpha}_{AC} \cdot \hat{\boldsymbol{u}}_c \times \hat{\boldsymbol{u}}_p = \omega_c\omega_p, \tag{5.2-31j}$$

in which the various quantities are obtained as explained in the section on these connections, page 228. This is one of the important points of this Example: the unknown $\boldsymbol{M}_C$ and this equation come together "in a brace," one embodying the *kinetic* ramification of the clevis, the other the *kinematic* one; *both* must be implemented. Had there been a ball-and-socket at C, on the other hand, *neither* a moment *nor* the kinematic condition (4.4-27) would have been included, as the previous example demonstrated.

This last equation closes the circle, and the system is soluble: all unknowns appear only *linearly*, and, although the system of equations is impressive, it *can*, "in theory," be solved. Just to show this, the desired $\boldsymbol{\alpha}_{AB} = \alpha_{AB}\hat{\boldsymbol{K}} \equiv \frac{d^2\theta}{dt^2}\hat{\boldsymbol{K}}$, where θ is the angle in the XY-plane between the X-axis and the vertical plane through OA and the Z-axis:

$$\alpha_{AB} = \frac{3}{4}\frac{num}{ma^2(32 + 32\cos^2\theta - 22\cos^4\theta + 3\cos^6\theta)}, \tag{5.2-32a}$$

in which

$$num = (48 - 24\cos^2\theta + 3\cos^4\theta)M_O + \\ ma^2\omega^2 \sin\theta\,(64\cos\theta - 31\cos^3\theta + 4\cos^5\theta) \quad (5.2\text{-}32b)$$

—and that's one of the *shorter* answers![20] |

The primary reason for getting *this* answer is the following:

Determination of the Motion of a System

We have pointed out that the equations of motion are *instantaneous* in nature, giving the *kinetic* forces and moments consistent with a given *kinematic* translational and rotational motion. And, like most presentations of this material, all the above examples have analyzed the motion *at a particular instant.*

But it is too easy to overlook a fundamental point: the classic problem in dynamics considers the kinetics as given and desires to find the kinematics, not at an instant, but over an *interval of time.* The equations of motion govern this motion; how would such instantaneous principles yield a solution over a finite time?

The answer, of course, is that these equations involve [translational and rotational] accelerations—*second [time] derivatives*—and thus determine the *second-order, ordinary differential equations* governing the motion. In particular, if we have *general* expressions giving the accelerations at *each* instant, these equations "can" be integrated to yield the subsequent motion. For instance, in Example 5.2.9 we found that

$$M_O = \frac{2}{9}ma^2\alpha \equiv \frac{2}{9}ma^2\frac{d^2\theta}{dt^2};$$

if we regarded the couple moment as being a known *constant,* then we would have the angular motion immediately:

$$\theta = \frac{9}{4ma^2}M_O t^2 + \dot{\theta}_o t + \theta_o,$$

in which $\dot{\theta}_o$ and θ_o were the values of the first and zeroth time derivatives of θ at $t = 0$.

Unfortunately, the differential equations are rarely so simple. In the preceding Example 5.2.11, even though there is only one component to $\boldsymbol{\alpha}$—and that one parallel to the applied $\boldsymbol{M}_O$—the equation of motion governing this rotation could most politely only be described as "daunting": $\alpha_{AB} \equiv \frac{d^2\theta}{dt^2}$ satisfies

[20]No, this *wasn't* done by hand! Like most of the examples in this book, it was solved with a MAPLE routine. MAPLE and its ilk become a boon to the analytic solution of a vast array of problems, but it is particularly useful in large systems of linear equations such as these.

equations (5.2-32), in which now "*num*" becomes

$$num = (48 - 24\cos^2\theta + 3\cos^4\theta)M_O + \\ ma^2\sin\theta\,(64\cos\theta - 31\cos^3\theta + 4\cos^5\theta)\left(\frac{d\theta}{dt}\right)^2$$

—an expression which, recall, is divided by another polynomial in $\cos\theta$, making it wildly non-linear! Even if M_O is constant, this is virtually impossible to solve analytically, and one must resort to a numerical integration. And this was a single equation for the single variable θ, which *happened* to be able to be decoupled from the other 27 unknowns in that Example; more generally, these differential equation are *coupled*, with variables, and their derivatives, appearing in the second-order equations for one another.

The Good News is that there are alternatives to this "brute force" approach to solving the fundamental differential equations directly—alternatives which involve already-integrated forms of these equations applicable in many, if not all, problems. We shall discuss these in Section 5.4

Homework:

1. Rod on frictionless horizontal surface, resting against frictionless vertical one, released from rest. Find $\boldsymbol{\alpha}$, normal forces.
2. [*Computer project*] Numerically solve the above problem for $\boldsymbol{a}$ of one end (2 components) and plot the path. Maybe indicate intervals of time with bigger dots?

Summary. Though this section has been extensive, much of its length results from the level of detail in the Examples. We ultimately developed the fundamental *equations of motion* (5.2-22) and (5.2-23)—*instantaneous* principles governing the motion of such bodies, and relating the *kinetic* forces and/or [couple] moments and *kinematic* translational and rotational accelerations. We discussed how these equations applied to the motion of both single rigid bodies and machines, in particular indicating how to deal with the forces and/or moments at *interconnections*, recommending that ball-and-sockets, in particular, be treated as separate bodies, and how to deal with the couple moments introduced by a clevis connection.

These forces, moments, and accelerations enter the equations of motion purely *linearly*; it is only the angular velocities which enter non-linearly through the $\boldsymbol{\omega}\times(\boldsymbol{\omega}\times\boldsymbol{r})$ term. To the extent the last are known, the balance can thus be found through the simultaneous solution of a purely *linear* system of equations embodying both the kinetics and the kinematics of the problem *at a particular instant*. However grand these systems might be—and even for the relatively simple case of just two rigid bodies above, they *were*—in principle they are all "merely" linear systems. To find the motion over an *interval*, however, it is necessary to *integrate* these instantaneous equations of motion—a task that is generally difficult.

5.2.3 A Special Case—the Gyroscope

We have obtained *general* equations of motion for a *general* rigid body. We shall now consider these equations in a particularly special type of rigid body, one which is *symmetric* about one axis. Though this section is entitled "The Gyroscope," in fact the results are more general than this particular application, however intrinsically important it might be. Ultimately, however, we get equations involving two separate components of the angular motion: one about the axis of symmetry of the body and the other measuring the motion *of* this axis—the "spin" and "precession/nutation" of the gyroscope, for example.

Gyroscope Coordinate Axes and Angular Velocities

To keep the identity of these two separate, it is necessary to set up a set of coordinates peculiar to the analysis of the "gyroscope": We define, relative to the fixed (X, Y, Z) axes, the z-axis to lie along the axis of symmetry of the body; this choice clearly isolates the spin to lie along that axis. In fact, it is normal to define that axis in the direction *of* the spin, so that $\hat{\boldsymbol{k}}$ is defined: $\boldsymbol{p} \equiv p\hat{\boldsymbol{k}}$ with $p > 0$. Given that, then we define the x axis to be perpendicular to the plane containing both the z- and Z-axes, defined *via* the right-hand rule: $\hat{\boldsymbol{\imath}} \equiv \hat{\boldsymbol{K}} \times \hat{\boldsymbol{k}}$ as shown; this choice is motivated by the fact that, assuming $\hat{\boldsymbol{K}}$ is vertical, the moment due to gravity will always lie in this direction. (We are implicitly assuming that the spin axis is not along $\hat{\boldsymbol{K}}$; otherwise, although $\hat{\boldsymbol{k}}$ would be defined, $\hat{\boldsymbol{\imath}}$ wouldn't!) Given $\hat{\boldsymbol{\imath}}$ and $\hat{\boldsymbol{k}}$, $\hat{\boldsymbol{\jmath}}$, and the y-axis, are then determined by the right-hand rule: $\hat{\boldsymbol{\jmath}} \equiv \hat{\boldsymbol{k}} \times \hat{\boldsymbol{\imath}}$. Note that our definition, particularly of the x- (and thus y-) axes, force the spin to take place *relative* to these axes, which are defined fundamentally in terms of the spin axis; these axes do *not* rotate with the gyroscope. We could, however, define yet another set of axes (x', y', z') which *do*, taking $\hat{\boldsymbol{k}}' \equiv \hat{\boldsymbol{k}}$ and defining the x'-axis [perpendicular to $\hat{\boldsymbol{k}}'$] to be fixed in the body, with $\hat{\boldsymbol{\jmath}}'$ following from the right-hand rule again.

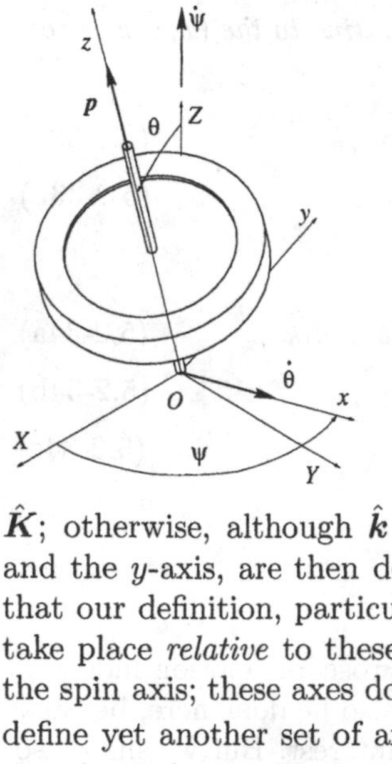

Having defined the appropriate axes, we can now define a set of angles measuring their position: we define θ to be the angle between $\hat{\boldsymbol{K}}$ and $\hat{\boldsymbol{k}}$—between the Z- and z-axes—measured positively about $\hat{\boldsymbol{\imath}}$; ψ is defined to be the angle between $\hat{\boldsymbol{I}}$ and $\hat{\boldsymbol{\imath}}$—between the X- and x-axes—measured positively about $\hat{\boldsymbol{K}}$. If we were to fix a set in the body, we could measure the angle ϕ between the x- and x'-axes, measured positively about $\hat{\boldsymbol{k}}$.[21] Once again, the first two angles completely specify the orientation of the spin axis, the third the actual orientation of the gyroscope itself. Note that these are precisely the *Euler angles*

[21]The reader is warned that the notation describing these angles is anything but uniform among authors; see, for example, the footnotes in Goldstein, *Classical Mechanics, Second Edition*, p. 145.

describing the orientation of axes, introduced in Section 4.2. In particular, [vector] points relative to the (x, y, z) axes could be expressed relative to the fixed axes through the pre-multiplication of the rotation matrices $\mathsf{C}^{-1}(-\psi)\mathsf{A}^{-1}(-\theta)$; matrices; see Example 3.6.3. Transforming the inertia matrix would also require post-multiplication by the inverse of their product. (Points relative to (x', y', z') would require an additional first $\mathsf{C}^{-1}(-\phi)$-rotation.)

The *motion* of the body is described through the time derivatives of these three angles. In particular, we can define *vector* angular velocities and the names used to describe them:

- $\boldsymbol{p} \equiv \dot{\phi}\hat{\boldsymbol{k}} \equiv p\hat{\boldsymbol{k}}, \quad p > 0$—the *spin*,
- $\dot{\boldsymbol{\theta}} \equiv \dot{\theta}\hat{\boldsymbol{\imath}}$—the *nutation*, and
- $\dot{\boldsymbol{\psi}} \equiv \dot{\psi}\hat{\boldsymbol{K}} = \dot{\psi}(\sin\theta\hat{\boldsymbol{\jmath}} + \cos\theta\hat{\boldsymbol{k}})$—the *precession.*

Note that the spin $p \equiv \dot{\phi}$ is defined to be measured *relative to the (x, y, z) axes. Absolute* motion of the spin axis is given by

$$\boldsymbol{\Omega} \equiv \dot{\boldsymbol{\psi}} + \dot{\boldsymbol{\theta}} = \dot{\theta}\hat{\boldsymbol{\imath}} + \dot{\psi}\sin\theta\hat{\boldsymbol{\jmath}} + \dot{\psi}\cos\theta\hat{\boldsymbol{k}} \tag{5.2-33a}$$

$$\equiv \Omega_x\hat{\boldsymbol{\imath}} + \Omega_y\hat{\boldsymbol{\jmath}} + \Omega_z\hat{\boldsymbol{k}}, \tag{5.2-33b}$$

while the total angular velocity of the body

$$\boldsymbol{\omega} = \boldsymbol{\Omega} + \boldsymbol{p} = \dot{\theta}\hat{\boldsymbol{\imath}} + \dot{\psi}\sin\theta\hat{\boldsymbol{\jmath}} + (\dot{\psi}\cos\theta + p)\hat{\boldsymbol{k}} \tag{5.2-34a}$$

$$= \Omega_x\hat{\boldsymbol{\imath}} + \Omega_y\hat{\boldsymbol{\jmath}} + (\Omega_z + p)\hat{\boldsymbol{k}} \tag{5.2-34b}$$

$$\equiv \omega_x\hat{\boldsymbol{\imath}} + \omega_y\hat{\boldsymbol{\jmath}} + \omega_z\hat{\boldsymbol{k}}. \tag{5.2-34c}$$

Equations of Motion

It is customary to derive the relevant equations of gyroscopic motion independently of the more general Euler equations; this will also be done here, because it serves better to identify the relevant quantities of interest. But we shall also show how these equations can be derived from the more general ones, just to emphasize that there's really nothing new—nothing *different*—about this case.

There is no change in the force equation (5.2-22): $\sum_i \boldsymbol{F}_i = m\overline{\boldsymbol{a}}$ no matter *what* the system is! But the *moment* equations (5.2-23) we have been using to this point were, recall, derived using rotating coordinates rotating *with* the body and its angular velocity $\boldsymbol{\omega}$.

The most general form of moment equation is given by Theorem 5.2.5:

$$\sum_i \boldsymbol{M}_{A_i} = \boldsymbol{r}_{G/A} \times m\overline{\boldsymbol{a}} + \dot{\overline{\boldsymbol{H}}}.$$

We shall focus attention on the $\dot{\overline{\boldsymbol{H}}}$ first, obtaining other variants of this equation from it.

Note that $\overline{\boldsymbol{H}} = \bar{\mathsf{I}}\boldsymbol{\omega}$, and it is useful to examine briefly the inertia matrix itself: The assumed symmetry makes *any* planes through the spin axis principal, and so the spin axis itself; thus the (x, y, z) axes themselves are principal, and the inertia matrix will be diagonalized relative to these axes from the start. The center of mass is clearly located on the spin axis, so $\bar{\mathsf{I}}$ will be diagonalized; and, as long as a fixed point O about which the gyroscope spins is located along one of these axes—in practice, it invariably is—so will I_O. Thus the "generic" form of the inertia matrix, with the coordinates chosen, is

$$\mathsf{I} = \begin{pmatrix} I_O & 0 & 0 \\ 0 & I_O & 0 \\ 0 & 0 & I \end{pmatrix}, \tag{5.2-35}$$

where, again, symmetry makes $I_{xx} = I_{yy} \equiv I_O$.

Direct Derivation of the Gyroscopic Equations of Motion. To compare these with the present ones, it is useful briefly to review the previous derivation: If we denote the time derivative measured *relative* to a coordinate system rotating with angular velocity $\boldsymbol{\omega}$ by "$\frac{d}{dt}\big|_{\boldsymbol{\omega}}$," as in Homeworks 1 on page 187 and 1 on page 305, the derivation of the moment equation, Euler's equation on page 305, uses Theorem 4.3.1:

$$\begin{aligned}\dot{\overline{\boldsymbol{H}}} &= \frac{d}{dt}\bigg|_{\boldsymbol{\omega}} \overline{\boldsymbol{H}} + \boldsymbol{\omega} \times \overline{\boldsymbol{H}} = \frac{d}{dt}\bigg|_{\boldsymbol{\omega}} \bar{\mathsf{I}}\boldsymbol{\omega} + \boldsymbol{\omega} \times \bar{\mathsf{I}}\boldsymbol{\omega} \\ &= \left(\frac{d}{dt}\bigg|_{\boldsymbol{\omega}} \bar{\mathsf{I}}\right)\boldsymbol{\omega} + \bar{\mathsf{I}}\frac{d}{dt}\bigg|_{\boldsymbol{\omega}} \boldsymbol{\omega} + \boldsymbol{\omega} \times \bar{\mathsf{I}}\boldsymbol{\omega} \\ &= \mathbf{0} + \bar{\mathsf{I}}\boldsymbol{\alpha} + \boldsymbol{\omega} \times \bar{\mathsf{I}}\boldsymbol{\omega}\end{aligned}$$

because the axes are rotating *with* the body, and, again by Theorem 4.3.1,

$$\boldsymbol{\alpha} \equiv \frac{d\boldsymbol{\omega}}{dt} = \frac{d}{dt}\bigg|_{\boldsymbol{\omega}} \boldsymbol{\omega} + \boldsymbol{\omega} \times \boldsymbol{\omega} = \frac{d}{dt}\bigg|_{\boldsymbol{\omega}} \boldsymbol{\omega}.$$

But consider the present system, axes rotating with the axis of symmetry—with $\boldsymbol{\Omega}$, *not* $\boldsymbol{\omega}$–inside which the gyroscope rotates with $\boldsymbol{p}$: Now

$$\dot{\overline{\boldsymbol{H}}} = \frac{d}{dt}\bigg|_{\boldsymbol{\Omega}} \bar{\mathsf{I}}\boldsymbol{\omega} + \boldsymbol{\Omega} \times \bar{\mathsf{I}}\boldsymbol{\omega}. \tag{5.2-36}$$

But

$$\frac{d}{dt}\bigg|_{\boldsymbol{\Omega}} \bar{\mathsf{I}}\boldsymbol{\omega} = \left(\frac{d}{dt}\bigg|_{\boldsymbol{\Omega}} \bar{\mathsf{I}}\right)\boldsymbol{\omega} + \bar{\mathsf{I}}\frac{d}{dt}\bigg|_{\boldsymbol{\Omega}} \boldsymbol{\omega},$$

in which the first term vanishes, *not* because the axes are rotating with the body, but rather because of the assumed *symmetry* of that body. Thus

$$\dot{\overline{\boldsymbol{H}}} = \boldsymbol{\Omega} \times \bar{\mathsf{I}}\boldsymbol{\omega} + \bar{\mathsf{I}}\frac{d}{dt}\bigg|_{\boldsymbol{\Omega}} \boldsymbol{\omega}.$$

Gyroscopic Equations of Motion from Euler's Equations. As was mentioned at the beginning of this section, the gyroscope equations are invariably developed *de nuvo* without any reference to the more general Euler equations. This is unfortunate, because it can lend a certain air of mystery about these equations, as if they are somehow "different" from the general ones. In fact, they're not, as the following shows:

Expressing Euler's equations using the fact that $\boldsymbol{\omega} = \boldsymbol{\Omega} + \boldsymbol{p}$,

$$\begin{aligned}\dot{\overline{\boldsymbol{H}}} &= \boldsymbol{\omega} \times \bar{\mathsf{I}}\boldsymbol{\omega} + \bar{\mathsf{I}}\boldsymbol{\alpha} \\ &= (\boldsymbol{\Omega} + \boldsymbol{p}) \times \bar{\mathsf{I}}\boldsymbol{\omega} + \bar{\mathsf{I}}\frac{d\boldsymbol{\omega}}{dt} \\ &= (\boldsymbol{\Omega} \times \bar{\mathsf{I}}\boldsymbol{\omega} + \boldsymbol{p} \times \bar{\mathsf{I}}\boldsymbol{\omega}) + \bar{\mathsf{I}}\left(\left.\frac{d}{dt}\right|_{\Omega} \boldsymbol{\omega} + \boldsymbol{\Omega} \times \boldsymbol{\omega}\right),\end{aligned} \tag{5.2-37}$$

where we have, more or less arbitrarily, expressed $\boldsymbol{\alpha}$ in terms of a time derivative relative to an $\boldsymbol{\Omega}$-rotating system. Now

$$\begin{aligned}\boldsymbol{p} \times \bar{\mathsf{I}}\boldsymbol{\omega} &= \boldsymbol{p} \times \bar{\mathsf{I}}(\boldsymbol{\Omega} + \boldsymbol{p}) \\ &= \boldsymbol{p} \times \bar{\mathsf{I}}\boldsymbol{\Omega} + \boldsymbol{p} \times \bar{\mathsf{I}}\boldsymbol{p} \\ &= \boldsymbol{p} \times \bar{\mathsf{I}}\boldsymbol{\Omega}\end{aligned}$$

because $\boldsymbol{p}$ *is defined to be along a principal axis*, so $\bar{\mathsf{I}}\boldsymbol{p} \parallel \boldsymbol{p}$ and the cross product vanishes; this is, in effect, analogous to the observation that $\left.\frac{d}{dt}\right|_{\Omega} \bar{\mathsf{I}} \equiv \mathbf{0}$ in the above classical derivation. And

$$\begin{aligned}\bar{\mathsf{I}}(\boldsymbol{\Omega} \times \boldsymbol{\omega}) &= \bar{\mathsf{I}}(\boldsymbol{\Omega} \times (\boldsymbol{\Omega} + \boldsymbol{p})) \\ &= \bar{\mathsf{I}}(\boldsymbol{\Omega} \times \boldsymbol{\Omega}) + \bar{\mathsf{I}}(\boldsymbol{\Omega} \times \boldsymbol{p}) = \bar{\mathsf{I}}(\boldsymbol{\Omega} \times \boldsymbol{p}) \\ &= (\bar{\mathsf{I}}\boldsymbol{\Omega}) \times \boldsymbol{p} = -\boldsymbol{p} \times \bar{\mathsf{I}}\boldsymbol{\Omega}.\end{aligned}$$

The passage from the second line to the last, $\bar{\mathsf{I}}(\boldsymbol{\Omega} \times \boldsymbol{p}) = (\bar{\mathsf{I}}\boldsymbol{\Omega}) \times \boldsymbol{p}$ (where the parentheses indicate the grouping of multiplication), requires some explanation, because this is not true in general. Here it is the result of the very special matrix and vectors involved: the form of $\bar{\mathsf{I}}$ (equation (5.2-35)), and the fact that $\boldsymbol{p} \equiv p\hat{\boldsymbol{k}}$, gives the result for arbitrary $\boldsymbol{\Omega}$.

Homework:

1. Show that, for $\bar{\mathsf{I}}$ satisfying (5.2-35) and $\boldsymbol{p} \equiv p\hat{\boldsymbol{k}}$, $\bar{\mathsf{I}}(\boldsymbol{\Omega} \times \boldsymbol{p}) = (\bar{\mathsf{I}}\boldsymbol{\Omega}) \times \boldsymbol{p}$ for arbitrary $\boldsymbol{\Omega}$.

Thus the second and fourth terms in (5.2-37) cancel out, and we end up with the same equation above:

$$\dot{\overline{\boldsymbol{H}}} = \boldsymbol{\Omega} \times \bar{\mathsf{I}}\boldsymbol{\omega} + \bar{\mathsf{I}}\left.\frac{d}{dt}\right|_{\Omega} \boldsymbol{\omega}.$$

Though this latter derivation makes direct reference to Euler's equations, the first, more classical, one is more in the spirit of the development of those fundamental equations and stresses the application of the $\mathbf{\Omega}$-rotation of the (x, y, z) axes, here introduced more or less formally. The present derivation *does*, however, point out the importance of the principal nature of these axes and the axial symmetry presumed in the rigid body.

However the expression for $\dot{\overline{\mathbf{H}}}$ is obtained, if we write the time derivative in its last term

$$\left.\frac{d}{dt}\right|_{\Omega} \boldsymbol{\omega} \equiv \boldsymbol{\alpha}_{rel}, \tag{5.2-38}$$

—the angular acceleration *relative to the* $\mathbf{\Omega}$*-rotating axes* (x, y, z)—then equation (5.2-36) becomes

$$\dot{\overline{\mathbf{H}}} = \mathbf{\Omega} \times \bar{\mathsf{I}}\boldsymbol{\omega} + \bar{\mathsf{I}}\boldsymbol{\alpha}_{rel}, \tag{5.2-39}$$

analogous to the result of Theorem 5.2.5, in which $\boldsymbol{\omega} \times \bar{\mathsf{I}}\boldsymbol{\omega} \to \mathbf{\Omega} \times \bar{\mathsf{I}}\boldsymbol{\omega}$ and $\boldsymbol{\alpha} \to \boldsymbol{\alpha}_{rel}$; thus the most general moment equation for the gyroscope becomes

$$\sum_i \mathbf{M}_{A_i} = \mathbf{r}_{G/A} \times m\bar{\mathbf{a}} + \dot{\overline{\mathbf{H}}} = \mathbf{r}_{G/A} \times m\bar{\mathbf{a}} + \mathbf{\Omega} \times \bar{\mathsf{I}}\boldsymbol{\omega} + \bar{\mathsf{I}}\boldsymbol{\alpha}_{rel}.$$

Although the derivation dealt with finding $\dot{\overline{\mathbf{H}}}$ in which $\overline{\mathbf{H}} = \bar{\mathsf{I}}\boldsymbol{\omega}$, if we instead had used angular momentum relative to a fixed point O, $\mathbf{H}_O = \mathsf{I}_O\boldsymbol{\omega}$, all the above would have carried through with $\bar{\mathsf{I}} \to \mathsf{I}_O$; thus we can conclude immediately that

$$\sum_i \mathbf{M}_{O_i} = \mathbf{\Omega} \times \mathsf{I}_O\boldsymbol{\omega} + \mathsf{I}_O\boldsymbol{\alpha}_{rel}.$$

And, of course, summing moments about the center of mass would give the $\dot{\overline{\mathbf{H}}}$ already obtained. Thus, bringing all the equations of motion for a gyroscope together in one place, we have, analogous to equations (5.2-22) and (5.2-23) the set

Translation:

$$\sum_i \mathbf{F}_i = m\bar{\mathbf{a}} \tag{5.2-40}$$

Rotation:

$$\sum_i \overline{\mathbf{M}}_i = \mathbf{\Omega} \times \bar{\mathsf{I}}\boldsymbol{\omega} + \bar{\mathsf{I}}\boldsymbol{\alpha}_{rel} \quad \text{(general)} \tag{5.2-41a}$$

$$\sum_i \mathbf{M}_{O_i} = \mathbf{\Omega} \times \mathsf{I}_O\boldsymbol{\omega} + \mathsf{I}_O\boldsymbol{\alpha}_{rel} \quad (\textit{only} \text{ if rotating about } \textit{fixed } O) \tag{5.2-41b}$$

$$\sum_i \mathbf{M}_{A_i} = \mathbf{r}_{G/A} \times m\bar{\mathbf{a}} + \mathbf{\Omega} \times \bar{\mathsf{I}}\boldsymbol{\omega} + \bar{\mathsf{I}}\boldsymbol{\alpha}_{rel} \quad (\textit{completely} \text{ general}) \tag{5.2-41c}$$

In the desire to obtain equations analogous to those we have been using for more general motion, we have been rather cavalier regarding precisely what the term "$\boldsymbol{\alpha}_{rel}$" actually is; it was merely defined, somewhat facilely, in equation (5.2-38), and then we barged ahead to get the above. We now examine that term more closely:

Again, what it stands for is the angular acceleration *relative to the $\boldsymbol{\Omega}$-rotating axes* (x, y, z). Note that we can no longer conclude that this term is the *absolute* $\boldsymbol{\alpha}$:

$$\boldsymbol{\alpha}_{rel} \equiv \left.\frac{d}{dt}\right|_{\boldsymbol{\Omega}} \boldsymbol{\omega} \neq \left.\frac{d}{dt}\right|_{\boldsymbol{\omega}} \boldsymbol{\omega} = \boldsymbol{\alpha}.$$

But it *is*

$$\left.\frac{d}{dt}\right|_{\boldsymbol{\Omega}} \boldsymbol{\omega} = \left.\frac{d}{dt}\right|_{\boldsymbol{\Omega}} (\boldsymbol{\Omega} + \boldsymbol{p}) = \left.\frac{d}{dt}\right|_{\boldsymbol{\Omega}} \boldsymbol{\Omega} + \left.\frac{d}{dt}\right|_{\boldsymbol{\Omega}} \boldsymbol{p} = \dot{\boldsymbol{\Omega}} + \left.\frac{d}{dt}\right|_{\boldsymbol{\Omega}} \boldsymbol{p},$$

by the same reasoning that led before to $\frac{d\omega}{dt} = \left.\frac{d}{dt}\right|_{\boldsymbol{\omega}} \boldsymbol{\omega}$. The first term $\dot{\boldsymbol{\Omega}}$ is precisely the *absolute* precessional acceleration. [By rights, this could be written as a capital $\boldsymbol{\alpha}$, but it is too easily confused with "$\boldsymbol{A}$"!] This leaves us with only the last term, $\left.\frac{d}{dt}\right|_{\boldsymbol{\Omega}} \boldsymbol{p}$, to interpret.

$\boldsymbol{p}$ is the spin which is, recall, always along the axis of symmetry and thus fixed to be in the $\hat{\boldsymbol{k}}$-direction: $\boldsymbol{p} \equiv p\hat{\boldsymbol{k}}$; thus, recalling that the "*rel*" time derivative operates only on the *components* of the expression, *relative to the $\boldsymbol{\Omega}$-rotating axes,*

$$\left.\frac{d}{dt}\right|_{\boldsymbol{\Omega}} \boldsymbol{p} = \dot{p}\hat{\boldsymbol{k}},$$

the spin acceleration, so that

$$\boldsymbol{\alpha}_{rel} = \left.\frac{d}{dt}\right|_{\boldsymbol{\Omega}} \boldsymbol{\omega} = \dot{\boldsymbol{\Omega}} + \dot{p}\hat{\boldsymbol{k}}. \tag{5.2-42a}$$

This can also be written in terms of the time derivatives of the [Euler] angles θ and ψ, in terms of which $\boldsymbol{\Omega}$ was expressed in equations (5.2-33): since $\left.\frac{d}{dt}\right|_{\boldsymbol{\Omega}} \boldsymbol{\omega}$ is the "*rel*" time derivative relative to the $\boldsymbol{\Omega}$-rotating axes, and since such time derivatives differentiate only the *components* of the vector representations in their [rotating] basis (equation (4.3-3))—here $\hat{\boldsymbol{\imath}}$, $\hat{\boldsymbol{\jmath}}$, and $\hat{\boldsymbol{k}}$—we can use equations (5.2-34) to obtain

$$= \dot{\omega}_x\hat{\boldsymbol{\imath}} + \dot{\omega}_y\hat{\boldsymbol{\jmath}} + \dot{\omega}_z\hat{\boldsymbol{k}} \tag{5.2-42b}$$

$$= \dot{\Omega}_x\hat{\boldsymbol{\imath}} + \dot{\Omega}_y\hat{\boldsymbol{\jmath}} + \frac{d}{dt}(\Omega_z + p)\hat{\boldsymbol{k}} \tag{5.2-42c}$$

$$= \ddot{\theta}\hat{\boldsymbol{\imath}} + (\ddot{\psi}\sin\theta + \dot{\psi}\dot{\theta}\cos\theta)\hat{\boldsymbol{\jmath}} + \frac{d}{dt}(\dot{\psi}\cos\theta + p)\hat{\boldsymbol{k}}. \tag{5.2-42d}$$

The form of the generic inertia matrix, equation (5.2-35), means that $\mathsf{I}\boldsymbol{\omega}$ and $\mathsf{I}\boldsymbol{\alpha}$ will both be of the form $I_{O\_\_x}\hat{\boldsymbol{\imath}} + I_{O\_\_y}\hat{\boldsymbol{\jmath}} + I_{O\_\_z}\hat{\boldsymbol{k}}$, where the underscores would be either ω or α. Substituting the expressions for $\boldsymbol{\Omega}$ and $\boldsymbol{\omega}$ from (5.2-33b) and (5.2-34b) into $\mathsf{I}\boldsymbol{\omega}$ in the last row of the cross product

$$\begin{aligned}\boldsymbol{\Omega} \times \mathsf{I}\boldsymbol{\omega} &= \begin{vmatrix} \hat{\imath} & \hat{\jmath} & \hat{\boldsymbol{k}} \\ \Omega_x & \Omega_y & \Omega_z \\ I_O\Omega_x & I_O\Omega_y & I(\Omega_z + p) \end{vmatrix} \\ &= (I\Omega_y(\Omega_z + p) - I_O\Omega_y\Omega_z)\,\hat{\imath} - (I\Omega_x(\Omega_z + p) - I_O\Omega_x\Omega_z)\,\hat{\jmath}\end{aligned}$$

—note that there is *no* $\hat{\boldsymbol{k}}$-component; writing

$$\mathsf{I}\boldsymbol{\alpha}_{rel} = I_O\dot{\Omega}_x\hat{\imath} + I_O\dot{\Omega}_y\hat{\jmath} + I\frac{d}{dt}(\Omega_z + p)\hat{\boldsymbol{k}};$$

and substituting the expressions for these in terms of the Euler angles, (5.2-34a) and (5.2-42d); we get ultimately

$$\begin{aligned}\boldsymbol{\Omega} \times \mathsf{I}\boldsymbol{\omega} + \mathsf{I}\boldsymbol{\alpha}_{rel} &= \Big(I\Omega_y(\Omega_z + p) - I_O(\Omega_y\Omega_z - \dot{\Omega}_x)\Big)\,\hat{\imath} \\ &\quad - \Big(I\Omega_x(\Omega_z + p) - I_O(\Omega_x\Omega_z + \dot{\Omega}_y)\Big)\,\hat{\jmath} \\ &\quad + I\frac{d}{dt}(\Omega_z + p)\hat{\boldsymbol{k}} \qquad (5.2\text{-}43a) \\ &= \Big(I_O(\ddot{\theta} - \dot{\psi}^2\sin\theta\cos\theta) + I\dot{\psi}\sin\theta(\dot{\psi}\cos\theta + p)\Big)\,\hat{\imath} \\ &\quad + \Big(I_O(\ddot{\psi}\sin\theta + 2\dot{\psi}\dot{\theta}\cos\theta) - I\dot{\theta}(\dot{\psi}\cos\theta + p)\Big)\,\hat{\jmath} \\ &\quad + I\frac{d}{dt}(\dot{\psi}\cos\theta + p)\hat{\boldsymbol{k}}. \qquad (5.2\text{-}43b)\end{aligned}$$

Note that, despite their separate development, *these equation are nothing more than a special case*—in the principal axes of an axially-symmetric body—*of Euler's equations* (5.2-23); this was the intent of the alternate derivation on page 330. What they do, however, is to break out the separate components of angular velocity due to spin, precession, and nutation. They should *not* be memorized! In the first place, they are written relative to only a specific, albeit useful, set of axes in the gyroscope. More importantly, however, the thrust of this text is to minimize the material to *be* memorized. In fact, this is the first time full-blown general equations have been expanded in components at all; the only reason for doing so here is to show the form the matrix products will assume in this particular formulation, and to enable a general analysis of the gyroscope in Example 5.4.5. Specific examples will rely on setting up specific representations of the general equations (5.2-40) and (5.2-41), as we shall see.

Special Case—Moment-free Gyroscopic Motion

We consider at first what might seem to be an altogether *too* special case: the motion of a "gyroscope" (in the generalized sense used in this section) free of any external moments. Any support located other than the center of mass of a spinning body will generate a moment. Yet such a situation can arise in practice: an axially-symmetric body—a football, for example—going through the air [in the absence of friction, at least] has only a gravitational force passing through its center of mass, and the same holds for bodies in space. In both these conditions, then, the moment about the center of mass vanishes. Almost invariably, then, we use the center-of-mass formulation of the moment equation, (5.2-41a),

$$\sum_i \overline{\boldsymbol{M}}_i = \boldsymbol{\Omega} \times \bar{\mathsf{I}}\boldsymbol{\omega} + \bar{\mathsf{I}}\boldsymbol{\alpha}_{rel}.$$

Before turning to a specific example, let us consider the general properties of bodies obeying this law of motion:

We must first establish a non-rotating reference system. To do so, we use the fact that: since there *is* no moment about the center of mass, relative to any such inertial basis $\overline{\boldsymbol{H}}$ must be a constant, in both magnitude and direction; thus define the $\hat{\boldsymbol{K}}$-direction to be parallel to this constant:

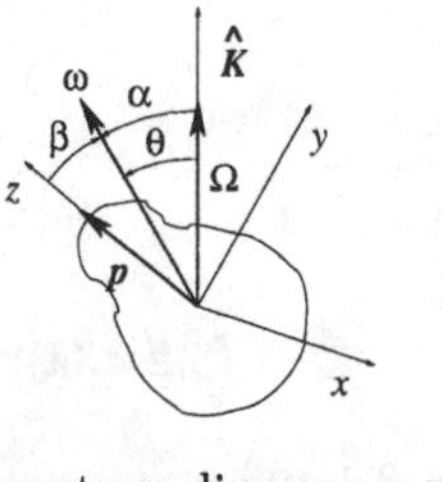

$$\overline{\boldsymbol{H}} \equiv \overline{H}\hat{\boldsymbol{K}}, \qquad \overline{H} \text{ constant.}$$

That done, we set up our gyroscopic axes as usual, taking $\hat{\boldsymbol{k}}$ along the body's axis of symmetry, making an angle $\theta \neq 0$ with $\hat{\boldsymbol{K}}$, and thence $\hat{\boldsymbol{\imath}} \equiv \hat{\boldsymbol{K}} \times \hat{\boldsymbol{k}}$ and $\hat{\boldsymbol{\jmath}} \equiv \hat{\boldsymbol{k}} \times \hat{\boldsymbol{\imath}}$. This is the system diagrammed at left.

Now consider $\overline{\boldsymbol{H}}$ represented both relative to the fixed and body-oriented basis vectors:

$$\begin{aligned}\overline{\boldsymbol{H}} &\equiv \overline{H}\hat{\boldsymbol{K}}\\ &= \overline{H}_x \quad \hat{\boldsymbol{\imath}} + \overline{H}_y \quad \hat{\boldsymbol{\jmath}} + \overline{H}_z \qquad \hat{\boldsymbol{k}}\\ &= 0 \qquad + \overline{H}\sin\theta\hat{\boldsymbol{\jmath}} + \overline{H}\cos\theta \quad \hat{\boldsymbol{k}}\end{aligned} \tag{5.2-44a}$$

—an equation relating $\overline{\boldsymbol{H}}$ and θ—or, multiplying out $\overline{\boldsymbol{H}} = \bar{\mathsf{I}}\boldsymbol{\omega}$ using (5.2-35) and (5.2-34b),

$$= I_O\Omega_x\hat{\boldsymbol{\imath}} + I_O\Omega_y \;\hat{\boldsymbol{\jmath}} + I(\Omega_z + p)\hat{\boldsymbol{k}}. \tag{5.2-44b}$$

Note that this tells us immediately that $\Omega_x = \omega_x \equiv 0$; *i.e.*, that $\boldsymbol{\omega}$ always lies in the yz-plane. This observation has an important side effect: it means that both $\boldsymbol{\Omega}$ and $\boldsymbol{\omega}$ have only y- and z-components; so that $\bar{\mathsf{I}}\boldsymbol{\omega}$ [expressed in principal axes] does also, and thus $\boldsymbol{\Omega} \times \bar{\mathsf{I}}\boldsymbol{\omega}$ *has only an* $\hat{\boldsymbol{\imath}}$*-component.*

To this point, θ could be varying: though $\hat{\boldsymbol{K}}$ is fixed, $\hat{\boldsymbol{\jmath}}$ and $\hat{\boldsymbol{k}}$ are rotating, recall. But we have just shown that $\Omega_x \equiv 0$, where $\Omega_x = \dot{\theta}$ from equations (5.2-33). Thus we see immediately that θ is a *constant*, so the (x, y, z)-axes rotate about $\hat{\boldsymbol{K}}$ at a constant angle—there is no nutation—and the spin axis' motion, equation (5.2-33a), specializes to

$$\boldsymbol{\Omega} = \Omega\hat{\boldsymbol{K}} = \dot{\psi}\sin\theta\hat{\boldsymbol{\jmath}} + \dot{\psi}\cos\theta\hat{\boldsymbol{k}} = \Omega_y\hat{\boldsymbol{\jmath}} + \Omega_z\hat{\boldsymbol{k}}$$

—there is only *precession*, $\boldsymbol{\Omega} = \dot{\boldsymbol{\psi}}$, of the spin axis.

Finally, since by equations (5.2-44) and (5.2-34),

$$\overline{H}\sin\theta = I_O\Omega_y = I_O\omega_y \quad \text{and} \quad \overline{H}\cos\theta = I(\Omega_z + p) = I\omega_z, \tag{5.2-45}$$

we can determine ω_y and ω_z—at least if we know $\overline{H}$. But even if we don't, we can find the [constant] angle β these components make with the [rotating] z-axis from $\boldsymbol{\omega}$:

$$\tan\theta = \frac{\overline{H}\sin\theta}{\overline{H}\cos\theta} = \frac{I_O\Omega_y}{I(\Omega_z + p)} = \frac{I_O\omega_y}{I\omega_z} = \frac{I_O}{I}\tan\beta.$$

[It is perhaps useful to refer to the above diagram to see the geometry.] These equations relate θ and β to the components of $\boldsymbol{\omega}$ and the physically characteristic moments of inertia of the body; in particular, they show that θ and β (both $< 90°$) must lie in the same quadrant.

Equations (5.2-45) giving ω_y and ω_z, since $\overline{H}$ and θ are constant, mean that these components of $\boldsymbol{\omega}$ are constant—$\omega_x \equiv 0$ already is! Alternatively, the constant y-component, $\overline{H}\sin\theta = I_O\Omega_y$, shows that $\Omega_y = \dot{\psi}\sin\theta$, so $\dot{\psi} = \|\Omega\|$, is constant—$\hat{\boldsymbol{\Omega}} \equiv \hat{\boldsymbol{K}}$ already is—and $\boldsymbol{\Omega}$ is constant; similar constancy of the z-component $\overline{H}\cos\theta = I(\Omega_z + p)$ shows that $p \equiv \|\boldsymbol{p}\|$ is constant. In either event, $\boldsymbol{\alpha}_{rel} = \dot{\boldsymbol{\Omega}} + \dot{p}\hat{\boldsymbol{k}}$ (see (5.2-42a-b)) in the equations of motion vanishes identically! Thus, ultimately, the moment equation of motion for the moment-free gyroscope, using the fact that $\dot{\theta} \equiv 0$, becomes merely a special case of equation (5.2-43b):

$$\sum_i \overline{\boldsymbol{M}}_i = \boldsymbol{\Omega} \times \bar{\mathbf{I}}\boldsymbol{\omega} = \Big((I - I_O)\dot{\psi}\cos\theta + Ip\Big)\,\dot{\psi}\sin\theta\hat{\boldsymbol{\imath}} = \mathbf{0}. \tag{5.2-46}$$

(Again, it needn't be memorized, coming out automatically as a result of the indicated multiplication!) This is effectively an equation relating the angle θ—still non-0 by hypothesis—and $\dot{\psi} \neq 0$, the precession—the only motion of the spin axis, recall:

$$\dot{\psi}\cos\theta = \Omega_z = \frac{Ip}{I_O - I}.$$

The equation is written in this form to avoid the apparently pathological problem if $\theta = 90°$: in this case [non-0] precession is taking place about an axis perpendicular to the axis of symmetry, and the above moment equation becomes $Ip\dot{\psi} = 0$—*i.e.*, $p = 0$ and *there can be no spin!*

We can easily find the angle θ^* between the spin axis $\boldsymbol{p}$ and the precession $\dot{\boldsymbol{\psi}}$ (generally different from θ between $\boldsymbol{p}$ and $\hat{\boldsymbol{K}} \parallel \overline{\boldsymbol{H}}$): since $\dot{\theta} \equiv 0$,

$$\begin{aligned}
\boldsymbol{p} \cdot \dot{\boldsymbol{\psi}} &= \|\boldsymbol{p}\|\,\|\dot{\boldsymbol{\psi}}\|\cos\theta^* \\
&= p\hat{\boldsymbol{k}} \cdot (\dot{\psi}(\sin\theta\hat{\boldsymbol{\jmath}} + \cos\theta\hat{\boldsymbol{k}})) = p\dot{\psi}\cos\theta \\
&= \frac{Ip^2}{I_O - I}.
\end{aligned}$$

In particular, then, the angle between the two axes will be less than, or greater than, 90° depending on the *sign* of $I_O - I$:

- if $I_O > I$, $\theta < 90°$; this case is referred to as "*direct*" precession, since it is "in the same direction" (more or less) as the spin. Similarly,
- if $I_O < I$, $\theta > 90°$, referred to as "*retrograde*" precession. Finally,
- if $I_O = I$ the moment equation above, 5.2-46, reduces to $Ip\dot{\psi}\sin\theta = 0$ and, assuming $p \neq 0$—$\theta \neq 0$ by definition of the entire problem, recall—$\dot{\psi}$ must vanish: there *is* no precession!

The utility of this analysis lies in being able to predict the easily-observable spin (symmetry) and precessional axis directions knowing the relative magnitudes of the two moments of inertia. In particular, if a body is long and thin—approaching a "slender rod," the moment about the spin axis, "I," will be smaller than that about a perpendicular axis, "I_O"; conversely, if the body is "flat"—more or less plate-shaped, $I > I_O$.

Above it was mentioned that θ and β must lie in the same quadrant. An approach similar to the above allows us to predict that quadrant, depending only on the relative sizes of the two moments of inertia: Again noting that $\boldsymbol{\omega}$ has no component due to nutation,

$$\begin{aligned}\boldsymbol{p}\cdot\boldsymbol{\omega} &= \|\boldsymbol{p}\|\,\|\boldsymbol{\omega}\|\cos\beta\\ &= p\hat{\boldsymbol{k}}\cdot(\dot{\psi}\sin\theta\hat{\boldsymbol{\jmath}} + (\dot{\psi}\cos\theta + p)\hat{\boldsymbol{k}}) = p\left(\frac{Ip}{I_O - I} + p\right)\\ &= \frac{I_O p^2}{I_O - I}.\end{aligned}$$

We see that this angle depends only the direct or retrograde nature of the precession.

Before continuing, let us briefly make a note regarding the explicit hypothesis that θ, between the fixed $\hat{\boldsymbol{K}}$ and symmetry axis $\hat{\boldsymbol{k}}$ be non-0: This was required initially in order to define $\hat{\boldsymbol{\imath}}$, and thence $\hat{\boldsymbol{\jmath}}$, from the cross product of these two. But if the angle between them happens to vanish, $\hat{\boldsymbol{K}}$ and $\hat{\boldsymbol{k}}$ are coincident and $\hat{\boldsymbol{\jmath}}$ can be arbitrarily fixed relative to $\hat{\boldsymbol{k}}/\hat{\boldsymbol{K}}$. Then the reader should confirm that there is no pathology in any of the above equations—$\sin\theta$ always entering in the numerators of the various expressions—except for (5.2-46), where the following expression for $\dot{\psi}\cos\theta$ presumes $\sin\theta \neq 0$. But if that angle vanishes, $\boldsymbol{p} \parallel \hat{\boldsymbol{K}}$ and $\dot{\psi} \equiv 0$, and $\dot{\theta}$ still vanishes by equations (5.2-44), which remain valid in this special case.

If $\boldsymbol{p}$ and $\boldsymbol{\Omega}$ are relatively easy to observe, the total angular velocity $\boldsymbol{\omega} = \boldsymbol{\Omega} + \boldsymbol{p}$ is less so. But it is nevertheless possibly interesting to note that, although the angle θ between $\boldsymbol{p}$ and $\boldsymbol{\Omega}$ can be anywhere between 0° and 180°, the angle $\alpha = \theta - \beta$ [again, see diagram] between $\boldsymbol{\omega}$ and $\boldsymbol{\Omega} \parallel \overline{\boldsymbol{H}}$ must always be *less* than a right angle:

$$\begin{aligned}\|\boldsymbol{\omega}\|\,\|\overline{\boldsymbol{H}}\|\cos\alpha &= \boldsymbol{\omega}\cdot\overline{\boldsymbol{H}}\\ &= (\omega_x\hat{\boldsymbol{\imath}} + \omega_y\hat{\boldsymbol{\jmath}} + \omega_z\hat{\boldsymbol{k}})\cdot(I_O\omega_x\hat{\boldsymbol{\imath}} + I_O\omega_y\hat{\boldsymbol{\jmath}} + I\omega_z\hat{\boldsymbol{k}})\\ &= I_O\omega_x^2 + I_O\omega_y^2 + I\omega_z^2 > 0.\end{aligned}$$

In summary, we see that moment-free rotation is characterized by a constant precession—no nutation—of the body about a fixed $\overline{\boldsymbol{H}}$ [|| Z-] axis, with the body's angular velocity being constant *relative to the body-oriented system*: $\omega_x \equiv 0$ and $\dot{\omega}_y \equiv 0 \equiv \dot{\omega}_z$, and making a constant angle β with the body's axis of symmetry: $\boldsymbol{\alpha}_{rel} \equiv \mathbf{0}$. In terms of the gyroscopic equations of motion (5.2-41), this means that only the term $\boldsymbol{\Omega} \times \mathsf{I}\boldsymbol{\omega}$ survives the moment equation, and that has only a non-vanishing $\hat{\imath}$-component! The upshot is that we can generally determine only a single scalar unknown from this analysis.

The above analysis really made no use of the precession, $\dot{\psi}$ and spin p, though we did show the nutation $\dot{\theta} \equiv 0$. In fact, they never made much use of the equations (5.2-43) we seem to have gone to such lengths to obtain! While the latter could have been used, it was much easier, and thus more transparent, to use the above analysis to determine the characteristics of the motion than to deal with the various time derivatives appearing in (5.2-43b). But if we *are* interested in precession, nutation, and spin, we must introduce their angular representations, as the next example, rather more characteristic of the means of approaching gyroscopic problems, shows:

Example 5.2.12 *Moment-free Gyroscopic Motion.* A football is thrown, rather badly, in the middle of a pick-up game one afternoon after classes. Its axis of symmetry is observed to "wobble" at a rate of s, making an angle of θ with a fixed axis. We are asked to find the *spin* rate p about its axis, assuming I about the symmetry axis and I_O about those perpendicular are known.

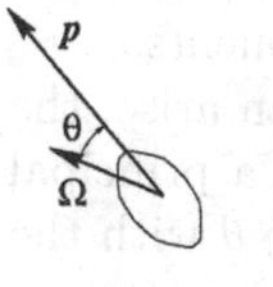

This is clearly a case of moment-free precession: there *is* no moment about the center of mass; thus we will use

$$\sum_i \overline{\boldsymbol{M}}_i = \boldsymbol{\Omega} \times \bar{\mathsf{I}}\boldsymbol{\omega} + \bar{\mathsf{I}}\boldsymbol{\alpha}_{rel} = \boldsymbol{\Omega} \times \bar{\mathsf{I}}\boldsymbol{\omega},$$

since $\boldsymbol{\alpha}_{rel} \equiv 0$ in this moment-free case. The "wobble" is precisely the precession $\dot{\psi}$, measured implicitly relative to some fixed axis [never specified]. In terms of its representation, we know from equations (5.2-33) that generally

$$\boldsymbol{\Omega} \equiv \Omega_x \hat{\boldsymbol{\imath}} + \Omega_y \hat{\boldsymbol{\jmath}} + \Omega_z \hat{\boldsymbol{k}} = \dot{\theta}\hat{\boldsymbol{\imath}} + \dot{\psi}\sin\theta\hat{\boldsymbol{\jmath}} + \dot{\psi}\cos\theta\hat{\boldsymbol{k}},$$

where, for moment-free motion, $\dot{\theta} \equiv 0$. Knowing θ, we immediately have

$$\boldsymbol{\Omega} = s\sin\theta\hat{\boldsymbol{\jmath}} + s\cos\theta\hat{\boldsymbol{k}}$$

and, up to the scalar unknown p,

$$\boldsymbol{\omega} = s\sin\theta\hat{\boldsymbol{\jmath}} + (s\cos\theta + p)\hat{\boldsymbol{k}},$$

with the expected only y- and z-components. From the generic form of the gyroscopic I—here corresponding to $\bar{\mathsf{I}}$—in equation (5.2-35), we have immediately that

$$\begin{aligned}\boldsymbol{\Omega} \times \bar{\mathsf{I}}\boldsymbol{\omega} &= \begin{vmatrix} \hat{\imath} & \hat{\jmath} & \hat{k} \\ 0 & s\sin\theta & s\cos\theta \\ 0 & I_O s\sin\theta & I(s\cos\theta + p) \end{vmatrix} \\ &= \left((I - I_O)s^2 \sin\theta\cos\theta + Is\sin\theta p\right)\hat{\imath} \equiv \mathbf{0},\end{aligned}$$

featuring only the expected $\hat{\imath}$-component. Solving, we obtain $p = \frac{(I_O - I)s\cos\theta}{I}$. Note that, since $p > 0$ by definition, and, for a football, $I_O > I$, $\cos\theta > 0$ so that $\theta < 90°$: precession must be *direct*! |

General Case—Gyroscope *with* Moment

We have seen that the moment-free gyroscope equations ultimately give a simple *precession*—no nutation—of the axis of symmetry about some fixed direction, ultimately leading to only a single non-vanishing component of the [vanishing] moment equation. Furthermore, the angular velocity $\boldsymbol{\omega}$ is *fixed* in the body-centered axes, forcing $\boldsymbol{\alpha}_{rel} \equiv \mathbf{0}$. More generally, however, there *will* be some moment, and we must return to the full-blown equations (5.2-40) and (5.2-41)—the latter of which, in particular, generally involves all three components.

Yet even in the presence of a moment, certain simplifications can arise, the most common involving the case in which $\boldsymbol{\Omega}$ happens to lie along a principal axis of the body: Note that, in general, $\boldsymbol{\Omega}$ makes an arbitrary angle θ with the spin axis; thus the term involving this quantity,

$$\boldsymbol{\Omega} \times \mathsf{I}\boldsymbol{\omega} = \boldsymbol{\Omega} \times \mathsf{I}(\boldsymbol{\Omega} + \boldsymbol{p}) = \boldsymbol{\Omega} \times \mathsf{I}\boldsymbol{\Omega} + \boldsymbol{\Omega} \times \mathsf{I}\boldsymbol{p},$$

will have a contribution due to $\boldsymbol{\Omega} \times \mathsf{I}\boldsymbol{\Omega}$. But in the special case that $\boldsymbol{\Omega}$ itself happens to be along a principal axis—*i.e.*, $\boldsymbol{\Omega} \perp \boldsymbol{p}$[22] (since $\boldsymbol{p}$ is *defined* to be along a principal axis, and the symmetry means any axis perpendicular to the spin will be principal)—then $\mathsf{I}\boldsymbol{\Omega} \parallel \boldsymbol{\Omega}$ and this term drops out; thus, for example,

$$\sum_i \overline{\boldsymbol{M}}_i = \boldsymbol{\Omega} \times \bar{\mathsf{I}}\boldsymbol{\omega} + \bar{\mathsf{I}}\boldsymbol{\alpha}_{rel} = \boldsymbol{\Omega} \times \bar{\mathsf{I}}\boldsymbol{p} + \bar{\mathsf{I}}\boldsymbol{\alpha}_{rel}.$$

If, in addition, the second term $\boldsymbol{\alpha}_{rel} = \dot{\boldsymbol{\Omega}} + \dot{p}\hat{\boldsymbol{k}}$ by (5.2-42), happens to vanish, this becomes simply

$$\sum_i \overline{\boldsymbol{M}}_i = \boldsymbol{\Omega} \times \bar{\mathsf{I}}\boldsymbol{p},$$

in which, because $\boldsymbol{p}$ is along a principal axis, $\bar{\mathsf{I}}\boldsymbol{p} \parallel \boldsymbol{p}$. This means that the set $(\boldsymbol{\Omega}, \boldsymbol{p}, \sum_i \overline{\boldsymbol{M}}_i)$ forms a right-handed triad, and some interesting qualitative information can be gleaned from just that. Even if $\boldsymbol{\alpha}_{rel}$ *doesn't* vanish,

[22] In the case of a *moment-free* gyroscope, this could not happen: we saw above that there can be no spin for a body precessing about an axis perpendicular to that of symmetry! But here there *is* a moment, and such precession can occur.

however—say in the case that the spin axis rotates with a constant angular velocity: $\dot{\boldsymbol{\Omega}} = \mathbf{0}$—the moment equation is still surprisingly simple in form:

$$\sum_i \overline{\boldsymbol{M}}_i = \boldsymbol{\Omega} \times \bar{\mathsf{I}}\boldsymbol{p} + \bar{\mathsf{I}}\dot{p}\hat{\boldsymbol{k}}.$$

Even these simple equations can lead to some surprising results:

Example 5.2.13. *Qualitative Analysis of "Gyroscopic" Motion.* A plane moving with constant speed is at the bottom of a vertical circle of constant radius moving in the direction shown. Find the effect of this maneuver on the plane.

We consider the "gyroscopic" component of the plane: the propeller. Taking $\hat{\boldsymbol{\imath}}$, $\hat{\boldsymbol{\jmath}}$, and $\hat{\boldsymbol{k}}$ in the customary directions, we see that, in the body-oriented axes, $\boldsymbol{p} = -p\hat{\boldsymbol{\imath}}$ and $\boldsymbol{\Omega} = -\Omega\hat{\boldsymbol{k}}$. Since both these quantities are constant, p in magnitude and $\boldsymbol{\Omega}$ in magnitude *and* direction, $\boldsymbol{\alpha}_{rel} \equiv \mathbf{0}$; thus the [couple] moment required to enforce this motion, predicted by $\sum_i \overline{\boldsymbol{M}}_i = \boldsymbol{\Omega} \times \bar{\mathsf{I}}\boldsymbol{p}$, is $\boldsymbol{M} \parallel \boldsymbol{\Omega} \times \boldsymbol{p} \parallel \hat{\boldsymbol{\jmath}}$ because $\mathsf{I}\boldsymbol{p} \parallel \boldsymbol{p}$. But this is the couple moment applied *to the propeller*; by Newton's Third, this applies an equal but opposite moment *to the plane.* Thus the plane feels a moment in the $-\hat{\boldsymbol{\jmath}}$-direction—*i.e.*, to the *right* as it is travelling. Thus the plane must "yaw" to the *left* in order to keep moving in this vertical circle!

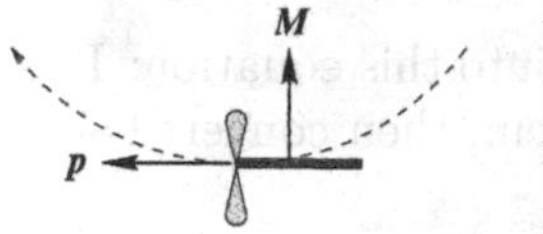

Clearly, some care is required in accounting for Newton's Third Law! And, although the above example was qualitative in nature, had we been given values for spin, precession, and the symmetric axis' moment of inertia, we could find the value of that required moment. |

Of course, $\boldsymbol{\Omega}$ *needn't* be perpendicular to the spin axis. The next example[23] is a rather more general example of the gyroscope equations

Example 5.2.14 A uniform cylinder with mass m, radius a, and length ℓ, spins with constant angular velocity $\boldsymbol{p}$ about its spin axis at the same time it is carried in a *light* frame about the vertical Z-axis through fixed point O with constant angular velocity $\boldsymbol{\Omega}$. Find the couple moment $\boldsymbol{M}$ about the vertical axis required to do this, assuming that the end of the cylinder's axis is a height b directly above O, and that the axis makes an angle ϕ with the horizontal as shown.[24]

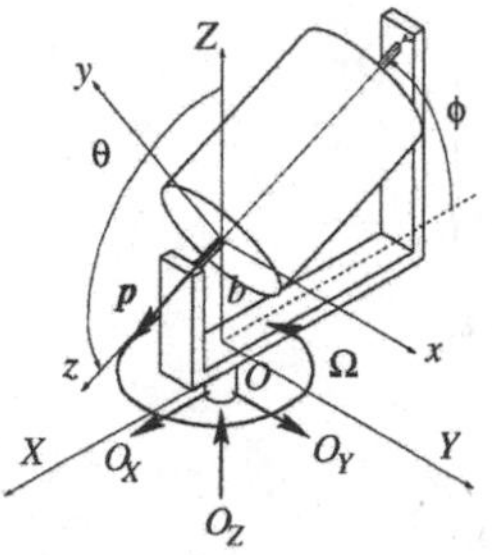

We note that the support for the frame itself consists of a full complement of three components of both forces $\boldsymbol{O}$ and [couple] moment $\boldsymbol{M}$ (though, to prevent clutter, only the former are diagrammed). We are given the two components of the total angular velocity $\boldsymbol{\omega} = \boldsymbol{\Omega} + \boldsymbol{p}$ of the cylinder, and this *is* supposed to exemplify gyroscopic motion, so we will use the equations of motion appropriate to this.

[23] From several editions of Meriam, *Engineering Mechanics: Dynamics.*

[24] Adapted from a problem in Meriam, *Dynamics, 2nd ed.*, 1986.

But this means we must set up the body-oriented axes utilized by these equations: First comes $\hat{\boldsymbol{k}}$ *in the direction of* $\boldsymbol{p}$—so $\theta = 90° + \phi > 90°$ as shown; this defines $\hat{\boldsymbol{\imath}} \equiv \hat{\boldsymbol{K}} \times \hat{\boldsymbol{k}}$ and thus $\hat{\boldsymbol{\jmath}} \equiv \hat{\boldsymbol{k}} \times \hat{\boldsymbol{\imath}}$.

Since we care only about $\boldsymbol{M}$, we can eliminate the $\boldsymbol{O}$ forces by summing moments about that point. But O is not a fixed point *about which the body is rotating* (to be fussy), nor is it the center of mass; thus we will use the generalized form of the gyroscope equations

$$\sum_i \boldsymbol{M}_{O_i} = \boldsymbol{r}_{G/O} \times m\overline{\boldsymbol{a}} + \boldsymbol{\Omega} \times \bar{\mathsf{I}}\boldsymbol{\omega} + \bar{\mathsf{I}}\boldsymbol{\alpha}_{rel}.$$

We note that, since both $\boldsymbol{p}$ and $\boldsymbol{\Omega}$ are constant, $\boldsymbol{\alpha}_{rel} \equiv \boldsymbol{0}$; any $\dot{\boldsymbol{\Omega}} = \dot{\Omega}\hat{\boldsymbol{K}}$ would enter here.

From here on out, we merely enter the known values into this equation: It is likely easiest to calculate the first term in the fixed system, then convert back:

$$\begin{aligned}\overline{\boldsymbol{a}} &= \boldsymbol{a}_O + \dot{\boldsymbol{\Omega}} \times \boldsymbol{r}_{G/O} + \boldsymbol{\Omega} \times (\boldsymbol{\Omega} \times \boldsymbol{r}_{G/O}) + \boldsymbol{0} + \boldsymbol{0}\\ &= \boldsymbol{0} + \boldsymbol{0} + \Omega\hat{\boldsymbol{K}} \times \left(\Omega\hat{\boldsymbol{K}} \times \left(-\frac{\ell}{2}\cos\phi\hat{\boldsymbol{I}} + \left(b + \frac{\ell}{2}\sin\phi\right)\hat{\boldsymbol{K}}\right)\right) = \frac{\ell\Omega^2}{2}\cos\phi\hat{\boldsymbol{I}},\end{aligned}$$

from which we get immediately that

$$\boldsymbol{r}_{G/O} \times m\overline{\boldsymbol{a}} = m\left(\frac{b}{2} + \frac{\ell}{4}\sin\phi\right)\ell\Omega^2\cos\phi\hat{\boldsymbol{J}} = m\left(\frac{b}{2} + \frac{\ell}{4}\sin\phi\right)\ell\Omega^2\cos\phi\hat{\boldsymbol{\imath}}.$$

Then, since in terms of the favored gyroscopic basis

$$\boldsymbol{\Omega} = \Omega(\cos\phi\hat{\boldsymbol{\jmath}} - \sin\phi\hat{\boldsymbol{k}})$$

so that

$$\boldsymbol{\omega} = \boldsymbol{\Omega} + \boldsymbol{p} = \Omega\cos\phi\hat{\boldsymbol{\jmath}} + (p - \Omega\sin\phi)\hat{\boldsymbol{k}},$$

it is easy, in the principal gyroscopic coordinates, to write down immediately the term

$$\begin{aligned}\boldsymbol{\Omega} \times \bar{\mathsf{I}}\boldsymbol{\omega} &= \begin{vmatrix} \hat{\boldsymbol{\imath}} & \hat{\boldsymbol{\jmath}} & \hat{\boldsymbol{k}} \\ 0 & \Omega\cos\phi & -\Omega\sin\phi \\ 0 & \bar{I}_O\Omega\cos\phi & \bar{I}(p - \Omega\sin\phi) \end{vmatrix}\\ &= \left(\bar{I}p + (\bar{I}_O - \bar{I})\Omega^2\sin\phi\cos\phi\right)\hat{\boldsymbol{\imath}},\end{aligned}$$

in which, from tables,

$$\bar{I}_O = \frac{1}{4}ma^2 + \frac{1}{12}m\ell^2 \qquad \text{and} \qquad \bar{I} = \frac{1}{2}ma^2.$$

Thus, in the end, we get that

$$\begin{aligned}\boldsymbol{M} &= m\left(\frac{1}{2}a^2p + \left(\frac{1}{12}\ell^2 - \frac{1}{4}a^2\right)\Omega^2\sin\phi\cos\phi + \left(\frac{b}{2} + \frac{\ell}{4}\sin\phi\right)\ell\Omega^2\cos\phi\right)\hat{\boldsymbol{\imath}}\\ &= m\left(\frac{1}{2}a^2p + \left(\left(\frac{1}{12}\ell^2 - \frac{1}{4}a^2\right)\sin\phi + \left(\frac{b\ell}{2} + \frac{\ell^2}{4}\sin\phi\right)\right)\Omega^2\cos\phi\right)\hat{\boldsymbol{J}}.\end{aligned}$$

$\mathbf{\Omega} \times \bar{\mathbf{I}}\boldsymbol{\omega}$ had only an $\hat{\imath}$-component here, just as with the moment-free gyroscope considered previously. But this is only because there was no nutation, only precession, in this problem. Similarly, it is only coincidence that the additional $\boldsymbol{r}_{G/O} \times m\bar{\boldsymbol{a}}$ was also in that direction; it could generally be in *any*. |

Homework:

1. Work the above example using *Euler's* equations, rather than the gyroscopic formulation used here.

In the above example, the motion—the *kinematics*—was known, and we found the forces/moments—the *kinetics*. Clearly the problem might be cast in the other direction, in which known forces and moments would generate an *unknown* motion. We now turn to this aspect of gyroscopic systems with moments.

In the first place, we note that it is generally possible to find either any $\dot{p}$ or $\dot{\mathbf{\Omega}}$, knowing the other and $\mathbf{\Omega}$; the situation is rather similar to the Euler equations, requiring knowing $\boldsymbol{\omega}$ to determine $\boldsymbol{\alpha}$. Indeed, since (in "generic" form for moment summation about the center of mass or a fixed point about which the body is rotating),

$$\begin{aligned}\sum_i \boldsymbol{M}_i &= \mathbf{\Omega} \times \mathbf{I}\boldsymbol{\omega} + \mathbf{I}\boldsymbol{\alpha}_{rel}\\ &= \mathbf{\Omega} \times \mathbf{I}(\mathbf{\Omega} + \boldsymbol{p}) + \mathbf{I}(\dot{\mathbf{\Omega}} + \dot{p}\hat{\boldsymbol{k}})\\ &= \mathbf{\Omega} \times \mathbf{I}\mathbf{\Omega} + \mathbf{\Omega} \times Ip\hat{\boldsymbol{k}} + \mathbf{I}\dot{\mathbf{\Omega}} + I\dot{p}\hat{\boldsymbol{k}}.\end{aligned}$$

[Note that we have observed the fact that $\mathbf{I}\boldsymbol{p} = Ip\hat{\boldsymbol{k}}$ in the last two lines.] In particular, then, $\dot{p}$ is isolated in the z-component of that equation:

$$\dot{p} = \frac{1}{I}\left(\sum_i \boldsymbol{M}_i - \mathbf{\Omega} \times \mathbf{I}\mathbf{\Omega} - \mathbf{\Omega} \times Ip\hat{\boldsymbol{k}} - \mathbf{I}\dot{\mathbf{\Omega}}\right)_z .$$

To find $\dot{\mathbf{\Omega}}$ is, in principal, little harder:

$$\dot{\mathbf{\Omega}} = \mathbf{I}^{-1}\left(\sum_i \boldsymbol{M}_i - \mathbf{\Omega} \times \mathbf{I}\mathbf{\Omega} - \mathbf{\Omega} \times \mathbf{I}(\mathbf{\Omega} + \boldsymbol{p}) - I\dot{p}\hat{\boldsymbol{k}}\right).$$

Again, these equations are merely meant to indicate how these quantities *can* be found from the general equations; they are not regarded as otherwise overly important!

We now consider a more general analysis of gyroscopic motion in the presence of a moment; the latter is assumed given, and it is the resulting motion we wish to ascertain. We have already observed the purely precessional motion of such a system in the absence of a moment; we here examine this case *with* one. Such motion is characterized by the absence of any nutation, a component which has, to this point, gotten rather short shrift! We *will* consider this, but in Section 5.4.5, where we view it as an application of conservation to rigid body motion in general.

Steady-state Precession of Gyroscope with Moment. By "steady-state precession," we enforce two primary conditions on the motion:

- *nutation vanishes*: this means that $\dot{\theta} = \Omega_x \equiv 0$
- *both precession and spin are constant in magnitude*: $\|\Omega\| = \|\dot{\boldsymbol{\psi}}\| = \dot{\psi}$ and $\|\boldsymbol{p}\| = p$ are constant, so $\ddot{\psi}$ and $\dot{p}$ both vanish.

In particular, the former means that

$$\boldsymbol{\omega} = \boldsymbol{\Omega} + \boldsymbol{p} = \dot{\boldsymbol{\psi}} + \boldsymbol{p},$$

while this and the latter implies $\boldsymbol{\alpha}_{rel} \equiv \mathbf{0}$ [see equation (5.2-42d)], so the [generic] moment equation reduces to

$$\sum_i \boldsymbol{M}_i = \boldsymbol{\Omega} \times \mathsf{I}\boldsymbol{\omega} = \begin{vmatrix} \hat{\boldsymbol{\imath}} & \hat{\boldsymbol{\jmath}} & \hat{\boldsymbol{k}} \\ 0 & \Omega_y & \Omega_z \\ 0 & I_O\Omega_y & I(\Omega_z + p) \end{vmatrix} = ((I - I_O)\Omega_y\Omega_z + I\Omega_y p)\,\hat{\boldsymbol{\imath}}$$

—*i.e.*, only the $\hat{\boldsymbol{\imath}}$-component of the moment equation survives, as in the moment-*free* case! (This also means that, in general, only a single scalar unknown can be determined in this circumstance.) Substituting in the values for the components of $\boldsymbol{\Omega}$ in terms of the non-vanishing Euler angles' time derivatives, this becomes, then,

$$\sum_i M_{i_x} = (I - I_O)\dot{\psi}^2 \sin\theta\cos\theta + I\dot{\psi}p\sin\theta, \tag{5.2-47}$$

involving $\dot{\psi}$ quadratically and p linearly.

Let us first consider the case in which p is known and we wish to determine the precession: Solving the above equation for $\dot{\psi}$, we get

$$\dot{\psi} = \frac{-Ip\sin\theta \pm \sqrt{(Ip\sin\theta)^2 + 4M(I - I_O)\sin\theta\cos\theta}}{2(I - I_O)\sin\theta\cos\theta} \tag{5.2-48}$$

Note that, in order to have real roots to this equation—*i.e.*, to have precession at all!—we must have

$$(Ip\sin\theta)^2 + 4M(I - I_O)\sin\theta\cos\theta \geq 0, \textit{i.e.}$$

$$p^2 \geq \frac{4M(I - I_O)\cot\theta}{I^2};$$

otherwise there is no precession at all!

For **large p**, the radical in (5.2-48) is approximated by $Ip\sin\theta$, and we have two solutions:

$$\dot{\psi}_1 \doteq 0 \qquad \text{and} \qquad \dot{\psi}_2 \doteq \frac{Ip}{(I_O - I)\cos\theta}.$$

Consider the latter solution first: since $p \gg 1$, so will be $\dot{\psi}_2$, corresponding to a high rate of precession. This "high energy" state tends to be mechanically unstable and hard to observe in practice. For example, if $p \approx \dot{\psi}$, it is possible for p and $\dot{\psi}$ to be [nearly] commensurate: $\frac{\dot{\psi}}{p} = \frac{m}{n}$ for relatively small integral values of m and n; in this case any imbalance in the rotor—there is always *some*!—will quickly cause the gyroscope to topple due to a "resonant" forcing.

Thus it is the first, "slow" precession we will examine more closely. Since $p \gg 1$ [still], equation (5.2-47) becomes approximated by

$$M \equiv \sum_i M_{i_x} = (I - I_O)\dot{\psi}^2 \sin\theta\cos\theta + I\dot{\psi}p\sin\theta \doteq I\dot{\psi}p\sin\theta;$$

thus

$$\dot{\psi}_1 \doteq \frac{M}{Ip\sin\theta}$$

varies *inversely* with p and is, for large such values, indeed "small."

As mentioned above, it is this "slow precession" most commonly observed as in, for example, the common gyroscope, which precesses at a far slower rate than its spin! If one considers such $\dot{\psi}_1$ in the standard body-oriented gyroscopic axes, the only moment about the fixed support, of magnitude $mg\bar{r}\sin\theta$ ($\bar{r}$ being the distance to the center of mass from the support point), will thus generate a precession of

$$\dot{\psi}_1 \doteq \frac{mg\bar{r}\sin\theta}{Ip\sin\theta} = \frac{g\bar{r}}{k^2 p},$$

where we have used the radius of gyration $I \equiv mk^2$ to eliminate m. All quantities are fixed except for p, and a simple experiment will verify that, the faster p is, the slower the [steady] precession.

Again, just to focus the thrust of the above discussion, all the results will fall out naturally from the moment equation; this is meant just to give an overview of the problem and point out that it is the *lesser* of the two $\dot{\psi}$'s one accepts as "the precession" when this is being solved for.

If, on the other hand, $\dot{\psi}$ is known, the single value of p which must have led to that is found immediately from equation (5.2-47). There is little left to be said!

We end this section with a *very* general analysis of a symmetric body *not* moving about a fixed point. Although, as always, it could be worked using Euler's equations, we treat it as a "gyroscope" because of this symmetry. More than an example of these equations, however, it also serves as one demonstrating acceleration under the non-slip condition.

Example 5.2.15 *Non-slip motion of a Thin Disk.* A thin disk rolls without

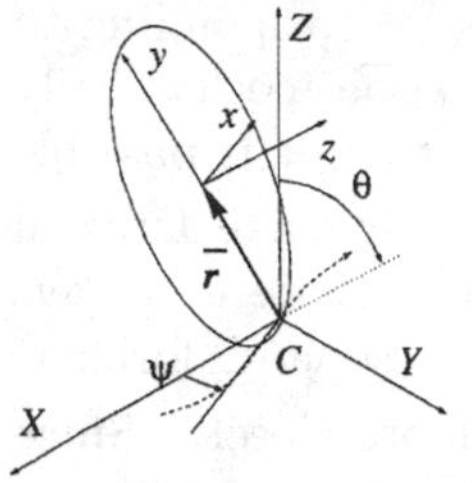

slipping on the horizontal XY-plane. At the instant represented, its z-axis of symmetry makes an angle θ with the vertical Z-axis, and the tangent to its path makes an angle ψ with the X-axis as shown. Find the equations of motion describing its angular motion.

We note that, in order to enforce the non-slip condition, there must be friction at the point of contact, C. We could sum forces and take moments about, here, the center of mass, obtaining a set of equations giving $\overline{\boldsymbol{a}}$ and $\boldsymbol{\alpha}$ (related through the non-slip condition), as well as the friction and normal forces. However, since we are interested only in the *angular* motion, we would like to sum moments about C to eliminate these forces, that we really don't care about! *But C is neither the center of mass nor a fixed point about which the body is rotating*; in fact, the point is generally *accelerating*. This is where the most general form of moment equation (5.2-41c) becomes useful: it allows moments to be taken about *any* point, fixed or not! That, then, is the route we shall follow here:

That moment equation is, in this case,

$$\sum_i \boldsymbol{M}_{C_i} = \boldsymbol{r}_{G/C} \times m\overline{\boldsymbol{a}} + \boldsymbol{\Omega} \times \bar{\mathbf{I}}\boldsymbol{\omega} + \bar{\mathbf{I}}\boldsymbol{\alpha}_{rel},$$

where we will denote $\boldsymbol{r}_{G/C} \equiv \overline{\boldsymbol{r}}$. The centroidal gyroscopic body-oriented (x, y, z) axes are as shown for the orientation given in the diagram. It remains only to determine the various parameters in this equation. We note that the precession

$$\begin{aligned}\boldsymbol{\Omega} &= \dot{\theta}\hat{\boldsymbol{\imath}} + \dot{\psi}\sin\theta\hat{\boldsymbol{\jmath}} + \dot{\psi}\cos\theta\hat{\boldsymbol{k}} \\ &\equiv \Omega_x\hat{\boldsymbol{\imath}} + \Omega_y\hat{\boldsymbol{\jmath}} + \Omega_z\hat{\boldsymbol{k}},\end{aligned}$$

while the total angular velocity of the body

$$\begin{aligned}\boldsymbol{\omega} &= \dot{\theta}\hat{\boldsymbol{\imath}} + \dot{\psi}\sin\theta\hat{\boldsymbol{\jmath}} + (\dot{\psi}\cos\theta + p)\hat{\boldsymbol{k}} \\ &= \Omega_x\hat{\boldsymbol{\imath}} + \Omega_y\hat{\boldsymbol{\jmath}} + (\Omega_z + p)\hat{\boldsymbol{k}} \\ &\equiv \omega_x\hat{\boldsymbol{\imath}} + \omega_y\hat{\boldsymbol{\jmath}} + \omega_z\hat{\boldsymbol{k}}.\end{aligned}$$

We shall work most of the problem in components of $\boldsymbol{\Omega}$ and $\boldsymbol{\omega}$ to keep the expressions tidy.

The first problem, of course, is determination of $\overline{\boldsymbol{a}}$ under the non-slip constraint. Something similar was done in Example 4.4.2, but there we used axes rotating with the disk $\boldsymbol{\omega}$; here we wish to use those rotating with the symmetry axis $\boldsymbol{\Omega}$. But that's not a major alteration:

$$\overline{\boldsymbol{v}} = \boldsymbol{\omega} \times \overline{\boldsymbol{r}} = (\omega_x\hat{\boldsymbol{\imath}} + \omega_y\hat{\boldsymbol{\jmath}} + \omega_z\hat{\boldsymbol{k}}) \times r\hat{\boldsymbol{\jmath}} = -r\omega_z\hat{\boldsymbol{\imath}} + r\omega_x\hat{\boldsymbol{k}}.$$

From that we obtain, just as in the Example cited [noting that the "*rel*" derivatives differentiate only the *components*],

$$\begin{aligned}\overline{\boldsymbol{a}} &= \dot{\overline{\boldsymbol{v}}}_{rel} + \boldsymbol{\Omega}\times\overline{\boldsymbol{v}}\\ &= (-r\dot{\omega}_z\hat{\boldsymbol{\imath}} + r\dot{\omega}_x\hat{\boldsymbol{k}}) + \begin{vmatrix}\hat{\boldsymbol{\imath}} & \hat{\boldsymbol{\jmath}} & \hat{\boldsymbol{k}}\\ \Omega_x & \Omega_y & \Omega_z\\ -r\omega_z & 0 & r\omega_x\end{vmatrix}\\ &= r\left((\omega_x\Omega_y - \dot{\omega}_z)\hat{\boldsymbol{\imath}} - (\omega_x\Omega_x + \omega_z\Omega_z)\hat{\boldsymbol{\jmath}} + (\omega_z\Omega_y + \dot{\omega}_x)\hat{\boldsymbol{k}}\right).\end{aligned}$$

Thus, finally,

$$\begin{aligned}\overline{\boldsymbol{r}}\times m\overline{\boldsymbol{a}} = r\hat{\boldsymbol{\jmath}}\times m\overline{\boldsymbol{a}} &= mr^2\left((\omega_z\Omega_y + \dot{\omega}_x)\hat{\boldsymbol{\imath}} - (\omega_x\Omega_y - \dot{\omega}_z)\hat{\boldsymbol{k}}\right)\\ &= mr^2\left(\left(\omega_z\dot{\psi}\sin\theta + \ddot{\theta}\right)\hat{\boldsymbol{\imath}} + (\dot{\omega}_z - \dot{\psi}\dot{\theta}\sin\theta)\hat{\boldsymbol{k}}\right).\end{aligned}$$

Calculation of the other two terms is straightforward: noting that $\boldsymbol{\omega}$ and $\boldsymbol{\Omega}$ are the same in the x- and y-directions,

$$\begin{aligned}\boldsymbol{\Omega}\times\overline{\mathsf{I}}\boldsymbol{\omega} &= \begin{vmatrix}\hat{\boldsymbol{\imath}} & \hat{\boldsymbol{\jmath}} & \hat{\boldsymbol{k}}\\ \Omega_x & \Omega_y & \Omega_z\\ I_O\Omega_x & I_O\Omega_y & I\omega_z\end{vmatrix}\\ &= (I\omega_z\Omega_y - I_O\Omega_y\Omega_z)\hat{\boldsymbol{\imath}} - (I\omega_z\Omega_x - I_O\Omega_x\Omega_z)\hat{\boldsymbol{\jmath}}\\ &= (I\omega_z\dot{\psi}\sin\theta - I_O\dot{\psi}^2\sin\theta\cos\theta)\hat{\boldsymbol{\imath}} + (I_O\dot{\psi}\dot{\theta}\cos\theta - I\omega_z\dot{\theta})\hat{\boldsymbol{\jmath}};\\ \overline{\mathsf{I}}\boldsymbol{\alpha}_{rel} &= I_O\ddot{\theta}\hat{\boldsymbol{\imath}} + I_O(\ddot{\psi}\sin\theta + \dot{\psi}\dot{\theta}\cos\theta)\hat{\boldsymbol{\jmath}} + I\dot{\omega}_z\hat{\boldsymbol{k}}.\end{aligned}$$

Note that we have not expanded ω_z or its time derivative. This is motivated by the fact that $\dot{\omega}_z$ appears only in the z-components here, and there is no moment in that direction. Maintaining the identity of ω_z will expedite its eventual integration; we do this in what follows. (By rights, we could substitute in the values of I and I_O for the thin disk under discussion, but the equations are completely general as they stand for *any* shape body. We retain that generality. Note, however, that if the disk were "thick," the non-slip would still hold, but with some adjustment in the value of $\overline{\boldsymbol{r}}$, which would no longer be in the $\hat{\boldsymbol{\jmath}}$-direction.) Collecting components in the above three terms, we get that

$$\begin{aligned}\sum_i \boldsymbol{M}_{C_i} &= \left((I+mr^2)\omega_z\dot{\psi}\sin\theta - I_O\dot{\psi}^2\sin\theta\cos\theta + (I_O+mr^2)\ddot{\theta}\right)\hat{\boldsymbol{\imath}}\\ &\quad + \left(I_O\ddot{\psi}\sin\theta + 2I_O\dot{\psi}\dot{\theta}\cos\theta - I\omega_z\dot{\theta}\right)\hat{\boldsymbol{\jmath}}\\ &\quad + \left((I+mr^2)\dot{\omega}_z - mr^2\dot{\psi}\dot{\theta}\sin\theta\right)\hat{\boldsymbol{k}}\\ &= -mg\overline{r}\cos\theta\hat{\boldsymbol{\imath}}. \qquad (5.2\text{-}49)\end{aligned}$$

Granted, it's not a pretty sight! But less important than the angular equations of motion themselves is the relatively mundane, even routine, calculation sufficient to get them. That is the major point to the example, as well as a review of how to deal with the non-slip condition kinematically. But the final equations provide an example of stability later. |

Summary. This section has treated a special formulation of the equations of motion to treat an axially-symmetric body, the *gyroscope*. In order to keep separate the angular velocity of the axis—itself consisting of *precession* and *nutation*—and the *spin* about that axis, a canonical set of body-oriented coordinates is set up, with $\hat{\boldsymbol{k}}$ along the symmetric axis and measured in the direction of the spin, and $\hat{\boldsymbol{\imath}} \equiv \hat{\boldsymbol{K}} \times \hat{\boldsymbol{k}}$. This forces these axes to be principal, so the inertia matrix is always diagonalized. The equations of motion leave the force equation unchanged, but the moment equation in these axes assumes a new form (5.2-41).

We then considered cases in which moments were actually applied or not. In the latter case, motion was characterized by constant precession of the symmetry about a fixed line (the direction of constant angular momentum); in the former, similar motion was examined, with more general motion forestalled until a later section.

Again, it must be emphasized that the gyroscope equations are merely a special formulation of the Euler equations for an axially-symmetric body. Their primary utility is precisely the ability to separate out the two components of angular motion of the body.

5.3 Dynamic Stability

We leave basic kinetics to examine briefly an application of the equations of motion we have developed to this point: dynamic stability. What do we mean by this? In fact, what do we really mean by the term "stability" at all?

In theory, a pencil standing on its tip on a horizontal surface should stay there: it satisfies all the equilibrium equations. Yet such a configuration is rarely observed in practice. Why is this? Life being what it is, the pencil will inevitably suffer some slight displacement from this position—someone walking by, or a puff of air (generally from a bothersome confederate). Once displaced from this position, the pencil falls: the [couple] moment about its tip due to gravity generates a moment which brings it down. On the other hand, if the pencil is *hung* by its tip and displaced from its equilibrium vertical position, a similar displacement from this orientation generates a moment which brings it *back* to the vertical. There is clearly an important qualitative difference between these two equilibrium "solutions": one is *unstable*, the other *stable*. A similar analysis is commonly applied to the stability of bodies floating in a fluid—*hydrostatic stability*; analogous conditions regarding the couple generated by the weight and, here, the buoyant force lead to the idea of "righting" moments and "tipping" moments.

Despite the existence of a [couple] moment in both the above examples, the concept of stability is not a kinetic idea, but a *kinematic* one: *if the body is displaced from a solution "a bit," will it return to, or diverge from, that solution.*

In our choice of words we have cleverly provided a bridge between statics and dynamics: the issue isn't whether it "remains in place" but whether it "returns to its *solution*"—the latter formulation being as applicable to Dynamics, where we ultimately consider the *solutions* to the differential equations resulting from Newton's Second, as it is to Statics.

That bridge, then, allows us to firm up the idea mathematically in the jargon of Differential Equations: It suffices to consider the *first*-order *vector* differential equation $\dot{\boldsymbol{x}} = \boldsymbol{f}(\boldsymbol{x}; t)$, since any higher-order equation can be cast in this form through the common device of introducing each order of differentiation as a component of the vector (see, *e.g.*, Section 2.4). Let us say, then, that we have a solution, $\boldsymbol{x}(t)$, to this equation; in the present context, this would represent the solution to Newton's Second or the Euler equations. A *displacement* from this solution could be represented by, say, $\boldsymbol{\xi}(t)$, where, to investigate stability, it suffices to regard this as *small*.[25] (The "small" $\boldsymbol{\xi}(t)$ is commonly described as being "of *first order*," in contrast to $\boldsymbol{x}(t)$ "of *zeroeth order*.") Thus the motion is now given, not by $\boldsymbol{x}(t)$, but by $\boldsymbol{x}(t) + \boldsymbol{\xi}(t)$. The question is, then, *how does* $\boldsymbol{\xi}(t)$ *behave*: does it remain small or even go to $\mathbf{0}$ so that $\boldsymbol{x}(t) + \boldsymbol{\xi}(t) \to \boldsymbol{x}(t)$—$\boldsymbol{x}(t)$ then being *stable*, or does it diverge—meaning that the original solution $\boldsymbol{x}(t)$ is *unstable*?

But this means that we must somehow determine $\boldsymbol{\xi}(t)$ itself. What handle do we have on this; in particular, what equation does it satisfy? The "perturbed" solution $\boldsymbol{x}(t) + \boldsymbol{\xi}(t)$ is, recall, a solution to the *original* differential equation $\dot{\boldsymbol{x}} = \boldsymbol{f}(\boldsymbol{x}; t)$; *i.e.*,

$$\frac{d}{dt}(\boldsymbol{x} + \boldsymbol{\xi}) = \dot{\boldsymbol{x}} + \dot{\boldsymbol{\xi}} = \boldsymbol{f}(\boldsymbol{x} + \boldsymbol{\xi}; t); \tag{5.3-1}$$

note that this reduces to the original "unperturbed" equation $\dot{\boldsymbol{x}} = \boldsymbol{f}(\boldsymbol{x}; t)$ when $\boldsymbol{\xi} \to \mathbf{0}$. $\boldsymbol{f}$ can be *any* function, arbitrarily complicated, depending on the original system, which makes the prospect of determining $\boldsymbol{f}(\boldsymbol{x} + \boldsymbol{\xi}; t)$ somewhat unappealing. But recall that a stability analysis admits $\boldsymbol{\xi}(t)$ to be *small*, suggesting immediately the technique of expanding $\boldsymbol{f}(\boldsymbol{x} + \boldsymbol{\xi}; t)$ in a *Taylor Series* about [the "unperturbed" solution] $\boldsymbol{x}$: if $\boldsymbol{x}$ (and thus $\boldsymbol{f}$) are n-vectors, the components of this expansion become

$$\begin{pmatrix} f_1(\boldsymbol{x} + \boldsymbol{\xi}; t) \\ f_2(\boldsymbol{x} + \boldsymbol{\xi}; t) \\ \vdots \\ f_n(\boldsymbol{x} + \boldsymbol{\xi}; t) \end{pmatrix} = \begin{pmatrix} f_1(\boldsymbol{x}; t) + \sum_{i=1}^n \frac{\partial f_1(\boldsymbol{x}; t)}{\partial x_i} \xi_i + \ldots \\ f_2(\boldsymbol{x}; t) + \sum_{i=1}^n \frac{\partial f_2(\boldsymbol{x}; t)}{\partial x_i} \xi_i + \ldots \\ \vdots \\ f_n(\boldsymbol{x}; t) + \sum_{i=1}^n \frac{\partial f_n(\boldsymbol{x}; t)}{\partial x_i} \xi_i + \ldots \end{pmatrix},$$

in which the higher-order terms in the ξ_i, and accompanying higher-order partial derivatives, are not written explicitly. *But the* ξ_i *are small*; thus the full, "perturbed," differential equation can be approximated by

$$\dot{\boldsymbol{x}} + \dot{\boldsymbol{\xi}} = \boldsymbol{f}(\boldsymbol{x} + \boldsymbol{\xi}; t) \doteq \boldsymbol{f}(\boldsymbol{x}; t) + \mathsf{D}_{\boldsymbol{f}(\boldsymbol{x};t)} \boldsymbol{\xi},$$

[25] Even stable solutions can be upset by a *large* displacement!

in which $\mathbf{D}_{\boldsymbol{f}(\boldsymbol{x};t)}$ is the *Jacobian matrix* of partial derivatives $\left(\frac{\partial f_i(\boldsymbol{x};t)}{\partial x_j}\right)$, i and j being the row and column indices, respectively, each partial being evaluated at the [*known*] *original* solution (see page 180). But, recall, $\boldsymbol{x}(t)$ satisfies the original differential equation $\dot{\boldsymbol{x}} = \boldsymbol{f}(\boldsymbol{x};t)$; thus the equation which the unknown $\boldsymbol{\xi}(t)$ must satisfy is "simply"

$$\dot{\boldsymbol{\xi}} = \mathbf{D}_{\boldsymbol{f}(\boldsymbol{x};t)}\boldsymbol{\xi} \tag{5.3-2}$$

—a *linear* equation, sometimes called the "variational equation," in that variable which can, in principle, be solved. In practice, it may or may not be necessary to form the Jacobian explicitly; the above is merely meant to show that, in general, this analysis always reduces to a linear equation.

In the particularly appealing case that the matrix multiplying $\boldsymbol{\xi}$ is a *constant*, we know the solutions for each component will be: exponentially decaying to, or expanding from, 0—stable or unstable respectively; purely periodic ("indifferently" stable); or one itself factored by t—this occurs when there are repeated eigenvalues—so unstable. But recall (Section 3.5.4) that the presumption that the solution of this linear system is of the form $\boldsymbol{\xi}(t) = \boldsymbol{\xi}_o e^{\lambda t}$ ultimately yields the eigenvalues λ_i of the matrix multiplying $\boldsymbol{\xi}$ (and the eigen*vectors* $\boldsymbol{\xi}_o$). Thus stability in such cases is tantamount to determining the eigenvalues of the [constant] matrix: *if any eigenvalue is* positive *(or, if complex, has a* positive real part*), then the overall solution is unstable.*

There is, in practice, one implicit working assumption invariably made which might require a little explicit elucidation: that if a quantity ξ is small, then so are its derivatives $\dot{\xi}$ and $\ddot{\xi}$. Intuitively, this makes sense: if a quantity is small, then its change $\Delta\xi$ in $\frac{d\xi}{dt} \equiv \lim_{\Delta t \to 0} \frac{\Delta\xi}{\Delta t}$ should be also. While this assumption is fairly easy to expect relative to $\dot{\xi}$, its veracity relative to $\ddot{\xi}$, which can be affected by external forces or moments of arbitrary magnitude (*e.g.*, *impact*; see page 370), is somewhat more tenuous. It is the assumption made nonetheless!

In order to "flesh out" all the above theory, we now consider a couple of examples:

Example 5.3.1. *Stability of Moment-free Motion.* Let us first examine the moment-free rotation of an arbitrary body for stability about each of its three principal axes:

We note first that this rotation remains *constant*: the Euler equation (5.2-23a)—this is *not* a gyroscope, since we don't know that any principal axis is one of *symmetry*—is simply

$$\sum_i \overline{\boldsymbol{M}}_i \equiv \mathbf{0} = \boldsymbol{\omega} \times \bar{\mathsf{I}}\boldsymbol{\omega} + \bar{\mathsf{I}}\boldsymbol{\alpha}.$$

But, for $\boldsymbol{\omega}$ about a *principal* axis, $\bar{\mathsf{I}}\boldsymbol{\omega} \parallel \boldsymbol{\omega}$, so the first term "crosses out" and we are left with $\bar{\mathsf{I}}\boldsymbol{\alpha} \equiv \mathbf{0}$—constant angular velocity.

Now let us consider a *perturbed* rotation about, say, the [arbitrary] z-axis: $\boldsymbol{\omega} + \boldsymbol{\xi} = \xi_x\hat{\imath} + \xi_y\hat{\jmath} + (\omega + \xi_z)\hat{k}$, in which the ξ's are presumed *small.* The above equation becomes, noting that $\boldsymbol{\alpha} = \boldsymbol{\alpha}_{rel}$ (differentiating only the *components*) for axes rotating with the body by Example 4.3.1,

$$
\begin{aligned}
\mathbf{0} &= \boldsymbol{\omega} \times \bar{\mathbf{I}}\boldsymbol{\omega} + \bar{\mathbf{I}}\boldsymbol{\alpha} \\
&= \begin{vmatrix} \hat{\imath} & \hat{\jmath} & \hat{k} \\ \xi_x & \xi_y & \omega + \xi_z \\ I_x\xi_x & I_y\xi_y & I_z(\omega+\xi_z) \end{vmatrix} + I_x\dot{\xi}_x\hat{\imath} + I_y\dot{\xi}_y\hat{\jmath} + I_z(\dot{\omega}+\dot{\xi}_z)\hat{k} \\
&\doteq \Big(I_x\dot{\xi}_x + \omega(I_z - I_y)\xi_y\Big)\,\hat{\imath} + \Big(I_y\dot{\xi}_y - \omega(I_z - I_x)\xi_x\Big)\,\hat{\jmath} + I_z(\dot{\omega}+\dot{\xi}_z)\hat{k},
\end{aligned}
$$

in which, following the ground rules for the "small" quantities, we have dropped second-order products in them. This is, to the order 1 retained, equation (5.3-1); note that, again, it reduces to the "unperturbed" Euler equation above when the $\xi_i \to 0$. We have carried $\dot{\xi}_z$ through in the last component to make a point: since it is "small" compared to the "0-th order" ω, it too is dropped in the *sum*, showing that, once again, ω is constant: $0 = I_z(\dot{\omega}+\dot{\xi}_z) \doteq I_z\dot{\omega} \Rightarrow \dot{\omega} = 0$.

The remaining two components, the first-order part of the above, can be put in the form of a matrix differential equation with constant coefficients:

$$
\frac{d}{dt}\begin{pmatrix} \xi_x \\ \xi_y \end{pmatrix} = \begin{pmatrix} 0 & -\frac{\omega(I_z-I_y)}{I_x} \\ \frac{\omega(I_z-I_x)}{I_y} & 0 \end{pmatrix}\begin{pmatrix} \xi_x \\ \xi_y \end{pmatrix};
$$

these are the equations the two small displacements must satisfy, corresponding to the variational equation (5.3-2), obtained here without recourse to calculation of the Jacobian. The solutions for this will involve exponentials with t times the eigenvalues of the matrix, the latter satisfying the equation

$$
\lambda^2 = -\frac{(I_z - I_y)(I_z - I_x)}{I_x I_y}\omega^2.
$$

If the right-hand side is *positive*—since the I's are all positive, this means that the differences $(I_z - I_y)$ and $(I_z - I_x)$ have different signs—then λ will be $\pm a$ for some constant a and, in particular, will have a positive eigenvalue.[26] Thus there is a divergent solution possible for $\boldsymbol{\omega} + \boldsymbol{\xi}$, and we regard the original system to be *unstable.* Conversely, if both have the *same* sign, λ is purely imaginary; $\xi_x\hat{\imath}$ and $\xi_y\hat{\jmath}$ will be oscillatory, but at least with their "small" amplitudes, about $\boldsymbol{\omega}$.

What this means physically is that if $I_x < I_z < I_y$—"z" is the axis with the *intermediate* moment of inertia—then motion is unstable; while as long as this axis has one of the extreme values, motion is stable. This is exactly what happens! (Try it with a book. Not this one.) |

For the second example, we consider the "thin disk" with which we concluded the previous section:

Example 5.3.2. *Stability of Thin Disk Rolling without Slipping.* We consider a thin disk rolling without slipping on a horizontal surface in a fixed vertical

[26]The general solution, $C_1e^{at} + C_2c^{-at}$, certainly admits of a stable solution if $C_1 \equiv 0$. But this solution is *singular* in the sense that any slight change in the initial conditions will force $C_1 \neq 0$ and the solution to go unstable. Thus it is dismissed from consideration here!

plane—no precession or nutation. Note that in this case, just as in the above example, $\sum \overline{\boldsymbol{M}} \equiv 0$ implies that $\boldsymbol{\alpha} \equiv \mathbf{0}$ so that $\boldsymbol{\omega}$ is constant; this is the "unperturbed" motion. In order to investigate the stability of the motion, we consider "small" departures—precession and nutation—*from* this constant rotation; as much as for as for intrinsic interest of the problem itself, this provides an example of how the "smallness" assumption is reflected in trigonometric functions. But this means we need the equations to account for the more general motion. In the previous Section, we obtained just such general equations for the disk, considered a "gyroscope." Unfortunately, we must modify them slightly to deal with the present case:

In the case of a gyroscope, we define the angle θ to be that between the fixed Z- and body-oriented z-axes, the latter *perpendicular* to the surface of the disk. But we wish to measure instead the "small" departures from a *vertical* plane—the angle between the Z-axis and the *parallel* to that surface—in our case, the y-axis. To prevent confusion, we will call that angle θ'. We observe that $\theta' = 90° - \theta$; this means that we must interchange the "sin" and "cos" in the previous equations when we use the new argument. But we must also change the sign of the derivatives: $\dot{\theta} \to -\dot{\theta}'$ and $\ddot{\theta} \to -\ddot{\theta}'$.

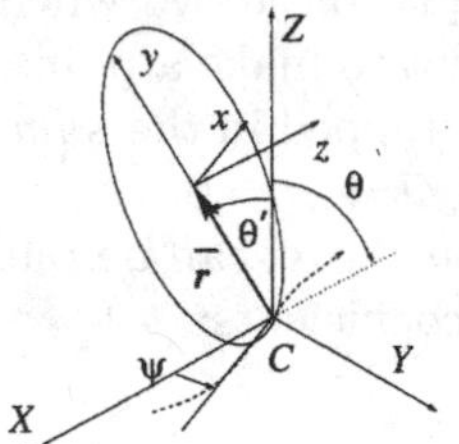

That done, the equations (5.2-49) now become

$$\sum_i \boldsymbol{M}_{C_i} = \Big((I + mr^2)\omega_z\dot{\psi}\cos\theta' - I_O\dot{\psi}^2\sin\theta'\cos\theta' - (I_O + mr^2)\ddot{\theta}'\Big)\,\hat{\boldsymbol{\imath}}$$
$$+ \Big(I_O\ddot{\psi}\cos\theta' - 2I_O\dot{\psi}\dot{\theta}'\sin\theta' + I\omega_z\dot{\theta}'\Big)\,\hat{\boldsymbol{\jmath}}$$
$$+ \Big((I + mr^2)\dot{\omega}_z + mr^2\dot{\psi}\dot{\theta}'\cos\theta'\Big)\,\hat{\boldsymbol{k}}$$
$$= -mgr\sin\theta'\hat{\boldsymbol{\imath}}.$$

Again, these are the full equations corresponding to equation (5.3-1) in the general discussion; no approximations or ordering have been made yet. To do this, we will consider ψ and θ', and thus their derivatives, the precession $\dot{\psi}$ and nutation $\dot{\theta}$, to be small [recall the working assumption that both the variables *and* their derivatives are of the same order]; thus their products above will be ignored. But we make the additional approximations that, to first order in the Taylor expansions of these functions, $\sin\theta' \doteq \theta'$ and $\cos\theta' \doteq 1$. Introducing these approximations into the equations, we get that

$$\Big((I + mr^2)\omega_z\dot{\psi} - (I_O + mr^2)\ddot{\theta}'\Big)\,\hat{\boldsymbol{\imath}} + (I_O\ddot{\psi} + I\omega_z\dot{\theta}')\hat{\boldsymbol{\jmath}}$$
$$+ \big((I + mr^2)\dot{\omega}_z\big)\,\hat{\boldsymbol{k}} \doteq -mgr\theta'\hat{\boldsymbol{\imath}}.$$

Though these equations could be solved explicitly for the variational equations (5.3-2) in ψ and θ' as we did before, we shall be content to solve for these variables by direct integration of the components: The vanishing of the $\hat{\boldsymbol{k}}$-component—of "order 0"—gives us immediately that ω_z is constant; this is the very reason we kept it separate in the original derivation of the equations! In point of fact, however, this is just a vestige of the original unperturbed solution, where $\boldsymbol{\omega} = \omega_z \hat{\boldsymbol{k}}$ is itself constant. We just didn't explicitly subtract that solution off here.

The other two components of this equation comprise the first-order variational equations (5.3-2): Similar integration of the $\hat{\jmath}$-component gives that $I_O\dot{\psi} + I\omega_z\theta' = C$, a constant; this shows that the "precession" $\dot{\psi}$ is linear in [small] θ'. Thus, finally, the $\hat{\imath}$-component becomes

$$(I_O + mr^2)\ddot{\theta}' + \left(\frac{I(I + mr^2)\omega_z^2}{I_O} - mgr\right)\theta' = \frac{C(I + mr^2)\omega_z}{I_O}$$

—a non-homogeneous linear differential equation in θ'. The particular part of its solution is just a constant, but the part that interests us is the homogeneous part: if the coefficient of θ' is positive, we will have periodic oscillation in θ' about the z-axis. But if that coefficient is negative, then we have exponential growth in θ'. The boundary between the two occurs when that coefficient vanishes; *i.e.*, when

$$\omega_z = \sqrt{\frac{I_O mgr}{I(I + mr^2)}}.$$

For ω_z greater than this value, the motion is stable—a quickly rolling coin (or bicycle wheel) will not keel over; if it's too small, the presumed motion is unstable and it will. For the thin disk announced, $I = \frac{1}{2}mr^2$ and $I_O = \frac{1}{4}mr^2$; then $\omega_z = \sqrt{\frac{g}{3r^3}}$, from which the [unperturbed] critical speed for stability would be, $\overline{\boldsymbol{r}}$ being *nearly* vertical, $\overline{v} = \|\boldsymbol{\omega} \times \overline{\boldsymbol{r}}\| \doteq \sqrt{\frac{g}{3r}}$. But—and this is the advantage of using general I and I_O—we could also use, for example, parameters for a "thin" *hoop* to approximate a bicycle wheel; then $\omega_z = \frac{1}{2}\sqrt{\frac{g}{r^3}}$—smaller than that for the disk, due to the greater concentration of mass at r, and thus greater angular momentum. |

The above has been cast purely in terms of the ramifications of small *physical* disturbances from a solution to the equations of motion. But they have a profound impact on the solution itself when the latter has been obtained *numerically*—a practice which must generally be resorted to: Numerical calculation is invariably limited by that fixed precision used to obtain a result. But these are forms of *numerical* "perturbation"—a variation from the "actual" solution taking place generally in the innards of a computer, "roundoff error." But it is also limited by the particular technique of numerical integration used to arrive at the final result, "truncation error." If a system is *physically unstable*, this instability will surely evidence itself in the numerical output; conversely, if it is *stable*, the numerical solution will be so as well (though likely not "exactly" the right one, due to the numerical issues). If the numerical and "actual" solutions are the same *qualitatively*, however, stability issues clearly make the solution for an inherently unstable system highly suspect.

Roundoff error can be mitigated by hardware/software techniques: modern computers, for example, will often *round* internal calculations before storing them for later steps; older machines tended merely to *truncate* the solution at a given fixed accuracy—systematically "rounding *down*." But truncation error is endemic, since numerical integration implicitly uses a Taylor-series approximation which, practically, must itself be stopped—"truncated"—at some order. In order to get some measure of confidence in a numerical answer, it is useful to continually make side calculations of, say, kinetic constants like energy and/or momentum which might be known to be conserved: these provide an independent check on the calculations and can be used to halt the process when the error in one of these quantities has exceeded an acceptable tolerance. Just because it comes out of a computer doesn't make it *right.* With all the advances in speed made in computer technology, one can merely generate garbage more efficiently.

Homework:

1. [*Computer project*] Numerically integrate the single-degree-of-freedom simple harmonic oscillator equation $\ddot{x} + \omega^2 x = 0$ using your favorite integrator, printing out the energy (which should be conserved) at each 1000 steps. See what happens.

Summary. Stability analysis mathematically introduces small variation from a known solution. By considering the general equations of motion for this altered case, we invariably obtain *linear* differential equations in these "perturbations." More important than their actual solution is their *qualitative* behavior, particularly whether they diverge or not from the original, unperturbed solution. If they do, the original motion is regarded as *unstable*; otherwise it is *stable* (or, in the case of purely periodic motion, "*indifferently* stable").

5.4 Alternatives to Direct Integration

We have seen that, just as with particles, the equations of motion for rigid bodies give rise to second-order ordinary differential equations in the *instantaneous* translational and rotational components. Unless such acceleration-type quantities are all that's desired, at some point we are going to have to confront the problem of *integrating* these—of obtaining information of how the system will behave over an *interval* of time. While it would be possible to do this directly on a case-by-case basis, it would be nice to be able to predict which such integration could be done, and even to obtain general expressions corresponding to just such integrated forms of the equations of motion.

In fact, this very possibility has already been hinted at: in Section 5.1, we saw recurring the "kinetic quantities" of *linear momentum*, *angular momentum*, and *energy.* At that time we alluded to conditions under which these quantities, or even just separate *components* of the vector momenta, would be "conserved" over an interval. In that Section, we formulated these very quantities initially for single particles, then extended their definitions to *systems* of particles. But rigid bodies are, at the end of the day, simply *systems* of particles constrained to move in such a way that their inter-particle distances remain constant, and thus Section 5.2 started by expressing linear and angular momentum in this special case. In fact, it was precisely the equations of motion for linear and angular momentum for such systems which ultimately led to the equations of motion for rigid bodies in that Section.

One is justified, then, in hoping that conservation laws analogous to those for particles and systems of them would hold for rigid bodies, albeit likely formulated to account for the special type of motion such bodies pursue. This is precisely the case. We shall consider first the scalar *energy*, the form [kinetic] energy assumes for rigid bodies, and what changes—or *doesn't* change—this energy. We then do the same for the two varieties of vector momenta (though their forms for rigid bodies have already been obtained above).

5.4.1 Energy

On page 258, we obtained the most fundamental form of energy equation, the *Principle of Work and [Kinetic] Energy*, equation (5.1-9); it was this on which the formulations of *Conservation of Energy*, (5.1-11) and the most general form of work and *total* energy, (5.1-14) were ultimately based. This will form the starting point here, too, with the later results following from it. But that means that we must consider precisely what might be different about *work* and *kinetic energy* in the special case of rigid bodies. This we do now (though in the opposite order):

Kinetic Energy

As might be expected from the above discussion, we must start the investigation of kinetic energy in rigid bodies by considering what it is for systems of particles: Though we *defined* T as simply the sum of the separate particles' kinetic energies in equation (5.1-16c), we ultimately expressed it in terms of the center-of-mass motion and motion *about* the center of mass in equations (5.1-21d-e), which are repeated here:

$$T = \frac{1}{2}m\overline{v}^2 + \overline{T}, \qquad \overline{T} \equiv \sum_i \frac{1}{2}m_i\dot{\rho}_i^2;$$

in view of the prominent role the center of mass has played to date, it is not surprising that this is the formulation we will focus on.

In particular, we desire to express $\overline{T}$ in the special case of rigid-body motion: Noting that the above $\dot{\rho}_i^2 = \dot{\boldsymbol{\rho}}_i \cdot \dot{\boldsymbol{\rho}}_i$, and that we have already obtained a general expression (5.2-2), $\dot{\boldsymbol{\rho}}_i = \boldsymbol{\omega} \times \boldsymbol{\rho}_i$, in the development of the equations of motion for the special case of rigid-body motion, we can immediately find

$$\dot{\rho}_i^2 = \dot{\boldsymbol{\rho}}_i \cdot \dot{\boldsymbol{\rho}}_i = (\boldsymbol{\omega} \times \boldsymbol{\rho}_i) \cdot (\boldsymbol{\omega} \times \boldsymbol{\rho}_i) = \boldsymbol{\omega} \times \boldsymbol{\rho}_i \cdot (\boldsymbol{\omega} \times \boldsymbol{\rho}_i)$$

(nothing magical about this: we are simply dropping the parentheses in the definition of the triple scalar product $\boldsymbol{A} \times \boldsymbol{B} \cdot \boldsymbol{C} \equiv (\boldsymbol{A} \times \boldsymbol{B}) \cdot \boldsymbol{C}$)

$$= \boldsymbol{\omega} \cdot \boldsymbol{\rho}_i \times (\boldsymbol{\omega} \times \boldsymbol{\rho}_i),$$

since we can interchange the dot and cross products in the triple scalar product. Thus

$$\overline{T} = \sum_i \frac{1}{2} m_i \boldsymbol{\omega} \cdot \boldsymbol{\rho}_i \times (\boldsymbol{\omega} \times \boldsymbol{\rho}_i) = \frac{1}{2} \boldsymbol{\omega} \cdot \sum_i m_i \boldsymbol{\rho}_i \times (\boldsymbol{\omega} \times \boldsymbol{\rho}_i),$$

factoring $\frac{1}{2}\boldsymbol{\omega}$, which doesn't depend on i, out of the summation. *But the term after the dot product is precisely the ["generic"]* $\boldsymbol{H}$, *equation* (5.2-3)*!* Accounting for the fact that all quantities are measured relative to the center of mass, then, we see that

$$\overline{T} = \frac{1}{2} \boldsymbol{\omega} \cdot \overline{\boldsymbol{H}} = \frac{1}{2} \boldsymbol{\omega} \cdot \bar{\mathsf{I}} \boldsymbol{\omega}. \tag{5.4-1}$$

Thus the general expression for kinetic energy is

$$T = \frac{1}{2} m \bar{v}^2 + \overline{T} = \frac{1}{2} m \bar{v}^2 + \frac{1}{2} \boldsymbol{\omega} \cdot \bar{\mathsf{I}} \boldsymbol{\omega}.$$

We note the marginally interesting fact that this can be written symmetric in *both* momenta:

$$T = \frac{1}{2} \overline{\boldsymbol{v}} \cdot \boldsymbol{G} + \frac{1}{2} \boldsymbol{\omega} \cdot \bar{\mathsf{I}} \boldsymbol{\omega}.$$

The reader will recall that in developing the equations of motion on page 307, we showed how the most general form of the moment equation (5.2-23c) reduced to that in the special case of motion about a fixed point, (5.2-23b), through the intercession of the Parallel Axis Theorem. We shall now show a similar circumstance holds for energy: *If* the body is rotating about a fixed point O, we can write $\overline{\boldsymbol{v}} = \boldsymbol{\omega} \times \overline{\boldsymbol{r}}$ where $\overline{\boldsymbol{r}}$ is measured from O, so

$$\begin{aligned}
\frac{1}{2} m \overline{\boldsymbol{v}}^2 &= \frac{1}{2} m (\boldsymbol{\omega} \times \overline{\boldsymbol{r}}) \cdot (\boldsymbol{\omega} \times \overline{\boldsymbol{r}}) \\
&= \frac{1}{2} m \boldsymbol{\omega} \times \overline{\boldsymbol{r}} \cdot (\boldsymbol{\omega} \times \overline{\boldsymbol{r}}) \\
&= \frac{1}{2} m \boldsymbol{\omega} \cdot \overline{\boldsymbol{r}} \times (\boldsymbol{\omega} \times \overline{\boldsymbol{r}}) \\
&= \frac{1}{2} \boldsymbol{\omega} \cdot m [\bar{r}^2 \boldsymbol{\omega} - (\boldsymbol{\omega} \cdot \overline{\boldsymbol{r}}) \overline{\boldsymbol{r}}],
\end{aligned}$$

using the ubiquitous equations (4.1-5) once again to expand the triple vector product. Thus

$$T = \frac{1}{2} m \bar{v}^2 + \frac{1}{2} \boldsymbol{\omega} \cdot \bar{\mathsf{I}} \boldsymbol{\omega} = \frac{1}{2} \boldsymbol{\omega} \cdot \left[m [\bar{r}^2 \boldsymbol{\omega} - (\boldsymbol{\omega} \cdot \overline{\boldsymbol{r}}) \overline{\boldsymbol{r}}] + \bar{\mathsf{I}} \boldsymbol{\omega} \right] = \frac{1}{2} \boldsymbol{\omega} \cdot \mathsf{I}_O \boldsymbol{\omega},$$

once again using Problem 2 on page 303 to show that the bracketed terms are nothing more than a matrix formulation of the Parallel Axis Theorem transferring the origin from the center of mass to O.

In developing the equations of motion, we noted at this point that, for radially-symmetric spherical bodies moving over a fixed, flat plate, we could employ the "trick" of summing moments about the contact point, treating it as if it were a fixed point about which the body rotates. We now show that, somewhat more generally, the same device can be applied to energy. Again, examination of how the *general* expression for energy, equation (5.4-2a), reduces to (5.4-2b) shows that this rests on the fact that $\overline{\boldsymbol{v}} = \boldsymbol{\omega} \times \overline{\boldsymbol{r}}$. But, *if the surface is fixed*, this is precisely the velocity non-slip condition (4.4-13), since, due to the mass symmetry, $\overline{\boldsymbol{v}} = \boldsymbol{v}_O$, the velocity of the *geometrical* center O, and $\overline{\boldsymbol{r}} = \boldsymbol{r}_{O/C'}$.

In sum, we have the kinetic energy expressed

$$\begin{aligned} T &= \frac{1}{2}m\overline{v}^2 + \frac{1}{2}\boldsymbol{\omega}\cdot\bar{\mathsf{I}}\boldsymbol{\omega} \quad \text{(completely general)} \qquad (5.4\text{-}2a)\\ &= \frac{1}{2}\boldsymbol{\omega}\cdot\mathsf{I}_O\boldsymbol{\omega} \quad (\textit{only} \text{ if rotating about } \textit{fixed } O) \quad (5.4\text{-}2b) \end{aligned}$$

Work

Recall that the *work* done by a *force* is defined through a *differential* quantity, equation (5.1-8):

$$dU \equiv \sum_i \boldsymbol{F}_i \cdot d\boldsymbol{r},$$

where $d\boldsymbol{r}$ is the *absolute* differential translational displacement of the point at which $\boldsymbol{F}_i$ acts; certainly the same definition would hold for work done by forces on a rigid body. That's all well and good when talking about particles, whose only motion *is* translational, and thus on which only forces show any effect, but on an extended rigid body, *couple* moments can also act, since the pair of forces generating such a couple can act over a non-vanishing moment arm. What, then, would be the work done by a couple?

Though, having introduced force couples, one immediately abstracts their effect to the "pure twist" of a couple, since work is *defined* in terms of forces, it makes sense to return to the basic definition of a couple as a pair of opposing forces of equal magnitude with non-vanishing moment arm.

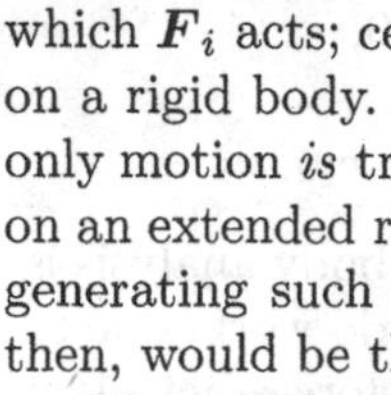

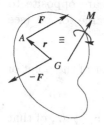

Thus, in determining the force due to a couple $\boldsymbol{M}$, we reintroduce a pair of forces, $\boldsymbol{F}$ and $-\boldsymbol{F}$ positioned with relative points of application separated by $\boldsymbol{r}$ such that $\boldsymbol{M} = \boldsymbol{r} \times \boldsymbol{F}$; the actual values of $\boldsymbol{F}$ and $\boldsymbol{r}$ are unimportant as long as they generate $\boldsymbol{M}$. In particular, so are their points of application on the body, since the couple moment's effect is independent of its placement; thus we will choose to place $-\boldsymbol{F}$ at the center of mass, G, with $\boldsymbol{r}$ being measured from that point to the other force's position A.

Using the [force] definition of work, the net work done by these two forces is $dU \equiv \boldsymbol{F} \cdot d\boldsymbol{r}_A + (-\boldsymbol{F}) \cdot d\boldsymbol{r}_G$. Now $d\boldsymbol{r}_A = d\boldsymbol{r}_G + d\boldsymbol{r}_{A/G}$, but we can get an expression for the relative displacement $d\boldsymbol{r}_{A/G}$ by invoking (4.1-9): $d\boldsymbol{r}_{A/G} = d\boldsymbol{\theta} \times \boldsymbol{r}$, where $d\boldsymbol{\theta}$ is the infinitesimal rotation of $\boldsymbol{r}$, here that of the body as a whole. Thus

$$\begin{aligned} dU &\equiv \boldsymbol{F} \cdot d\boldsymbol{r}_A + (-\boldsymbol{F}) \cdot d\boldsymbol{r}_G = \boldsymbol{F} \cdot (d\boldsymbol{r}_G + d\boldsymbol{\theta} \times \boldsymbol{r}) + (-\boldsymbol{F}) \cdot d\boldsymbol{r}_G \\ &= \boldsymbol{F} \cdot d\boldsymbol{\theta} \times \boldsymbol{r} = -\boldsymbol{F} \cdot \boldsymbol{r} \times d\boldsymbol{\theta} = -\boldsymbol{F} \times \boldsymbol{r} \cdot d\boldsymbol{\theta} \\ &= +\boldsymbol{r} \times \boldsymbol{F} \cdot d\boldsymbol{\theta} \equiv \boldsymbol{M} \cdot d\boldsymbol{\theta} \end{aligned}$$

Thus, corresponding to the [definition of the] work done by a *force* under a *translation*, we see that the work done by a *couple* under a *rotation* is simply

$$dU = \boldsymbol{M} \cdot d\boldsymbol{\theta}. \tag{5.4-3}$$

Note that θ here must be measured in [unitless] *radians*.

We now know how to calculate the work done by both forces and couples, but which of those forces and couples acting on a rigid body actually *do* work? We shall answer this question in the negative giving a brief (though hardly exhaustive) list of those which *don't*:

- *internal forces and moments*: For an action/reaction pair of *couples*, this is immediate: $dU = \boldsymbol{M} \cdot d\boldsymbol{\theta} + (-\boldsymbol{M}) \cdot d\boldsymbol{\theta} \equiv 0$ because the $d\boldsymbol{\theta}$ is the *same* for both. And for such a pair of *forces*, acting at, say, $\boldsymbol{r}_1$ and $\boldsymbol{r}_2$,

  $$\begin{aligned} dU &= \boldsymbol{f} \cdot d\boldsymbol{r}_1 + (-\boldsymbol{f}) \cdot d\boldsymbol{r}_2 = \boldsymbol{f} \cdot (d\boldsymbol{r}_1 - d\boldsymbol{r}_2) \\ &= \boldsymbol{f} \cdot d(\boldsymbol{r}_1 - \boldsymbol{r}_2) = \boldsymbol{f} \cdot d\boldsymbol{\theta} \times (\boldsymbol{r}_1 - \boldsymbol{r}_2) \equiv 0, \end{aligned}$$

  since the last is a triple scalar product in which $\boldsymbol{f} \parallel (\boldsymbol{r}_1 - \boldsymbol{r}_2)$.

  Contrast this with the systems of particles on which rigid-body analysis is based: for systems, both external *and* internal forces can do work, due to the independent motion possible for points at which the forces act. But for rigid bodies this is precluded.

- *frictionless machine connections*: Forces on the "ball" are opposite to those on the members it connects, and the $d\boldsymbol{r}$ is the same for all (see page 312).

- *normal forces*: They are *perpendicular* to the displacement of the point of contact.

- *friction imposing a non-slip condition at a fixed surface*: $\boldsymbol{v}$, so $d\boldsymbol{r}$, of that point vanishes altogether (equation (4.4-14b)).

This greatly simplifies the problem by reducing the number of forces which need to be considered.

Energy Principles

From here on out, everything we have said regarding energy, its conservation, and the generalized principle of work and energy, equation (5.1-14)

$$(U_N)_{1\to 2} = E_2 - E_1,$$

holds in the case of rigid bodies: if all forces doing work are *conservative*—the criteria for which being given by equation (5.1-12), then E is conserved, the *work* done these forces being accounted for in the *potential energy* V in the *total energy* $E \equiv T + V$.

Since we are now going to make use of energy, it is appropriate to make several remarks regarding it:

- This is a *generally* integrated form of the original Newton equations for a particle, now generalized to rigid bodies; thus, unlike the latter's *instantaneous* nature, these hold over an *interval* of time. In particular, if E is conserved, that *function* has the constant *value* over the interval.
- One need account only for those forces *actually doing work*, be they in the work term or potential energy, *over the entire interval.*
- It is a *scalar* formulation; this has both good and bad points:
  - Only a *single* scalar unknown can be found; in particular, absent any other equations, it can be applied only to "single degree of freedom" systems—those in which the entire orientation of the system can be specified with a single variable (that's maybe *bad*).
  - $\overline{v}$, in particular, is the *speed* of the center of mass: $\overline{v}^2 = \overline{v}_x^2 + \overline{v}_y^2 + \overline{v}_z^2$; the same considerations apply to ω. *It is not possible to "break out" components of this* as it is with *vector* formulations! (*Very* bad if you try it!) And *directions* can't be found, either.
  - As a scalar, its expression is *independent of the vector coordinate system in terms of which forces and velocities are specified*; in particular, it is admissible to "mix" coordinates in terms of which we express the various quantities (that's *good*).
  - In line with the above observation, $\overline{T} = \frac{1}{2}\boldsymbol{\omega} \cdot \overline{\mathsf{I}}\boldsymbol{\omega}$, *despite the vector nature of its constituents*, can be expressed in *any* basis (that's *very* good).
- It is a relation between *speed*, linear and angular, (through T) and *position* (through U and/or V); this distinguishes it from the equations of motion which relate forces and accelerations. Now that we are beginning to build a repertoire of principles to solve problems, observations such as this can often guide one in the choice of which to use.

The observant reader will recall that, in the discussion of systems of particles (on which the analysis of rigid bodies is based), we also referred kinetic energy to center-of-mass coordinates. In particular, we obtained work-[kinetic] energy relations (5.1-22d) and (5.1-22e):

$$\int_{\overline{r}_1}^{\overline{r}_2} \sum_i \boldsymbol{F}_i \cdot d\overline{\boldsymbol{r}} = \frac{1}{2}m\overline{v}_2^2 - \frac{1}{2}m\overline{v}_1^2,$$

$$\sum_i \int_{\rho_{i_1}}^{\rho_{i_2}} (\boldsymbol{F}_i + \sum_j \boldsymbol{f}_{ij}) \cdot d\boldsymbol{\rho}_i = \overline{T}_2 - \overline{T}_1.$$

But we have already argued that the internal forces $\boldsymbol{f}_{ij}$ *do* no work in a rigid body; thus, in that case, this reduces to just

$$\sum_i \int_{\rho_{i_1}}^{\rho_{i_2}} \boldsymbol{F}_i \cdot d\boldsymbol{\rho}_i = \overline{T}_2 - \overline{T}_1.$$

The center-of-mass formulation has been fundamental to the equations of motion, yet we seem to ignored the above relations altogether. We now correct that apparent oversight:

It is, in principle, certainly possible to reformulate all the above discussion regarding the energy of a rigid body in terms of these relations. The first is effectively the energy principle for a *particle*—the center-of-mass particle—on which *all* the applied forces act, and separates out the *translational* motion of the rigid body. The last, on the other hand, is a measure of purely *rotational* motion: $\overline{T} = \frac{1}{2}\boldsymbol{\omega} \cdot \overline{\mathsf{I}}\boldsymbol{\omega}$. So far, so good. But now consider the nature of the *work* terms in the above relations:

$d\overline{\boldsymbol{r}}$ is the *absolute* displacement of the center of mass. $d\boldsymbol{\rho}_i$, on the other hand, measures the displacement of the point at which $\boldsymbol{F}_i$ is applied *relative to the center of mass*; this, too, will only occur if the body *rotates*: by equation (4.1-9), $d\boldsymbol{\rho}_i = d\boldsymbol{\theta} \times \boldsymbol{\rho}_i$. We consider a couple of simple examples to illustrate the differences between these and the previous "absolute" work:

1. *a flat plate drawn over a horizontal plane with friction*: Here *both* forces, $\boldsymbol{P}$ and $\boldsymbol{F}_f$, appear in $\sum \boldsymbol{F} \cdot d\boldsymbol{r}$, but *neither* enters $\sum \boldsymbol{F} \cdot d\boldsymbol{\rho}$ because there is no rotation!

2. *a cylinder drawn without slipping on a horizontal plane*: Here only $\boldsymbol{P}$ appears in $\sum \boldsymbol{F} \cdot d\boldsymbol{r}$ because of the non-slip at the point of contact, but only $\boldsymbol{F}_f$—which actually *generates* the rotation—enters $\sum \boldsymbol{F} \cdot d\boldsymbol{\rho}$ because $\boldsymbol{P}$ passes *through* the center of mass.

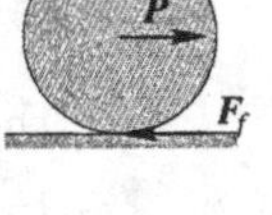

In *both* cases, however, *both* forces enter $\sum \boldsymbol{F} \cdot d\overline{\boldsymbol{r}}$!

With this background in hand, it is possible to apply all the above principles including the concept of "conservative" forces doing work, now limited to the above considerations. There is some minor advantage to the translational form, but one must involve *all* the forces, whereas one appealing feature of energy is the ability to *eliminate* forces! The rotational work-energy form, however, seems to be little used in practice.

We now provide a couple of examples of the use of energy:

Example 5.4.1 *A Simple Example—Single Rigid Body.* Consider a *thin* uni-

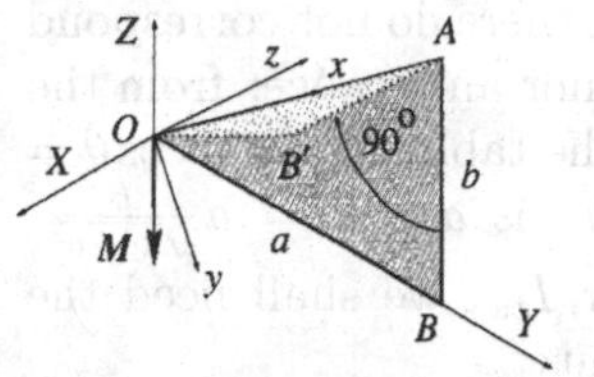

form right-triangular plate of mass m and sides a and b (OA the hypotenuse!) as shown, free to pivot about the fixed OA-axis relative to the vertical Z-axis. It is released from rest in a position at which the edge AB has been rotated 90° about the OA-axis upward from the vertical; during the time it falls, a constant couple moment $\boldsymbol{M} = M(-\hat{\boldsymbol{K}})$ is applied at O. We are asked to determine its angular velocity when it passes the vertical position shown.

This is clearly an instance in which energy would be useful: we are asked to determine its angular *velocity* after its position has changed, but this "system" has only a single degree of freedom—its complete orientation can be specified knowing only the angle it assumes about the OA-axis—and thus, up to its sense, the velocity is determined by its *speed*. In this case, we are interested in the *value* of the energy at the endpoints of the interval.

The forces and moments acting on it are whatever constrains it along the OA-axis, its weight, and the moment. But the first does no work, since there is no displacement of its hypotenuse, leaving only the weight and the couple. The first of these is known to be conservative: $V = mg\overline{Z}$, but we're not sure about the couple. In fact, since the criteria for "conservative" applies only to *forces*, the issue becomes moot; thus we choose to put only the work done by gravity into a potential, leaving that done by the couple in a work term:

$$U_{1\to 2}(\boldsymbol{M}) = E_2 - E_1,$$

in which $E = T + V = T + mg\overline{Z}$. This leaves us with T to determine; since the body rotates about the fixed point O, we shall use $T = \frac{1}{2}\boldsymbol{\omega} \cdot \mathbf{I}_O\boldsymbol{\omega}$.

Such "thin triangular plates" are not necessarily common in tables of moments of inertia, but we have craftily determined the inertia matrix for such a plate about the origin in Homework Problem 2 on page 277; if they were not given, we would have to determine them from first principles as in that problem. We note that the axes given there are not principal, or even through the center of mass; but as noted above, this does not matter: we can use *any* convenient system because T is a scalar. We're given the appropriate matrix in that set, so they qualify as "convenient." Noting that, the first impulse is simply to start writing, filling in the entries in the inertia matrix and start calculating. But a little foresight can save us a great deal of work:

We note in the first place that, in the chosen axes, $\boldsymbol{\omega}$ is represented $\boldsymbol{\omega} = \omega\hat{\boldsymbol{\imath}}$, so ω_y and ω_z both vanish. This means that the only terms in the inertia matrix which survive the multiplication $\mathbf{I}_O\boldsymbol{\omega}$ are those in the first column, so all we need are I_{xx}, I_{xy}, and I_{xz}: $\mathbf{I}_O\boldsymbol{\omega} = I_{xx}\omega_x\hat{\boldsymbol{\imath}} + I_{xy}\omega_x\hat{\boldsymbol{\jmath}} + I_{xz}\omega_x\hat{\boldsymbol{k}}$. But, when we form the dot product with $\boldsymbol{\omega}$, the only term which remains after *this* is $I_{xx}\omega_x^2$. Thus all we really need is I_{xx}, and $T = \frac{1}{2}I_{xx}\omega^2$!

We're not out of the woods yet, however: as seems usual with tables, the

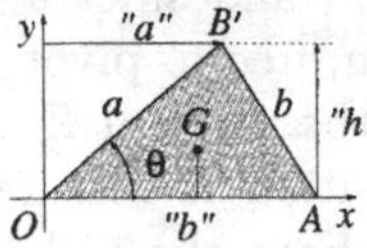

parameters "a," "b," and "h" given there do not correspond to those here. But that's just a minor annoyance: from the diagram at left we see that "a" in the tables is just $a\cos\theta = a\frac{a}{\sqrt{a^2+b^2}}$, "$b$" is $\sqrt{a^2+b^2}$, and "h" is $a\sin\theta = a\frac{b}{\sqrt{a^2+b^2}}$. (Though only "$h$" is necessary for I_{xx}, we shall need the others to find $\overline{Z}$ in the potential energy presently.) Thus

$$E = T + V = \frac{1}{2}I_{xx}\omega^2 + mg\overline{Z} = \frac{1}{12}\frac{ma^2b^2}{a^2+b^2}\omega^2 + mg\overline{Z}.$$

Now we must find the work due to $\boldsymbol{M}$: by equation (5.4-3), and noting that $d\boldsymbol{\phi} = (\cos\theta\hat{\boldsymbol{J}} + \sin\theta\hat{\boldsymbol{K}})\,d\phi$ as the plate falls, we have that

$$dU(\boldsymbol{M}) = \boldsymbol{M}\cdot d\boldsymbol{\phi} = M(-\hat{\boldsymbol{K}})\cdot(\cos\theta\hat{\boldsymbol{J}} + \sin\theta\hat{\boldsymbol{K}})\,d\phi = -M\sin\theta\,d\phi$$

(using "$d\phi$" so as not to confuse with "θ" above), in which M (and θ!) are *constant*. Thus, finally, we have that

$$U_{1\to2}(\boldsymbol{M}) = -\frac{Mb}{\sqrt{a^2+b^2}}\frac{\pi}{2}.$$

All that's left is to calculate are the two values of $\overline{Z}$ for the potential energy: $\overline{Z}_1$ is the height of the centroid when the plate is rotated up, $\overline{Z}_2$ when it reaches its final position. Though we could find $\overline{\boldsymbol{r}}$ in the (x, y, z) coordinates and then use rotation matrices to find its Z-components, we will use a more elementary means by observing that the difference in vertical heights of the center of mass is simply $\overline{y}\cos\theta = \frac{\text{“}h\text{”}}{3}\cos\theta$—"$\overline{y}$" and "$\theta$" the same as in the above Table diagram, the latter also being the angle between OA and the horizontal:

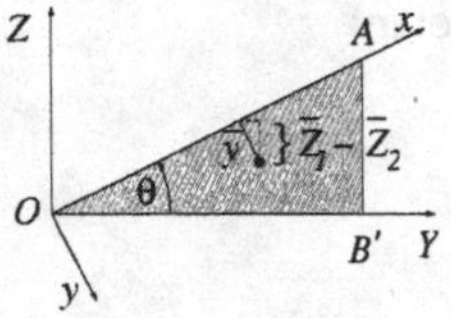

$$\overline{Z}_1 - \overline{Z}_2 = \frac{\text{“}h\text{”}}{3}\cos\theta = \frac{ab}{3\sqrt{a^2+b^2}}\frac{a}{\sqrt{a^2+b^2}} = \frac{a^2b}{3(a^2+b^2)}$$

(note the sign).

Finally, substituting into the energy equation,

$$\begin{aligned}
U_{1\to2}(\boldsymbol{M}) &= -\frac{Mb}{\sqrt{a^2+b^2}}\frac{\pi}{2}\\
&= E_2 - E_1 = \left(\frac{1}{12}\frac{ma^2b^2}{a^2+b^2}\omega_2^2 + mg\overline{Z}_2\right) - (0 + mg\overline{Z}_1)\\
&= \frac{1}{12}\frac{ma^2b^2}{a^2+b^2}\omega_2^2 - mg\frac{a^2b}{3(a^2+b^2)}
\end{aligned}$$

or

$$\omega_2 = \sqrt{\frac{12(a^2+b^2)}{ma^2b^2}\left(mg\frac{a^2b}{3(a^2+b^2)} - \frac{\pi Mb}{2\sqrt{a^2+b^2}}\right)}.$$

Note that this places a condition on M: it must be small enough so that the radicand is non-negative, a zero value corresponding to the maximum value allowing the plate to fall completely. Furthermore, we cannot determine the *sense* in which the plate is rotating from this analysis, since energy involves only the speeds—the *magnitudes* of the desired velocities—rather than their vector components. But this can be determined from "intuition" (hand-waving).

Although we presumed $\boldsymbol{M}$ known and determined $\boldsymbol{\omega}$, this same approach could also be used to find the $\boldsymbol{M}$ required to raise the plate from rest: now $\omega_2 = 0$ would be known, and M unknown, but still determinable from a single equation.

As much as anything else, this example demonstrates how to coerce tables into doing what you want them to! |

The above example was a relatively simple one in which there was only a single body whose direction of rotation was known. In the next one, we consider a *machine* in which the rotations are *not* known:

Example 5.4.2 *A Machine, and Kinematical Considerations.* A pair of uniform slender rods, AB of length 310 mm, and BC of length 260 mm, and of masses—we shall assume them to be of the same stock—say, .31 and .26 kg, respectively, are connected together by a "small" 50-g collar at B. Point A is fixed in the YZ-plane, and B supported by the horizontal XY-plane. End C has another 50-g collar sliding along a spiral track whose radius is given by $r = \frac{.24}{\pi}\theta$ m, where the angle is measured from the negative Z-axis as shown. Applied to the collar is a force $\boldsymbol{F}$ of magnitude $F = \|\boldsymbol{F}\|$, directed radially outward from O, and to AB a couple $\boldsymbol{M} = 2\hat{\boldsymbol{K}}$ N·m. If the collar C is moving with a speed $v_C = 5$ m/s when C is on the X-axis $(\theta = \frac{\pi}{2})$, we are asked to find F to just bring that collar to the positive Z-axis $(\theta = \pi)$, ignoring any friction.

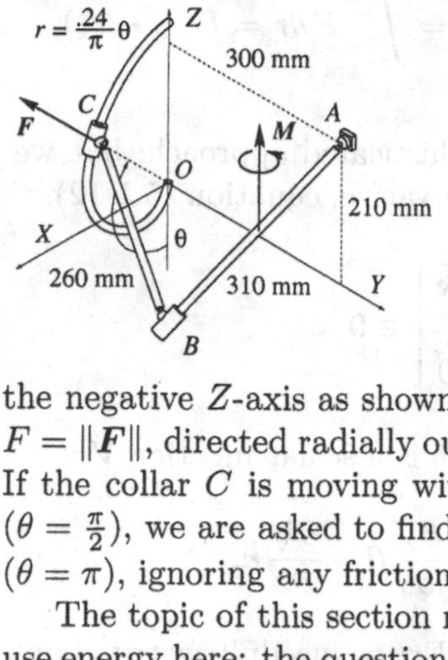

The topic of this section notwithstanding, it is fairly clear that we desire to use energy here: the question asks to find a relation between speed and position: we know v_C initially and wish v_C to vanish at the end, and we are once again interested in the *values* of the energy at its endpoints. Furthermore, the system has only a single degree of freedom: since the lengths of the rods are fixed, once we know the position of C—and thus the length AC—the size of the resulting triangle is known, and its orientation is determined by C and the fact that B must lie on the given plane. What is not so obvious is that we shall consider the energy of the *system* of the two bodies together:

The agents doing work are the weights, $\boldsymbol{F}$, and $\boldsymbol{M}$: the support at A does none because that point is fixed, while that acting on the collar C due to the spiral rod does none because it is always perpendicular to $d\boldsymbol{r}_C$—at least as long as the collar doesn't break the spiral rod! Nor, despite the fact it moves, is work done by the ball-and-socket forces at B, since the opposite forces it exerts on each rod move through the same $d\boldsymbol{r}_B$; the desire to "internalize" this work is precisely the reason we considered the system as a whole. We note further, however, that weight does no work on AB, since its center of mass moves perpendicularly to the weight, remaining at a constant height of 210 mm above the horizontal plane; similarly, the collar at B remains at $Z_B \equiv 0$. Thus we consider only the weight acting on BC and the collar at C, choosing to put them into a potential energy [of the system]. The couple moment isn't a force at all, so we put it in a work term.

This leaves us with the issue of how to deal with $\boldsymbol{F}$ which, although constant in magnitude, is not constant in direction. We actually have a couple of options, depending on whether we account for the work due to $\boldsymbol{F}$ as:

- *work*: Since $\boldsymbol{F} = F\hat{\boldsymbol{r}}$ in polar coordinates, we can find the "line element" $d\boldsymbol{r} = \frac{d\boldsymbol{r}}{dt}dt = dr\hat{\boldsymbol{r}} + r\,d\theta\hat{\boldsymbol{\theta}}$:

$$U_{1\to 2}(\boldsymbol{F}) \equiv \int_{r_1}^{r_2} \boldsymbol{F}\cdot d\boldsymbol{r}$$
$$= \int_{r_1}^{r_2} F\hat{\boldsymbol{r}}\cdot(dr\hat{\boldsymbol{r}} + r\,d\theta\hat{\boldsymbol{\theta}}) = \int_{r_1}^{r_2} F\,dr = F(r_2 - r_1).$$

- *potential energy*: This is a slightly more sophisticated approach, but we note that the criterion for a force to be conservative, equation (5.1-12),

$$\nabla\times\boldsymbol{F} = \frac{1}{r}\begin{vmatrix} \hat{\boldsymbol{r}} & r\hat{\boldsymbol{\theta}} & \hat{\boldsymbol{k}} \\ \frac{\partial}{\partial r} & \frac{\partial}{\partial \theta} & \frac{\partial}{\partial z} \\ F & 0 & 0 \end{vmatrix} \equiv \boldsymbol{0}$$

  is satisfied. Thus, by the same criterion, there is a scalar function V:

$$\boldsymbol{F} = F\hat{\boldsymbol{r}} = -\nabla V = -\frac{\partial V}{\partial r}\hat{\boldsymbol{r}} - \frac{1}{r}\frac{\partial V}{\partial \theta}\hat{\boldsymbol{\theta}} - \frac{\partial V}{\partial z}\hat{\boldsymbol{k}},$$

  from which we get immediately that $V = -Fr$; we needn't worry about any additive constant, since we consider only the *difference* of any potential energy.

Note that, in the first case we will have a $+F(r_2 - r_1)$ on the work side, in the second we will get, in $E_2 - E_1$, $-F(r_2 - r_1)$ on the energy side; the approaches are equivalent. We chose the second option in the following, so that, including the potential energy of the collar,

$$V = m_{BC}g\overline{Z}_{BC} + m_C g Z_C - Fr.$$

Observe that we use three different coordinates, $\overline{Z}_{BC}$, Z_C, and r, to measure the potential energy; this is an instance of the "mixing" of coordinates possible with the scalar principle of energy mentioned above in the notes.

This leaves us with T to find: Since AB moves about the fixed point A while

BC doesn't, we express the *system*

$$\begin{aligned} T &= T_{AB} + T_{BC} + T_C \\ &= \frac{1}{2}\boldsymbol{\omega}_{AB} \cdot \mathsf{I}_{AB_A}\boldsymbol{\omega}_{AB} + \\ &\left(\frac{1}{2}\boldsymbol{\omega}_{BC} \cdot \overline{\mathsf{I}}_{BC}\boldsymbol{\omega}_{BC} + \frac{1}{2}m_{BC}\overline{v}_{BC}^2\right) + \frac{1}{2}m_C v_C^2, \end{aligned} \tag{5.4-4}$$

where we choose to leave the "small collar" at B part of body AB itself. (Note that "small" bodies—*particles*—such as the collar at C have no rotational kinetic energy, only translational. Putting collar B in body AB does not violate this; it merely "reconstructs" the kinetic energy of the particle B from the inertia matrix of the rigid body AB when the latter is multiplied by $\boldsymbol{\omega}$!) But we don't know *either* angular velocity, suggesting a kinematic analysis. And there is the further problem to deal with that, in such analysis, we will only be able to determine that component of angular velocity *perpendicular* to each rod, $\boldsymbol{\omega}_\perp$; the parallel component $\boldsymbol{\omega}_\parallel$ is indeterminate (see page 237). While the first has its own wrinkles, the second can be dispatched immediately: we note that the moment of inertia *along* the rod—that which would multiply $\boldsymbol{\omega}_\parallel$ in principal axes—itself vanishes, so that component of angular velocity never even enters the energy. This suggests that it might be necessary explicitly to *find* the principal axes to evaluate the kinetic energy, but, carrying this principal axis concept one step further, we note that we can *imagine* that we have obtained these with z along the rod and the perpendicular principal x-axis *along* $\boldsymbol{\omega}_\perp$; then the expression for kinetic energy reduces to simply $\frac{1}{2}I_{xx}\omega_\perp^2$. Once again we see how helpful the scalar nature of energy becomes: we can, and here *do*, utilize entirely different bases in the evaluation of the I's and $\boldsymbol{\omega}$'s!

We are finally ready to calculate. We start with the two rods' [*scalar*] moments of inertia about a perpendicular axis: since the collar B is attached to AB, and noting that a point ("small") mass has moment of inertia mr^2—this was the original formulation of the inertia tensor, recall—we get that

$$\begin{aligned} I_{AB_A} &= \frac{1}{3}(0.31)(0.31)^2 + (0.05)(0.31)^2 = 1.474 \times 10^{-2}\,\text{kg·m}^2 \\ \overline{I}_{BC} &= \frac{1}{12}(0.26)(0.26)^2 = 1.465 \times 10^{-3}\,\text{kg·m}^2. \end{aligned}$$

These are necessary to find T_1; but so is quite a bit more, since we don't know the angular velocities at this point. But we do have the kinematical constraints relating the motions of A, B, and C:

$$\boldsymbol{v}_B = \boldsymbol{v}_A + \boldsymbol{\omega}_{AB} \times \boldsymbol{r}_{B/A} + \mathbf{0} = \boldsymbol{\omega}_{AB} \times \boldsymbol{r}_{B/A} \tag{5.4-5a}$$

$$\boldsymbol{v}_C = \boldsymbol{v}_B + \boldsymbol{\omega}_{BC} \times \boldsymbol{r}_{C/B} + \mathbf{0} \tag{5.4-5b}$$

as well as the constraints on the angular velocities arising from the ball-and-socket connections:

$$0 \equiv \boldsymbol{\omega}_{AB} \cdot \boldsymbol{r}_{B/A} \tag{5.4-5c}$$

$$0 \equiv \boldsymbol{\omega}_{BC} \cdot \boldsymbol{r}_{C/B}. \tag{5.4-5d}$$

[Even though the angular velocities will each have three components, the last two conditions ensure that these will be perpendicular to the corresponding positions.] We note that $v_{B_Z} \equiv 0$; and $\boldsymbol{v}_{C1} = v_{C1}\hat{\boldsymbol{v}}_{C_1}$, where we can find the direction $\hat{\boldsymbol{v}}_{C_1}$ from the polar coordinate representation of the spiral in the XZ-plane:

$$\boldsymbol{v}_{C_1} = \dot{r}_1\hat{\boldsymbol{r}}_1 + r_1\dot{\theta}_1\hat{\boldsymbol{\theta}}_1 = \frac{.24}{\pi}\dot{\theta}_1\hat{\boldsymbol{r}}_1 + (.12)\dot{\theta}_1\hat{\boldsymbol{\theta}}_1 = .12\dot{\theta}_1\left(\frac{2}{\pi}\hat{\boldsymbol{I}} + (1)\hat{\boldsymbol{K}}\right),$$

so

$$\hat{\boldsymbol{v}}_{C_1} = \frac{\frac{2}{\pi}\hat{\boldsymbol{I}} + \hat{\boldsymbol{K}}}{\sqrt{\left(\frac{2}{\pi}\right)^2 + 1}} = 0.5370\hat{\boldsymbol{I}} + 0.8436\hat{\boldsymbol{K}}.$$

Thus, since $v_{C1} = 5\,\mathrm{m/s}$, $\boldsymbol{v}_{C_1} = v_{C_1}\hat{\boldsymbol{v}}_{C_1} = 2.685\hat{\boldsymbol{I}} + 4.218\hat{\boldsymbol{K}}\,\mathrm{m/s}$.

But we still need to know the two $\boldsymbol{r}$'s. Using the same approach we did in Example 4.4.7 to find the position of B (assumed to have positive X-components), we get that $\boldsymbol{r}_{B_1} = 0.22\hat{\boldsymbol{I}} + 0.24\hat{\boldsymbol{J}}\,\mathrm{m}$ so that $\boldsymbol{r}_{B/A} = 0.22\hat{\boldsymbol{I}} - 0.06\hat{\boldsymbol{J}} - 0.21\hat{\boldsymbol{K}}\,\mathrm{m}$ while $\boldsymbol{r}_{C/B} = -0.10\hat{\boldsymbol{I}} - 0.24\hat{\boldsymbol{J}}\,\mathrm{m}$. Substituting these values and the $\boldsymbol{v}_C$ found before into the above equations (5.4-5), we get ultimately the set corresponding to those:

$$\begin{aligned} v_{B_X} &= -.21\omega_{AB_Y} + .06\omega_{AB_Z} \\ v_{B_Y} &= .21\omega_{AB_X} + .22\omega_{AB_Z} \\ 0 &= -.06\omega_{AB_X} - .22\omega_{AB_Y} \end{aligned} \tag{5.4-6a}$$

$$\begin{aligned} 2.685 &= v_{B_X} + .24\omega_{BC_Z} \\ 0 &= v_{B_Y} - 0.1\omega_{BC_Z} \\ 4.218 &= -.24\omega_{BC_X} + 0.1\omega_{BC_Y} \end{aligned} \tag{5.4-6b}$$

$$0 = .22\omega_{AB_X} - .06\omega_{AB_Y} - .21\omega_{AB_Z} \tag{5.4-6c}$$

$$0 = -0.1\omega_{BC_X} - .24\omega_{BC_Y}. \tag{5.4-6d}$$

Solving [we don't bother checking rank and all, just seeing if a solution is possible], we get

$$\begin{aligned} \boldsymbol{\omega}_{AB} &= 2.195\hat{\boldsymbol{I}} - 0.5987\hat{\boldsymbol{J}} + 2.471\hat{\boldsymbol{K}}\,\mathrm{s}^{-1} \\ \boldsymbol{\omega}_{BC} &= -14.97\hat{\boldsymbol{I}} + 6.239\hat{\boldsymbol{J}} + 10.05\hat{\boldsymbol{K}}\,\mathrm{s}^{-1} \\ \boldsymbol{v}_B &= 0.2740\hat{\boldsymbol{I}} + 1.005\hat{\boldsymbol{J}}\,\mathrm{m/s}, \end{aligned}$$

allowing us to find the requisite

$$\overline{\boldsymbol{v}}_{BC} = \boldsymbol{v}_B + \boldsymbol{\omega}_{BC} \times \overline{\boldsymbol{r}}_{BC} = 1.480\hat{\boldsymbol{I}} + 0.5023\hat{\boldsymbol{J}} + 2.109\hat{\boldsymbol{K}}\,\mathrm{m/s},$$

which, when substituted into equation (5.4-4), give $T_1 = 1.8703\,\mathrm{J}$.

After all that, V_1 is easy: since $\overline{Z}_{BC} = Z_C \equiv 0$ initially, the only contribution to that term is the potential energy due to $\boldsymbol{F}$, and $V_1 = -.12F$. Thus, finally,

$$E_1 = T_1 + V_1 = 1.8703 - .12F\,\mathrm{J}.$$

E_2 is essentially trivial: since nothing is moving, $T_2 \equiv 0$, $\overline{Z}_{BC} = .12\,\mathrm{m}$, $Z_C = .24\,\mathrm{m}$, and "r" in the potential energy of $\boldsymbol{F}$ is $.24\,\mathrm{m}$, so

$$E_2 = T_2 + V_2 = 0 + (((0.12)(0.26) + (0.24)(0.05))\,g - (0.24)F) = 0.4238 - .24F.$$

Finally, we have the effect of the moment to find. Since

$$dU_{1\to 2}(\boldsymbol{M}) = \boldsymbol{M} \cdot d\boldsymbol{\theta}_{AB} = M\hat{\boldsymbol{K}} \cdot (d\theta_{AB_X}\hat{\boldsymbol{I}} + d\theta_{AB_Y}\hat{\boldsymbol{J}} + d\theta_{AB_Z}(-\hat{\boldsymbol{K}}))$$

where M is constant, $U_{1\to 2} = -M\,\Delta\theta_{AB_Z}$ and we need $\Delta\theta_{AB_Z}$. But this is the angle between the projections of $\boldsymbol{r}_{B/A}$ in the XY-plane $\boldsymbol{t}_1 = 0.22\hat{\boldsymbol{I}} - 0.06\hat{\boldsymbol{J}}$ and $\boldsymbol{t}_2 = 0.06\hat{\boldsymbol{I}} - 0.22\hat{\boldsymbol{J}}$; solving

$$\boldsymbol{t}_1 \cdot \boldsymbol{t}_2 = 2.640 \times 10^{-2} = \|\boldsymbol{t}_1\|\|\boldsymbol{t}_2\| \cos\Delta\theta_{AB_Z} = (2.280 \times 10^{-1})^2 \Delta\theta_{AB_Z},$$

we get finally that $\Delta\theta_{AB_Z} = 1.455\,[\mathrm{rad}]$, so that $U_{1\to 2} = -2.910\,\mathrm{N{\cdot}m}$.

The generalized work-energy equation thus becomes

$$\begin{aligned} U_{1\to 2} = -2.910 = E_2 - E_1 &= (0.4238 - .24F) - (1.8703 - .12F) \\ &= -1.447 - .12F, \end{aligned}$$

solving which we get $F = 19.12\,\mathrm{N}$.

This example has been rather more "busy" than normal: in addition to the energy principle it was meant to illustrate, there were also details regarding the kinetic energy of slender rods, a reminder regarding the moment of inertia of point masses, determination of the direction of motion from a path, calculation of the relevant angle in the moment's work term, and ultimately the kinematics necessary to reduce the various variables in a single-degree-of-freedom system to a single independent one. Though the last *is* a general technique often required in even single-body problems, one should not lose sight of the basic point regarding the implementation of energy; the rest *are* just details! |

The above examples have utilized the *values* of the energy at its endpoints: finding information about a system's motion *after* it has changed its orientation. While this is its typical application, there is another useful aspect to energy: the fact that it is an *already*-integrated form of the equations of motion, whether this integration is done numerically or analytically, and that this integrated form—a *function*, remember—holds over the interval in question. In particular, then, this means that what start out as generally *second*-order equations of motion will, upon integration to an energy form, become *first*-order ones involving $\boldsymbol{v}$ and $\boldsymbol{\omega}$. And one would much rather integrate the latter than the former![27]

[27] Note, however, that solution for $\boldsymbol{v}$ or $\boldsymbol{\omega}$ will invariably involve a square root, making the resulting first-order equations non-linear.

There remains, of course, the fact that there is still only a *single* component to such an energy equation, and it is thus necessary to reduce everything to a *single* variable to be found from this procedure. This important aspect is exemplified presently in Example 5.4.5, regarding general motion of the gyroscope.

Summary. Energy, a generally-integrated form of the equations of motion which relates *speed* and *position*, is as useful in three-dimensional systems—perhaps even *more* useful—as it is in planar problems: the fact that it is a *scalar* formulation makes it very forgiving of both signs and coordinate systems, and the elimination of forces which do no work greatly reduces the number to be considered. At the same time, its application is limited by the fact that it *is* only a scalar, so separate components of the motion cannot be found, and its ability to determine even speed is limited to single-degree-of-freedom systems in the absence of other equations. Furthermore, it is generally necessary to use a kinematical analysis to relate the various variables in such systems to a single one.

It is most often used in finding the *value* of various quantities at the end-points of the interval over which it is examined. But it can be used more generally to find the *first-order differential equations* involving the quantities entering the energy, which are at the root of this integrated form.

5.4.2 Momentum

In Section 5.1, we pointed out that the *impulse* was *defined* to be just the time-integrated form of the equations of motion. Applying this approach to the case of rigid bodies, we get immediately that the *linear impulse*

$$\int_{t_1}^{t_2} \sum_i \boldsymbol{F}_i \, dt = \boldsymbol{G}_2 - \boldsymbol{G}_1,$$

in which, introducing the linear momentum for a system/rigid body, equation (5.1-21a),

$$\boxed{\boldsymbol{G} \equiv m\overline{\boldsymbol{v}} \quad (5.4\text{-}7)}$$

There is only one such. But there are *two* forms of angular momentum, that relative to a fixed point O, $\boldsymbol{H}_O$, and that relative to the center of mass, $\overline{\boldsymbol{H}}$; integrating their equations of motion, (5.1-17b) and (5.1-22c), we get the respective *angular impulses*

$$\int_{t_1}^{t_2} \sum_i \overline{\boldsymbol{M}}_i \, dt = \overline{\boldsymbol{H}}_2 - \overline{\boldsymbol{H}}_1, \qquad \overline{\boldsymbol{H}} = \bar{\mathsf{I}}\boldsymbol{\omega}$$

$$\int_{t_1}^{t_2} \sum_i \boldsymbol{M}_{O_i} \, dt = \boldsymbol{H}_{O_2} - \boldsymbol{H}_{O_1}, \qquad \boldsymbol{H}_O = \overline{\boldsymbol{r}} \times m\overline{\boldsymbol{v}} + \bar{\mathsf{I}}\boldsymbol{\omega},$$

where we have introduced $\overline{\boldsymbol{H}} = \bar{\mathsf{I}}\boldsymbol{\omega}$ and (5.2-16) (see page 303). Note that the above expression for $\boldsymbol{H}_O$ is the *general* one; in the special case that the point O is a *fixed* point about which the body is rotating, this reduces, just as it has for the equations of motion and energy through the intermediary of the Parallel Axis Theorem, to simply $\mathsf{I}_O\boldsymbol{\omega}$: see Homework Problem 2 on page 303. Collecting these results for the expression of the angular momentum of rigid bodies:

$$\overline{\boldsymbol{H}} = \bar{\mathsf{I}}\boldsymbol{\omega} \quad \text{(general)} \quad (5.4\text{-}8a)$$

$$\boldsymbol{H}_O = \mathsf{I}_O\boldsymbol{\omega} \quad \text{(\textit{only} if rotating about \textit{fixed} } O) \quad (5.4\text{-}8b)$$

$$= \bar{\boldsymbol{r}} \times m\bar{\boldsymbol{v}} + \bar{\mathsf{I}}\boldsymbol{\omega} \quad \text{(\textit{completely} general)} \quad (5.4\text{-}8c)$$

We make, as usual, a couple of notes regarding the above impulse-momentum equations:

- Unlike the energy principles of the previous section, these are *vector* principles; thus rather than a single equation, we have in general *three* components to each.
- These relate the *vector velocity*—and, for angular impulse, the associated vector *angular velocity*—to the *vector* forces and/or moments and the *time*. To the extent we are interested in these quantities—or their *components*, this is the principle to use (an observation made, again, to guide its application).

As we noted in Section 5.1, there is really nothing special or profound in these equation: they are simply direct first integrations of the equations of motion—the very thing we are allegedly trying to *avoid*!

Nonetheless, their application, particularly the angular impulse, can be very useful: Recall that one of the classic techniques utilized in Statics is the ability to sum moments about an arbitrary point, since the equilibrium equations hold about *any* point. This fact is used to *eliminate* forces from the moment equation, and thus make the resulting system easier to solve. The same thing can occur in Dynamics, as the next example illustrates:

Example 5.4.3 *Linear and Angular Impulse.* Two light inextensible cords, OB and AB, of lengths 65 cm and 61 cm respectively, are fixed at their ends along the X-axis as shown. They are pulled taut to the position B in the vertical XZ-plane indicated,[28] and then wrapped carefully around a uniform cylinder of radius 10 cm and length 50 cm, keeping one end against the YZ-plane and its axis horizontal, until it is against the X-axis. If the cylinder is released from rest in this position, find its angular velocity after 0.2 s.

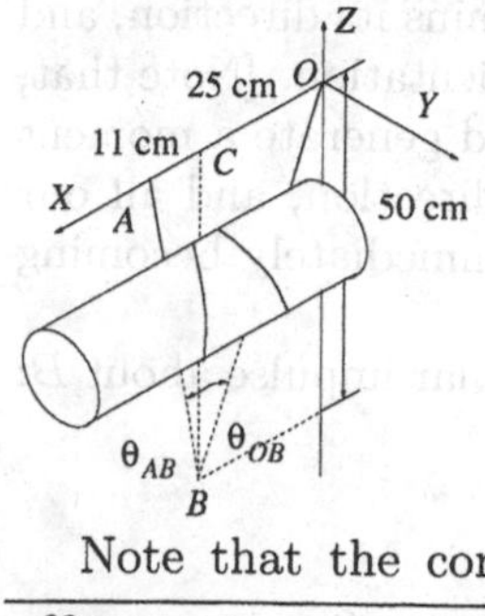

Note that the conditions regarding maintaining the horizontal aspect of

[28]This information is given, but it could be found easily using trigonometry.

the cylinder's axis is not unreasonable: if the center of the cylinder is rolled against the line BC, then the amount of each cord wrapped around the cylinder is the same because the tangents make the angles θ_{OB} and θ_{AB} with the cylinder: $ds_{OB}\cos\theta_{OB} = d\overline{r} = ds_{AB}\cos\theta_{AB}$:

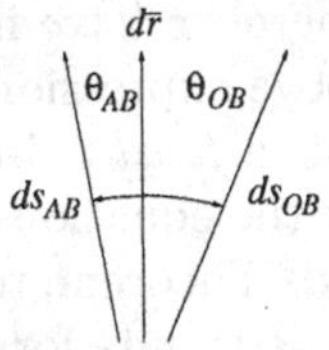

This is not the sort of problem in which one would like to integrate the equations of motion directly! These do provide us with a couple of useful facts however: The only forces acting on the cylinder are the tensions in the cords and gravity. Since the cylinder is initially against the XZ-plane, the cord forces lie in the XZ-plane at that instant; thus *all* the forces are initially in vertical planes. There is no horizontal force—the normal force the XZ-plane *could* generate has nothing to react against—or velocity in the $\hat{\boldsymbol{J}}$-direction to change that, so the forces *remain* in this planes, and we see that the center of mass moves purely vertically. In fact, the *kinematical* condition imposed by the wrapped cord shows that there is no acceleration in the $\hat{\boldsymbol{I}}$-direction, so the center of mass moves purely vertically parallel to BC and the X-components of the two tensions must vanish.

This deals with the *translational* motion of the center of mass, but what about the *rotational* motion of the body? Here we come to the key point about this problem: from the above, *the forces* $\boldsymbol{T}_{OB}$ *and* $\boldsymbol{T}_{AB}$ *will always pass through the [fixed] point* B; thus, if we sum moments about that point, both forces will have been *eliminated* from that moment equation. The process is precisely the same as in Statics. This leaves us with only the *known* gravitational force to deal with. Let us tentatively assume that $\boldsymbol{\omega} = -\omega\hat{\boldsymbol{I}}$; from that moment equation—about neither a fixed point about which the body is rotating nor the center of mass—

$$\begin{aligned}\sum_i \boldsymbol{M}_{B_i} &= \boldsymbol{r}_{G/B}\times m\boldsymbol{g} = (0.1\hat{\boldsymbol{J}} + Z\hat{\boldsymbol{K}})\times(-mg\hat{\boldsymbol{K}}) = -0.1mg\hat{\boldsymbol{I}}\\ &= \boldsymbol{r}_{G/B}\times m\overline{\boldsymbol{a}} + \boldsymbol{\omega}\times\bar{\mathsf{I}}\boldsymbol{\omega} + \bar{\mathsf{I}}\boldsymbol{\alpha}\\ &= (0.1\hat{\boldsymbol{J}} + Z\hat{\boldsymbol{K}})\times m(-\overline{a}\hat{\boldsymbol{K}}) + \boldsymbol{0} + \bar{\mathsf{I}}\boldsymbol{\alpha}\\ &= -0.1m\overline{a}\hat{\boldsymbol{I}} + \bar{\mathsf{I}}\boldsymbol{\alpha},\end{aligned} \tag{5.4-9}$$

the term $\boldsymbol{\omega}\times\bar{\mathsf{I}}\boldsymbol{\omega}$ vanishing because $\boldsymbol{\omega}$ is along a principal axis, so $\bar{\mathsf{I}}\boldsymbol{\omega} \parallel \boldsymbol{\omega}$. But this shows that the only $\boldsymbol{\alpha}$ is in the $\hat{\boldsymbol{I}}$-direction; thus $\boldsymbol{\omega}$ retains its direction, and there is also no rotation of the cylinder out if its initial orientation. [Note that, had the center of mass lain *off* the line BC, then it would generate a moment and resulting angular acceleration component in the $\hat{\boldsymbol{J}}$-direction, and all our kinematic assumptions would be invalid, with one cord immediately becoming slack.]

Thus the central equation we will use involves the angular impulse about B:

$$\begin{aligned}\int_0^{0.2\,\mathrm{s}} \overline{\boldsymbol{r}}\times m\boldsymbol{g}\,dt &= -0.02mg\hat{\boldsymbol{I}}\\ &= \boldsymbol{H}_{B_2} - \boldsymbol{H}_{B_1} = (\boldsymbol{r}_{G/B}\times m\overline{\boldsymbol{v}} + \bar{\mathsf{I}}\boldsymbol{\omega}) - \boldsymbol{0}\end{aligned}$$

$$= (0.1\hat{\boldsymbol{J}} + Z\hat{\boldsymbol{K}}) \times m(-\omega\hat{\boldsymbol{I}} \times 0.1\hat{\boldsymbol{J}}) + \overline{I}_{xx}(-\omega\hat{\boldsymbol{I}}),$$

$\overline{\boldsymbol{v}}$ replaced by the "non-slip" motion along the cords, and the principal nature of these axes used to find

$$\overline{\mathsf{I}}\omega = \overline{I}_{xx}(-\omega\hat{\boldsymbol{I}} = -\left(0.01m\omega + \frac{1}{2}m(0.1)^2\omega\right)\hat{\boldsymbol{I}},$$

from which we conclude that $\omega = 13.08\,[\text{rad}]/\text{s}$.

Note that, in the above, we *assumed* that $\boldsymbol{\omega} \equiv -\omega\hat{\boldsymbol{I}}$, $\boldsymbol{\alpha} \equiv -\alpha\hat{\boldsymbol{I}}$, and $\overline{\boldsymbol{a}} \equiv -\overline{a}\hat{\boldsymbol{K}}$, so that the non-slip conditions $\overline{\boldsymbol{v}} = \boldsymbol{\omega} \times \overline{\boldsymbol{r}}$ and $\overline{\boldsymbol{a}} = \boldsymbol{\alpha} \times \overline{\boldsymbol{r}}$ gave that $\overline{v} = +\overline{r}\omega$ and $\overline{a} = +\overline{r}\alpha$. We could just as easily (and just a validly) presumed $\boldsymbol{\omega} \equiv +\omega\hat{\boldsymbol{I}}$ and/or $\boldsymbol{\alpha} \equiv +\alpha\hat{\boldsymbol{I}}$—$\omega$ and α are *unknowns*, recall—in which case we would obtain $\overline{v} = -\overline{r}\omega$ and $\overline{a} = -\overline{r}\alpha$. But whichever is chosen, one must remain *consistent.*

Clearly, the impulse approach is nothing more than an integration of the moment equation of motion, (5.4-9). But it does tend to emphasize the *kinetic* aspect a bit more and is useful to look for in cases in which it can be applied.

There is an interesting sidelight to this example: The above analysis was based on the assumption that the cords, enforcing both the non-slip motion and the moment itself, remained *taut.* But *do* they? Or, as the cylinder spins faster, does the cylinder spin so fast that the cords become slack? (They do this simultaneously, since $\sum F_X = T_{OB_X} + T_{AB_X} = 0$.)

In fact, they don't: Summing *forces* in the vertical direction,

$$\sum F_Z = T_{OB_Z} + T_{AB_Z} - mg = -m\overline{a} \quad \text{or} \quad m(g - \overline{a}) = T_{OB_Z} + T_{AB_Z} \geq 0;$$

(where we have once again taken $\overline{a} > 0$), so that $\overline{a} \leq g$, equality occurring when the tensions go to zero. But, again by the non-slip condition, assuming that $\boldsymbol{\alpha} = -\alpha\hat{\boldsymbol{I}}$, $\overline{a} = +0.1\alpha$, so (5.4-9)—which holds whether the cords are taut or not—tells us that $\alpha = \frac{20}{3}g$, a *constant.* In particular, then, $\overline{a} = \frac{2}{3}g < g$, *also* constant, so never allowing the cords to slacken. |

Homework:

1. For Example 5.4.3,
   (a) find the radius of the cylinder for which the cords would *not* exert a moment on it as it fell, and
   (b) describe the motion of the cylinder under these conditions.

There is one other useful aspect to the impulse equations: just as in Section 5.1.1, when an impulse *vanishes*, then the corresponding momentum is *conserved.* While, as pointed out there, this generally requires the very sort of integration we are trying to avoid, there are easy sufficiency conditions which obviate this: if the integrand, or a *component* of the integrand, happens to vanish, then the corresponding kinetic quantity, or *component* of that quantity, is conserved—at least subject to the constraints of the *caveat* on page 262. In particular, if there is no [net] *force* in a certain [fixed] direction, that component of *linear* momentum is conserved, while if there is no *moment* in such a direction, the *angular* momentum component is conserved.

In the above example, use of angular impulse was an optional means of eliminating unknown forces from an equation of motion; had we chosen to sum moments about, say, the mass center, the resulting equations would have been equivalent (see the remark regarding this on page 309), but not so convenient to solve, generally requiring the use of the force equations, too. But in certain cases, use of impulse/momentum—particularly its angular form—becomes absolutely essential:

Impactive forces are large forces acting over a very short period of time. Over that period Δt, such forces, $\boldsymbol{F}_I$, say, exert a far greater effect than more general forces, say $\boldsymbol{F}_G$: $\boldsymbol{F}_I \gg \boldsymbol{F}_G$. But this means the *impulse* these forces generate is also much greater:

$$\boldsymbol{G}_2 - \boldsymbol{G}_1 = \int_t^{t+\Delta t} \sum \boldsymbol{F}_I\, dt + \int_t^{t+\Delta t} \sum \boldsymbol{F}_G\, dt \doteq \int_t^{t+\Delta t} \sum \boldsymbol{F}_I\, dt;$$

in fact, because of their "small" magnitude, the latter effectively produce *no* impact over the short Δt (though they would over a longer interval of time). A similar equation would hold for the angular impulse. The upshot is that, in the presence of such impactive forces, we can effectively consider *only* impactive forces, and any moments they might generate.

We need consider only those impactive forces *external* to the rigid body—the "system"—because of the equations of motion for systems. But these external forces can be either *applied* or *reaction* forces. Thus, for example, when one hammers a block of wood resting atop a concrete floor, the block jumps *up*, due to the *impactive, reaction* normal force exerted by the floor against it!

For a rigid body, these forces result in qualitatively *discontinuous* jumps in the velocity $\overline{\boldsymbol{v}} = \boldsymbol{G}/m$, as can be see from the above equation. Yet they are also considered to take place over so short a period of time that the *position* remains unchanged. (If this seems a little far-fetched, consider what happens when a bat hits a baseball: the ball *instantaneously* changes its velocity, in fact, reversing its direction, all while the ball "remains" at home plate during the encounter.) This situation is schematically diagrammed at left.

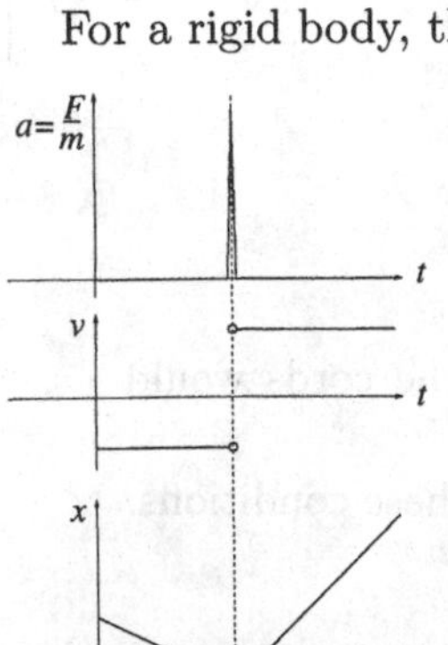

The nature of impact—a "δ-function," having negligible "width" but non-vanishing "area"—make its integration mathematically difficult to deal with explicitly, and mechanicians go to great length to avoid this if at all possible. Thus, for example, in the analysis of *frictionless particle impact* covered in an elementary

Dynamics course, the linear momentum of the *system* is considered in order to internalize this force.[29] In the case of rigid body motion, if such an impactive force is localized to a *point*, it can be eliminated by saying that *angular momentum is conserved about that point*: we need consider only the impactive forces and their moments during impact, and summing [impactive] moments *about* this point of impact eliminates it even there. We should note, however, that although momentum may be conserved during impact, *energy virtually never is*: the *mechanical* energy inherent in its motion is invariably converted to other, non-mechanical forms. This is the thrust of the next example:

Example 5.4.4 *Impact—Angular Momentum Conservation.* A uniform thin rectangular plate, 1 m×2 m in size, falls against the stationary corner of the horizontal XY-plane at the origin with center-of-mass velocity $\overline{\boldsymbol{v}} = -1\hat{\boldsymbol{K}}$ m/s and angular velocity $\omega = 2\hat{\boldsymbol{K}}$ [rad]/s as shown. Assuming no rebound of the corner during the encounter, find the angular velocity of the plate after impact.

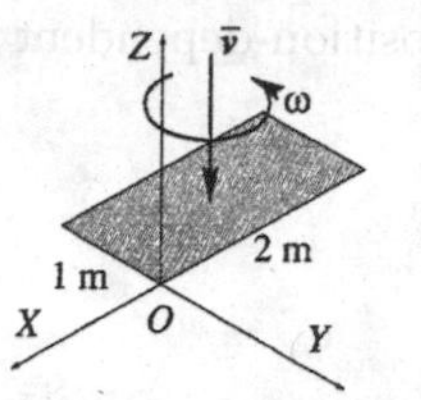

Even though gravity is operative in this example, we never consider it: in the presence of impact, such normal-magnitude forces are ignored compared with the effect of the large impactive forces. We observe that that force (one of reaction) is concentrated at the corner O; this means that the sum of the *impactive* moments—that's all we have to consider during impact—about O vanishes, and thus angular momentum is conserved: $\boldsymbol{H}_{O_2} = \boldsymbol{H}_{O_1}$.

It remains only to calculate the initial and final angular momentum. Since the plate is *not* initially rotating about O, we must use the general form of angular momentum; the orientation of the plate is gratifyingly principal relative to the reference axes:

$$\begin{aligned}\boldsymbol{H}_{O_1} &= \boldsymbol{r}_{G/O} \times m\overline{\boldsymbol{v}} + \bar{\mathsf{I}}\boldsymbol{\omega}\\ &= (-1\hat{\boldsymbol{I}} - 0.5\hat{\boldsymbol{J}}) \times m(-1\hat{\boldsymbol{K}})\\ &\quad+ \begin{pmatrix}\frac{1}{12}m(1)^2 & 0 & 0\\ 0 & \frac{1}{12}m(2)^2 & 0\\ 0 & 0 & \frac{1}{12}m(1^2+2^2)\end{pmatrix}\begin{pmatrix}0\\0\\2\end{pmatrix}\\ &= \left(\frac{1}{2}\hat{\boldsymbol{I}} - 1\hat{\boldsymbol{J}} + \frac{5}{6}\hat{\boldsymbol{K}}\right)m.\end{aligned}$$

After the impact, however, the plate *is* rotating about O, due to the lack of rebound; these axes are *not* principal, though the Z-axis is:

$$\begin{aligned}\boldsymbol{H}_{O_2} &= \mathsf{I}_O\boldsymbol{\omega}'\\ &= m\begin{pmatrix}\frac{1}{12}(1)^2+(0.5)^2 & -(0.5)(1) & 0\\ -(0.5)(1) & \frac{1}{12}(2)^2+(1)^2 & 0\\ 0 & 0 & \frac{1}{12}(1^2+2^2)+(0.5^2+1^2)\end{pmatrix}\begin{pmatrix}\omega'_X\\ \omega'_Y\\ \omega'_Z\end{pmatrix}\end{aligned}$$

[29]Then, of course, the presentation turns right around and introduces it explicitly, cleverly eliminating it through the linear impulse relation.

$$= \left(\left(\frac{1}{3}\omega'_X - \frac{1}{2}\omega'_Y\right)\hat{\boldsymbol{I}} + \left(\frac{4}{3}\omega'_Y - \frac{1}{2}\omega'_X\right)\hat{\boldsymbol{J}} + \frac{5}{3}\omega'_Z\hat{\boldsymbol{K}}\right)m.$$

Solving the above three vector component equations in the three unknown components of $\boldsymbol{\omega}'$, we get that

$$\boldsymbol{\omega}' = 0.857\hat{\boldsymbol{I}} - 0.429\hat{\boldsymbol{J}} + 0.5\hat{\boldsymbol{K}}.$$

We mentioned above that energy is almost never conserved during impact; from previous experience with *particle* impact we would expect it not to be here, since there is no rebound (the coefficient of restitution $e = 0$), while energy in the particle case is conserved if and only if $e = 1$. This is precisely the case: since the position is posited not to change during impact, the position-dependent potential energy does not change, and we need find only

$$T_1 = \frac{1}{2}m\overline{v}^2 + \frac{1}{2}\boldsymbol{\omega}\cdot\overline{\mathbf{I}}\boldsymbol{\omega} = 1.333m$$

and

$$T_2 = \frac{1}{2}\boldsymbol{\omega}'\cdot\mathbf{I}_O\boldsymbol{\omega}' = \frac{1}{2}\boldsymbol{\omega}'\cdot\boldsymbol{H}_{O_2} = 0.637m.$$

Thus we have an energy change $\frac{\Delta T}{T_1} = (-)52.2\%$. In fact, the vertical motion of the center of mass of the plate is stopped "dead in its tracks": if we calculate

$$\overline{\boldsymbol{v}}' = \boldsymbol{\omega}' \times \overline{\boldsymbol{r}} = -0.25\hat{\boldsymbol{I}} - 0.5\hat{\boldsymbol{J}},$$

the Z-component vanishes identically! |

Summary. Just as for particles and systems, we can conceptually integrate the linear and angular momentum equations for a rigid body to obtain the *linear* and *angular impulse*, giving the change in the respective momenta over an interval. However appealing these are to write, there is really nothing new here, since they are just first time-integrations of those equations of motion. But there are *vector* quantities, so can give up to *three* scalar equations each to assist in the solution of the original equations of motion. In particular, if any of these quantities is *conserved*—the associated impulse *vanishes*—in a *fixed* direction (relative to the oft-cited *caveat* on page 262), we can conclude immediately that that quantity is a constant.

5.4.3 Conservation Application in General

In the two previous sections, we have carefully used only *one* conservation principle in the examples; this is a routine pedagogical device to exemplify the concepts being presented. But, more generally, *more* than a single kinetic quantity may be conserved. In the hopes of forestalling integration of the equations of motion as long as possible, *all* possible quantities should be examined for possible conservation: energy, and *each component*—at least subject to the considerations of the "*caveat*" on page 262—of linear and angular momentum. Doing so will uncover already-integrated forms of the equations of motion.

In some problems, the desired information can be found without recourse to the equations of motion at all; thus, in Example 5.4.4, the results of angular momentum were used to calculate the [kinetic] energy and post-impact center-of-mass velocity, which themselves could be used as input to an energy analysis—all without ever invoking the full equations of motion. We conclude this section, however, with an example not only of intrinsic interest, but also demonstrating how already-integrated, conserved quantities can be useful in the analysis of the equations of motion from which they were derived:

Example 5.4.5 *General Motion of the Gyroscope—Nutation.* In the previous Section, we made an analysis of the motion of a standard gyroscope supported at one fixed point, under the action of its own weight. But this was explicitly limited to the case of "steady-state precession." We now consider the most general case of that system, diagrammed again at left.

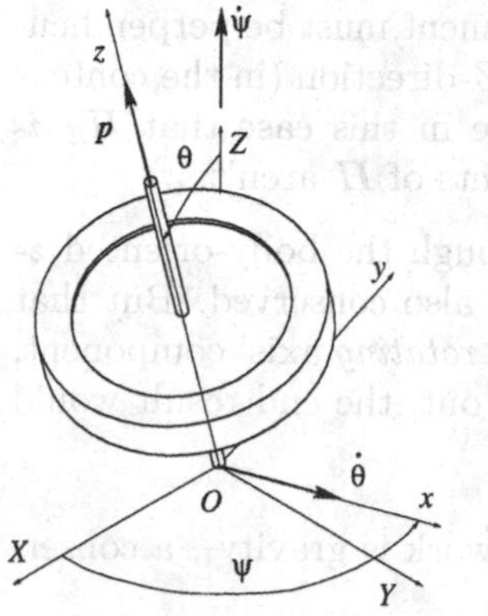

Being primarily interested in the angular motion of the gyroscope, we will try to effect the analysis using only the moment equation. Once again, we will take moments about the fixed support point at the origin of coordinates—call it O. And, since we are interested in an analysis of precession and nutation, we shall also utilize those equations introducing those specific quantities, equations (5.2-43b).

Now, however, we have no simplifications resulting from the previously-presumed steady-state motion, and our equations become

$$\sum_i \mathbf{M}_{O_i} = \overline{r}mg\sin\theta\hat{\imath}$$

$$= \mathbf{\Omega}\times\mathbf{I}_O\boldsymbol{\omega} + \mathbf{I}_O\boldsymbol{\alpha}_{rel}$$

$$= \Big(I_O(\ddot{\theta} - \dot{\psi}^2\sin\theta\cos\theta) + I\dot{\psi}\sin\theta(\dot{\psi}\cos\theta + p)\Big)\,\hat{\imath} \qquad (5.4\text{-}10a)$$

$$+ \Big(I_O(\ddot{\psi}\sin\theta + 2\dot{\psi}\dot{\theta}\cos\theta) - I\dot{\theta}(\dot{\psi}\cos\theta + p)\Big)\,\hat{\jmath} \qquad (5.4\text{-}10b)$$

$$+ I\frac{d}{dt}(\dot{\psi}\cos\theta + p)\hat{\mathbf{k}}, \qquad (5.4\text{-}10c)$$

in which $\overline{r}$ represents the distance to the center of mass of the gyroscope from O. The last of these equations tells us immediately that

$$\dot{\psi}\cos\theta + p = \omega_z, \text{ a constant.} \qquad (5.4\text{-}11)$$

While vanishing of the second, (5.4-10b), *looks* promising, unlike (5.4-10c), it is not the total time derivative of a function; the best that can be done is to write it in the form

$$I_O \frac{d}{dt}(\dot{\psi}\sin\theta) + (I_O - I)\dot{\theta}(\dot{\psi}\cos\theta + p) - Ip(\dot{\psi}\cos\theta + p) = 0,$$

which *cannot* be integrated as it stands.

Recalling that conservations of energy and momenta are already-integrated forms of these equations, we now consider which of these might be conserved:

- *Linear Momentum.* Not conserved: [external] applied forces due to the fixed support and gravity appear generally in all three components.

- *Angular Momentum.* Not conserved: there is a moment generated about O. *But*: relative to the *fixed* (X, Y, Z)-system, that moment is generated by the weight, in the $\hat{\boldsymbol{K}}$-direction; thus that moment must be perpendicular to $\hat{\boldsymbol{K}}$. Since $\dot{H}_Z = \sum M_{O_Z} \equiv 0$ in the *fixed* Z-direction (in the context of the "*caveat*" on page 262), we *can* conclude in this case that H_Z *is* conserved, even though the other two components of $\boldsymbol{H}$ aren't.

  It is certainly the case that *all* forces pass through the body-oriented z-axis, and it is tempting to conclude that H_z is also conserved. But that would be attempting to apply conservation to a *rotating* axis' component, so would be specious (even though, as it turns out, the end result would be the same)!

- *Energy.* Since O is fixed, the only force doing work is gravity—a conservative force. Thus energy *is* conserved.

Now let us implement the conserved quantities we have discovered above:

$$\begin{aligned} H_Z &= H_y \sin\theta + H_z \cos\theta = (I_O\dot{\psi}\sin\theta)\sin\theta + (I\omega_z)\cos\theta \\ &= I_O\dot{\psi}\sin^2\theta + I\omega_z\cos\theta \end{aligned} \tag{5.4-12}$$

—where he have introduced (5.4-11)—is constant. [Note that this is *not* what we did above, where we concluded *directly from the differential equation* $I\dot{\omega}_z = 0$ that ω_z was constant, rather than saying $\dot{H}_z = \sum M_{O_z} \equiv 0$ implies that $H_z = I\omega_z$ is constant.] This illustrates the advantages of considering the possible conservation of kinetic quantities in this problem allegedly dealing with the equations of motion: even though the former are derived from the latter, it is unlikely that a mere examination of the equations of motion would uncover the particular combination (5.4-12) as being one of their integrals! Equations (5.4-11) and (5.4-12) provide us with two *first*-order equations in the *first* derivatives $\dot{\theta}$, $\dot{\psi}$, and $p = \dot{\phi}$, but we need a third in order to be able to solve for all three.

That, of course [given the conservation analysis above] is energy. Kinetic energy in principal gyroscopic coordinates is easy: $T = T_O = \frac{1}{2}(I_O\omega_x^2 + I_O\omega_y^2 + I\omega_z^2) = \frac{1}{2}(I_O\dot{\theta}^2 + I_O(\dot{\psi}\sin\theta)^2 + I\omega_z^2)$; if we define the potential energy to vanish when $\theta = 0$ and eliminate $\dot{\psi}$ using H_Z from equation (5.4-12), we get that

$$E = T + V = \frac{1}{2}\left(I_O\dot{\theta}^2 + I_O\left(\frac{H_Z - I\omega_z\cos\theta}{I_O\sin^2\theta}\right)^2\sin^2\theta + I\omega_z^2\right) - mg\bar{r}(1-\cos\theta) \tag{5.4-13}$$

—an equation in θ alone—is also constant. This is where energy has been particularly useful: it is still a *first*-order equation in the variables, particularly here, $\dot{\theta}$. Note that, unlike the previous examples of energy, we are applying its conservation *throughout the interval*, rather than just at the endpoints. Writing this in slightly different form to *solve* for that derivative, we get

$$\sin^2\theta\,\dot{\theta}^2 = \left(\frac{2(E+mg\bar{r}) - I\omega_z^2}{I_O}\right)\sin^2\theta - \frac{2mg\bar{r}}{I_O}\cos\theta\sin^2\theta - \left(\frac{H_Z - I\omega_z\cos\theta}{I_O}\right)^2, \tag{5.4-14}$$

which doesn't look too promising until we note that $\frac{d\cos\theta}{dt} = -\sin\theta\dot{\theta}$. This suggests writing the above equation entirely in terms of $\cos\theta$; introducing the identity $\sin^2\theta = 1-\cos^2\theta$, in particular into the second term on the right-hand side, demonstrates that this is a cubic equation in $\cos\theta$:

$$\begin{aligned}\left(\frac{d\cos\theta}{dt}\right)^2 &= \left(\frac{2(E+mg\bar{r}) - I\omega_z^2}{I_O} - \frac{H_Z^2}{I_O^2}\right) + 2\left(\frac{H_Z I\omega_z}{I_O^2} - \frac{mg\bar{r}}{I_O}\right)\cos\theta \\ &\quad - \left(\frac{2(E+mg\bar{r}) - I\omega_z^2}{I_O} + \frac{I^2\omega_z^2}{I_O^2}\right)\cos^2\theta + \frac{2mg\bar{r}}{I_O}\cos^3\theta \\ &\equiv f(\cos\theta),\end{aligned} \tag{5.4-15}$$

in which all coefficients are constants depending on the initial conditions. If we can integrate this equation to find $\cos\theta(t)$, then equations (5.4-12) and (5.4-11) determine $\dot{\psi}$ and $p = \dot{\phi}$, which are no longer constant but depend on $\theta(t)$:

$$\dot{\psi}(t) = \frac{H_Z - I\omega_z\cos\theta(t)}{I_O\sin^2\theta(t)} = \frac{H_Z - I\omega_z\cos\theta(t)}{I_O(1-\cos^2\theta(t))} \tag{5.4-16}$$

$$\therefore p(t) = \omega_z - \dot{\psi}(t)\cos\theta(t). \tag{5.4-17}$$

We can, in theory, actually *find* $\theta(t)$: equation (5.4-15) is a first order equation, and thus we have, as the saying goes, "reduced the problem to quadratures," which is generally a phrase to indicate that we just don't now *how* to find the solution! In this case, however, we actually *can*: writing $u \equiv \cos\theta$ and simply separating variables, we get that

$$t - t_o = \int_{u_o}^{u}\frac{du}{\sqrt{f(u)}},$$

where $\cos\theta(t_o) = u_o$. But $f(u)$ is a *cubic* in u, not a quadratic such as would lead to the "elementary" trig substitutions familiar from a second Calculus course. In fact, however, there is a whole area of so-called "elliptic functions"—having nothing, incidentally, to do with ellipses—which allow such integration of square roots of cubic and even quartic functions. These involve the introduction of new functions "sn" and "cn" which are, like their more elementary counterparts sin and cos, *periodic* in nature.[30] In fact, in this case the solution would involve the sn *of* the sin of a constant times t, so the solution is assured to be periodic! That's all we'll say about the subject, but the periodicity will be important.

Rather than finding the actual solution, however, we shall be content with a rather more qualitative analysis of it: Clearly, to be real—and it is *real* mechanics we're interested in, after all!—$\left(\frac{d\cos\theta}{dt}\right)^2 = f(\cos\theta)$ must be non-negative. In particular, when $u \equiv \cos\theta = -1$, equation (5.4-15) shows that

$$f(-1) = -\frac{H_Z^2}{I_O^2} - \frac{I^2\omega_z^2}{I_O^2} + 2\frac{H_Z I\omega_z}{I_O^2} = -\frac{(H_Z - I\omega_z)^2}{I_O^2} < 0, \tag{5.4-18a}$$

and, similarly,

$$f(+1) = -\frac{H_Z^2}{I_O^2} - \frac{I^2\omega_z^2}{I_O^2} - 2\frac{H_Z I\omega_z}{I_O^2} = -\frac{(H_Z + I\omega_z)^2}{I_O^2} < f(-1) < 0; \tag{5.4-18b}$$

differentiation of the same equation also yields

$$\begin{aligned} f'(-1) &= 2\frac{H_Z I\omega_z + 4I_O mg\bar{r} + 2I_O E + I(I - I_O)\omega_z^2}{I_O^2} \\ &= 2\frac{H_Z I\omega_z + I_O(2E + 4mg\bar{r} - I\omega_z^2) + I^2\omega_z^2}{I_O^2} \\ &> 0, \end{aligned} \tag{5.4-18c}$$

the only negative term $II_O\omega_z^2$ being balanced through multiplication of the inequalities resulting from (5.4-13)

$$2E + 4mg\bar{r} \geq 2E + 2mg\bar{r}(1 - \cos\theta) = 2T \geq I\omega_z^2$$

by I_O. Furthermore we have that

$$\lim_{u\to-\infty} f(u) = -\infty \quad \text{while} \quad \lim_{u\to+\infty} f(u) = +\infty. \tag{5.4-18d}$$

(5.4-18d) shows immediately that the cubic $f(u)$ goes from $-\infty$ on the left to $+\infty$ on the right; while (5.4-18a-b) (particularly that $f(+1) < f(-1)$), and the fact that $f(u)$ is still rising at (-1) by (5.4-18c) (and, of course, continuity of its derivative from the polynomial nature of $f(u)$), mean that the function assumes a maximum somewhere on the interval $(-1, +1)$.

[30] In fact, they are *doubly* periodic: there are *two* distinct periods. The classic reference on the subject is by Byrd and Friedman, *Handbook of Elliptic Integrals for Engineers and Physicists*.

Thus we see that the cubic $f(u)$ is qualitatively as at left: assuming that $f_{max}(u) > 0$ for some $u \equiv \cos\theta \in (-1, 1)$, there are three roots, two inside the open interval $(-1, 1)$ bounding u itself—call them u_1 and u_2—and the other outside.[31] Those two inside the interval are the points at which $\frac{d\cos\theta}{dt} = -\sin\theta\dot{\theta}$ must vanish—*i.e.*, at which $\dot{\theta} = 0$; they bound the region in which $\dot{u}^2 \geq 0$ and thus of admissible motion. This means that $\theta(t)$, periodic due to its elliptic functional nature, varies between the two values $\theta_i = \arccos u_i$—the nutation observed of a gyroscope.

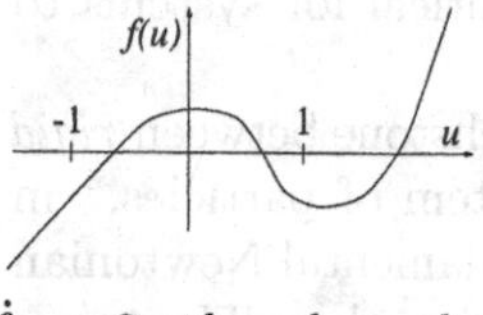

It is, of course, possible that the two distinct roots above coalesce into a single, repeated root if the initial conditions make the coefficients of $f(u)$ give such; then there is only a *single* angle at which such precession takes place. But, then, by equations (5.4-12) and (5.4-16), $\dot{\psi}$ and p are also constant since θ is. This is precisely the case of *steady-state precession* studied at the end of Section 5.2.3, where we *started* with the conditions that nutation vanished while spin and precession were constant in magnitude; here we investigate the origins of *those* conditions.

The above is meant as much to describe a method of analysis and to verify mathematically a phenomenon familiar physically as to provide an example of "conservation." But what the analysis also shows is something likely *not* expected from physical intuition—that the precession is related to the nutation by equation (5.4-16). In particular, depending on the values of the various parameters in the problem, the periods of θ—recall that the elliptic functions are also periodic—and ψ can be in any ratio to one another, generating "Lissajous"-like figures in the trace of θ *vs.* ψ if the two variables are plotted (mod 2π). |

Summary. Though we tend to compartmentalize linear momentum, angular momentum, energy, and the equations of motion as being separate, in point of fact they are all different facets of the *same* basic equations of motion. Thus each of the "kinetic quantities" can be brought to bear on the analysis of the system, including the equations of motion themselves.

Summary

This is likely the longest chapter in the book, since it is essentially the culmination of *all* that has gone before: We first reviewed *particles*, defining *linear momentum*, *angular momentum*—obtaining their "equations of motion" and *impulse*, and *energy*—obtaining the *work-[kinetic] energy principle.* Then we considered *systems of particles*, where we defined the same [*single*] "kinetic quantities" for such a system, both in terms of their definition relative to a *fixed*, inertial system, and then relative to the *non*-inertial center-of-mass system.

[31] It is not immediately obvious from what has been said thus far that, in fact, $f(u)$ actually *goes* positive in the interval $(-1, 1)$. But, on physical grounds, this must be so: equation (5.4-15) demonstrates that, as a square [of a presumably real number!], $f(u) \geq 0$.

Rather remarkably, it turned out that the equations of motion for momenta, and work-energy principle for energy, were essentially identical for systems to those for particles.

Systems were important, because they provided the touchstone between *rigid body* on the one hand—the last being just a special "system of particles," in which the inter-point distance remains fixed—and the fundamental Newtonian equations of motion, formulated only for *single particles* on the other. The *translational* motion of a rigid body was given by the system equation of motion for linear momentum: $\sum \boldsymbol{F} = \dot{\boldsymbol{G}} = m\overline{\boldsymbol{a}}$. Similarly, the *rotational* motion resulted from the system equation of motion for angular momentum, in which now that assumed (relative to the center of mass) the form $\overline{\boldsymbol{H}} = \bar{\mathsf{I}}\boldsymbol{\omega}$. $\bar{\mathsf{I}}$ is the *inertia tensor* or [its *representation*] *matrix*—a real, symmetric matrix. The latter depends not only on the basis vectors, but also the coordinates in terms of which the individual entries in the matrix are represented; like any other linear transformation, it could be transformed to other bases through *rotation matrices*; but it could also be translated to other coordinates—other *origins*—through the *Parallel Axis Theorem.* In particular, as a real symmetric matrix, it was possible to diagonalize this matrix through a simple proper rotation; the axes in which it was diagonalized were called *principal axes*, and the inertia matrix could then be represented only in terms of its *moments of inertia*, the off-diagonal *products of inertia* vanishing because of the diagonalization. That done, we then considered the equations of motion themselves, comprising most generally equations (5.2-22) and (5.2-23), the latter being forms of *Euler's equations.* These could be applied to single rigid bodies or systems of rigid bodies—*machines*—demanding special care at the interconnections. In the special case the body had an axis of symmetry about which spin could take place—a *gyroscope*, these assumed a special form, particularly the moment equation (5.2-41), separating out the *spin* component of the gyroscope's angular velocity about its axis from the *precession* and *nutation* of that axis.

We briefly discussed *stability*, showing that the stability of motion could be investigated mathematically the same way it is conceptually: by taking small disturbances—"*perturbations*"—from the motion and examining their solutions to see whether they amplified, died away, or merely oscillated about the original solution. Fortuitously, the *variational equations* of motion for such perturbations are invariably linear, as a result of considering only first-order departures from the unperturbed solution.

Finally, we considered alternatives to direct integration of the second-order equations of motion—generally-integrated forms of those equations. These came back to the original kinetic quantities of *energy* and *momenta*, both linear and angular; the former is purely scalar, allowing its representation in arbitrary "mixed" coordinates, while the latter are vectors, demanding their representation relative to a particular basis. As with particles, we can talk of the possible *conservation* of these quantities, energy begin conserved if all the forces [doing work] are themselves conservative, and the momenta being conserved along *fixed* basis directions if the corresponding impulse vanishes. In general, we can profitably bring *all* these tools—equations of motion *and* conservation—to bear on a given problem, with conservation often uncovering integrals of the motion which would elude discovery by examining the equations of motion alone.

Epilogue

Given its length, comments on this Part are going to be uncharacteristically brief! But we have considered both the *kinematics* and *kinetics* of the motion of rigid bodies, showing how the first is invariably required just to *describe* the second, which then *determines* the first.

The kinetics, in particular, is founded on Newton's Laws, particularly the Second and (in the case of interconnected rigid bodies) the Third. These laws are a *vector* formulation, so must be expressed relative to a given basis—a given *coordinate system.* In a typical introductory course on planar motion, one is introduced to three or four of these—Cartesian, cylindrical, tangential-normal, and perhaps spherical coordinates. It is pointed out there that it is expeditious to select a coordinate system best suited to a particular problem: for motion in a straight line, for example, a basis similarly fixed in direction, Cartesian coordinates, are most useful, while for motion in a circle, cylindrical or tangential-normal are best. One can envision more complicated problems in which it might be nice to have a coordinate system particularly well-suited to that problem; but there just aren't many more than those already enumerated, and, in such cases, one invariably defaults to the "system of last resort," Cartesian coordinates.

One's choices are limited with regard to vector coordinate systems.

The next Part removes those limitations, allowing us to define *any* coordinates convenient to a given problem, free of the constraints imposed by the classic vector coordinate systems. The situation is similar to that observed in energy, where point was made of the freedom to "mix" coordinates due to the scalar nature of energy itself: *scalar* quantities, unlike *vector* quantities, are *invariant* under vector transformations. In fact, we have already let the cat out of the bag: The next Part deals with an *energy*-based formulation of Mechanics, rather than the *vector*-based formulation of Newton. Yet we can't just *ignore* Newton; Nature seems inherently vector-based, and his Laws are, after all, that foundation on which all motion rests. How, then, can we separate out the various *components* of motion from a *scalar*?

Read on....

Part III

Analytical Dynamics

Prologue

Newton's laws, on which all of mechanics is based, are a *vector* formulation: the forces and moments, and resulting linear and angular accelerations, are all *vectors*. Thus, in order to implement them, it is necessary to express—to *represent*—these laws in a given *vector* coordinate system. This is the reason that courses in a given area will typically start with a discussion of *kinematics*—the *mathematical* expression of position, velocity, and acceleration (hence the vector coordinate systems), and the relations among them, before introducing *kinetics*—the *physical* laws governing motion and their ramifications.

In Statics, of course, the kinematics is trivial: $\overline{a} \equiv 0 \equiv \alpha$; thus Cartesian coordinates are totally adequate to the description of the *non*-motion, and the primary focus of that field is the kinetics. But when one studies Dynamics, it is necessary to express the various accelerations and velocities, so a great deal of time is spent developing vector expressions for these, too. And other coordinates systems are introduced in addition to the Cartesian: polar, tangential-normal, and often spherical coordinates are presented as alternatives to the "old standby". Though as [absolute] *vector* formulations, the physical laws should not need such a variety, it turns out to be expeditious to use one or another of these systems in a given problem, the choice being guided typically by the *type of motion*—the *kinematical constraints*—imposed on the movement of the system. Yet even when this can be implemented, the [kinetic] constraint forces still appear in the equations of motion, despite the fact these constraints are already implicit in the [kinematical] expressions of velocity and acceleration.

While the use of these fundamental vector laws turns out to be altogether adequate to analyze the motion of a *single* particle or rigid body, when it comes to *systems* of such objects, this approach soon becomes burdensome: Different parts have different motions generally not amenable to the special features of alternative coordinates systems; one must usually doggedly default to the "coordinate system of last resort", Cartesian, expressing the kinematic relations among the parts using *rotating* coordinate systems. And the *forces* interconnecting parts must be included in the separate equations of motion for each, resulting in a ponderous system of equations in which both the accelerations and forces as unknowns—a system endemically *coupled* by the latter.

The same problem holds in Statics, of course, though the system of equations is simpler, due to the absence of accelerations. But an alternative approach is presented there, *virtual work*, which eliminates not only the coupling pin forces (assuming they're frictionless), but also the external supports; the only forces remaining in the equations of equilibrium are the "active" forces—the ones *doing [virtual] work* under a "virtual displacement" ("consistent with the constraints") from the equilibrium position assumed. In a very real sense, the "redundancy" inherent in accounting for constraint forces in both the kinetics and the kinematics in equations of motion is eliminated: the constraint forces never appear, the constraint being implicit in selecting a [kinematical] virtual displacement "consistent with the constraints". In addition to this elimination of constraint forces, however, there is also a subtle advantage to be gained from the use of virtual work: work is a *scalar*, so its description can be made in *any* coordinates, since scalars retain their value regardless of coordinate system. Thus it is common to select any convenient coordinates in terms of which to express work, and even to *mix* coordinates in its evaluation. The other side of this coin, however, is the necessity ultimately to *relate* the various coordinates in the final scalar expression, which can only be solved for a single unknown. And, of course, since the constraint forces never appear in the virtual work expression, this approach cannot be used to find the support or pin forces.

"Work" is a fundamental quantity in dynamics, too, and it seems reasonable that some form of Principle of Virtual Work should be applicable to that field. Presumably, once again, constraint forces would be eliminated from consideration, and one would have the freedom to select any set of convenient coordinates in terms of which to describe the system. Yet, to this point, Dynamics has had little to say on the possibility. Clearly this would be a somewhat delicate matter, for since the body is moving, any "virtual displacements" would involve the *motion* of the bodies in question, and somehow we would have to conceptualize displacements from this [actual] motion. And there's still the issue of how one might relate any "mixed" coordinates introduced for convenience. But if we could somehow overcome these obstacles, the major advantage of such a formulation would be the ability to express the motion of systems in *any* coordinates, chosen at convenience, bursting unshackled from the customary sets one *must* use in a vector formulation.

This is precisely the approach taken by analytical dynamics: It considers a virtual displacement, from the actual motion satisfying Newton's laws of motion, consistent with the constraints imposed by supports and frictionless interconnecting pins. But these constraints, originally *kinetic* in nature, only evidence themselves through *kinematic* constraints on the motion: they never enter the equations themselves, since they do no work! Such kinematic constraints on the motion, in turn, are reflected in the choice of coordinates we make, just as they were in the vector coordinate systems exemplified above. But here the similarity ends: Because the foundation of this formulation is *work*, and since work is a *scalar* able to be expressed in *any* convenient coordinates, such coordinates can be *completely arbitrary*—chosen at will—subject only to the stricture that they be able to describe the configuration of the system. Such *generalized coordinates*, rather than being chosen from the standard set of vector coordinate systems, are completely at our discretion, and the most "natural" set according to personal taste. Only the non-constraint, active, forces enter the equations of motion; the constraint forces are implicit in the coordinates themselves.

To develop this subject, we first briefly review how constraints are treated in the classical vector formulation, and the concepts fundamental to Virtual Work in Statics. In order to motivate the balance of the discussion, several simple examples of constraints and how they affect the form of the position, $\mathbf{r}$, are presented.

We then focus on the kinematics, discussing precisely what a "virtual displacement" is in the context of dynamics: a *differential* displacement from the actual motion *at a given time*. This is motivated by a detailed examination of constraints as *differential* quantities, describing how they relate the components of such a virtual displacement, *regardless of what coordinates we use to describe them*, categorizing them according to integrability and time dependence, and finally relating them to the kinetic forces generating them in the first place. We then come to the central idea in Lagrangian dynamics, *generalized coordinates*—an arbitrary set of *independent* coordinates, from which all relations have been explicitly removed, describing the configuration of the system; it is these in terms of which the motion is ultimately described.

With this in hand, we discuss the kinetics, obtaining the general form of the equations of motion in these variables, the *Euler-Lagrange equations*, utilizing virtual displacements from the actual motion satisfying Newton's laws. and the accompanying virtual work done. Since they are based on work, it is not surprising to see that this form is strongly reminiscent of the general work/energy relation from dynamics; just as in the more classical field, there is a special form of of these equation holding if all forces are conservative, *Lagrange's equations*. And in the same way that energy conservation can be put in a form generalized to include non-conservative forces, Lagrange's equations are so generalized. We shall then briefly touch on the case in which one might choose to use a set other than a canonical one of independent coordinates, and how *Lagrange multipliers* might be used to determine "constraint forces" imposing those constraints not explicitly eliminated.

We then introduce a perhaps new perspective on the solution of problems in dynamics, "integrals of the motion". As a particular instance of these, we introduce a generalization of the energy integral familiar from dynamics, *Jacobi's integral*. We then show how the solution to the system is, in some ways, tantamount to the selection of the "right set" of coordinates.

Finally we provide an introduction to an alternative form of Lagrange's equations, the *Hamilton equations*, pointing out how this concept of the "right" coordinates can be used to solve the complete problem.

A word of advice here: The student will likely recall the reception given to Virtual Work in Statics. Some of this might result from its typically unfortunate placement at the end of the course. But some is inherent in the subject itself: if forces and moments are easy to visualize, work, particularly *virtual* work, is far less so; furthermore, the fundamental idea seems almost *philosophical*, rather than *physical*, in nature—something "done with mirrors" rather than "common sense". Both of these impressions are, unfortunately, somewhat justified, but the only way to overcome the stigma attached to virtual work is to understand it and look back to its foundations after some rudimentary facility has been gained with the procedure.

If this is true with Statics, it is possibly even more so in the context of dynamics. Analytical dynamics (the name given to these scalar, work-based formulations of mechanics) *is* initially somewhat arcane: there is a lengthy discourse with a lot of mathematics, some philosophy, and apparently unrelated facts; then, suddenly, there appears the *deus ex machina*—a remarkably simple formulation almost routine in application, precisely as back in Statics! But, as with virtual work, any time taken to review the development, particularly after working some problems, will result in a far less jaundiced view of the subject and the feeling that, in fact, it *does* rest on a solid foundation conceptually. This approach is encouraged.

Chapter 6

Analytical Dynamics: Perspective

Introduction

In order to provide a perspective on the issues dealt with by *scalar* Analytical Dynamics, we briefly discuss them from the viewpoint of *vector* Classical Dynamics. Central is the subject of *constraint*: a given motion is reflected both in the *forces*—the *kinetics* and the [vector] coordinates used to describe this—the *kinetics.* Yet Virtual Work in Statics manages to *eliminate* the forces of constraint (as well as those of interconnection in frictionless machines) because they *do no work.* And Virtual Work somehow manages, from a *scalar* principle (work), to get "components" of the balance of the forces. These issues become central to that generalization of Virtual Work to Dynamics that we call Analytical Dynamics.

6.1 Vector Formulations and Constraints

Newton's Second Law, the foundation of all of mechanics, is a *vector* principle; one must express it—in particular, the various forces, moments, and accelerations—in a *vector* coordinate system. Thus, in the kinematics of dynamics, one typically studies Cartesian, polar, tangential-normal and possibly spherical coordinates systems, giving expressions for $\boldsymbol{r}$ (at least in all but tangential-normal coordinates, which *start* with velocity), $\boldsymbol{v}$, and $\boldsymbol{a}$ in these systems. There is immediately a justifiable question: Why is it *necessary* to have all these systems? In a very real sense, it isn't; physical laws involve *physical* quantities, not *mathematical* ones, and one should be able to express a physical principle in *any* coordinate system.

A body *free* to move in two- or three-dimensional space requires coordinates equal in number to the corresponding dimension to specify its motion. But motion which is *not* free—which is constrained to move in a certain direction—imposes relations among its coordinates. The implementation of such *kinematical constraints* can be facilitated by an appropriate choice of coordinate system: If a particle is moving in a straight line the constraint imposed on this motion can easily be imposed by taking Cartesian basis vectors with one parallel to the motion, and saying that the coordinate[s] in the direction[s] *perpendicular* to that basis vector are constant; thus if we take $\hat{\boldsymbol{\imath}}$ in the direction of motion, the kinematical constraints are that $y = y_o, z = z_o$, and

$$\boldsymbol{r} = x\hat{\boldsymbol{\imath}} + y_o\hat{\boldsymbol{\jmath}} + z_o\hat{\boldsymbol{k}}, \quad \boldsymbol{v} = \dot{x}\hat{\boldsymbol{\imath}}, \quad \boldsymbol{a} = \ddot{x}\hat{\boldsymbol{\imath}}$$

provide reasonably simple expressions for these quantities.

We *could* use, say, polar coordinates (evidencing at least *some* wisdom by taking the reference plane through the polar origin and the line, so that $z \equiv 0$), but then our constraint imposing straight-line motion would be that $r\cos\theta = d$, where d is the perpendicular distance between the chosen origin of the polar coordinates and the motion's trajectory; this would have to be solved for $r(\theta)$; then differentiated, and explicitly solved for, say, $\dot{r}(\theta, \dot{\theta})$; then differentiated *again* and solved for, say $\ddot{r}(\theta, \dot{\theta}, \ddot{\theta})$; finally expressing

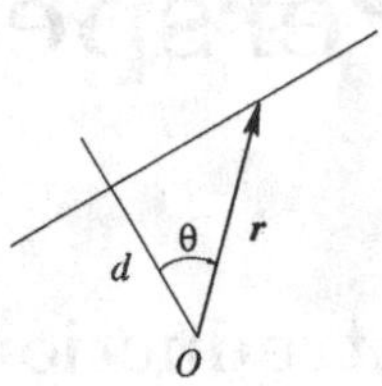

$$\boldsymbol{r} = r(\theta)\hat{\boldsymbol{r}}, \quad \boldsymbol{v} = \dot{r}(\theta, \dot{\theta})\hat{\boldsymbol{r}} + r(\theta)\dot{\theta}\hat{\boldsymbol{\theta}},$$
$$\boldsymbol{a} = (\ddot{r}(\theta, \dot{\theta}, \ddot{\theta}) - r(\theta)\dot{\theta}^2)\hat{\boldsymbol{r}} + (2\dot{r}(\theta, \dot{\theta})\dot{\theta} + r(\theta)\ddot{\theta})\hat{\boldsymbol{\theta}}$$

—not a pleasant sight!

Conversely, if motion were in a circle, the constraint that $r = a$ is easy to implement in polar coordinates; but the above process would have to be carried out on the corresponding constraint $x^2 + y^2 = a^2$, expressing, say, $y(x)$, $\dot{y}(x, \dot{x})$, and $\ddot{y}(x, \dot{x}, \ddot{x})$ if one used Cartesian coordinates. (If, of course, the motion is neither straight nor circular—along a curve $y = f(x)$, say—there is no recourse but to use [here] Cartesian coordinates, utilizing this constraint explicitly by taking its time derivatives as discussed.)

The above deals only with the *kinematics* of the problem. When we come to the *kinetics*, it is necessary to express the *forces* imposing the above constraints in whichever coordinate system has been chosen, as motivated by the kinematics. But the best choice for the kinematic constraints is mirrored in that for the kinetics enforcing these: Again in the straight-line case, if Cartesian coordinates were selected, the *normal* force would have components in the $\hat{\boldsymbol{\jmath}}$- and $\hat{\boldsymbol{k}}$- directions, and friction would have simply an $\hat{\boldsymbol{\imath}}$ component. But if polar coordinates were invoked, each of these would have to be resolved in terms of θ along the basis vectors $\hat{\boldsymbol{r}}$ and $\hat{\boldsymbol{\theta}}$ (with at least the $\hat{\boldsymbol{k}}$-component cooperating!)[1]

[1] A feature bearing on the choice of coordinates, rarely emphasized, is the fact that the polar coordinate $\hat{\boldsymbol{r}}$ and $\hat{\boldsymbol{\theta}}$ are undefined at the origin. Thus these presumably should be avoided in cases in which such motion is possible.

There is an important observation regarding these examples: a *kinetic* constraint (the normal forces enforcing the above motions) reflects itself in a *kinematic* constraint imposed on the expressions for position, velocity, and acceleration. In this regard, the two are equivalent—a constraint force leads to a certain type of motion, and a given motion must result from some force; this reflexive nature will play a pivotal role in the present Part. But *both* must be accounted for in Newton's Second Law, one in the expression for $\boldsymbol{a}$ in $m\boldsymbol{a}$, and the other in the $\sum \boldsymbol{F}$ side; it almost seems redundant.

If the above dealt with a single particle, motion of a single rigid body would involve similar considerations. But what if we were given a *system* of rigid bodies—a "*machine*" in engineering parlance? Because of the different motions of the various centers of mass, it would be necessary to break up the system into its constituent parts, writing equations of motion for each. While one body might possess an axis about which fixed rotation occurs, admitting the use of polar coordinates to describe *its* motion, others would not. For the latter we typically resort to "the coordinate system of last resort", Cartesian coordinates; now the kinematical constraints are enforced through the use of *rotating* coordinate systems, while the kinetic forces generating these constraints *including the supports and pin forces holding the machine together* must still appear in the force side of the equation of motion. There results an impressive system of linear equations in the accelerations and these forces, *coupled* with one another through both the pin forces and accelerations, which must be solved simultaneously.

6.2 Scalar Formulations and Constraints

"Machines" are also treated in what is typically the first engineering course, Statics. The problem there is simplified by the fact that all accelerations vanish, though the mathematical nature of the linear system of equation resulting from such problems remains. But an alternative approach to the analysis of machines is generally presented (typically the last topic in the course—"Is this going to be on the final?"): *virtual work*. There one considers, not work in the normal sense of dynamics, but *virtual* work—work *thought* about but not done.[2] Rather than considering the [vector] *forces/moments* acting on a body or system, virtual work considers the *work* these do under a *virtual displacement* from the equilibrium configuration. Further, since it is based on work, one need consider only those forces/moments *doing* [virtual] work, unlike the Newtonian formulation, where one must consider *all* agents acting on a body; those doing work are generally only a subset of the totality of all forces and couples. In particular, the supports do no work as long as displacements are "consistent" with such constraints; neither do frictionless interconnecting pins in a machine. The only forces which end up doing work are the applied, "active" forces and/or couples.

[2]Given the propensity for this practice among students, it is surprising virtual work is not more popular with them.

Work, of course, is a scalar. There is an important ramification of this fact which transcends merely the distinction between "vectors" and "scalars": *Scalar quantities are independent of vector coordinate systems*—they retain their value regardless of the *vector* system in which they are embedded. For example, the distance between two points is the same, whether those points' positions are measured in Cartesian coordinates or polar. This means that we can use any convenient [scalar] coordinates to measure them. This evidences itself in dynamics, where work/energy problems will often "mix" various scalar coordinates in the measurement of the work and/or potential energy in a given system—though it is necessary, either explicitly or implicitly, to relate these coordinates to one another.

Analytical dynamics, particularly *Lagrangian* dynamics, is *based* on the principle of virtual work; we shall see this in the derivation of the Lagrange equations. But now we consider "virtual displacements" from the actual *motion* of the dynamical system, rather than from its static *equilibrium.* (Still left open is the question of just what such a "virtual displacement" actually entails in a moving body.) Like the Statics Principle of Virtual Work, constraint forces of support will do no work (as long as they're not violated), so they won't enter the equations of motion. Like the Principle of Virtual Work, interconnecting pin forces in a system will similarly not enter (as long as they're *frictionless*); it is merely the *active* forces on which we focus our attention. And like the Principle of Virtual Work, its scalar nature allow us to use *any* convenient coordinates in terms of which to formulate our problems; in fact, *the entire dynamics of a system is contained in a scalar function.* This is likely the major advantage of analytical dynamics: we are no longer shackled by the relatively small number of vector coordinate systems at our disposal and can select *any* coordinates of convenience, or preference, in terms of which to describe the problem. There is an additional advantage to the analytical formulations of dynamics: the equations of motion generally depend on whether or not a given coordinate (or, in Hamiltonian dynamics, a *momentum*) actually *appears* in the scalar function describing the system, alluded to above. This fact makes it possible to *solve* the system (at least implicitly) without resorting to the solution of differential equations! We shall see this in Chapter 9.

In view of this conceptual heritage common to both virtual work in Statics and the present subject, it is useful briefly to review the former, since these ideas will recur continually in what is to follow.

6.3 Concepts from Virtual Work in Statics

For a body in static equilibrium the *actual* displacement, $d\boldsymbol{r}$, of every point is zero: $d\boldsymbol{r} \equiv \mathbf{0}$; thus no *actual* work is done by any force which might act at that point: $dU = \boldsymbol{F}\cdot d\boldsymbol{r} \equiv 0$. But one can consider an arbitrary *virtual displacement*, denoted $\delta\boldsymbol{r}$, from that actual [non]-motion: $\delta\boldsymbol{r} \neq d\boldsymbol{r} \equiv \mathbf{0}$. If a force $\boldsymbol{F}$ acts at $\boldsymbol{r}$, then $\boldsymbol{F}$ generally does *virtual* work $\delta U \equiv \boldsymbol{F}\cdot\delta\boldsymbol{r} \neq 0$. Similarly, any couple moment, $\boldsymbol{M}$, does work $\delta U \equiv \boldsymbol{M}\cdot\delta\boldsymbol{\theta}$, where $\delta\boldsymbol{\theta}$ is the [virtual] *angular* displacement of the body to which $\boldsymbol{M}$ is applied, its direction defined *via* the right-hand rule along the axis about which such rotation takes place, and its magnitude measured in a plane perpendicular to that axis. The *Principle of Virtual Work* then states that: *a body/system is in equilibrium if and only if the net work done by all forces and couples, under* any *virtual displacement from the alleged equilibrium configuration, is zero.*

The method of proving this is important. The original system of forces and couples acting on a rigid body is equivalent to a *single* force, $\boldsymbol{R} = \sum \boldsymbol{F}_i$ acting at, say, the center of mass G of the body, plus a *single* couple, $\overline{\boldsymbol{M}} \equiv \sum \boldsymbol{\rho}_i \times \boldsymbol{F}_i + \sum \boldsymbol{M}_i$ (where the $\boldsymbol{\rho}_i$ are then measured from the center of mass), acting on the body as in the diagram at left. Since these two systems are equivalent, we can utilize the simpler, reduced system to prove the desired result.

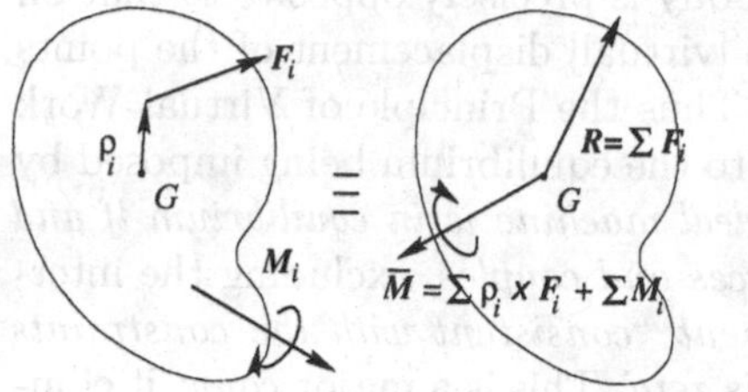

The net virtual work done under an arbitrary virtual displacement of the body is then

$$\delta U \equiv \boldsymbol{R} \cdot \delta \overline{\boldsymbol{r}} + \overline{\boldsymbol{M}} \cdot \delta \boldsymbol{\theta} \tag{6.3-1}$$

—in which $\delta \overline{\boldsymbol{r}}$ is the absolute [virtual] translational displacement of G and $\delta \boldsymbol{\theta}$ the absolute [virtual] rotational displacement of the body. If the body is in equilibrium, then $\boldsymbol{R}$ and $\overline{\boldsymbol{M}}$, as the sums of the forces and moments, respectively, must both vanish; thus $\delta U \equiv 0$; this is easy to swallow.

But proof of the converse is particularly significant, since it seems to rely on an almost *semantic* argument, hinging on the single word "any": If $\delta U \equiv 0$ for *any* virtual displacement, we can consider one in which $\delta \boldsymbol{\theta} \equiv \boldsymbol{0}$, but $\delta \overline{\boldsymbol{r}} \neq \boldsymbol{0}$ (and, in particular, not perpendicular to $\boldsymbol{R}$); the only way δU can vanish is for $\boldsymbol{R} \equiv \boldsymbol{0} = \sum \boldsymbol{F}_i$. Similarly, considering $\delta \overline{\boldsymbol{r}} \equiv \boldsymbol{0}$ but $\delta \boldsymbol{\theta} \neq \boldsymbol{0}$, $\overline{\boldsymbol{M}} \equiv \sum \boldsymbol{r}_i \times \boldsymbol{F}_i + \sum \boldsymbol{M}_i \equiv \boldsymbol{0}$. But the last two results are precisely the force and moment equilibrium equations, so the vanishing of δU under "*any*" virtual displacement implies equilibrium, as desired. The argument implicitly takes $\delta \overline{\boldsymbol{r}}$ and $\delta \boldsymbol{\theta}$ to be *independent*: we can vary each independently of the other. It is this independence which allows us to conclude that vanishing of the *summation* (6.3-1) implies vanishing of the *individual terms.* This is at the heart of the argument in Statics, and this same independence, somewhat formalized, will play a critical role in the development of Lagrangian Dynamics. Indeed, it is precisely the explicit or implicit implementation of constraints relating the coordinates which renders those remaining to *be* independent.

This basic Principle involves *all* forces and couples, among them the support *constraints*, say $\boldsymbol{F}^{(c)}$ and $\boldsymbol{M}^{(c)}$. But it turns out that *constraint forces and couples do no work* (as long as they are not "violated"): a pin constraint does no work because $\delta \boldsymbol{r} \equiv 0$, a constraining couple does no work because $\delta \boldsymbol{\theta} \equiv 0$, and a normal force $\boldsymbol{N}$ does no work because $\delta \boldsymbol{r} \perp \boldsymbol{N}$; so for *all* constraints we can conclude that $\boldsymbol{F}^{(c)} \cdot \delta \boldsymbol{r} \equiv 0$ and $\boldsymbol{M}^{(c)} \cdot \delta \boldsymbol{\theta} \equiv 0$, as least as long as these constraints are observed, *i.e.*, not "violated". Thus the above principle specializes to: *a body/system is in equilibrium if and only if the net work done by all* applied *forces and couples, under a virtual displacement* "consistent with the constraints" *from the alleged equilibrium configuration, is zero.*

While this Principle can be applied to a single rigid body, it turns out to be more involved to do this than merely to invoke the standard vector conditions for equilibrium. But if one considers a "machine", virtual work comes to the fore: assuming the interconnecting pins are *frictionless* (as one invariably does), they do no net work: the pin force on one body is precisely opposite to that on the other (Newton's Third Law), while the [virtual] displacement of the points at which this force is applied is the same. Thus the Principle of Virtual Work applied to such "ideal machines" simplifies to the equilibrium being imposed by only the *applied* forces and moments: *an ideal machine is in equilibrium if and only if the net work done by all applied forces and couples*, excluding the interconnecting pins, *under a virtual displacement "consistent with the constraints from the alleged equilibrium configuration, is zero.* This is a major *coup*: it eliminates from consideration the interconnecting pins—the very forces coupling the *vector* equilibrium conditions in that approach.

Work is a scalar; this means that we can use any convenient [scalar] coordinates to measure it. Recall, in this regard, how the [scalar] potential energy due to gravity is customarily measured utilizing [Cartesian] height y, while that due to a spring typically uses its [polar] length $r - r_o$ measured from its fixed end (or some reasonable facsimile)—*both in the same expression for energy*: it is not necessary to hold doggedly to a particular set of coordinates as one must do in a vector coordinate system. Virtual work, based on work, is thus also scalar in nature, enjoying the advantages of this freedom. There is a down side to this fact, however: Since it *is* a scalar principle, *we get only a single equation* by invoking it; thus it is really only useful when the virtual displacements considered can be parameterized by a single scalar quantity—when the system has only a "single degree of freedom"—and all these variables we have chosen to express the work must ultimately be expressed in term of this parameter. Furthermore, since the constraint forces—often the very forces one is interested in finding in Statics!—do no work, *they can't be found* using this Principle! Nonetheless, having information regarding the system that virtual work gives us, we can often return to the vector formulation with an easier system to solve, now that some of the unknown quantities have been determined independently of the vector form.

Summary

Vector formulations of mechanics demand that one express each equation of motion in terms of a *vector* coordinate system. Further, *all* forces and couples acting on the bodies must be accounted for, both in the *kinetic* force/moment side of the equations, and in the *kinematic* acceleration side.

Yet Statics proposes a *scalar* alternative based on energy, the *Principle of Virtual Work*. There entire classes of forces and moments—constraining supports and interconnecting frictionless pins in a machine—do no work as long as they are not "violated", and therefore do not appear in the expression for the virtual work done under any virtual displacement of the system one considers. In this formulation, such "constraints" are merely *kinematical*, implicit in the allowable choice of virtual displacement. As a scalar principle, this can be expressed in terms of *any* scalar variables. The demonstration of this principle rests on the conceptual *independence* of the virtual displacements: though the net [virtual] work done under such displacements is a [scalar] summation, this independence allows us to conclude that the vanishing of the *sum* implies the vanishing of the *individual terms* in the sum.

Chapter 7

Lagrangian Dynamics: Kinematics

Introduction

As is done with classical dynamics, we shall start by examining the kinematical considerations in analytical dynamics. The primary focus of this chapter will be a discussion of *virtual displacements* and what they are in the context of dynamics, as well as *generalized coordinates* and what might qualify as an appropriate set of such coordinates. But, in view of the relationship between kinetic constraints and their kinematical manifestation, we also see how the first effect the latter, particularly with regard to possible [virtual] displacements. In order to put this all into some perspective, however, we start with a brief discussion of how constraints affect the specification of of position—used to describe the *configuration* of a system—in the first place.

7.1 Background: Position and Constraints

Newton's Second Law for a particle is $\sum \boldsymbol{F} = m\boldsymbol{a}$.[1] This involves the *acceleration* of the particle, but we are almost invariably less interested in the acceleration that we are in the *position*, $\boldsymbol{r}$. Nonetheless, we can get an equation for $\boldsymbol{a}$ by merely differentiating—getting the second *total* time derivative, $\ddot{\boldsymbol{r}} \equiv \boldsymbol{a}$. But this derivative depends on the form which $\boldsymbol{r}$ assumes; thus we will examine this question more closely through some simple, even prosaic, 2-D examples, in order to motivate the discussion to follow.

Example 7.1.1. *Unconstrained Motion of a Particle.* For motion in two dimensions, it requires generally *two*, *independent* coordinates to specify r. Thus,

[1] Actually, it isn't. See Section 5.1.1.

for example, we could use Cartesian coordinates

$$\boldsymbol{r} \equiv \boldsymbol{r}(x, y) = x\hat{\boldsymbol{\imath}} + y\hat{\boldsymbol{\jmath}}.$$

The expression for $\boldsymbol{r}$ in polar coordinates

$$\boldsymbol{r} \equiv \boldsymbol{r}(r, \theta) = r\hat{\boldsymbol{r}}$$

seems to belie the above statement regarding the number of coordinates until we realize that $\hat{\boldsymbol{r}} = \hat{\boldsymbol{r}}(\theta)$; thus the second coordinate is implicit in the basis itself, and two coordinates *are* still required. |

However elementary this example has been, it does demonstrate several profoundly important points regarding position vectors and their expression using coordinates:

- Specification of the position requires [scalar] coordinates equal in number to the [linearly independent] basis vectors—the *dimension* of the space.

- For unconstrained motion, that position depends *only* on the coordinates, not time or the coordinates' time derivatives.[2]But that doesn't mean that the *total* $\frac{d\boldsymbol{r}}{dt} \equiv 0$; the latter is, after all, the velocity, which generally doesn't vanish. Rather, time enters $\boldsymbol{r}$ *implicitly* through the coordinates, which *do* depend on time: if, as we shall be doing a great deal of, we denote by $\boldsymbol{q}$ an *arbitrary* choice of coordinates (Cartesian, polar, *etc.*),

$$\frac{d\boldsymbol{r}}{dt} = \sum_i^N \frac{\partial \boldsymbol{r}}{\partial q_i}\dot{q}_i + \frac{\partial \boldsymbol{r}}{\partial t}$$

  (where N is the number of coordinates); that $\boldsymbol{r}$ is independent of time *explicitly* means that $\frac{\partial \boldsymbol{r}}{\partial t} = 0$. Similarly, $\frac{\partial \boldsymbol{r}}{\partial \dot{q}_i} \equiv 0$ because $\boldsymbol{r}$ depends only on the coordinates, not their time derivatives.

The next two examples will consider the motion of the above particle to be *constrained* by similar conditions: in both cases it is to move along a line with slope m. The motion is no longer "free", but the x- and y-components of $\boldsymbol{r}$ will be *related* through some equation involving the line. The primary difference between them is the fact that, in the first example, the line is *fixed*; in the second (shaded, and with parenthesized symbols in the diagram at left) the line will be *moving*.

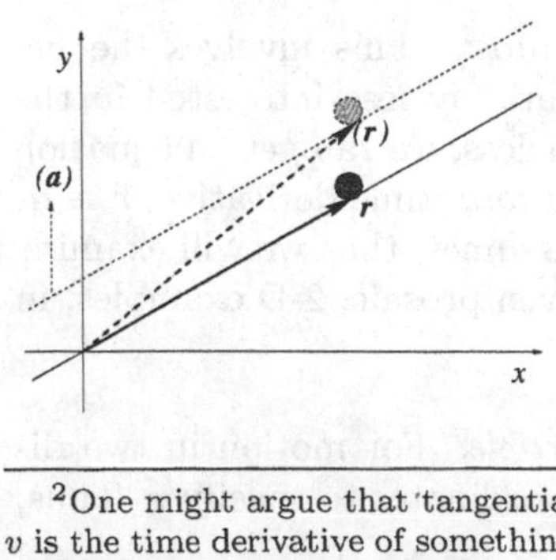

[2]One might argue that tangential-normal coordinates violate this maxim—that in $\boldsymbol{v} = v\hat{\boldsymbol{t}}$, v is the time derivative of something. But, in fact, this set of coordinates *starts* with $\boldsymbol{v}$; there *is* no [vector] expression for $\boldsymbol{r}$!.

Example 7.1.2. *Motion with Fixed Constraint.* Assume the particle is constrained to move along the *fixed* line passing through the origin of coordinates. The x- and y-components of $\boldsymbol{r}$ are related through the equation $y = mx$. Thus we can either represent $\boldsymbol{r}$ as above, with the side condition that this constraint be satisfied:

$$\begin{aligned} \boldsymbol{r} \equiv \boldsymbol{r}(x,y) &= x\hat{\imath} + y\hat{\jmath}, \text{ where } \phi(x,y) \equiv y - mx = 0 \\ \therefore \boldsymbol{v} &= \dot{x}\hat{\imath} + \dot{y}\hat{\jmath}, \text{ where } \dot{\phi}(x,y) = \dot{y} - m\dot{x} = 0 \\ \boldsymbol{a} &= \ddot{x}\hat{\imath} + \ddot{y}\hat{\jmath}, \text{ where } \ddot{\phi}(x,y) = \ddot{y} - m\ddot{x} = 0 \end{aligned} \tag{7.1-1}$$

—in which case the two coordinates are no longer independent, due to the additional constraint; or we can explicitly *eliminate* the constraint:

$$\begin{aligned} \boldsymbol{r} \equiv \boldsymbol{r}(x) &= x\hat{\imath} + mx\hat{\jmath} \\ \therefore \boldsymbol{v} = \dot{x}\hat{\imath} + m\dot{x}\hat{\jmath}, &\quad \boldsymbol{a} = \ddot{x}\hat{\imath} + m\ddot{x}\hat{\jmath} \end{aligned} \tag{7.1-2}$$

In this latter case, the position is no longer a function of x and y, but only x, due to the explicit elimination of the constraint: only one coordinate is required to specify the position of the particle, even though motion allegedly takes place in 2-D space.

We note in passing that $\phi(x, y) = 0$ in equation (7.1-1) can be differentiated to give a relation between the *differentials* of these coordinates:

$$d\phi(x,y) = dy - m\,dx = 0;$$

conversely, *knowing* the expression for $\boldsymbol{v}$ in (7.1-2), the same relation could be obtained by multiplying the $\hat{\jmath}$ component of $\boldsymbol{v} = \dot{x}\hat{\imath} + m\dot{x}\hat{\jmath} = \dot{x}\hat{\imath} + \dot{y}\hat{\jmath}$—that component in which the constraint is introduced above—by dt. The significance of this fact will become clear below. |

The first example showed that the number of coordinates required to specify an *unconstrained* $\boldsymbol{r}$ is equal to the dimension of the space. This one shows that a [kinematic] constraint *reduces* that number (by one for each such constraint). Note, however, that time has yet to enter the $\boldsymbol{r}$ expression.

Example 7.1.3. *Motion with Moving Constraint.* Now let us assume that the above constraining line, though still having a fixed slope m, is *moving* in the y-direction with constant acceleration a; thus, assuming that the line passes through the origin at $t = 0$, now $y = mx + \frac{1}{2}at^2$, which can be expressed as a scalar function $\phi(x, y; t) = y - (mx + \frac{1}{2}at^2) = 0$. We could get an equation like (7.1-1) by simply carrying this constraint along; or, analogously to (7.1-2), we could eliminate it. But now, unlike any of the previous cases, *the time t is introduced explicitly*:

$$\begin{aligned} \boldsymbol{r} \equiv \boldsymbol{r}(x;t) &= x\hat{\imath} + (mx + \frac{1}{2}at^2)\hat{\jmath} \\ \therefore \boldsymbol{v} = \dot{x}\hat{\imath} + (m\dot{x} + at)\hat{\jmath}, &\quad \boldsymbol{a} = \ddot{x}\hat{\imath} + (m\ddot{x} + a)\hat{\jmath} \end{aligned}$$

Observe that, in contrast to (7.1-2), there is an additional term corresponding to $\frac{\partial \boldsymbol{r}(x,t)}{\partial t}$ in $\boldsymbol{r}$, and an additional term in the *total* time derivative $\boldsymbol{a} = \ddot{\boldsymbol{r}}$.

Once more, we can differentiate $\phi(x, y; t) = 0$:

$$d\phi(x, y; t) = dy - m\,dx - at\,dt = 0,$$

yielding a relation involving not only x and y, but now again also t, due to the moving nature of the constraint in this case. As in the previous example, *knowing* $\boldsymbol{v}$ we could merely multiply the $\hat{\jmath}$ component of $\boldsymbol{v}$ by dt to get this same result. |

Even in the presence of a time-*in*dependent constraint relating the coordinates, $\boldsymbol{r} = \boldsymbol{r}(\boldsymbol{q})$ and $\frac{\partial \boldsymbol{r}}{\partial t} \equiv 0$. But if the constraint is a *moving* one, $\boldsymbol{r}$ depends on the time, and $\frac{\partial \boldsymbol{r}}{\partial t} \neq 0$ for the first time. This is the *only* way t can enter the expression for $\boldsymbol{r}$ explicitly.

These two examples have been intentionally straightforward. In both we were able to embody the kinematic constraint in a function, $\phi = 0$, expressed in terms of the coordinates themselves (and, in the case of a *moving* constraint, also involving the time explicitly). The next, example, however, cannot be so written:

Example 7.1.4 *Motion with Differential Constraints.* We consider a surprisingly prosaic example: A disk, with its diameter always vertical, rolls without slipping on a horizontal plane. We can completely specify its configuration with four coordinates: x and y, such that (x, y) is the position of the disk's center O; α, the orientation of the disk's plane (measured between that plane and the yz-plane); and θ, the rotation of the disk relative to the horizontal. We allege that, despite the relative simplicity of this system, constraints relating the above variables satisfy only non-integrable differential constraints.

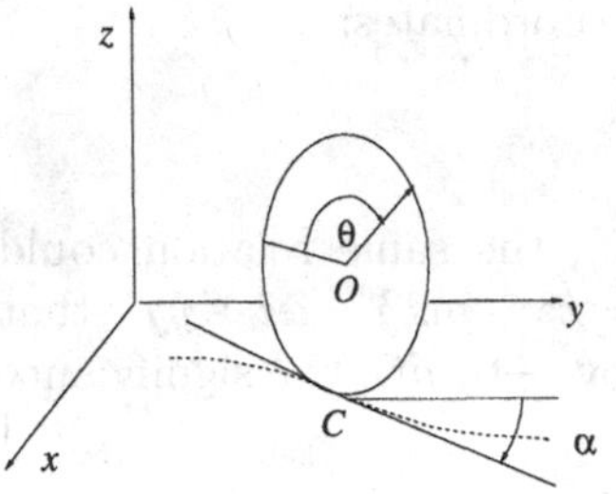

Before getting to the bulk of the problem, it is interesting to see just *why* only four coordinates are required to specify the motion of this system: As a rigid body in three dimensions, we would normally expect there to be required six coordinates—three for the center, say, plus three Euler angles. Clearly, however, O will always have a z-coordinate equal to the radius of the disk, say r, with associated constraint $\phi_1 = z - r = 0$, leaving only the (x, y) cited above to specify the position of the center. And the orientation of the disk requires only *two* Euler angles: axes with, say, z' along that radius making the angle θ with the horizontal, and x' perpendicular to the disk (y' then determined by the right-hand rule) can be brought coincident with the (x, y, z) system by only two rotations, one around x' through $\theta - 90°$ (bringing $\hat{\boldsymbol{k}}'$ and $\hat{\boldsymbol{k}}$ coincident), followed by a second around the [*new*] z'-axis through α, generating a composite rotation matrix $\mathsf{R} = \mathsf{A}(\theta - 90)\mathsf{C}(\alpha)$. Yet there are generally *three* Euler angles, with rotations through each, first about one axis, then about a second, and finally one about the first again. Where is that third angle in this case, and how can it be specified by a constraint?

The answer to this comes from the observation that each of the three rotation matrices reduces to the identity matrix **1** when the rotation angle is 0. The pattern of rotations "first/second/first", then, will be maintained if we simply "dummy in" an initial rotation angle of $\psi_{z'} = 0$ about the z'-axis, *or* a final one of $\psi_{x'} = 0$ about x', the associated constraint then becoming either $\phi_2 = \psi_{z'} = 0$ or $\phi_2 = \psi_{x'} = 0$. This constraint, along with the first, then reduces the number of coordinates to four.[3] Note that both of these constraints are integrable; we now return to the problem to exemplify two others which *aren't*:

In the previous problems, we obtained constraints relating the coordinates, then used time derivatives of these constraints to obtain velocity and acceleration. Here we reverse that procedure: By the non-slip condition, we can relate the velocity of its center to its [*absolute*] *angular* velocity, $\boldsymbol{\Omega}$: $\boldsymbol{v}_O = \boldsymbol{\Omega} \times \boldsymbol{r}$, where $\boldsymbol{r}$ is measured from the point of contact to the center and, in this case, is always vertical. Here $\boldsymbol{\Omega} = \dot{\boldsymbol{\theta}} + \dot{\boldsymbol{\alpha}}$, where $\dot{\boldsymbol{\theta}}$ is measured about an axis $\hat{\boldsymbol{\theta}}$, always horizontal, through O, and $\dot{\boldsymbol{\alpha}}$ about a vertical one; because of the latter, however, $\dot{\boldsymbol{\alpha}} \times \boldsymbol{r} \equiv 0$. Thus

$$\boldsymbol{v}_O = \dot{\boldsymbol{\theta}} \times \boldsymbol{r},$$

relating the translational velocity $\boldsymbol{v}_O = \dot{x}\hat{\boldsymbol{\imath}} + \dot{y}\hat{\boldsymbol{\jmath}}$ and the angular $\dot{\boldsymbol{\theta}} = \dot{\theta}\hat{\boldsymbol{\theta}}$, and thus the *time derivatives* of these coordinates.

We can relate the differentials of the coordinates by multiplying $\boldsymbol{v}_O = \dot{\boldsymbol{\theta}} \times \boldsymbol{r}$ by dt—the same device noted at the end of the previous two examples:

$$d\boldsymbol{r}_O = d\boldsymbol{\theta} \times \boldsymbol{r}.^4$$

But, since the direction of rotation is given by $(-\cos\alpha\hat{\boldsymbol{\imath}} + \sin\alpha\hat{\boldsymbol{\jmath}})$, we have immediately that $d\boldsymbol{\theta} = d\theta(-\cos\alpha\hat{\boldsymbol{\imath}} + \sin\alpha\hat{\boldsymbol{\jmath}})$, so that

$$\begin{aligned} d\boldsymbol{r}_O &= dx\hat{\boldsymbol{\imath}} + dy\hat{\boldsymbol{\jmath}} \\ &= d\boldsymbol{\theta} \times \boldsymbol{r} \\ &= d\theta(-\cos\alpha\hat{\boldsymbol{\imath}} + \sin\alpha\hat{\boldsymbol{\jmath}}) \times r\hat{\boldsymbol{k}} \\ &= r\sin\alpha\, d\theta\hat{\boldsymbol{\imath}} + r\cos\alpha\, d\theta\hat{\boldsymbol{\jmath}}. \end{aligned}$$

Thus we have *two* relations among the differentials of the chosen coordinates:

$$dx - r\sin\alpha\, d\theta = 0$$
$$dy - r\cos\alpha\, d\theta = 0.$$

[3]Though these are the customary sets of coordinates—position of one point plus a rotation—another pair of conditions is possible: for example, that the point of contact always have $z_C \equiv 0$, and that the diameter be vertical so that $z_O = r$.

[4]It might appear that we have "forced" a differential form here, by taking an already integrated form and differentiating it! But note that, without knowing the solution $\boldsymbol{r}_O(t)$—and at this stage, we don't—we cannot relate θ to x and y. [On a more subtle level, the relation $\boldsymbol{\omega} \equiv d\boldsymbol{\theta}/dt$ involves the *vector* $\boldsymbol{\omega}$, which can't be integrated to a *vector* "$\boldsymbol{\theta}$": *finite* rotations are not vectors! See the remark on page 154.]

Recall that, in the previous examples, we were able to write such relations in terms of a [scalar] relation among the coordinates $\boldsymbol{q}$: $\phi(\boldsymbol{q};t) = 0$. It is natural to ask whether we can do the same in the present case. To be able to do so would require that each of the above two relations be *exact differentials* of a function—that each would be $d\phi_i$ for some functions ϕ_i. But this is true if and only if the "mixed partials" of the above sums be the same; *e.g.*, for the first relation

$$(1)dx + (0)dy + (-r\sin\alpha)\,d\theta + (0)d\alpha \equiv a_{11}dx + a_{12}dy + a_{13}d\theta + a_{14}d\alpha,$$

it would be necessary that, for example

$$\frac{\partial a_{13}}{\partial\alpha} = \frac{\partial a_{14}}{\partial\theta}.$$

But if we try to verify this:

$$\frac{\partial a_{13}}{\partial\alpha} = -r\cos\alpha \neq 0 = \frac{\partial a_{14}}{\partial\theta},$$

so this differential constraint is *not* an exact differential.

Thus we see that, in the spirit of the previous examples, it is possible to relate velocities and accelerations. Yet, even in this relatively simple, non-general example, we cannot directly relate the four *coordinates* (x, y, θ, α) or express any relation of the form $\phi(x, y, \theta, \alpha) = 0$. Instead we get a constraint involving the *differentials* of the coordinates. |

The above are meant to exemplify constraints, how they enter the expressions for $\boldsymbol{r}$, $\boldsymbol{v}$, and $\boldsymbol{a}$ in Newtonian dynamics, and the effect of time dependency in the constraint on these expressions. Most importantly, however, we note that the most general form of constraint is a *differential* one: though we have obtained such a differential form from the differentiation of scalar constraints relating the coordinates, the last example demonstrates that, in general, we cannot go in the other direction!

Constraints, in some sense, relate the *coordinates* in a problem. In the first examples above, this could be done explicitly because we had an explicit [scalar] relation among the coordinates, and the latter could be used to *eliminate* the "superfluous" ones and reduce the number of actual coordinates required. In such a case it is possible to express $\boldsymbol{r} = \boldsymbol{r}(\boldsymbol{q};t)$, where $\boldsymbol{q}$ is some reduced set of coordinates, written for convenience as a vector. (Use of the semicolon to separate the two arguments is meant to emphasize that t, rather than being simply another parameter in the problem, is that parameter on which $\boldsymbol{q}$ generally depends.) We see that t enters explicitly on the presence of *moving*—*i.e.*, *time-dependent*—constraints which have been explicitly eliminated from the expression for $\boldsymbol{r}$. In the last example, however, such elimination couldn't be done, since the relation was only on the *differential* level. Yet even such a differential relation *somehow* relates the coordinates themselves, if only implicitly.

Categorization of Differential Constraints

If the above demonstrates how a function relating the coordinates can be used to obtain one relating the differentials, it points out more significantly that, in general, we have a relation among the *differentials* of both the coordinates and time

$$\sum_{i=1} a_i(\boldsymbol{q};t)\,dq_i + a_t(\boldsymbol{q};t)\,dt \equiv \sum_{i=0} a_i(\boldsymbol{q};t)\,dq_i = 0 \qquad (7.1\text{-}3)$$

(where, for convenience, we take $q_0 \equiv t$). This most general form of constraint is called its *Pfaffian form* and is the ultimate relation among differentials of the coordinates and time. Although we can always obtain such relations by taking the differential of a scalar relation among the variables themselves, this is really just a very special case of the Pfaffian form; indeed, as we saw in Example 7.1.4, only if the mixed partials of the coefficients in (7.1-3) are equal:

$$\frac{\partial a_i(\boldsymbol{q};t)}{\partial q_j} = \frac{\partial a_j(\boldsymbol{q};t)}{\partial q_i} \qquad (7.1\text{-}4)$$

can we conclude that the Pfaffian form reduces to the *exact differential* of some scalar function ϕ:

$$\sum_{i=1} a_i(\boldsymbol{q};t)\,dq_i = d\phi(\boldsymbol{q};t) = 0. \qquad (7.1\text{-}5)$$

This distinction between integrable and non-integrable constraints is actually quite fundamental: In general, we *start* with a non-integrable form (Example 7.1.4). If there *is* such a ϕ—if the constraint is *integrable*—then we can recover a relation among the coordinates themselves, which is part of the solution involving these variables; otherwise we can't. This observation, if nothing else, justifies the classification of differential constraints according to such integrability:

Definition 7.1.1. A [Pfaffian] constraint (7.1-3) satisfying (7.1-4) (so there *is* a $\phi(\boldsymbol{q};t)$, (7.1-5)), is called *holonomic*; otherwise, it is described as being *non-holonomic.*

There is an important fact regarding these two categories of constraints: If the constraint is holonomic, it might be possible to express one coordinate in terms of the others and thus reduce the number of differentials *and* coordinates; but if it is *non*-holonomic, we can't even do that. Even if the constraint *is* holonomic, however, the relation among the coordinates is generally *non-linear*; so, even though solution for some variable in terms of the others *should* be possible in principle, this is often "inconvenient," or even mathematically impossible.[5]

The general differential constraint, on the other hand, is always *linear* in the differentials, as above, *always* allowing solution for one differential in terms of the others. One might argue that this merely suppresses the problem—that the

[5] The Implicit Function Theorem says this *can* be done under reasonable conditions, but only locally. And Galois Theory in Modern Algebra demonstrates that, in general, fifth-degree polynomials are *in*soluble.

$a_i(\boldsymbol{q}; t)$ in (7.1-3) still involve *all* the coordinates (even though one coordinate's *differential* has been eliminated), that these *still* generally enter non-linearly, and that we are no better off than we were before: though we have eliminated a differential, the *coordinates themselves* still remain! The result: a general Pfaffian constraint yields a dq_i expressed in terms of not only the remaining $n-1$ differentials, but also in terms of *all* n "independent" coordinates! But this was not a problem in energy, where—since it is a *scalar*—potential energy can be expressed in any number of, generally, related coordinates. In the end, it is assumed that we know the values of *all* the coordinates, [differentially] related or not, and we can substitute in these values at the end.

Even if a constraint *is* holonomic, however, the overall situation is still murky: there might exist symmetries in the problem, such as $x^2 + y^2 + z^2 =$ *constant*. It is difficult to choose which variable should be eliminated here; indeed, such symmetries are intrinsically important, and one might prefer they be protected rather than eliminated! We shall return to these issues in Section 8.3.

If non-integrable constraints are the norm, some integrable ones are quite important:

Example 7.1.5. *Fixed Distance Constraint.* We provide another example of constraint, this one holonomic, of fundamental importance to the study of rigid-body motion: Two particles of masses m_1 and m_2 are fixed to the end of a rigid rod of length d. We shall obtain the constraint that fixed distance imposes on the system.

Taking the positions of the two masses to be at $\boldsymbol{r}_i = x_i\hat{\boldsymbol{\imath}} + y_i\hat{\boldsymbol{\jmath}} + z_i\hat{\boldsymbol{k}}$, the vector from, say, m_1 to m_2 is just $\boldsymbol{r}_{12} \equiv (x_2 - x_1)\hat{\boldsymbol{\imath}} + (y_2 - y_1)\hat{\boldsymbol{\jmath}} + (z_2 - z_1)\hat{\boldsymbol{k}}$. Thus the condition that the two particles be a fixed distance $r_{12} \equiv |\boldsymbol{r}_{12}| = d$ apart is just that $r_{12}^2 = (x_2 - x_1)^2 + (y_2 - y_1)^2 + (z_2 - z_1)^2 = d^2$. While this is a holonomic constraint, we are interested here in how it affects the virtual displacement. Since the latter follows the same rules as differentials, we find that

$$d(r_{12})^2 = d(\boldsymbol{r}_{12} \cdot \boldsymbol{r}_{12}) = 2\boldsymbol{r}_{12} \cdot d\boldsymbol{r}_{12} \equiv 0 \qquad (7.1\text{-}6)$$

where $d\boldsymbol{r}_{12} = (dx_2 - dx_1)\hat{\boldsymbol{\imath}} + (dy_2 - dy_1)\hat{\boldsymbol{\jmath}} + (dz_2 - dz_1)\hat{\boldsymbol{k}}$. Note that this demonstrates that any $d\boldsymbol{r}_{12}$ consistent with the fixed distance must be *orthogonal* to $\boldsymbol{r}_{12}$ itself, already used in the proof of Theorem 4.1.1. |

The above classification described the integrability of the [differential] constraint. We conclude this topic with one more categorization, now dealing with whether or not the constraint depends on *time*:

Definition 7.1.2. If a constraint is *independent of time*, it is called *scleronomic*; if it is *time-dependent*, it is described as *rheonomic*.

When it comes to problems which actually have to be solved, this classification seems rather academic, compared with the "practical" holonomic character of the constraint. But it has profound ramifications relative to the process described in Example 7.1.6, particularly equation (7.1-8)—which, recall, involves only *coordinate* differentials, not the time. This will be exemplified in Example 7.1.7.

Constraints and Linear Independence

There is an important interpretation to be given to any form of differential constraint: in the language of vector spaces, (7.1-3) says that *the differentials of the various coordinates are linearly dependent.* In Section 1.2.2 we introduced the idea of linear dependence among a set of vectors $\{\boldsymbol{v}_i\}$: that there are $\{c_1, c_2, \ldots, c_n\}$ such that $c_1\boldsymbol{v}_1 + c_2\boldsymbol{v}_2 + \ldots + c_n\boldsymbol{v}_n = \mathbf{0}$—a definition essentially saying that one $\boldsymbol{v}_i$ can be solved for in terms of the others. While this concept has been applied uniformly to *vectors* (and is stated here in terms of vectors), it was first motivated by being able to express one vector in terms of the others in the set—an idea which applies equally well to scalar coordinates (or their differentials). From this perspective, (7.1-3) is precisely Definition 1.2.3 cast in terms of differentials.

This means that certain of these differentials are, in some sense, *redundant* and can be eliminated in favor of the others. When such "redundant" differentials have been eliminated, however, those remaining are linearly *in*dependent, so that we can conclude of any equation involving the balance that

$$\sum c_i(\boldsymbol{q};t)\,dq_i = 0 \Rightarrow c_i(\boldsymbol{q};t) \equiv 0 \text{ for all } i \tag{7.1-7}$$

—precisely the definition of linear independence, Definition 1.2.3. In general, we would not be justified in concluding that the coefficients of the various dq_i vanish just because their *sum* vanishes; it is the *independence* of the dq_i which admits that result. Just to illustrate the situation, we consider an example:

Example 7.1.6. *Explicit and Implicit Introduction of Constraint.* Assume we have a set of n coordinates $\{q_1, q_2, \ldots, q_n\}$ *linearly independent* in their differentials in the above sense: that

$$\sum_{i=1}^{n} c_i(\boldsymbol{q};t)\,dq_i = 0$$

means that *all* $c_i \equiv 0$. If there is imposed on this system a constraint of the form (7.1-3) in which, although time appears, its *differential* does *not*,

$$\sum_{i=1}^{n} a_i(\boldsymbol{q};t)\,dq_i = 0, \tag{7.1-8}$$

this equation states that the original differentials are now linearly dependent, by definition: certainly not all a_i can vanish, so this is a linear combination of the dq_i which vanishes even though not all coefficients (the a_i) themselves vanish.

Thus far the constraint has not been eliminated; it must be carried as a "side condition" imposed after all coordinate manipulations have been performed. We might try to introduce it explicitly: Since not all coefficients can vanish, say that $a_{i_o} \neq 0$. Then, solving for

$$dq_{i_o} = -\frac{1}{a_{i_o}(\boldsymbol{q};t)} \sum_{i \neq i_o}^{n} a_i(\boldsymbol{q};t)\,dq_i \tag{7.1-9}$$

and substituting this into the left-hand side of the linear combination criterion (7.1-7),

$$\begin{aligned}\sum_{i=1}^{n} c_i(\boldsymbol{q};t)\,dq_i &= \sum_{i\neq i_o}^{n} c_i(\boldsymbol{q};t)\,dq_i + c_{i_o}(\boldsymbol{q};t)\,dq_{i_o}\\ &= \sum_{i\neq i_o}^{n} c_i(\boldsymbol{q};t)\,dq_i + c_{i_o}(\boldsymbol{q};t)\sum_{i\neq i_o}^{n}\Big(-\frac{a_i(\boldsymbol{q};t)}{a_{i_o}(\boldsymbol{q};t)}\Big)\,dq_i\\ &= \sum_{i\neq i_o}^{n}\Big(c_i(\boldsymbol{q};t) - \frac{a_i(\boldsymbol{q};t)c_{i_o}(\boldsymbol{q};t)}{a_{i_o}(\boldsymbol{q};t)}\Big)\,dq_i.\end{aligned}$$

By Theorem 1.2.2, this subset of a linearly independent set is also linearly independent, so if the sum vanishes, the coefficients of the dq_i must also; in particular, for $i \neq i_o$,

$$c_i(\boldsymbol{q};t) = \frac{a_i(\boldsymbol{q};t)c_{i_o}(\boldsymbol{q};t)}{a_{i_o}(\boldsymbol{q};t)} \neq 0$$

in terms of [arbitrary] $c_{i_o}(\boldsymbol{q};t)$, and we have a linear combination of the dq_i with non-0 coefficients which vanishes; *i.e.*, a linearly *de*pendent set. *But this is merely* $\frac{c_{i_o}(\boldsymbol{q};t)}{a_{i_o}(\boldsymbol{q};t)}$ *times the constraint equation*; in testing linear dependence of the set of differentials, we have actually recovered the constraint condition! Alternatively, we can eliminate the associated differential *implicitly*, restricting our consideration to the *sub*set of differentials $\{dq_1, dq_2, \ldots, dq_n, i \neq i_o\}$ ignoring q_{i_o}—dq_{i_o} can be expressed using (7.1-9); this subset *is* linearly independent.

Observe that the $a_i(\boldsymbol{q};t)$ generally depend on *all* the q_i, so though we may have eliminated dq_{i_o} using (7.1-9), the associated coordinate q_{i_o} itself will generally remain in the a_i. |

It is important to note that these two concepts—independence of the coordinates and [linear] independence of the *differentials* of these coordinates—are distinct: The differential of a holonomic constraint will yield a Pfaffian form, but consider a non-holonomic one; by definition, there is no relation among the variables, just among the differentials of these variables. The *differentials* are linearly dependent, but the coordinates themselves *aren't*, linearly or otherwise; else there would be some function relating them, and, by hypotheses, there isn't! *Q.E.D.* A purely semantic argument, but one showing it possible to have linearly dependent coordinate *differentials* without a corresponding dependence among the coordinates themselves!

Example 7.1.7. *Time-dependent Constraint.* Consider a single particle in 3-D. In Cartesian coordinates (x, y, z), linear independence of coordinates (actually, the coordinates' *differentials*), equation (7.1-7), becomes

$$c_1\,dx + c_2\,dy + c_3\,dz = 0. \tag{7.1-10a}$$

The presumed linear independence of the three coordinates means that the $c_i \equiv 0$ above.

Now assume the particle is constrained to move on the surface of a parabolic sheet $z = ax^2$—one moving with velocity $v_o\hat{k}$ which has passed through the origin at $t = 0$; this imposes the constraint [function] $\phi(x, y, z) = z - (ax^2 + v_o t) = 0$. Put in terms of *actual* coordinate differentials, this becomes

$$dz = 2ax\,dx + v_o\,dt. \tag{7.1-10b}$$

While this is the solution for dz, let us (somewhat more perversely) solve instead for $dx = (dz - v_o\,dt)/(2ax)$ [assuming $x \neq 0$], just to illustrate that such constraints generally depend on the coordinates—even those being "eliminated." Now the linear independence criterion, equation (7.1-10a), becomes

$$c_1\frac{dz - v_o\,dt}{2ax} + c_2 dy + c_3 dz = c_2 dy + \left(\frac{c_1}{2ax} + c_3\right) dz - \frac{c_1 v_o}{2ax} dt = 0. \tag{7.1-10c}$$

If the coefficients of dx, dy, and dz must vanish due to their assumed linear independence, what of those of dy, dz, and dt? Those of dy and dt force c_2 and c_1, respectively, to vanish; the latter (from the coefficient of dz) then makes $c_3 \equiv 0$. That these are the same in terms of the new equation is not surprising: it is merely a restatement of the linear independence of the original variables, now recast (through the constraint) in terms of the new.

But if we *implicitly* choose dx to have been eliminated by the constraint, we say from the start that only dy and dz are linearly independent; the constraint itself never even appears, and the linear independence criterion becomes

$$c_2\,dy + c_3\,dz = 0,$$

which, to vanish, requires $c_2 \equiv 0 \equiv c_3$.

Note that it was presence of the dt term in (7.1-10c) that forced c_1, and thus c_3, to vanish. That term, in turn, resulted from the [differential] relation resulting from a *time-dependent*, "rheonomic," constraint. Had the constraint been time-*in*dependent—$v_o \equiv 0$, say—then (7.1-10c) would have been

$$c_2\,dy + \left(\frac{c_1}{2ax} + c_3\right) dz = 0.$$

In particular, choosing any $c_1 \neq 0$, $c_2 \equiv 0$ and $c_3 \equiv -\frac{c_1}{2ax} \neq 0$—*note that this still depends on that variable, x, whose* differential *has been eliminated*—will allow the linear *de*pendence criterion to be satisfied for *non*-0 coefficients; indeed,

$$0 = c_1\,dx + c_2\,dy + c_3\,dz = c_1(dx) + 0 + (-\frac{c_1}{2ax})dz = c_1\left(dx - \frac{dz}{2ax}\right),$$

which is precisely the constraint equation (7.1-10b) for $v_o \equiv 0$, just as predicted by Example 7.1.6. But such recovery of the constraint equation—indeed, of the linear *de*pendence itself—holds *only for time-independent constraints*! This is the thrust of the concept of *virtual displacements* considered next. |

Independence of the coordinate differentials is fundamental: The ability, *independently*, to vary these was what enabled the conclusion in the [Statics] Principle of Virtual Work in Section 6.3, and their linear independence is what will ultimately allow us to obtain multiple "component" equations from a *scalar* differential in Section 8.1. In the short term, we will make the assumption that, in fact, superfluous differentials have been eliminated from all expressions; thus the remaining differentials are independent. We subsequently show how to deal with *non*-holonomic constraints (with a smaller number of independent differentials than coordinates) using "Lagrange Multipliers" (Section 8.3.1).

Summary. Constraints are *scalar* relations among the *differentials* of the coordinate variables. When such a relation exists, the differentials are *linearly dependent*; only if all such differential constraints have been explicitly eliminated can we regard the remaining ones as linearly *in*dependent, and can conclude that the vanishing of a linear combination of such differentials implies that the coefficients of the differentials must vanish separately.

In general such relations are *not exact differentials—integrable* functions of the coordinates. If they *are*, they are termed *holonomic*; otherwise, they are *non-holonomic.* Similarly, constraints are categorized as *scleronomic* or *rheonomic*, according as to whether they are time-independent or not; the latter result from *moving* constraints. The elimination of redundant variables utilizing holonomic constraints *before* differentiation renders the differentials of the remaining variables linearly independent.

7.2 Virtual Displacements

As discussed in the discussion of virtual displacements in Statics, we categorize *infinitesimal* displacements of points in a body according to whether they are the real, *actual* displacements, $d\mathbf{r}$, or imagined, *virtual* displacements, $\delta\mathbf{r}$. That seems an awfully great burden to place on notation, but there it is! In particular, in Statics, $d\mathbf{r} \equiv \mathbf{0}$; there is *no* actual displacement whatsoever, and *any* displacement from the equilibrium position of a point is a virtual one. That's all well and good for Statics, but how are we to treat such virtual displacements in dynamics, where things *do* move?

This is where things begin to get a little "philosophical". We craftily introduced, in the previous section, the fact that the time t plays a distinguished role in mechanics: it is that variable on which all other variables ultimately depend. It may enter the expression for $\mathbf{r}$ *explicitly* (if one goes to the trouble of eliminating a *moving* constraint relating the variables) or it may not, as the examples above demonstrate. But whether it does or not, the *coordinates ultimately depend on* t, so t enters the expression for $\mathbf{r}$ *implicitly* through $\mathbf{q}$.

Recall that Newton's Second Law is an *instantaneous* one: when we write $\sum \mathbf{F} = m\mathbf{a}$, we write it *at a particular instant in time.* If we are lucky, and can manage to integrate up the resulting second-order differential equation, we will get the solution $\boldsymbol{r}(t)$ over an *interval* in t, but it must be remembered that this results from an equation that holds, generally, at each *instant*.

If we are going to try to uncover an alternative formulation of mechanics utilizing [virtual] displacements from the actual, we can't just ignore Newton; we must ultimately refer back to his fundamental Law! But that Law holds at a particular [fixed] instant t; thus it makes sense that, in considering them, *virtual displacements from the actual motion take place at a fixed* t. Thus we are finally in a position to define what me mean by a virtual displacement in the context of dynamics:

Definition 7.2.1. A *virtual displacement* $\delta\boldsymbol{r}$ is any displacement different from the actual displacement $d\boldsymbol{r}$ *at a fixed time.*

The distinction between the *real* and *virtual* displacements becomes clear when we consider what this means mathematically, but such a demonstration becomes merely formal—even arbitrary—unless we understand the above background. We saw in the previous section that, most generally, $\boldsymbol{r} = \boldsymbol{r}(\boldsymbol{q}; t)$ (where, again, the semicolon emphasizes the special nature of the variable t). Thus the *actual* infinitesimal displacement, $d\boldsymbol{r}$, is simply the total differential of $\boldsymbol{r}$:

$$d\boldsymbol{r} = d\boldsymbol{r}(\boldsymbol{q}; t) = \sum_i \frac{\partial \boldsymbol{r}}{\partial q_i} dq_i + \frac{\partial \boldsymbol{r}}{\partial t} dt. \tag{7.2-1}$$

On the other hand, the *virtual* displacement, $\delta\boldsymbol{r}$, *holds at a particular instant*; thus in the above, $dt \equiv 0$:

$$\delta\boldsymbol{r} = \delta\boldsymbol{r}(\boldsymbol{q}; t) \equiv \sum_i \frac{\partial \boldsymbol{r}}{\partial q_i} \delta q_i \tag{7.2-2}$$

—where we have used "δq_i" rather than "dq_i" to emphasize the virtual nature of the displacement of each coordinate. It is tempting to indicate the distinction between the two by writing simply

$$d\boldsymbol{r} = \delta\boldsymbol{r} + \frac{\partial \boldsymbol{r}}{\partial t} dt.$$

To the extent that this ignores the difference between δq_i in $\delta\boldsymbol{r}$ and dq_i in $d\boldsymbol{r}$, it is misleading; but, at least informally, it does suggest the contrast we are trying to draw.

The distinction between "real" and virtual [differential] displacements can be applied directly to constraints of the previous Section. There Example 7.1.7 showed how differential constraints could be introduced. The series of equations (7.1-10) involving *true* differentials could be expressed in terms of *virtual* ones: Equation (7.1-10a) would read

$$c_1\delta x + c_2\delta y + c_3\delta z = 0$$

—almost the same as its corresponding number. But now $\delta x = \delta z/(2ax)$ (accounting for the fact that $\delta t \equiv 0$ for *virtual* displacements), and Equation (7.1-10b) becomes

$$\delta z = 2ax\, \delta x.$$

In the same way, (7.1-10c) is now

$$c_2\delta y + \left(\frac{c_1}{2ax} + c_3\right)\delta z = 0,$$

while, assuming we have only *implicitly* eliminated dx (δx), (7.1-10d) is just

$$c_2\delta y + c_3\delta z = 0.$$

This is precisely what was pointed out in Example 7.1.7: by forcing dt (or δt) to vanish—here through the definition of virtual displacements—we similarly force the *constraint* among the coordinates to be reflected in *linear dependence* among their differentials. Had $dt \neq 0$, the need to make the term involving it to vanish would have resulted in linear *in*dependence among the differentials, just as it did in the Example—certainly not what we want!

Note that we have blithely calculated $\delta \boldsymbol{r}$ above as if it were a *differential.* In fact it is, and there is an important point to be noted regarding this practice: Like any other differential, the derivative part must be evaluated at a particular point. In this case, because $\delta \boldsymbol{r}$ is taken from the actual motion, all the $\frac{\partial \boldsymbol{r}}{\partial q_i}$ above are evaluated at the "actual" position $\boldsymbol{r}$. In particular then, *the* $\boldsymbol{q}$ *in* $\boldsymbol{r}(\boldsymbol{q};t)$ *above corresponds to the solution for these variables.*

To serve as examples, we can calculate $d\boldsymbol{r}$ and $\delta \boldsymbol{r}$ for those in the previous section. We only briefly cite the results in order to show the differences between the real and virtual displacements in the three cases:

Example 7.1.1: $$d\boldsymbol{r} = dx\hat{\imath} + dy\hat{\imath} \quad (7.2\text{-}3a)$$
$$\delta\boldsymbol{r} = \delta x\hat{\imath} + \delta y\hat{\jmath} \quad (7.2\text{-}3b)$$

Example 7.1.2: $$d\boldsymbol{r} = dx\hat{\imath} + m\,dx\hat{\jmath} \quad (7.2\text{-}3c)$$
$$\delta\boldsymbol{r} = \delta x\hat{\imath} + m\,\delta x\hat{\jmath} \quad (7.2\text{-}3d)$$

Example 7.1.3: $$d\boldsymbol{r} = dx\hat{\imath} + (m\,dx + at\,dt)\hat{\jmath} \quad (7.2\text{-}3e)$$
$$\delta\boldsymbol{r} = \delta x\hat{\imath} + m\,\delta x\hat{\jmath} \quad (7.2\text{-}3f)$$

Example 7.1.4: $$d\boldsymbol{r}_O = dx\hat{\imath} + dy\hat{\jmath} \quad (7.2\text{-}3g)$$
$$= r\sin\alpha\,d\theta\hat{\imath} - r\cos\alpha\,d\theta\hat{\jmath} \quad (7.2\text{-}3h)$$
$$\delta\boldsymbol{r}_O = r\sin\alpha\,\delta\theta\hat{\imath} - r\cos\alpha\,\delta\theta\hat{\jmath} \quad (7.2\text{-}3i)$$

The reader should be certain these examples are understood. In particular, it is only when $\boldsymbol{r}$ is *explicitly* dependent on t, such as occurs when there is a "moving", time-dependent constraint, that there is a critical distinction—more than one of mere notation—between $d\boldsymbol{r}$ and $\delta\boldsymbol{r}$.

These ideas can be schematized diagrametrically: Let us consider the motion taking place in two dimensions. In order to emphasize the "fixed t" nature of virtual displacements, it is useful initially to imagine the system evolving in time, as shown at left, where a point starting at $t = 0$ moves in a *three*-dimensional space $(r_x(t), r_y(t); t)$. At each value of t there will be a two-dimensional

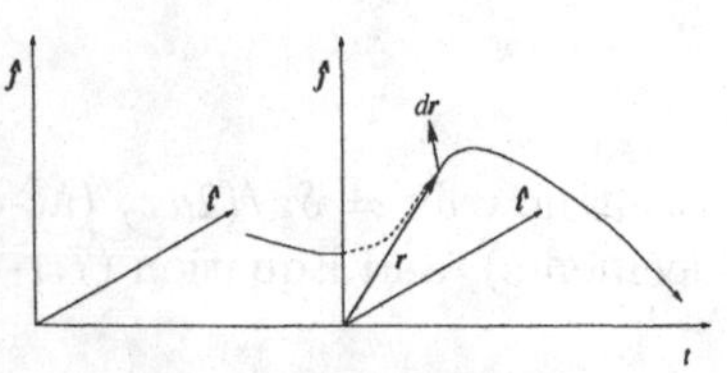

"slice" describing the position, $\boldsymbol{r} = r_x\hat{\imath} + r_y\hat{\jmath}$, of the point *at that instant*. Despite the "three-dimensional" nature of the motion, *this is the plane in which we actually observe* $\boldsymbol{r}$ and $\boldsymbol{v}$, and thus, particularly, $d\boldsymbol{r} = \boldsymbol{v}\,dt$: we don't see the "third" dimension of t any more than we do the "fourth dimension" t advocated by relativity in our own [3-D] world.

Now we examine such a "slice", with the basis vectors $\hat{\imath}$ and $\hat{\jmath}$. In that plane we see the *actual*, *two*-dimensional displacement $d\boldsymbol{r}$, along with a typical *virtual* displacement $\delta\boldsymbol{r} \neq d\boldsymbol{r}$, at that instant. Assuming $\boldsymbol{r}$ depends on the two coordinates q_1 and q_2—not necessarily the Cartesian coordinates "x" and "y"—as well as t, the former is expressed $d\boldsymbol{r} = (d\boldsymbol{r})_t + (d\boldsymbol{r})_q$, where, using the notation common to the description of partial derivatives, subscripts indicate which variables are held constant while partial differentiation is being done with the others, here $\boldsymbol{q}$ and t, respectively:

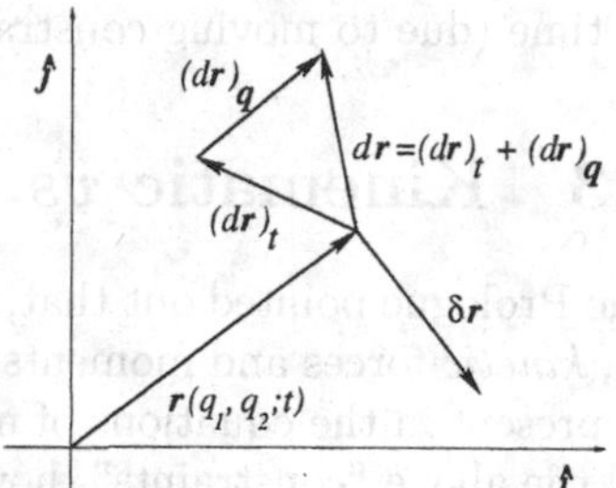

$$(d\boldsymbol{r})_t \equiv \frac{\partial \boldsymbol{r}}{\partial q_1}dq_1 + \frac{\partial \boldsymbol{r}}{\partial q_2}dq_2$$
$$(d\boldsymbol{r})_q \equiv \frac{\partial \boldsymbol{r}}{\partial t}dt.$$

Note here that the dq_i are the *actual* [infinitesimal] changes in the two coordinates during the *actual* motion; the *virtual* displacement

$$\delta\boldsymbol{r} \equiv \frac{\partial \boldsymbol{r}}{\partial q_1}\delta q_1 + \frac{\partial \boldsymbol{r}}{\partial q_2}\delta q_2$$

looks like $(d\boldsymbol{r})_t$ (since $dt \equiv 0$) but involves *arbitrary* δq_i. Were t to have been absent *explicitly* from the expression for $\boldsymbol{r}$, $(d\boldsymbol{r})_q$ would vanish identically and $(d\boldsymbol{r})_t$ would be the same as $d\boldsymbol{r}$. The virtual displacement is one *different* from $(d\boldsymbol{r})_t$ in this case, and, in any event, generally different from $d\boldsymbol{r}$ itself.

[Differential] constraints can be holonomic or non-holonomic. We have seen above that non-holonomic constraints—at least thus far—are not all that inconvenient: due to the linearity in the differentials of such constraints, we can still solve for dependent variables' differentials and express the differential $\delta\boldsymbol{r}$ in terms of a reduced set. There *is*, however, a major problem with such constraints, dealing not with the differential $\delta\boldsymbol{r}$, but rather with quantities which depend on the coordinate variables themselves, particularly forces (which, in general, depend on position). And any solutions obtained will presumably give only the reduced set, yet still expressed in the "redundant" set which cannot be related to the smaller one.

We can get some feel for the problem even at this point: In point of fact, even for a general differential constraint of form (7.1-3), the $\dot{q}_i$ *are* related:

$$\sum_{i=1} a_i(\boldsymbol{q};t)\,\dot{q}_i + a_t(\boldsymbol{q};t) \equiv \sum_{i=0} a_i(\boldsymbol{q};t)\,\dot{q}_i = 0;$$

The difficulty is the fact that we can't similarly relate the *coordinates*. We shall see this later in the derivation of the equations of motion (Section 8.1), and ultimately develop a technique to deal with this situation (Section 8.3).

Summary. A *virtual* displacement $\delta\boldsymbol{r}$ (as opposed to the *actual* displacement $d\boldsymbol{r}$), is any from $\boldsymbol{r}$ *at a specific instant in time—i.e.*, holding t *fixed*. Operationally, it is a *differential* quantity, but in cases in which $\boldsymbol{r}$ depends explicitly on time (due to moving constraints), we must set $dt \equiv 0$.

7.3 Kinematic *vs.* Kinetic Constraints

The Prologue pointed out that, in the vector Newtonian formulation of mechanics, *kinetic* forces and moments enforced *kinematic* constraints, yet both had to be present in the equations of motion; this was described as "redundant" there. All the above "constraints", however, have been *kinematic* in nature; and, when it comes to examining virtual work, the forces generating such constraints will not appear precisely because they *do* no work! What we shall briefly describe here is just how, even though kinetic constraints don't enter explicitly in the ultimate equations, they are *implicit* in the [differential] constraints themselves.

Let us consider such a constraint force, $\boldsymbol{F}^{(c)}$ (similar considerations holding for constraint *couples*). In the Statics formulation of Virtual Work, these do no work either because $\delta\boldsymbol{r} \equiv \boldsymbol{0}$ (in a fixed pin, for example), or because $\delta\boldsymbol{r} \perp \boldsymbol{F}^{(c)}$ (with normal forces); in either case $\boldsymbol{F}^{(c)} \cdot \delta\boldsymbol{r} \equiv \boldsymbol{0}$ and such forces don't enter the virtual work equation.

In the present context, we turn that argument around, effectively *defining* a [kinetic] constraint force as that *enforcing* a [kinematical] constraint:

Definition 7.3.1. Given an arbitrary virtual displacement $\delta\boldsymbol{r} \neq \boldsymbol{0}$ *consistent with a scalar constraint*, we associate a *constraint force* $\boldsymbol{F}^{(c)}$ to be any force doing no work under that virtual displacement: $\boldsymbol{F}^{(c)} \cdot \delta\boldsymbol{r} \equiv 0$.

In effect, what we are saying is that, given any constraint, *there must be a [kinetic] constraint force* $\boldsymbol{F}^{(c)}$ *generating that [kinematical] constraint!*

Though we don't know its *magnitude* (any more than we know the magnitude of the associated normal force in Statics), its *direction* must satisfy $\boldsymbol{F}^{(c)} \cdot \delta\boldsymbol{r} \equiv \boldsymbol{0}$. Thus, for example, if a constraint states that motion of a given point is restricted to a particular surface, the associated constraint force must be *normal* to that surface in order that it do no work under *any* virtual displacement along the surface. And in Example 7.1.5, equation (7.1-6) (particularly the part that $\boldsymbol{r}_{12} \cdot d\boldsymbol{r}_{12} \equiv 0$) says that the force associated with the constraint maintaining m_1 and m_2 at a fixed distance must be along $\boldsymbol{r}_{12}$.

In the particular case that a constraint is holonomic—that it is associated with a [scalar] function $\phi = 0$ of the coordinates—this observation makes it particularly easy to determine the direction of the force: taking the differential of this function

$$d\phi = \sum \frac{\partial \phi}{\partial q_i}\delta q_i + \frac{\partial \phi}{\partial t}dt = 0.$$

But at *fixed t*, using the defining property of the gradient of a scalar (which, recall, takes derivatives only with respect to the *coordinates*), this reduces to

$$\delta\phi \equiv \nabla\phi \cdot \delta\boldsymbol{r} = 0,$$

and we see that $\nabla\phi$ *is in the direction of the constraint force generating that constraint.* We note, however, that in the event the constraint, even though holonomic, is *scleronomic*, work *is* done under an *actual* displacement: since we can write $\boldsymbol{F}^{(c)} = F^*\nabla\phi$ for some scalar F^*,

$$d\phi(\boldsymbol{q};t) = \nabla\phi \cdot d\boldsymbol{r} + \frac{\partial\phi}{\partial t}dt = 0 \Rightarrow \boldsymbol{F}^{(c)} \cdot d\boldsymbol{r} = F^*\nabla\phi \cdot d\boldsymbol{r} = -F^*\frac{\partial\phi}{\partial t}dt \neq 0.$$

The constraint still does no work, however, under a *virtual* displacement, since $\nabla\phi \cdot \delta\boldsymbol{r}$ still vanishes identically.

To drive these points home, let us consider three examples, with holonomic, non-holonomic, and rheonomic constraints, respectively:

Example 7.3.1. *Holonomic Constraint.* Consider a body constrained to move on the surface of a hemispherical bowl of radius a. Though this would most appropriately be worked using spherical coordinates, just to demonstrate the effect of constraints, let us consider Cartesian ones. The most convenient set of such coordinates would have their origin at the center of the hemisphere; thus the coordinates (x, y, z) satisfy the constraint $\phi(x, y, z) = x^2 + y^2 + z^2 - a^2 = 0$. (Note the "standard form" in which such holonomic constraints are customarily written, so we can conclude immediately that $\delta\phi = 0$.) We can use this relation to eliminate, say, δx from the general form for $\delta\boldsymbol{r} = \delta x\hat{\boldsymbol{\imath}} + \delta y\hat{\boldsymbol{\jmath}} + \delta z\hat{\boldsymbol{k}}$:

$$\delta\boldsymbol{r} = \Big(-\frac{y\,\delta y + z\,\delta z}{x(y,z)}\Big)\hat{\boldsymbol{\imath}} + \delta y\hat{\boldsymbol{\jmath}} + \delta z\hat{\boldsymbol{k}}. \tag{7.3-1}$$

We note that this constraint has reduced the number of coordinates from three to two. This makes sense: if we had been rational and chosen to use spherical coordinates in the first place, only the two angular variables would have been necessary. Even there, however, we would be imposing the constraint that the radial coordinate ρ was a constant.

If, now, we take the gradient of ϕ, $\nabla\phi = 2(x\hat{\boldsymbol{\imath}} + y\hat{\boldsymbol{\jmath}} + z\hat{\boldsymbol{k}})$, we see that this is in the direction radially outward from the origin—normal to the surface of the hemisphere—as expected. This is not actually the force *or* the direction, since the units are wrong for both; but we can at least find the unit vector in this direction by dividing by its magnitude:

$$\hat{\boldsymbol{F}} = \frac{x\hat{\boldsymbol{\imath}} + y\hat{\boldsymbol{\jmath}} + z\hat{\boldsymbol{k}}}{a}.$$

A direct calculation verifies that $\nabla\phi \cdot \delta\boldsymbol{r} = 2(-(y\,\delta y + z\,\delta z) + y\,\delta y + z\,\delta z) \equiv 0$, as expected. |

Example 7.3.2. *Non-holonomic Constraint.* Let us consider an arbitrary non-holonomic (though scleronomic) constraint $a_x(x,y,z)\delta x + a_z(x,y,z)\delta z = 0$. Subject to this constraint, the allowable virtual displacement

$$\delta \boldsymbol{r} = (-\frac{a_z}{a_x}\delta z)\hat{\boldsymbol{\imath}} + \delta y\hat{\boldsymbol{\jmath}} + \delta z\hat{\boldsymbol{k}}.$$

We can associate with this constraint the force $\boldsymbol{F}^{(c)} \equiv F_x\hat{\boldsymbol{\imath}} + F_y\hat{\boldsymbol{\jmath}} + F_z\hat{\boldsymbol{k}}$ by enforcing the fact that, for any $\delta\boldsymbol{r}$, $\boldsymbol{F}^{(c)}\cdot\delta\boldsymbol{r} \equiv 0$:

$$\boldsymbol{F}^{(c)}\cdot\delta\boldsymbol{r} = F_y\delta y + (F_z - \frac{a_z}{a_x}F_x)\delta z \equiv 0$$

which must hold for *any* δy and δz. (Note that y and z are, at this stage, *independent.*) The only way this can happen is for the coefficients of δy and δz to vanish identically:

$$F_y \equiv 0, \quad F_z = \frac{a_z}{a_x}F_x$$

or

$$\boldsymbol{F}^{(c)} = F_x\hat{\boldsymbol{\imath}} + \frac{a_z}{a_x}F_x\hat{\boldsymbol{k}}.$$

Non-holonomic constraints are clearly more general than holonomic ones; thus we expect that we should be able to use this approach to get the $\boldsymbol{F}^{(c)}$ even for holonomic constraints. This is, in fact, true, as can be seen by applying the present methodology to the previous, holonomic example. In that case, using $\delta\boldsymbol{r}$ from (7.3-1),

$$\boldsymbol{F}^{(c)}\cdot\delta\boldsymbol{r} = (F_y - \frac{y}{x}F_x)\delta y + (F_z - \frac{z}{x}F_x)\delta z \equiv 0$$

for *any* δy and δz. Thus

$$F_y = \frac{y}{x}F_x, \quad F_z = \frac{z}{x}F_x$$

or

$$\boldsymbol{F}^{(c)} = F_x(\hat{\boldsymbol{\imath}} + \frac{y}{x}\hat{\boldsymbol{\jmath}}) + \frac{z}{x}F_x\hat{\boldsymbol{k}} = \frac{F_x}{x}(x\hat{\boldsymbol{\imath}} + y\hat{\boldsymbol{\jmath}} + z\hat{\boldsymbol{k}})$$

—again a force (the units *do* work out in this case) in the direction of the result in the previous example. |

Example 7.3.3. *Rheonomic Constraint.* We conclude this section with an example which manages to bring together "virtually" all the concepts discussed to date: A particle is constrained to move along a line, moving with velocity $\boldsymbol{v}_o = v_o\hat{\boldsymbol{\jmath}}$ and having a [constant] slope m; the line initially passes through the origin at time $t = 0$. At a particular instant of time, the point is located at $\boldsymbol{r}$ relative to the origin as shown. We shall determine expressions for the actual displacement $d\boldsymbol{r}$, the virtual displacement $\delta\boldsymbol{r}$, and the constraint force $\boldsymbol{N}$ enforcing the given constraint.

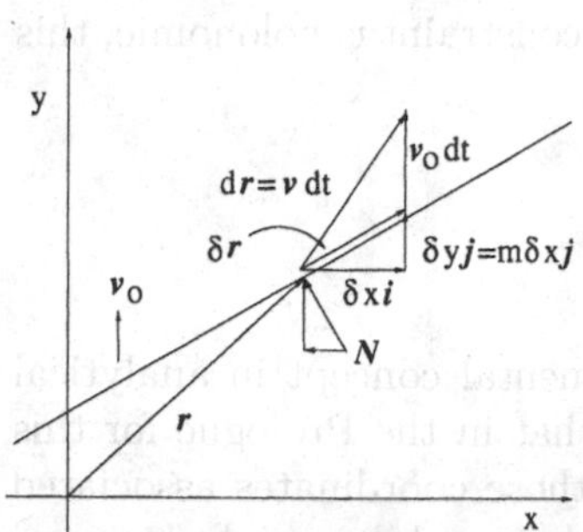

Diagrammed at left are the various vectors of interest in this example. In particular, we see the two components of $d\boldsymbol{r}$, one ($\delta\boldsymbol{r}$) along the constraining plane and the other ($\boldsymbol{v}_o dt$) in the direction of motion, as well as the Cartesian components of $\delta\boldsymbol{r}$ which, since $dt \equiv 0$, will always be a vector parallel to the plane. (Though not altogether general, for purposes of illustration, $\delta\boldsymbol{r}$ has been chosen to be the same as that component of $d\boldsymbol{r}$ parallel to the line.) Also indicated are the two Cartesian components of the constraining $\boldsymbol{N}$.

We start by determining the explicit form for the constraint, here a rheonomic one (since the line is moving) which happens to be holonomic: from the initial conditions on the line's position, we have immediately that $\phi(x, y; t) \equiv y - (mx + v_o t) = 0$. Substituting this into the *un*constrained $\boldsymbol{r} = x\hat{\imath} + y\hat{\jmath}$ gives us an expression for the constrained $\boldsymbol{r}$, and thence the real and virtual displacements:

$$\begin{aligned}
\boldsymbol{r} &= x\hat{\imath} + (mx + v_0 t)\hat{\jmath} \\
d\boldsymbol{r} &= (dx\hat{\imath} + m\,dx\hat{\jmath}) + v_0\,dt\hat{\jmath} \\
\delta\boldsymbol{r} &= \delta x\hat{\imath} + m\,\delta x\hat{\jmath}.
\end{aligned}$$

(The grouping in the expression for $d\boldsymbol{r}$ is meant to facilitate comparison of these analytical expressions with the diagram, which is strongly recommended at this point!)

We now determine the constraint force. Because the constraint is holonomic, the *direction* of this force can be determined by taking the gradient of that function:

$$\nabla\phi = -m\hat{\imath} + \hat{\jmath};$$

thus the unit vector in this direction is just $(-m\hat{\imath} + \hat{\jmath})/\sqrt{1+m^2}$, and the normal force can be expressed

$$\boldsymbol{N} = \frac{N}{\sqrt{1+m^2}}(-m\hat{\imath} + \hat{\jmath}).$$

As expected, this $\boldsymbol{N}$ is perpendicular to the line: its slope is $(-1/m)$—the negative reciprocal of the slope of the line itself.

As a side note, we observe that $\boldsymbol{N} \cdot \delta\boldsymbol{r} = N(-m+m)/\sqrt{1+m^2}\,\delta x = 0$, but $\boldsymbol{N} \cdot d\boldsymbol{r} = N((-m+m)\,dx + v_o\,dt)/\sqrt{1+m^2} = Nv_o/\sqrt{1+m^2}\,dt \neq 0$! Thus, although no [virtual] work will be done under a *virtual* displacement, the rheonomic nature of the constraint results in its doing work under an *actual* displacement, as expected. |

Summary. Although the forces imposing these constraints never enter any virtual work (at least under virtual displacements consistent with the constraints), they are implicitly defined in direction through the [infinitesimal] constraints; this, in fact, effectively becomes the *definition* of a constraint force in the current context. In the particular case that the constraint is holonomic, this direction becomes merely that of $\nabla\phi$.

7.4 Generalized Coordinates

We finally come to what is likely the most fundamental concept in analytical dynamics, that of *generalized coordinates.* Recall that in the Prologue for this Part, note was made of the obligation to hold to those coordinates associated with a *vector* coordinate system. This can be a bother, and downright inconvenient in cases involving multiple bodies, as was alluded to there. The campaign is ultimately to allow *any* choice of coordinates, most particularly those best suited to a given problem.

Recall that, at the beginning of Section 7.1, we pointed out that, once a form for the position $\boldsymbol{r}$ has been chosen, then $\boldsymbol{a} \equiv \ddot{\boldsymbol{r}}$ is determined; thus we shall focus our attention on the specification of position. In finally obtaining the equations of motion for Lagrangian dynamics, using virtual work (as has been mentioned continually), we shall follow the lead of classical dynamics, doing this first for a *system of particles*, then (unfortunately, somewhat informally) passing to the limit of *continuous* systems—*i.e., rigid bodies.* But this means that we shall generally require more than a single position, and we shall denote the position vector to a given particle of mass m_i as $\boldsymbol{r}_i$.

But the specification of $\boldsymbol{r}_i$ requires a set of *coordinates*, and it is here that the *generalized* coordinates enter: we allege that we can pick *any* such set, as long as we can completely specify the orientation—the *configuration*—of the system. But how many such coordinates are required? For a set of N particles in two dimensions we generally require $2N$ such coordinates, $3N$ in three. But recall that we typically have *constraints* on the system, certain points which are fixed, or prescribed to move in a certain direction due to external supports. Such constraints evidence themselves through scalar relations among the coordinates or, more generally, the coordinates' *differentials* (see Section 7.1); such constraints mean that the original coordinates are, in fact, *linearly dependent. If these constraints are holonomic*, we can use these relations successively to express each coordinate in terms of the balance, ultimately ending up with a reduced set of linearly *in*dependent coordinates in terms of which we can describe the problem. In particular, if there are k scalar constraints among the coordinates, only $2N-k$ or $3N-k$ remain after all the constraint dependencies have been explicitly eliminated. Those which do are precisely the generalized coordinates:

Definition 7.4.1. *Generalized coordinates* are any *independent* set completely specifying the configuration of the system.

Note, in this regard, that by "independent" we mean independent in the sense of equation (7.1-7): that the vanishing of any [scalar] linear combination of the dq_i or δq_i means that we can conclude that the coefficient of *each* such differential must vanish. Thus, even though the discussion leading to this Definition entailed *holonomic* constraints, it would hold even for *non*-holonomic ones.

The number of such coordinates, $2N - k$ or $3N - k$ in 2- or 3-dimensional space subject to k constraints, is also given a name:

Definition 7.4.2. The number of generalized coordinates required completely to specify the configuration of a system is called the number of *degrees of freedom.*

In most of the [*particle*] examples to this point, there has been only one constraint; thus all 2-D problems required only a single coordinate, while Example 7.3.1, a 3-D problem, required two (x and z chosen there). On the other hand Example 7.1.4 started with *six* coordinates (the position, $\overline{\mathbf{r}}$, of O, and three Euler angles, say), but two holonomic and two non-holonomic constraints reduced the number of *independent* coordinates to two. Each of these describes the number of degrees of freedom for the corresponding system. We shall consider one more example of this idea of fundamental importance:

Example 7.4.1. *Degrees of Freedom in Rigid Body Motion.* The constraints required to describe the orientation of a body will be reduced by the number of constraints on the system. Generally we conceive of such constraints as limiting the motion of specific *points* in the body—those sliding over a surface or connected to other bodies with pin connections; there are no others if one is discussing the motion of a "particle"—a "point" in its own right. But merely to qualify as a "rigid body"—in this example meaning a *system* of particles moving as a unit—implicitly imposes further constraints on the system; these, in turn, are precisely of the type discussed in Example 7.1.5: constraints maintaining the inter-particle distance. It is this type of constraint, and the resulting number of generalized coordinates required to describe the motion of a rigid body independent of the former "point constraints", which we shall examine here.

As noted above, that number will be either $2N - k$ or $3N - k$, depending on whether we are addressing 2- or 3-D dynamics; thus it is not surprising that we must treat these two cases separately. But the real issue is what k is; *i.e.*, how many constraints are required to enforce the condition of being a "rigid body". We now examine this equation for the two cases.

Planar Motion. The simplest rigid system consists of *two* particles connected by a "light rigid rod" with fixed length. Here there are four coordinates required to specify the *un*constrained motion of the particles; this number is reduced to three generalized coordinates by the single constraint.

Now consider *three* particles. These would form a triangle—itself an inherently rigid structure[6]—connected by *three* rods; thus the six unconstrained coordinates are reduced again to three by the three inter-particle distances.

When we go to *four* particles, however, simply specifying the four lengths of

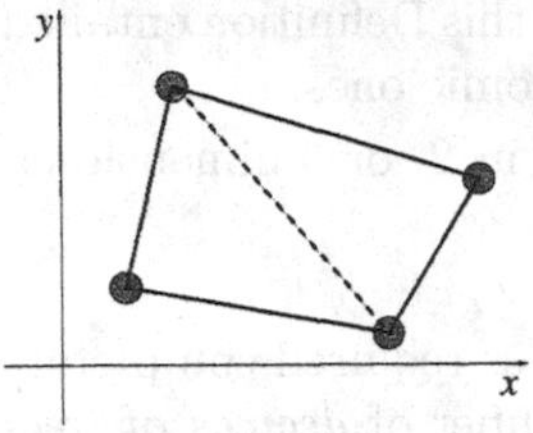

the sides of the resulting quadrilateral is not enough, as anyone who has tried to construct a bookcase by merely nailing four boards together knows: the resulting structure is not rigid! To *make* it rigid, one must specify an *additional* inter-particle distance—a *diagonal* as it were—in effect forming two [rigid] triangles with a common side. This suggests the general inductive mode for building up a planar rigid structure with more than three particles: one attaches the ith particle to the *rigid* system of $(i-1)$ particles using *two* rods of fixed length. The two additional [unconstrained] coordinates required to specify the position of that particle are then eliminated by the additional two constraints imposed by the lengths of the connecting rods, and the net number of *generalized* coordinates remains at three.

Thus in all these cases, *three generalized coordinates are sufficient*—the coordinates of one particle (or the center of mass of the system) plus the angle specifying the orientation of the entire rigid body relative to some fixed line, say—and planar rigid bodies have *three degrees of freedom.*

"Space" Motion. Now we consider motion in three dimensions, generally requiring *three* coordinates to specify the position of each particle. Again we start with the pair of *two* particles connected by a rigid rod: here the *six* unconstrained coordinates are reduced by the single constraint maintaining the inter-particle distance to *five* generalized coordinates.

Three particles, however, require *six* generalized coordinates: nine [unconstrained] coordinates to specify the positions of each of the three particles, reduced by the three inter-particle distances.

Now consider *four* particles. Our experience with the planar model suggests the approach to attach the fourth to the previous three: to maintain rigidity, we must connect it with *three* additional rods to form a *tetrahedron.* Thus we now have *six* constraints—the original three for the base triangle, plus the additional three to fix the fourth particle—reducing the number of generalized coordinates from *twelve* to *six* again. Further points will be added similarly: each additional requires three additional coordinates, reduced by three constraints.

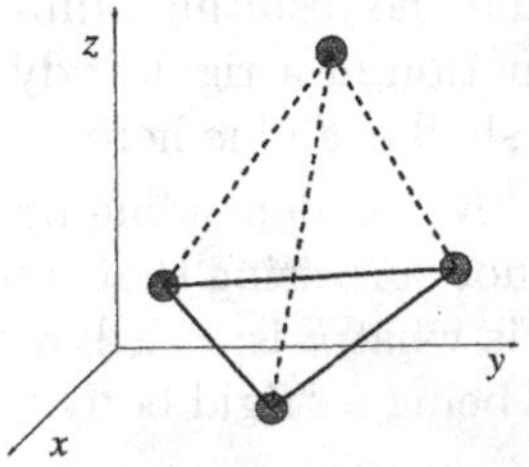

So the number of degrees of freedom remains at *six*: three coordinates of one particle (or the center of mass) plus *three Euler angles* (see Section 4.2) specifying the rotations required to bring the body coordinates from the reference system to its instantaneous orientation. This holds for all systems of $N \geq 3$. But why does it require only *five* coordinates for two particles? This is a degenerate case, insensitive to rotation about the interconnecting rod itself. The angular orientation of the rod, measuring the position of one particle relative to the other, will effectively trace out a sphere as that first particle moves around

[6] Note that triangles *are* rigid because any two with equal sides are congruent.

its companion; but *this* orientation requires only *two* angles, say "latitude" and "longitude", in addition to the three coordinates specifying the position of the reference particle.

The student might perceive a familiar ring in the above arguments. In point of fact, they are precisely the ones used to describe the geometry of *trusses*, both 2- and 3-D, in Statics (where they are also often categorized as "planar" and "space"). The number of members in such structures required to maintain rigidity lead to a relation which must hold between the number of joints in the truss (corresponding here to the masses) and the number of members (here the interconnecting "rods").

We are thus finally ready to describe the various positions in a system. Each $\boldsymbol{r}_i$ will generally depend on its [arbitrary] generalized coordinates, equal in number to the number of degrees of freedom. But, to cover all possible eventualities, we must allow for the possibility that some of these constraints be *rheonomic*, time-dependent, in character; in that case, $\boldsymbol{r}_i$ will generally depend on the time t. Thus in the end, $\boldsymbol{r}_i = \boldsymbol{r}_i(\boldsymbol{q}; t)$, where each $q_i = q_i(t)$. *Note that, although $\boldsymbol{r}_i$ depends on $\boldsymbol{q}$ and t, it does not on $\dot{\boldsymbol{q}}$.*

If we denote the number of degrees of freedom (and thus the number of independent q_i) by f, we can express various quantities of interest in terms of these q's: corresponding to $\boldsymbol{r}_i$ is the *velocity*

$$\boldsymbol{v}_i \equiv \frac{d\boldsymbol{r}_i(\boldsymbol{q};t)}{dt} = \sum_{j=1}^{f} \frac{\partial \boldsymbol{r}_i}{\partial q_j}\dot{q}_j + \frac{\partial \boldsymbol{r}_i}{\partial t}. \tag{7.4-1}$$

(We could similarly find $\boldsymbol{a}_i$, but it will turn out not to be needed.) We observe that $\boldsymbol{v}_i = \boldsymbol{v}_i(\boldsymbol{q}, \dot{\boldsymbol{q}}; t)$; in particular, $\boldsymbol{v}_i$ is *linear* in the $\dot{q}_i$. And, with an eye to the distinction between the infinitesimal *real* and *virtual* displacements, we note that

$$d\boldsymbol{r}_i = \sum_{j=1}^{f} \frac{\partial \boldsymbol{r}_i}{\partial q_j}\,dq_j + \frac{\partial \boldsymbol{r}_i}{\partial t}\,dt$$

while, since $dt \equiv 0$ for virtual displacements,

$$\delta\boldsymbol{r}_i = \sum_{j=1}^{f} \frac{\partial \boldsymbol{r}_i}{\partial q_j}\,\delta q_j. \tag{7.4-2}$$

Both to illustrate these results and to set up further examples in this section, we consider a brief example:

Example 7.4.2. Consider "Cartesian" coordinates $\boldsymbol{r} = x\hat{\boldsymbol{\imath}} + y\hat{\boldsymbol{\jmath}}$ expressed in terms of "polar" variables (r, θ): since $x = r\cos\theta$ and $y = r\sin\theta$, we have immediately that $\boldsymbol{r} = r\cos\theta\hat{\boldsymbol{\imath}} + r\sin\theta\hat{\boldsymbol{\jmath}}$; thus

$$\boldsymbol{v} = \frac{\partial \boldsymbol{r}}{\partial r}\dot{r} + \frac{\partial \boldsymbol{r}}{\partial \theta}\dot{\theta} = (\cos\theta\hat{\boldsymbol{\imath}} + \sin\theta\hat{\boldsymbol{\jmath}})\,\dot{r} + (-r\sin\theta\hat{\boldsymbol{\imath}} + r\cos\theta\hat{\boldsymbol{\jmath}})\,\dot{\theta}$$

and

$$\delta\boldsymbol{r} = \frac{\partial \boldsymbol{r}}{\partial r}\,\delta r + \frac{\partial \boldsymbol{r}}{\partial \theta}\,\delta\theta = (\cos\theta\hat{\boldsymbol{\imath}} + \sin\theta\hat{\boldsymbol{\jmath}})\,\delta r + (-r\sin\theta\hat{\boldsymbol{\imath}} + r\cos\theta\hat{\boldsymbol{\jmath}})\,\delta\theta.$$

(Here, since $\boldsymbol{r}$ is independent of t, no extra terms appear in $\boldsymbol{v}$ due to partial derivatives of that variable.) |

The time derivatives of generalized coordinates, the $\dot{q}_j$, are called "*generalized velocities*" (even though they are scalars!) Note that although the $\boldsymbol{v}_i$ will always have units $[L/T]$, the generalized velocities need not: $\dot{\theta}$ above, for example, has units simply $[T]^{-1}$.

Derivatives of $\boldsymbol{r}$ and $\boldsymbol{v}$ with respect to Generalized Coordinates and Velocities

We now prove two important properties of generalized coordinates and their velocities:

Theorem 7.4.1. For $\boldsymbol{r}_i(\boldsymbol{q};t)$,

$$\frac{\partial \dot{\boldsymbol{r}}_i}{\partial \dot{q}_k} = \frac{\partial \boldsymbol{r}_i}{\partial q_k}.$$

This property relates derivatives of the position with respect to coordinates to those of the velocity with respect to the *generalized* velocities; it is sometimes informally described by saying "dots cancel"!

Proof. This property follows immediately from (7.4-1):

$$\frac{\partial \dot{\boldsymbol{r}}_i}{\partial \dot{q}_k} = \frac{\partial \boldsymbol{v}}{\partial \dot{q}_k} = \frac{\partial}{\partial \dot{q}_k}\Big(\sum_{j=1}^{f} \frac{\partial \boldsymbol{r}_i}{\partial q_j}\dot{q}_j + \frac{\partial \boldsymbol{r}_i}{\partial t}\Big) = \frac{\partial \boldsymbol{r}_i}{\partial q_k}$$

since $\boldsymbol{r}$ is explicitly a function only of $\boldsymbol{q}$ and t—not $\dot{\boldsymbol{q}}$—so the $\dot{q}_j$ enter only explicitly as above. □

This result can be exemplified by returning to the above Example 7.4.2; in particular

$$\frac{\partial \boldsymbol{v}}{\partial \dot{\theta}} = -r\sin\theta\hat{\boldsymbol{\imath}} + r\cos\theta\hat{\boldsymbol{\jmath}} = \frac{\partial}{\partial \theta}(r\cos\theta\hat{\boldsymbol{\imath}} + r\sin\theta\hat{\boldsymbol{\jmath}}) = \frac{\partial \boldsymbol{r}}{\partial \theta}.$$

Theorem 7.4.2. For $\boldsymbol{r}_i(\boldsymbol{q};t)$,

$$\frac{\partial}{\partial q_k}\Big(\frac{d\boldsymbol{r}_i}{dt}\Big) = \frac{d}{dt}\Big(\frac{\partial \boldsymbol{r}_i}{\partial q_k}\Big).$$

The essence of this property is that "$\frac{\partial}{\partial q_k}$ and $\frac{d}{dt}$ commute" on $\boldsymbol{r}$. It is really a non-trivial result: one side of the equation is the *partial* of a *total* derivative, the other is the *total* of a *partial*.

Proof. Again referring to (7.4-1),

$$\frac{\partial \dot{\mathbf{r}}_i}{\partial \dot{q}_k} = \frac{\partial \mathbf{v}}{\partial \dot{q}_k} = \frac{\partial}{\partial q_k}\Big(\sum_{j=1}^{f} \frac{\partial \mathbf{r}_i}{\partial q_j}\dot{q}_j + \frac{\partial \mathbf{r}_i}{\partial t}\Big)$$

$$= \sum_{j=1}^{f} \frac{\partial}{\partial q_k}\Big(\frac{\partial \mathbf{r}_i}{\partial q_j}\Big)\dot{q}_j + \frac{\partial}{\partial q_k}\Big(\frac{\partial \mathbf{r}_i}{\partial t}\Big)$$

($\frac{\partial \dot{q}_k}{\partial q_k} \equiv 0$ since $q_k = q_k(t)$ only)

$$= \sum_{j=1}^{f} \frac{\partial}{\partial q_k}\Big(\frac{\partial \mathbf{r}_i}{\partial q_j}\Big)\dot{q}_j + \frac{\partial}{\partial t}\Big(\frac{\partial \mathbf{r}_i}{\partial q_k}\Big)$$

(assuming the partials commute on the last term)

$$= \frac{d}{dt}\Big(\frac{\partial \mathbf{r}_i}{\partial q_k}\Big) \qquad \square$$

Again this is verified in Example 7.4.2:

$$\begin{aligned}
\frac{\partial}{\partial \theta}\Big(\frac{d\mathbf{r}}{dt}\Big) &= \frac{\partial}{\partial \theta}((\cos\theta\hat{\imath} + \sin\theta\hat{\jmath})\,\dot{r} + (-r\sin\theta\hat{\imath} + r\cos\theta\hat{\jmath})\,\dot{\theta}) \\
&= (-\sin\theta\hat{\imath} + \cos\theta\hat{\jmath})\,\dot{r} + (-r\cos\theta\hat{\imath} + r\sin\theta\hat{\jmath})\,\dot{\theta}) \\
\frac{d}{dt}\Big(\frac{\partial \mathbf{r}}{\partial \theta}\Big) &= \frac{d}{dt}(-r\sin\theta\hat{\imath} + r\cos\theta\hat{\jmath})) \\
&= -(\dot{r}\sin\theta + r\cos\theta\dot{\theta})\hat{\imath} + (\dot{r}\cos\theta - r\sin\theta\dot{\theta})\hat{\jmath}
\end{aligned}$$

Summary. Generalized coordinates are an *arbitrary* set of *independent* variables sufficient completely to describe the orientation of the system; the number of such coordinates is called the number of *degrees of freedom.* The independence of these coordinates is important: it means that all constraints among the variables have been explicitly eliminated—something conceptually requiring the constraints to be *holonomic*, though we allow such constraints to be even *non*-holonomic in practice—and that the vanishing of any linear combination of [the differentials of] these variables allows us to conclude that the coefficients of those variables must vanish individually. The time derivatives of these generalized coordinates, even though they are scalars, are referred to as *generalized velocities.*

We have obtained, in terms of such generalized coordinates and velocities, expressions for velocity, as well as differentials of the actual and virtual motions. Finally we have determined two important properties of derivatives of $\mathbf{r}$ with respect to these variables. Virtually all of these will return in the next section, where we determine the equations of motion of an arbitrary system in terms of this arbitrary set of coordinates.

Summary

Constraints on the motion of a particle generally reduce the number of [scalar] coordinates required to specify its position. Such constraints can most generally expressed in the form of *scalar* relations among the *differentials* of the coordinates and the time (*Pfaffian form*). In the event such differential relations are *integrable* to get a [scalar] relation among the coordinates themselves, they are called *holonomic*; otherwise they are *non-holonomic*. Similarly, if they are time-*de*pendent, generally arising from constraints which are *moving*, they are called *rheonomic*; otherwise *scleronomic*. The number of *independent* coordinates—those from which all constraints have been eliminated—is called the number of *degrees of freedom*.

A *virtual displacement* $\delta \boldsymbol{r}$ is one away from the *actual* displacement $d\boldsymbol{r}$, conceived to occur at the *same time* as the actual one; thus, in terms of differentials, $dt \equiv 0$; conceptually, they are *differential* in nature, so one can can obtain $\delta \boldsymbol{r}$ from $d\boldsymbol{r}$ by simply taking $dt = 0$. But this same contemporaneous character must be observed with regard to *constraints*; in particular, if $\boldsymbol{r} = \boldsymbol{r}(\boldsymbol{q}; t)$, the differential constraint on the *virtual* δq_i arising from a *rheonomic* constraint on the dq_i must set $dt \equiv 0$, too.

Since constraints are imposed on the coordinates, they are *kinematic* in nature. But such constraints must be generated by *kinetic* forces (or couples). In fact, they are *equivalent* in the sense that a constraint force $\boldsymbol{F}^{(c)}$ imposing a *virtual* displacement $\delta \boldsymbol{r}$ at a point must satisfy $\boldsymbol{F}^{(c)} \cdot \delta \boldsymbol{r} \equiv 0$—that constraint forces *do no work* under displacements consistent with them; this condition can, in fact, be used as a *definition* of such constraint forces, which otherwise will never appear in any expressions involving work.

Chapter 8

Lagrangian Dynamics: Kinetics

Introduction

The previous chapter discussed the fundamental *kinematical* concepts associated with Lagrangian dynamics: virtual displacements and generalized coordinates, are and their relation to constraints. The present chapter now moves to the *kinetics*—the introduction of *forces* and how they effect the motion.

Given the fact that analytical dynamics is based on virtual *work*, itself a concept from *energy*, it is not surprising that features of the latter pervade the current discussion: in particular, in cases in which the forces are *conservative*, certain simplifications to the general Principle of Work and [Kinetic] Energy arise in dynamics, and a similar simplification will occur here. But one must always have the more general Principle to fall back on in the event all forces *aren't* conservative, and thus we start with the most general formulation of Lagrangian dynamics, the *Euler-Lagrange Equations*. We then discuss the special case of conservative forces, obtaining the *Lagrange Equations*, ultimately generalizing them to a form which allows even non-conservative forces to be included.

The derivations will demonstrate the critical role that independence of the [differentials of the] coordinates plays in the present context: just as in the derivation of the Principle of Virtual work in Section 6.3, we ultimately conclude that the vanishing of a *sum* of terms in the virtual work—done by the applied forces under an arbitrary virtual displacement from the actual motion—implies that *each* term must vanish *separately*. These demonstrations presume initially that any constraints are *holonomic*. But we have already seen (Example 7.1.4) that *non*-holonomic constraints can arise quite naturally; thus the subsequent section discusses just how these could be implemented, culminating in a discussion of *Lagrange multipliers* and their interpretation.

The chapter concludes with a discussion of how even *time*—that *ultimate* variable on which all others depend—can be regarded as just another coordinate! Yet the special character of this variable requires correspondingly special treatment, as we shall see. Aside from the perhaps philosophical merit of the subject, it is introduced primarily to allow us to treat the same topic in a subsequent chapter on Hamiltonian dynamics.

8.1 Arbitrary Forces: Euler-Lagrange Equations

We are finally in a position to obtain the equations of motion in the present formulation. As mentioned above, we shall do this considering a system of N particles, ultimately generalizing to the "continuous systems" of rigid bodies. We shall also assume, for the time being, that *all constraints have been eliminated explicitly*, leaving the system configuration completely describable in terms of f [the number of degrees of freedom] generalized coordinates. Recall, in this regard, the remark on page 411: while even non-holonomic constraints can be used to eliminate relations among the differentials in $\delta \boldsymbol{r}$, if these constraints cannot be integrated, corresponding relations among the variables themselves cannot be found; this will confound the elimination of dependencies in the expressions of such coordinate-dependent quantities as force and velocity. Thus we must also assume for the nonce that *all constraints are holonomic*; we shall see in a later section (8.3) how to circumvent this assumption. In any event, the f generalized coordinates will be *linearly independent*—a fact we shall have to invoke.

Consider the *net* force $\boldsymbol{F}_i = \boldsymbol{F}_i^{(a)} + \boldsymbol{F}_i^{(c)}$ acting on the ith mass m_i, where the $\boldsymbol{F}_i^{(a)}$ is the [net] *active* force, and $\boldsymbol{F}_i^{(c)}$ the net *constraint* force acting on that mass. Although the latter will ultimately vanish when we consider the virtual work, we must include them here because we initially resort to Newton's Second Law applied to this mass:

$$\boldsymbol{F}_i = \boldsymbol{F}_i^{(a)} + \boldsymbol{F}_i^{(c)} = \frac{d}{dt}(m_i \dot{\boldsymbol{r}}_i)$$

to describe the *actual* motion of m_i. Under a virtual displacement $\delta \boldsymbol{r}_i$ *from* this motion, these forces do *virtual work* on the mass; this can be related to the resulting motion:

$$\delta U_i \equiv \boldsymbol{F}_i \cdot \delta \boldsymbol{r}_i = \boldsymbol{F}_i^{(a)} \cdot \delta \boldsymbol{r}_i + \boldsymbol{F}_i^{(c)} \cdot \delta \boldsymbol{r}_i = \frac{d}{dt}(m_i \dot{\boldsymbol{r}}_i) \cdot \delta \boldsymbol{r}_i;$$

thus the *net* virtual work done on the entire system, under similar virtual displacements from their actual motion for *all* the m_i, is

$$\delta U \equiv \sum_{i=1}^{n} \delta U_i = \sum_{i=1}^{n} \boldsymbol{F}_i \cdot \delta \boldsymbol{r}_i =$$

$$\sum_{i=1}^{n} \boldsymbol{F}_i^{(a)} \cdot \delta \boldsymbol{r}_i + \sum_{i=1}^{n} \boldsymbol{F}_i^{(c)} \cdot \delta \boldsymbol{r}_i = \sum_{i=1}^{n} \frac{d}{dt}(m_i \dot{\boldsymbol{r}}_i) \cdot \delta \boldsymbol{r}_i. \quad (8.1\text{-}1)$$

Even just at this stage in the development, it is important to take note of a point likely to get lost in the subsequent formalism: The q_i satisfying the above equations—entering them through the $\boldsymbol{F}_i$ generally, and the $\dot{\boldsymbol{r}}_i$ always—describe the actual motion of the particles, since they satisfy Newton's Second Law. Thus *the q_i appearing in any results of these equations will themselves represent the actual motion.*

Consider first the right-hand, last term of this equation: By (7.4-2),

$$\sum_{i=1}^{n}\frac{d}{dt}(m_i\dot{\boldsymbol{r}}_i)\cdot\delta\boldsymbol{r}_i=\sum_{i=1}^{n}\frac{d}{dt}(m_i\dot{\boldsymbol{r}}_i)\cdot\sum_{j=1}^{f}\frac{\partial\boldsymbol{r}_i}{\partial q_j}\,\delta q_j$$
$$=\sum_{j=1}^{f}\Big(\sum_{i=1}^{n}\frac{d}{dt}(m_i\dot{\boldsymbol{r}}_i)\cdot\frac{\partial\boldsymbol{r}_i}{\partial q_j}\Big)\,\delta q_j \qquad (8.1\text{-}2)$$

But we can write

$$\frac{d}{dt}(m_i\dot{\boldsymbol{r}}_i)\cdot\frac{\partial\boldsymbol{r}_i}{\partial q_j}=\frac{d}{dt}\Big(m_i\dot{\boldsymbol{r}}_i\cdot\frac{\partial\boldsymbol{r}_i}{\partial q_j}\Big)-m_i\dot{\boldsymbol{r}}_i\cdot\frac{d}{dt}\Big(\frac{\partial\boldsymbol{r}_i}{\partial q_j}\Big)$$
$$=\frac{d}{dt}\Big(m_i\dot{\boldsymbol{r}}_i\cdot\frac{\partial\dot{\boldsymbol{r}}_i}{\partial\dot{q}_j}\Big)-m_i\dot{\boldsymbol{r}}_i\cdot\frac{\partial}{\partial q_j}\Big(\frac{d\boldsymbol{r}_i}{dt}\Big)$$

—using Theorems 7.4.1 and 7.4.2 in the first and second terms, respectively, of this equation—

$$=\frac{d}{dt}\Big(\frac{\partial}{\partial\dot{q}_j}\big(m_i\frac{\dot{\boldsymbol{r}}_i\cdot\dot{\boldsymbol{r}}_i}{2}\big)\Big)-\frac{\partial}{\partial q_j}\big(m_i\frac{\dot{\boldsymbol{r}}_i\cdot\dot{\boldsymbol{r}}_i}{2}\big)$$
$$=\frac{d}{dt}\Big(\frac{\partial}{\partial\dot{q}_j}\big(m_i\frac{v_i^2}{2}\big)\Big)-\frac{\partial}{\partial q_j}\big(m_i\frac{v_i^2}{2}\big)$$
$$\equiv\frac{d}{dt}\Big(\frac{\partial}{\partial\dot{q}_j}(T_i)\Big)-\frac{\partial}{\partial q_j}(T_i)$$

—where T_i is just the kinetic energy of the ith particle! Thus the last term in (8.1-2)

$$\sum_{j=1}^{f}\Big(\sum_{i=1}^{n}\frac{d}{dt}(m_i\dot{\boldsymbol{r}}_i)\cdot\frac{\partial\boldsymbol{r}_i}{\partial q_j}\Big)\,\delta q_j=\sum_{j=1}^{f}\Big(\sum_{i=1}^{n}\frac{d}{dt}\Big(\frac{\partial}{\partial\dot{q}_j}(T_i)\Big)-\frac{\partial}{\partial q_j}(T_i)\Big)\,\delta q_j$$
$$=\sum_{j=1}^{f}\Big(\frac{d}{dt}\Big(\frac{\partial}{\partial\dot{q}_j}\big(\sum_{i=1}^{n}T_i\big)\Big)-\frac{\partial}{\partial q_j}\big(\sum_{i=1}^{n}T_i\big)\Big)\,\delta q_j$$
$$\equiv\sum_{j=1}^{f}\Big(\frac{d}{dt}\Big(\frac{\partial}{\partial\dot{q}_j}(T)\Big)-\frac{\partial}{\partial q_j}(T)\Big)\,\delta q_j$$

—where now $T \equiv \sum_{i=1}^{n} T_i$ is the total *kinetic energy of the system.* Thus equation (8.1-1) reads that:

$$\delta U = \sum_{i=1}^{n} \frac{d}{dt}(m_i \dot{\boldsymbol{r}}_i) \cdot \delta \boldsymbol{r}_i = \sum_{j=1}^{f} \Big(\frac{d}{dt}\Big(\frac{\partial T}{\partial \dot{q}_j}\Big) - \frac{\partial T}{\partial q_j} \Big) \delta q_j \tag{8.1-3}$$

We now consider the penultimate term in (8.1-1), involving the virtual work done by the applied and constraint forces. That done by the constraint forces *vanishes*, however, *as long as we take the* $\delta \boldsymbol{r}_i$ *consistent with these constraints*; this is precisely the definition 7.3.1! *Assuming* such virtual displacements, then,

$$\delta U = \sum_{i=1}^{n} \boldsymbol{F}_i^{(a)} \cdot \delta \boldsymbol{r}_i. \tag{8.1-4}$$

We once again invoke (7.4-2):

$$\begin{aligned} \sum_{i=1}^{n} \boldsymbol{F}_i^{(a)} \cdot \delta \boldsymbol{r}_i &= \sum_{i=1}^{n} \boldsymbol{F}_i^{(a)} \cdot \sum_{j=1}^{f} \frac{\partial \boldsymbol{r}_i}{\partial q_j} \delta q_j \\ &= \sum_{j=1}^{f} \Big(\sum_{i=1}^{n} \boldsymbol{F}_i^{(a)} \cdot \frac{\partial \boldsymbol{r}_i}{\partial q_j} \Big) \delta q_j \\ &\equiv \sum_{j=1}^{f} Q_j \, \delta q_j \end{aligned}$$

i.e.,

$$\delta U = \sum_{i=1}^{n} \boldsymbol{F}_i^{(a)} \cdot \delta \boldsymbol{r}_i = \sum_{j=1}^{f} Q_j \, \delta q_j \tag{8.1-5}$$

—where the

$$Q_j \equiv \sum_{i=1}^{n} \boldsymbol{F}_i^{(a)} \cdot \frac{\partial \boldsymbol{r}_i}{\partial q_j} \tag{8.1-6}$$

are called *generalized forces.* Like the generalized velocities, these, too, are scalars! Note that, while the $\boldsymbol{F}_i^{(a)}$ always have units $[F]$, and the $\boldsymbol{r}_i$ always have units $[L]$, the q_j, and thus the δq_j, can have any units whatsoever. Thus, for example, if q_j happens to be an angle, the corresponding Q_j by (8.1-6) would have units of $[F][L]$—*i.e.*, units of *moment*; only if $[q_j] = [L]$ will the partial $\frac{\partial \boldsymbol{r}_i}{\partial q_j}$ be unitless, making $[Q_j] = [F]$.

We now set the two expressions for δU, (8.1-3) and (8.1-5), equal to one another:

$$\delta U = \sum_{j=1}^{f} \Big(\frac{d}{dt}\Big(\frac{\partial T}{\partial \dot{q}_j}\Big) - \frac{\partial T}{\partial q_j} \Big) \delta q_j = \sum_{j=1}^{f} Q_j \, \delta q_j$$

or, just to be fussy,

$$\sum_{j=1}^{f}\left\{\left(\frac{d}{dt}\left(\frac{\partial T}{\partial \dot{q}_j}\right)-\frac{\partial T}{\partial q_j}\right)-Q_j\right\}\delta q_j=0. \tag{8.1-7}$$

This says that the above *sum* must vanish. But recall that the generalized coordinates $\{q_1, q_2, \ldots, q_f\}$ *have had all constraints eliminated from among them*; thus they are *linearly independent*; *i.e.*, by equation (7.1-7), we can conclude that the coefficients of the δq_j must vanish *separately*, not just in the above sum. There results a *set* of f equations

$$\boxed{\frac{d}{dt}\left(\frac{\partial T}{\partial \dot{q}_j}\right)-\frac{\partial T}{\partial q_j}=Q_j, \quad j=1,2,\ldots,f \quad (8.1\text{-}8)}$$

These are referred to as the *Euler-Lagrange equations*. They are *the* fundamental equations, and, in view of their rather remarkable character, some discussion of them is warranted:

Notes on the Euler-Lagrange Equations

First a few conceptual observations regarding these equations:

- The virtual displacements above were presumed consistent with all the constraints; that is the reason only the $\boldsymbol{F}^{(a)}$ entered the virtual work term. But since the $\boldsymbol{F}^{(c)}$ don't enter, *constraint forces cannot be determined.*[1] This is precisely what obtained in the Statics presentation of virtual work.
- The above development considered virtual displacements *from the actual motion*; Newton's Second Law was used right from the start. Thus the q_j satisfying (8.1-8) *satisfy the actual motion of the system.*
- The equations were derived through the use of [virtual] *work*; thus it is not surprising that what results involves the *energy*, particularly the kinetic energy, T. It is otherwise completely general; there is no restriction placed on the type of [applied] forces which can be considered. In all these regards, (8.1-8) is somewhat analogous to the Principle of Work and [Kinetic] Energy in classical dynamics.

We now consider the *form* of these equations in some detail:

- There is an interesting division of labor between the kinetic energy and generalized forces: The former involves the *speed* of the particles; the expression for this, in turn, depends purely on the coordinates chosen, *as*

[1] Well, actually, they *can* be found using this technique, but only if one selects a virtual displacement *violating* a given constraint.

well as the constraints. The latter, though, also dependent on coordinates, are where any active forces are described. Thus, in some sense, the *kinematics* of the problem is contained in T, while the *kinetics* relies on the Q's.

- The kinetic energy T for a system, is defined by

$$T \equiv \sum_{i=1}^{N} \frac{1}{2} m_i v_i^2;$$

"v_i" in this equation is the *speed*—the magnitude of the *actual* velocity—of m_i. But, using equation (7.4-1),

$$\begin{aligned} v_i^2 = \boldsymbol{v}_i \cdot \boldsymbol{v}_i &= \left(\sum_{j=1}^{f} \frac{\partial \boldsymbol{r}_i}{\partial q_j} \dot{q}_j + \frac{\partial \boldsymbol{r}_i}{\partial t} \right) \cdot \left(\sum_{k=1}^{f} \frac{\partial \boldsymbol{r}_i}{\partial q_k} \dot{q}_k + \frac{\partial \boldsymbol{r}_i}{\partial t} \right) \\ &= \sum_j^f \sum_k^f \frac{\partial \boldsymbol{r}_i}{\partial q_j} \cdot \frac{\partial \boldsymbol{r}_i}{\partial q_k} \dot{q}_j \dot{q}_k + 2 \sum_j^f \frac{\partial \boldsymbol{r}_i}{\partial q_j} \cdot \frac{\partial \boldsymbol{r}_i}{\partial t} \dot{q}_j + \frac{\partial \boldsymbol{r}_i}{\partial t} \cdot \frac{\partial \boldsymbol{r}_i}{\partial t}. \end{aligned} \tag{8.1-9}$$

(Note that the last two terms in this somewhat obnoxious sum result from *time-dependent*—i.e., *rheonomic*—*constraints*; in particular they will vanish if the constraints are *scleronomic*.) Thus we can write

$$\begin{aligned} T = &\sum_i^n \frac{1}{2} m_i \sum_j^f \sum_k^f \frac{\partial \boldsymbol{r}_i}{\partial q_j} \cdot \frac{\partial \boldsymbol{r}_i}{\partial q_k} \dot{q}_j \dot{q}_k + \\ &\sum_i^n m_i \sum_j^f \frac{\partial \boldsymbol{r}_i}{\partial q_j} \cdot \frac{\partial \boldsymbol{r}_i}{\partial t} \dot{q}_j + \sum_i^n \frac{1}{2} m_i \frac{\partial \boldsymbol{r}_i}{\partial t} \cdot \frac{\partial \boldsymbol{r}_i}{\partial t} \\ = &\sum_j^f \sum_k^f \left(\sum_i^n \frac{1}{2} m_i \frac{\partial \boldsymbol{r}_i}{\partial q_j} \cdot \frac{\partial \boldsymbol{r}_i}{\partial q_k} \right) \dot{q}_j \dot{q}_k + \\ &\sum_j^f \left(\sum_i^n m_i \frac{\partial \boldsymbol{r}_i}{\partial q_j} \cdot \frac{\partial \boldsymbol{r}_i}{\partial t} \right) \dot{q}_j + \sum_i^n \frac{1}{2} m_i \frac{\partial \boldsymbol{r}_i}{\partial t} \cdot \frac{\partial \boldsymbol{r}_i}{\partial t} \end{aligned} \tag{8.1-10}$$

Note that the terms multiplying the $\dot{q}_j$ involve derivatives of $\boldsymbol{r}_i(\boldsymbol{q};t)$ so are themselves functions only of $\boldsymbol{q}$ and t. We see, then, that T is generally *quadratic* in the $\dot{q}_j$; thus $\frac{\partial T}{\partial \dot{q}_j}$ will be *linear* in the $\dot{q}_j$; thus $\frac{d}{dt}(\frac{\partial T}{\partial \dot{q}_j})$ will give an expression in $\boldsymbol{q}$ times $\ddot{q}_j$—a *second order differential equation for* q_j (which, however, will generally be non-linear).

It will prove useful for later development to write this as a summation of *matrix* products.

$$T \equiv \dot{\mathbf{q}}^{\mathsf{T}} \mathsf{T}_2 \dot{\mathbf{q}} + \dot{\mathbf{q}}^{\mathsf{T}} \mathsf{T}_1 + \mathsf{T}_0 \tag{8.1-11}$$

—in which $\mathbf{q}$ is the $f \times 1$ column vector of the q_i, T_2 is an $f \times f$ matrix (of sums of the individual terms for each m_i)—note that this is *symmetric*—T_1 is a simple $f \times 1$ column matrix (also a sum), and T_0 a scalar. Again, all these are functions of q and t only, just as in (8.1-10).

- The *generalized forces*, (8.1-6),

$$Q_j \equiv \sum_{i=1}^{n} \boldsymbol{F}_i^{(a)} \cdot \frac{\partial \boldsymbol{r}_i}{\partial q_j}$$

are, if generally straightforward to calculate, admittedly a little involved. Note that one deals with vectors, so it is necessary to introduce basis vectors to express the $\boldsymbol{F}_i^{(a)}$ and $\boldsymbol{r}_i$, though these can be expressed in whatever q_j have been selected for the particular problem.

But there are actually a couple of alternatives: The fact that $Q_j \delta q_j$ is just the *work* done under a virtual displacement δq_j suggests that this might be an alternative method for determining the Q_j: simply take the virtual δq_j—*at fixed t, and holding all other generalized coordinates fixed*—and, "by inspection", determine what work is done; the coefficient of the δq_j is then the desired Q_j. In this method, it is necessary to be consistent with regard to signs of both the δq_j and the resulting $\delta U_j \equiv Q_j \delta q_j$. To help enforce this, it is wise to take δq_j *positive*; this will assure that the δU_j, and thus the corresponding Q_j, will have appropriate signs. While such an approach has a certain simplistic appeal, and is certainly useful in simple problems (see first example below), it is quite possible that varying one coordinate alone will result in rather intense "inspection" for the associated work (see second example).

Still another alternative, likely most popular among those who remember their virtual work from Statics, is merely to take the same [vector] virtual displacements used there—simple differentials, as they have been all along—of the $\boldsymbol{r}_i$ and calculate the *total* virtual work

$$\delta U = \sum_{i=1}^{n} \boldsymbol{F}_i^{(a)} \cdot \delta \boldsymbol{r}_i = \sum_{j=1}^{f} Q_j \, \delta q_j,$$

equation (8.1-5), directly. The [net] coefficient of δq_j is then just Q_j. It is important in this regard, to note that we are calculating the "absolute" work done under the $\delta \boldsymbol{r}_i$, be that virtual or not. This means that the generating $\boldsymbol{r}_i$ *must be measured from a fixed point*; otherwise the resulting δU would be different, depending on which [moving] point $\boldsymbol{r}_i$ is chosen to be measured from! By virtue of the fact that

$$\delta \boldsymbol{r}_i = \sum_{j=1}^{f} \frac{\partial \boldsymbol{r}_i}{\partial q_j} \, \delta q_j.$$

(equation (7.4-2)), this is altogether equivalent to the first approach (which simply circumvents introduction of the δq_j), though it might have a greater appeal of familiarity. But this means that the same *caveat*, regarding the measurement of $\boldsymbol{r}_i$ from a fixed point, which applies here must also be observed in the first. In any event, both the first and third are more "automatic" (so less prone to error) than the second. Regardless of which is chosen, it is recommended the student pick one and stick by it!

Each of these approaches, referred to as "a)", "b)", and "c)", respectively, is exemplified in the following examples, which also, for the first time, introduce generalized coordinates themselves.

The first example is a simple one—one likely encountered in a more elementary classical dynamics course:

Example 8.1.1 *A Single Rigid Body.* We consider a uniform "slender rod" of mass m and length $2a$, suspended from a fixed, frictionless pin O, free to swing in a vertical plane. This example could be approached using the planar equations for a rigid body, probably utilizing polar coordinates centered at O. But it is easy to see that this system possesses only a single degree of freedom (under the constraint that $\bar{r}$, the distance from O to the center of mass, is constant—the same constraint which would be imposed in polar coordinates)—and we choose to utilize, as our [single] generalized coordinate, q, the angle from the fixed horizontal line to the axis of the rod. Note that, since it is measured from the horizontal, q increases counterclockwise; this observation is necessary to get the sign of δq correct in both approaches. Of the forces acting on the body—the pin force $\boldsymbol{O}$ and the gravitational force acting at the center of mass (here the centroid, due to uniformity)—the former is a constraint force, so won't enter the equations of motion; thus all we need consider in the calculation of the Q_j is the active weight: $\boldsymbol{F}^{(a)} \equiv -mg\hat{\jmath}$, where $\hat{\jmath}$ points conventionally upward.

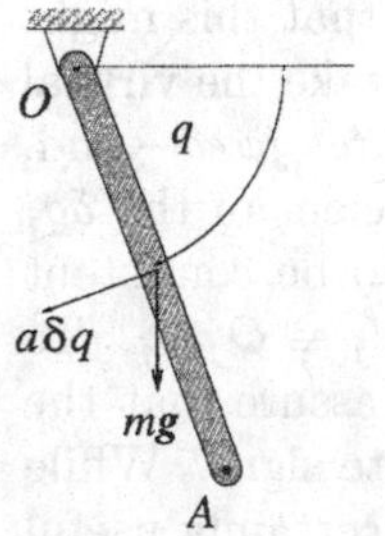

- a) (definition of the Q_j, (8.1-6)): Since $\boldsymbol{mg}$ acts at the centroid, we can introduce standard Cartesian coordinates, expressed in terms of the generalized coordinate q to obtain $\bar{\boldsymbol{r}} = a\cos q\hat{\imath} - a\sin q\hat{\jmath}$; thus

$$Q_j \equiv \sum_{i=1}^{n} \boldsymbol{F}_i^{(a)} \cdot \frac{\partial \boldsymbol{r}_i}{\partial q_j} = (-mg\hat{\jmath}) \cdot \frac{\partial}{\partial q}(a\cos q\hat{\imath} - a\sin q\hat{\jmath})$$

$$= (-mg\hat{\jmath}) \cdot (-a\sin q\hat{\imath} - a\cos q\hat{\jmath}) = +mga\cos q$$

- b) (calculating work due to δq_j): We observe that $+\delta q$ generates a $\delta\bar{\boldsymbol{r}}$ perpendicular to the rod and of magnitude $a\,\delta q$ as indicated in the diagram (note direction!) This results in a $\delta U = +(mg)(a\,\delta q)\cos q$.

- c) (calculating virtual work directly): Referring to the $\overline{r}$ in a), the net virtual work

$$\begin{aligned}\delta U \equiv \sum_{i=1}^{n} \boldsymbol{F}_i^{(a)} \cdot \delta \boldsymbol{r}_i &= (-mg\hat{\jmath}) \cdot \delta(a\cos q\hat{\imath} - a\sin q\hat{\jmath})\\ &= (-mg\hat{\jmath}) \cdot (-a\sin q\hat{\imath} - a\cos q\hat{\jmath})\,\delta q\\ &= +mga\cos q\,\delta q \equiv Q\,\delta q\end{aligned}$$

  we see, once again, that $Q = mga\cos q$

All three approaches are equivalent and, as expected, lead to precisely the same results.

Note in the two vector formulations a) and c), that, because we ultimately take a dot product of the vector derivatives/differentials with a force purely in the $\hat{\jmath}$, we really didn't need to find their $\hat{\imath}$ components at all! Observing this from the start can save a little work, as the next example will demonstrate. |

We now apply this virtual work approach to a proper "machine":

Example 8.1.2 *A Machine.* We consider a *double* pendulum, consisting of two uniform "slender rods", each of mass m and length $2a$. The first, OA suspended from the fixed frictionless pin O, being connected to the second through a similar frictionless pin at A; both pins admit only of motion in a vertical plane. We shall once again calculate the generalized forces doing work under an admissible virtual displacement.

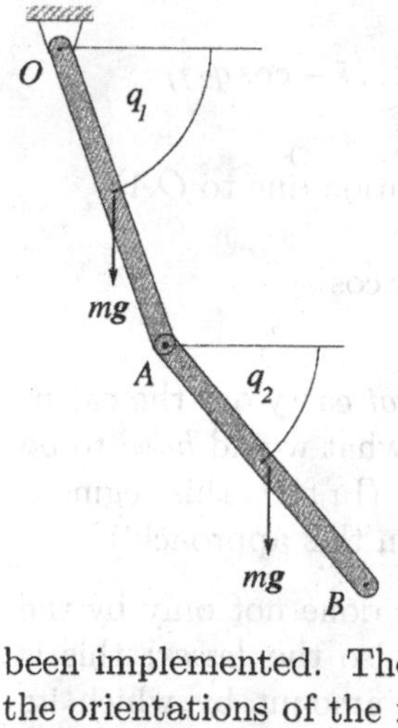

The first issue is clearly to select generalized coordinates. Two *free* rigid bodies in a plane would require six coordinates (two for a point, plus one for an angle for each). But the end point of OA is fixed at O; while, knowing the angle q_1 and length $2a$ for OA, the coordinates of A in AB are determined. Thus only *two* coordinates are required once all constraints have been implemented. The "obvious" choice here is the pair of angles describing the orientations of the respective bodies, diagrammed at left. While, as in the previous example, q_1 corresponds to the standard polar coordinate angle θ about the fixed point O, *polar coordinates are not applicable to the analysis of AB*: the point A is not fixed, and the existence of a fixed point relative to which the position is measured is built into polar coordinates! Thus, with classical dynamics, there would be no choice but to utilize Cartesian coordinates to describe the motion of AB, imposing kinematic constraints on the motion of that body utilizing rotating coordinates, for example; this will be done when we mount a frontal attack on a similar problem utilizing the Lagrange equations. For the time being, however, one can at least appreciate the ability to utilize such a "natural" set as the q's.

As in the previous example, the pin at O does no work under an admissible displacement. But neither does that at A: this can be seen either by appeal to its "constraint" nature, or by noting that the forces the pin exerts on each member are opposite, but that the [virtual] displacements of the points in the respective rods are the same, resulting in a net virtual work of zero. Thus the only forces doing work under a displacement [from the actual] not violating the constraints are the two weights. The work done by the upper rod's has already been obtained above; allowing for the change in notation, this is just $mga \cos q_1$—a term which will have to be added to any additional contribution to Q_1 we obtain from the lower rod. Thus we concentrate our attention on work done by gravity on AB.

- a) (definition of the Q_j): Again $m\mathbf{g}$ acts at the centroid of the lower rod—a point whose [vector] position we must now express. The first inclination might be to measure this from the point A, *but that would be wrong*: in order to measure the *absolute* displacement of the point, it must be measured from a *fixed* point. O provides an obvious choice; utilizing a Cartesian basis in the customary directions, and noting once again that the force is purely in the $\hat{\jmath}$ direction, we get that $\overline{\mathbf{r}}_2 = 2a(\ldots\hat{\imath} - \sin q_1\hat{\jmath}) + a(\ldots\hat{\imath} - \sin q_2\hat{\jmath})$. Then, since

$$\frac{\partial \overline{\mathbf{r}}_2}{\partial q_1} = 2a(\ldots\hat{\imath} - \cos q_1\hat{\jmath}) \quad \text{and} \quad \frac{\partial \overline{\mathbf{r}}_2}{\partial q_2} = a(\ldots\hat{\imath} - \cos q_2\hat{\jmath})$$

  we obtain immediately (accounting for the contribution due to OA)

$$Q_1 = 3mga \cos q_1 \quad \text{and} \quad Q_2 = mga \cos q_2$$

- b) (calculating work due to q_j): Actually, we will *not* carry out the calculations for this method; rather we will only sketch what would *have* to be done to find the generalized forces "by inspection". (In fact, this segment is included primarily to show why *not* to depend on this approach!)

  When q_1 changes *holding* q_2 *fixed* [virtual] work is done not only by the upper bar (whose orientation q_1 measures), but also the *lower*; this is precisely the additional term to Q_1 above. The amount by which its center is displaced is going to be $\rho\, \delta q_1$, where "ρ" is the distance from O to the lower bar's center. That is not an easy thing to calculate: it is the third side of a triangle whose other two have lengths $2a$ and a, and we don't have explicitly any of the angles in that triangle! Likely the easiest way to find *them* is to resort to finding the horizontal and vertical distances from O to the rod's center—*but this is essentially* all *we had to do in a)!* It is not a pleasant thing to contemplate, particularly given the relatively straightforward—even automatic—techniques available.

- c) (calculating virtual work directly): We shall approach this as if we never had the previous example's results, since this is typically the situation:

$$\begin{aligned}
\delta U &= (-mg\hat{\jmath})\cdot\delta\overline{r}_1 + (-mg\hat{\jmath})\cdot\delta\overline{r}_2 \\
&= (-mg\hat{\jmath})\cdot\delta(\ldots\hat{\imath} - a\sin q_1\hat{\jmath}) + \\
&\qquad (-mg\hat{\jmath})\cdot\delta(\ldots\hat{\imath} - (2a\sin q_1 - a\sin q_2)\hat{\jmath}) \\
&= (-mg\hat{\jmath})\cdot(\ldots\hat{\imath} - a\cos q_1\,\delta q_1\hat{\jmath}) + \\
&\qquad (-mg\hat{\jmath})\cdot(\ldots\hat{\imath} - (2a\cos q_1\,\delta q_1 - a\cos q_2\,\delta q_2)\hat{\jmath}) \\
&= mga\cos q_1\,\delta q_1 + 2mga\cos q_1\,\delta q_1 + mga\cos q_2\,\delta q_2 \\
&\equiv Q_1\,\delta q_1 + Q_2\,\delta q_2
\end{aligned}$$

—from which we get the same results as in a).

We leave whatever conclusions are to be drawn to the reader. |

There is one final note to be made regarding the above derivation in general. We have considered a *system of particles*—a "gas" of particles subject to *some* constraints, but otherwise free to move relative to one another. This is a stepping stone on the way to developing the rigid body equations of motion in classical dynamics, where, in presenting "systems", the distinction is made between *external* forces—those arising from *outside* the system, and *internal* ones—those due to forces particles in the system exert on one another. It is shown there that, while only the *external* forces affect the linear and angular momentum of a system, *both external and internal forces do work*; thus, in the "two-body problem"—two point masses subject only to their mutual gravitation—there are no external forces (so the center of mass of the system moves with constant velocity), but the internal gravitational force appears in the energy of the system.

We have been focusing here on the *applied/constraint* division between forces; but these, too, can each be either "external" or "internal" (though the classification for constraint forces becomes moot, since they do no work anyway). Considering the applied forces, however, *both internal and external applied forces do [virtual] work.* And even though the internal $\boldsymbol{f}_{ij}$, the force on m_i due to m_j, say, is equal in magnitude but opposite in direction to $\boldsymbol{f}_{ji}$ (by Newton's Third Law), the net virtual work done by these two forces on their respective masses will generally not vanish, due to the unconstrained nature of this "gas" of particles (see the remark on page 266).

But now consider such a system of particles in which they *move as a unit*, having their inter-particles distances fixed; *i.e.*, as a *rigid body.* This is precisely the situation examined in Example 7.1.5, where we observed that the imposition of such fixed distances is tantamount to putting *constraints*—holonomic ones, at that—on the system. These [internal] forces, rather than being active, now become *constraint* forces *which do no work.* Thus, *in considering the motion of a rigid body, we can ignore internal forces.* This is, in the context of constraints in Analytical Mechanics, precisely the result obtained for rigid bodies on page 356.

Finally, with regard to the motion of rigid bodies, we note that "T" in the above equations was simply the kinetic energy of the *system* [of particles]. But, in classical dynamics, rigid-body expressions for T have already been obtained passing from particles to the limit of continuous mass distribution. Thus the above equations hold for these systems, too, and the expressions (5.4-2) for kinetic energy of rigid bodies can be implemented directly.

We conclude with several examples of the general Euler-Lagrange equations. The first of these is, once again, one which would be a relatively straightforward problem in classical dynamics; it is included to show how the analytical mechanics would work the problem. The second, however, involves a *machine*, which is where the present approach really shows its worth. The last is also routine in the Newtonian formulation, but it is included to indicate how to deal with time-dependent forces.

In all three, comparison is made with the same problem worked using the classical vector approach. This is done for two reasons: In the first place, the answers are the same; that's encouraging, but not surprising, since the current approach is based on Newton's Laws anyway! But, more importantly, it is to stress that *the problem can be worked using either the Newton equations or the Euler-Lagrange ones.* Students have a tendency to "type cast" problems: "Oh, this is an *energy* problem;" "This is a *virtual work* problem." The fact is that, at the end of the day, *all problems are* Newton *problems*: Newton's laws are the foundation on which all other things rest. The alternatives are merely general conclusions which can be derived from Newton's laws (so one doesn't, for example, have to rederive the energy equation each time), or approaches with special features making them more amenable to a given problem. In studying the field, one is learning a repertoire of techniques, as well as the wisdom of which might be most appropriate to find the desired information in a given instance.

Example 8.1.3 *Frictional Force.* We consider a small mass—a "particle"—of mass m, attached to a [linear] spring whose other end is fixed, as shown. The mass moves on a horizontal table with which it has [kinetic] coefficient of friction μ_k. We shall determine the equations of motion for this mass.

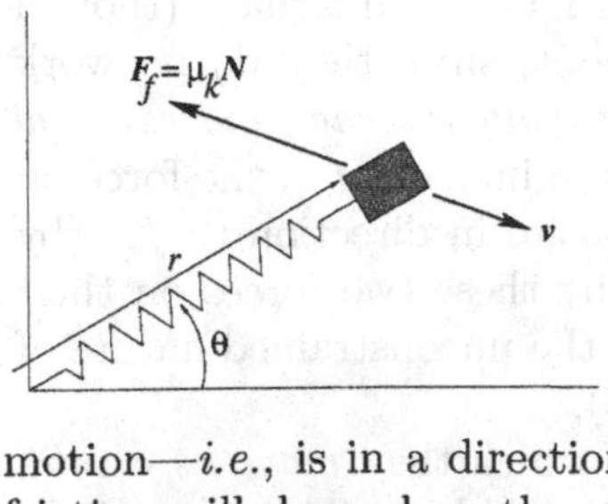

There are two forces doing work here, the spring and the friction. The former is always along the spring, the latter opposes the motion—*i.e.*, is in a direction opposite to the velocity $\boldsymbol{v}$. Specification of the friction will depend on the expression for velocity in whatever coordinates we choose. With an eye to determining the generalized forces using one of the vector methods above, it would be nice to use a set amenable to easy expression of the velocity. The obvious choice would be tangential-normal coordinates, but these do not admit of an expression for position, which we need for the spring; furthermore, we have no idea of what the radius of curvature of the motion might be in case we need it. Thus the friction force provides little guidance on a "convenient" set of coordinates. But the spring, whose direction changes and whose magnitude depends on the amount by which the spring has been stretched, virtually begs the use of classical polar coordinates; thus we will use those.

In order to calculate the generalized forces, we will express both these forces as vectors:

$$\text{spring:} \qquad \boldsymbol{F}_s = -k(r - r_o)\hat{\boldsymbol{r}}$$
$$\text{friction:} \qquad \boldsymbol{F}_f = -\mu_k mg\hat{\boldsymbol{v}}$$

(r_o being the unstretched length of the spring) where, in polar coordinates

$$\hat{\boldsymbol{v}} = \frac{\boldsymbol{v}}{|\boldsymbol{v}|} = \frac{\dot{r}\hat{\boldsymbol{r}} + r\dot{\theta}\hat{\boldsymbol{\theta}}}{\sqrt{\dot{r}^2 + (r\dot{\theta})^2}}.$$

In terms of these coordinates, the Euler-Lagrange equations can be written

$$\frac{d}{dt}\left(\frac{\partial T}{\partial \dot{r}}\right) - \frac{\partial T}{\partial r} = Q_r, \quad \frac{d}{dt}\left(\frac{\partial T}{\partial \dot{\theta}}\right) - \frac{\partial T}{\partial \theta} = Q_\theta;$$

we must determine, then, T, Q_r and Q_θ.

T is easy: either dotting the polar coordinate $\boldsymbol{v} = \dot{r}\hat{\boldsymbol{r}} + r\dot{\theta}\hat{\boldsymbol{\theta}}$ with itself, or utilizing the orthogonal character of these [vector] coordinates, we get that

$$T = \frac{1}{2}mv^2 = \frac{1}{2}m(\dot{r}^2 + (r\dot{\theta})^2)$$

We now find the generalized forces, for example, by calculating the virtual work done under a virtual displacement: The expression for $\delta\boldsymbol{r}$ can most easily be found noting that, since $\dot{\boldsymbol{r}} = \boldsymbol{v} = \dot{r}\hat{\boldsymbol{r}} + r\dot{\theta}\hat{\boldsymbol{\theta}}$, the differential $\delta\boldsymbol{r} = \delta r\hat{\boldsymbol{r}} + r\delta\theta\hat{\boldsymbol{\theta}}$, so

$$\begin{aligned}
\delta U &= \delta U(\boldsymbol{F}_s) + \delta U(\boldsymbol{F}_f) \\
&= -k(r - r_o)\hat{\boldsymbol{r}} \cdot (\delta r\hat{\boldsymbol{r}} + r\delta\theta\hat{\boldsymbol{\theta}}) + \frac{-\mu_k mg(\dot{r}\hat{\boldsymbol{r}} + r\dot{\theta}\hat{\boldsymbol{\theta}})}{v} \cdot (\delta r\hat{\boldsymbol{r}} + r\delta\theta\hat{\boldsymbol{\theta}}) \\
&= -k(r - r_o)\,\delta r - \frac{\mu_k mg\dot{r}}{v}\,\delta r - \frac{\mu_k mgr\dot{\theta}}{v} r\,\delta\theta \\
&\equiv Q_r\,\delta r + Q_\theta\,\delta\theta
\end{aligned}$$

—in which, for brevity, we write $v \equiv |\boldsymbol{v}| = \sqrt{\dot{r}^2 + (r\dot{\theta})^2}$. Thus

$$\begin{aligned}
\frac{d}{dt}\left(\frac{\partial T}{\partial \dot{r}}\right) - \frac{\partial T}{\partial r} &= \frac{d}{dt}(m\dot{r}) - mr\dot{\theta}^2 = m\ddot{r} - mr\dot{\theta}^2 \\
&= Q_r = -k(r - r_o) - \frac{\mu_k mg\dot{r}}{v} \\
\frac{d}{dt}\left(\frac{\partial T}{\partial \dot{\theta}}\right) - \frac{\partial T}{\partial \theta} &= \frac{d}{dt}(mr^2\dot{\theta}) - 0 = 2mr\dot{r}\dot{\theta} - mr^2\ddot{\theta} \\
&= Q_\theta = -\frac{\mu_k mgr^2\dot{\theta}}{v}.
\end{aligned}$$

Note that above we made direct appeal to the polar coordinates we chose as our "generalized" ones in order to get expressions for $\boldsymbol{v}$ and $\delta\boldsymbol{r}$. A more general approach, and one which would work without foreknowledge of the necessary quantities, would be to set up *quasi*-Cartesian coordinates in, say, the directions in the diagrams. Then, expressing the Cartesian coordinates in terms of our choice of generalized coordinates,

$$\boldsymbol{r} = x\hat{\boldsymbol{\imath}} + y\hat{\boldsymbol{\jmath}} = r\cos\theta\hat{\boldsymbol{\imath}} + r\sin\theta\hat{\boldsymbol{\jmath}};$$

thus

$$\delta\boldsymbol{r} = (\cos\theta\hat{\boldsymbol{\imath}} + \sin\theta\hat{\boldsymbol{\jmath}})\,\delta r + (-r\sin\theta\hat{\boldsymbol{\imath}} + r\cos\theta\hat{\boldsymbol{\jmath}})\,\delta\theta$$

and

$$\begin{aligned} v^2 &= \dot{x}^2 + \dot{y}^2 = (\dot{r}\cos\theta - r\sin\theta\dot{\theta})^2 + (\dot{r}\sin\theta + r\cos\theta\dot{\theta})^2 \\ &= \dot{r}^2 + r^2\dot{\theta}^2. \end{aligned}$$

—the cross terms in $\sin\theta\cos\theta$ cancelling, and the others both multiplying $(\cos^2\theta + \sin^2\theta) \equiv 1$.

Two things are notable about this example: In the first place, after all the ponderous development leading up to this point, the final procedure is surprisingly straightforward; the most involved thing above was calculation of the Q's!

The other point refers to the classical vector approach to this problem: As noted above, we would likely pick polar coordinates in terms of which to work it; then the resulting equation would be

$$\sum F_r = -k(r - r_o) - \frac{\mu_k mg\dot{r}}{v} = ma_r = m(\ddot{r} - r\dot{\theta}^2)$$

$$\sum F_\theta = -\frac{\mu_k mgr\dot{\theta}}{v} = ma_\theta = m(2\dot{r}\dot{\theta} + r\ddot{\theta})$$

—precisely the equations obtained above (*mod* the extra factor of r in the θ equation). Comparing the proposed approach with the classical one, we see we've got to find the vector forces to find the Q's anyhow, and the Newtonian formulation avoids the extra step of determining these. In point of fact, this example has done little to demonstrate any advantage! The next one probably does. |

The next example involves a system—a "machine" of sorts:

Example 8.1.4 *System of Particles.* We return to the double pendulum of Example 8.1.2, with some generalizations, but the significant simplification that, rather than using slender rods ("rigid bodies"), we consider two point masses, m_1 and m_2. Each is fixed to the end of a "light" rigid rod, of lengths l_1 and l_2 respectively; the first is suspended from the fixed point O while the second attached to m_1, both connections using those "frictionless pins". (We choose different generalized coordinates, too—four free [two planar particles] reduced by two fixed-distance constraints—but that is of little consequence.)

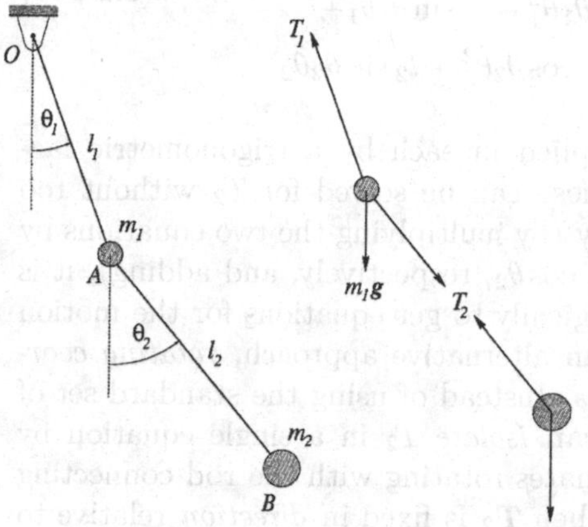

Included in the picture are the free body diagrams of the two masses (don't confuse the tensional forces "T_i" with kinetic energy!); this is done because, in order to drive home the point that there is nothing "special" about the Euler-Lagrange equations, as well as to compare the effort and features involved in both the classical the present approaches, the problem will first be at least set up using Newtonian dynamics, but measured using the generalized coordinates θ_1 and θ_2 we shall be using later, just to facilitate comparison.

Newtonian Equations We note first that, since the rods are light, the forces exerted on them, and thus which they exert on the two masses, are along the respective rods. Further, it will be necessary to write individual equations of motion for each mass.

The first body's motion is amenable to the use of polar coordinates, using O as the fixed origin, and choosing to have θ increasing in the same direction as θ_1. That done, the equations of motion for m_1 are relatively straightforward:

$$\begin{aligned}\sum F_r &= -T_1 + T_2\cos(\theta_2-\theta_1) + m_1 g\cos\theta_1 \\ &= m_1 a_r = m_1(\ddot{r}_1 - r_1\dot{\theta}_1^2) = -m_1 l_1 \dot{\theta}_1^2\end{aligned} \tag{8.1-12a}$$

$$\begin{aligned}\sum F_\theta &= T_2\sin(\theta_2-\theta_1) - m_1 g\sin\theta_1 \\ &= m_1 a_\theta = m_1(2\dot{r}_1\dot{\theta}_1 + r_1\ddot{\theta}_1) = m_1 l_1 \ddot{\theta}_1.\end{aligned} \tag{8.1-12b}$$

Description of the second mass's motion becomes rather more interesting. Since it doesn't move in a circle, and we have no clue regarding its radius of curvature, we default to the "coordinates of last resort", Cartesian.[2] Taking the customary direction for such coordinates, we get that

$$\begin{aligned}\sum F_X &= -T_2\sin\theta_2 = m_2\ddot{x}_2 = m_2\frac{d^2}{dt^2}(l_1\sin\theta_1 + l_2\sin\theta_2) \\ &= m_2(-l_1\sin\theta_1\dot{\theta}_1^2 + l_1\cos\theta_1\ddot{\theta}_1 - \\ &\qquad l_2\sin\theta_2\dot{\theta}_2^2 + l_2\cos\theta_2\ddot{\theta}_2)\end{aligned} \tag{8.1-13a}$$

[2]Lest we be accused of "mixing coordinates", note that we will ultimately be getting *scalar* component equations in the various unknowns; how those are obtained then becomes moot.

$$\sum F_Y = T_2 \cos\theta_2 - m_2 g = m_2 \ddot{y}_2 = m_2 \frac{d^2}{dt^2}(-l_1 \cos\theta_1 - l_2 \cos\theta_2)$$
$$= m_2(l_1 \cos\theta_1 \dot{\theta}_1^2 + l_1 \sin\theta_1 \ddot{\theta}_1 + l_2 \cos\theta_2 \dot{\theta}_2^2 + l_2 \sin\theta_2 \ddot{\theta}_2). \tag{8.1-13b}$$

The above equations have T_2 multiplied in each by a trigonometric factor. While these can be solved for T_2 without too much difficulty (by multiplying the two equations by $(-\sin\theta_2)$ and $\cos\theta_2$, respectively, and adding), it is useful pedagogically to get equations for the motion of m_2 using an alternative approach, *rotating coordinate systems.* Instead of using the standard set of $\hat{\imath}$ and $\hat{\jmath}$, we can *isolate* T_2 in a single equation by taking coordinates rotating with the rod connecting m_1 and m_2; then $\boldsymbol{T}_2$ is fixed in *direction* relative to these axes, and we can select one, say $\hat{\jmath}$, to be parallel to that. This set of axes is diagrammed at left—note that this is not a free-body diagram, but rather a vector diagram to obtain the components—where we see the rotating $\hat{\imath}$ and $\hat{\jmath}$, generally making an angle of $(\theta_2 - \theta_1)$ relative to the m_1 directions $\hat{r}$ and $\hat{\boldsymbol{\theta}}$.

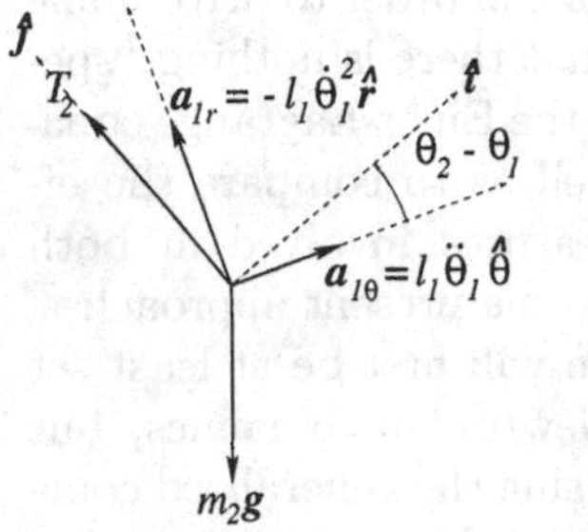

Further, since we know the motion of m_1, we can use that point as the origin of the rotating set; relative to these axes and origin, then, m_2 is itself *fixed*, so the terms "$\boldsymbol{v}_{rel}$" and "$\boldsymbol{a}_{rel}$" for that point both vanish. Thus, finally, we get that

$$\begin{aligned}\sum \boldsymbol{F} &= T_2\hat{\jmath} - m_2 g(\sin\theta_2\hat{\imath} + \cos\theta_2\hat{\jmath})\\ &= m_2\boldsymbol{a}_2 = m_2(\boldsymbol{a}_1 + \boldsymbol{\alpha}\times\boldsymbol{r} + \boldsymbol{\omega}\times(\boldsymbol{\omega}\times\boldsymbol{r}))\\ &= m_2(\ell_1\ddot{\theta}_1(\cos(\theta_2-\theta_1)\hat{\imath} - \sin(\theta_2-\theta_1)\hat{\jmath})\\ &\quad + \ell_1\dot{\theta}_1^2(\sin(\theta_2-\theta_1)\hat{\imath} + \cos(\theta_2-\theta_1)\hat{\jmath})\\ &\quad + \ddot{\theta}_2\hat{\boldsymbol{k}}\times(-\ell_2\hat{\jmath}) + \dot{\theta}\hat{\boldsymbol{k}}\times(\dot{\theta}\hat{\boldsymbol{k}}\times(-\ell_2\hat{\jmath}))).\end{aligned}$$

The $\hat{\imath}$- and $\hat{\jmath}$-components of this equation, respectively, then become

$$-m_2 g\sin\theta_2 = m_2(\ell_1\ddot{\theta}_1\cos(\theta_2-\theta_1) + \ell_1\dot{\theta}_1^2\sin(\theta_2-\theta_1) + \ell_2\ddot{\theta}_2) \tag{8.1-14a}$$

$$T_2 - m_2 g\cos\theta_2 = m_2(-\ell_1\ddot{\theta}_1\sin(\theta_2-\theta_1) + \ell_1\dot{\theta}_1^2\cos(\theta_2-\theta_1) + \ell_2\dot{\theta}_2^2). \tag{8.1-14b}$$

Having done this, then, four of these equations—the two (8.1-12) for m_1 and either of the above pairs (8.1-13) or (8.1-14) for m_2—determine the four unknowns T_1, T_2, $\ddot{\theta}_1$ and $\ddot{\theta}_2$, which "can be" solved: they are linear in these variables (though multiplied by trigonometric functions) and expressed in terms of θ_1, θ_2, $\dot{\theta}_1$ and $\dot{\theta}_2$. (The greater convenience of the latter for m_2 is obvious: its second equation gives T_2 immediately; this replaced in the $\hat{r}$ equation for m_1

then gives T_1.) In particular, we obtain differential equations for the θ's which must be solved.

Note that, as is necessary in Newtonian mechanics, T_1 and T_2 had to be included in these equations, even though we might not be so interested in them as in the motion, $\theta_i(t)$. In any event, the resulting four unknowns *had* to be solved for from the four equations to obtain the differential equations for the θ's.

Euler-Lagrange Equations We shall use the equations

$$\frac{d}{dt}\Big(\frac{\partial T}{\partial \dot\theta_1}\Big) - \frac{\partial T}{\partial \theta_1} = Q_1, \quad \frac{d}{dt}\Big(\frac{\partial T}{\partial \dot\theta_2}\Big) - \frac{\partial T}{\partial \theta_2} = Q_2$$

for the θ_i, where T is the kinetic energy *of the system*; we get that first. Recall that, for a system of particles,

$$T \equiv \sum T_i = \frac{1}{2}m_1 v_1^2 + \frac{1}{2}m_2 v_2^2$$

The v's are scalars here; we can use any coordinates we choose to determine them. v_1 is easy: since m_1 moves in a circle about O, we could use the results from polar coordinates [based on m_1] we have invoked previously:

$$\boldsymbol{v}_1 = \dot r_1 \hat{\boldsymbol{r}}_1 + r_1\dot\theta_1\hat{\boldsymbol{\theta}}_1 = l_1\dot\theta_1\hat{\boldsymbol{\theta}}_1 \quad \Rightarrow \quad v_1^2 \equiv \|\boldsymbol{v}_1\|^2 = l_1^2\dot\theta_1^2$$

—$\hat{\boldsymbol{r}}_1$ and $\hat{\boldsymbol{\theta}}_1$ being the polar coordinate unit vectors for mass m_1—or use the more general approach utilized above of expressing $\boldsymbol{v}_1$ in Cartesian coordinates but using the present variables: taking $\hat{\boldsymbol{I}}$ and $\hat{\boldsymbol{J}}$ in the customary directions,

$$\begin{aligned}
v_1^2 \equiv \|\dot{\boldsymbol{r}}_1\|^2 &= \|\frac{d}{dt}(l_1\sin\theta_1\hat{\boldsymbol{I}} - l_1\cos\theta_1\hat{\boldsymbol{J}})\|^2 \\
&= \|l_1\cos\theta_1\dot\theta_1\hat{\boldsymbol{I}} + l_1\sin\theta_1\dot\theta_1\hat{\boldsymbol{J}}\|^2 \\
&= (l_1\cos\theta_1\dot\theta_1)^2 + (l_1\sin\theta_1\dot\theta_1)^2 \\
&= l_1^2\dot\theta_1^2.
\end{aligned}$$

While cited for purpose of completeness, the latter approach is included here because it *must* be used if we choose to find v_2 using Cartesian coordinates:

$$\begin{aligned}
v_2^2 \equiv \|\dot{\boldsymbol{r}}_2\|^2 &= \|\frac{d}{dt}([l_1\sin\theta_1 + l_2\sin\theta_2]\hat{\boldsymbol{I}} - [l_1\cos\theta_1 + l_2\cos\theta_2]\hat{\boldsymbol{J}})\|^2 \\
&= \|[l_1\cos\theta_1\dot\theta_1 + l_2\cos\theta_2\dot\theta_2]\hat{\boldsymbol{I}} + [l_1\sin\theta_1\dot\theta_1 + l_2\sin\theta_2\dot\theta_2]\hat{\boldsymbol{J}}\|^2 \\
&= (l_1\cos\theta_1\dot\theta_1 + l_2\cos\theta_2\dot\theta_2)^2 + (l_1\sin\theta_1\dot\theta_1 + l_2\sin\theta_2\dot\theta_2)^2 \\
&= l_1^2\dot\theta_1^2 + l_2^2\dot\theta_2^2 + 2l_1l_2(\cos\theta_1\cos\theta_2 + \sin\theta_1\sin\theta_2)\dot\theta_1\dot\theta_2 \\
&= l_1^2\dot\theta_1^2 + l_2^2\dot\theta_2^2 + 2l_1l_2\cos(\theta_2 - \theta_1)\dot\theta_1\dot\theta_2.
\end{aligned}$$

(The excruciating detail in the above is included merely to indicate the type of calculations one might expect, as well as the simplification in form afforded by the use of elementary trigonometric identities.)

Once again, however, rotating coordinates provide an alternative, arguably more convenient, means of finding v_2: using the rotating $(\hat{\imath},\hat{\jmath})$ basis above and finding components of $\boldsymbol{v}_1$ with respect to this,

$$\begin{aligned}\boldsymbol{v}_2 &= \boldsymbol{v}_1 + \boldsymbol{\omega}_2 \times \boldsymbol{r}_2\\ &= \ell_1\dot{\theta}_1((\cos(\theta_2-\theta_1)\hat{\imath} - \sin(\theta_2-\theta_1)\hat{\jmath}) + \dot{\theta}_2\hat{\boldsymbol{k}} \times (-\ell_2\hat{\jmath})\\ &= (\ell_1\dot{\theta}_1\cos(\theta_2-\theta_1) + \ell_2\dot{\theta}_2)\hat{\imath} - \ell_1\dot{\theta}_1\sin(\theta_2-\theta_1)\hat{\jmath},\end{aligned}$$

whose components squared give the above v_2^2. In either event, note that this is *not* merely $(\ell_1\dot{\theta}_1)^2 + (\ell_2\dot{\theta}_2)^2$; the latter would hold if and only if the angle between the two rods was 90°!

With the above two expression for the v's—compare them with the general $v_i^2(\boldsymbol{q}, \dot{\boldsymbol{q}}; t)$, equation (8.1-9)—it is an easy matter to find

$$T = \frac{1}{2}(m_1+m_2)l_1^2\dot{\theta}_1^2 + \frac{1}{2}m_2 l_2^2\dot{\theta}_2^2 + m_2 l_1 l_2\cos(\theta_2-\theta_1)\dot{\theta}_1\dot{\theta}_2.$$

Now it is necessary to determine the Q_i which do work under a virtual displacement not violating the constraints. Under such a displacement, the pin force at O ("T_1" above) and the connection between the lower rod and m_1 ("T_2" above) are both constraint forces which do no work; thus the only forces doing "active" work are the two $m_i\boldsymbol{g}$'s, both in the vertical direction. We find the Q_i using the standard virtual work approach:

$$\delta U = \delta U(m_1\boldsymbol{g}) + \delta U(m_2\boldsymbol{g}) \equiv Q_1\,\delta\theta_1 + Q_2\,\delta\theta_1$$

in which [noting we need only the $\hat{\boldsymbol{J}}$-component of $\delta\boldsymbol{r}$]

$$\begin{aligned}\delta U(m_1\boldsymbol{g}) &= -m_1 g\hat{\boldsymbol{J}}\cdot\delta(\cdots - l_1\cos\theta_1\hat{\boldsymbol{J}})\\ &= -m_1 g\hat{\boldsymbol{J}}\cdot(\cdots - l_1\sin\theta_1\,\delta\theta_1\hat{\boldsymbol{J}})\\ &= -m_1 g l_1\sin\theta_1\,\delta\theta_1\end{aligned}$$

and, *noting that we must measure the position at which $m_2\boldsymbol{g}$ acts from a fixed point,*

$$\begin{aligned}\delta U(m_2\boldsymbol{g}) &= -m_2 g\hat{\boldsymbol{J}}\cdot\delta(\cdots - (l_1\cos\theta_1 + l_2\cos\theta_2)\hat{\boldsymbol{J}})\\ &= -m_2 g\hat{\boldsymbol{J}}\cdot(\cdots + (l_1\sin\theta_1\,\delta\theta_1 + l_2\sin\theta_2\,\delta\theta_2)\hat{\boldsymbol{J}})\\ &= -m_2 g(l_1\sin\theta_1\,\delta\theta_1 + l_2\sin\theta_2\,\delta\theta_2);\end{aligned}$$

thus

$$\begin{aligned}\delta U &= -(m_1+m_2)g l_1\sin\theta_1\,\delta\theta_1 - m_2 g l_2\sin\theta_2\,\delta\theta_2\\ &\equiv Q_1\,\delta\theta_1 + Q_2\,\delta\theta_2.\end{aligned}$$

Finally we calculate the two equations of motion: For θ_1,

$$
\begin{aligned}
\frac{d}{dt}\left(\frac{\partial T}{\partial \dot\theta_1}\right) - \frac{\partial T}{\partial \theta_1} &= \frac{d}{dt}\left[(m_1+m_2)l_1^2\dot\theta_1 + m_2 l_1 l_2 \cos(\theta_2-\theta_1)\dot\theta_2\right] \\
&\quad - m_2 l_1 l_2 \sin(\theta_2-\theta_1)\dot\theta_1\dot\theta_2 \\
&= [(m_1+m_2)l_1^2\ddot\theta_1 + m_2 l_1 l_2 \cos(\theta_2-\theta_1)\ddot\theta_2 \\
&\qquad - m_2 l_1 l_2 \sin(\theta_2-\theta_1)(\dot\theta_2-\dot\theta_1)\dot\theta_2] \\
&\quad - m_2 l_1 l_2 \sin(\theta_2-\theta_1)\dot\theta_1\dot\theta_2 \\
&= (m_1+m_2)l_1^2\ddot\theta_1 + m_2 l_1 l_2 \cos(\theta_2-\theta_1)\ddot\theta_2 \\
&\quad - m_2 l_1 l_2 \sin(\theta_2-\theta_1)\dot\theta_2^2 \\
&= Q_1 = -(m_1+m_2) g l_1 \sin\theta_1.
\end{aligned}
$$

In a similar fashion, for θ_2,

$$
\begin{aligned}
\frac{d}{dt}\left(\frac{\partial T}{\partial \dot\theta_2}\right) - \frac{\partial T}{\partial \theta_2} &= \frac{d}{dt}\left[m_2 l_2^2\dot\theta_2 + m_2 l_1 l_2 \cos(\theta_2-\theta_1)\dot\theta_1\right] \\
&\quad + m_2 l_1 l_2 \sin(\theta_2-\theta_1)\dot\theta_1\dot\theta_2 \\
&= m_2 l_2^2\ddot\theta_2 + m_2 l_1 l_2 \cos(\theta_2-\theta_1)\ddot\theta_1 + m_2 l_1 l_2 \sin(\theta_2-\theta_1)\dot\theta_1^2 \\
&= Q_2 = -m_2 g l_2 \sin\theta_2.
\end{aligned}
$$

Here we have only *two* equations to solve for the *two* quantities $\ddot\theta_1$ and $\ddot\theta_2$; the solution is not trivial, but the unknowns still appear only linearly (albeit again multiplied by trigonometric functions) and at least it's easier than in the Newtonian system! These equations are altogether equivalent to the equations for $\theta_i(t)$ obtained above using the Newtonian approach: the θ_2 equation is precisely the $\hat{\imath}$-component of the *rotating* coordinate set's (8.1-14) for m_2, and the θ_1 equation can be shown to be a linear combination of the rotating $\hat{\imath}$- and $\hat{\jmath}$-equations (8.1-13), and the $\hat{\boldsymbol\theta}$-equation for m_1, (8.1-12b). The T_i, forces at the connections, do not appear because they do no [virtual] work; thus they couldn't be found using this method in which such "constraints" are not violated. But oftentimes one is merely interested in the motion of the system, and these equations give it; if one really wants the T_i, they can now be found with the Newtonian [polar coordinate] equations, but now in only *two* unknowns. |

Example 8.1.5 *Time-dependent Force.* In order to indicate how time-dependent forces can be dealt with using the Euler-Lagrange equations, we include a brief example of such a system. We consider one similar to Example 8.1.3, but now, rather than a [kinetic] frictional force—we will assume the horizontal table to be "frictionless"—we apply a radial time-dependent force. T will remain the same, but the generalized forces will be dif-

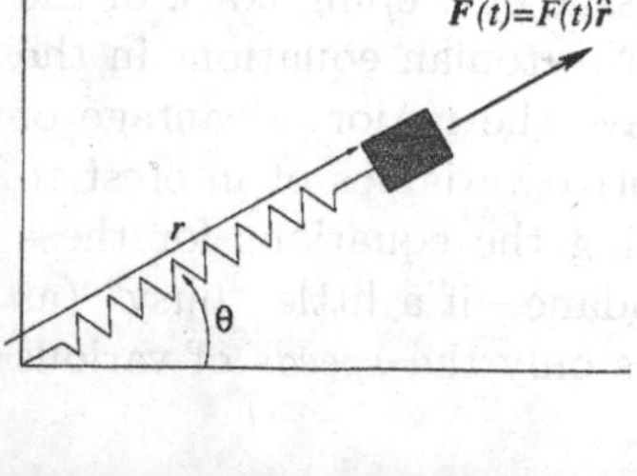

ferent. Thus the first step required is calculation of these forces.

We already have that associated with the spring, so we here find that resulting from $\boldsymbol{F}(t)$. Again, it is necessary to introduce a temporary vector coordinate system in order to evaluate the virtual work. Given the "polar" nature of the problem, the easiest choice is likely to express $\boldsymbol{F}(t) \equiv F_r(t)\hat{\boldsymbol{r}}$. In terms of these (and again obtaining $\delta\boldsymbol{r}$ from $d\boldsymbol{r} = \frac{d\boldsymbol{r}}{dt}\,dt = \boldsymbol{v}\,dt$ in polar coordinates),

$$\delta U \equiv \boldsymbol{F}(t) \cdot \delta\boldsymbol{r} = F_r(t)\,\delta r \equiv Q_r \delta r$$

—which will be added onto the corresponding terms for the spring in Example 8.1.3. From here, application of the Euler-Lagrange equations is routine:

$$\frac{d}{dt}\left(\frac{\partial T}{\partial \dot{r}}\right) - \frac{\partial T}{\partial r} = m\ddot{r} - mr\dot{\theta}^2 = Q_r = -k(r - r_o) + F_r(t)$$
$$\frac{d}{dt}\left(\frac{\partial T}{\partial \dot{\theta}}\right) - \frac{\partial T}{\partial \theta} = 2mr\dot{r}\dot{\theta} - mr^2\ddot{\theta} = Q_\theta = 0$$

—again equations obtained by writing the Newtonian equations of motion in polar coordinates. |

Of the above examples, Example 8.1.4 is likely more typical of the types of problems where the features of analytical dynamics becomes a powerful technique for analysis: *systems* of interconnected bodies—"machines"—for which one wishes to determine the *motion* (as opposed to the connection forces required to support this motion). Actually, this is not too surprising: even in Statics, using virtual work to solve a problem involving a *single* rigid body/particle involves an order of magnitude more work than simply writing out the equilibrium equations; it is in the analysis of *machines* that virtual work really shows its stuff! And if one *does* require the internal connection or support forces, it is always possible to revert *afterwards* to the Newtonian formulation to find them, where now the [kinematic] *motion is known*; then there are only the kinetic forces or moments to find, instead of both: we get two independent *pairs*, say, of equations in two unknowns each, rather than a single system of *four* equations in four unknowns for the two-degree-of-freedom system of Example 8.1.4.

As is typical in mechanics, the interconnecting pins were assumed to be frictionless. But friction in pins manifests itself through a couple moment $\boldsymbol{M}$, and, in the presence of such an effect, it could be accounted for by merely including a term $-\boldsymbol{M} \cdot \delta\boldsymbol{\theta}$ under a [virtual] rotation $\delta\boldsymbol{\theta}$.

Clearly the complexity of the equations would become greater as the number of bodies increased, but it would increase *doubly* if Newtonian mechanics was used (and that doesn't even begin to address the issue of keeping track of the vector coordinate systems required to express the Newtonian equations in the first place). But the Euler-Lagrange equations have the major advantage of being *scalar* in character and allowing the descriptive variables of interest to be chosen at will. As a practical matter, obtaining the equations for these variables is straightforward—almost tediously mundane—if a little "busy" (as demonstrated in the second example): one requires only the *speeds* of various points, and this means that only the first time derivative, instead of the second required to obtain accelerations in Newtonian mechanics, need be found. The only other bookkeeping required is determination of the generalized forces, but this, too, is relatively routine.

It is an important feature of these equations that they deal with *all* types of forces, including dissipative ones—a characteristic they share with their counterpart from classical dynamics, the work-[kinetic] energy principle. In that, too, calculation of the work done by various forces over an interval was somewhat tedious, at least when compared with the relative simplicity of energy conservation, where the forces doing work entered through their potential energies rather than work. Given the energy nature of the Euler-Lagrange equations, and their correspondence with the Principle of Work and [Kinetic] Energy, it might be suspected that analytical mechanics has a similar formulation, with similar simplicities, in the special case that all forces [doing virtual work] are conservative. This is in fact true, and it will be dealt with in the next section.

Summary. We have developed the Euler-Lagrange equations

$$\frac{d}{dt}\left(\frac{\partial T}{\partial \dot{q}_i}\right) - \frac{\partial T}{\partial q_i} = Q_i$$

for systems in which all constraints are *holonomic* and have been eliminated among the generalized coordinates; the latter was required in order that linear independence among the resulting f differentials, equal in number to that of the degrees of freedom, could be assumed. The equations are otherwise completely general, applicable to any system of forces.

T in this equation is the total energy of the system, expressed in terms of the generalized coordinates and their [generalized] velocities; this turns out to be a purely quadratic function of the latter. The Q_i are the *active* "generalized forces", themselves calculated using virtual work. *Constraint* forces never appear, at least as long as they are not violated; thus the Euler-Lagrange equations cannot be used to obtain the forces of constraint. (These can be found, however, by solving for the $q_i(t)$ obtained from the Euler-Lagrange equations, then substituting these solutions back into the Newtonian formulation; this procedure *decouples* the solution of the motion from that of solution for constraints.) The resulting equations yield f second-order ordinary differential equations in the q_i; the q_i satisfying these equations are then precisely those giving the actual motion of the system according to Newton's Laws, from which these equations have been derived in the first place.

8.2 Conservative Forces: Lagrange Equations

It has been mentioned several times that the Euler-Lagrange Equations are reminiscent of the Principle of Work and [Kinetic] Energy in classical dynamics. To explain that, aid in the development of a special case of the Euler-Lagrange Equations, and show the correspondence between that new formulation and the conservation of energy, let us briefly review the development of energy in Section 5.1:

We start by integrating the differential [definition] of *work*, $dU \equiv \boldsymbol{F}\cdot d\boldsymbol{r}$ (equation 5.1-8) to get the Principle of Work and [Kinetic] Energy, equation (5.1-9)[3]

$$U_{1\to 2} \equiv \int_{\boldsymbol{r}_1}^{\boldsymbol{r}_2} \sum \boldsymbol{F}_i \cdot d\boldsymbol{r}_i = T_2 - T_1. \tag{8.2-1}$$

The net work $U_{1\to 2}$ is thus a *line integral*, generally dependent on the *path* to get from $\boldsymbol{r}_1$ to $\boldsymbol{r}_2$, but we consider the possibility that this work might be *independent of the path—i.e.*, dependent only on [positions of] the endpoint, $\boldsymbol{r}_1$ and $\boldsymbol{r}_2$. This would be the case if the integrand of the line integral were an *exact differential*: $\sum \boldsymbol{F}_i \cdot d\boldsymbol{r}_i = -dV(\boldsymbol{r})$, say, for then

$$\begin{aligned} U_{1\to 2} &\equiv \int_{\boldsymbol{r}_1}^{\boldsymbol{r}_2} \sum \boldsymbol{F}_i \cdot d\boldsymbol{r}_i \\ &= \int_{\boldsymbol{r}_1}^{\boldsymbol{r}_2} (-dV(\boldsymbol{r})) = -V(\boldsymbol{r}_2) + V(\boldsymbol{r}_1) = T_2 - T_1, \end{aligned} \tag{8.2-2}$$

or, for $V_i \equiv V(\boldsymbol{r}_i)$, simply

$$-V_2 + V_1 = T_2 - T_1.$$

But this says that $T_2 + V_2 = T_1 + V_1$, or defining the *[total] energy* $E \equiv T + V$, that (equation (5.1-11))

$$E_2 = E_1, \tag{8.2-3}$$

the *conservation of energy.* Forces whose work *is* path-independent are called *conservative*, and we have a set of equivalences, (5.1-12), furnishing both a criterion for $\boldsymbol{F}$ to be conservative, and giving an equation for the above $V(\boldsymbol{r})$, the *potential energy*. This has the major advantage over the Work/[Kinetic] Energy Principle that the work can be calculated from a *scalar* function $V(\boldsymbol{r})$, rather that by evaluating a line integral. We even point out that this can be applied to time-dependent forces to give a time-dependent "potential energy", thus making E itself be time-dependent (see page 261).

But conservation of energy demands that *all* forces do path-independent work; if any doesn't, we must revert to the Work/[Kinetic] Energy Principle with its line integrals. To keep the advantages of the scalar potential energy while allowing certain forces to be *non*-conservative, we ultimately obtain a "hybrid" form of Work and Energy,

$$(U_N)_{1\to 2} = E_2 - E_1, \tag{8.2-4}$$

—where E includes *conservative* forces (the only ones for which there is a V in the first place), and $(U_N)_{1\to 2}$ involves *all non*-conservative forces and any *conservative* forces that one, by choice or ignorance, doesn't put into a potential energy.

[3]We renumber these equations in this Section for convenience.

What has all this got to do with analytical dynamics? The Euler-Lagrange equations in the previous section put *all* [active] forces into *generalized* forces, Q_i, forces which required a separate step to obtain. But these equations are based on energy—virtual work is used to get them in the first place, recall—and it seems plausible that, at least in the case of *conservative* forces, we might be able to bypass this separate step, perhaps even using potential energy. This is indeed the case, and the above review was included to show the parallels between classical energy arguments and the ones to follow, where we consider first the case in which *all* forces are conservative, and then obtain the most general form of the equations with "mixed" forces.

Recall that the Euler-Lagrange equations resulted from equation (8.1-3), reproduced here:

$$\delta U = \sum_{i=1}^{n} \frac{d}{dt}(m_i \dot{\boldsymbol{r}}_i) \cdot \delta \boldsymbol{r}_i = \sum_{j=1}^{f} \left(\frac{d}{dt}\left(\frac{\partial T}{\partial \dot{q}_j}\right) - \frac{\partial T}{\partial q_j} \right) \delta q_j,$$

in which the virtual work, assuming done through virtual displacements consistent with the constraints,

$$\delta U = \sum_{i=1}^{n} \boldsymbol{F}_i^{(a)} \cdot \delta \boldsymbol{r}_i.$$

Assume, now that the $\boldsymbol{F}_i^{(a)}$ are [all] *conservative.* By the conditions (5.1-12) for this to hold, just as in the classical analysis, this means that

$$\delta U = \sum_{i=1}^{n} (-\nabla V_i^{(a)}) \cdot \delta \boldsymbol{r}_i \equiv -\delta V = -\sum_{j=1}^{f} \frac{\partial V}{\partial q_j} \delta q_j.^{4}$$

Substituting this into the above [reproduced equation (8.1-3)] and putting all terms on one side, we get that

$$\sum_{j=1}^{f} \left\{ \left(\frac{d}{dt}\left(\frac{\partial T}{\partial \dot{q}_j}\right) - \frac{\partial T}{\partial q_j} \right) + \frac{\partial V}{\partial q_j} \right\} \delta q_j = 0. \tag{8.2-5}$$

Again we invoke the linear independence of the δq's, allowing us to conclude that, *for each* j [separately]

$$\left(\frac{d}{dt}\left(\frac{\partial T}{\partial \dot{q}_j}\right) - \frac{\partial T}{\partial q_j} \right) + \frac{\partial V}{\partial q_j} = 0.$$

But recall that $V = V(\boldsymbol{r}; t)$, and $\boldsymbol{r} = \boldsymbol{r}(\boldsymbol{q}; t)$; in particular, both are *independent of the* $\dot{q}_j$; thus $\frac{\partial V}{\partial \dot{q}_j} \equiv 0$ and we can write, at least formally, that

$$\left(\frac{d}{dt}\left(\frac{\partial T}{\partial \dot{q}_j}\right) - \frac{\partial T}{\partial q_j} \right) + \frac{\partial V}{\partial q_j} = \left(\frac{d}{dt}\left(\frac{\partial T}{\partial \dot{q}_j}\right) - \frac{\partial T}{\partial q_j} \right) - \left(\frac{d}{dt}\left(\frac{\partial V}{\partial \dot{q}_j}\right) - \frac{\partial V}{\partial q_j} \right) = 0.$$

[4] Note that, because the "δ" varies only the *coordinates*, this equation holds even if $\frac{\partial V}{\partial t} \neq 0$, *i.e.*, if $dV \neq \delta V$.

This motivates our tidying things up a bit by defining the *Lagrangian*

$$L \equiv T - V, \tag{8.2-6}$$

resulting in the final form

$$\frac{d}{dt}\left(\frac{\partial L}{\partial \dot{q}_j}\right) - \frac{\partial L}{\partial q_j} = 0 \quad j = 1, 2, \ldots, f \tag{8.2-7}$$

—the *Lagrange Equations.*

The Lagrange equations—merely the Euler-Lagrange equations applied to a conservative system—can be obtained by invoking the *Principle of Least Action*: that the *variation*, δ, of the *action*, $A \equiv \int_{t_1}^{t_2} L\,dt$, vanish:

$$\delta \int_{t_1}^{t_2} L(q, \dot{q}; t)\,dt = 0,$$

effectively "minimizing" this integral (more appropriately, *extremizing* it).[5] "Variation" here is precisely the variation we have been considering: it involves varying the path by varying the values of the coordinates, and hence the value of the integral, *at a fixed time.* But the integral itself imposes the further stricture that, whatever various paths are chosen, *they must have the same starting and ending points*: $\boldsymbol{q}(t_1)$ and $\boldsymbol{q}(t_2)$ are held *constant.* The Euler-Lagrange equations, in this context, are simply the conditions on the integrand that the integral be extremized.

This is a common approach to developing the Lagrange equations, but it is not utilized here for several reasons:

- It requires a long time just to develop the theory of the *calculus of variations*, that field of mathematics dealing with the minimization of such integrals. However important this is intrinsically, students invariably never see the topic again, making it a sort of "hanger" in the course.
- Having developed this theory, one apparently simply *posits* the Principle of Least Action as a fundamental law—an approach many find unsatisfying.
- Since the Lagrangian is defined only for *conservative* systems, the introduction of *non*-conservative forces, endemic in the Real World, becomes problematic! In fact, many books starting from the Lagrangian itself can be, perhaps justifiably, accused of a somewhat *ad hoc* approach to general forces.

The approach used here does not require any more than the most rudimentary knowledge of variations, no variational calculus, no *deus ex machina* in the form of curious philosophical propositions, and immediate application to *all* forces. Nonetheless, the reader should be aware of the alternative, since many of the classic texts in the field *do* develop the subject this way.

[5]The first sentence in C.L. Siegel's book *Lectures on Celestial Mechanics* (1971), as translated by Moser, reads "Ours, according to Leibniz, is the best of all possible worlds, and the laws of nature can therefore be described in terms of extremal principles."

Equation (8.2-7) presumes that *all* forces are conservative, so that we can write (8.2-5); but, as with the classical development, if some forces are conservative and others aren't (or aren't *chosen* to be put into V), we can get the most general form of Lagrange's equations,

$$\frac{d}{dt}\left(\frac{\partial L}{\partial \dot{q}_j}\right) - \frac{\partial L}{\partial q_j} = Q_j \quad j = 1, 2, \ldots, f \quad (8.2\text{-}8)$$

—*generalized Lagrange equations.* In a very real sense, the Euler-Lagrange equations (8.1-8), Lagrange equations (8.2-7), and generalized Lagrange equations (8.2-8) stand to one another in exactly the same way the classical work-[kinetic] energy equation (8.2-1), conservation of energy (8.2-3), and generalized work-energy (8.2-4) do: (8.2-1) and (8.1-8) are the original, general forms from which their respective special cases in which *all* forces are conservative, equations (8.2-3) and (8.2-7), are derived; both of the latter can be generalized to their most general forms (8.2-4) and (8.2-8), allowing use of both work and potential energy in the same equation. As with the classical equations, $V \equiv 0$ in (8.2-8) reduces to the Euler-Lagrange equations, while $Q_j \equiv 0$ for all j reduces to the standard Lagrange equations. Like the classical generalized energy equations, this allows one to use potential energy functions for those forces which are conservative, obviating the need to calculate *generalized* forces in such cases, while admitting of their use for non-conservative forces.

Note that the equations (8.2-8) are just an alternative form of the Euler-Lagrange equations: we have merely introduced the potential energy where possible. Thus, as with the earlier formulation, these will give us a set of f second-order ordinary differential equations in the q_j. (See the remarks related to T in the Euler-Lagrange equations, page 428.)

We illustrate these equations, first by briefly returning to the previous examples of the Euler-Lagrange equations, merely indicating the modifications in the procedure; the point here is that, just as in the classical case of the work-energy principle, *the Euler-Lagrange equations always work*; it's just usually simpler to utilize potential energy wherever possible. We then present an entirely new problem, exemplifying (for the first time) rigid body motion; as with previous examples, the analysis required classically will be presented, in order that the advantages of the present approach can be appreciated.

Example 8.2.1. *Example 8.1.3 using Lagrange Equations.* We return to the cited example, which the reader is recommended to review: only the modifications to that example are given here.

We shall now use the generalized Lagrange equations, (8.2-8). Here we need $L \equiv T - V$ and any necessary Q_j. $T = \frac{1}{2}mv^2 = \frac{1}{2}m(\dot{r}^2 + (r\dot{\theta})^2)$ is easy, using polar coordinates or a more general approach. In this case, of the two active forces (the spring and friction), only the spring is conservative; thus we can put its effect into a potential energy $V = \frac{1}{2}k(r - r_o)^2$. But friction is *not* conservative; thus it must remain the generalized force terms as in that example, and be determined in the same way. There results the Lagrangian

$$L = T - V = \frac{1}{2}m(\dot{r}^2 + (r\dot{\theta})^2) - \frac{1}{2}k(r - r_o)^2.$$

Substitution into the Lagrange equations yield

for r:

$$\frac{d}{dt}\left(\frac{\partial L}{\partial \dot{r}}\right) - \frac{\partial L}{\partial r} = \frac{d}{dt}(m\dot{r}) - (mr\dot{\theta}^2 - k(r - r_o)) = m\ddot{r} - mr\dot{\theta}^2 + k(r - r_o)$$

$$= Q_r = -\frac{\mu_k mg\dot{r}}{v}$$

while for θ:

$$\frac{d}{dt}\left(\frac{\partial L}{\partial \dot{\theta}}\right) - \frac{\partial L}{\partial \theta} = \frac{d}{dt}(mr^2\dot{\theta}) - 0 = 2mr\dot{r}\dot{\theta} - mr^2\ddot{\theta}$$

$$= Q_\theta = -\frac{\mu_k mgr^2\dot{\theta}}{v}.$$

The θ equation is exactly the same as before: V is independent of θ, so it never enters. The r equation has the spring force transferred to the left-hand side with opposite sign; this is precisely what happens when potential energy, rather than work, is used in energy problems in classical dynamics. |

Example 8.2.2. *Example 8.1.4 using Lagrange Equations.* Again, we need $L = T - V$ and any Q's. T must still be found in the same way it was before

$$T = \frac{1}{2}(m_1 + m_2)l_1^2\dot{\theta}_1^2 + \frac{1}{2}m_2 l_2^2\dot{\theta}_2^2 + m_2 l_1 l_2 \cos(\theta_2 - \theta_1)\dot{\theta}_1\dot{\theta}_2;$$

it is merely cited here (see the previous incantation of this example). All active forces are conservative; this suggests putting them into potential energy, leaving the $Q_j \equiv 0$ ("Lagrange's equations"):

$$V = mg\bar{y}_1 + mg\bar{y}_2 = -m_1 g l_1 \cos\theta_1 - m_2 g(l_1 \cos\theta_1 + l_2 \cos\theta_2)$$

Thus the Lagrangian in this problem is just

$$L \equiv T - V = \frac{1}{2}(m_1 + m_2)l_1^2\dot{\theta}_1^2 + \frac{1}{2}m_2 l_2^2\dot{\theta}_2^2 + m_2 l_1 l_2 \cos(\theta_2 - \theta_1)\dot{\theta}_1\dot{\theta}_2$$
$$- (-m_1 g l_1 \cos\theta_1 - m_2 g(l_1 \cos\theta_1 + l_2 \cos\theta_2).$$

Calculating the Lagrange equations:

$$\frac{d}{dt}\left(\frac{\partial L}{\partial \dot{\theta}_1}\right) - \frac{\partial L}{\partial \theta_1} = (m_1 + m_2)l_1^2\ddot{\theta}_1 + m_2 l_1 l_2 \cos(\theta_2 - \theta_1)\ddot{\theta}_2$$
$$- m_2 l_1 l_2 \sin(\theta_2 - \theta_1)\dot{\theta}_2^2 + (m_1 + m_2)g l_1 \sin\theta_1 = 0$$

$$\frac{d}{dt}\left(\frac{\partial L}{\partial \dot{\theta}_2}\right) - \frac{\partial L}{\partial \theta_2} = m_2 l_2 \ddot{\theta}_2 + m_2 l_1 l_2 \cos(\theta_2 - \theta_1)\ddot{\theta}_1$$
$$+ m_2 l_1 l_2 \sin(\theta_2 - \theta_1)\dot{\theta}_1^2 + m_2 g l_2 \sin\theta_2 = 0$$

—again, the same as before, with the force terms transposed.

Example 8.2.3 *Rigid Body Motion.* We introduce, for the first time, rigid body motion into the Lagrangian formalism. Recall, in this regard, the remarks on page 433: that, although we have derived the Lagrangian equations for a system of particles, T represents the kinetic energy of the system, which has already been determined for rigid bodies in classical mechanics. We shall use that here.

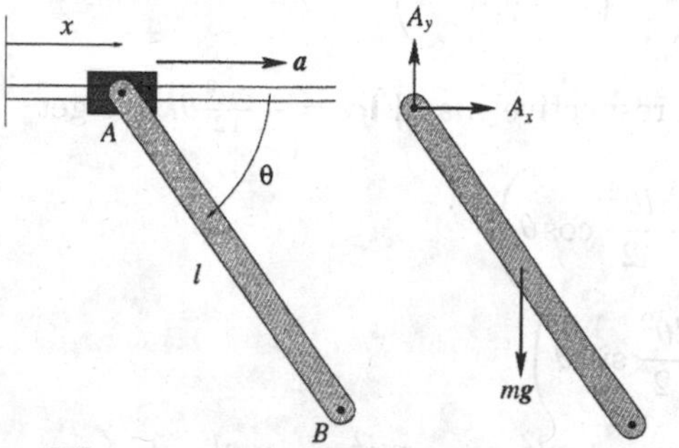

We consider a "uniform slender rod" AB, of mass m and length l. It is attached by a frictionless pin at A to a light collar which, in turn, moves along a horizontal rod with acceleration $\boldsymbol{a}$ as shown. We desire to determine the motion of AB, which is assumed to be constrained to be in the vertical plane.

Newtonian Equations. As was mentioned in the introduction to these examples, we will work this first using classical dynamics, in order that comparison with the Lagrangian approach can be made; thus there is a free body diagram above. Now matter *how* it is worked, however, it must be noted that, since the point A is accelerating, there must be some force applied to the rod at A, even though it is not drawn in the original statement's diagram.

There is no available fixed point, due to the acceleration of A; thus we will utilize rotating Cartesian coordinates in the customary directions. The equations of motion for this planar problem are

$$\sum F_x = A_x = m\overline{a}_x$$
$$\sum F_y = A_y - mg = m\overline{a}_y$$
$$\sum \boldsymbol{M}_A = -\frac{mgl}{2}\cos\theta\hat{\boldsymbol{k}} = \boldsymbol{r}_{G/A} \times m\overline{\boldsymbol{a}} + \overline{I}\boldsymbol{\alpha}$$

—where we choose to take moments about A in order to eliminate the pin force there; but, since this is not the center of mass or a fixed point about which the body is rotating, it is necessary to use the moment transfer to get the moment equation. Unlike the others, this is written in vector form for later calculation.

Though these are three equations in the five unknowns A_x, A_y, $\overline{a}_x$, $\overline{a}_y$, and α, we can get two more equations by relating $\overline{\boldsymbol{a}}$ and $\boldsymbol{\alpha}$ through the rotating coordinate equations, using the fact that $\boldsymbol{a}_A = a\hat{\boldsymbol{\imath}}$: defining $\boldsymbol{\alpha} \equiv -\ddot{\theta}\hat{\boldsymbol{k}}$, and $\boldsymbol{\omega} \equiv -\dot{\theta}\hat{\boldsymbol{k}}$, we finally get (after tedious but straightforward calculation!) that

$$\overline{\boldsymbol{a}} = \boldsymbol{a}_A + \boldsymbol{\alpha} \times \overline{\boldsymbol{r}} + \boldsymbol{\omega} \times (\boldsymbol{\omega} \times \overline{\boldsymbol{r}})$$
$$= \left(a - \frac{\alpha l}{2}\sin\theta - \frac{l\omega^2}{2}\right)\hat{\boldsymbol{\imath}} + \left(-\frac{\alpha l}{2}\cos\theta + \frac{l\omega^2}{2}\right)\hat{\boldsymbol{\jmath}}.$$

But this means that (after *more* calculation!)

$$\boldsymbol{r}_{G/A} \times m\overline{\boldsymbol{a}} = m\left(-\frac{\alpha l^2}{4} + \frac{al}{2}\sin\theta\right)\hat{\boldsymbol{k}}.$$

Finally, replacing α and ω by $\ddot{\theta}$ and $\dot{\theta}$, respectively, and $\bar{\mathsf{I}}\boldsymbol{\alpha} = -\frac{ml^2}{12}\ddot{\theta}\hat{\boldsymbol{k}}$ we get

$$A_x = m\left(a - \frac{l\ddot{\theta}}{2}\sin\theta - \frac{l\dot{\theta}^2}{2}\cos\theta\right)$$

$$A_y - mg = m\left(-\frac{l\ddot{\theta}}{2}\cos\theta + \frac{l\dot{\theta}^2}{2}\sin\theta\right)$$

$$-\frac{mgl}{2}\cos\theta = -\frac{ml^2}{4}\ddot{\theta} + \frac{mal}{2}\sin\theta - \frac{ml^2}{12}\ddot{\theta} = -\frac{ml^2}{3}\ddot{\theta} + \frac{mal}{2}\sin\theta$$

—three equations in A_x, A_y, and $\ddot{\theta}$, which must be solved simultaneously just to get the last. (They are *linear* in those three unknowns, so they *can* be solved. And, of course, these *do* give $\boldsymbol{A}$.)

Lagrange Equations. We observe that there are *two* degrees of freedom in this problem: the normal two (for the collar "particle") plus three (for rigid AB) being reduced by the fact that the positions of the collar and A are the same (two constraints) and that the y-coordinate of both is constant. knowing, for example, where A is, we can measure the orientation of the rod from that point. Thus we choose to use, as our generalized coordinates, the variables x and θ shown on the diagram. (Note that these do not correspond to any vector coordinate system whatsoever. But they are convenient, and we are at liberty to pick convenient coordinates!)

As usual, we first determine the Lagrangian function itself. This requires expressions for T and V (if any) in terms of our chosen coordinates. For general planar rigid body motion $T = \frac{1}{2}m\overline{v}^2 + \frac{1}{2}\overline{I}\omega^2$; we must now find the two component parts. Again using Cartesian coordinates relative to the *fixed* point at the end of the horizontal rod,

$$\begin{aligned}\overline{\boldsymbol{v}} &= \frac{d}{dt}((x + \frac{l}{2}\cos\theta)\hat{\boldsymbol{\imath}} - \frac{l}{2}\sin\theta\hat{\boldsymbol{\jmath}})\\ &= (\dot{x} - \frac{l}{2}\sin\theta\dot{\theta})\hat{\boldsymbol{\imath}} - \frac{l}{2}\cos\theta\dot{\theta}\hat{\boldsymbol{\jmath}}.\end{aligned}$$

(The same result could have been obtained using *rotating* coordinates, centered at A and rotating with the rod, having *instantaneously* the same $\hat{\boldsymbol{\imath}}$- and $\hat{\boldsymbol{\jmath}}$-directions:

$$\begin{aligned}\overline{\boldsymbol{v}} &= \boldsymbol{v}_A + \boldsymbol{\omega} \times \overline{\boldsymbol{r}}\\ &= \dot{x}\hat{\boldsymbol{\imath}} + (-\dot{\theta}\hat{\boldsymbol{k}}) \times \frac{\ell}{2}(\cos\theta\hat{\boldsymbol{\imath}} - \sin\theta\hat{\boldsymbol{\jmath}});\end{aligned}$$

note the sign given to $\boldsymbol{\omega}(\dot{\theta})$, due to the direction of positive θ.) Thus, after slight simplification,

$$\overline{v}^2 = \dot{x}^2 - l\sin\theta\dot{x}\dot{\theta} + \frac{l^2\dot{\theta}^2}{4},$$

so that, finally, multiplying by $\frac{1}{2}m$ and adding on $\frac{1}{2}\bar{I}\omega^2 = \frac{ml^2}{24}\dot{\theta}^2$,

$$T = \frac{1}{2}m\dot{x}^2 - \frac{1}{2}ml\sin\theta\dot{x}\dot{\theta} + \frac{1}{6}ml^2\dot{\theta}^2.$$

(The presence of $\dot{x}$ is somewhat troubling: we are only given $\boldsymbol{a}$, not $\boldsymbol{v}$. And, although we normally wouldn't have it available, the above Newtonian solution didn't need it!)

We now consider the active forces. The weight is conservative, with $V = -mg\frac{l}{2}\sin\theta$ (in these coordinates), and, though the vertical component of the pin force at A does no work for a virtual displacement consistent with the motion, A_x *does*; thus we need its generalized force expression. Just to be fussy, we include both components of $\boldsymbol{A}$ and note that the only possible virtual displacement which does not violate the pin constraint at A is a horizontal one:

$$\delta U = (A_x\hat{\imath} + A_y\hat{\jmath}) \cdot \delta x\hat{\imath} = A_x\delta x \equiv Q_x\delta x$$

so that $Q_x = A_x$ itself in this case.

Thus we finally have our system: with the above V,

$$L \equiv T - V = \frac{1}{2}m\dot{x}^2 - \frac{1}{2}ml\sin\theta\dot{x}\dot{\theta} + \frac{1}{6}ml^2\dot{\theta}^2 + mg\frac{l}{2}\sin\theta$$

and, with the above Q_x, we get our final [generalized] Lagrangian equations of motion:

for x:

$$\begin{aligned}\frac{d}{dt}\left(\frac{\partial L}{\partial \dot{x}}\right) - \frac{\partial L}{\partial x} &= \frac{d}{dt}(m\dot{x} - \frac{1}{2}ml\dot{\theta}\sin\theta) - 0 = m\ddot{x} - \frac{1}{2}ml(\sin\theta\ddot{\theta} + \cos\theta\dot{\theta}^2)\\ &= ma - \frac{1}{2}ml(\sin\theta\ddot{\theta} + \cos\theta\dot{\theta}^2)\\ &= Q_x = A_x\end{aligned}$$

while, after intermediate simplification, for θ:

$$\begin{aligned}\frac{d}{dt}\left(\frac{\partial L}{\partial \dot{\theta}}\right) - \frac{\partial L}{\partial \theta} &= \frac{d}{dt}(-\frac{1}{2}ml\sin\theta\dot{x} + \frac{1}{3}ml^2\dot{\theta}) - (-\frac{1}{2}ml\cos\theta\dot{x}\dot{\theta} + \frac{mgl}{2}\cos\theta)\\ &= -\frac{1}{2}ml\sin\theta\ddot{x} + \frac{1}{3}ml^2\ddot{\theta} - \frac{mgl}{2}\cos\theta\\ &= -\frac{1}{2}ml\sin\theta + \frac{1}{3}ml^2\ddot{\theta} - \frac{mgl}{2}\cos\theta = 0.\end{aligned}$$

(The time derivative generates a term $-\frac{1}{2}ml\cos\theta\dot{x}\dot{\theta}$ when $\sin\theta$ is differentiated; this is what eliminates the awkward $\dot{x}$ term we complained about above.) The x-equation here is precisely the x-component of the Newtonian example, while the θ-equation is just the Newtonian moment. (A_y, of course, doesn't enter the Lagrange equations, since it is a constraint force.) |

Example 8.2.4. *Example 8.1.5 using Lagrange Equations.* Once again we return to a previous problem, but now the emphasis is on the "time-dependent potential energy" mentioned above. The spring's potential energy is "clean": $V = \frac{1}{2}k(r - r_o)^2$. But we note of $\boldsymbol{F}(t) = F(t)\hat{\boldsymbol{r}}$ that, since it is independent of coordinates altogether *and lies only in the* $\hat{\boldsymbol{r}}$*-direction*, it trivially satisfies the condition of equations (5.1-12) that $\nabla \times \boldsymbol{F} \equiv 0$. Thus, by those equations, there is a V $(= V(t))$ such that $\boldsymbol{F} = -\nabla V$; the appropriate V expressed in the generalized "polar" coordinates of choice, either by solving the partial differential equation

$$\boldsymbol{F}(t) = F_r(t)\hat{\boldsymbol{r}} = -\nabla V = -\frac{\partial V}{\partial r}\hat{\boldsymbol{r}} - \frac{1}{r}\frac{\partial V}{\partial \theta}\hat{\boldsymbol{\theta}},$$

or the equivalent approach using the *definition* of ∇V in polar coordinates

$$-\nabla V \cdot d\boldsymbol{r} \equiv -dV = -\frac{\partial V}{\partial r}\,dr - \frac{\partial V}{\partial \theta}\,d\theta = \boldsymbol{F}(t)\cdot d\boldsymbol{r} = F_r(t)\,dr,$$

we get a "potential energy" $V(t) = -F_r(t)r$; thus the Lagrangian

$$L \equiv T - V = \frac{1}{2}m(\dot{r}^2 + (r\dot{\theta})^2) - \frac{1}{2}k(r - r_o)^2 + rF_r(t).$$

Taking the Lagrange equations—the system now being "conservative"—

$$\begin{aligned}
\frac{d}{dt}\left(\frac{\partial L}{\partial \dot{r}}\right) - \frac{\partial L}{\partial r} &= \frac{d}{dt}(m\dot{r}) - (mr\dot{\theta}^2 - k(r - r_o) + F_r(t)) \\
&= m\ddot{r} - mr\dot{\theta}^2 + k(r - r_o) - F_r(t) = 0 \\
\frac{d}{dt}\left(\frac{\partial L}{\partial \dot{\theta}}\right) - \frac{\partial L}{\partial \theta} &= \frac{d}{dt}(mr^2\dot{\theta}) - 0 \\
&= 2mr\dot{r}\dot{\theta} - r^2\ddot{\theta} = 0
\end{aligned}$$

Again, these are precisely the same equations resulting previously. |

Having obtained the above generalized Lagrange equations and assessed their advantages, we shall supplant the earlier Euler-Lagrange set with them: the latter can be obtained from the Lagrange form; and the ability to be able to use potential energy (where applicable), rather than having continually to rederive the generalized forces, is just too good to pass up. The choice to do so warrants examining these Lagrange equations, and the Lagrangian function itself, in a little more detail:

Properties of the Lagrangian

Clearly, if one selects a different set of generalized coordinates in terms of which to express the Lagrangian, the form of L will change. But one might expect that, given a choice of coordinates, that form would be determined—that L would be *unique*. The following asserts that this is not quite the case, at least insofar as the equations of motion are concerned:

Theorem 8.2.1. Given a set of coordinates $\{q_1, q_2, \ldots, q_n\}$, and an arbitrary function $W = W(\mathbf{q}; t)$, the equations of motion for a Lagrangian $L(\mathbf{q}, \dot{\mathbf{q}}; t)$ and another Lagrangian $L'(\mathbf{q}, \dot{\mathbf{q}}; t) = L(\mathbf{q}, \dot{\mathbf{q}}; t) + \frac{dW(\mathbf{q};t)}{dt}$ are the same.

Note that W does *not* depend on $\dot{\mathbf{q}}$, since otherwise $\frac{dW}{dt}$ would introduce terms in $\ddot{\mathbf{q}}$, affecting the equations of motion!

Proof. This is so if and only if

$$\frac{d}{dt}\Big(\frac{\partial}{\partial \dot{q}_j}\Big(\frac{dW(\mathbf{q};t)}{dt}\Big)\Big) - \frac{\partial}{\partial q_j}\Big(\frac{dW(\mathbf{q};t)}{dt}\Big) \equiv 0$$

—*i.e.*, $\frac{dW}{dt}$ satisfies the Lagrange equations *identically*—which is straightforward to demonstrate:

$$\frac{dW(\mathbf{q};t)}{dt} = \sum_k \frac{\partial W(\mathbf{q};t)}{\partial q_k}\dot{q}_k + \frac{\partial W(\mathbf{q};t)}{\partial t}$$

$$\therefore \frac{\partial}{\partial \dot{q}_j}\Big(\frac{dW(\mathbf{q};t)}{dt}\Big) = \frac{\partial W(\mathbf{q};t)}{\partial q_j},$$

and the rest follows from Theorem 7.4.2. □

Homework:

1. Prove this.
2. Show that a Lagrangian L and another $L' = L + q_1\dot{q}_1 + \frac{1}{2}q_2^2\dot{q}_2$ give the same equations of motion by
   (a) showing the added terms are the total time derivative of some function, W, and
   (b) taking the operation

$$\frac{d}{dt}\Big(\frac{\partial}{\partial \dot{q}_j}\Big(\frac{dW(\mathbf{q};t)}{dt}\Big)\Big) - \frac{\partial}{\partial q_j}\Big(\frac{dW(\mathbf{q};t)}{dt}\Big) \equiv 0.$$

   Hint: use Theorems 7.4.1 and 7.4.2.

The next result holds even if one goes to a *different* set of generalized coordinates, at least for *conservative* systems (where the "generalized" Lagrange equations go to *the* Lagrange equations):

Theorem 8.2.2. The Lagrange equations of motion are *invariant in form under an arbitrary coordinate transformation* $q_i \to q_i(\boldsymbol{Q};t)$; *i.e.*, under such a transformation, the equations of motion in terms of the new variables for the *transformed* Lagrangian $L^*(\boldsymbol{Q},\dot{\boldsymbol{Q}};t) \equiv L(\boldsymbol{q}(\boldsymbol{Q};t),\dot{\boldsymbol{q}}(\boldsymbol{Q},\dot{\boldsymbol{Q}};t);t)$ are

$$\frac{d}{dt}\Big(\frac{\partial L^*}{\partial \dot{Q}_j}\Big) - \frac{\partial L^*}{\partial Q_j} = 0.$$

Again, note that the transformation depends only on the Q's, not their time derivatives; further, this result applies only to Lagrange's equations, *not* the *generalized* form in which the Q's do not vanish.

Proof. Assume that $L(\boldsymbol{q},\dot{\boldsymbol{q}};t)$ satisfies

$$\frac{d}{dt}\Big(\frac{\partial L}{\partial \dot{q}_j}\Big) - \frac{\partial L}{\partial q_j} = 0.$$

Then

$$\begin{aligned}\frac{d}{dt}\Big(\frac{\partial L^*}{\partial \dot{Q}_j}\Big) &\equiv \frac{d}{dt}\Big(\frac{\partial}{\partial \dot{Q}_j}L(\boldsymbol{q}(\boldsymbol{Q};t),\dot{\boldsymbol{q}}(\boldsymbol{Q},\dot{\boldsymbol{Q}};t);t)\Big)\\ &= \frac{d}{dt}\Big(\sum_k \frac{\partial L}{\partial \dot{q}_k}\frac{\partial \dot{q}_k}{\partial \dot{Q}_j}\Big) = \frac{d}{dt}\Big(\sum_k \frac{\partial L}{\partial \dot{q}_k}\frac{\partial q_k}{\partial Q_j}\Big)\end{aligned}$$

(since "dots cancel", by Theorem 7.4.1)

$$= \sum_k \frac{d}{dt}\Big(\frac{\partial L}{\partial \dot{q}_k}\Big)\frac{\partial q_k}{\partial Q_j} + \sum_k \frac{\partial L}{\partial \dot{q}_k}\frac{d}{dt}\Big(\frac{\partial q_k}{\partial Q_j}\Big);$$

Similarly,

$$\begin{aligned}\frac{\partial L^*}{\partial Q_j} &\equiv \frac{\partial}{\partial Q_j}L(\boldsymbol{q}(\boldsymbol{Q};t),\dot{\boldsymbol{q}}(\boldsymbol{Q},\dot{\boldsymbol{Q}};t);t)\Big)\\ &= \sum_k \frac{\partial L}{\partial q_k}\frac{\partial q_k}{\partial Q_j} + \sum_k \frac{\partial L}{\partial \dot{q}_k}\frac{\partial \dot{q}_k}{\partial Q_j}\\ &= \sum_k \frac{\partial L}{\partial q_k}\frac{\partial q_k}{\partial Q_j} + \sum_k \frac{\partial L}{\partial \dot{q}_k}\frac{d}{dt}\Big(\frac{\partial q_k}{\partial Q_j}\Big)\end{aligned}$$

(since "$\frac{\partial}{\partial Q}$ and $\frac{d}{dt}$ commute", by Theorem 7.4.2). Subtracting these two subexpressions, the last sums in each cancel, leaving

$$\frac{d}{dt}\Big(\frac{\partial L^*}{\partial \dot{Q}_j}\Big) - \frac{\partial L^*}{\partial Q_j} = \sum_k \Big(\frac{d}{dt}\Big(\frac{\partial L}{\partial \dot{q}_k}\Big) - \frac{\partial L}{\partial q_k}\Big)\frac{\partial q_k}{\partial Q_j} \equiv 0$$

—since the term in parentheses satisfies the Lagrange equation (for each k). □

In a very real sense, the above theorem and proof are superfluous: it is clear that, had we started with either the $\{q_1, q_2, \ldots, q_n\}$ or the $\{Q_1, Q_2, \ldots, Q_n\}$, we would have ended up with equations of motion in standard Lagrangian [non-generalized] form. But this is included to stress the idea of *transformation* of coordinates—a concept which will return.

Summary. We have seen that, in the same way that the Euler-Lagrange equations are related to the Principle of Work and Energy, we can obtain, in the case that the active forces are conservative, a specialized form related to the Conservation of Energy, the *Lagrange equations.* And in the same way that energy conservation can be cast in a "mixed" form, utilizing both work and potential energy, the Lagrange equations can be put in a form using both generalized forces and potential energy. The latter preserves the advantages of potential energy (obviating the need to determine the associated generalized forces), while allowing it to be applied to cases in which the forces are not *all* conservative. Given the importance of the Lagrangian, we also obtained a couple of results pertaining to that function.

8.3 Differential Constraints

Though we have introduced both holonomic and non-holonomic constraints in Section 7.1, we have limited our discussion to those of the former variety, particularly in the derivation of the Euler-Lagrange and Lagrange equations of motion. But where was that limitation actually imposed in an important way? A review of the derivations shows that it was invoked only at the last step in each, where equations (8.1-7) and (8.2-5), involving the vanishing of a *sum* in the δq_i, implied the ultimate equations (8.1-8) and (8.2-8), respectively, the *separate* vanishing of coefficients of the δq_i. These conclusions were justified because of the assumed *linear independence of the [differential]* δq_i, *not* the integrability of the constraints generating these differentials!

And, aside from the psychological security which might arise from dealing with coordinates rather than their differentials, holonomic constraints are not necessarily such a Good Thing: As a practical issue, even though there might be a relation among the coordinates, this relation is generally going to be nonlinear, and the process required to express some of these in terms of the others introduces its own difficulties. But, on an almost aesthetic level, certain constraints may introduce symmetries into the problem—for example, the [holonomic] constraint $\phi(x, y, z) = x^2 + y^2 + z^2 - a^2 = 0$ of Example 7.3.1, forcing the point to move on the surface of a hemisphere, is symmetric in all three variables—a symmetry which is broken when one chooses, more or less arbitrarily, which to solve for. Such symmetries are often an important feature of a system, and one would like to preserve them, not destroy them!

Thus the present section will examine these issues more closely, focussing on the *differential* nature of the constraints, independent of whether they are holonomic or not; its results, then, will be applicable to *any* constraint, holonomic or non-holonomic. And, although most of the examples thus far have demonstrated single constraints, we shall put this in the more general context of problems involving *systems* of constraints. In particular, we will consider a system describable with N [unconstrained] coordinates $\{q_1, q_2, \ldots, q_N\}$, subject to a set of m differential constraints:

$$\sum_{j=1}^{N} a_{ij}(\boldsymbol{q};t)\,dq_j + a_{it}(\boldsymbol{q};t)\,dt \equiv \sum_{j=0}^{N} a_{ij}(\boldsymbol{q};t)\,dq_j = 0, \quad \text{i=1,\ldots,m}$$

—in which, for compactness, we write $q_0 \equiv t$. (Though these are in *Pfaffian form*, (7.1-3), they might have resulted from the differential of a holonomic constraint.) Clearly $m \leq N$, for otherwise the system is *over*constrained; generally $m < N$, or all coordinates are already determined by the constraints, independent of any equations of motion! We will assume (reasonably) that these constraints (if not the coordinates themselves) are [linearly] *independent*—that none of them can be written as a linear combination of the others. This set of constraints immediately yields a similar set of constraints on the *virtual displacements*:

$$\sum_{j=1}^{N} a_{ij}(\boldsymbol{q};t)\,\delta q_j = 0, \quad \text{i=1,\ldots,m}, \tag{8.3-1}$$

also independent.

We shall first consider these from a purely *algebraic* viewpoint, showing how one might explicitly solve for the dependent differentials, making the balance independent. We shall then introduce the technique of "*Lagrange multipliers*", which obviates this solution process and admits of a rather elegant interpretation.

8.3.1 Algebraic Approach to Differential Constraints

Recall that, in Example 7.1.7, we indicated how explicit introduction of the *differential* constraints among the variables rendered the resulting virtual differentials of the coordinates independent; this is the procedure to be followed here, generalized to multiple constraints. The above relations (8.3-1) are *linear*; thus we can write these as a [matrix] system of algebraic equations, with the N-vector unknown $\delta\boldsymbol{q}$ is multiplied by an $m \times N$ matrix A (whose elements are themselves generally functions of $\boldsymbol{q}$ and t):

$$\mathsf{A}\delta\boldsymbol{q} \equiv \begin{pmatrix} a_{11} & a_{12} & \cdots & a_{1N} \\ a_{21} & a_{22} & \cdots & a_{2N} \\ \vdots & \vdots & & \vdots \\ a_{m1} & a_{m2} & \cdots & a_{mN} \end{pmatrix} \begin{pmatrix} \delta q_1 \\ \delta q_2 \\ \vdots \\ \delta q_N \end{pmatrix} = 0.$$

The presumed independence of the coordinates means that the rank of A is m; thus (see page 84) we can solve for m of the δq_i, say $\{\delta q_i, i = (N-m+1), \ldots, N\}$, as linear combinations of the $(N-m)$ remaining ones:

$$\delta q_j = \delta q_j(q_1, q_2, \ldots, q_N; t) \equiv \sum_{k=1}^{N-m} c_{jk}(q;t)\,\delta q_k, \quad j = N-m+1, \ldots, N. \tag{8.3-2}$$

Now consider the derivation of the Euler-Lagrange equations, on which that for the generalized Lagrange equations was based: The various summations $\sum_{j=1}^{f}$ there arose because we had eliminated all dependent variables from the start, leaving us with f degrees of freedom, and such summations resulted from the substitution of equation (7.4-2) into equations (8.1-1) and (8.1-4). But the upper limit f in equation (7.4-2) came from the posited number of degrees of freedom; more generally, it would extend to N for the general differential form (here on N variables) on which the equation is based. In this case, equation (8.1-7) becomes

$$\sum_{j=1}^{N}\left\{\left(\frac{d}{dt}\left(\frac{\partial T}{\partial \dot{q}_j}\right) - \frac{\partial T}{\partial q_j}\right) - Q_j\right\}\delta q_j = 0 \tag{8.3-3}$$

and still holds, having come from manipulations among the various quantities; it's just that we cannot conclude that the separate coefficients of the δq_i vanish.

But this sum can be broken up into two parts:

$$\begin{aligned}
&\sum_{j=1}^{N}\left\{\left(\frac{d}{dt}\left(\frac{\partial T}{\partial \dot{q}_j}\right) - \frac{\partial T}{\partial q_j}\right) - Q_j\right\}\delta q_j \\
&= \sum_{j=1}^{N-m}\left\{\left(\frac{d}{dt}\left(\frac{\partial T}{\partial \dot{q}_j}\right) - \frac{\partial T}{\partial q_j}\right) - Q_j\right\}\delta q_j \\
&\qquad + \sum_{j=N-m+1}^{N}\left\{\left(\frac{d}{dt}\left(\frac{\partial T}{\partial \dot{q}_j}\right) - \frac{\partial T}{\partial q_j}\right) - Q_j\right\}\delta q_j \\
&= \sum_{j=1}^{N-m}\left\{\left(\frac{d}{dt}\left(\frac{\partial T}{\partial \dot{q}_j}\right) - \frac{\partial T}{\partial q_j}\right) - Q_j\right\}\delta q_j \\
&\qquad + \sum_{j=N-m+1}^{N}\left\{\left(\frac{d}{dt}\left(\frac{\partial T}{\partial \dot{q}_j}\right) - \frac{\partial T}{\partial q_j}\right) - Q_j\right\}\sum_{k=1}^{N-m} c_{jk}\,\delta q_k \\
&= \sum_{j=1}^{N-m}\left\{\left(\frac{d}{dt}\left(\frac{\partial T}{\partial \dot{q}_j}\right) - \frac{\partial T}{\partial q_j}\right) - Q_j\right\}\delta q_j \\
&\qquad + \sum_{j=1}^{N-m}\sum_{k=N-m+1}^{N} c_{kj}\left\{\left(\frac{d}{dt}\left(\frac{\partial T}{\partial \dot{q}_k}\right) - \frac{\partial T}{\partial q_k}\right) - Q_k\right\}\delta q_j \\
&= \sum_{j=1}^{N-m}\left\{\left(\frac{d}{dt}\left(\frac{\partial T}{\partial \dot{q}_j}\right) - \frac{\partial T}{\partial q_j}\right) - Q_j\right.
\end{aligned}$$

$$+\sum_{k=N-m+1}^{N} c_{kj}\left(\left(\frac{d}{dt}\left(\frac{\partial T}{\partial \dot{q}_k}\right)-\frac{\partial T}{\partial q_k}\right)-Q_k\right)\Bigg\}\,\delta q_j=0.$$

(Note that we have merely switched indices from the introduction of (8.3-2) in the third, going to the fourth, equation above to make "folding" into the last equation more transparent.) These δq_j *are* linearly independent, and we can conclude that each coefficient vanishes identically; this is precisely what happened in the simpler Example 7.1.7. Clearly the result would obtain for the generalized Lagrange equations, since the only difference is to include those Q_k which are conservative in a potential energy in L.

This procedure demonstrates that what is important to the [derivation of the] fundamental equations of analytical dynamics is linear independence of the *differentials* of the coordinates, rather than of the coordinates themselves. We observe that the equations are expressed in terms of all N coordinates, rather than just the $f = N - m$ remaining after all constraints have been eliminated, so that any symmetries among the variables would be preserved. But the technique, if conceptually straightforward, requires explicit solution for the m differentials as well as explicit substitution into the original set of equations, and lacks a certain elegance, particularly when compared to the alternative:

8.3.2 Lagrange Multipliers

The reader has likely noted that all constraints, both holonomic and non-holonomic, are typically written in a "standard form", with the scalar function, or linear combination of differentials, always being set equal to 0. The reason for this will become clear here.

Though we will return to the *systems* of constraints (8.3-1), in order to clarify the procedure, let us once again consider a *single* [Pfaffian] constraint

$$\sum_{j=1}^{N} a_j(\boldsymbol{q};t)\,\delta q_j=0$$

for the time being, and return to the place in the derivations, equation (8.3-3),

$$\sum_{j=1}^{N}\left\{\left(\frac{d}{dt}\left(\frac{\partial T}{\partial \dot{q}_j}\right)-\frac{\partial T}{\partial q_j}\right)-Q_j\right\}\delta q_j=0$$

at which we must be assured of linear independence in order to get separate equations. We know that the constraint vanishes; thus we can subtract 0, in the clever guise of λ times that constraint, from the above equation:

$$\sum_{j=1}^{N}\left\{\left(\frac{d}{dt}\left(\frac{\partial T}{\partial \dot{q}_j}\right)-\frac{\partial T}{\partial q_j}\right)-Q_j\right\}\delta q_j-\lambda\sum_{j=1}^{N}a_j\,\delta q_j$$
$$=\sum_{j=1}^{N}\left\{\left(\frac{d}{dt}\left(\frac{\partial T}{\partial \dot{q}_j}\right)-\frac{\partial T}{\partial q_j}\right)-Q_j-\lambda a_j\right\}\delta q_j=0.$$

Note that it is λ times the coefficient of δq_j which enters each term in brackets. The N variables are still linearly dependent, since no elimination to reduce the set to the constrained set of $(N-1)$ variables has been done, explicitly or implicitly; in particular, we cannot conclude that the coefficients of δq_j vanishes for all j.

λ is a free variable, apparently introduced capriciously. At least two of the $a_j \neq 0$ in the constraint, for otherwise there is no relation among the differentials! Assume, say, that $a_N \neq 0$ in the constraint; then let us determine λ so that the $j = N$ term of this sum vanishes:

$$\frac{d}{dt}\left(\frac{\partial T}{\partial \dot{q}_N}\right) - \frac{\partial T}{\partial q_N} - Q_N - \lambda a_N \equiv 0;$$

note that this will yield a λ depending generally on the q_i, $\dot{q}_i$, and the $\ddot{q}_i$! Then the above sum reduces to

$$\sum_{j=1}^{N-1}\left\{\left(\frac{d}{dt}\left(\frac{\partial T}{\partial \dot{q}_j}\right) - \frac{\partial T}{\partial q_j}\right) - Q_j - \lambda a_j\right\}\delta q_j = 0$$

using the above-determined value for λ. *These* variables *can* be considered linearly independent, since they are only $(N-1)$ in number; in effect, we have eliminated the constraint *implicitly* by using λ to force the $j = N$ term vanish, rather than any assumption on linear independence. But that equation (used to determine λ initially) then becomes the equation of motion for q_N, while the previous $(N-1)$ equations (the coefficients of the δq_j vanishing *separately* because they are linearly independent) become those for the $\delta q_j, j < N$. *All have the same form*, though one is used to find λ. In effect we have $(N+1)$ equations—the $(N-1)$ [independent] equations for $j < N$, the equation forcing the coefficient of q_N to vanish, and the constraint itself—in $(N+1)$ unknowns—the N equations of motion, plus λ.

Observe that the equations of motion in this case assume the form

$$\frac{d}{dt}\left(\frac{\partial T}{\partial \dot{q}_j}\right) - \frac{\partial T}{\partial q_j} - \lambda a_j = Q_j$$

—the standard Euler-Lagrange equations, but with λa_j on the derivative side. Again, these can be recast immediately in terms of the generalized Lagrange equations

$$\frac{d}{dt}\left(\frac{\partial L}{\partial \dot{q}_j}\right) - \frac{\partial L}{\partial q_j} - \lambda a_j = Q_j$$

—once more, with λa_j on the left-hand side.

If all this has been presented in differential form, useful for non-holonomic constraints, it is immediately applicable to holonomic ones, too, since the a_j are merely the partial derivatives of the constraint ϕ:

$$a_j \equiv \frac{\partial \phi}{\partial q_j}.$$

But rather than carry out that differentiation and then substitute into the above equations, it is conceptually easier simply to subtract $\lambda\phi$ from the Lagrangian itself: since $\phi = \phi(\boldsymbol{q}; t)$ (and not $\dot{\boldsymbol{q}}$),

$$\frac{d}{dt}\left(\frac{\partial L}{\partial \dot{q}_j}\right) - \frac{\partial L}{\partial q_j} - \lambda a_j = \frac{d}{dt}\left(\frac{\partial L}{\partial \dot{q}_j}\right) - \frac{\partial L}{\partial q_j} - \lambda\frac{\partial \phi}{\partial q_j} = \frac{d}{dt}\left(\frac{\partial (L-\lambda\phi)}{\partial \dot{q}_j}\right) - \frac{\partial (L-\lambda\phi)}{\partial q_j} = Q_j$$

—tantamount to using a "modified Lagrangian" $L^* \equiv L - \lambda\phi$.

Before moving to the case in which there is more than a single constraint, it is useful to see how the Lagrange multiplier technique and the earlier algebraic approach relate to one another in this simpler case: Assuming that $a_N \neq 0$, we can solve for that variable from the [single] constraint equation above:

$$\delta q_N = \sum_{j=1}^{N-1}\left(-\frac{a_j}{a_N}\right)\delta q_j,$$

corresponding to (8.3-2) in the *algebraic* method. Substituting this into (8.3-3), we get the linear combination of the $N-1$, now linearly *in*dependent, δq_j:

$$\begin{aligned}
\sum_{j=1}^{N}&\left\{\left(\frac{d}{dt}\left(\frac{\partial T}{\partial \dot{q}_j}\right) - \frac{\partial T}{\partial q_j}\right) - Q_j\right\}\delta q_j \\
&= \sum_{j=1}^{N-1}\left\{\left(\frac{d}{dt}\left(\frac{\partial T}{\partial \dot{q}_j}\right) - \frac{\partial T}{\partial q_j}\right) - Q_j\right\}\delta q_j \\
&\qquad + \left\{\left(\frac{d}{dt}\left(\frac{\partial T}{\partial \dot{q}_N}\right) - \frac{\partial T}{\partial q_N}\right) - Q_N\right\}\sum_{j=1}^{N-1}\left(-\frac{a_j}{a_N}\right)\delta q_j \\
&= \sum_{j=1}^{N-1}\left(\left\{\left(\frac{d}{dt}\left(\frac{\partial T}{\partial \dot{q}_j}\right) - \frac{\partial T}{\partial q_j}\right) - Q_j\right\} - \frac{a_j}{a_N}\left\{\left(\frac{d}{dt}\left(\frac{\partial T}{\partial \dot{q}_N}\right) - \frac{\partial T}{\partial q_N}\right) - Q_N\right\}\right)\delta q_j,
\end{aligned} \tag{8.3-4}$$

which must vanish for *each* $j = 1, \ldots, N-1$.

If the above has given the result from the algebraic perspective, the *Lagrange multiplier* approach would determine, using a_N from the [single] constraint in the *form* of the coefficient of δq_N to determine the [single] λ,

$$\frac{d}{dt}\left(\frac{\partial T}{\partial \dot{q}_N}\right) - \frac{\partial T}{\partial q_N} - Q_N - \lambda a_N = 0$$

$$\text{or} \qquad \lambda = \frac{1}{a_N}\left\{\frac{d}{dt}\left(\frac{\partial T}{\partial \dot{q}_N}\right) - \frac{\partial T}{\partial q_N}\right) - Q_N\right\}.$$

This value of λ is then used in the [now, linearly *in*dependent] coefficients of $\delta q_j, j = 1, \ldots, N-1$ to get the equations of motion:

$$\left(\frac{d}{dt}\left(\frac{\partial T}{\partial \dot{q}_j}\right) - \frac{\partial T}{\partial q_j}\right) - Q_j - \lambda a_j =$$
$$\left(\frac{d}{dt}\left(\frac{\partial T}{\partial \dot{q}_j}\right) - \frac{\partial T}{\partial q_j}\right) - Q_j - \frac{a_j}{a_N}\left\{\left(\frac{d}{dt}\left(\frac{\partial T}{\partial \dot{q}_N}\right) - \frac{\partial T}{\partial q_N}\right) - Q_N\right\} = 0$$

—precisely the result obtained above in (8.3-4).

With this special case of a single constraint in hand, it is easy to generalize to the *system* of m constraints, (8.3-1): Now the "clever disguise" for 0 is not just the single Pfaffian form of constraint multiplied by the single λ, but the *sum* of such terms:

$$\sum_{k=1}^{m} \lambda_k \sum_{j=1}^{N} a_{kj}(\boldsymbol{q};t)\,\delta q_j = 0. \tag{8.3-5}$$

The N equations we use, coefficients of the δq_j in the sum, now become

$$\frac{d}{dt}\left(\frac{\partial T}{\partial \dot{q}_j}\right) - \frac{\partial T}{\partial q_j} - \sum_{k=1}^{m} \lambda_k a_{kj}(\boldsymbol{q};t) = Q_j;$$

once again it is the coefficients of δq_j, each multiplied by its own λ_k, which enter the equation for the jth coordinate. Of these, we use, say, the last m to determine the λ_k:

$$\frac{d}{dt}\left(\frac{\partial T}{\partial \dot{q}_{(N-m+1)}}\right) - \frac{\partial T}{\partial q_{(N-m+1)}} - \sum_{k=1}^{m} \lambda_k a_{k(N-m+1)}(\boldsymbol{q};t) = Q_{(N-m+1)}$$
$$\frac{d}{dt}\left(\frac{\partial T}{\partial \dot{q}_{(N-m+2)}}\right) - \frac{\partial T}{\partial q_{(N-m+2)}} - \sum_{k=1}^{m} \lambda_k a_{k(N-m+2)}(\boldsymbol{q};t) = Q_{(N-m+2)}$$
$$\vdots$$
$$\frac{d}{dt}\left(\frac{\partial T}{\partial \dot{q}_N}\right) - \frac{\partial T}{\partial q_N} - \sum_{k=1}^{m} \lambda_k a_{kN}(\boldsymbol{q};t) = Q_N$$

—itself a system of m linear equations in the m λ_k. The remaining $(N-m)$ equations *with the above values of* λ_k,

$$\frac{d}{dt}\left(\frac{\partial T}{\partial \dot{q}_j}\right) - \frac{\partial T}{\partial q_j} - \sum_{k=1}^{m} \lambda_k a_{kj}(\boldsymbol{q};t) = Q_j, \quad j = 1, \ldots, (N-m),$$

in the now linearly *in*dependent $q_1, q_2, \ldots, q_{(N-m)}$—note that $(N-m) = f$, the number of degrees of freedom once again—are then the equations of motion for *these* variables. We get the N equations of motion and the m λ_k from the above N "Euler-Lagrange" equations and m constraints. These can be cast in terms of the Lagrangian, using the corresponding N [generalized] "Lagrange" equations

$$\frac{d}{dt}\left(\frac{\partial L}{\partial \dot{q}_j}\right) - \frac{\partial L}{\partial q_j} - \sum_{k=1}^{m} \lambda_k a_{kj}(q;t) = Q_j. \qquad (8.3\text{-}6)$$

In the special case that the differential constraints can be determined from *holonomic* constraints, these can more easily be recovered by utilizing the modified Lagrangian

$$L^* = L - \sum_{k=1}^{m} \lambda_k \phi_k.$$

Recall that when we were laying the groundwork for the kinematics of analytical dynamics in Section 7.1, "side conditions" were mentioned when constraints had not been explicitly eliminated from a set of coordinates (see, *e.g.*, Example 7.1.2). It was certainly not made clear at that time just how such conditions would be imposed explicitly, but we now see a means of doing just that: use Lagrange multipliers! The constraints maintain their identity, yet no explicit elimination is effected until *after* the equations of motion have been obtained.

It's time for an example:

Example 8.3.1 *Lagrange Multipliers.* We consider a relatively simple problem: a "small" body (read "particle") of mass m moves along a smooth surface making an angle of θ with the horizontal as shown. We shall determine its motion.

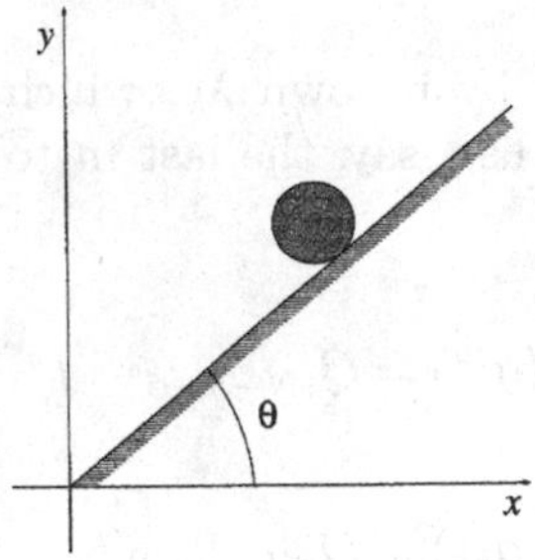

As we have done so often, we will work this in two ways: first using the approach of eliminating the constraint imposed on the body's motion explicitly; then using the Lagrange multipliers of this section. We note that the only active force here is gravity: the normal force acting on the mass is one of constraint, and does no work under a virtual displacement (along the incline) which doesn't violate it. In order to illustrate constraints, we will use *un*constrained Cartesian coordinates (x, y), along with the explicit holonomic constraint $\phi(x, y) = y - (\tan\theta)x = 0$.

For both approaches we need the Lagrangian. In the original Cartesian coordinates, this is trivially

$$T - V = \frac{1}{2}m(\dot{x}^2 + \dot{y}^2) - mgy.$$

Explicit Elimination. The constraint, along with giving us $y(x)$, also gives us $\dot{y}(x, \dot{x})$ (in general it depends on both):

$$\dot{y} = (\tan\theta)\dot{x}$$

Substituting these into the above general form for $T - V$, we get the Lagrangian

$$L = \frac{1}{2}(1 + \tan^2\theta)m\dot{x}^2 - mg(\tan\theta)x$$

—depending on the single variable x. The Lagrange equation for this conservative force is thus

$$\frac{d}{dt}\left(\frac{\partial L}{\partial \dot{x}}\right)-\frac{\partial L}{\partial x}=(1+\tan^2\theta)m\ddot{x}+mg(\tan\theta)=0 \quad \text{or} \quad \ddot{x}=-\frac{g\tan\theta}{1+\tan^2\theta}.$$

Lagrange Multipliers. In terms of the *two* [unconstrained] variables, we have now *two* Lagrange equations using the modified Lagrangian $L-\lambda\phi$:

$$\frac{d}{dt}\left(\frac{\partial L}{\partial \dot{x}}\right)-\frac{\partial L}{\partial x}-\lambda\frac{\partial\phi}{\partial x}=m\ddot{x}+\lambda\tan\theta=0$$
$$\frac{d}{dt}\left(\frac{\partial L}{\partial \dot{y}}\right)-\frac{\partial L}{\partial y}-\lambda\frac{\partial\phi}{\partial y}=m\ddot{y}+mg-\lambda=0$$

These two, along with ϕ itself, constitute three equations in $\ddot{x}$, $\ddot{y}$ (their equations of motion, really), and λ.

Given the presence of a constraint, we know that one of the two original variables can be eliminated; we (more or less arbitrarily) pick y to replace. Thus we will use the y equation to determine λ by *forcing* that to vanish; we get $\lambda=m(g+\ddot{y})$. This is where the [general] constraint equation ϕ comes in; it can be differentiated twice to give $\ddot{y}=(\tan\theta)\ddot{x}$, eliminating y in favor of x: $\lambda=m(g+(\tan\theta)\ddot{x})$. Substituting this into the x Lagrange equation

$$m\ddot{x}+\lambda\tan\theta=m\ddot{x}+m(g+\tan\theta\ddot{x})tan\theta=m(1+\tan^2\theta)\ddot{x}+mg(\tan\theta)=0$$

—the same result as explicit elimination. |

Example 8.3.2. *Lagrange Multipliers—Non-holonomic Constraint.* We have already seen a physical example of non-holonomic constraints in Example 7.1.4; we now return to that example, both to exemplify how they could be treated using Lagrange Multipliers, and to show the implementation of multiple constraints. We shall also see the need generally to have, not only the *virtual* displacement constraint, but the *actual* differential constraint relating the coordinates and, if rheonomic, the time.

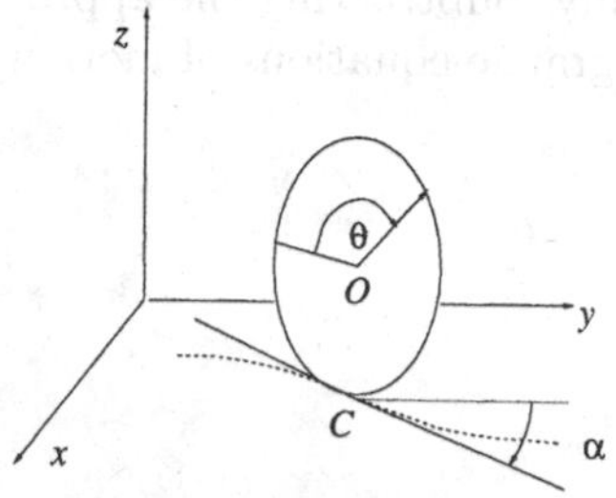

The system's diagram is repeated at left. Briefly to review, we have a disk, always with its diameter vertical, rolling without slipping on a horizontal surface. We saw there that, including the constraints imposed through these restrictions, we needed only four coordinates, (x, y), the center of the disk, α, the orientation of the disk's plane relative to the xy-plane, and θ, the rotation angle of the disk itself in its own plane. Though we didn't need it for the purely kinematical example cited, we must now make the further prescription that the disk is, say, "thin" and uniform, with mass m to account for *kinetics*.

In the previous kinematical example, we have already determined the [actual] constraints on the motion, that

$$\begin{aligned} d\mathbf{r}_O &= r\sin\alpha\, d\theta\hat{\mathbf{i}} + r\cos\alpha\, d\theta\hat{\mathbf{j}} \\ &= dx\hat{\mathbf{i}} + dy\hat{\mathbf{j}}, \end{aligned} \tag{8.3-7}$$

from which we obtained the virtual constraints

$$\begin{aligned} \delta x - r\sin\alpha\,\delta\theta &= 0 \\ \delta y - r\cos\alpha\,\delta\theta &= 0. \end{aligned}$$

But now we must also determine the Lagrangian of the system. This requires the calculation of T; this is three-dimensional, with $\mathbf{\Omega} = \dot{\boldsymbol{\theta}} + \dot{\boldsymbol{\alpha}}$, but, since we are dealing with scalars here, we can take any set of convenient coordinates—that means *principal* axes (x', y', z'), rotated from the reference system by α—to express the vector quantities:

$$\begin{aligned} T &= \frac{1}{2}m\overline{v}^2 + \frac{1}{2}\mathbf{\Omega}\cdot\bar{\mathsf{I}}\mathbf{\Omega} \\ &= \frac{1}{2}m(\dot{x}^2 + \dot{y}^2) + \frac{1}{2}(\dot{\theta}\hat{\mathbf{i}}' + \dot{\alpha}\hat{\mathbf{k}}')\cdot(\frac{1}{2}mr^2\dot{\theta}\hat{\mathbf{i}}' + \frac{1}{4}mr^2\dot{\alpha}\hat{\mathbf{k}}') \\ &= \frac{1}{2}m(\dot{x}^2 + \dot{y}^2) + \frac{1}{2}(\frac{1}{2}mr^2\dot{\theta}^2 + \frac{1}{4}mr^2\dot{\alpha}^2) \end{aligned}$$

We note that, under the strictures in the problem, $V = mgr$ is constant, so it will not affect the equations; in fact, if we took the height datum for gravitational potential energy to be 0 at the center of the disk, potential energy would vanish altogether! Thus the above T is all there is to the Lagrangian function L. The only other force in the problem is the frictional force at the point of contact; but, by the non-slip condition, $\delta\mathbf{r} \equiv \mathbf{0}$ for this point, so it doesn't enter, either.

Corresponding to the above differential *virtual* constraints are the "clever forms"

$$\begin{aligned} \lambda_1(\delta x - r\sin\alpha\,\delta\theta) &\equiv \lambda_1(a_{1x}\delta x + a_{1\theta}\delta\theta) = 0 \\ \lambda_2(\delta y - r\cos\alpha\,\delta\theta) &\equiv \lambda_2(a_{2y}\delta y + a_{2\theta}\delta\theta) = 0, \end{aligned}$$

with the balance of the a_{kj} terms vanishing identically. Subtracting the appropriate portion of these relations from each of the Lagrange equations of motion gives the four equations for x, y, θ, and α:

$$\begin{aligned} \frac{d}{dt}\Big(\frac{\partial L}{\partial \dot{x}}\Big) - \frac{\partial L}{\partial x} - \sum_{k=1}^{2}\lambda_k a_{kx} &= m\ddot{x} - \lambda_1 = 0 \\ \frac{d}{dt}\Big(\frac{\partial L}{\partial \dot{y}}\Big) - \frac{\partial L}{\partial y} - \sum_{k=1}^{2}\lambda_k a_{ky} &= m\ddot{y} - \lambda_2 = 0 \\ \frac{d}{dt}\Big(\frac{\partial L}{\partial \dot{\theta}}\Big) - \frac{\partial L}{\partial \theta} - \sum_{k=1}^{2}\lambda_k a_{k\theta} &= \frac{1}{2}mr^2\ddot{\theta} + \lambda_1 r\sin\alpha + \lambda_2 r\cos\alpha = 0 \end{aligned}$$

$$\frac{d}{dt}\left(\frac{\partial L}{\partial \dot{\alpha}}\right) - \frac{\partial L}{\partial \alpha} - \sum_{k=1}^{2} \lambda_k a_{k\alpha} = \frac{1}{4} m r^2 \ddot{\alpha} = 0.$$

It is clear which of these we will use to solve for the λ's! We get that $\lambda_1 = m\ddot{x}$ and $\lambda_2 = m\ddot{y}$, implicitly choosing to eliminate x and y. The last equation, already free of these variables, immediately gives us that $\dot{\alpha}$ is constant, while the third becomes

$$\frac{1}{2} m r^2 \ddot{\theta} + r m \ddot{x} \sin\alpha + m \ddot{y} r \cos\alpha = 0.$$

But this equation still involves the variables x and y, which we allegedly eliminated through our choice of equations used to determine the λ_i! This is where the knowledge of the *actual* (rather than *virtual*) differential constraint becomes important. Using the differentials (8.3-7), we get that

$$\begin{aligned} \dot{x} - r\sin\alpha\,\dot{\theta} &= 0 \\ \dot{y} - r\cos\alpha\,\dot{\theta} &= 0 \end{aligned}$$

from which

$$\begin{aligned} \ddot{x} - r\cos\alpha\,\dot{\alpha}\dot{\theta} - r\sin\alpha\,\ddot{\theta} &= 0 \\ \ddot{y} + r\sin\alpha\,\dot{\alpha}\dot{\theta} - r\cos\alpha\,\ddot{\theta} &= 0 \end{aligned}$$

so that the equation for θ is just

$$\frac{1}{2} m r^2 \ddot{\theta} + m r^2 \left[(\cos\alpha\sin\alpha - \sin\alpha\cos\alpha)\,\dot{\alpha}\dot{\theta} + (\sin^2\alpha + \cos^2\alpha)\,\ddot{\theta} \right] = \frac{3}{2} m r^2 \ddot{\theta} = 0.$$

That's an awful lot of work to get what we could have by consideration of the torqueless system (and a little hand-waving), but it's really meant to indicate how such non-holonomic constraints could be implemented! Note, however, that knowledge of the *actual* differentials in the variables was required, since, for a rheonomic constraint, there would generally be an additional term arising from the differential in t. This term, absent from the *virtual* displacement in which $dt \equiv 0$, could not otherwise be determined. |

Note that, in both of the above examples, the "equations of motion" (8.3-6) *maintain their form for all N [*dependent*] variables.* We merely select m of these equation to determine the λ's (implicitly solving for the associated variables' virtual differentials in terms of the rest), then the balance (with these λ values) to obtain equations of motion for the now *in*dependent variables.

Interpretation of the Lagrange Multipliers

Thus far, Lagrange multipliers have been introduced merely as a means of accounting for differential (or, in the case of holonomic constraints, even *non*-differential) constraints. However, they admit of a rather elegant interpretation:

We observe that, in the above equations of motion where they are employed, the terms $\lambda_k a_{kj}$ (or $\lambda_k \frac{\partial \phi}{\partial q_k}$ for holonomic constraints) *must have the same units as the generalized force* associated with q_j, since, wherever the latter appear, those term enters additively. Furthermore, the kth constraint in equation (8.3-5),

$$\sum_{j=1}^{N} \lambda_k a_{kj}(\boldsymbol{q};t)\,\delta q_j = 0,$$

can be interpreted as saying that the "*vector* generalized force", with components $\lambda_k a_{kj}(\boldsymbol{q};t)$ (whose units may be different from one another, depending on those of δq_j!), *does no work.* But this is precisely the criterion defining constraint forces in Section 7.3! Thus, in effect, *the Lagrange multipliers (times the terms in the differential constraints) are just the [generalized] constraint forces.*

We can see this where we used Lagrange multipliers in the above Example 8.3.1: There we determine

$$\lambda = m(g + \tan\theta\ddot{x}) = mg(1 - \frac{\tan^2\theta}{1+\tan^2\theta}) = mg\cos^2\theta;$$

thus, since the constraint in this problem is the holonomic $\phi = y - (tan\theta)x = 0$,

$$\lambda a_1 \to \lambda\frac{\partial\phi}{\partial x} = (mg\cos^2\theta)(-\tan\theta) = -mg\sin\theta\cos\theta$$
$$\lambda a_2 \to \lambda\frac{\partial\phi}{\partial y} = (mg\cos^2\theta)(1).$$

But these can be recognized as just the x- and y-components of the normal force $\boldsymbol{N} = mg\cos\theta(-\cos\theta\hat{\boldsymbol{\imath}} + \sin\theta\hat{\boldsymbol{\jmath}})$ producing that constraint in the first place. (In this case the q_j both have $[L]$, so their *generalized* forces will, in fact, have $[F]$.) Similarly, the two multipliers in Example 8.3.2 are just $m\ddot{x}$ and $m\ddot{y}$, the friction enforcing non-slip—the only unbalanced force, giving the center of mass motion.

Summary. We have noted that what is important in the derivation of the Euler-Lagrange/generalized Lagrange equations of motion is not so much the ability to find a set of independent *coordinates* as it is to find such a set whose *differentials* are independent; it is this which allows us to conclude that the various *summations* $\sum c_i\,\delta q_i = 0$ implies that the *separate* $c_i \equiv 0$. More to show conceptually that this is so than for its practical importance, we first used a purely algebraic approach to eliminate dependent differentials.

We then introduced the technique of *Lagrange multipliers* to effect the same thing: By factoring each of the m [differential] constraints by such a term, and subtracting the results from the Lagrangian, we got a full set of generalized Lagrange equations in all N variables; though only $(N - m)$ of these were independent, we used the remaining equations to *determine* the multipliers. The latter had the added advantage of giving the [generalized] *constraint* forces themselves. In the particular case that the constraints were holonomic, we showed that the same result could be achieved by simply modifying the

Lagrangian itself, subtracting the constraint functions, times their respective multipliers, from the unconstrained Lagrangian and applying the equations of motion to the altered function.

But ultimately we have obtained a technique which can be applied to *any* system, holonomic or not—one which *implicitly* reduces the original set of N [unconstrained] coordinates to $(N - m) = f$ whose differentials are linearly independent. Once again, we recover a set in number equal to the number of degrees of freedom.

8.4 Time as a *Coordinate*

This section will obtain a result possibly interesting in its own right, but more important conceptually for what is to follow than for its practical significance. We shall devise a means to consider time, up to now the fundamental variable in mechanics, as merely another coordinate.

In this regard, we shall consider it, as well as all the other generalized coordinates, to be functions of some arbitrary variable—a *parameter*—τ. In particular, for $t = t(\tau)$,

$$t' \equiv \frac{dt}{d\tau} = t'(\tau)$$

so that derivatives with respect to t can be expressed in terms of those with respect to τ:

$$\frac{d}{dt} = \frac{d\tau}{dt}\frac{d}{d\tau} = \frac{1}{t'}\frac{d}{d\tau}$$

and, for $q_i = q_i(t(\tau))$,

$$q_i' \equiv \frac{dq_i}{d\tau} = \dot{q}_i t',$$

and thus the Lagrangian

$$L(\mathbf{q}, \dot{\mathbf{q}}; t) = L(\mathbf{q}, \frac{\mathbf{q}'}{t'}; t).$$

(Note that we have admitted the possibility that there are rheonomic constraints or a time-dependent potential in L.)

We shall first obtain the Lagrange equations, originally derived from Newton's Laws, and thus tied to t, now expressed rather in terms of τ. We have

$$\frac{\partial L(\mathbf{q}, \dot{\mathbf{q}}; t)}{\partial \dot{q}_i} = \frac{\partial L(\mathbf{q}, \frac{\mathbf{q}'}{t'}; t)}{\partial \dot{q}_i} = \frac{\partial L(\mathbf{q}, \frac{\mathbf{q}'}{t'}; t)}{\partial q_i'} t' = \frac{\partial (L(\mathbf{q}, \frac{\mathbf{q}'}{t'}; t)\, t')}{\partial q_i'}$$

—since $t' = t'(\tau)$ only. Thus

$$\frac{d}{dt}\Big(\frac{\partial L(\mathbf{q}, \dot{\mathbf{q}}; t)}{\partial \dot{q}_i}\Big) = \frac{1}{t'}\frac{d}{d\tau}\Big(\frac{\partial (L(\mathbf{q}, \frac{\mathbf{q}'}{t'}; t)\, t')}{\partial q_i'}\Big).$$

Furthermore, we can write

$$\frac{\partial}{\partial q_i} L(\mathbf{q}, \dot{\mathbf{q}}; t) = \frac{\partial}{\partial q_i} L(\mathbf{q}, \frac{\mathbf{q}'}{t'}; t) = \frac{1}{t'}\frac{\partial}{\partial q_i}(L(\mathbf{q}, \frac{\mathbf{q}'}{t'}; t)\, t')$$

for the same reason. Thus, finally,

$$\frac{d}{dt}\Big(\frac{\partial L(\boldsymbol{q},\dot{\boldsymbol{q}};t)}{\partial \dot{q}_i}\Big)-\frac{\partial}{\partial q_i}L(\boldsymbol{q},\dot{\boldsymbol{q}};t)=$$
$$\frac{1}{t'}\Big[\frac{d}{d\tau}\Big(\frac{\partial(L(\boldsymbol{q},\frac{\boldsymbol{q}'}{t'};t)\,t')}{\partial q'_i}\Big)-\frac{\partial}{\partial q_i}(L(\boldsymbol{q},\frac{\boldsymbol{q}'}{t'};t)\,t')\Big] \qquad (8.4\text{-}1)$$

and we see that Lagrange equations in t for L vanish if and only if the Lagrange equations in τ for the *modified* Lagrangian Lt' do. Thus under such a change of variable from t to $t(\tau)$, the Lagrangian transforms from L to Lt' (expressed in the new variables and derivatives).

Now consider a Lagrangian like that above, but also regard t as a *coordinate*: $t = t(\tau) \equiv q_o$. Then we can write the modified Lagrangian

$$L(\boldsymbol{q},\dot{\boldsymbol{q}};t)t' \equiv L(\boldsymbol{q},\frac{\boldsymbol{q}'}{t'};q_o)t' \equiv \overline{L}(\overline{\boldsymbol{q}},\overline{\boldsymbol{q}}'),$$

—*note that* $\overline{L}$ *is independent of the variable* τ *itself*—in which the generalized coordinates and their velocities are now

$$\overline{\boldsymbol{q}} \equiv (q_o,\ldots,q_f)^{\mathsf{T}}$$
$$\overline{\boldsymbol{q}}' \equiv (q'_o,\ldots,q'_f)^{\mathsf{T}}, \quad q'_o \equiv t'.$$

These will satisfy the Lagrange equations (8.4-1) in τ for the modified Lagrangian $\overline{L}(\overline{\boldsymbol{q}},\overline{\boldsymbol{q}}') = L(\boldsymbol{q},\frac{\boldsymbol{q}'}{t'};q_o)\,t'$.

Though we have not yet encountered the term, it is traditional to assign a name to the derivative of L with respect to the generalized velocities: we define the *generalized momentum* to be $p_i \equiv \frac{\partial L}{\partial \dot{q}_i}$, and it will become of increasing importance in the sequel. For present purposes, however, we note merely that there is a generalized momentum associated with the "coordinate" $t = q_o$; namely, because the t' enter L only through the f arguments $\boldsymbol{q}'/t'$,

$$\begin{aligned}\overline{p}_o \equiv \frac{\partial \overline{L}(\overline{\boldsymbol{q}},\overline{\boldsymbol{q}}')}{\partial \overline{q}'_o} = \frac{\partial(L(\boldsymbol{q},\frac{\boldsymbol{q}'}{t'};t)t')}{\partial t'} &= L + \Big(\sum_{i=1}^{f}\frac{\partial L}{\partial q'_i}\frac{\partial q'_i}{\partial t'}\Big)t' \\ &= L + \sum_{i=1}^{f}\frac{\partial L}{\partial \dot{q}_i}(-\frac{q'_i}{t'^2})t' \qquad (8.4\text{-}2)\\ &\equiv L - \sum_{i=1}^{f} p_i\dot{q}_i\end{aligned}$$

since

$$\frac{q'_i}{t'} \equiv \frac{\frac{dq_i}{d\tau}}{\frac{dt}{d\tau}} = \dot{q}_i.$$

In particular, if t—*i.e.*, q_o—happens to be absent from the Lagrangian $\overline{L}$, then the Lagrange equation for q_o,

$$\frac{d}{d\tau}\Big(\frac{\partial(Lt')}{\partial t'}\Big)-\frac{\partial(Lt')}{\partial q_o} \equiv \frac{dp_o}{d\tau}-0=0;$$

i.e., $p_o = L - \sum p_i\dot{q}_i$ is a *constant* in τ (whatever that might happen to be).

Both the argument used above regarding the constancy of p_o, and the function $L-\sum p_i \dot{q}_i$ (or, more exactly, its *negative*) will return in the next chapter. The function, in particular, happens to be precisely what we will define there as *Jacobi's function*, which will also reappear as the *Hamiltonian* in a later chapter!

Summary. We have seen that it is possible to regard time itself as a generalized coordinate, of no greater significance than any other coordinate, if we view them all as functions of yet another *independent* variable, here called τ. From that perspective, then, there is a generalized *momentum* associated with t, the function $L - \sum p_i \dot{q}_i$. In particular, if t is absent from the Lagrangian, this function is *constant.*

Summary

We have ultimately obtained the most general form of *Lagrange equations*—ones which encourage use of potential energy where possible, while admitting of use of generalized forces where necessary:

$$\frac{d}{dt}\left(\frac{\partial L}{\partial \dot{q}_i}\right) - \frac{\partial L}{\partial q_i} = Q_i, \quad i = 1, 2, \ldots, f,$$

in which $L \equiv T - V$ and f is the number of *degrees of freedom*, so the number of [independent] generalized coordinates available. V includes only potential energies due to *conservative* forces $\boldsymbol{F}$:

$$\boldsymbol{F} \text{ conservative} \quad \textit{iff} \quad \boldsymbol{F} = -\nabla V \quad \textit{iff} \quad \nabla \times \boldsymbol{F} \equiv 0.$$

(the only ones for which it is defined in the first place); Q_i is that generalized force arising from *non*-conservative forces

$$Q_j \equiv \sum_{i=1}^{n} \boldsymbol{F}_i^{(a)} \cdot \frac{\partial \boldsymbol{r}_i}{\partial q_j} \quad \text{or} \quad \delta U = \sum_{i=1}^{n} \boldsymbol{F}_i^{(a)} \cdot \delta \boldsymbol{r}_i = \sum_{j=1}^{f} Q_j \, \delta q_j$$

(or any conservative ones not, by choice or ignorance, included in V). In character, it is very much like the most general form of energy principle: that $U_{1\to2} = E_2 - E_1$, in which E includes the potential forces and U the rest. The only forces which need be included in this formulation are those doing virtual work (on which the principle is based); namely, the *active* forces—assuming that the virtual displacement from the actual motion is *consistent* with the constraints, which then do no work.

The function L is not unique: the equations of motion remain the same if a total time derivative of an arbitrary function $W(\boldsymbol{q}; t)$ is added onto L, because $\dot{W}$ satisfies the Lagrange equations with $Q_i \equiv 0$. Further, the equations—for $Q_i \equiv 0$—retain their form under an arbitrary transformation to new coordinates, $q_i \rightarrow q_i(\boldsymbol{Q}; t)$.

Though the equations are initially derived assuming all constraints to be *holonomic* and eliminated among the original set of coordinates, all that is really required to obtain them is linear independence among the *virtual differentials*, δq_i. The latter can be imposed either explicitly through algebraic solution, or (more elegantly) *implicitly* through the introduce of *Lagrange multipliers*; in the latter case, the multiplier, times its associated [kinematic] constraint, is nothing more than the [kinetic] constraint enforcing that constraint. In fact, the [generalized] Lagrange equations appear in exactly the same form as for independent coordinates; one just uses a number, equal to the number of constraints, to *determine* the multipliers, with the balance of the equations (using those multiplier values) then holding as independent coordinates. Though, for purposes of generality, *non*-holonomic constraints are used in the derivation, *holonomic* constraints $\phi_k(\mathbf{q};t)$ can be accounted for automatically by using a modified Lagrangian $L^* = L - \sum_{k=1}^{m} \lambda_k \phi_k$.

Despite its distinguished character, even *time* can be regarded as just another variable: it is necessary only to introduce another independent variable τ on which all variables, including t, depend, and utilize a slightly different Lagrangian $\overline{L} \equiv Lt' \equiv L\frac{dt}{d\tau}$.

Chapter 9

Integrals of Motion

Introduction

Recall that, for each degree of freedom, the [generalized] Lagrange equations (8.2-8) give us a *second order ordinary differential equation* for each q_j (see pages 447 and 428). Though we have been focussing on *obtaining* the equations, at some point they must be *solved*! The same thing occurred in rigid-body motion in the previous Part; there we ultimately invoked *conservation laws* to glean information about the system at hand. This chapter will introduce a new perspective on that process, "integrals of the motion". We will then obtain a particular such integral, energy-like in character, *Jacobi's integral*, which holds generally for conservative systems and which reduces to the classical energy in a certain case. Finally, we will discuss an approach to "solution" of the system, resulting not from solving a given set of differential equations, but rather from finding a set of coordinates in which the Lagrangian function assumes a particular form in which the problem is already half-solved *without any integration of differential equations*!

9.1 Integrals of the Motion

What is meant by the "solution" to a problem in dynamics? We have already seen that the Lagrange equations, just like the Newtonian equations on which they are based, result in a system of second-order ordinary differential equations. Though such a system can always be solved if they are linear, and though the $\ddot{q}_j$ certainly enter linearly, the balance of the derivatives generally aren't; thus, unless he is lucky, one must usually resort to approximations to get an analytical solution, or just give up and get a numerical one.

But any of these techniques, because of the second-order nature of the system, must ultimately somehow effect two integrations, *each of which introduces an [additive] constant* generally used to match the conditions of the system at a given time. In particular, for a system with f degrees of freedom, we require

$2f$ such integrations introducing $2f$ associated constants.

The integrations themselves rarely give an *explicit* solution $q_i = q_i(t)$, any more than they do in the analytical integration of differential equations; rather they yield an *implicit* solution—a *relation* among the variables and their [first] derivatives, generally of the form $f_i(q_1, q_2, \ldots, q_n, \dot{q}_1, \dot{q}_2, \ldots, \dot{q}_n; t) = C_i$. It's not that this is the best system, or certainly the most desirable one (that would likely be an *explicit* solution), but, with differential equations, one is just happy with what he gets!

In the present context, such a relation is referred to as an "*integral of the motion*". Aside from the jargon introduced here, such [general] integrals have already been encountered in the previous Part; in particular, for particles, they include

- *energy*: $E = T + V = T(\dot{\boldsymbol{r}}) + V(\boldsymbol{r})$ (one constant), holding if all forces are conservative
- *angular momentum*: $\boldsymbol{H}_O = m\boldsymbol{r} \times \boldsymbol{v}$ (three constants), holding if $\frac{d\boldsymbol{H}_O}{dt} = \sum \boldsymbol{M}_O \equiv 0$
- *linear momentum*: $\boldsymbol{G} = m\boldsymbol{v}$ (three constants), holding if $\frac{d\boldsymbol{G}}{dt} = \sum \boldsymbol{F} \equiv 0$

—"conserved" under the given conditions, and yielding the indicated number of [scalar] "constants", the *constants of integration* alluded to above. And there are corresponding integrals of the motion for *systems* of particles and rigid bodies, too.

It is significant that the last two of these integrals result from the vanishing of the *time* derivatives—the "*equations of motion*"—of the corresponding quantities $\boldsymbol{H}_O$ and $\boldsymbol{G}$; the energy principle, on the other hand, has no such time derivative $\dot{E}$ (though, to be conserved, quite clearly $\dot{E} \equiv 0$, and we have obtained an expression for $\dot{E}$ in equation (5.1-13)). Equations of motion are generally regarded as *time* derivatives.

But no matter what they're called, to completely solve a system with f degrees of freedom, *one must obtain* $2f$ *integrals of the motion*, explicit or implicit. In some sense, then, the solution to a problem in [Lagrangian] dynamics is tantamount to discovering a full set of such integrals. Even if they are all *implicit*, one can presumably solve the resulting system in order to obtain the values of the q_i and $\dot{q}_i$ at any (or each) time t.

The classical integrals of classical dynamics cited above are known to hold under the conditions given; in particular, if one knows those conditions hold, then the corresponding functions of $\boldsymbol{r}$ and $\boldsymbol{v}$ are known to be constant *without explicitly solving the differential equations.* Given the initial conditions on the system, the *values* of the "constants" E and/or $\boldsymbol{H}_O$ and/or $\boldsymbol{G}$ are known, and we are already on our way to a full solution of the problem. (Sometimes only one or two *components* of the vector integrals are known to be conserved; that's still better than nothing.) Oftentimes, one of these integrals is enough to get the information desired; that's the approach taken in typical undergraduate problems in dynamics, which necessarily present these general integrals piecemeal.

But, at the end of the day, we ultimately want to know the *full* solution to the *full* problem, and the number of available integrals of the motion allows us to measure how far along the process we are.

All the examples of integrals of the motion thus far have been taken from the realm of classical dynamics. We shall now obtain such an integral in the present context of Lagrangian dynamics.

Summary. "Integrals of the motion" are simply relations among the unknowns which are constant in *time*; *i.e.*, if the time derivative of a certain quantity vanishes, that quantity is one of the requisite integrals. In particular, if we have as many [independent] integrals for each degree of freedom in the system as the order of the differential equation describing its motion, the problem is solved, albeit *implicitly.*

9.2 Jacobi's Integral—an Energy-like Integral

We shall *assume all active forces are conservative* and have been put into a potential energy. Then Lagrange's equations hold, and, for $L \equiv T - V = L(q, \dot{q}; t)$, we have that

$$\frac{d}{dt}\left(\frac{\partial L}{\partial \dot{q}_j}\right) - \frac{\partial L}{\partial q_j} = 0$$

for all $j = 1, \ldots, f$.

We shall now examine the *[total] time derivative* of the function L—its *equation of motion*:

$$\frac{dL}{dt} = \sum_{i=1}^{f} \frac{\partial L}{\partial q_i}\dot{q}_i + \sum_{i=1}^{f} \frac{\partial L}{\partial \dot{q}_i}\ddot{q}_i + \frac{\partial L}{\partial t}.$$

By the Lagrange equations, however, $\frac{\partial L}{\partial q_i}$ in the first term can be replaced by $\frac{d}{dt}(\frac{\partial L}{\partial \dot{q}_i})$; furthermore, $\ddot{q}_i \equiv \frac{d\dot{q}_i}{dt}$ in the second; thus

$$\begin{aligned}\frac{dL}{dt} &= \sum_{i=1}^{f}\left(\frac{d}{dt}\left(\frac{\partial L}{\partial \dot{q}_i}\right)\dot{q}_i + \frac{\partial L}{\partial \dot{q}_i}\frac{d\dot{q}_i}{dt}\right) + \frac{\partial L}{\partial t}\\ &= \sum_{i=1}^{f}\frac{d}{dt}\left(\frac{\partial L}{\partial \dot{q}_i}\dot{q}_i\right) + \frac{\partial L}{\partial t} = \frac{d}{dt}\left(\sum_{i=1}^{f}\frac{\partial L}{\partial \dot{q}_i}\dot{q}_i\right) + \frac{\partial L}{\partial t}.\end{aligned}$$

Thus we get finally that

$$\frac{d}{dt}\left(\sum_{i=1}^{f}\frac{\partial L}{\partial \dot{q}_i}\dot{q}_i - L\right) = -\frac{\partial L}{\partial t}.$$

We shall now further *assume that* $\frac{\partial L}{\partial t} \equiv 0$, *i.e.*, that L does not depend on time *explicitly* (it still does *implicitly* through the q_i and $\dot{q}_i$). Then, just as occurred in the classical conservations of linear and angular momentum, where the corresponding equation of motion vanished, we get that

$$h = h(\boldsymbol{q}, \dot{\boldsymbol{q}}) \equiv \sum_{i=1}^{f} \frac{\partial L}{\partial \dot{q}_i} \dot{q}_i - L \qquad (9.2\text{-}1)$$

is a *constant* (in time). Thus we have uncovered what appears to be another integral of the motion—*Jacobi's integral.* The function h could be evaluated at a particular time, and the resulting form—a relation among the q_i and $\dot{q}_i$—would then always assume that value. Note the perhaps overly cautious remark that this "appears to be another integral of the motion". The emphasis here is on the word "another": *is* this truly a new integral, or is it merely contained in the above classical ones? The facts that this is a scalar, and it requires all active forces to be conservative—both characteristics of the *energy* integral—make us even more suspicious. We now examine that issue.

Remember that $L = L(\boldsymbol{q}, \dot{\boldsymbol{q}})$—*not t!*—in the above function. But how can t enter $L = T - V$? Consider first V: We have seen (page 261) that V *can* be an explicit function of t if there is a time-dependent $\boldsymbol{F}(t)$ such that $\nabla \times \boldsymbol{F} \equiv 0$. In this case, even if T were to be time-dependent, it would be impossible for t-dependent terms between T and V to cancel because $V = V(\boldsymbol{r}; t)$ doesn't depend on $\dot{\boldsymbol{q}}$ but T does; thus any accompanying t in these terms arising from time-dependent constraints would remain (see equation (8.1-10)). So L would be dependent on t, violating the conditions for Jacobi's Integral.

Thus let us assume that $V = V(\boldsymbol{r})$ only. Recall that, in general, $\boldsymbol{r}_i = \boldsymbol{r}_i(\boldsymbol{q}; t)$. so $\boldsymbol{v}_i = \boldsymbol{v}_i(\boldsymbol{q}, \dot{\boldsymbol{q}}; t)$, and thus $T \equiv \sum \frac{1}{2} m_i \boldsymbol{v}_i \cdot \boldsymbol{v}_i = T(\boldsymbol{q}, \dot{\boldsymbol{q}}; t)$; furthermore, $V = V(\boldsymbol{r}_i) = V(\boldsymbol{r}_i(\boldsymbol{q}; t))$. Now T depends on the v_i^2, and we have obtained a general expression for T, equation (8.1-10), which we can examine to see where t enters explicitly. Doing so reveals that this can happen only if $\boldsymbol{r}_i$ *explicitly* depends on time; that, in turn, occurs only if $\boldsymbol{r}_i$ is subject to time-dependent—*i.e.*, *rheonomic*—constraints; if the constraints are *scleronomic*, on the other hand, the discussion surrounding (8.1-9) points out that T is a purely quadratic function of the $\dot{q}_i$. And certainly if $\boldsymbol{r}_i$ depends on time, so does $V(\boldsymbol{r}_i(\boldsymbol{q}; t))$. This suggests consideration of the special case that $\boldsymbol{r} = \boldsymbol{r}(\boldsymbol{q})$ alone to see whether *it* might lead to one of the classical integrals; this turns out to be precisely the case:

Theorem 9.2.1. If $\boldsymbol{r} = \boldsymbol{r}(\boldsymbol{q})$, then Jacobi's integral (9.2-1) reduces to the energy integral.

Proof. We note that $V = V(\boldsymbol{r}_i(\boldsymbol{q}))$ alone (for "time-dependent potentials" it might depend on t, too, but it *doesn't* depend on $\dot{\boldsymbol{q}}$!); thus the $\dot{q}_i$ enter L only through T, which we will now examine.

We have already cited the general expression for T, equation (8.1-10). In the hypothesized case that $\frac{\partial \mathbf{r}_i}{\partial t} \equiv 0$, that equation reduces to

$$T \equiv \sum_i^n \frac{1}{2} m_i \sum_j^f \sum_k^f \frac{\partial \mathbf{r}_i}{\partial q_j} \cdot \frac{\partial \mathbf{r}_i}{\partial q_k} \dot{q}_j \dot{q}_k$$

$$= \sum_j^f \sum_k^f \left(\sum_i^n \frac{1}{2} m_i \frac{\partial \mathbf{r}_i}{\partial q_j} \cdot \frac{\partial \mathbf{r}_i}{\partial q_k} \right) \dot{q}_j \dot{q}_k \equiv \sum_j^f \sum_k^f t_{2_{jk}} \dot{q}_j \dot{q}_k$$

We can write T in a form isolating the $\dot{q}_i$ terms:

$$T = \sum_{j \neq i}^f \left(\sum_{k \neq i}^f t_{2_{jk}} \dot{q}_j \dot{q}_k + t_{2_{ji}} \dot{q}_j \dot{q}_i \right) + \left(\sum_{k \neq i}^f t_{2_{ik}} \dot{q}_i \dot{q}_k + t_{2_{ii}} \dot{q}_i^{\,2} \right)$$

(where the second term in parentheses is just the $j = i$ term left out in the first such). Thus

$$\frac{\partial L}{\partial \dot{q}_i} = \frac{\partial T}{\partial \dot{q}_i} = \sum_{j \neq i}^f (0 + t_{2_{ji}} \dot{q}_j) + \left(\sum_{k \neq i}^f t_{2_{ik}} \dot{q}_k + 2t_{2_{ii}} \dot{q}_i \right)$$

$$= \sum_j^f t_{2_{ji}} \dot{q}_j + \sum_k^f t_{2_{ik}} \dot{q}_k$$

(where we have allocated the $2t_{2_{ii}} \dot{q}_i$ between the two summations). Thus,[1]

$$\sum_i^f \frac{\partial L}{\partial \dot{q}_i} \dot{q}_i = \sum_i^f \sum_j^f t_{2_{ji}} \dot{q}_i \dot{q}_j + \sum_i^f \sum_k^f t_{2_{ik}} \dot{q}_i \dot{q}_k = 2T$$

It is instructive to carry out the above manipulations with the *matrix* expression for T, equation (8.1-11): We note that we can write

$$\sum_{i=1}^f \frac{\partial L}{\partial \dot{q}_i} \dot{q}_i = \left(\frac{\partial L}{\partial \dot{q}_1}, \frac{\partial L}{\partial \dot{q}_2}, \ldots, \frac{\partial L}{\partial \dot{q}_f} \right) \begin{pmatrix} \dot{q}_1 \\ \dot{q}_2 \\ \vdots \\ \dot{q}_f \end{pmatrix}.$$

[1]This is a special case of "Euler's Theorem". If there are nearly as many "Euler's Theorems" as there are fields of mathematics, this one deals with *homogeneous functions*—functions satisfying the condition

$$f(\lambda x_1, \lambda x_2, \ldots, \lambda x_n) = \lambda^n f(x_1, x_2, \ldots, x_n).$$

Loosely, this merely states that the exponents of each term in a polynomial sum to the same value; thus $x^2 y^{\frac{1}{2}} + y^{\frac{5}{2}}$ is homogeneous [of degree $\frac{5}{2}$], while $x^2 y + y^{\frac{5}{2}}$ isn't. In this context, Euler's Theorem states that

$$\sum x_i \frac{\partial f}{\partial x_i} = n f(x_1, x_2, \ldots, x_n),$$

and is proven by simply differentiating the homogeneity condition with respect to [the general] λ and setting that parameter to 1 in the result.

Again, T is a purely quadratic function of the $\dot{q}_j$; thus (8.1-11) reduces to

$$T = \dot{\mathbf{q}}^{\mathsf{T}}\mathbf{T}_2\dot{\mathbf{q}}$$

—where, recall, $\mathbf{T}_2$ is symmetric (see discussion relating to (8.1-11)). Differentiating this with respect to $\dot{q}_i$ (see Homework 2, Section 2.1.1),

$$\begin{aligned}\frac{\partial L}{\partial \dot{q}_i} = \frac{\partial T}{\partial \dot{q}_i} &= \mathbf{e}_i^{\mathsf{T}}\mathbf{T}_2\dot{\mathbf{q}} + \dot{\mathbf{q}}^{\mathsf{T}}\mathbf{T}_2\mathbf{e}_i \\ &= 2\dot{\mathbf{q}}^{\mathsf{T}}\mathbf{T}_2\mathbf{e}_i,\end{aligned} \tag{9.2-2}$$

since $(\mathbf{e}_i^{\mathsf{T}}\mathbf{T}_2\dot{\mathbf{q}})^{\mathsf{T}} = \dot{\mathbf{q}}^{\mathsf{T}}\mathbf{T}_2^{\mathsf{T}}\mathbf{e}_i$ (by Theorem 2.1.2), which is just $\dot{\mathbf{q}}^{\mathsf{T}}\mathbf{T}_2\mathbf{e}_i$, by the symmetry of $\mathbf{T}_2$. Thus

$$\sum_{i=1}^{f}\frac{\partial L}{\partial \dot{q}_i}\dot{q}_i = 2\dot{\mathbf{q}}^{\mathsf{T}}\mathbf{T}_2(\mathbf{e}_1, \mathbf{e}_2, \ldots, \mathbf{e}_f)\dot{\mathbf{q}} = 2\dot{\mathbf{q}}^{\mathsf{T}}\mathbf{T}_2\mathbf{I}\dot{\mathbf{q}} = 2T$$

(using block multiplication)—just as before. Note that, aside from the appealing compactness of this approach, it avoids all the annoying index bookkeeping necessary when one uses the above explicit summations.

In either event, Jacobi's integral

$$h \equiv \sum_{i=1}^{f}\frac{\partial L}{\partial \dot{q}_i}\dot{q}_i - L = 2T - (T - V) = T + V = E$$

in this case, as desired. □

We can collect all these results into a single theorem:

Theorem 9.2.2 (Jacobi's Integral). If all active forces in a system are *conservative*, and if the resulting *Lagrangian function is [explicitly] independent of time*, then the quantity

$$h(q, \dot{q}) \equiv \sum_{i=1}^{f}\frac{\partial L}{\partial \dot{q}_i}\dot{q}_i - L$$

is conserved. Furthermore, if the system is subject to only *scleronomic constraints*, h is nothing more than the energy function.

Actually, the above can be stated in a slightly more general form: if we define Jacobi's *function*

$$h(q, \dot{q}; t) \equiv \sum_{i=1}^{f}\frac{\partial L(q, \dot{q}; t)}{\partial \dot{q}_i}\dot{q}_i - L(q, \dot{q}; t), \tag{9.2-3}$$

the conditions of this theorem become that all forces are conservative and $\frac{\partial h}{\partial t} = 0$; under those conditions Jacobi's *function* becomes a constant of the motion, reducing to the energy [function] if all constraints are scleronomic. This function will return in the next chapter.

So we have shown that Jacobi's integral reduces to the classical energy integral *if* $\mathbf{r} = \mathbf{r}(q)$ *alone*; *i.e.*, if all constraints are *scleronomic*. But it holds in the more general case of rheonomic constraints, too; it is a *generalization* of the energy integral, not strictly one of the classical ones. Note, however, that we have expressed L in terms of its f generalized coordinates (and their generalized velocities); in particular, all constraints have already been eliminated. Thus, at least as presented here, this demands that the constraints be *holonomic* in character; the devices in the previous chapter cannot be applied.

Example 9.2.1. *Jacobi's Integral.* To illustrate Jacobi's integral in a nontrivial fashion, it is necessary to pick an example embodying *rheonomic* [and holonomic] constraints; as shown in the above theorem, scleronomic constraints reduce to simply the energy integral. Furthermore, the active forces must themselves be conservative in order to satisfy the hypotheses of the general Theorem. Yet the Lagrangian must be independent of time. It would seem to be a somewhat delicate matter to bring all these restrictions together in a problem! In point of fact, however, Jacobi's integral arose originally in connection with a problem in celestial mechanics, the "restricted three-body problem",[2] in which variables are referred to a rotating coordinate system. Thus we shall present a similar example.

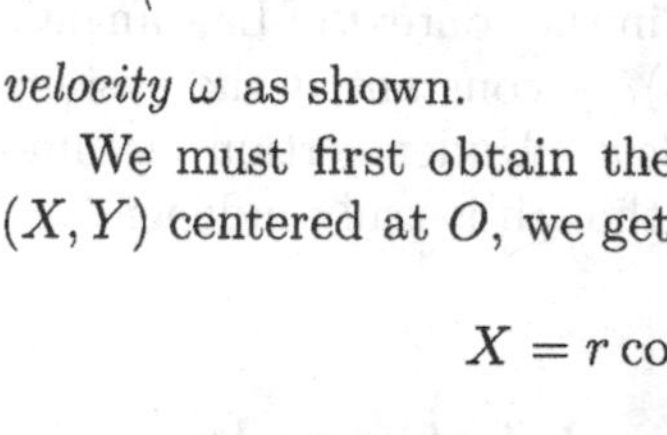

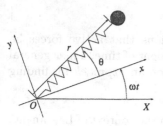

We consider a [linear] spring with unextended length r_o, with one end connected to a fixed point O, and the other to a particle of mass m moving on a horizontal plane without friction (so that the only active force is conservative). We shall obtain an integral of the motion involving the two generalized coordinates r and θ *measured relative to a coordinate system rotating with constant angular velocity* ω as shown.

We must first obtain the Lagrangian. Taking *fixed* Cartesian coordinates (X, Y) centered at O, we get that the components of the mass are

$$X = r\cos(\theta + \omega t), \quad Y = r\sin(\theta + \omega t).$$

Note that the rotating coordinates introduce a "rheonomic constraint" due not to a constraint force *per se*, but rather to the rotation itself. This is not surprising in retrospect: we are observing the motion relative to a *non-inertial* coordinate system, introducing intrinsic accelerations due to the rotation. From the above, we obtain immediately that

$$\dot{X} = \dot{r}\cos(\theta + \omega t) - r\sin(\theta + \omega t)(\dot{\theta} + \omega)$$
$$\dot{Y} = \dot{r}\sin(\theta + \omega t) + r\cos(\theta + \omega t)(\dot{\theta} + \omega);$$

[2]See, *e.g.*, Danby, *Fundamentals of Celestial Mechanics*, 1962, §8.1.

substituting these into $v^2 = \dot{X}^2 + \dot{Y}^2$ and subtracting off the potential energy due to the spring, we get finally that

$$L \equiv T - V = \frac{1}{2}m(\dot{r}^2 + r^2(\dot{\theta} + \omega)^2) - \frac{1}{2}k(r - r_o)^2. \tag{9.2-4}$$

(The same result could be obtained with polar or rotating coordinates, where $\dot{\theta} \rightarrow \dot{\theta} + \omega$.) Note that, even though $\mathbf{r} = \mathbf{r}(q; t)$, the time dependence vanishes in the Lagrangian! Then calculating the two terms in the Jacobi function

$$\frac{\partial L}{\partial \dot{r}}\dot{r} = m\dot{r}^2$$
$$\frac{\partial L}{\partial \dot{\theta}}\dot{\theta} = m(r^2(\dot{\theta} + \omega))\dot{\theta} = m(r^2\dot{\theta}^2 + r^2\omega\dot{\theta}),$$

we get that

$$h \equiv \frac{\partial L}{\partial \dot{r}}\dot{r} + \frac{\partial L}{\partial \dot{\theta}}\dot{\theta} - L = \frac{1}{2}m(\dot{r}^2 + r^2\dot{\theta}^2 - r^2\omega^2) + \frac{1}{2}k(r - r_o)^2.$$

This is *not* the energy integral

$$E = T + V = \frac{1}{2}m(\dot{r}^2 + r^2(\dot{\theta} + \omega)^2) + \frac{1}{2}k(r - r_o)^2$$

but an independent one. |

Thus we have obtained, under only the conditions that active forces be conservative and the resulting Lagrangian be independent of time, a *new* general integral of the motion. We consider in the next section an *approach* for finding others.

Summary. We have uncovered a general integral in the context of Lagrangian dynamics—that *Jacobi's function*, equation (9.2-3), is constant if the system is *conservative* and the Lagrangian *time-independent.* Under certain circumstances, it reduces to the classical energy integral, though is truly independent of it.

9.3 "Ignorable Coordinates" and Integrals

We shall once again, from the start, *assume all forces are conservative*, so that the [non-generalized] form of Lagrange's equations holds. Let us say that, when the Lagrangian is expressed in a certain set of coordinates, it is *independent* of, say, q_l; then

$$0 = \frac{\partial L}{\partial q_l} = \frac{d}{dt}\left(\frac{\partial L}{\partial \dot{q}_l}\right) \tag{9.3-1}$$

(where the second part of this results from Lagrange's equations). This is not so uncommon a result as might be expected; in fact, it occurred in the previous

Example 9.2.1, in which the Lagrangian $L = \frac{1}{2}m(\dot{r}^2 + r^2(\dot{\theta}+\omega)^2) - \frac{1}{2}k(r - r_o)^2$, equation (9.2-4), happens to be free of the variable θ! But the last term above is itself an "equation of motion"—the time derivative of $\frac{\partial L}{\partial \dot{q}_l}$. Note that, since $L = L(\boldsymbol{q}, \dot{\boldsymbol{q}}_l; t)$, this partial derivative will itself be a *function* of the various coordinates and their time derivatives; the fact its time derivative vanishes, just as in the case of the classical and Jacobi integrals, means that *this function is an integral of the motion.* Thus in the Lagrangian of the previous example,

$$\frac{\partial L}{\partial \dot{\theta}} = mr^2(\dot{\theta} + \omega)$$

is constant throughout the motion. Such a coordinate which does not appear in the Lagrangian of a system is called an "*ignorable coordinate*", and we see that its absence in a *conservative* system immediately implies the existence of an integral of the motion. (Before going any farther, it should be noted that the same does *not* apply to "ignorable velocities": the absence of a generalized *velocity* in a conservative Lagrangian means that its corresponding coordinate must be absent, too, by (9.3-1)—"you can't have a coordinate without its velocity"!)

In the same way that analytical dynamics talks about *scalar* "generalized coordinates", "generalized velocities", and "generalized forces"—these are often not "real" velocities and forces, recall—one can define the partial derivative

$$p_i \equiv \frac{\partial L}{\partial \dot{q}_i} \tag{9.3-2}$$

to be the *generalized momentum* corresponding to q_i. The terminology is motivated by the observation that, in Cartesian systems, in which the kinetic energy (in conservative systems, the only term containing the $\dot{q}_i$) is of the form $T = \frac{1}{2}m(\dot{x}^2 + \dot{y}^2)$, so that

$$p_x \equiv \frac{\partial L}{\partial \dot{x}} = m\dot{x} \quad \text{and} \quad p_y \equiv \frac{\partial L}{\partial \dot{y}} = m\dot{y}.$$

Note, however, that once again in polar coordinates, where $T = \frac{1}{2}m(\dot{r}^2 + (r\dot{\theta})^2)$,

$$p_r \equiv \frac{\partial L}{\partial \dot{r}} = m\dot{r} \quad \text{and} \quad p_\theta \equiv \frac{\partial L}{\partial \dot{\theta}} = mr^2\dot{\theta},$$

the latter is the *angular* momentum: just as with its counterparts, these quantities are *scalars*, and do not necessarily have the expected units. In terms of these, the above discussion regarding ignorable coordinates can be expressed by saying that *if a coordinate is ignorable, then its corresponding momentum is constant.*

We have already noted one example of this above. We now provide another:

Example 9.3.1. *Ignorable Coordinate.* Consider the "one body problem" of celestial mechanics: A mass m moves under an inverse-square law towards a fixed focus; in terms of its position $\boldsymbol{r}$ relative to this point, the force law is

$$\boldsymbol{F} = -\frac{\mu}{r^2}\hat{\boldsymbol{r}} = -\frac{\mu}{r^3}\boldsymbol{r}$$

where $r \equiv |\boldsymbol{r}|$. Using the criteria collected in equation (5.1-12), $\nabla \times \boldsymbol{F} \equiv 0$, so this is a conservative force; thus there is an associated potential energy V such that $\boldsymbol{F} = -\nabla V$. In fact,

$$V = -\frac{\mu}{r}.$$

If we consider only planar motion, and express the Lagrangian in terms of Cartesian coordinates, we get that

$$L = \frac{1}{2}m(\dot{x}^2 + \dot{y}^2) + \frac{\mu}{\sqrt{x^2 + y^2}}$$

and no coordinates are ignorable. But if we use *polar* coordinates to write it,

$$L = \frac{1}{2}m(\dot{r}^2 + (r\dot{\theta})^2) + \frac{\mu}{r}$$

—in which, as above, θ is ignorable; thus we conclude that $p_\theta = mr^2\dot{\theta}$, the angular momentum, is conserved. A not overly perspicacious choice of coordinate system has generated an integral, without any integration! |

What is the importance of this? We note that

- whether a coordinate is ignorable or not depends on the *form* of the Lagrangian, and
- the form of the Lagrangian depends on the *coordinates* used to express it.

And now we pull together all of the ideas in this chapter: We desire to find *integrals of the motion.* If coordinates are ignorable, their corresponding momenta *are* integrals of the motion. But this depends on the *form* of the Lagrangian, and the form depends on the *coordinates* chosen. In particular, *if we could find a set of coordinates in which the Lagrangian [form] renders them all ignorable, we would immediately have half the required integrals of the motion*! We wouldn't have to find such integrals by integration; rather we find them by seeking the "right" coordinates in terms of which to cast the Lagrangian. The onus is shifted from solving differential equations to discovering coordinates. This is where Theorem 8.2.2 comes in: it says that, given a Lagrangian in one set of coordinates, we can *transform* to any other set and *the form of the equations of motion will be the same in the new coordinates.* Thus we can search for such a transformation, concentrating only on the form the *Lagrangian* assumes in the new coordinates, in the assurance that the equations of motion will still be of the same form.

Examination of the above arguments explains why we are restricting our attention to *conservative* systems: The existence of integrals of the motion according to this criterion depends on only the presence or absence of given generalized coordinates in the *Lagrangian*, not on where they might appear in any generalized forces; this, in turn, relies on the right-hand side of equations (8.2-8) being 0. A coordinate transformation would likely change the *form* of any Q_j there, but it generally won't make it *vanish*.

Unfortunately, there is no practical means of uncovering just the "right" set of coordinates in Lagrangian problems. But the possibility of avoiding the explicit solution of differential equations, even after little experience with them, does have some appeal. And, although there isn't a general means of finding coordinates to get half the integrals in Lagrangian problems, there *is* one in the allied field of Hamiltonian dynamics—one which gets them *all*! This will be the subject of the next, and final, section in this Part.

Summary. If a generalized coordinate is absent from the Lagrangian—if it is an "*ignorable coordinate*"—the "generalized momentum" corresponding to that variable is an integral of the motion. But this depends on the *form* of the Lagrangian, which, in turn, depends on the coordinates chosen to describe it. Thus, if we could find a set of coordinates in which the Lagrangian depends only on the generalized velocities, we would immediate have f of the $2f$ integrals of motion required to solve the system [*implicitly*].

Summary

This chapter has generally been involved with uncovering *integrals of the motion*—functions of the q_i and $\dot{q}_i$ which are constant in time. We have produced a function, the *Jacobi integral*, which is generally known to be such a constant under the *proviso* that all active forces be conservative and the Lagrangian is free of time; this function is shown to reduce to the energy integral under scleronomic constraint. Finally we discussed how the discovery of up to half the required integrals of the motion was tantamount to selection of the appropriate coordinates—an achievement signalled by the presence (or *absence*) of *ignorable coordinates*.

Chapter 10

Hamiltonian Dynamics

Introduction

We shall, in this chapter, develop to Lagrangian dynamics a companion formulation, *Hamiltonian dynamics.* If the former utilizes the Lagrangian function, the latter makes fundamental use of the *Hamiltonian* [function], which enjoys a sort of reflexive relationship with the Lagrangian; in fact, much of the development relies on results from Lagrangian dynamics. Like the Lagrangian, the Hamiltonian equations of motion depend on the *form* of that function. Hamiltonian dynamics, however, also has a number of unique characteristics, as we shall see, among them the ability to *find* coordinates in which the Hamiltonian assumes a desired form—a property not shared by the Lagrangian—a form in which the *entire* problem is solved.

10.1 The Variables

The Lagrangian, recall, utilizes f generalized coordinates which, along with their time derivatives (the generalized *velocities*) and the time itself, appear in the Lagrangian:

$$L = L(\boldsymbol{q}, \dot{\boldsymbol{q}}; t) \equiv T - V.$$

In the previous section we introduced, almost offhandedly, the f *generalized momenta* associated with generalized coordinates:

$$p_i = p_i(\boldsymbol{q}, \dot{\boldsymbol{q}}; t) \equiv \frac{\partial L(\boldsymbol{q}, \dot{\boldsymbol{q}}; t)}{\partial \dot{q}_i}.$$

These quantities, merely a sidelight previously, will become fundamental to Hamiltonian dynamics. In fact, the *Hamiltonian*, a function in this formulation playing the same role as the Lagrangian in the previous one, is expressed not in terms of the $\boldsymbol{q}$ and $\dot{\boldsymbol{q}}$, but rather the [same] $\boldsymbol{q}$ and these $\boldsymbol{p}$.[1] These sets are

[1]This choice is possibly the origin of the phrase that "You must watch your p's and q's"?

called *conjugate variables*, with each q_i said to be [the variable] *conjugate* to its p_i (and conversely).

The above definition of the p_i is, effectively, a set of f equations in the $3f$ variables $\boldsymbol{p}$, $\boldsymbol{q}$, and $\dot{\boldsymbol{q}}$ *explicitly* defining the p_i: given values of $\boldsymbol{q}$ and $\dot{\boldsymbol{q}}$, we immediately have a corresponding value for p_i. But we might then also expect that, given values of the $\boldsymbol{q}$ and $\boldsymbol{p}$, say, the value for $\dot{\boldsymbol{q}}$ would be determined—that, in essence, $\dot{\boldsymbol{q}}$ is defined *implicitly* by this same set of equations, and that we might be able to "solve" this system for $\dot{\boldsymbol{q}}(\boldsymbol{q},\boldsymbol{p};\mathrm{t})$—at least under certain conditions. This is indeed the case, and we now examine the conditions required to ensure the ability to find this solution.

Solution for $\dot{q}(\boldsymbol{q},\boldsymbol{p};t)$

Once more we note that the $\dot{q}$'s enter only through T and return to the matrix expression for that quantity, equation (8.1-11) (repeated here):

$$T \equiv \dot{\mathbf{q}}^{\mathsf{T}}\mathsf{T}_2\dot{\mathbf{q}} + \dot{\mathbf{q}}^{\mathsf{T}}\mathsf{T}_1 + \mathsf{T}_0$$

—in which the various terms, except for the $\dot{\mathbf{q}}$, are all functions only of $\boldsymbol{q}$ and t. Thus each

$$\begin{aligned} p_i \equiv \frac{\partial L}{\partial \dot{q}_i} = \frac{\partial T}{\partial \dot{q}_i} &= \frac{\partial}{\partial \dot{q}_i}(\dot{\mathbf{q}}^{\mathsf{T}}\mathsf{T}_2\dot{\mathbf{q}} + \dot{\mathbf{q}}^{\mathsf{T}}\mathsf{T}_1 + \mathsf{T}_0) \\ &= 2\mathbf{e}_i^{\mathsf{T}}\mathsf{T}_2\dot{\mathbf{q}} + \mathbf{e}_i^{\mathsf{T}}\mathsf{T}_1, \end{aligned} \tag{10.1-1}$$

using the results of (9.2-2). Note that p_i is a *linear* function in the $\dot{q}_j$ (though its dependence on the q_j, through $\mathsf{T}_i(q,\dot{q};t)$, in generally *non*-linear). These can be "stacked" into a column f-vector (more block multiplication):

$$\mathsf{p} = \begin{pmatrix} \mathbf{e}_1^{\mathsf{T}} \\ \mathbf{e}_2^{\mathsf{T}} \\ \vdots \\ \mathbf{e}_f^{\mathsf{T}} \end{pmatrix} (2\mathsf{T}_2\dot{\mathbf{q}} + \mathsf{T}_1) = 2\mathsf{T}_2\dot{\mathbf{q}} + \mathsf{T}_1,$$

the "stacked" $\mathbf{e}_i^{\mathsf{T}}$ being just $\mathbf{1}$, the identity matrix. But this is a linear equation in $\dot{\mathbf{q}}$, which can be solved

$$\dot{\mathbf{q}} = \frac{1}{2}\mathsf{T}_2{}^{-1}(\mathsf{p} - \mathsf{T}_1) \tag{10.1-2}$$

as long as T_2^{-1} *exists*—*i.e.*, as long as $|\mathsf{T}_2| \neq 0$. We note, in this regard, that (10.1-1) gives us

$$\frac{\partial^2 L}{\partial \dot{q}_i \partial \dot{q}_j} = \frac{\partial^2 T}{\partial \dot{q}_i \partial \dot{q}_j} = 2\mathbf{e}_i^{\mathsf{T}}\mathsf{T}_2\mathbf{e}_j = 2t_{2_{ij}}$$

where $\mathsf{T}_2 = (t_{2_{ij}})$; *i.e.*, $2\mathsf{T}_2$ is just the matrix of the $\frac{\partial^2 L}{\partial \dot{q}_i \partial \dot{q}_j}$, the so-called *Hessian matrix* of L with respect to the $\dot{q}_i$. This is the form in which the condition for

$\dot{q}$ solution is usually cast:

$$\left|\left(\frac{\partial^2 L}{\partial \dot{q}_i \partial \dot{q}_j}\right)\right| \neq 0 \tag{10.1-3}$$

—since it makes direct reference to the definition of $p_i \equiv \frac{\partial L}{\partial \dot{q}_i}$. Equivalently, it also shows that the p_i, regarded as functions of the $\dot{q}_j$, are linearly independent of one another:

$$\left|\left(\frac{\partial^2 L}{\partial \dot{q}_i \partial \dot{q}_j}\right)\right| = \left|\left(\frac{\partial p_j}{\partial \dot{q}_i}\right)\right| \neq 0$$

(actually, that their partial *derivatives*, so their *directional derivatives*, are linearly independent). Regardless of how this condition is expressed, however, we note that $\dot{\mathbf{q}}$ will itself be *linear* in $\mathbf{p}$ (though, again, arbitrarily complicated in the q's), by equation (10.1-2).

It's all well and good to obtain this criterion for solution, but we have to show that the condition is *met* for the Lagrangian systems of interest. To do this, we will return to the topic of *matrix diagonalization*, Section 3.7.1, not so much for the diagonalization itself, as to reduce $\mathbf{T}_2$ to a diagonalized form, $\mathbf{T}_2'$, whose determinant we can find easily. Then using the fact that this diagonalization is effected through a similarity transformation, which leaves the determinant of the matrix to which it is applied unchanged, we have that of $\mathbf{T}_2$ (which we are, recall, trying to find).

We note first that, in fact, $\mathbf{T}_2$ above *can* be diagonalized: its entries are real, and it is symmetric, so its eigenvectors will be linearly independent (Theorem 3.5.2). We must be careful about what basis we are diagonalizing with respect to, however: confusion can arise because this matrix, as defined in equation (8.1-11), was just the representation of equation (8.1-10), shown above to be the Hessian,

$$\mathbf{T}_2 \equiv (t_{2_{ij}}) \equiv \left(\sum_k^n \frac{1}{2} m_k \frac{\partial \boldsymbol{r}_k}{\partial q_i} \cdot \frac{\partial \boldsymbol{r}_k}{\partial q_j}\right) = \left(\frac{1}{2} \frac{\partial^2 T}{\partial \dot{q}_i \partial \dot{q}_j}\right),$$

and, in some sense, we've got *two* sets of vectors, $\boldsymbol{r}_i$ and $\boldsymbol{q}$ (whose basis is that of the $\dot{\boldsymbol{q}}$), in play here. But, since $\mathbf{T}_2$ is applied to the latter, it is with respect to *their* basis (the natural basis of column f-tuples) we will diagonalize.

Carrying out the procedure in Section 3.7.1, we see that, through a change of basis, represented here by the matrix $\mathbf{S}$ (so as not to confuse it with the $\mathbf{T}$'s), $\mathbf{T}_2$ and $\mathbf{q}$ assume the new representations

$$\mathbf{T}_2' = \mathbf{S}^{-1}\mathbf{T}_2\mathbf{S} \quad \text{and} \quad \mathbf{q}' = \mathbf{S}^{-1}\mathbf{q}.$$

But this represents the same $\mathbf{T}_2$ relative to the new $\mathbf{q}$ basis:

$$\mathbf{T}_2' = \left(\sum_i^n \frac{1}{2} m_i \frac{\partial \boldsymbol{r}_i}{\partial q_j'} \cdot \frac{\partial \boldsymbol{r}_i}{\partial q_k'}\right) = \left(\sum_i^n \frac{1}{2} m_i \frac{\partial \boldsymbol{r}_i}{\partial q_j'} \cdot \frac{\partial \boldsymbol{r}_i}{\partial q_j'}\right) = \left(\sum_i^n \frac{1}{2} m_i \left|\frac{\partial \boldsymbol{r}_i}{\partial q_j'}\right|^2\right)$$

—since the matrix *is* diagonalized; thus each of its diagonal terms is *non-negative*, and there must be at least *one* non-0 $\frac{\partial \mathbf{r}_i}{\partial q'_j}$, or the $\mathbf{r}_i$ don't depend on the q's at all! Thus, since the determinant of a diagonalized matrix is simply the product of its diagonal terms, we have that

$$0 < |\mathsf{T}_2'| = |\mathsf{S}^{-1}||\mathsf{T}_2||\mathsf{S}| = |\mathsf{T}_2|,$$

by Theorems 3.2.8 and 3.2.6. Thus $|\mathsf{T}_2| \neq 0$, and it is always possible to solve for $\dot{\boldsymbol{q}}(\boldsymbol{q}, \boldsymbol{p}; t)$.

Before leaving this topic, there is one interesting comparison to be made regarding the variables $\boldsymbol{q}$ and $\dot{\boldsymbol{q}}$ in the Lagrangian formulation, and those $\boldsymbol{q}$ and $\boldsymbol{p}$ in the Hamiltonian one: In the former, selection of $\boldsymbol{q}$ immediately determines the $\dot{\boldsymbol{q}}$; in effect, specification of the Lagrangian is determined completely by choice of the $\boldsymbol{q}$. In the Hamiltonian systems, on the other hand, $\boldsymbol{p}$ is defined as the partial derivatives of $L(\boldsymbol{q}, \dot{\boldsymbol{q}}; t)$, determined not only by $\boldsymbol{q}$, *but also by the system itself*! Conceptually, $\boldsymbol{p}$ is linked to the *system*. To the extent that, in any Lagrangians of physical interest, $\dot{\boldsymbol{q}}$ enters only through the kinetic energy, and potential energy depends purely on the *coordinates* chosen—*i.e.*, $\boldsymbol{q}$—this distinction is moot, but it interesting nonetheless.

Summary. The *generalized momenta* $p_i = p_i(\boldsymbol{q}, \dot{\boldsymbol{q}}; t) \equiv \frac{\partial L}{\partial \dot{q}_i}$ are variables fundamental to the Hamiltonian formulation of dynamics, which utilizes the $2f$ variables $\boldsymbol{q}$ and $\boldsymbol{p}$. We can always invert this relation to find $\dot{\boldsymbol{q}}(\boldsymbol{q}, \boldsymbol{p}; t)$.

10.2 The Equations of Motion

Just as in obtaining the Euler-Lagrange and Lagrange equations, we will assume, at least initially, that we have eliminated all constraints explicitly from the start; *i.e.*, that the system is *holonomic*. (We shall consider a technique analogous to that for non-holonomic constraints in Lagrangian systems in Section 10.8.) Having obtained $\dot{\boldsymbol{q}}(\boldsymbol{q}, \boldsymbol{p}; t)$, we now form, from the Lagrangian, the *Hamiltonian*:

$$H(\boldsymbol{q}, \boldsymbol{p}; t) \equiv \boldsymbol{p} \cdot \dot{\boldsymbol{q}}(\boldsymbol{q}, \boldsymbol{p}; t) - L(\boldsymbol{q}, \dot{\boldsymbol{q}}(\boldsymbol{q}, \boldsymbol{p}; t); t). \tag{10.2-1}$$

This might seem somewhat arbitrary, but in point of fact, we have already encountered this function: the first term in this definition of H is just

$$\boldsymbol{p} \cdot \dot{\boldsymbol{q}} = \sum_j^f p_j \dot{q}_j \equiv \sum_j^f \frac{\partial L}{\partial \dot{q}_j} \dot{q}_j;$$

thus, comparing the definition of the Hamiltonian with equation (9.2-3), we see that *the Hamiltonian is just Jacobi's function*, merely expressed in terms of $\boldsymbol{q}$, $\boldsymbol{p}$, and t, rather than $\boldsymbol{q}$, $\dot{\boldsymbol{q}}$ and t! In particular, then, under the conditions that this function be an integral of the motion—namely, that all active forces be conservative, and that $\frac{\partial L}{\partial t} \equiv 0$ (so $\frac{\partial H}{\partial t} \equiv 0$)—the Hamiltonian itself will be a constant; if constraints are scleronomic, H is nothing more than the energy of the system (Theorem 9.2.2).

We will now obtain the equations of motion for this Hamiltonian formulation. Differentiating H with respect to each of its $2f+1$ arguments:

$$\frac{\partial H}{\partial q_i} = \sum_j^f p_j \frac{\partial \dot{q}_j}{\partial q_i} - \frac{\partial L}{\partial q_i} - \sum_j^f \frac{\partial L}{\partial \dot{q}_j}\frac{\partial \dot{q}_j}{\partial q_i} = -\frac{\partial L}{\partial q_i}$$

$$\frac{\partial H}{\partial p_i} = \dot{q}_i + \sum_j^f p_j \frac{\partial \dot{q}_j}{\partial p_i} - \sum_j^f \frac{\partial L}{\partial \dot{q}_j}\frac{\partial \dot{q}_j}{\partial p_i} = \dot{q}_i$$

$$\frac{\partial H}{\partial t} = \sum_j^f p_j \frac{\partial \dot{q}_j}{\partial t} - \sum_j^f \frac{\partial L}{\partial \dot{q}_j}\frac{\partial \dot{q}_j}{\partial t} - \frac{\partial L}{\partial t} = -\frac{\partial L}{\partial t}$$

—where the final simplification in each case results from the definition of the $p_j \equiv \frac{\partial L}{\partial \dot{q}_j}$. But, in the first equation, the [generalized] Lagrange equations tell us that

$$-\frac{\partial L}{\partial q_i} = Q_i - \frac{d}{dt}\left(\frac{\partial L}{\partial \dot{q}_i}\right) \equiv Q_i - \frac{d}{dt}p_i$$

Thus, we get finally the equations of motion with a slight reordering,

$$\dot{q}_i = \frac{\partial H}{\partial p_i} \qquad (10.2\text{-}2a)$$

$$\dot{p}_i = -\frac{\partial H}{\partial q_i} + Q_i \qquad (10.2\text{-}2b)$$

$$\frac{\partial H}{\partial t} = -\frac{\partial L}{\partial t} \qquad (10.2\text{-}2c)$$

(the last being simply a restatement of the time dependence of the Jacobi function).

We see that, rather than getting f *second*-order differential equations in the q_i as in the Lagrangian formulation, we get $2f$ *first*-order equations in each the q_i and p_i. Actually, this fact is not really all that remarkable by itself: employing the standard technique of defining a new variable

$$\mathbf{x} \equiv \begin{pmatrix} \dot{\mathbf{q}} \\ \mathbf{q} \end{pmatrix},$$

we can formally represent even the second-order Lagrangian equations as a set of *first*-order ones:

$$\dot{\mathbf{x}} = \begin{pmatrix} \ddot{\mathbf{q}} \\ \dot{\mathbf{q}} \end{pmatrix} = \mathbf{f}(\mathbf{x})$$

—where the top half of $\mathbf{f}$ is obtained from the Lagrange equations (expressed in terms of the components of $\mathbf{x}$) and the bottom is just the top half of $\mathbf{x}$ itself.

What *is* notable, however, is that each equation of motion depends on the partial derivative with respect to its "conjugate" variable—on the *form* of the

Hamiltonian—just as occurred in the Lagrangian formulation, but now holding with respect to *both* sets of variables. That is a significant observation, as we shall see in Section 10.3.

Clearly the Hamiltonian depends on the Lagrangian—it's *defined* in *terms* of the Lagrangian! But in the introduction to this section, it was mentioned that there was a "reflexive relationship" between these two functions. We briefly examine this.

10.2.1 Legendre Transformations

Rather than talking about either Lagrangians or Hamiltonians, consider for the time being an *arbitrary* scalar function $F(\boldsymbol{u}, \boldsymbol{x})$ of the $2n$ variables $u_1, u_2, \ldots, u_n$ and $x_1, x_2, \ldots, x_n$. Now introduce two new sets of variables $v_1, v_2, \ldots, v_n$ and $y_1, y_2, \ldots, y_n$ defined by

$$v_i = v_i(\boldsymbol{u}, \boldsymbol{x}) \equiv \frac{\partial F}{\partial u_i}, \quad y_i = y_i(\boldsymbol{u}, \boldsymbol{x}) \equiv \frac{\partial F}{\partial x_i}; \tag{10.2-3}$$

subject to the condition that the Hessian of F,

$$\left| \frac{\partial^2 F}{\partial u_i \partial u_j} \right| \neq 0, \tag{10.2-4}$$

the resulting v_i will be linearly independent, and the inversion giving $\boldsymbol{u}(\boldsymbol{v}, \boldsymbol{x})$ can be carried out.

Now consider a function

$$G(\boldsymbol{v}, \boldsymbol{x}) \equiv \boldsymbol{u}(\boldsymbol{v}, \boldsymbol{x}) \cdot \boldsymbol{v} - F(\boldsymbol{u}(\boldsymbol{v}, \boldsymbol{x}), \boldsymbol{x}). \tag{10.2-5}$$

By direct calculation,

$$\begin{aligned}\frac{\partial G}{\partial v_i} &= \sum_j^n \frac{\partial u_j}{\partial v_i} v_j + u_i - \sum_j^n \frac{\partial F}{\partial u_j} \frac{\partial u_j}{\partial v_i} \\ &\equiv \sum_j^n \frac{\partial u_j}{\partial v_i} v_j + u_i - \sum_j^n v_j \frac{\partial u_j}{\partial v_i}\end{aligned}$$

and

$$\begin{aligned}\frac{\partial G}{\partial x_i} &= \sum_j^n \frac{\partial u_j}{\partial x_i} v_j - \sum_j^n \frac{\partial F}{\partial u_j} \frac{\partial u_j}{\partial x_i} - \frac{\partial F}{\partial x_i} \\ &\equiv \sum_j^n \frac{\partial u_j}{\partial x_i} v_j - \sum_j^n v_j \frac{\partial u_j}{\partial x_i} - \frac{\partial F}{\partial x_i},\end{aligned}$$

or

$$\frac{\partial G}{\partial v_i} = u_i, \quad \frac{\partial G}{\partial x_i} = -y_i. \tag{10.2-6}$$

The symmetry in equations (10.2-3) and (10.2-6), as well as the fact that (10.2-5) is itself symmetric in form:

$$F(\boldsymbol{u}, \boldsymbol{x}) = \boldsymbol{u} \cdot \boldsymbol{v}(\boldsymbol{u}, \boldsymbol{x}) - G(\boldsymbol{v}(\boldsymbol{u}, \boldsymbol{x}))$$

exhibit an almost incestuous relation—a *duality*—between the functions and the variables in terms of which they are expressed. We note the parallels in form between the above and the definitions of $\boldsymbol{p}$ and H; we return to this shortly.

Examples of such relations, other than the above transformation from Lagrangian to Hamiltonian, are not so far-fetched as might initially be thought. In classical thermodynamics, the *enthalpy* $H(S,P)$ (not to be confused with the *Hamiltonian* H!) is defined in terms of *entropy* S and pressure P such that

$$T = \frac{\partial H(S,P)}{\partial S} = T(S,P), \quad \text{and} \quad V = \frac{\partial H(S,P)}{\partial P} = V(S,P),$$

where T (not to be confused with *kinetic energy* T) and V (not to be confused with *potential* energy V) are temperature and volume, respectively. Again, we assume we can invert T to get $S = S(P,T)$. The *Gibbs free energy* (not to be confused with G above!)

$$G(P,T) \equiv H(S(P,T),P) - TS(P,T)$$

so that

$$S(P,T) = -\frac{\partial G(P,T)}{\partial T} \quad \text{and} \quad V(P,T) = \frac{\partial G(P,T)}{\partial P},$$

(V playing the role of the parameter "x" above) is the [sign-modified] dual function, as can be verified directly.

What does this have to do with Lagrangians and Hamiltonians? The situation is somewhat muddied here due to the presence of additional variables, but in point of fact, the transformation between the Lagrangian and Hamiltonian is just a Legendre transformation *insofar as the $\dot{q}_i$ and p_i are concerned.* In order to dramatize this, we will briefly notationally suppress the variables $\boldsymbol{q}$ and t by merely keeping their places; then the above four equations become

$$(10.2\text{-}3) \rightarrow (9.3\text{-}2): \quad p_i(\_, \dot{\boldsymbol{q}}; \_) \equiv \frac{\partial L}{\partial \dot{q}_i}$$

$$(10.2\text{-}4) \rightarrow (10.1\text{-}3): \quad \left| \frac{\partial^2 L}{\partial \dot{q}_i \partial \dot{q}_j} \right| \neq 0$$

$$(10.2\text{-}5) \rightarrow (10.2\text{-}1): \quad H(\_, \boldsymbol{p}; \_) \equiv \boldsymbol{p} \cdot \dot{\boldsymbol{q}}(\_, \boldsymbol{p}; \_) - L(\_, \dot{\boldsymbol{q}}(\_, \boldsymbol{p}; \_); \_)$$

$$(10.2\text{-}6) \rightarrow (10.2\text{-}2a): \quad \dot{q}_i = \frac{\partial H}{\partial p_i}.$$

In particular, then, the last equation above, (10.2-2a) in the set of Hamiltonian equations of motion, is merely a result of the Legendre transformation from L to H. *There is no dynamics involved*—L could be anything!—and the fact that we get $\dot{\boldsymbol{q}}$ is the result of having chosen to use that variable as one of the Legendre ones, nothing deeper. This set is almost tautological in character, with $\boldsymbol{q}$ and t playing the role of "free parameters". Equation (10.2-2b) for $\dot{\boldsymbol{p}}$, on the other hand, *is* an equation of motion, obtained *independently* of the other one; here $\boldsymbol{p}$ *does* play an important role, being a variable used in favor of, but to account for the effects of, $\dot{\boldsymbol{q}}$ in the Lagrangian. Equation (10.2-2c), still different, is just a result of the calculus. But since the first of these equations is obtained independently of the second, *in Hamiltonian dynamics,* $\boldsymbol{q}$ *and* $\boldsymbol{p}$ *are treated as independent variables.* The Hamilton equations (10.2-2) are truly a set of $2f+1$ first-order equations in $2f+1$ *independent* variables $\boldsymbol{q}$, $\boldsymbol{p}$, and t.

We remark that this duality goes in the other direction, too: defining the Legendre "variable" $\dot{\boldsymbol{q}}$:

$$\dot{q}_i(\boldsymbol{q},\boldsymbol{p};t) \equiv \frac{\partial H}{\partial p_i}, \tag{10.2-7}$$

subject to the condition that

$$\left|\frac{\partial^2 H}{\partial p_i \partial p_j}\right| \neq 0,$$

and defining

$$L(\boldsymbol{q},\dot{\boldsymbol{q}};t) \equiv \boldsymbol{p}(\boldsymbol{q},\dot{\boldsymbol{q}};t)\cdot\dot{\boldsymbol{q}} - H(\boldsymbol{q},\boldsymbol{p}(\boldsymbol{q},\dot{\boldsymbol{q}};t);t)$$

we get that

$$p_i = \frac{\partial L}{\partial \dot{q}_i}.$$

We can pass freely between the two functions and their corresponding formulations (as long as the appropriate equations of motion are observed).

10.2.2 q and p as *Lagrangian* Variables

Above we pointed out that the equations of motion for $\boldsymbol{q}$ and $\boldsymbol{p}$ were derived independently—one merely from the Legendre transformation, the other from the *Lagrangian* equations of motion—so could be regarded as "independent variables". In view of the foundation that Lagrangian dynamics provides for Hamiltonian dynamics, it is reasonable to return to that function briefly.

In reinforcing the idea of the "duality" between the Lagrangian and Hamiltonian systems, we showed how L could be obtained from H through the Legendre transformation: given $H(\boldsymbol{q},\boldsymbol{p};t)$, we could obtain the Lagrangian by "reinverting" to find $\boldsymbol{p}(\boldsymbol{q},\dot{\boldsymbol{q}};t)$ and then substituting into the Legendre transformation

$$L(\boldsymbol{q},\dot{\boldsymbol{q}};t) \equiv \boldsymbol{p}(\boldsymbol{q},\dot{\boldsymbol{q}};t)\cdot\dot{\boldsymbol{q}} - H(\boldsymbol{q},\boldsymbol{p}(\boldsymbol{q},\dot{\boldsymbol{q}};t);t).$$

But, for sake of argument, say we *didn't* bother to solve for $\boldsymbol{p}$ in terms of the other variables—that we merely expressed the Legendre transformation in terms of $\boldsymbol{q}$, $\boldsymbol{p}$, *and* $\dot{\boldsymbol{q}}$:

$$\tilde{L}(\boldsymbol{q},\dot{\boldsymbol{q}},\boldsymbol{p};t) \equiv \boldsymbol{p}\cdot\dot{\boldsymbol{q}} - H(\boldsymbol{q},\boldsymbol{p};t). \tag{10.2-8}$$

Then a couple of remarkable things happen: In the first place, $\tilde{L}$ *is independent of the value of* $\mathbf{p}$, since

$$\frac{\partial \tilde{L}}{\partial p_i} = \dot{q}_i - \frac{\partial H}{\partial p_i} \equiv 0 \qquad (10.2\text{-}9)$$

by virtue of nothing more than the Legendre transformation, (10.2-7); but this means that $\tilde{L}$ *is truly independent of* $\mathbf{p}$!

But this result leads to an equally remarkable second one: from (10.2-9), and the fact that $\dot{p}$ doesn't occur in $\tilde{L}$ at all,

$$\frac{d}{dt}\left(\frac{\partial \tilde{L}}{\partial \dot{p}_i}\right) - \frac{\partial \tilde{L}}{\partial p_i} \equiv 0; \qquad (10.2\text{-}10)$$

i.e., $\tilde{L}$ *satisfies the Lagrange equations in* $\mathbf{p}$! There's no dynamics here—this is really just $\frac{d}{dt}0 - 0 = 0$!—and *no* δp; in fact, there's not even much left of the Legendre transformation, aside from the definition of the Hamiltonian and the fact that $\dot{q}_i \equiv \frac{\partial H}{\partial p_i}$ used to obtain equation (10.2-9), and those could be viewed as independent definitions. That's all well and good for $\mathbf{p}$, but do the Lagrange equations for $\tilde{L}$ still hold for the original $\mathbf{q}$, or, in going from L to $\tilde{L}$, have we "thrown out the baby with the bathwater"? In point of fact, they *do*: going back to the definition of $\tilde{L}$, equation (10.2-8),

$$\frac{\partial \tilde{L}}{\partial \dot{q}_i} = p_i \equiv \frac{\partial L}{\partial \dot{q}_i}$$

$$\frac{\partial \tilde{L}}{\partial q_i} = -\frac{\partial H}{\partial q_i} = +\frac{\partial L}{\partial q_i}$$

(the last equation resulting from the definition of H itself), so that

$$\frac{d}{dt}\left(\frac{\partial \tilde{L}}{\partial \dot{q}_i}\right) - \frac{\partial \tilde{L}}{\partial q_i} = \frac{d}{dt}\left(\frac{\partial L}{\partial \dot{q}_i}\right) - \frac{\partial L}{\partial q_i} = Q_i.$$

Thus, in fact, $\tilde{L}$ *satisfies the Lagrange equations in* both $\mathbf{q}$ *and* $\mathbf{p}$!

And *this* result provides an even more significant one: Now returning to the definition of $\tilde{L}$ and applying to it the Lagrange equations in *both* $\mathbf{q}$ *and* $\mathbf{p}$, we obtain the $2f$ equations

$$\frac{d}{dt}\left(\frac{\partial \tilde{L}}{\partial \dot{q}_i}\right) - \frac{\partial \tilde{L}}{\partial q_i} = \frac{d}{dt}(p_i) - \left(-\frac{\partial H}{\partial q_i}\right) = Q_i$$

$$\frac{d}{dt}\left(\frac{\partial \tilde{L}}{\partial \dot{p}_i}\right) - \frac{\partial \tilde{L}}{\partial p_i} = \frac{d}{dt}(0) - \left(\dot{q}_i - \frac{\partial H}{\partial p_i}\right) = 0$$

—precisely the Hamiltonian equations of motion (10.2-2a and b). (The last of those, equation (10.2-2c), follows as before from the Jacobi function results.) Note that there are no generalized forces associated with the p_i.

10.2.3 An Important Property of the Hamiltonian

When the Lagrangian was first introduced in Section 8.2, we presented two theorems: Theorem 8.2.1 said that an arbitrary time derivative of the coordinates [only] could be added to the Lagrangian without affecting its equations of motion; the other, Theorem 8.2.2, said these equations were invariant under a coordinate transformation. We shall obtain similar results in this Chapter, but the latter warrants its own section (10.4), so we restrict our attention here to an analog to the former.

In Section 10.2.2, we showed that if we didn't explicitly eliminate $\boldsymbol{p}$ in going from the Hamiltonian back to its dual Lagrangian with a Legendre transformation, we could obtain the Hamiltonian equations of motion by applying the Lagrange equations in *both* $\boldsymbol{q}$ *and* $\boldsymbol{p}$ to that Lagrangian we called there $\tilde{L}(\boldsymbol{q},\dot{\boldsymbol{q}},\boldsymbol{p};t)$. Just as $\dot{W}(\boldsymbol{q};t)$ satisfies the Lagrange equations [in $\boldsymbol{q}$] *identically*, a $\dot{W}(\boldsymbol{q},\boldsymbol{p};t)$ will satisfy the Lagrange equations *in* $\boldsymbol{q}$ *and* $\boldsymbol{p}$ identically. But this means that, in exactly the same way that $\dot{W}(\boldsymbol{q};t)$ could be added onto L without changing its equations of motion in $\boldsymbol{q}$, $\dot{W}(\boldsymbol{q},\boldsymbol{p};t)$ can be added onto $\tilde{L}$ without changing *its* equations of motion *in* $\boldsymbol{q}$ *and* $\boldsymbol{p}$. This result is formalized in the following:

Theorem 10.2.1. Given a set of coordinates $\{q_1,q_2,\ldots,q_f\}$ and their conjugate momenta $\{p_1,p_2,\ldots,p_f\}$, and an arbitrary $W=W(\boldsymbol{q},\boldsymbol{p};t)$, the equations of motion *in these two sets of variables* for a Lagrangian $\tilde{L}(\boldsymbol{q},\dot{\boldsymbol{q}},\boldsymbol{p};t)$ and another Lagrangian $\tilde{L}'(\boldsymbol{q},\dot{\boldsymbol{q}},\boldsymbol{p};t)\equiv\tilde{L}(\boldsymbol{q},\dot{\boldsymbol{q}},\boldsymbol{p};t)+\frac{dW(\boldsymbol{q},\boldsymbol{p};t)}{dt}$ are the same.

The reason this result is *notable* is the observation made with regard to Theorem 8.2.1: that W there *could not depend on* $\dot{\boldsymbol{q}}$! Yet in Hamiltonian systems, $\boldsymbol{p}=\boldsymbol{p}(\boldsymbol{q},\dot{\boldsymbol{q}};t)$, so any $W(\boldsymbol{q},\boldsymbol{p};t)$ will generally introduce [implicitly] not only $\dot{\boldsymbol{q}}$, but also $\ddot{\boldsymbol{q}}$ into the expression for $\dot{W}$.

The reason this result is *significant* is the fact that the *Hamiltonian* equations of motion result from application of the *Lagrange* equations of motion (in both $\boldsymbol{q}$ and $\boldsymbol{p}$) to $\tilde{L}$. The Hamiltonian equations, in turn, apply to the *original* H, *not* to $H+\dot{W}$. We demonstrate what this theorem says (and what it *doesn't*!) with a simple example:

Example 10.2.1. We consider a one-degree-of-freedom system $H(q,p;t)$. Take $W(q,p)\equiv pq$; then $\dot{W}=p\dot{q}+\dot{p}q$, so $\tilde{L}=p\dot{q}-H(q,p;t)$, and $\tilde{L}'\equiv\tilde{L}+\dot{W}=2p\dot{q}+q\dot{p}-H$. Then the *Lagrangian* equations become

$$\frac{d}{dt}\Big(\frac{\partial\tilde{L}'}{\partial\dot{q}}\Big)-\frac{\partial\tilde{L}'}{\partial q}=\frac{d}{dt}(2p)-(\dot{p}-\frac{\partial H}{\partial q})=\dot{p}+\frac{\partial H}{\partial q}=0$$
$$\frac{d}{dt}\Big(\frac{\partial\tilde{L}'}{\partial\dot{p}}\Big)-\frac{\partial\tilde{L}'}{\partial p}=-\frac{d}{dt}(q)-(2\dot{q}-\frac{\partial H}{\partial p})=-\dot{q}+\frac{\partial H}{\partial p}=0,$$

exactly as we would get with the Hamilton equations for H. Note, however, that had we tried to apply the *Hamiltonian* equations to $H'=H+\dot{W}=H+p\dot{q}+\dot{p}q$, we would get

$$\dot{q} = \frac{\partial H'}{\partial p} = \frac{\partial H}{\partial p} + \dot{q}$$
$$\dot{p} = -\frac{\partial H'}{\partial q} = -\frac{\partial H}{\partial q} - \dot{p},$$

which are very peculiar, indeed! |

We end this section with an example:

Example of Hamiltonian Dynamics

To show the correspondence between these two formulations, we consider an example worked previously in Lagrangian dynamics, Example 8.1.3, now using the Hamiltonian approach. This is one involving a non-conservative force, friction, to which Hamiltonian's equations are *not* generally applied, for reasons to be cited later. But it is introduced here just to show a) that Hamiltonian *can* be used in such problems, and b) that the results are the same as obtained with the Lagrangian approach (which, in turn, jibes with the classical one).

Example 10.2.2 *Hamiltonian Dynamics—Non-conservative system.* To review the problem briefly: A particle of mass m is attached to a linear spring with constant k and unextended length r_o, whose other end is fixed. The mass moves on a horizontal table with which it has friction, given by a kinetic friction coefficient of μ_k. We take, as generalized coordinates in terms of which to formulate the Lagrangian (required by the above discussion to express the Hamiltonian) the same variables r and θ we did previously. Unlike that example, however, meant to exemplify the Euler-Lagrange Equations, we will now introduce the potential energy to account for the spring (though the friction *must* be left in generalized force terms).

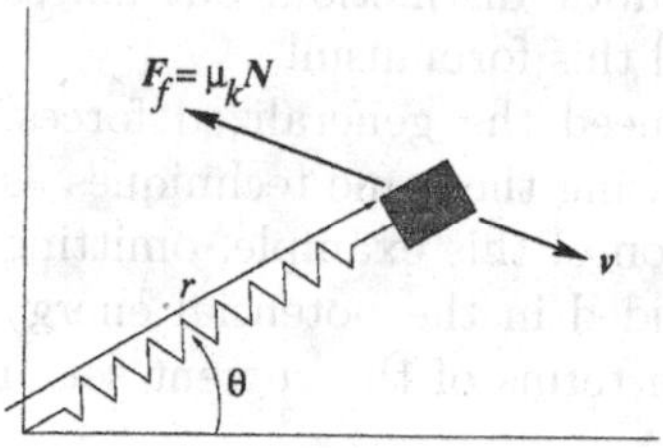

Our first task is to define the Hamiltonian: The Lagrangian should, by this time, have become almost routine:

$$L = T - V = \frac{1}{2}m(\dot{r}^2 + (r\dot{\theta})^2) - \frac{1}{2}k(r - r_o)^2.$$

This defines the generalized momenta

$$p_r \equiv \frac{\partial L}{\partial \dot{r}} = m\dot{r} \quad \text{and} \quad p_\theta \equiv \frac{\partial L}{\partial \dot{\theta}} = mr^2\dot{\theta}$$

(linear, as expected, in the $\dot{q}_i$) which can easily be inverted to express the $\dot{\boldsymbol{q}}$ in terms of the $\boldsymbol{p}$:

$$\dot{r} = \frac{p_r}{m} \quad \text{and} \quad \dot{\theta} = \frac{p_\theta}{mr^2};$$

thus, finally, the Hamiltonian is found by replacing the $\dot{\boldsymbol{q}}$ by these expressions in terms of $\boldsymbol{p}$ in

$$H \equiv \boldsymbol{p} \cdot \dot{\boldsymbol{q}} - L = \frac{p_r^2}{m} + \frac{p_\theta^2}{mr^2} - \frac{1}{2}m\left(\left(\frac{p_r}{m}\right)^2 + r^2\left(\frac{p_\theta}{mr^2}\right)^2\right) + \frac{1}{2}k(r - r_o)^2$$

$$= \frac{1}{2}\left(\frac{p_r^2}{m} + \frac{p_\theta^2}{mr^2}\right) + \frac{1}{2}k(r - r_o)^2. \tag{10.2-11}$$

Observe that, in the absence of friction, this *would* be the time-independent Hamiltonian for a *conservative* system with scleronomic constraints—the Jacobi function equal to the energy, but expressed in terms of $\boldsymbol{q}$ and $\boldsymbol{p}$.

We now take the equations of motion: For the q's,

$$\dot{r} = \frac{\partial H}{\partial p_r} = \frac{p_r}{m} \tag{10.2-12a}$$

$$\dot{\theta} = \frac{\partial H}{\partial p_\theta} = \frac{p_\theta}{mr^2} \tag{10.2-12b}$$

—simply the relation between the $\dot{\boldsymbol{q}}$ and $\boldsymbol{p}$, just as expected from the Legendre duality; nothing new here. But, in the present context, they must be *interpreted* as being a relation, not between the original Lagrangian $\dot{\boldsymbol{q}}$ and the new Hamiltonian $\boldsymbol{p}$ (which they are, after all), but between the Hamiltonian $\boldsymbol{p}$ and time derivatives of the *Hamiltonian* $\boldsymbol{q}$. That's a subtle distinction, but unless it's understood, one feels he's chasing his tail in all this formalism!

For the equations of motion for the p's, we need the generalized forces. These would normally have to be found *de nuvo*, using the same techniques as for Lagrangians. But, from the previous incantation of this example, omitting the terms due to the spring (which are now included in the potential energy in the Lagrangian/Hamiltonian) and expressing in terms of the current set of variables, they are

$$Q_r = -\frac{\mu_k m g \dot{r}}{v} = -\frac{\mu_k g p_r}{v}$$

and

$$Q_\theta = -\frac{\mu_k m g r^2 \dot{\theta}}{v} = -\frac{\mu_k g p_\theta}{v},$$

where, recall, v is the speed, here expressed in $\boldsymbol{q}$ and $\boldsymbol{p}$:

$$v = \sqrt{\left(\frac{p_r}{m}\right)^2 + \left(\frac{p_\theta}{mr}\right)^2}.$$

Thus

$$\dot{p}_r = -\frac{\partial H}{\partial r} + Q_r = +\frac{p_\theta^2}{mr^3} - k(r - r_o) - \frac{\mu_k g p_r}{v} \tag{10.2-13a}$$

$$\dot{p}_\theta = -\frac{\partial H}{\partial \theta} + Q_\theta = 0 - \frac{\mu_k g p_\theta}{v}. \tag{10.2-13b}$$

Right! We've got these first-order equations in the four variables r, θ, p_r, and p_θ. How do we begin to get a solution? Equations (10.2-12) give a relation between the Hamiltonian p's and the time derivatives of the [*Hamiltonian*!] q's. Differentiating them,

$$\dot{p}_r = m\ddot{r} \quad \text{and} \quad \dot{p}_\theta = m(2r\dot{r}\dot{\theta} + r^2\ddot{\theta})$$

and substituting both these and the $\boldsymbol{p}(\dot{\boldsymbol{q}})$ from (10.2-12) into equations (10.2-13), returns a pair of *second*-order differential equations for the [Hamiltonian] q's:

$$\begin{aligned} m\ddot{r} &= \frac{(mr^2\dot{\theta})^2}{mr^3} - k(r - r_o) - \frac{\mu_k g(m\dot{r})}{v} \\ &= mr\dot{\theta}^2 - k(r - r_o) - \frac{\mu_k m g\dot{r}}{v} \end{aligned}$$

and

$$m(2r\dot{r}\dot{\theta} + r^2\ddot{\theta}) = -\frac{\mu_k g m r^2\dot{\theta}}{v}.$$

These are precisely the equations obtained in the Euler-Lagrange treatment of this problem, Example 8.1.3. |

Summary. The Hamiltonian equations of motion, equation (10.2-2), are a set of $2f$ first-order differential equations for the $\boldsymbol{q}$ and $\boldsymbol{p}$. They result from two independent lines: the *Legendre transformation*, in the variables $\dot{\boldsymbol{q}}$ and $\boldsymbol{p}$, from the Lagrangian to its "dual" Hamiltonian, plus the Lagrangian equations of motion. In this sense, we regard the "conjugate" variables $\boldsymbol{q}$ and $\boldsymbol{p}$, in terms of which the Hamiltonian is defined, to be independent insofar as the equations of motion are concerned. But this independence is more than merely technical: in fact, the Lagrangian associated with a Hamiltonian, expressed in terms of $\boldsymbol{q}$, $\boldsymbol{p}$, t *and* $\dot{\boldsymbol{q}}$, is *insensitive* to the value of $\boldsymbol{p}$. Thus it turns out that this form of the Lagrangian formally satisfies the Lagrangian equations of motion in *both* $\boldsymbol{q}$ and $\boldsymbol{p}$, resulting in the *Hamiltonian* equations. In turn, it also allows the time derivative of a function in *both* sets of variables to be added onto this Lagrangian without affecting its *Lagrangian* equations of motion [in both variables], and thus the *Hamiltonian* ones.

10.3 Integrals of the Motion

Integrals of the motion were introduced in the previous chapter, where we also saw that "ignorable coordinates" signalled the existence of such integrals in *conservative* Lagrangian systems. In particular, if a given coordinate were absent from the Lagrangian—if the partial derivative of L with respect to that coordinate vanished—its corresponding "generalized momentum" was constant—*i.e.*, an integral. But the presence or absence of coordinates in the Lagrangian depended on which coordinates were chosen to express it: if we could find a set of coordinates whose Lagrangian for a system was free of *all* coordinates, then we would immediately have f of the $2f$ integrals required to obtain a complete solution [implicitly]. Though we showed that the Lagrangian equations of motion were invariant under a coordinate transformation from $\boldsymbol{q}$ to $\boldsymbol{q}(\boldsymbol{Q})$, unfortunately, as was noted there, there is no practical systematic means of discovering the appropriate transformation to such sets of coordinates.

We now apply the same type of argument to *conservative* Hamiltonian systems. Like Lagrangian ones, the Hamiltonian equations of motion depend on the presence or absence of the variables. In fact, although it was not pointed out then, if there were no friction in Example 10.2.2—*i.e.*, $\mu_k \equiv 0$ in that problem—the Hamiltonian (10.2-11) would be independent of θ; thus we could conclude immediately that $p_\theta = mr^2\dot{\theta}$ is a constant of the motion. (Of course it *isn't* in the problem as stated: the generalized force Q_θ arising from the friction invalidates any such result.) But, unlike Lagrangians, where only the time derivatives of the *momenta* depended on the coordinates in the problem, *in Hamiltonian dynamics, the equations of motion for* all *variables depend on their conjugate variables being in the Hamiltonian function.* In Lagrangian dynamics, only the absence of a *coordinate* can signal the existence of an integral; in Hamiltonian dynamics, the absence of *either* a coordinate *or* a momentum implies its conjugate is an integral of the motion.

Thus, in the same way that we can at least imagine a set of coordinates in which the Lagrangian is half-solved, we can consider the possibility that there might be a set of coordinates *and their conjugate momenta* which are totally absent from the Hamiltonian—that the Hamiltonian is *constant*:

$$\dot{q}_i = \frac{\partial H}{\partial p_i} \equiv 0 \Rightarrow q_i = Q_{io}, \text{ constant}$$
$$\dot{p}_i = -\frac{\partial H}{\partial q_i} \equiv 0 \Rightarrow p_i = P_{io}, \text{ constant;}$$

in this case *all* $2f$ *integrals of the motion would be available*, again, not through explicit solution of the differential equations, but only by virtue of the coordinates (and associated momenta) in terms of which the Hamiltonian is expressed. Actually, we don't even have to aim for an ideal form $H \equiv$ *constant*: if H depends on only *one* set of conjugate variables, $\boldsymbol{q}$, say, the Hamiltonian equations of motion tell us that

$$\dot{q}_i = \frac{\partial H(\boldsymbol{q}, \_)}{\partial p_i} \equiv 0 \Rightarrow q_i = Q_{io} \text{ constant} \tag{10.3-1}$$
$$\dot{p}_i = -\frac{\partial H(\boldsymbol{q}, \_)}{\partial q_i} \Rightarrow p_i(t) = -\frac{\partial H(\boldsymbol{Q}_o, \_)}{\partial q_i} t + P_{io}; \tag{10.3-2}$$

i.e., the p_i are linear in the time, *since the partial derivatives in* $p_i(t)$ *are evaluated at constant values.* In either event, *one focuses on the* form *of the Hamiltonian to solve the problem.*

From this discussion, we see why we are continually restricting our Hamiltonians to those of *conservative* systems: like Lagrangian systems, any generalized force in the equations of motion (10.2-2) could certainly be expressed in terms of the new variables, but these would be *added* to the partial derivatives we are counting on to give us the integrals, and the above arguments, relying on merely the presence or absence of variables in the Hamiltonian itself, would no longer be valid.

The reader may have noticed that the concept of "transformation" has been stressed in this discussion. Unless one is *very* lucky, it is unlikely that the first set of coordinates selected to express the Hamiltonian will yield one of the desired [constant] form. Thus generally one starts with a particular set of coordinates in which the Lagrangian/Hamiltonian is *not* of "ideal form" and *transforms* to a set in which it is. With the Lagrangian, whose form depends only on the generalized coordinates, this is straightforward: *any* transformation to new coordinates—which implicitly transforms the generalized velocities on which the Lagrangian also depends—will leave the form of the Lagrangian equations of motion unchanged (Theorem 8.2.2), so the form of the transformed Lagrangian is all we need. Of course, as has been mentioned repeatedly, there is no general means of uncovering the desired transformation. And, even if it's *found*, we still have only f integrals—half the number we need.

With Hamiltonians, if such a transformation can be found, *all* $2f$ integrals are manifest immediately. And it turns out that there *is* a systematic means of ascertaining it. On the other hand, the situation is rather more convoluted: Here one has *two* sets of variables, coordinates and momenta, which must be transformed; the equations of motion depend on *both*, and it is not at all obvious that an *arbitrary* transformation of these variables will leave the *form* of the equations of motion invariant. And it is maintenance of the form of the *equations of motion* on which the desired form of the *Hamiltonian* is predicated. In point of fact, the form of transformation to new variables in which the Hamiltonian equations of motion preserve their form is *not* arbitrary; this will be discussed in the next section.

Summary. As with Lagrangians, the *form* of the Hamiltonian function in *conservative* systems enables us to uncover integrals of the motion. Unlike the former, however, "ignorable" coordinates *or* momenta will each correspond to integrals. In particular, in that case in which the Hamiltonian is entirely free of either set of conjugate variables, or free of the both, the system is already "solved".

10.4 Canonical Transformations

If we seek to *transform* the [*conservative*] Hamiltonian to one amenable in form to one of the above trivial solutions, we must be assured that the form of the equations of motion remains unchanged under the transformation. We saw (Theorem 8.2.2) that this was true for *any* transformation $\boldsymbol{q} \rightarrow \boldsymbol{Q} : \boldsymbol{q} = \boldsymbol{q}(\boldsymbol{Q}; t)$

in Lagrangian systems; we now investigate what conditions, if any, might be necessary for the form to be maintained under a transformation in Hamiltonian ones. We will assume such transformations are of similar form $\boldsymbol{q} = \boldsymbol{q}(\boldsymbol{Q}, \boldsymbol{P})$ and $\boldsymbol{p} = \boldsymbol{p}(\boldsymbol{Q}, \boldsymbol{P})$—*i.e.*, that they are *time independent.* (It turns out that this condition can be relaxed, but forestall consideration of that until the next section.) We start with a bit of jargon to describe such transformations:

Definition 10.4.1. A transformation $(\boldsymbol{q}, \boldsymbol{p}) \rightarrow (\boldsymbol{Q}, \boldsymbol{P})$ which preserves the form of the Hamiltonian equations of motion—*i.e.*

$$\begin{aligned} \dot{q}_i &= \frac{\partial H(\boldsymbol{q}, \boldsymbol{p}; t)}{\partial p_i} \quad \text{and} \quad \dot{p}_i = -\frac{\partial H(\boldsymbol{q}, \boldsymbol{p}; t)}{\partial q_i} \\ \Rightarrow \dot{Q}_i &= \frac{\partial H^*(\boldsymbol{Q}, \boldsymbol{P}; t)}{\partial P_i} \quad \text{and} \quad \dot{P}_i = -\frac{\partial H^*(\boldsymbol{Q}, \boldsymbol{P}; t)}{\partial Q_i} \end{aligned} \tag{10.4-1}$$

in which $H^*(\boldsymbol{Q}, \boldsymbol{P}; t) \equiv H(\boldsymbol{q}(\boldsymbol{Q}, \boldsymbol{P}), \boldsymbol{p}(\boldsymbol{Q}, \boldsymbol{P}); t)$—is called *canonical.* Note that the notation "H^*" is used merely to express the difference in functional *form* from H (one doesn't just replace $\boldsymbol{q}$ and $\boldsymbol{p}$ by $\boldsymbol{Q}$ and $\boldsymbol{P}$, which would be $H(\boldsymbol{Q}, \boldsymbol{P}; t)$); they are, however, *equal* "in value".

Thus we shall find in this section the criterion that a transformation $(\boldsymbol{q}, \boldsymbol{p}) \rightarrow (\boldsymbol{Q}, \boldsymbol{P})$ be canonical.

It will turn out to be convenient to express this transformation in a matrix formulation, but this means we must express the equations of motion in such a form. We shall use a single $2f$-tuple to express the entire set of conjugate variables

$$\boldsymbol{\rho} \equiv \begin{pmatrix} \boldsymbol{q} \\ \boldsymbol{p} \end{pmatrix}$$

and denote the derivative of a scalar function with respect to an n-vector variable by

$$\frac{\partial f}{\partial \boldsymbol{x}} \equiv \begin{pmatrix} \frac{\partial f}{\partial x_1} \\ \vdots \\ \frac{\partial f}{\partial x_n} \end{pmatrix}. \tag{10.4-2}$$

In terms of these, then, we can write the Hamiltonian equations of motion in the form

$$\dot{\boldsymbol{\rho}} = \begin{pmatrix} \dot{\boldsymbol{q}} \\ \dot{\boldsymbol{p}} \end{pmatrix} = \begin{pmatrix} \frac{\partial H}{\partial \boldsymbol{p}} \\ -\frac{\partial H}{\partial \boldsymbol{q}} \end{pmatrix} = \begin{pmatrix} \mathbf{0} & \mathbf{1} \\ -\mathbf{1} & \mathbf{0} \end{pmatrix} \begin{pmatrix} \frac{\partial H}{\partial \boldsymbol{q}} \\ \frac{\partial H}{\partial \boldsymbol{p}} \end{pmatrix} \equiv \mathsf{J} \begin{pmatrix} \frac{\partial H}{\partial \boldsymbol{q}} \\ \frac{\partial H}{\partial \boldsymbol{p}} \end{pmatrix} \tag{10.4-3}$$

—in which the matrix J, consisting of $f \times f$ blocks of 0's and diagonal blocks of ± 1, is called the *symplectic matrix.* We wish to find the conditions that, under the above transformation to $\boldsymbol{Q}$ and $\boldsymbol{P}$, the new equations of motion are

$$\dot{\boldsymbol{R}} \equiv \begin{pmatrix} \dot{\boldsymbol{Q}} \\ \dot{\boldsymbol{P}} \end{pmatrix} = \mathsf{J} \begin{pmatrix} \frac{\partial H^*}{\partial \boldsymbol{Q}} \\ \frac{\partial H^*}{\partial \boldsymbol{P}} \end{pmatrix}. \tag{10.4-4}$$

Since we are dealing with [partial] derivatives, here, it is not surprising that we shall make use of the *Jacobian matrix* of the transformation[2]

$$\mathsf{T} \equiv \begin{pmatrix} \frac{\partial \boldsymbol{Q}}{\partial \boldsymbol{q}} & \frac{\partial \boldsymbol{Q}}{\partial \boldsymbol{p}} \\ \frac{\partial \boldsymbol{P}}{\partial \boldsymbol{q}} & \frac{\partial \boldsymbol{P}}{\partial \boldsymbol{p}} \end{pmatrix} \equiv \begin{pmatrix} \frac{\partial Q_1}{\partial q_1} & \frac{\partial Q_1}{\partial q_2} & \cdots & \frac{\partial Q_1}{\partial p_f} \\ \frac{\partial Q_2}{\partial q_1} & \frac{\partial Q_2}{\partial q_2} & \cdots & \frac{\partial Q_2}{\partial p_f} \\ \vdots & \vdots & \cdots & \vdots \\ \frac{\partial P_f}{\partial q_1} & \frac{\partial P_f}{\partial q_2} & \cdots & \frac{\partial P_f}{\partial p_f} \end{pmatrix} \tag{10.4-5}$$

[which seems to run counter to the above notation (10.4-2)—we'll see why this is chosen below], assumed to have non-vanishing determinant so that the transformation is, in fact, invertible. Using this in the expression of (10.4-3) in terms of the new variables (using block multiplication again),

$$\begin{pmatrix} \dot{\boldsymbol{Q}} \\ \dot{\boldsymbol{P}} \end{pmatrix} = \begin{pmatrix} \frac{\partial \boldsymbol{Q}}{\partial \boldsymbol{q}}\dot{\boldsymbol{q}} + \frac{\partial \boldsymbol{Q}}{\partial \boldsymbol{p}}\dot{\boldsymbol{p}} \\ \frac{\partial \boldsymbol{P}}{\partial \boldsymbol{q}}\dot{\boldsymbol{q}} + \frac{\partial \boldsymbol{P}}{\partial \boldsymbol{p}}\dot{\boldsymbol{p}} \end{pmatrix} = \begin{pmatrix} \frac{\partial \boldsymbol{Q}}{\partial \boldsymbol{q}} & \frac{\partial \boldsymbol{Q}}{\partial \boldsymbol{p}} \\ \frac{\partial \boldsymbol{P}}{\partial \boldsymbol{q}} & \frac{\partial \boldsymbol{P}}{\partial \boldsymbol{p}} \end{pmatrix} \begin{pmatrix} \dot{\boldsymbol{q}} \\ \dot{\boldsymbol{p}} \end{pmatrix} \equiv \mathsf{TJ} \begin{pmatrix} \frac{\partial H}{\partial \boldsymbol{q}} \\ \frac{\partial H}{\partial \boldsymbol{p}} \end{pmatrix}$$

(by (10.4-4)) [now we see the reason for notation (10.4-5)!]

$$= \mathsf{TJ} \begin{pmatrix} \frac{\partial H^*}{\partial \boldsymbol{Q}}\left(\frac{\partial \boldsymbol{Q}}{\partial \boldsymbol{q}}\right)^{\mathsf{T}} + \frac{\partial H^*}{\partial \boldsymbol{P}}\left(\frac{\partial \boldsymbol{P}}{\partial \boldsymbol{q}}\right)^{\mathsf{T}} \\ \frac{\partial H^*}{\partial \boldsymbol{Q}}\left(\frac{\partial \boldsymbol{Q}}{\partial \boldsymbol{p}}\right)^{\mathsf{T}} + \frac{\partial H^*}{\partial \boldsymbol{P}}\left(\frac{\partial \boldsymbol{P}}{\partial \boldsymbol{p}}\right)^{\mathsf{T}} \end{pmatrix} = \mathsf{TJ} \begin{pmatrix} \left(\frac{\partial \boldsymbol{Q}}{\partial \boldsymbol{q}}\right)^{\mathsf{T}} & \left(\frac{\partial \boldsymbol{P}}{\partial \boldsymbol{q}}\right)^{\mathsf{T}} \\ \left(\frac{\partial \boldsymbol{Q}}{\partial \boldsymbol{p}}\right)^{\mathsf{T}} & \left(\frac{\partial \boldsymbol{P}}{\partial \boldsymbol{p}}\right)^{\mathsf{T}} \end{pmatrix} \begin{pmatrix} \frac{\partial H^*}{\partial \boldsymbol{Q}} \\ \frac{\partial H^*}{\partial \boldsymbol{P}} \end{pmatrix}$$

(where the transposes enter because we are summing on the Q's and P's—the *rows* of T)

$$\equiv \mathsf{TJT}^{\mathsf{T}} \begin{pmatrix} \frac{\partial H^*}{\partial \boldsymbol{Q}} \\ \frac{\partial H^*}{\partial \boldsymbol{P}} \end{pmatrix} \equiv \mathsf{J} \begin{pmatrix} \frac{\partial H^*}{\partial \boldsymbol{Q}} \\ \frac{\partial H^*}{\partial \boldsymbol{P}} \end{pmatrix}.$$

Thus we have proven the

Theorem 10.4.1 (Symplectic Condition). A *time-independent* transformation from $(\boldsymbol{q}, \boldsymbol{p})$ to $(\boldsymbol{Q}, \boldsymbol{P})$ with Jacobian matrix J, (10.4-5), is *canonical* if and only if the *symplectic condition*, that

$$\mathsf{TJT}^{\mathsf{T}} = \mathsf{J}, \tag{10.4-6}$$

is satisfied.

[2]Note that, unlike previous transformations, the Jacobian presumes *new* variables expressed in terms of the *old*.

Actually carrying out the indicated [block] multiplications in this condition,

$$\mathsf{TJT}^\mathsf{T} = \begin{pmatrix} \frac{\partial \boldsymbol{Q}}{\partial \boldsymbol{q}} & \frac{\partial \boldsymbol{Q}}{\partial \boldsymbol{p}} \\ \frac{\partial \boldsymbol{P}}{\partial \boldsymbol{q}} & \frac{\partial \boldsymbol{P}}{\partial \boldsymbol{p}} \end{pmatrix} \begin{pmatrix} \mathbf{0} & \mathbf{1} \\ -\mathbf{1} & \mathbf{0} \end{pmatrix} \begin{pmatrix} (\frac{\partial \boldsymbol{Q}}{\partial \boldsymbol{q}})^\mathsf{T} & (\frac{\partial \boldsymbol{P}}{\partial \boldsymbol{q}})^\mathsf{T} \\ (\frac{\partial \boldsymbol{Q}}{\partial \boldsymbol{p}})^\mathsf{T} & (\frac{\partial \boldsymbol{P}}{\partial \boldsymbol{p}})^\mathsf{T} \end{pmatrix}$$

$$= \begin{pmatrix} \frac{\partial \boldsymbol{Q}}{\partial \boldsymbol{q}} & \frac{\partial \boldsymbol{Q}}{\partial \boldsymbol{p}} \\ \frac{\partial \boldsymbol{P}}{\partial \boldsymbol{q}} & \frac{\partial \boldsymbol{P}}{\partial \boldsymbol{p}} \end{pmatrix} \begin{pmatrix} (\frac{\partial \boldsymbol{Q}}{\partial \boldsymbol{p}})^\mathsf{T} & (\frac{\partial \boldsymbol{P}}{\partial \boldsymbol{p}})^\mathsf{T} \\ -(\frac{\partial \boldsymbol{Q}}{\partial \boldsymbol{q}})^\mathsf{T} & -(\frac{\partial \boldsymbol{P}}{\partial \boldsymbol{q}})^\mathsf{T} \end{pmatrix}$$

$$= \begin{pmatrix} (\frac{\partial \boldsymbol{Q}}{\partial \boldsymbol{q}})(\frac{\partial \boldsymbol{Q}}{\partial \boldsymbol{p}})^\mathsf{T} - (\frac{\partial \boldsymbol{Q}}{\partial \boldsymbol{p}})(\frac{\partial \boldsymbol{Q}}{\partial \boldsymbol{q}})^\mathsf{T} & (\frac{\partial \boldsymbol{Q}}{\partial \boldsymbol{q}})(\frac{\partial \boldsymbol{P}}{\partial \boldsymbol{p}})^\mathsf{T} - (\frac{\partial \boldsymbol{Q}}{\partial \boldsymbol{p}})(\frac{\partial \boldsymbol{P}}{\partial \boldsymbol{q}})^\mathsf{T} \\ (\frac{\partial \boldsymbol{P}}{\partial \boldsymbol{q}})(\frac{\partial \boldsymbol{Q}}{\partial \boldsymbol{p}})^\mathsf{T} - (\frac{\partial \boldsymbol{P}}{\partial \boldsymbol{p}})(\frac{\partial \boldsymbol{Q}}{\partial \boldsymbol{q}})^\mathsf{T} & (\frac{\partial \boldsymbol{P}}{\partial \boldsymbol{q}})(\frac{\partial \boldsymbol{P}}{\partial \boldsymbol{p}})^\mathsf{T} - (\frac{\partial \boldsymbol{P}}{\partial \boldsymbol{p}})(\frac{\partial \boldsymbol{P}}{\partial \boldsymbol{q}})^\mathsf{T} \end{pmatrix} = \begin{pmatrix} \mathbf{0} & \mathbf{1} \\ -\mathbf{1} & \mathbf{0} \end{pmatrix}$$

we see that the upper right and lower left blocks are just negative transposes of each other, so identical in content. There remain three different block equations:

$$(\frac{\partial \boldsymbol{Q}}{\partial \boldsymbol{q}})(\frac{\partial \boldsymbol{Q}}{\partial \boldsymbol{p}})^\mathsf{T} - (\frac{\partial \boldsymbol{Q}}{\partial \boldsymbol{p}})(\frac{\partial \boldsymbol{Q}}{\partial \boldsymbol{q}})^\mathsf{T} = \mathbf{0} \tag{10.4-7a}$$

$$(\frac{\partial \boldsymbol{Q}}{\partial \boldsymbol{q}})(\frac{\partial \boldsymbol{P}}{\partial \boldsymbol{p}})^\mathsf{T} - (\frac{\partial \boldsymbol{Q}}{\partial \boldsymbol{p}})(\frac{\partial \boldsymbol{P}}{\partial \boldsymbol{q}})^\mathsf{T} = \mathbf{1} \tag{10.4-7b}$$

$$(\frac{\partial \boldsymbol{P}}{\partial \boldsymbol{q}})(\frac{\partial \boldsymbol{P}}{\partial \boldsymbol{p}})^\mathsf{T} - (\frac{\partial \boldsymbol{P}}{\partial \boldsymbol{p}})(\frac{\partial \boldsymbol{P}}{\partial \boldsymbol{q}})^\mathsf{T} = \mathbf{0} \tag{10.4-7c}$$

—the so-called *Poisson conditions* that a transformation be canonical. Expanding these, noting that the second factor in each involves the transpose, so that summation will be on p's and q's (which we must watch!), we get a corresponding set

$$\sum_k^f \left(\frac{\partial Q_i}{\partial q_k} \frac{\partial Q_j}{\partial p_k} - \frac{\partial Q_i}{\partial p_k} \frac{\partial Q_j}{\partial q_k} \right) = 0 \tag{10.4-8a}$$

$$\sum_k^f \left(\frac{\partial Q_i}{\partial q_k} \frac{\partial P_j}{\partial p_k} - \frac{\partial Q_i}{\partial p_k} \frac{\partial P_j}{\partial q_k} \right) = \delta_{ij} \tag{10.4-8b}$$

$$\sum_k^f \left(\frac{\partial P_i}{\partial q_k} \frac{\partial P_j}{\partial p_k} - \frac{\partial P_i}{\partial p_k} \frac{\partial P_j}{\partial q_k} \right) = 0 \tag{10.4-8c}$$

—where δ_{ij} is the Kronecker delta (see equation (4.2-3)). We point out that the above equations (10.4-8) are nothing more than the $(ij)^{th}$ element in the matrix TJT^T. In fact, this form of differentiation

$$\{u, v\}_{\boldsymbol{q},\boldsymbol{p}} \equiv \sum_k^f \left(\frac{\partial u}{\partial q_k} \frac{\partial v}{\partial p_k} - \frac{\partial u}{\partial p_k} \frac{\partial v}{\partial q_k} \right)$$

is called the *Poisson bracket* of $u(\boldsymbol{q}, \boldsymbol{p})$ and $v(\boldsymbol{q}, \boldsymbol{p})$. Aside from mere name-calling, however, this operation actually is quite useful: $\{u, H\} = \frac{du}{dt}$! Thus the determination of integrals is equivalent to finding functions whose Poisson brackets with H vanish (though we will not pursue this approach here).

There is an equivalent form for the symplectic condition which will prove useful below. If we take the transpose of equation (10.4-6), and using the fact (from homework 2 on page 502) that $\mathsf{J}^{\mathsf{T}} = -\mathsf{J}$,

$$\mathsf{T}^{\mathsf{T}}\mathsf{J}\mathsf{T} = \mathsf{J}.$$

Expanding this as we did leading up to equations (10.4-7), we get now, corresponding to those equations, the set

$$\begin{aligned}
(\frac{\partial \boldsymbol{Q}}{\partial \boldsymbol{q}})^{\mathsf{T}}(\frac{\partial \boldsymbol{P}}{\partial \boldsymbol{q}}) - (\frac{\partial \boldsymbol{P}}{\partial \boldsymbol{q}})^{\mathsf{T}}(\frac{\partial \boldsymbol{Q}}{\partial \boldsymbol{q}}) &= \mathbf{0}\\
(\frac{\partial \boldsymbol{Q}}{\partial \boldsymbol{q}})^{\mathsf{T}}(\frac{\partial \boldsymbol{P}}{\partial \boldsymbol{p}}) - (\frac{\partial \boldsymbol{P}}{\partial \boldsymbol{q}})^{\mathsf{T}}(\frac{\partial \boldsymbol{Q}}{\partial \boldsymbol{p}}) &= \mathbf{I}\\
(\frac{\partial \boldsymbol{Q}}{\partial \boldsymbol{p}})^{\mathsf{T}}(\frac{\partial \boldsymbol{P}}{\partial \boldsymbol{p}}) - (\frac{\partial \boldsymbol{P}}{\partial \boldsymbol{p}})^{\mathsf{T}}(\frac{\partial \boldsymbol{Q}}{\partial \boldsymbol{p}}) &= \mathbf{0}.
\end{aligned}$$

Now the expansions corresponding to (10.4-8), due to transposes on the *first* factors, and the resultant summation on the Q's and P's, become

$$\sum_k^f \left(\frac{\partial Q_k}{\partial q_i}\frac{\partial P_k}{\partial q_j} - \frac{\partial P_k}{\partial q_i}\frac{\partial Q_k}{\partial q_j} \right) = 0 \qquad (10.4\text{-}10a)$$

$$\sum_k^f \left(\frac{\partial Q_k}{\partial q_i}\frac{\partial P_k}{\partial p_j} - \frac{\partial P_k}{\partial q_i}\frac{\partial Q_k}{\partial p_j} \right) = \delta_{ij} \qquad (10.4\text{-}10b)$$

$$\sum_k^f \left(\frac{\partial Q_k}{\partial q_k}\frac{\partial P_k}{\partial p_j} - \frac{\partial P_k}{\partial p_i}\frac{\partial Q_k}{\partial p_j} \right) = 0. \qquad (10.4\text{-}10c)$$

Again, this is an equivalent criterion to the previous one, each holding if and only if the transformation is canonical. The corresponding forms,

$$[u,v]_{q,p} \equiv \sum_k^f \left(\frac{\partial q_k}{\partial u}\frac{\partial p_k}{\partial v} - \frac{\partial p_k}{\partial u}\frac{\partial q_k}{\partial v} \right)$$

are called *Lagrange brackets*, the above criteria for canonicity called the *Lagrange conditions.*

We observe that, in the not uncommon case of a single degree of freedom (f, i, and j all 1), both statements of the symplectic condition, Equations 10.4-8 and 10.4-10, simplify: conditions a) and c) are satisfied identically, and the b) forms each reduce to

$$\frac{\partial Q}{\partial q}\frac{\partial P}{\partial p} - \frac{\partial Q}{\partial p}\frac{\partial P}{\partial q} = 1.$$

It must be emphasized that *only transformations satisfying these equations will preserve the form of the Hamiltonian equations.* And while the symplectic condition is useful to determine whether a given transformation is canonical or not, it does not actually help us to *find* such a transformation. That is the topic of the next section.

Summary. Unlike Lagrangian systems, whose equations of motion remain invariant under an arbitrary coordinate transformation, $\boldsymbol{q}$ and $\boldsymbol{p}$ must transform *together* in order that their equations of motion retain Hamiltonian form. We obtain a pair of *symplectic conditions* enabling us to evaluate the canonical nature of a given transformation.

Homework:

1. Show that $\mathsf{J}^{-1} = -\mathsf{J}$.
2. Show that $\mathsf{J}^{\mathsf{T}} = -\mathsf{J}$.
3. Show that $\{u, H\} = \frac{du}{dt}$.
4. Show that J is an orthogonal matrix.
5. Show that $|-J| = |\mathsf{J}|$, *not* $-|\mathsf{J}|$! (Hint: use Theorem 3.2.3.)
6. Show that $|\mathsf{J}| = 1$.
7. In Example 10.6.1, we shall show that

$$H = \frac{1}{2m}p^2 + \frac{1}{2}k(x - x_o)^2$$

is the Hamiltonian for the "simple harmonic oscillator" (the undamped spring-mass system).

(a) Find the [Hamiltonian] equations of motion for this system.

A transformation $(x, p) \leftrightarrow (X, P)$ which brings $H(x, p)$ to a "simple form" for $H^*(X, P)$ is given by

$$x - x_o = \sqrt{\frac{2X}{k}} \sin P$$

$$p = \sqrt{2mX} \cos P.$$

Verify this allegation by

(b) explicitly calculating H^* and thence

(c) finding the Hamiltonian equations for $\dot{X}$ and $\dot{P}$.

Unfortunately, these results are incorrect. Show this by

(d) solving—"inverting"—the above transformation for $X(x, p)$ and $P(x, p)$ and then differentiating these directly to find $\dot{X}(x, p, \dot{x}, \dot{p})$ and $\dot{P}(x, p, \dot{x}, \dot{p})$, finally

(e) substituting in the expressions for $\dot{x}$ and $\dot{p}$ found above in part (a). These are the *correct* $\dot{X}$ and $\dot{P}$, but *not* those found in (c). The given transformation, then, is *not* canonical.

(f) Now show this transformation to be non-canonical using the Poisson conditions.

10.5 Generating Functions

Above we examined an already-existing transformation to determine whether it was canonical; this section will develop a means of *producing* a transformation in the assurance that it *will be* canonical. If the symplectic condition is both necessary and sufficient ("if and only if"), the method presented here will be only *sufficient*; in that regard, then, it does not deal with the entire class of *all* canonical transformations, but only a subset of them. Notwithstanding, it will give us a means of *finding* those variables in which the Hamiltonian assumes an "ideal form"—*i.e.*, of *solving* the system. The method, if appealing in its generality, is not without fault: in particular, it will give the canonical transformation only *implicitly*, and the issue of how we get an *explicit* transformation will be conveniently brushed aside. But we will show that an arbitrary function of "mixed"—"old" and "new"—variables will, in fact *generate* a canonical transformation.

We first obtain a result on which the balance of this section is based:

Theorem 10.5.1. A transformations $(\boldsymbol{q},\boldsymbol{p}) \rightarrow (\boldsymbol{Q},\boldsymbol{P})$ is canonical if and only if the differential $\sum(p_k\,dq_k - P_k\,dQ_k)$ is exact; *i.e.*, if and only if

$$\sum_k^f (p_k\,dq_k - P_k\,dQ_k) = dW \tag{10.5-1}$$

for some function W.

Before proving this, let us make a couple of observations regarding it:

In the first place, no arguments are stated for W. At first blush, we might expect that $W = W(\boldsymbol{q},\boldsymbol{Q})$, in order that the differentials on the left "match" those on the right when dW is taken. But what of $\boldsymbol{p}$ and $\boldsymbol{P}$ on the left: where are *they* to come from? Actually, we are assuming implicitly that these, too, be functions of the same variables W is. But, in point of fact, the above relation could be expressed in terms of *any* set of $2f$ conjugate variables (see the example below). In any event, we see already the "mixed" nature of the variables on which W depends.

Let us also make an observation to motivate the significance of the particular form (10.5-1): Though we are dealing with the Hamiltonian formulation in this chapter, its legacy ultimately traces back to *Lagrangians*, from which Hamiltonians are derived (equation (10.2-1)). In fact, in Section 10.2.2, we showed how the Hamiltonian equations of motion could be obtained by consideration of the Lagrangian, applying the Lagrange equations *in both the independent $\boldsymbol{q}$ and $\boldsymbol{p}$* to $\tilde{L} = \boldsymbol{p}\cdot\dot{\boldsymbol{q}} - H$. [We are purposely suppressing arguments here, for the same reasons alluded to in the above paragraph.]

Now let us consider something we did not there, the transformation of variables. If such a transformation $(\boldsymbol{q},\boldsymbol{p}) \rightarrow (\boldsymbol{Q},\boldsymbol{P})$ is applied, H will assume a new form H^*; clearly $H = H^*$ ["in value"]. This will generate a new Lagrangian $L^* \equiv \boldsymbol{P}\cdot\dot{\boldsymbol{Q}} - H^*$. At first blush, one might expect that it is necessary that $L^* = L$, which would then also require that $\boldsymbol{p}\cdot\dot{\boldsymbol{q}} = \boldsymbol{P}\cdot\dot{\boldsymbol{Q}}$. *But we desire* *only that the equations of motion for L^* remain invariant*; we don't really *care* about its "value"! And Theorem 10.2.1 says that the *Lagrangian* equations of motion remain invariant if one adds the time derivative of a function $W(\boldsymbol{Q},\boldsymbol{P};t)$ onto the Lagrangian; thus the equations of motion will be the same for L^* and $L^*+\dot{W}$. But *this* says that we can transform the *Lagrangians*:

$$L = \boldsymbol{p}\cdot\dot{\boldsymbol{q}} - H = L^* \equiv \boldsymbol{P}\cdot\dot{\boldsymbol{Q}} - H^* + \dot{W}$$

or, equivalently, using the fact that $H = H^*$,

$$\boldsymbol{p}\cdot\dot{\boldsymbol{q}} - \boldsymbol{P}\cdot\dot{\boldsymbol{Q}} = \dot{W};$$

this is, except in [total] differential form, precisely what (10.5-1) says—at least for $W = W(\boldsymbol{Q},\boldsymbol{P})$ (We treat the more general case in which W depends explicitly on t below). From the new L^*, we can obtain its equations of motion—the equivalent *Hamiltonian* equations of motion in the new variables. But those equations depend only on the *form* of L^*, which is (except for the irrelevant $\dot{W}$) precisely the same as that for L. Thus the *Hamiltonian* equations of motion will retain their form, and (10.5-1) is nothing more than the condition that this hold.

Having observed that, let us finally get on with the formal proof of the above theorem:

Proof. Let us assume that we have expressed $\boldsymbol{Q}$ and $\boldsymbol{P}$ in terms of $(\boldsymbol{q},\boldsymbol{p})$. This means, in particular, that

$$dQ_k = \sum_l^f \frac{\partial Q_k}{\partial q_l}\,dq_l + \sum_l^f \frac{\partial Q_k}{\partial p_l}\,dp_l$$

so that (10.5-1) becomes

$$\sum_l^f \Big(p_l - \sum_k^f P_k \frac{\partial Q_k}{\partial q_l}\Big)\,dq_l - \sum_l^f \Big(\sum_k^f P_k \frac{\partial Q_k}{\partial p_l}\Big)\,dp_l \equiv \sum_l^f A_l\,dq_l + \sum_l^f B_l\,dp_l.$$

The condition that a differential form be exact is that its "crossed partial derivatives" are the same. Here we have two sets of variables with respect to which to find these partials:

$$\frac{\partial A_i}{\partial q_j} = \frac{\partial A_j}{\partial q_i}: \qquad -\sum_k \frac{\partial P_k}{\partial q_j}\frac{\partial Q_k}{\partial q_i} - \sum_k^f P_k \frac{\partial^2 Q_k}{\partial q_j \partial q_i} =$$

$$-\sum_k \frac{\partial P_k}{\partial q_i}\frac{\partial Q_k}{\partial q_j} - \sum_k^f P_k \frac{\partial^2 Q_k}{\partial q_i \partial q_j};$$

the second sum on each side drops out, leaving terms which satisfy this equation if and only if (10.4-10a) holds. Similarly

$$\frac{\partial A_j}{\partial p_i} = \frac{\partial B_i}{\partial q_j}: \qquad \delta_{ij} - \sum_k \frac{\partial P_k}{\partial p_i}\frac{\partial Q_k}{\partial q_j} - \sum_k^f P_k \frac{\partial^2 Q_k}{\partial p_i \partial q_j} =$$

$$-\sum_k \frac{\partial P_k}{\partial q_j}\frac{\partial Q_k}{\partial p_i} - \sum_k^f P_k \frac{\partial^2 Q_k}{\partial q_j \partial p_i}$$

(the δ_{ij} entering because p_i only appears in A_j if $i = j$); again, the second partials cancel, leaving a condition true if and only if (10.4-10b) is. (The condition that $\frac{\partial B_j}{\partial p_i} = \frac{\partial B_i}{\partial q_j}$ is similar to the first of the above, so is omitted.) □

We note that we can write

$$\sum_k^f (p_k\, dq_k - P_k\, dQ_k) = -\sum_k^f (q_k\, dp_k + P_k\, dQ_k) + d\sum_k^f p_k q_k \tag{10.5-2a}$$

$$= \sum_k^f (p_k\, dq_k + Q_k\, dP_k) - d\sum_k^f P_k Q_k \tag{10.5-2b}$$

$$= -\sum_k^f (q_k\, dp_k - Q_k\, dP_k) + d\sum_k^f (p_k q_k - P_k Q_k); \tag{10.5-2c}$$

since each of the first differential forms on the right-hand side of the above equations differs from the original (10.5-1) by an exact differential, by the above theorem, the transformation is canonical if and only if any *one* of them is an exact differential. This is the form in which we shall apply the Theorem.

Pollard[3] makes an important point—"a subtlety, often overlooked" he describes it—which, along with his example, is paraphrased here: For sake of argument, we shall take $f = 2$. To say that, for example, the differential form in (10.5-2a), $q_1\, dp_1 + q_2\, dp_2 + P_1\, dQ_1 + P_2\, dQ_2$, is exact does *not* mean that there is a function $W(p_1, p_2, Q_1, Q_2)$ such that

$$q_1\, dp_1 + q_2\, dp_2 + P_1\, dQ_1 + P_2\, dQ_2 = \frac{\partial W}{\partial p_1}\, dp_1 + \frac{\partial W}{\partial p_2}\, dp_2 + \frac{\partial W}{\partial Q_1}\, dQ_1 + \frac{\partial W}{\partial Q_2}\, dQ_2$$

in the sense we can simply equate coefficients:

$$\frac{\partial W}{\partial p_1} = q_1, \quad \frac{\partial W}{\partial p_2} = q_2, \quad \frac{\partial W}{\partial Q_1} = P_1 \quad \frac{\partial W}{\partial Q_2} = P_2. \tag{10.5-3}$$

[3] *Mathematical Introduction to Celestial Mechanics* (1966), p. 67

For example, consider the transformation $(\boldsymbol{q},\boldsymbol{p}) \to (\boldsymbol{Q},\boldsymbol{P})$ given by

$$(Q_1, Q_2, P_1, P_2) \equiv (q_1, p_2, p_1, -q_2);$$

expressing the $\boldsymbol{Q}$ and $\boldsymbol{P}$ in terms of the $\boldsymbol{q}$ and $\boldsymbol{p}$ in the differential form cited above,

$$\sum_k^f (q_k\,dp_k + P_k\,dQ_k) = q_1\,dp_1 + q_2\,dp_2 + p_1\,dq_1 - q_2\,dp_2 = d(q_1p_1)$$

is, indeed, exact, so this transformation is canonical, by the above Theorem. Note, however, that this exact differential involves q_1, a variable not appearing [explicitly] in $W(p_1, p_2, Q_1, Q_2)$; in fact, this differential depends on both the *old* coordinates and momenta. But there is *no* W *such that* (10.5-3) *is satisfied.* Consider for example, the second component of that:

$$\frac{\partial W(p_1, p_2, Q_1, Q_2)}{\partial p_2} = q_2 \equiv -P_2;$$

this is impossible since P_2, allegedly the result of this differentiation, doesn't even appear in the function being differentiated! In fact, as Goldstein points out regarding this example,[4] this transformation can be derived from the function $q_1P_1 + q_2Q_2 = W(q_1, q_2, P_1, Q_2)$—a "mixed function" of the mixed coordinates!

But if there *is* such a function—a "*generating function*"—for a given canonical transformation, then the above equations (10.5-1) and (10.5-2) allow us to conclude, respectively, that

$$p_k = \frac{\partial W(\boldsymbol{q},\boldsymbol{Q})}{\partial q_k} \qquad P_k = -\frac{\partial W(\boldsymbol{q},\boldsymbol{Q})}{\partial Q_k} \tag{10.5-4a}$$

$$q_k = -\frac{\partial W(\boldsymbol{p},\boldsymbol{Q})}{\partial p_k} \qquad P_k = -\frac{\partial W(\boldsymbol{p},\boldsymbol{Q})}{\partial Q_k} \tag{10.5-4b}$$

$$p_k = \frac{\partial W(\boldsymbol{q},\boldsymbol{P})}{\partial q_k} \qquad Q_k = \frac{\partial W(\boldsymbol{q},\boldsymbol{P})}{\partial P_k} \tag{10.5-4c}$$

$$q_k = -\frac{\partial W(\boldsymbol{p},\boldsymbol{P})}{\partial p_k} \qquad Q_k = +\frac{\partial W(\boldsymbol{p},\boldsymbol{P})}{\partial P_k}; \tag{10.5-4d}$$

conversely, for *any* function W of *mixed* variables, the transformations *implicitly* defined by these equations will be canonical. It is the latter fact we utilize in order to transform the Hamiltonian to an "ideal" form in the next section. Thus, even though all the conditions thus far have been of the "if and only if" variety, to the extent that we use only functions of this type, we are dealing with only a *subset* of all canonical transformations.

Clearly the precise signs on the above equations are less important than whether those for a given transformation have the same or different ones: *both* would be switched simply replacing W by $-W$. To remember whether a given

[4] *Classical Mechanics*, 2nd ed. (1980), p. 388

generating function's transformations should have alternating signs, examination of these equations verifies the mnemonic: "*Like* arguments of the generating function [both coordinates or both momenta] have *unlike* signs in the transformations" (and conversely).

Once again, note the "mixed" nature of these functions W, all of which involve conjugate variables from both the original and transformed sets. But that means that each of those variables resulting from the indicated partial derivatives are also in terms of both old and new variables; *i.e.*, the transformation is only *implicitly* defined. In order to get an *explicit* transformation, some form of inversion is required. Thus, though we are assured such transformations are *canonical*, they can hardly be described as *convenient*!

Example 10.5.1. *Generating Functions.* As a very simple example of the above ideas, consider a function $W \equiv pQ + p^2 = W(p, Q)$. Since the arguments are "different", we know the corresponding transformation equations will involve the partials of this function with "like" signs:

$$q = \frac{\partial W}{\partial p} = Q + 2p \qquad P = \frac{\partial W}{\partial Q} = p.$$

The transformation is, indeed, only implicit, but in this case it is a trivial matter to solve for the explicit one:

$$Q = q - 2p \qquad P = p.$$

We note that, in this case, the fundamental differential form (10.5-1)

$$p\,dq - P\,dQ = p\,dq - p(dq - 2\,dp) = 2pdp = d(p^2)$$

is verified. |

Homework:

1. Show that the other three differential forms in equations (10.5-2) are verified in the above Example.

There is one more point to be made regarding (10.5-1): Theorem 10.2.1 admits the possibility that W depend on *time*, so that there would generally be an additional term in dW corresponding to $\frac{\partial W}{\partial t}\,dt$. And there is no "$dt$" on the left-hand side to "match" with this! What to do? Again we return to the *Lagrangian* foundation of the Hamiltonian formulation: there we regarded *virtual* displacements, *holding t fixed.* Thus in the same way that non-holonomic constraints—themselves differential forms—were implemented using their *virtual* differentials in the equations of motion for Lagrangians, (10.5-1) will, most generally, be cast in the form of *virtual* differentials:

$$\sum_k^f (p_k\,\delta q_k - P_k\,\delta Q_k) = \delta W.$$

Equivalently, in terms of the generalized velocities appearing in the Hamiltonian, the derivative form can be expressed

$$\sum_k^f (p_k\dot{q}_k - P_k\dot{Q}_k) = \frac{dW}{dt} - \frac{\partial W}{\partial t}.$$

We can now summarize our results for this most general case that W depends explicitly on t: Given a Hamiltonian $H(\boldsymbol{q},\boldsymbol{p};t)$ and a transformation $(\boldsymbol{q},\boldsymbol{p}) \to (\boldsymbol{Q},\boldsymbol{P};t)$, yielding the transformed $H^*(\boldsymbol{Q},\boldsymbol{P};t)$, the associated *Lagrangian* functions (from which we obtain the *Hamiltonian* equations of motion for these Hamiltonians) are

$$L(\boldsymbol{q},\boldsymbol{p};t) = \boldsymbol{p}\cdot\dot{\boldsymbol{q}}(\boldsymbol{q},\boldsymbol{p};t) - H(\boldsymbol{q},\boldsymbol{p};t)$$

and

$$L^*(\boldsymbol{Q},\boldsymbol{P};t) = \boldsymbol{P}\cdot\dot{\boldsymbol{Q}}(\boldsymbol{Q},\boldsymbol{P};t) - H^*(\boldsymbol{Q},\boldsymbol{P};t)$$

which yield the *Hamiltonian* equations of motion (as in the above discussion) of the *same form*

$$\dot{q}_i = \frac{\partial H}{\partial p_i} \qquad \dot{p}_i = -\frac{\partial H}{\partial q_i}$$
$$\dot{Q}_i = \frac{\partial H^*}{\partial P_i} \qquad \dot{P}_i = -\frac{\partial H^*}{\partial Q_i}.$$

But the *Lagrangian* equations of motion for $\boldsymbol{Q}$ and $\boldsymbol{P}$ *are exactly the same if we add* $\dot{W}$ *to* L^*; *i.e.*, rather than using the above L^*, these equations remain unchanged if we use

$$L^{*'} \equiv \boldsymbol{P}\cdot\dot{\boldsymbol{Q}} - H^* + \frac{dW}{dt} = \boldsymbol{P}\cdot\dot{\boldsymbol{Q}} - H^* + \left(\frac{dW}{dt} - \frac{\partial W}{\partial t}\right) + \frac{\partial W}{\partial t}.$$

Thus the *Hamiltonian* equations of motion remain unchanged under this addition!

It is at this point that the *canonical* nature of the transformation comes in: if such a transformation transforms L above to $L^{*'} = L$ ["in value"]

$$L = \boldsymbol{p}\cdot\dot{\boldsymbol{q}} - H = L^{*'} \equiv \boldsymbol{P}\cdot\dot{\boldsymbol{Q}} - H^* + \left(\frac{dW}{dt} - \frac{\partial W}{\partial t}\right) + \frac{\partial W}{\partial t}$$

in the particular way that

$$\boldsymbol{p}\cdot\dot{\boldsymbol{q}} = \boldsymbol{P}\cdot\dot{\boldsymbol{Q}} + \left(\frac{dW}{dt} - \frac{\partial W}{\partial t}\right) \tag{10.5-5}$$

(condition (10.5-1), modified as above to account for t-dependent W's), then we see the canonical property is preserved. *But we also get the form for the new* H^*:

$$-H = -H^* + \frac{\partial W}{\partial t}$$

or

$$H^* = H + \frac{\partial W}{\partial t}.$$

This, then, is the form of the new H^* in that case in which the generating function W is explicitly dependent on t. Note that now $H \neq H^*$: since W depends explicitly on t, evaluation of H^* is more than a matter of merely substituting in the old *variables* expressed in terms of the new.

All the above discussion has been cast in the language that "W depends on t". But if it does, since the condition (10.5-5) is equivalent to the condition (10.5-1), the transformation to new coordinates will satisfy the same equations derived from the latter, namely equations (10.5-4). But these involve partial derivatives of W, which now *depends on t*; thus the transformations themselves are time-dependent; *i.e., a time-dependent generating function is precisely that defining a time-dependent transformation* $(\boldsymbol{q}, \boldsymbol{p}) \rightarrow (\boldsymbol{Q}, \boldsymbol{P}; t)$.

As a final remark, we note the critical role that Theorem 10.2.1 has played in the above discussion of generating function, both depending, or not, on t. But that theorem relies on the *independence* of $\boldsymbol{q}$ and $\boldsymbol{p}$ in applying the Lagrange equations to [the Legendre transformation] $L = \boldsymbol{p} \cdot \dot{\boldsymbol{q}} - H$. Without this conceptual foundation, the entire structure of generating functions collapses.

Summary. We have seen that *any* function of "old" and "new" variables—effectively corresponding to that function in these variables whose time derivative can be added to the Hamiltonian—generates an *implicit* canonical transformation. These are endemically "mixed" in nature (requiring some form of inversion to recover the *explicit* transformation), and do not encompass *all* canonical transformations.

10.6 Transformation Solution of Hamiltonians

We finally have the formalism developed to give a general approach to the solution of *conservative* Hamiltonian systems: Rather than attempting to solve the *original* system, we solve a *transformed* system whose solution is easy to find—one of the "ideal" systems referred to above. This is determined merely by the *form* of the Hamiltonian *as long as the transformation preserves the form of the equations of motion*; *i.e.*, as long as the transformation is *canonical*. The solution of the original system is then found "merely" by expressing its variables in terms of the new, soluble system's. In a very real sense, this implements the philosophy that "If you don't like the system you have, convert it into one you *do*!"

This is where the previous section's topic comes in: In order to enforce the canonical nature of the transformation, we utilize *generating functions*, in the assurance that the result will *be* canonical; *e.g.*, for $W = W(\boldsymbol{q}, \boldsymbol{P}; t)$, substituting the appropriate transformations from equations (10.5-4) ["unlike" variables], this process results in

$$H(\boldsymbol{q}, \frac{\partial W}{\partial q_i}; t) + \frac{\partial W}{\partial t} = H^*(\frac{\partial W}{\partial P_i}, \boldsymbol{P}; t)$$

—where $\frac{\partial W}{\partial t}$ results from the time dependence of W, as at the end of the last section.

The process of finding the solution of the original system is thus supplanted by that of finding the transformation—the generating function—to bring the Hamiltonian into a soluble form. There are some problems surrounding just that: since the transformations are expressed as partial derivatives of the generating function, we end up with *partial differential equations* to solve. But the major difficulty is at the final step, where we express the original variables in terms of the soluble ones—the "inversion" problem arising from the mixed nature of the generating function. We illustrate this process with an example:

Example 10.6.1. We consider a classic example of the transformation technique, the single-degree-of-freedom undamped linear spring-mass system. We must first find the Hamiltonian of the system. It is time-independent and conservative, and there are no rheonomic constraints; thus we know that the Hamiltonian will be nothing more than the energy. But this must be expressed in terms of the generalized momentum of the problem, and for that we still need the Lagrangian. That is trivial to find: for unextended length x_o,

$$L \equiv T - V = \frac{1}{2}m\dot{x}^2 - \frac{1}{2}k(x - x_o)^2;$$

thus

$$p \equiv \frac{\partial L}{\partial \dot{x}} = m\dot{x} \Rightarrow \dot{x} = \frac{p}{m}.$$

In terms of this we find the Hamiltonian

$$\begin{aligned} H = E = T + V &= \frac{1}{2}m\Big(\frac{p}{m}\Big)^2 + \frac{1}{2}k(x - x_o)^2 \\ &= \frac{1}{2m}p^2 + \frac{1}{2}k(x - x_o)^2. \end{aligned}$$

Now we must decide on the desired "ideal form" for the transformed Hamiltonian, H^*, to assume. Noting (though this is more or less "cheating"!) that H *is* constant suggests setting it equal to a new variable whose solution will also be constant; let us take $H^*(X, P) \equiv X$ (See equation 10.3-1; we could just as easily have set it equal to P.) And H is independent of t; this suggests a generating function also independent of t. These two observations motivate a $W = W(\_, X)$, but we must pick the "old" variable for W. The rather generic expression for T, and the fact that V depends on x, suggest we take the latter: $W \equiv W(x, X)$. In terms of these "like" arguments for W, we have immediately that the signs for the generating function transformation will be of the form

$$p = -\frac{\partial W}{\partial x} \qquad P = \frac{\partial W}{\partial X}$$

(or their opposites). Substituting this into the equation saying that $H = H^*$ (again, because of the t-independence of W), we get

$$H = \frac{1}{2m}\Big(-\frac{\partial W}{\partial x}\Big)^2 + \frac{1}{2}k(x - x_o)^2 = H^* \equiv X,$$

noting that $X \geq 0$.

We must now solve the system for W, satisfying the equation

$$\frac{\partial W}{\partial x} = \sqrt{2mX - mk(x - x_o)^2} = \sqrt{mk}\sqrt{\frac{2X}{k} - (x - x_o)^2}; \qquad (10.6\text{-}1)$$

the solution is

$$W = \sqrt{mk}\left\{\frac{x - x_o}{2}\sqrt{\frac{2X}{k} - (x - x_o)^2} + \frac{X}{k}\arcsin\left(\frac{x - x_o}{\sqrt{\frac{2X}{k}}}\right)\right\} \qquad (10.6\text{-}2)$$

—a function of x and X, as expected. (We can take the indicated square root in the argument of the inverse trig function, since, as noted above, $X \geq 0$. Further, since the transformation depends only on the *derivatives* of W, we dispense with any additive constants.)

We need at least one more relation among the variables to solve for x and p: we only have $p = \frac{\partial W}{\partial x}$ from (10.6-2), which is precisely (10.6-1)! We *could* solve for $P = -\frac{\partial W}{\partial X}$ by direct differentiation, but W is notably ugly, and we invoke the useful dodge of *Leibniz' rule*:

$$P = \frac{\partial W}{\partial X} = \frac{\partial}{\partial X}\int \frac{\partial W}{\partial x}\,dx = \int \frac{\partial}{\partial X}\frac{\partial W}{\partial x}\,dx = \int 2\sqrt{\frac{m}{k}}\frac{1}{\sqrt{\frac{2X}{k} - (x - x_o)^2}}\,dx$$

$$= \sqrt{\frac{m}{k}}\arcsin\left(\frac{x - x_o}{\sqrt{\frac{2X}{k}}}\right). \qquad (10.6\text{-}3)$$

This is immediately soluble for—this is the "inversion" step cited above—

$$x = x_o + \sqrt{\frac{2X}{k}}\sin\sqrt{\frac{k}{m}}P,$$

which, when substituted back into (10.6-1)—another inversion—yields

$$p = -\frac{\partial W}{\partial x} = -\sqrt{mk}\sqrt{\frac{2X}{k} - \frac{2X}{k}\sin^2\sqrt{\frac{k}{m}}P} = -\sqrt{2mX}\cos\sqrt{\frac{k}{m}}P.$$ [5]

It is easy to lose sight of the solutions for X and P in the above manipulations, but, from the form for H^*, the former is constant, while the latter is just $P =$

[5] The minus sign here could be avoided by taking the negative square root in (10.6-1).

$-t + P_o$. Thus, in fact, the above solutions for x and p show a fundamental [circular] frequency of $\sqrt{\frac{k}{m}}$, as expected from this system. But, often ignored, there is the fact that we must know the *values* of X and P_o in order to have the full solution! These are certainly defined implicitly by the above expressions for x and p at, say, $t = 0$:

$$x(0) - x_o = \sqrt{\frac{2X}{k}} \sin \sqrt{\frac{k}{m}} P_o$$
$$p(0) = m\dot{x}(0) = -\sqrt{2mX} \cos \sqrt{\frac{k}{m}} P_o.$$

Equivalently, they could be found *explicitly* by performing the last inversion, expressing X and P in terms of x and p:

$$\frac{1}{2}k(x - x_o)^2 + \frac{1}{2m}p^2 = X = H;$$

substituting this back into (10.6-3) gives

$$P = \sqrt{\frac{m}{k}} \arcsin\left(\frac{x - x_o}{\sqrt{(x - x_o)^2 + \frac{p^2}{mk}}}\right).$$

The values of $x(0)$ and $\dot{x}(0) = P_o/m$ will then give the corresponding values for X and P_o. In either event, some form of inversion is required to match initial conditions. |

Homework:

1. Use the Poisson conditions to show that the transformation $(x, p) \leftrightarrow (X, P)$ used in Example 10.6.1 *is* canonical.

The above was meant to show the general idea of implementing canonical transformations through generating functions. It doesn't really go very far to demonstrate the benefits of the approach: the solution of the ordinary differential equation $m\ddot{x} + k(x - x_o) = 0$ can effectively be found by inspection, not with two pages of calculus! In a similar vein, the next example is meant merely to show how these canonical transformations could be applied to *time-dependent* Hamiltonians.

Example 10.6.2. *Time-dependent Generating Function.* Again, we consider a single-degree-of-freedom system subject to a time-dependent forcing function $\boldsymbol{F}(t) = At$. As was noted on page 261, and utilized in Example 8.2.4, if such forces satisfy $\nabla \times \boldsymbol{F} \equiv 0$, then they can be included in a time-dependent potential; in this case $V = -Axt$. Thus the Lagrangian is $L(x, t) = \frac{1}{2}m\dot{x}^2 + Axt$, the corresponding generalized momentum $p = m\dot{x}$, so the Hamiltonian—effectively the *time-dependent* "energy" for this "conservative" system—

$$H \equiv p\dot{x} - L = \frac{p^2}{2m} - Axt.$$

As is generally done in time-dependent problems, we are going to use a canonical transformation to transform this Hamiltonian to the new one $H^* \equiv 0$; thus *all* new coordinates and momenta will be constant. The explicit appearance of t in the Hamiltonian suggests that our generating function W will depend on t, but what other variables will it depend on? Though above we used the transformation to eliminate the momentum from the Hamiltonian, the fact that the coordinate x is only *linear*, while the momentum quadratic, suggests that we eliminate x: $W = W(p, \_; t)$. Since $H^* \equiv 0$, this doesn't give any guidance for the choice of the new variable in terms of which to express the variables; we more or less arbitrarily pick X as the other variable; thus the transformations, now including the Hamiltonian's, will be given by [note "unlike" arguments of W]

$$x = \frac{\partial W(p, X; t)}{\partial p} \qquad P = \frac{\partial W(p, X; t)}{\partial X}$$
$$H^* = H + \frac{\partial W(p, X; t)}{\partial t}.$$

We then substitute these into the expression

$$H^* = H + \frac{\partial W}{\partial t} = \frac{p^2}{2m} - At\frac{\partial W(p, X; t)}{\partial p} + \frac{\partial W(p, X; t)}{\partial t} \equiv 0 \qquad (10.6\text{-}4)$$

to get an equation for W.

So now we've got to *find* W to solve this problem. Despite the simple linear nature of the above equation, it is, like all partial differential equations, a somewhat delicate matter to solve. This, fortunately, is only of first order; we will use probably the most general approach to such partial differential equations, *Lagrange's method.* We note that X doesn't appear in this equation at all; thus we shall not carry this superfluous variable in what follows.

Setting up the *Lagrange auxiliary equations*:

$$\frac{dW}{p^2/2m} = -\frac{dp}{At} = dt,$$

the last equation gives immediately $p = -At^2 + C_1$; thus the solution can be expressed as an arbitrary function of the function $\phi_1 \equiv p + At^2$. Unfortunately, this doesn't help very much:

$$\frac{p^2}{2m} - At\frac{\partial \phi_1}{\partial p} + \frac{\partial \phi_1(p, X; t)}{\partial t} \equiv 0;$$

i.e., ϕ_1 is the *homogeneous* solution! The first of the auxiliary equations above must contain the *non*-homogeneous one.

More unfortunately, that one contains, in addition to W and p, the term At, and there seems to be no linear combination of the differentials which eliminates this without reducing to the equation already solved. In the hopes that the next solution we [hope to] obtain, ϕ_2, is independent of ϕ_1, we resort to eliminating t in favor of p through the solution relating those variables above: $t = \sqrt{2(C_1 - p)/A}$. Then the first equation, writing W as "W_p" for reasons which will become clear presently, becomes

$$dW_p = -\frac{p^2\,dp}{2mAt} = -\frac{p^2\,dp}{2\sqrt{2A}m\sqrt{C_1 - p}};$$

therefore

$$\begin{aligned} W_p &= -\frac{1}{2\sqrt{2A}m}\int \frac{p^2\,dp}{\sqrt{C_1 - p}} \\ &= -\frac{1}{2\sqrt{2A}m}\frac{2(8C_1^2 + 4C_1 p + 3p^2)}{15}\sqrt{C_1 - p} + C_2 \\ &= -\frac{p^2 t}{2m} - \frac{Apt^3}{3m} - \frac{A^2 t^5}{15m} + C_2 \end{aligned}$$

—in which C_1 is eliminated in favor of $C_1 = p + At^2$. Though the final solution depends on an arbitrary function of $\phi_2 = W_p + \frac{p^2 t}{2m} + \frac{Apt^3}{3m} + \frac{A^2 t^5}{15m}$, direct substitution verifies that, in fact, W_p is the solution to the *non*-homogeneous equation (10.6-4), also verifying its independence of ϕ_1. *This* is what we were looking for! But we're not out of the woods yet:

W_p is independent of X; we *forced* that when we noted above that it doesn't appear in the equation equating H and H^*. But as things stand now, we have that $P = \frac{\partial W_p}{\partial X} \equiv 0$—a somewhat awkward result! Somehow, we've got to introduce X into W *explicitly*.

We might try to patch this up by assuming that W is *separable* in p:

$$W(p, X; t) = W_p(p; t) + W_X(p, X; t)$$

—we have already implicitly done that above by finding that part of W dependent on p (and t) only—and then noting that, since P doesn't appear in H^*, it is constant, concluding that

$$P = \frac{\partial W_X}{\partial X} \equiv P_o \quad \Rightarrow \quad W_X = P_o X.$$

But this would be an unproductive tack, for two reasons: In the first place, the whole argument flies in the face of the philosophy that the transformation is, in some sense, *independent* of the solution for H^*—the solution arising from H^* is *substituted* into the transformation in order to find $x(t)$ and $p(t)$—whereas here, we have *used* the solution—in particular, the fact that P is constant—in order to *determine* the transformation. In the second, it doesn't satisfy the original differential equation (10.6-4) which defines W in the first place!

The way out of this is to note that the W_p we have determined above is the solution to the *non*-homogeneous [partial] differential equation (10.6-4); on the other hand, if the additive part W_X happens to be a solution to the *homogeneous* equation

$$-At\frac{\partial W_X(p,X;t)}{\partial p}+\frac{\partial W_X(p,X;t)}{\partial t}\equiv 0,$$

then our hard-won W_p will still give the non-homogeneity in (10.6-4), so that $W = W_p + W_X$ will satisfy that equation. Yet the solution to the homogeneous part obtained above:

$$\phi_1 = p + \frac{At^2}{2}$$

still doesn't contain X! But if ϕ is a solution to the *homogeneous* equation, then so is $f(X)\phi$, for *any* $f(X)$. Thus, *finally*, we have a generating function of appropriate form:

$$W(p,X;t) = W_p(p;t) + W_X(p,X;t) = \\ -\frac{p^2t}{2m} - \frac{Apt^3}{3m} - \frac{A^2t^5}{15m} + f(X)\Big(p+\frac{At^2}{2}\Big),$$

where, for pedagogical reasons, we leave $f(X)$ unspecified for the moment.

We are now presumably at the point we can recover the solution for $x(t)$ and $p(t)$. We have that, from the original canonical transformation,

$$x = \frac{\partial W}{\partial p} = -\frac{pt}{m} - \frac{At^3}{3m} + f(X)$$

and

$$P = \frac{\partial W}{\partial X} = f'(X)\Big(p+\frac{At^2}{2}\Big)$$

When we finally get to the equations of motion for H^*, we will have immediately that X is constant; this is why $f(X)$ was left unspecified: *regardless* of what it is, x will merely be changed by an additive constant, while P will have $p - At^2/2$ times another constant added onto it. There is no particular virtue in taking f to be complicated; let us, more or less arbitrarily, select $f(X) \equiv X$. Then the specific transformation is

$$x = -\frac{pt}{m} - \frac{At^3}{3m} + X$$

and

$$P = p + \frac{At^2}{2}.$$

"Inverting" these—*i.e.*, solving for x and p in terms of X, P, and t—we get that

$$p = P - \frac{At^2}{2} \quad\Rightarrow\quad x = -\frac{t}{m}\Big(P - \frac{At^2}{2}\Big) - \frac{At^3}{3m} + X = \frac{At^3}{6m} - \frac{Pt}{m} + X;$$

conversely, getting X in x, p, and t (P already *is*, above!) to match initial conditions,

$$X = x + \frac{pt}{m} - \frac{At^3}{3m}.$$

At this point we introduce the fact that, from the equations of motion for H^*, $X = X_o$ and $P = P_o$, both constant, in the above transformations, giving the solution to the problem. Recognizing that $p = m\dot{x}$ in this case, we have precisely (with allowance for the values of the constants) those for $m\ddot{x} = At$. |

Neither of the above examples has done much to press the case for Hamiltonian dynamics: either could be solved much more simply in the Newtonian formulation—typically in one *line* rather than multiple *pages*! Notwithstanding, one can at least recognize that *we have complete control over the form of the transformed Hamiltonian*, and the above machinations are, in some sense, just the price to be paid for such freedom. It should also be pointed out that, however simpler the Lagrangian material in this Part might have seemed, the thrust there was merely to obtain the equations of motion: *we never solved the resulting differential equations.* What has occupied most of the volume above has been the *solution* of the Hamiltonian equations of motion, not obtaining them in the first place.

Having defended the study to that extent, however, there is an additional point which must be made: The *method* of solution above is qualitatively quite distinct from that in Lagrangian dynamics. In the latter, we had to resort to brute-force solution of the *system* of second-order ordinary differential equations resulting from the equations of motion: even though we showed these are *invariant* under a coordinate transformation, there is no practical, systematic means of uncovering the appropriate set of *ignorable* coordinates in Lagrangian dynamics. And even if we *could* find such coordinates, they would only give us *half* the number of integrals of motion required to solve the problem. In the present formulation, on the other hand, *canonical transformations provide a systematic means of discovering the appropriate set of coordinates and momenta*—of solving the *entire* problem. After having mentioned continually that the *form* of the Lagrangian or Hamiltonian functions can be used to obtain integrals of the motion, we finally have a means of implementing that approach. Given that fact, it is not too surprising that we must contend with a more powerful formalism and the resulting *partial* differential equations.

But there is even an observation to be made with regard to *that*. To a certain extent, the difficulty of solution, particularly in multi-degree-of-freedom systems, depends very much on how the transformed Hamiltonian is "broken up" in the partial differential equations resulting from the canonical transformations generated by the function. If this can be done mathematically in a way conforming to the *physical* system, we presumably have a greater chance of obtaining a mathematically tractable system—and conversely, as we shall see in further examples. Some of this is, frankly, just *insight*, and that's something which comes with brute *experience*: it can't merely be taught. But a great deal rests on the truism that mathematical simplicity generally results from picking the "right" system to solve in the first place, and that is often guided by the *physics* of the problem!

Summary. Hamiltonians can be expressed in any desired form through a canonical transformation resulting from a generating function, used to ensure its canonicity. Mere substitution of the generating function transformations into the original and transformed Hamiltonians results in a generally non-linear partial differential equation. Having the solution in the new variables, the solution of the original system is embodied in the transformation between the sets of variables themselves. Since that transformation is only *implicit* in the generating function formulation, the mixed-variable relations must be inverted explicitly, both to get the solution of the original variables, and to match initial conditions on the original variables to values of the new ones.

10.7 Separability

In Example 10.6.2 we had to resort to assuming—*hoping*—that the generating function we sought was *separable*—that it could be written as a *sum* of terms, some of which depended on only a single variable (and, in that case, because of the time-dependence, on t). We could do it there with $W_p(p;t)$, but W_X was still a function of both p and X. In such circumstances, p is referred to as a *separable variable.* Though we were able to find W_X even though it depended on both variables, this was primarily because its p-dependence was a solution to the homogeneous partial differential equation, which we could multiply by *any* function of X while retaining that property. Clearly, this is just a special situation and does not hold in general.

As a matter of practicality, one normally counts on such separability just to be able to solve the partial differential equations resulting from canonical transformation generated by an unknown function. In fact, one generally hopes for *complete* separability among, say, the various [old] coordinates and [new] momenta:

$$W = \sum_i W_i(q_i, \boldsymbol{P}; t)$$

—in which it is implicitly assumed the new momenta will be constants, due to the form of the desired H^*. Though there are, in fact, criteria to *predict* when such separability can be effected (the *Staeckel conditions*; see, *e.g.*, Goldstein, *Classical Mechanics*, 2nd ed, p. 453), these are somewhat technical, even a little "fussy". There are, however, fairly general situations in which separability can be assured relative to a particular variable.

10.7.1 The Hamilton-Jacobi Equation

Recall that the first example in the previous section used a time-independent generating function and transformed to an H^* depending on the coordinates, while the second's allowed W to depend on t and chose $H^* \equiv 0$. Equations determining the generating functions to effect a vanishing transformed Hamiltonian are generally referred to as the *Hamilton-Jacobi Equation*:

$$H\Big(\mathbf{q}, \frac{\partial W}{\partial q_i}; t\Big) + \frac{\partial W}{\partial t} \equiv 0.$$

What we intend to show here is that, if H is *independent of time* (*i.e.*, the type in Example 10.6.1), then

- this equation still holds, reducing to the form we utilized in the first example, and
- that the equation is, in fact, *separable* in t.

The two features are interrelated, and we shall actually consider them in the opposite order:

If $H = H(\mathbf{q}, \mathbf{p})$ is independent of t, then the Hamilton-Jacobi equation for a time-*de*pendent generating function, $W = W(\mathbf{q}, \mathbf{P}; t)$, say,

$$H\Big(\mathbf{q}, \frac{\partial W}{\partial q_i}\Big) + \frac{\partial W}{\partial t} \equiv 0$$

can be solved assuming W is separable in t:

$$W(\mathbf{q}, \mathbf{P}; t) = W_o(\mathbf{q}, \mathbf{P}) + W_t(t).$$

Goldstein refers to the time-*in*dependent $W_o(\mathbf{q}, \mathbf{P})$ as *Hamilton's characteristic function*. Indeed, under this assumption,

$$H\Big(\mathbf{q}, \frac{\partial W}{\partial q_i}\Big) + \frac{\partial W}{\partial t} = H\Big(\mathbf{q}, \frac{\partial W_o}{\partial q_i}\Big) + \frac{\partial W_t}{\partial t} \equiv 0,$$

where the first term is independent of t altogether. Thus, using a separation of variables argument, the most general solution is

$$H\Big(\mathbf{q}, \frac{\partial W_o}{\partial q_i}\Big) = constant = -\frac{\partial W_t}{\partial t}.$$

The right-hand equation gives us immediately that $W_t = (constant)\,t$, while the left-hand simply says that H is itself constant—just the form used in Example 10.6.1. The argument is reminiscent of that demonstrating constancy of the Jacobi function—precisely the Hamiltonian, recall. In a very real sense, then, W can *always* be regarded as a function of t; it's just that, when $\frac{\partial H}{\partial t} \equiv 0$, the generating function includes an additive term *linear* in the time. This blurs the distinction between the time-dependent and time-independent cases, and makes the Hamilton-Jacobi equation conceptually general in application.

10.7.2 Separable Variables

The reasoning applied above can be put in a more general context: Let us say that a given variable q_j *and its conjugate* p_j appear together in a group, $f_j(q_j, p_j)$, *independent* of the other variables. Then, writing $\boldsymbol{q}' \equiv (q_k, k \neq j)$ and $\boldsymbol{p}' \equiv (p_k, k \neq j)$, and assuming that H doesn't depend explicitly on t—*i.e.*, that t is itself separable—the Hamilton-Jacobi equation becomes

$$H\Big(\boldsymbol{q}, \frac{\partial W_o}{\partial q_i}\Big) = H'\bigg(\boldsymbol{q}', \frac{\partial W_o'}{\partial q_i}, f_j\Big(q_j, \frac{\partial W_j}{\partial q_j}\Big)\bigg) = constant$$

—in which we seek a *separable* generating function

$$W_o(\boldsymbol{q}, \boldsymbol{P}) = W_o'(\boldsymbol{q}', \boldsymbol{P}) + W_j(q_j, \boldsymbol{P}).$$

As Goldstein [*op. cit.*] argues, the above Hamilton-Jacobi equation can, at least in principle, be inverted to give

$$f_j\Big(q_j, \frac{\partial W_j}{\partial q_j}\Big) = F\Big(\boldsymbol{q}', \frac{\partial W_o'}{\partial q_i}\Big).$$

But the left-hand side depends only on q_j while the right depends only on $\boldsymbol{q}'$; thus, most generally, both must equal some constant. We have, indeed, an independent Hamilton-Jacobi equation for q_j, and separability has, in fact, been effected.

Special Case—Ignorable Coordinates

If q_j is altogether *absent* from H, then the $f_j = f_j(p_j)$ alone, and the above inversion of separability reduces to the Hamilton-Jacobi equation

$$f_j\Big(\frac{\partial W_j}{\partial q_j}\Big) = constant. \tag{10.7-1}$$

As a result of the above, such ignorable coordinates in the Hamiltonian can always be separated. (Presumably the same would hold for "ignorable *momenta*", too, but these are unlikely to arise without a corresponding coordinate!) This is, of course, just another way of looking at the conservation of the corresponding momentum.

The above Hamilton-Jacobi equation has, as its immediate solution, $W_j = (constant)\,q_j$; in particular, we can identify the constant with P_j, the [new] momentum, yielding $W_j = P_j q_j$. This is precisely the form, $P_o X$, we had when we tried a "cheap" means of working X into W_X on page 514; there it was regarded as a Bad Thing! Why is it valid in the present context? In Example 10.6.2, recall, we attempted to use the *constant* solution of the transformed Hamiltonian to justify such a term in the generating function; consequently, the *variable* P never appeared in W, and, in fact, the trial W didn't even satisfy the partial differential equation defining it. Here the variable is absent *from the start*, and it's the *original* system we utilize to obtain the [part of the] Hamilton-Jacobi equation (10.7-1) which *is* satisfied; in effect, we are merely trying not to tamper with a variable that's *already* constant!

Finally, just to bring everything together, the similarity of all the above arguments, particularly with regard to t-separability in the Hamilton-Jacobi equation for time-independent Hamiltonians, suggests that one could almost regard t itself as an "ignorable coordinate". In view of the special character t plays in dynamics, viewing it as "just another coordinate" seems mildly heretical, but we shall justify it in [optional] Section 10.9.

We conclude *this* section with an example of the above ideas:

Example 10.7.1. *Three-dimensional "One-body Problem"* We shall examine the problem of a body, subject to a force of magnitude $\frac{\mu}{r^2}$ when it is distance r from a fixed point, in three dimensions. The problem is conservative: $V = -\frac{\mu}{r}$. Using spherical coordinates (r, θ, ϕ), the Lagrangian (from which we must obtain the Hamiltonian) is

$$L \equiv T - V = \frac{1}{2}mv^2 - V = \frac{1}{2}m(\dot{r}^2 + (r\dot{\phi})^2 + (r\sin\phi\dot{\theta})^2) + \frac{\mu}{r}.$$

We obtain immediately

$$p_r \equiv \frac{\partial L}{\partial \dot{r}} = m\dot{r}, \qquad p_\phi \equiv \frac{\partial L}{\partial \dot{\phi}} = mr^2\dot{\phi}, \qquad p_\theta \equiv \frac{\partial L}{\partial \dot{\theta}} = mr^2\sin^2\phi\,\dot{\theta},$$

from which, since this is conservative and time-independent, so H equals the energy,

$$H = \frac{1}{2m}\left(p_r^2 + \frac{p_\phi^2}{r^2} + \frac{p_\theta^2}{r^2\sin^2\phi}\right) - \frac{\mu}{r}.$$

We note immediately that θ is an ignorable coordinate (though ϕ isn't); thus we know p_θ will correspond to an integral—a *constant.* Once that is done, we see that we can write the last terms in the parentheses as

$$\frac{p_\phi^2}{r^2} + \frac{p_\theta^2}{r^2\sin^2\phi} = \frac{1}{r^2}\left(p_\phi^2 + \frac{p_\theta^2}{\sin^2\phi}\right);$$

thus, accounting for the constancy of p_θ, this means that the (ϕ, p_ϕ) pair will be separable, and the above group itself a constant. The remaining terms will then depend on r alone, as well as the terms shown above to be constants.

Having examined the separability of this system, it is time to commit to the variables we will use, as well as the transformed form of the Hamiltonian, H^*. Let us use $W \equiv W(r, \theta, \phi, P_R, P_\Theta, P_\Phi)$ this time, where the P's will be constant; from the above strategy detailing the separation, we shall take

$$W(r, \theta, \phi, P_R, P_\Theta, P_\Phi) = W_r(r, \boldsymbol{P}) + W_\theta(\theta, \boldsymbol{P}) + W_\phi(\phi, \boldsymbol{P}).$$

From the transformations generated by this function [of "unlike" variables], we get that

$$P_\Theta^2 \equiv \left(\frac{\partial W}{\partial \theta}\right)^2$$
$$P_\Phi^2 \equiv \left(\frac{\partial W}{\partial \phi}\right)^2 + \frac{P_\Theta^2}{\sin^2\phi}$$
$$R \equiv \frac{1}{2m}\left(\left(\frac{\partial W}{\partial r}\right)^2 + \frac{P_\Phi^2}{r^2} - \frac{\mu}{r}\right)$$

or

$$\frac{\partial W}{\partial \theta} = \frac{\partial W_\theta}{\partial \theta} = P_\Theta$$
$$\frac{\partial W}{\partial \phi} = \frac{\partial W_\phi}{\partial \phi} = \sqrt{P_\Phi^2 - \frac{P_\Theta^2}{sin^2\phi}}$$
$$\frac{\partial W}{\partial r} = \frac{\partial W_r}{\partial r} = \sqrt{2mR - \frac{P_\Phi^2}{r^2} + \frac{\mu}{r}}.$$

We have craftily restrained from predicting the form for the transformed H^* to this point; we see now that the above transformations will result in just $H^* \equiv R$, so that P_R is linear in time. On the other hand, the variables P_Θ and P_Φ are altogether *absent* from this, but that's no problem: they are still constant, as are their associated conjugate coordinates, Θ and Φ! The entire solution process has been "reduced to quadratures"—the determination of first integrals. Of course, we will ignore the details of inversion of this transformation! But it should be pointed out that the device utilized in Example 10.6.1—of integrating the [partial] derivatives of W, rather than finding W explicitly and *then* differentiating—is of major benefit in actually *finding* the explicit transformations themselves. |

This example *does* begin to show the power of the Hamiltonian approach, as well as the use of generating functions resulting in what we know to be canonical transformations. It also demonstrates just how critical the separability was, and reinforces the maxim that one should *examine* the system before starting to *write*!

Summary. Actual solution for the generating function is expedited—usually *necessitated*!—by the ability to separate the function into a sum of terms, each of which depends on only a single coordinate/momentum pair. Although there are general conditions for such separability, it is assured if the Hamiltonian *groups* such pairs naturally, particularly if a particular coordinate is ignorable altogether.

10.8 Constraints in Hamiltonian Systems

The reader will recall how much attention was paid to constraints in the Lagrangian chapters; it seemed that every other sentence somehow had that word in it! Yet we have been strangely silent regarding this topic in the present context. In some ways this is understandable: the implementation of constraints in the Lagrangian context arose from consideration of the work done under *virtual displacements*, yet here there has been nary a word regarding these. At the same time, since the Hamiltonian is derivable from the Lagrangian, an understanding of how constraints enter the latter will explain their introduction into the former. That is precisely the approach we will take in this brief discussion of constraints in Hamiltonian systems, which is included for sake of completeness as much as anything else.

Recall that when we considered these in Section 8.3, particularly with the introduction of Lagrange multipliers, the independence of the δq_i played a critical role: because constraints are most generally non-holonomic—*differential* relations among the coordinates—we had to return to the differential level to introduce them. Yet we have never used virtual displacements directly in derivation of the Hamilton equations, so how can we introduce them now? The key lies in the observation that independence of the δq_i were used merely to obtain the [Euler-] Lagrange equations, and we have seen, in the discussion surrounding Theorem 10.2.1, that we can utilize the Lagrange equations *directly* to obtain the Hamiltonian ones, *if we regard both* $\mathbf{q}$ *and* $\mathbf{p}$ *as independent* in the function $\tilde{L}(\mathbf{q}, \dot{\mathbf{q}}, \mathbf{p}; t)$.

Thus we return once again to the Lagrangian

$$\tilde{L} = \mathbf{p} \cdot \dot{\mathbf{q}} - H$$

and the general form for the Lagrangian equations of motion, equation (8.3-6), in the special case that all generalized forces have been derivable from a potential: $Q_j \equiv 0$. Then, given general constraints of the form

$$\sum_{j=1}^{f} a_{kj}(\mathbf{q}; t)\, \delta q_j = 0, \qquad k = 1, \ldots, m$$

and applying these Lagrange equations, *now in both* $\mathbf{q}$ *and* $\mathbf{p}$ (regarding them as independent again), we get

$$\frac{d}{dt}\Big(\frac{\partial \tilde{L}}{\partial \dot{q}_j}\Big) - \frac{\partial \tilde{L}}{\partial q_j} - \sum_{k=1}^{m} \lambda_k a_{kj}(\mathbf{q}; t) = \dot{p}_j + \frac{\partial H}{\partial q_j} + \sum_{k=1}^{m} \lambda_k a_{kj} = 0$$

$$\frac{d}{dt}\Big(\frac{\partial \tilde{L}}{\partial \dot{p}_j}\Big) - \frac{\partial \tilde{L}}{\partial p_j} = 0 - \Big(\dot{q}_j - \frac{\partial H}{\partial p_j}\Big) = 0,$$

or

$$\dot{p}_j = -\frac{\partial H}{\partial q_j} - \sum_{k=1}^{m} \lambda_k a_{kj}$$
$$\dot{q}_j = \frac{\partial H}{\partial p_j}.$$

Just as in Section 8.3, in the special case that the above constraints are *holonomic*

$$\phi_k(\boldsymbol{q}; t) = 0, \qquad k = 1, \ldots, m,$$

the above reduce to the special form

$$\dot{p}_j = -\frac{\partial H}{\partial q_j} - \sum_{k=1}^{m} \lambda_k \frac{\partial \phi_k}{\partial q_j} = -\frac{\partial (H + \sum_{k=1}^{m} \lambda_k \phi_k)}{\partial q_j}$$
$$\dot{q}_j = \frac{\partial H}{\partial p_j}$$

—*i.e.*, we can use the modified function $H' \equiv H + \sum_{k=1}^{m} \lambda_k \phi_k$ as our Hamiltonian, similar to the modified Lagrangian we used before.

The above assumed constraints among the coordinates alone, as discussed in connection with Lagrangian dynamics. But, to the extent that we are viewing $\boldsymbol{q}$ and $\boldsymbol{p}$ as independent, there is no reason that we couldn't at least conceive of a similar set of constraints involving $\boldsymbol{p}$, too:

$$\sum_{j=1}^{f} a_{kj}(\boldsymbol{q}, \boldsymbol{p}; t)\, \delta q_j + \sum_{j=1}^{f} b_{kj}(\boldsymbol{q}, \boldsymbol{p}; t)\, \delta p_j = 0, \qquad k = 1, \ldots, m;$$

then the Hamiltonian equations of motion would include terms in the b's

$$\dot{p}_j = -\frac{\partial H}{\partial q_j} - \sum_{k=1}^{m} \lambda_k a_{kj}$$
$$\dot{q}_j = \frac{\partial H}{\partial p_j} + \sum_{k=1}^{m} \lambda_k b_{kj}.$$

If these were holonomic, $\phi_k(\boldsymbol{q}, \boldsymbol{p}; t) = 0$, we would get

$$\dot{p}_j = -\frac{\partial (H + \sum_{k=1}^{m} \lambda_k \phi_k)}{\partial q_j}$$
$$\dot{q}_j = \frac{\partial (H + \sum_{k=1}^{m} \lambda_k \phi_k)}{\partial p_j}.$$

While we have merrily obtained equations, we must not forget that, at some point, it is necessary actually to *obtain* the λ_k! But, just as in the Lagrangian presentation in Section 8.3—the above are obtained from the *Lagrangian* formulation, after all—we will use m of the above equations to determine the m λ_k; then the balance actually to derive their equations of motion.

We note a significant advantage to holonomic constraints in the Hamiltonian formulation: Recall that the major feature of Hamiltonian dynamics is the ability to obtain systematically those variables in which the Hamiltonian function assumes an "ideal" form. This rests on the observation that the Hamiltonian equations of motion depend on the presence or absence of variables in the Hamiltonian function itself, and the generating function approach to solution looks only to the form of the Hamiltonian under a canonical transformation; indeed, "canonical transformations" themselves are *defined* [Definition 10.4.1] only in terms of what they do to the Hamiltonian, not to any additional terms. Though non-holonomic constraints can be included in the Hamiltonian equations, even canonical transformations will not generally eliminate their additive nature from the transformed equations of motion. But *holonomic* constraints satisfy the same equations of motion the Hamiltonians do; thus one uses canonical transformations on the modified Hamiltonian $H' = H + \sum_{k=1}^{m} \lambda_k \phi_k$, effectively *ignoring* the fact that constraints are a part of that function, and retaining the canonical approach to solution.

As was mentioned in the opening to this section, it is included for completeness, as much because it was considered in the context of Lagrangian dynamics as anything else, though clearly the advantages of leaving constraints in the Hamiltonian rather than eliminating them are the same as in the Lagrangian: symmetries, in particular, are preserved. But it turns out that we will use these very ideas in the *next* section, where, again analogous to the Lagrangian presentation, we deal with time as a coordinate.

Summary. It is possible to introduce constraints into the Hamiltonian formulation in much the same way as in the Lagrangian one: non-holonomic constraints merely get appended to the Hamiltonian equations of motion, while holonomic ones can be simply *added* to the Hamiltonian function itself; the latter circumstance is then amenable to canonical transformations applied to the *modified* Hamiltonian. In both cases, the constraints are generally multiplied by appropriate Lagrange multipliers, just as for the Lagrangian formulation.

10.9 Time as a Coordinate in Hamiltonians

Recall that, in Section 8.4, we introduced the possibility that we might regard t as a coordinate, effectively reducing its otherwise special nature as the Ultimate Independent Variable to one of no greater significance than any other. This required the introduction of some other Ultimate Variable, τ, in terms of which all others were determined; the exact nature of τ has been left purposely vague, and will remain so here.

With regard to Lagrangians, this fact was more an oddity than anything else: there was no major advantage to adopting this perspective in Lagrangian dynamics, and, even from the point of view of the present Chapter, this was really done there only to introduce the generalized momentum conjugate to t. But viewing t as "just another variable" *could* have a practical advantage in the Hamiltonian formulation: Canonical transformations resulting from generating functions are presented as the primary means of solution of an arbitrary Hamiltonian system. But if H depends on t, so will the generating function, and this requires special treatment (as pointed out at the end of Section 10.5); this is just another manifestation of the special status of time. But if time could be reduced to the status of an "ordinary variable", this would be unnecessary, and there *might* be advantages to this approach.

And we seem to have all the formalism at hand to do this: we know that

$$H \equiv \mathbf{p} \cdot \dot{\mathbf{q}} - L(\mathbf{q}, \mathbf{p}; t) = \sum_{i=1}^{f} p_i \dot{q}_i - L(\mathbf{q}, \mathbf{p}; t),$$

can call t "q_o", and even have a momentum conjugate to t, equation (8.4-2). Blithely substituting this into the above definition for the Hamiltonian to obtain one in which t is a dependent variable, however, one obtains

$$\begin{aligned}\overline{H} &\equiv \sum_{i=0}^{f} p_i \dot{q}_i - L(\mathbf{q}, \mathbf{p}; t) \\ &= p_o \dot{q}_o + \sum_{i=1}^{f} p_i \dot{q}_i - L(\mathbf{q}, \mathbf{p}; t) \\ &\equiv (L - \sum_{i=1}^{f} p_i \dot{q}_i)(1) + \sum_{i=1}^{f} p_i \dot{q}_i - L(\mathbf{q}, \mathbf{p}; t)\end{aligned}$$

—in which we have substituted (8.4-2) for p_o and noted that $\dot{q}_o \equiv 1$—

$$\equiv 0$$

—a result bizarre beyond words: simply by renaming t, we have suddenly caused the Hamiltonian to vanish *identically*! While this is just what we want in, say, a *transformed* Hamiltonian, in the present context it says that $\mathbf{q}$ and $\mathbf{p}$ are all *constant* in the *original* one. Clearly something is dreadfully, dreadfully wrong here.

Actually, there are *two* things wrong here. In the first place, this naive, even mindless, manipulation blissfully ignored the fact that $\overline{H}$ is that *relative to* τ, not t. And, as noted above, τ is a somewhat shadowy variable, whose identity remains unspecified even to now. But that's precisely the point: τ *is an arbitrary parameter*, yet the equations of motion relative to that variable must remain valid regardless of what it is. In some sense, the power of τ is precisely its anonymity!

The other error above takes a little more digging, but is once again connected to the identity of τ as *the* independent variable: The above value for p_o came from the *Lagrangian* in τ, shown in Section 8.4 to be the *modified* Lagrangian Lt', where $t' \equiv \frac{dt}{d\tau} \equiv q_o'$. Thus the above Hamiltonian should *really* be

$$\overline{H} \equiv \sum_{i=0}^{f} p_i q_i' - \overline{L}(\overline{\mathbf{q}}, \overline{\mathbf{q}}') = \sum_{i=0}^{f} p_i q_i' - L(\mathbf{q}, \frac{\mathbf{q}'}{q_o'}; q_o) q_o'$$

in which $\overline{\mathbf{q}}$ and $\overline{\mathbf{q}}'$ are the $n+1$ vectors including $t \equiv q_o$ and $t' \equiv q_o'$,

$$q_i' \equiv \frac{dq_i}{d\tau} \quad \text{and} \quad p_i' \equiv \frac{\partial L(\mathbf{q}, \frac{\mathbf{q}'}{t'}; t) t'}{\partial q_i'},$$

The latter equation is solved for $q'(\boldsymbol{q},\boldsymbol{p}';t)$ in order to express $\overline{H}$ in terms of $\boldsymbol{p}'$. $\overline{H}$ still vanishes: this is effectively no more than the Jacobi function (Section 9.2) in $f+1$ coordinates $\overline{\boldsymbol{q}}$, in which the Lagrangian is independent of "time", τ.

But *we can't solve for the* q_i': Recall that, at the end of Section 10.1, we obtained the condition to be able to do this, namely that the Hessian of L with respect to the $\dot{\boldsymbol{q}}$ not vanish, equation (10.1-3); we claim that that for $\overline{L}$ relative to the $\overline{\boldsymbol{q}}'$ it *does.* This is shown by direct calculation. In the new "extended" variables $\overline{\boldsymbol{q}}$ and $\overline{\boldsymbol{p}}$, the new momenta satisfy, for $i\neq 0$,

$$\overline{p}_i \equiv \frac{\partial \overline{L}(\overline{\boldsymbol{q}},\overline{\boldsymbol{q}}')}{\partial \overline{q}_i'} = \frac{\partial (L(\boldsymbol{q},\frac{\boldsymbol{q}'}{t'};t)t')}{\partial q_i'} = \frac{\partial L}{\partial \dot{q}_i}\left(\frac{1}{t'}\right)t' = \frac{\partial L}{\partial \dot{q}_i}.$$

Thus we have the derivatives in the Hessian (for $i\neq 0$!), for $j=0$:

$$\frac{\partial^2 \overline{L}}{\partial \overline{q}_i'\partial \overline{q}_o'} = \frac{\partial}{\partial t'}\left(\frac{\partial L(\boldsymbol{q},\frac{\boldsymbol{q}'}{t'};t)}{\partial \dot{q}_i}\right) = \sum_{j=1}^{f}\frac{\partial^2 L}{\partial \dot{q}_i\partial \dot{q}_j}\left(-\frac{q_j'}{t'^2}\right) = -\sum_{j=1}^{f}\frac{\partial^2 L}{\partial \dot{q}_i\partial \dot{q}_j}\frac{\dot{q}_j}{t'}, \tag{10.9-1a}$$

while, for $j\neq 0$:

$$\frac{\partial^2 \overline{L}}{\partial \overline{q}_i'\partial \overline{q}_j'} = \frac{\partial}{\partial q_j'}\left(\frac{\partial L(\boldsymbol{q},\frac{\boldsymbol{q}'}{t'};t)}{\partial \dot{q}_i}\right) = \frac{\partial^2 L}{\partial \dot{q}_i\partial \dot{q}_j}\left(\frac{1}{t'}\right) \tag{10.9-1b}$$

—where, again, $\frac{q_j'}{t''} = \dot{q}_j$. Similarly, for $i=0$, writing the penultimate expression for $\overline{p}_o$ in equation (8.4-2) as

$$\overline{p}_o \equiv \frac{\partial \overline{L}(\overline{\boldsymbol{q}},\overline{\boldsymbol{q}}')}{\partial \overline{q}_o'} = L - \sum_{i=1}^{f}\frac{\partial L}{\partial \dot{q}_i}\left(\frac{q_i'}{t'}\right), \tag{10.9-2}$$

we get, for $j=0$:

$$\begin{aligned}
\frac{\partial^2 \overline{L}(\overline{\boldsymbol{q}},\overline{\boldsymbol{q}}')}{\partial \overline{q}_o'^2} &= \frac{\partial}{\partial t'}\left(L(\boldsymbol{q},\frac{\boldsymbol{q}'}{t'};t) - \sum_{i=1}^{f}\frac{\partial L(\boldsymbol{q},\frac{\boldsymbol{q}'}{t'};t)}{\partial \dot{q}_i}\left(\frac{q_i'}{t'}\right)\right)\\
&= \sum_{i=1}^{f}\frac{\partial L}{\partial \dot{q}_i}\left(-\frac{q_i'}{t'^2}\right) + \sum_{j=1}^{f}\sum_{i=1}^{f}\left(\frac{\partial^2 L}{\partial \dot{q}_i\partial \dot{q}_j}\frac{q_j'}{t'^2}\right)\left(\frac{q_i'}{t'}\right) + \sum_{i=1}^{f}\frac{\partial L}{\partial \dot{q}_i}\left(\frac{q_i'}{t'^2}\right) \qquad (10.9\text{-}3a)\\
&= \sum_{j=1}^{f}\sum_{i=1}^{f}\frac{\partial^2 L}{\partial \dot{q}_i\partial \dot{q}_j}\left(\frac{\dot{q}_i\dot{q}_j}{t'}\right) = \sum_{i=1}^{f}\dot{q}_i\sum_{j=1}^{f}\frac{\partial^2 L}{\partial \dot{q}_i\partial \dot{q}_j}\frac{\dot{q}_j}{t'},
\end{aligned}$$

while, for $j\neq 0$:

$$\begin{aligned}
\frac{\partial^2 \overline{L}}{\partial \overline{q}_i'\partial \overline{q}_o'} &= \frac{\partial}{\partial q_j'}\left(L(\boldsymbol{q},\frac{\boldsymbol{q}'}{t'};t) - \sum_{i=1}^{f}\frac{\partial L(\boldsymbol{q},\frac{\boldsymbol{q}'}{t'};t)}{\partial \dot{q}_i}\left(\frac{q_i'}{t'}\right)\right)\\
&= \frac{\partial L}{\partial \dot{q}_j}\left(\frac{1}{t'}\right) - \sum_{i=1}^{f}\left(\frac{\partial^2 L}{\partial \dot{q}_i\partial \dot{q}_j}\frac{1}{t'}\right)\left(\frac{q_i'}{t'}\right) - \frac{\partial L}{\partial \dot{q}_j}\left(\frac{1}{t'}\right) \qquad (10.9\text{-}3b)\\
&= -\sum_{i=1}^{f}\frac{\partial^2 L}{\partial \dot{q}_i\partial \dot{q}_j}\left(\frac{\dot{q}_i}{t'}\right) = -\sum_{i=1}^{f}\dot{q}_i\frac{\partial^2 L}{\partial \dot{q}_i\partial \dot{q}_j}\left(\frac{1}{t'}\right).
\end{aligned}$$

We observe that the equations in (10.9-3) are just linear combinations of the respective ones in (10.9-1), each of the latter multiplied by its corresponding $-\dot{q}_i$; thus the 0th row in the Hessian for $\overline{L}$ is a linear combination of the others, and its determinant vanishes (Theorem 3.2.6). We *cannot* solve for *all* $f+1$ $\overline{q}_i'$ in this case. But the $f \times f$ *sub*matrix with entries given by equation (10.9-1b) is just the *normal* Hessian matrix divided by t', and we have shown (at the end of Section 10.1) that *that* matrix has a non-vanishing determinant; thus the rank of the Hessian for $\overline{L}$ is f, and the above linear combination is the *only* one among the entries in this matrix.

The original f p_i are linearly independent; the $f+1$ $\overline{p}_i$ aren't. Thus there is some *relation* among them, particularly between p_o and the others, since it is the introduction of that variable which forces the entire set to go dependent. That relation is precisely equation (10.9-2), defining $\overline{p}_o$ and differentiated to get the above results, that

$$p_o \equiv L - \sum_{i=1}^{f} p_i \dot{q}_i = -H(\boldsymbol{q}, \boldsymbol{p}; t);$$

i.e.—and this is something we recognize explicitly for the first time—*the momentum conjugate to t in the τ equations of motion is nothing more than the [negative] Hamiltonian in terms of the* original *$\boldsymbol{q}$ and $\boldsymbol{p}$ itself!* This relation among the $\overline{q}$ and $\overline{p}$ is thus a *constraint* on these variables, and we are ultimately implementing Hamiltonian equations of motion *relative to the constraint*

$$\phi = p_o + H(\boldsymbol{q}, \boldsymbol{p}; t) = 0$$

—precisely the situation discussed in the previous section. (Note that this is a case in which the constraint depends on $\boldsymbol{p}$, too!) Since this constraint is holonomic, rather than eliminating it explicitly, we can include it in the Hamiltonian

$$\overline{H}' \equiv \overline{H} + \lambda\phi = 0 + \lambda(p_o + H(\boldsymbol{q}, \boldsymbol{p}; q_0)),$$

in which λ is the Lagrange multiplier. The Hamiltonian equations of motion in the $2f+2$ variables $\overline{q}$ and $\overline{p}$ are just:

$$\begin{aligned}
i = 0: \qquad & q_o' = \frac{\partial \overline{H}'}{\partial p_o} = \lambda \\
& p_o' = -\frac{\partial \overline{H}'}{\partial q_o} = -\lambda \frac{\partial H}{\partial q_o} \\
i \geq 1: \qquad & q_i' = \frac{\partial \overline{H}'}{\partial p_i} = \lambda \frac{\partial H}{\partial p_i} \\
& p_i' = -\frac{\partial \overline{H}'}{\partial q_i} = -\lambda \frac{\partial H}{\partial q_i}.
\end{aligned}$$

And now the flash of the wand: $q_o \equiv t$, so $q_o' = \lambda = t'$. And τ is, recall, *arbitrary*; in particular, we can select it to make $t' = \lambda \equiv 1$. Doing so, using the fact that $\frac{d}{d\tau} = \frac{1}{t'}\frac{d}{dt} = \frac{d}{dt}$, and substituting H for p_o explicitly at this stage, the above equations become

$$t' \equiv \frac{dt}{d\tau} = \frac{dt}{dt} = 1$$
$$H' \equiv \frac{dH}{d\tau} = \frac{dH}{dt} = -\frac{\partial H}{\partial t}$$
$$q_i' \equiv \frac{dq_i}{d\tau} = \frac{dq_i}{dt} = \frac{\partial H}{\partial p_i}$$
$$p_i' \equiv \frac{\partial p_i}{\partial \tau} = \frac{dp_i}{dt} = -\frac{\partial H}{\partial q_i}.$$

The first equation is merely a reflection of our choice of τ, the second is just equation (10.2-2c), while the last two are just equations (10.2-2a–b) (in the special case that $Q_i \equiv 0$)! In effect, the Hamiltonian $\overline{H}' \equiv p_o + H(\mathbf{q}, \mathbf{p}; q_o)$ in the $2f+2$ variables $\overline{\mathbf{q}}$ and $\overline{\mathbf{p}}$ in which $q_o \equiv t$—note that H is a function of only the *original* $2f$ variables $\mathbf{q}$ and $\mathbf{p}$—give precisely the equations of motion [in τ] of the original system [in t] when $\frac{dt}{d\tau} \equiv 1$! And $\overline{H}'$ is *always conservative*, since it doesn't depend on τ.

Now for the test to see what, if any, advantages might accrue from this approach:

Example 10.9.1. *Hamiltonian with t as Coordinate.* For sake of comparison, we shall again treat the problem in Example 10.6.2, with Hamiltonian

$$H = \frac{p^2}{2m} - Axt.$$

In the previous presentation of this example, we took advantage of the linear nature of the coordinate x, choosing to eliminate this term through a transformation $W(p, X; t)$ in which $x = \frac{\partial W}{\partial p}$. This put the Hamilton-Jacobi equation in the form (10.6-4):

$$\frac{p^2}{2m} - At\frac{\partial W}{\partial p} + \frac{\partial W}{\partial t} \equiv 0$$

Alternatively, we could (though perhaps not "just as easily") have used, say, $W(x, X; t)$ in which now $p = \frac{\partial W}{\partial x}$; this would have led to the equation

$$\frac{1}{2m}\left(\frac{\partial W}{\partial x}\right)^2 - Atx + \frac{\partial W}{\partial t} \equiv 0. \tag{10.9-4}$$

Now let us implement the approach of this section: We conceptually introduce the independent variable τ, making t a function of that variable. As we have observed, the new Hamiltonian $\overline{H} \equiv 0$ in this approach, but we get around this by viewing the equation for the momentum conjugate to t as a *constraint* $\phi \equiv p_o + H = 0$. Adding this with its Lagrange multiplier onto $\overline{H}$ we get what we called above

$$\overline{H}' \equiv \overline{H} + \lambda\phi = \lambda(p_o + H(x, p; q_o)) = \lambda\Big(p_o + \frac{p^2}{2m} - Atx\Big),$$

where, recall, we will ultimately take $\lambda \equiv 1$. This is now truly a two-degree-of-freedom system in the conjugate pairs (x, p) and (q_o, p_o).

For purposes of comparison, let us transform this to $H^* \equiv 0$ and use a generating function of the form $W(x, q_o, X, Q_o)$; then the Hamilton-Jacobi equation becomes

$$\frac{\partial W}{\partial q_o} + \frac{1}{2m}\left(\frac{\partial W}{\partial x}\right)^2 - Aq_o x = 0$$

—*precisely the form* (10.9-4) *we get from the normal Hamilton-Jacobi equation* for this form of function. We note, however, that this device does allow an alternative, not available to the standard Hamilton-Jacobi equation: if, rather, we use $W(p, p_o, X, Q_o)$, for example, then the equation becomes different in form:

$$p_o + \frac{p^2}{2m} - A\frac{\partial W}{\partial q_o}\frac{\partial W}{\partial p} = 0$$

—a flexibility which might be appealing in certain cases. |

After all this work, there seems to be little if any practical advantage to interpreting t as "just another coordinate" in the Hamiltonian formulation, either. But there is a *conceptual* advantage[6] we won't go into here.

Summary. Much like with Lagrangians, time can be regarded as a *de*pendent variable, if one considers it and all the other variables to be functions of an arbitrary parameter; the momentum conjugate to t is then nothing more than the *original* Hamiltonian itself. Doing so makes the Hamiltonian in the resulting $2f + 2$ variables *vanish identically*, and if one tries to express the q' in terms of the $f + 1$ p's, the defining equations are *dependent.* In fact, this dependency is contained precisely in the equating of the time's momentum with the Hamiltonian—a relation which, however, can be introduced as a *constraint* without explicitly eliminating it. Doing so, and taking $\frac{dt}{d\tau} \equiv 1$ (which makes the Lagrange multiplier $\lambda \equiv 1$), recovers the original $2f$ Hamiltonian equations in the original variables, as well as the time derivative of that Hamiltonian.

Summary

Given a Lagrangian $L(\boldsymbol{q}, \dot{\boldsymbol{q}}; t)$, we define the *generalized momentum conjugate to* q_i

$$p_i \equiv \frac{\partial L}{\partial \dot{q}_i}$$

—a relation which can always be inverted to obtain $\dot{\boldsymbol{q}} = \dot{\boldsymbol{q}}(\boldsymbol{q}, \boldsymbol{p}; t)$. This done, we form the *Hamiltonian* (the *Legendre transform* of L)

$$H(\boldsymbol{q}, \boldsymbol{p}; t) \equiv \boldsymbol{p} \cdot \dot{\boldsymbol{q}}(\boldsymbol{q}, \boldsymbol{p}; t) - L(\boldsymbol{q}, \dot{\boldsymbol{q}}(\boldsymbol{q}, \boldsymbol{p}; t); t)$$

which enjoys equations of motion

$$\begin{aligned} \dot{q}_i &= \frac{\partial H}{\partial p_i} \\ \dot{p}_i &= -\frac{\partial H}{\partial q_i} + Q_i \\ \frac{\partial H}{\partial t} &= \frac{\partial L}{\partial t} \end{aligned}$$

[6]See Lanczos, *The Variational Principles of Mechanics*, 3rd ed., pp 189-192.

—a set of $2f$ first-order equations in the variables $\boldsymbol{q}$ and $\boldsymbol{p}$, along with an equation describing the time variation of H. Despite the fact that $\boldsymbol{q} \rightarrow \dot{\boldsymbol{q}}$ in Lagrangian dynamics, the variables $\boldsymbol{q}$ and $\boldsymbol{p}$ are truly *independent*; in fact, the equations of motion can be obtained by applying the Lagrange equations to $\tilde{L}(\boldsymbol{q}, \dot{\boldsymbol{q}}, \boldsymbol{p}; t) \equiv \boldsymbol{p} \cdot \dot{\boldsymbol{q}} - H(\boldsymbol{q}, \boldsymbol{p}; t)$ in *both* sets of variables.

The form of these equations of motion is significant: it demonstrates that, like the Lagrangian, integrals of the motion depend only on the *form* of the function; but, unlike the Lagrangians, that form affects the equations for *all* $2f$ *variables.* In particular, if H is free of both sets of variables, they are constant; if it depends on only one set, that set is constant and its conjugate variables are linear in the time. In either event, the complete solution of the system is known.

Since the form of the Hamiltonian depends on the variables chosen, this suggests the possibility that we might *find* such a set—effect a *transformation of coordinates*—to bring H to a soluble form H^*. But such an approach is predicated on the form of the equations of motion being invariant under such a transformation. This is true if and only if the transformation satisfies the *symplectic conditions*; transformations satisfying this stricture are called *canonical.*

It turns out that these conditions are satisfied by that transformation resulting from partial derivatives of an *arbitrary* function—the *generating function*—of "*mixed*", old and new, variables. Unfortunately, the transformation is defined only *implicitly* and must be inverted to get the explicit transformation. And such transformations are only a [proper] subset of the class of all such canonical transformations.

Nonetheless, implementation of this approach is a powerful tool for the solution of such Hamiltonian systems—at least as long as the systems are *conservative*—since, unlike Lagrangian systems, they provide a systematic means of uncovering the "right variables" to make the Hamiltonian assume a soluble form. This approach generally requires the solution of a partial differential equation, the *Hamilton-Jacobi equation*, and ultimate inversion of the resulting transformation, both to get the solution and to match initial conditions.

Solution of the Hamilton-Jacobi equation is expedited if the variables are *separable*: if, most generally, they appear in *groups* independent of all the others. In that case, solution of the governing equation reduces to simple quadratures—integrals in a single variable.

In the same way that *constraints* in Lagrangian systems can be implemented by either explicit or implicit elimination—the latter utilizing *Lagrange multipliers*, Hamiltonian systems can be. Like Lagrangian systems, *holonomic* constraints are particularly appealing, since it is possible simply to apply the standard Hamiltonian equations of motion to the modified Hamiltonian $H' \equiv H + \sum_{k=1}^{m} \lambda_k \phi_k$.

Like Lagrangian systems, time can be considered to be merely another coordinate, itself dependent on yet another, arbitrary parameter. This, of course, requires a momentum conjugate to t; it turns out that momentum is nothing more than the negative of the original Hamiltonian itself! Appropriate conditions on the nature of the "arbitrary parameter" ultimately lead to precisely the equations of motion for the original Hamiltonian.

Epilogue

We have considered in some detail two energy-based formulations of mechanics, Lagrangian and Hamiltonian dynamics. The former is explicitly based on the Principle of Virtual Work, where virtual displacements are taken from the *motion* of the particle being considered; the latter, in major degree, is a by-product of the former. Both enjoy the heritage of virtual work: Constraint forces never enter the equations (as long as they're not violated). Pins connecting members of a machine never enter the equations (as long as they're frictionless). And, as least as important, *any* coordinates can be used (as long as the orientation of the system can be specified in terms of those chosen).

So how does one decide which of these two formulations to utilize in a given problem? To a certain extent, this depends on what one intends to do with the result. Though one might argue simplistically that the Hamiltonian formulation gives only first-order equations, even the Lagrangian ones can be put in a formally first-order form through the simple device of doubling the number of variables by introducing the *first* derivatives themselves as variables. Yet the Hamilton equations are *true* first-order equations, while doubling the number of variables to solve the Lagrangian systems merely disguises their truly second-order nature. In particular, one likely has a better chance at solving the Hamiltonian system *analytically* than the Lagrangian one. In fact, that is where the Hamiltonian formulation shows true superiority over the Lagrangian: in solution—the determination of "integrals of the motion". Hamiltonian dynamics has a means of *determining* those variables in which its fundamental function assumes a soluble form, however cumbersome it might appear to be on the surface. But that mechanism endemically presumes the system to be *conservative*: if there are any generalized forces which cannot be included in the Hamiltonian function itself, this approach fails, and one might just as well revert to the Lagrangian formalism.

In fact, if ultimately one is interested in the equations of motion merely to obtain a *numerical* solution, Lagrangian equations likely have the edge: one get immediately equations in the variables q_i and *their* time derivatives, rather than having to go through the intermediary $\boldsymbol{p}$ calculations. And, as pointed out above, the ability of Lagrangian equations of motion to deal with *any* type of force, conservative or non-conservative, gives them a generality not enjoyed—even viewing canonical transformations as "practical"—by Hamiltonian systems.

But the appealing symmetry of the Hamiltonian equations, and the ability to utilize the canonical formalism, make them the formulation of choice for the examination of general non-linear equations, of current major interest. Those with an interest in this field should have at least a nodding acquaintance with this approach, one of the primary reasons it was included in this text.

But these are *details*. The major advantage of analytical dynamics is the ability to deal with *systems* of bodies, and to introduce *arbitrary* coordinates of choice. The first of these is tedious in Newtonian mechanics, the second impossible. It stands to Dynamics as the Principle of Virtual Work (on which it is based) does to Statics, and should be co-equal in the repertoire of dynamical techniques to the Newtonian formulation.

Index

Italicized pages indicate primary references or definitions; "Hw" denotes Homework problems, "Ex" examples, and "Thrm" theorems.

Mechanical Engineering Series *(continued from page ii)*

K.C. Gupta, **Mechanics and Control of Robots**

J. Ida and J.P.A. Bastos, **Electromagnetics and Calculations of Fields**

M. Kaviany, **Principles of Convective Heat Transfer, 2nd ed.**

M. Kaviany, **Principles of Heat Transfer in Porous Media, 2nd ed.**

E.N. Kuznetsov, **Underconstrained Structural Systems**

P. Ladevèze, **Nonlinear Computational Structural Mechanics: New Approaches and Non-Incremental Methods of Calculation**

P. Ladevèze and J.-P. Pelle, **Mastering Calculations in Linear and Nonlinear Mechanics**

A. Lawrence, **Modern Inertial Technology: Navigation, Guidance, and Control, 2nd ed.**

R.A. Layton, **Principles of Analytical System Dynamics**

F.F. Ling, W.M. Lai, D.A. Lucca, **Fundamentals of Surface Mechanics: With Applications, 2nd ed.**

C.V. Madhusudana, **Thermal Contact Conductance**

D.P. Miannay, **Fracture Mechanics**

D.P. Miannay, **Time-Dependent Fracture Mechanics**

D.K. Miu, **Mechatronics: Electromechanics and Contromechanics**

D. Post, B. Han, and P. Ifju, **High Sensitivity and Moiré: Experimental Analysis for Mechanics and Materials**

F.P. Rimrott, **Introductory Attitude Dynamics**

S.S. Sadhal, P.S. Ayyaswamy, and J.N. Chung, **Transport Phenomena with Drops and Bubbles**

A.A. Shabana, **Theory of Vibration: An Introduction, 2nd ed.**

A.A. Shabana, **Theory of Vibration: Discrete and Continuous Systems, 2nd ed.**